Mechanical Measurements
Sixth Edition

Thomas G. Beckwith

Roy D. Marangoni
University of Pittsburgh

John H. Lienhard V
Massachusetts Institute of Technology

PEARSON

Prentice Hall

Upper Saddle River, NJ 07458

Library of Congress Catalog-in-Publication Data on File

Vice President and Editorial Director: *Marcia J. Horton*
Editorial Assistant: *Dolores Mars*
Executive Managing Editor: *Vince O'Brien*
Managing Editor: *David A. George*
Production Editor: *Rose Kernan*
Director for Creative Services: *Paul Belfanti*
Creative Director: *Juan Lopez*
Cover Designer: *Bruce Kenselaar*
Managing Editor, AV Management and Production: *Patricia Burns*
Art Editor: *Gregory Dulles*
Manufacturing Manager: *Alexis Heydt-Long*
Manufacturing Buyer: *Lisa McDowell*
Executive Marketing Manager: *Tim Galligan*

© 2007 Pearson Education, Inc.
Pearson Prentice Hall
Pearson Education, Inc.
Upper Saddle River, NJ 07458

Printed in the United States of America

10 9 8 7 6 5 4

ISBN 0-201-84765-5

Pearson Education, Inc., *Upper Saddle River, New Jersey*
Pearson Education Ltd., *London*
Pearson Education Australia Pty. Ltd., *Sydney*
Pearson Education Singapore, Pte. Ltd.
Pearson Education North Asia Ltd., *Hong Kong*
Pearson Education Canada, Inc., *Toronto*
Pearson Educación de Mexico, S.A. de C.V.
Pearson Education—Japan, *Tokyo*
Pearson Education—Malaysia, Pte. Ltd.

Contents

Preface

After more than 45 years, the basic purpose of this book can still be expressed by the following three paragraphs extracted from the Preface of the first edition of *Mechanical Measurements*, published in 1961.

> Experimental development has become a very important aspect of mechanical design procedure. In years past the necessity for "ironing out the bugs" was looked upon as an unfortunate turn of events, casting serious doubts on the abilities of a design staff. With the ever-increasing complexity and speed of machinery, a changed design philosophy has been forced on both the engineering profession and industrial management alike. An experimental development period is now looked upon, not as a problem to avoid, but as an *integral phase* of the whole design procedure. Evidence supporting this contention is provided by the continuing growth of research and development companies, subsidiaries, teams, and armed services R&D programs.

> At the same time, it should not be construed that the experimental development (design) approach reduces the responsibilities attending the preliminary planning phases of a new device or process. In fact, knowledge gained through experimental programs continually strengthens and supports the theoretical phases of design.

> *Measurement* and the correct interpretation thereof are necessary parts of any engineering research and development program. Naturally, the measurements must supply reliable information and their meanings must be correctly comprehended and interpreted. *It is the primary purpose of this book to supply a basis for such measurements.*

When Tom Beckwith introduced the first edition of this pioneer text in 1961, all sensors and recording equipment were of analog nature. From the first edition to the present sixth edition, we have seen a marked transition from analog measurement to digital measurement. During the time between the first edition and the fourth edition, Tom came to recognize and appreciate the change from analog techniques to digital techniques and this was reflected in the technical material in those editions. However, from the fourth edition to the present edition enormous changes in digital instrumentation have occurred.

In his wildest dreams, we doubt that Tom envisioned how the computer would change the area of mechanical measurements. The personal computer and software such as Lab-VIEW have revolutionized the measurement process. Instead of cabinet after cabinet of specialized signal conditioning equipment, students may now use a personal computer, with a dedicated interface and appropriate software, to drive array of sensors suitable for most measurements. Although Tom is no longer with us, we are sure that he looks down and smiles when he sees how that which he originally started has progressed.

The authors do not suggest that the sequence of materials as presented need be strictly adhered to. Wide flexibility of course contents should be possible, with text assignments tailored to fit a variety of basic requirements or intents. For example, the authors have found that, if desired, Chapters 1 and 2 can simply be made a reading assignment. Greater or lesser emphasis may be placed on certain chapters as the instructor wishes. Should a course consist of a lecture/recitation section plus a laboratory, available laboratory equipment may also dictate areas to be emphasized. Quite generally, as a text, the book can easily accommodate a two-semester sequence.

ACKNOWLEDGMENTS

Roy Marangoni thanks his wife, Lavonne, for her patience throughout several editions of this textbook. Roy also expresses his gratitude to his son, Gary, for sharing his expertise in the field of computer technology. John Lienhard would like to thank Lori Hyke and Suzanne Bunker for their assistance with various aspects of the manuscript. John also thanks his family, Theresa, Jasper, and Hannah, for their loving support over the years.

ROY D. MARANGONI
JOHN H. LIENHARD V

PART ONE

FUNDAMENTALS
OF MECHANICAL
MEASUREMENT

CHAPTER 1

The Process of Measurement: An Overview

1.1   INTRODUCTION

It has been said, "Whatever exists, exists in some amount." The determination of the amount is what measurement is all about. If those things that exist are related to the practice of mechanical engineering, then the determination of their amounts constitutes the subject of *mechanical measurements*.[1]

The process or the act of measurement consists of obtaining a quantitative comparison between a predefined *standard* and a *measurand*. The word *measurand* is used to designate the particular physical parameter being observed and quantified; that is, the input quantity to the measuring process. The act of measurement produces a *result* (see Fig. 1.1).

The standard of comparison must be of the same character as the measurand, and usually, but not always, is prescribed and defined by a legal or recognized agency or organization—for example, the National Institute of Standards and Technology (NIST), formerly called the National Bureau of Standards (NBS), the International Organization for Standardization (ISO), or the American National Standards Institute (ANSI). The meter, for example, is a clearly defined standard of length.

Such quantities as temperature, strain, and the parameters associated with fluid flow, acoustics, and motion, in addition to the fundamental quantities of mass, length, time, and so on, are typical of those within the scope of mechanical measurements. Unavoidably,

---

[1] *Mechanical measurements* are not necessarily accomplished by mechanical means: rather, it is to the measured quantity itself that the term *mechanical* is directed. The phrase *measurement of mechanical quantities*, or *of parameters*, would perhaps express more completely the meaning intended. In the interest of brevity, however, the subject is simply called *mechanical measurements*.

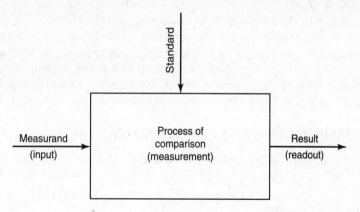

FIGURE 1.1: Fundamental measuring process.

the measurement of mechanical quantities also involves consideration of things electrical, since it is often convenient or necessary to change, or *transduce*, a mechanical measurand into a corresponding electrical quantity.

1.2 THE SIGNIFICANCE OF MECHANICAL MEASUREMENT

Measurement provides quantitative information on the actual state of physical variables and processes that otherwise could only be estimated. As such, measurement is both the vehicle for new understanding of the physical world and the ultimate test of any theory or design. Measurement is the fundamental basis for all research, design, and development, and its role is prominent in many engineering activities.

All mechanical design of any complexity involves three elements: experience, the rational element, and the experimental element. The element of experience is based on previous exposure to similar systems and on an engineer's common sense. The rational element relies on quantitative engineering principles, the laws of physics, and so on. The experimental element is based on measurement—that is, on measurement of the various quantities pertaining to the operation and performance of the device or process being developed. Measurement provides a comparison between what was intended and what was actually achieved.

Measurement is also a fundamental element of any control process. The concept of control *requires* the measured discrepancy between the actual and the desired performances. The controlling portion of the system must know the magnitude and direction of the difference in order to react intelligently.

In addition, many daily operations require measurement for proper performance. An example is in the central power station. Temperatures, flows, pressures, and vibrational amplitudes must be constantly monitored by measurement to ensure proper performance of the system. Moreover, measurement is vital to commerce. Costs are established on the basis of *amounts* of materials, power, expenditure of time and labor, and other constraints.

To be useful, measurement must be reliable. Having incorrect information is potentially more damaging than having no information. The situation, of course, raises the question of the accuracy or *uncertainty* of a measurement. Arnold O. Beckman, founder of Beckman Instruments, once stated, "One thing you learn in science is that there is no *perfect* answer, no *perfect* measure."[2] It is quite important that engineers interpreting the results of measurement have some basis for evaluating the likely uncertainty. Engineers should *never* simply read a scale or printout and blindly accept the numbers. They must carefully place realistic tolerances on each of the measured values, and not only should have a doubting mind but also should attempt to quantify their doubts. We will discuss uncertainty in more detail in Section 1.8 and as the subject of Chapter 3.

1.3   FUNDAMENTAL METHODS OF MEASUREMENT

There are two basic methods of measurement: (1) *direct comparison* with either a primary or a secondary standard and (2) *indirect comparison* through the use of a calibrated system.

1.3.1   Direct Comparison

How would you measure the length of a bar of steel? If you were to be satisfied with a measurement to within, let us say, $\frac{1}{8}$ in. (approximately 3 mm), you would probably use a steel tape measure. You would compare the length of the bar with a *standard* and would find that the bar is so many inches long because that many inch-units on your standard are the same length as the bar. Thus you would have determined the length by *direct comparison*. The standard that you have used is called a secondary standard. No doubt you could trace its ancestry back through no more than four generations to the primary length standard, which is related to the speed of light (Section 2.4).

Although to measure by direct comparison is to strip the measurement process to its barest essentials, the method is not always adequate. The human senses are not equipped to make direct comparisons of all quantities with equal facility. In many cases they are not sensitive enough. We can make direct comparisons of small distances using a steel rule, with a precision of about 1 mm (approximately 0.04 in.). Often we require greater accuracy. Then we must call for additional assistance from some more complex form of measuring system. Measurement by direct comparison is thus less common than is measurement by *indirect comparison*.

1.3.2   Using a Calibrated System

Indirect comparison makes use of some form of transducing device coupled to a chain of connecting apparatus, which we shall call, in toto, the *measuring system*. This chain of devices converts the basic form of input into an analogous form, which it then processes and presents at the output as a known function of the original input. Such a conversion is often necessary so that the desired information will be intelligible. The human senses are simply not designed to detect the strain in a machine member, for instance. Assistance is required from a system that senses, converts, and finally presents an analogous output in the form of a displacement on a scale or chart or as a digital readout.

Processing of the analogous signal may take many forms. Often it is necessary to increase an amplitude or a power through some form of amplification. Or in another case it may be necessary to extract the desired information from a mass of extraneous input by a process of filtering. A remote reading or recording may be needed, such as ground recording of a temperature or pressure within a rocket in flight. In this case the pressure or

---

[2]Emphasis added by the authors.

temperature measurement must be combined with a radio-frequency signal for transmission to the ground.

In each of the various cases requiring amplification, or filtering, or remote recording, electrical methods suggest themselves. In fact, the majority of transducers in use, *particularly for dynamic mechanical measurements*, convert the mechanical input into an analogous electrical form for processing.

1.4 THE GENERALIZED MEASURING SYSTEM

Most measuring systems fall within the framework of a general arrangement consisting of three phases or stages:

Stage 1. A detection-transduction, or *sensor-transducer*, stage
Stage 2. An intermediate stage, which we shall call the *signal-conditioning* stage
Stage 3. A terminating, or *readout-recording*, stage

Each stage consists of a distinct component or group of components that performs required and definite steps in the measurement. These are called *basic elements*; their scope is determined by their function rather than by their construction. Figure 1.2 and Table 1.1 outline the significance of each of these stages.

1.4.1 First, or Sensor-Transducer, Stage

The primary function of the first stage is to detect or to sense the measurand. At the same time, ideally, this stage should be insensitive to every other possible input. For instance, if it is a pressure pickup, it should be insensitive to, say, acceleration; if it is a strain gage, it should be insensitive to temperature; if a linear accelerometer, it should be insensitive to angular acceleration; and so on. Unfortunately, it is rare indeed to find a detecting device that is completely selective. Unwanted sensitivity is a measuring error, called *noise* when it varies rapidly and *drift* when it varies very slowly.

Frequently one finds more than a single transduction (change in signal character) in the first stage, particularly if the first-stage output is electrical (see Section 6.3).

1.4.2 Second, or Signal-Conditioning, Stage

The purpose of the second stage of the general system is to modify the transduced information so that it is acceptable to the third, or terminating, stage. In addition, it may perform one or more basic operations, such as selective filtering to remove noise, integration, dif-

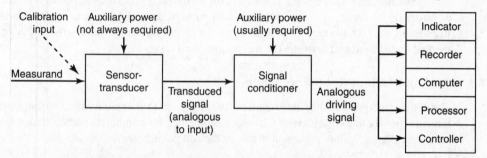

FIGURE 1.2: Block diagram of the generalized measuring system.

TABLE 1.1: Stages of the General Measurement System

Stage 1: Sensor-Transducer	Stage 2: Signal Conditioning	Stage 3: Readout-Recording
Senses desired input to exclusion of all others and provides analogous output	Modifies transduced signal into form usable by final stage. Usually increases amplitude and/or power, depending on requirement. May also selectively filter unwanted components or convert signal into pulsed form	Provides an indication or recording in form that can be evaluated by an unaided human sense or by a controller. Records data digitally on a computer
Types and Examples	*Types and Examples*	*Types and Examples*
Mechanical: Contacting spindle, spring-mass, elastic devices (e.g., Bourdon tube for pressure, proving ring for force), gyro	*Mechanical*: Gearing, cranks, slides, connecting links, cams, etc.	*Indicators (displacement type)*: Moving pointer and scale, moving scale and index, light beam and scale, electron beam and scale (oscilloscope), liquid column
Hydraulic-pneumatic: Buoyant float, orifice, venturi, vane, propeller	*Hydraulic-pneumatic*: Piping, valving, dashpots, plenum chambers	*Indicators (digital type)*: Direct alphanumeric readout
Optical: Photographic film, photoelectric diodes and transistors, photomultiplier tubes, holographic plates	*Optical*: Mirrors, lenses, optical filters, optical fibers, spatial filters (pinhole, slit)	*Recorders*: Digital printing, inked pen and chart, direct photography, magnetic recording (hard disk)
Electrical: Contacts, resistance, capacitance, inductance, piezoelectric crystals and polymers, thermocouple, semiconductor junction	*Electrical*: Amplifying or attenuating systems, bridges, filters, telemetering systems, various special-purpose integrated-circuit devices	*Processors and computers*: Various types of computing systems, either special-purpose or general, used to feed readout/recording devices and/or controlling systems
		Controllers: All types

ferentiation, or telemetering, as may be required.

Probably the most common function of the second stage is to increase either amplitude or power of the signal, or both, to the level required to drive the final terminating device. In addition, the second stage must be designed for proper matching characteristics between the first and second and between the second and third stages.

1.4.3 Third, or Readout-Recording, Stage

The third stage provides the information sought in a form comprehensible to one of the human senses or to a controller. If the output is intended for immediate human recognition, it is, with rare exception, presented in one of the following forms:

1. As a *relative displacement*, such as movement of an indicating hand or displacement of oscilloscope trace

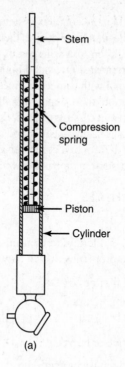

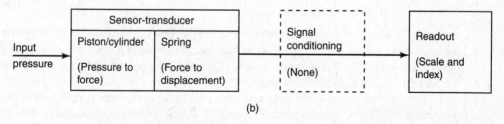

FIGURE 1.3: (a) Gage for measuring pressure in automobile tires. (b) Block diagram of tire-gage functions. In this example, the spring serves as a secondary transducer (see Section 6.3).

2. In *digital* form, as presented by a counter such as an automobile odometer, or by a liquid crystal display (LCD) or light-emitting diode (LED) display as on a digital voltmeter

To illustrate a very simple measuring system, let us consider the familiar tire gage used for checking automobile tire pressure. Such a device is shown in Fig. 1.3(a). It consists of a cylinder and piston, a spring resisting the piston movement, and a stem with scale divisions. As the air pressure bears against the piston, the resulting force compresses the spring until the spring and air forces balance. The calibrated stem, which remains in place after the spring returns the piston, indicates the applied pressure.

The piston-cylinder combination constitutes a force-summing apparatus, sensing and transducing pressure to force. As a secondary transducer (see Section 6.3), the spring converts the force to a displacement. Finally, the transduced input is transferred *without* signal conditioning to the scale and index for readout [see Fig. 1.3(b)].

As an example of a more complex system, let us say that a velocity is to be measured, as shown in Fig. 1.4. The *first-stage* device, the accelerometer, provides a voltage analogous to acceleration.[3] In addition to a voltage amplifier, the *second* stage may also include a filter that selectively attenuates unwanted high-frequency noise components. It may also integrate the analog signal with respect to time, thereby providing a velocity–time relation, rather than an acceleration–time signal. Finally, the signal voltage will probably need to be increased to the level necessary to be sensed by the *third*, or *recording and readout, stage*, which may consist of a data-acquisition computer (Section 8.8) and printer. The final record will then be in the form of a computer-generated graph; with the proper calibration, an accurate velocity-versus-time measurement should be the result.

1.5   TYPES OF INPUT QUANTITIES

1.5.1   Time Dependence

Mechanical quantities, in addition to their inherent defining characteristics, also have distinctive time-amplitude properties, which may be classified as follows:

1. Static—constant in time
2. Dynamic—varying in time

    (a) Steady-state periodic
    (b) Nonrepetitive or transient

        i. Single pulse or aperiodic
        ii. Continuing or random

Of course, the unchanging, static measurand is the most easily measured. If the system is terminated by some form of meter-type indicator, the meter's pointer has no difficulty in eventually reaching a definite indication. The rapidly changing, dynamic measurand presents the real measurement challenge.

Two general forms of dynamic input are possible: steady-state periodic input and transient input. The steady-state periodic quantity is one whose magnitude has a definite repeating time cycle, whereas the time variation of a transient quantity does not repeat. "Sixty-cycle" line voltage is an example of a steady-state periodic signal. So also are many mechanical vibrations, after a balance has been reached between a constant input exciting energy and energy dissipated by damping.

An example of a pulsed transient quantity is the acceleration–time relationship accompanying an isolated mechanical impact. The magnitude is temporary, being completed in a matter of milliseconds, with the portions of interest existing perhaps for only a few microseconds. The presence of extremely high rates of change, or wavefronts, can place severe demands on the measuring system. The nature of these inputs is discussed in detail in Chapter 4, and the response of the measuring system to such inputs is covered in Chapter 5.

1.5.2   Analog and Digital Signals

Most measurands of interest vary with time in a continuous manner over a range of magnitudes. For instance, the speed of an automobile, as it starts from rest, has some magnitude

---

[3]Although the accelerometer may be susceptible to an analysis of "stages" within itself, we shall forgo such an analysis in this example.

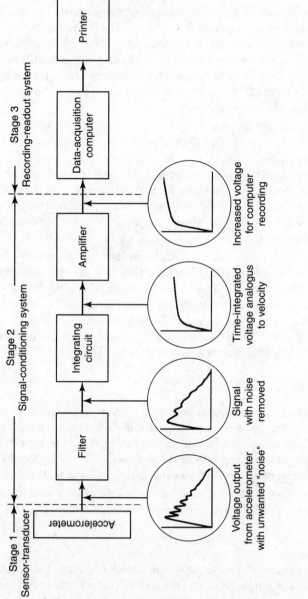

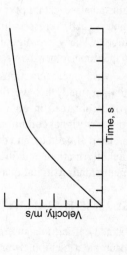

FIGURE 1.4: Block diagram of a relatively complex measuring system.

9

at every instant during its motion. A sensor that responds to velocity will produce an output signal having a time variation analogous to the time change in the auto's speed. We refer to such a signal as an *analog* signal because it is *analogous* to a continuous physical process. An analog signal has a value at every instant in time, and it usually varies smoothly in magnitude.

Some quantities, however, may change in a stepwise manner between two distinct magnitudes: a high and low voltage or on and off, for instance. The revolutions of a shaft could be counted with a cam-actuated electrical switch that is open or closed, depending on the position of the cam. If the switch controls current from a battery, current either flows with a given magnitude or does not flow. The current flow varies discretely between two values, which we could represent as single digits: 1 (flowing) and 0 (not flowing). The amplitude of such a signal may thus be called *digital*.

Many electronic circuits store numbers as sets of digits—strings of 1s and 0s—with each string held in a separate memory register. When digital circuits, such as those in computers, are used to record an analog signal, they do so only at discrete points in time because they have only a fixed number of memory registers. The analog signal, which has a value at every instant of time, becomes a *digital signal*. A digital signal is a set of discrete numbers, each corresponding to the value of the analog signal at a single specific instant of time. Clearly, the digital signal contains no information about the value of the analog signal at times other than sample times.

Mechanical quantities—such as temperatures, fluid-flow rates, pressure, stress, and strain—normally behave timewise in an analog manner. However, distinct advantages are often obtained in converting an analog signal to an equivalent digital signal for the purposes of signal conditioning and/or readout. Noise problems are reduced or sometimes eliminated altogether, and data transmission is simpler. Computers are designed to process digital information, and direct numerical display or recording is more easily accomplished by manipulating digital quantities. Digital techniques are discussed at length in Chapter 8.

1.6 MEASUREMENT STANDARDS

As stated earlier, measurement is a process of comparison. Therefore, regardless of our measurement method, we must employ a basis of comparison—*standardized units*. The standards must be precisely defined, and, because different systems of units exist, the method of conversion from system to system must be mutually agreed upon. Chapter 2, "Standards and Dimensional Units of Measurement," provides a detailed discussion of this subject.

Most importantly, a relationship between the standards and the readout scale of each measuring system must be established through a process known as *calibration*.

1.7 CALIBRATION

At some point during the preparation of a measuring system, *known* magnitudes of the input quantity must be fed into the sensor-transducer, and the system's output behavior must be observed. Such a comparison allows the magnitude of the output to be correctly interpreted in terms of the magnitude of the input. This *calibration* procedure establishes the correct output scale for the measuring system.

By performing such a test on an instrument, we both calibrate its scale and prove its ability to measure reliably. In this sense, we sometimes speak of *proving* an instrument. Of course, if the calibration is to be meaningful, the known input must itself be derived from a defined standard.

If the output is exactly proportional to the input (output = constant × input), then a single simultaneous observation of input and output will suffice to fix the constant of proportionality. This is called *single-point calibration*. More often, however, *multipoint calibrations* are used, wherein a number of different input values are applied. Multipoint calibration works when the output is not simply proportional, and, more generally, improves the accuracy of the calibration.

If a measuring system will be used to detect a time-varying input, then the calibration should ideally be made using a time-dependent input standard. Such *dynamic* calibration can be difficult, however, and a *static* calibration, using a constant input signal, is frequently substituted. Naturally, this procedure is not optimal; the more nearly the calibration standard corresponds to the measurand in all its characteristics, the better the resulting measurements.

Occasionally, the nature of the system or one of its components makes the introduction of a sample of the basic input quantity difficult or impossible. One of the important characteristics of the bonded resistance-type strain gage is the fact that, through quality control at the time of manufacture, *spot* calibration may be applied to a complete lot of gages. As a result, an indirect calibration of a strain-measuring system may be provided through the gage factor supplied by the manufacturer. Instead of attempting to apply a known unit strain to the gage installed on the test structure—which, if possible, would often result in an ambiguous situation—a resistance change is substituted. Through the predetermined gage factor, the system's strain response may thereby be obtained (see Section 12.4).

1.8   UNCERTAINTY: ACCURACY OF RESULTS

Error may be defined as the difference between the *measured* result and the *true* value of the quantity being measured (see Section 3.1). While we do know the measured value, we do not know the true value, and so we do not know the error either. If we estimate a likely upper bound on the magnitude of the error, that bound is called the *uncertainty*: We estimate, with some level of odds, that the error will be no larger than the uncertainty.

To estimate the size of errors, we must have some understanding of their causes and classifications. Errors can be of two basic types: *bias*, or *systematic, error* and *precision*, or *random, error*.

Should an unscrupulous butcher place a ball of putty under the scale pan, the scale readouts would be consistently in error. The scale would indicate a weight of product too great by the weight of the putty. This *zero offset* represents one type of systematic error.

Shrink rules are used to make patterns for the casting of metals. Cast steel shrinks in cooling by about 2%; hence the patterns used for preparing the molds are oversized by the proper percentage amounts. The pattern maker uses a shrink rule on which the dimensional units are increased by that amount. Should a pattern maker's shrink rule for cast steel be inadvertently used for ordinary length measurements, the readouts would be consistently undersized by $\frac{1}{50}$ in one (that is, by 2%). This is an example of *scale error*.

In each of the foregoing examples the errors are constant and of a systematic nature. Such errors *cannot* be uncovered by statistical analysis; however, they may be estimated by methods that we discuss in Chapter 3.

An inexpensive frequency counter may use the 60-Hz power-line frequency as a comparison standard. Power-line frequency is held very close to the 60-Hz standard. Although it does wander slowly above and below the average value, over a period of time—say, a day—the *average* is very close to 60 Hz. The wandering is random and the moment-to-moment error in the frequency meter readout (from this source) is called *precision*, or *random, error*.

Randomness may also be introduced by variations in the measurand itself. If a number of hardness readings is made on a given sample of steel, a range of readings will be obtained. An average hardness may be calculated and presented as the actual hardness. Single readings will deviate from the average, some higher and some lower. Of course, the primary reason for the variations in this case is the nonhomogeneity of the crystalline structure of the test specimen. The deviations will be random and are due to variations in the measurand. Random error may be estimated by statistical methods and is considered in more detail in Chapter 3.

1.9 REPORTING RESULTS

When experimental setups are made and time and effort are expended to obtain results, it normally follows that some form of written record or report is to be made. The purpose of such a record will determine its form. In fact, in some cases, several versions will be prepared. Reports may be categorized as follows:

1. Executive summary
2. Laboratory note or technical memo
3. Progress report
4. Full technical report
5. Technical paper

Very briefly, an executive summary is directed at a busy overseer who wants only the key features of the work: what was done and what was concluded, outlined in a few paragraphs. A laboratory note is written to be read by someone thoroughly familiar with the project, such as an immediate supervisor or the experimentalist himself or herself. A full report tells the complete story to one who is interested in the subject but who has not been in direct touch with the specific work—perhaps top officials of a large company or a review committee of a sponsoring agency. A progress report is just that—one of possibly several interim reports describing the current status of an ongoing project, which will eventually be incorporated in a full report. Ordinarily, a technical paper is a brief summary of a project, the extent of which must be tailored to fit either a time allotment at a meeting or space in a publication.

Several factors are common to all the various forms. With each type, the first priority is *to make sure* that the *problem or project* that has been tackled *is clearly stated*. There is nothing quite so frustrating as reading details in a technical report while never being certain of the raison d'être. It is extremely important to make certain that the reader is quickly clued in on the *why* before one attempts to explain the *how* and the results. A clearly stated objective can be considered the most important part of the report. The entire report should be written in simple language. A rule stated by Samuel Clemens is not inappropriate: "Omit unnecessary adverbs and adjectives."

1.9.1 Laboratory Note or Technical Memo

The laboratory note is written for a very limited audience, possibly even only as a memory jogger for the experimenter or, perhaps more often, for the information of an immediate supervisor who is thoroughly familiar with the work. In some cases, a single page may be sufficient, including a sentence or two stating the problem, a block diagram of the experimental setup, and some data presented either in tabular form or as a plotted diagram. Any pertinent observations not directly evident from the data should also be included. Sufficient information should be included so that the experimenter can mentally reconstruct the situation and results 1 year or even 5 years hence. A date and signature should always be included and, if there is a possibility of important developments stemming from the work, such as a patent, a second witnessing signature should be included and dated.

1.9.2   Full Report

The full report must relate all the facts pertinent to the project. It is even more important in this case to make the purpose of the project completely clear, for the report will be read by persons not closely associated with the work. The full report should also include enough detail to allow another professional to repeat the measurements and calculations.

One format that has much merit is to make the report proper—the main body—short and to the point, relegating to appendices the supporting materials, such as data, detailed descriptions of equipment, review of literature, sample calculations, and so on. Frequent reference to these materials can be made throughout the report proper, but the option to peruse the details is left to the reader. This scheme also provides a good basis for the technical paper, should it be planned.

1.9.3   Technical Paper

A primary purpose of a technical paper is to make known (to advertise) the work of the writer. For this reason, two particularly important portions of the writing are the *problem statement* and the *results*. Adequately done, these two items will attract the attention of other workers interested in the particular field, who can then make direct contact with the writer(s) for additional details and discussion.

Space, number of words, limits on illustrations, and perhaps time are all factors making the preparation of a technical paper particularly challenging. Once the problem statement and the primary results have been adequately established, the remaining available space may be used to summarize procedures, test setups, and the like.

1.10   FINAL REMARKS

An attempt has been made in this chapter to provide an overall preview of the problems of mechanical measurement. In conformance with Section 1.9, we have tried to state the problem as fully as possible in only a few pages. In the remainder of the book, we will expand on the topics introduced in this chapter. Figure 1.5 illustrates the interrelation of these topics and their organization within this book.

PROBLEMS

1.1.   Write an *executive summary* of this chapter.

1.2.   Consider a mercury-in-glass thermometer as a temperature-measuring system. Discuss the various stages of this measuring system in detail.

1.3.   For the thermometer of Problem 1.2, specify how practical single point calibration may be obtained.

1.4.   Set up test procedures you would use to estimate, with the aid only of your present judgment and experience, the magnitudes of the common quantities listed.

(**a**)   Distance between the centerlines of two holes in a machined part

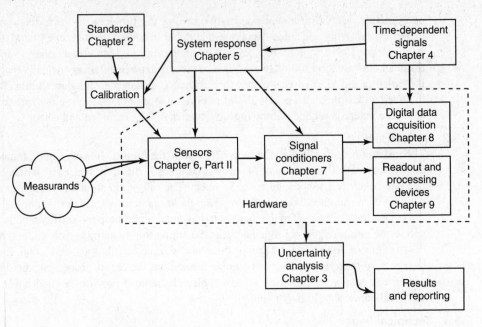

FIGURE 1.5: Conceptual organization of this book.

(b) Weight of two small objects of different densities

(c) Time intervals

(d) Temperature of water

(e) Frequency of pure tones

1.5. Consider the impact frame shown in Figure 1.6. Mass M, which travels along guide rails, is raised to an initial height H and released from rest. Discuss how you would measure the mass velocity just prior to impact with the test item in order to account for friction between mass M and the guide rails.

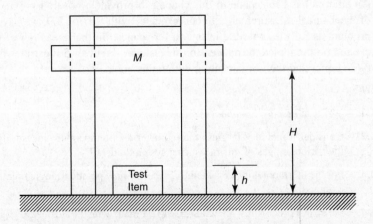

FIGURE 1.6: Impact test frame for Problem 1.5.

CHAPTER 2

Standards and Dimensional Units of Measurement

2.1   INTRODUCTION

The basis of measurement was outlined in Section 1.3: It is the comparison between a measurand and a suitable *standard*. In this chapter, we will take a closer look at the establishment of standards.

The term *dimension* connotes the defining characteristics of an entity (measurand), and the *dimensional unit* is the basis for quantification of the entity. For example, *length* is a dimension, whereas *centimeter* is a unit of length; *time* is a dimension, and the *second* is a unit of time. A dimension is unique; however, a particular dimension—say, length—may be measured in various units, such as feet, meters, inches, or miles. Systems of units must be established and agreed to; that is, the systems must be standardized. Because there are various systems, there must also be agreement on the basis for conversion from system to system. It is clear, then, that standards of measurement apply to units, to systems of units, and to unitary conversion between such systems.

In general terms, standards are ubiquitous. There are standards governing food preparation, marketing, professional behavior, and so on. Many are established and governed by either federal or state laws. So that we may avoid chaos, it is especially important that the basic measurement standards carry the authority of not only federal, but also international, laws.

In the following sections, we will discuss those standards, systems of units, and problems of conversion that are fundamental to mechanical measurement.

2.2   HISTORICAL BACKGROUND OF MEASUREMENT IN THE UNITED STATES

The legal authority to control measurement standards in the United States was assigned by the U.S. Constitution. Quoting from Article 1, Section 8, Paragraph 5, of the U.S.

Constitution: "The congress shall have power to ... fix the standard of weights and measures." Although Congress was given the power, considerable time elapsed before anything was done about it. In 1832, the Treasury Department introduced a uniform system of weights and measures to assist the customs service; in 1836, these standards were approved by Congress [1]. In 1866, the Revised Statutes of the United States, Section 3569, added the stipulation that "It shall be lawful throughout the United States of America to employ the weights and measures of the metric system." This simply makes it clear that the metric system *may* be used. In addition, this act established the following (and now obsolete) relation for conversion:

$$1 \text{ meter} = 39.37 \text{ inches}$$

An international convention help in Paris in 1875 resulted in an agreement signed for the United States by the U.S. ambassador to France. The following is quoted therefrom: "The high contracting parties engage to establish and maintain, at their expense, a scientific and permanent international bureau of weights and measures, the location of which shall be Paris" [2,3]. Although this established a central bureau of standards, which set up at Sèvres, a suburb of Paris, it did not, of course, bind the United States to make use of or adopt such standards.

On April 5, 1893, in all absence of further congressional action, Superintendent Mendenhall of the Coast and Geodetic Survey issued the following order [2,4]:

> The Office of Weights and Measures with the approval of the Secretary of the Treasury, will in the future regard the international prototype meter and the kilogram as *fundamental standards*, and the customary units, the yard and pound, will be derived therefrom in accordance with the Act of July 28, 1866.

The Mendenhall Order turned out to be a very important action. First, it recognized the meter and the kilogram as being fundamental units on which all other units of length and mass should be based. Second, it tied together the metric and English systems of length and mass in a definite relationship, thereby making possible international exchange on an exact basis.

In response to requests from scientific and industrial sources, and to a great degree influenced by the establishment of like institutions in Great Britain and Germany,[1] Congress on March 3, 1901 passed an act providing that "The Office of Standard Weights and Measures shall hereafter be known as 'The National Bureau of Standards'" [5]. Expanded functions of the new bureau were set forth and included development of standards, research basic to standards, and the calibration of standards and devices. The National Bureau of Standards (NBS) was formally established in July 1910, and its functions were considerably expanded by an amendment passed in 1950. In 1988, Congress changed the name of the bureau to "The National Institute of Standards and Technology" (NIST) [6].

Commercial standards are largely regulated by state laws; to maintain uniformity, regular meetings (National Conferences on Weights and Measures) are held by officials of NIST and officers of state governments. Essentially all state standards of weights and measures are in accordance with the Conference's standards and codes. International uniformity is maintained through regularly scheduled meetings (held at about 6-year intervals),

---

[1]The National Physical Laboratory, Teddington, Middlesex and Physikalisch-Technische Reichsanstalt, Braunschweig.

called the *General Conference on Weights and Measures* and attended by representatives from most of the industrial countries of the world. In addition, numerous interim meetings are held to consider solutions to more specific problems, for later action by the General Conference.

2.3   THE SI SYSTEM

2.3.1   Establishment of the SI System

The International System of Units, or SI System, has its origins in the Decimal Metric System that was introduced at the time of the French Revolution. During the next two centuries, metric systems of measurement continued to evolve, and they came to encompass both mechanical and electrical dimensions. Finally, in 1960, the Eleventh General Conference on Weights and Measures formally established the SI System, consisting of dimensional standards for length, mass, time, electric current, thermodynamic temperature, and luminous intensity. The Fourteenth General Conference on Weights and Measures (1971) added the mole as the unit for amount of substance, completing the seven dimensional system in use today [7].

The seven *base units* of the SI System are listed in Table 2.1. Other dimensions can be derived from these base units by multiplying or dividing them. A few such *derived units* are assigned special names; others are not. For example, the unit of force, the *newton*, is obtained from the kilogram, the meter, and the second as

$$1 \text{ newton} = 1 \text{ kg/m} \cdot \text{s}^2$$

In contrast, area is simply meters squared (m^2). Work and energy are expressed in *joules* ($\text{kg} \cdot \text{m}^2/\text{s}$). The term *hertz* is used for frequency (s^{-1}) and the term *pascal* is used for pressure (N/m^2). Some derived units carrying special names are listed in Table 2.2, and some without special names are given in Table 2.3. Note that, whereas those assigned special names that originate from proper names are not capitalized, the corresponding abbreviations are capitalized.

It should be clear that all *the various derived units can be expressed in terms of base units.* In certain instances when a unit balance is attempted for a given equation, it may be desirable, or necessary, to convert all variables to base units.

TABLE 2.1: Base Units in the SI System

Quantity	Unit	
	Name	Symbol
length	meter	m
mass	kilogram	kg
time	second	s
electric current	ampere	A
temperature	kelvin	K
amount of substance	mole	mol
luminous intensity	candela	cd

TABLE 2.2: SI-Derived Units with Special Names and Symbols

Quantity	Unit	Symbol	Expressed in Other Units
plane angle	radian	rad	$m \cdot m^{-1}$
solid angle	steradian	sr	$m^2 \cdot m^{-2}$
frequency	hertz	Hz	s^{-1}
force	newton	N	$kg \cdot m/s^2$
pressure, stress	pascal	Pa	N/m^2
energy	joule	J	$N \cdot m$
power	watt	W	J/s
electric charge	coulomb	C	$A \cdot s$
electric potential difference	volt	V	W/A
electric capacitance	farad	F	C/V
electric resistance	ohm	Ω	V/A
magnetic flux	weber	Wb	$V \cdot s$
magnetic flux density	tesla	T	Wb/m^2
inductance	henry	H	Wb/A

To accommodate the writing of very large or very small values, the SI System defines the multiplying prefixes shown in Table 2.4. For example, 2,500,000 Hz may be written as 2.5 MHz (megahertz), and 0.000 000 000 005 farad as 5 pF (picofarad). Only one prefix should be used with a given dimension; thus, it would be incorrect to write 2.5 kkHz in place of 2.5 MHz. Likewise, for units of mass, 1000 kg might be written as 1 Mg (megagram).

In the following sections, we discuss the SI standards of length, mass, time, and current in greater detail. The standards of luminous intensity and amount of substance are described in reference [7].

TABLE 2.3: Some SI-Derived Units

Derived Quantity	Symbol
area	m^2
acceleration	m/s^2
angular acceleration	rad/s^2
angular velocity	rad/s
density	kg/m^3
dynamic viscosity	$Pa \cdot s$
heat flux	W/m^2
moment of force	$N \cdot m$
specific heat capacity	$J/kg \cdot K$
velocity	m/s
volume	m^3

TABLE 2.4: Multiplying Factors

Multiple	Prefix	Symbol	Multiple	Prefix	Symbol
10^1	deka	da	10^{-1}	deci	d
10^2	hecto	h	10^{-2}	centi	c
10^3	kilo	k	10^{-3}	milli	m
10^6	mega	M	10^{-6}	micro	μ
10^9	giga	G	10^{-9}	nano	n
10^{12}	tera	T	10^{-12}	pico	p
10^{15}	peta	P	10^{-15}	femto	f
10^{18}	exa	E	10^{-18}	atto	a
10^{21}	zetta	Z	10^{-21}	zepto	z
10^{24}	yotta	Y	10^{-24}	yocto	y

2.3.2   Metric Conversion in the United States

In May 1965, the United States announced its intention of adopting the SI system. In 1968, the passage of Public Law 90-472 authorized the Secretary of Commerce to make a "U.S. Metric Study" to be reported by August 1971. After prolonged debates, studies, and public pronouncements of 10-year conversion plans, on December 23, 1975, the 94th Congress approved Public Law 94-168, called the Metric Conversion Act of 1975. Its stated purpose was as follows: "To declare a national policy of coordinating the increasing use of the metric system in the United States, and to establish a United States Metric Board to coordinate the voluntary conversion to the metric system." Note especially that the conversion was to be *voluntary* and that no time limit was set. The Act made clear that, in using the term *metric*, the SI System of units was intended.

In 1981, the U.S. Metric Board reported to Congress that it lacked the clear Congressional mandate needed to effectively bring about national conversion to the metric system; funding for the Board was eliminated after fiscal year 1982 [8].

Congress subsequently amended the Metric Conversion Act with the *Omnibus Trade and Competitiveness Act of 1988*, Public Law 100-418 [6]. These amendments provided strong incentives for industrial conversion to SI units. The amended act declares that the metric system is "the preferred system of weights and measures for United States trade and commerce." It further requires that all federal agencies use the metric system in procurement, grants, and business-related activities; this requirement was to be met by the end of fiscal year 1992, except in cases where conversion would harm international competitiveness.

Metrication in the United States has progressed, especially in the automotive industry and certain parts of the food and drink industries. Classroom use has increased to the point that most engineering courses rely primarily on SI units. Throughout this book, we shall use both the SI system and the English Engineering system,[2] with the hope of encouraging the complete conversion to SI units.[3]

---

[2]This term may appear to be incongruous given that the United Kingdom has adopted the SI System. However, this usage is so well established that the term has outlived its origins.

[3]Attention is directed to reference [9], which is an excellent guide for applying the metric system.

2.4 THE STANDARD OF LENGTH

The meter was originally intended to be one ten-millionth of the earth's quadrant. In 1889, the First General Conference on Weights and Measures defined the meter as the length of the International Prototype Meter, the distance between two finely scribed lines on a platinum-iridium bar when subject to certain specified conditions. On October 14, 1960, the Eleventh General Conference on Weights and Measures adopted a new definition of the meter as 1,650,763.73 wavelengths in vacuum of the radiation corresponding to the transition between the levels $2p_{10}$ and $5d_5$ of the krypton-86 atom. The National Bureau of Standards of the United States also adopted this standard, and the inch became 41,929.398 54 wavelengths of the krypton light.

As it turned out, the wavelength of krypton light could only be determined to about 4 parts per billion, limiting the accuracy of the meter to a similar level. During the 1960s and early 1970s, laser-based measurements of frequency and wavelength evolved to such accuracy that the uncertainty in the meter became the limiting uncertainty in determining the speed of light [10,11]. This limitation was of serious concern in both atomic and cosmological physics, and on October 20, 1983, the Seventeenth General Conference on Weights and Measures redefined the meter directly in terms of the speed of light:

> The meter is the length of the path travelled by light in vacuum during a time interval of 1/299,792,458 of a second.

This definition has the profound effect of *defining* the speed of light to be 299,792,458 m/s, which had been the accepted experimental value since 1975 [12].

2.4.1 Relationship of the Meter to the Inch

The 1866 U.S. Statute had specified that 1 m = 39.37 inches, resulting in the relationship

$$1 \text{ in.} = 2.540\,005\,08 \text{ cm}   \text{(approximately)}$$

In 1959, the National Bureau of Standards made a small adjustment to this relationship to ensure international agreement on the definition of the inch [13]:

$$1 \text{ in.} = 2.54 \text{ cm}   \text{(exactly)}$$

This simpler relationship had already been used as an approximation by engineers for years. The difference between these two standards may be written as

$$2.54\,005\,08/2.54 - 1 = 0.000\,002$$

or 0.0002%, which is about $\frac{1}{8}$ in. per mile.

We gain a sense of the significance of the difference by considering the following situation. In 1959, the work of the United States National Geodetic Survey was based on the 39.37 in./m relationship and a coordinate system with its origin located in Kansas. Changing the relationship from 39.37 in./m (exactly) to 2.54 in./cm (exactly) would have caused discrepancies of almost 16 ft at a distance of 1500 miles. One can only imagine the confusion over property lines if such a change had been made! This problem was resolved by defining separately the *U.S. survey foot* (12/39.37 m) and the *international foot* (12 × 2.54 cm). The survey foot is still used with U.S. geodetic data and U.S. statute miles [14].

2.5   THE STANDARD OF MASS

The *kilogram* is defined as the mass of the International Prototype Kilogram, a platinum-iridium weight kept at the International Bureau of Weights and Measures near Paris. Of the basic standards, this remains the only one established by a prototype (by which is meant *the* original model or pattern, *the* unique example, to which all others are referred for comparison). Various National Prototype Kilogram masses have been calibrated by comparison to the International Prototype Kilogram. These masses are in turn used by the standards agencies of various countries to calibrate other standard masses, and so on, until one reaches masses or weights of day-to-day goods and services.

Apart from the inconvenience of maintaining this chain of calibration, the definition of the kilogram by an international prototype leads to several very fundamental problems: The prototype can be damaged or destroyed; the mass of the prototype fluctuates by about one part in 10^8 owing to gas absorption and cleaning; and the prototype ages in an unknown manner, perhaps having resulted in 50 μg of variation during the past century [15].

In recent years, considerable effort has been given to developing a new mass standard that can be reproduced in any suitably equipped lab, without the use of a prototype. One approach being considered is to precisely determine Avogadro's number by mass and density measurements of silicon crystals. This value of Avogadro's number could then be used with an atomic unit of mass to define the kilogram as the mass of a specific number of atoms [15,16]. An alternative approach, which promises somewhat better accuracy, uses a "moving coil watt balance" to compare the mechanical and electrical power exerted on a current-carrying conductor that moves against gravity in a magnetic field. This technique leads to a definition of the kilogram in terms of fundamental physical quantities [17,18].

The pound was defined in terms of the kilogram by the Mendenhall Order of 1893. In 1959, the definition was slightly adjusted [13], giving the relationship still in use today:

$$1 \text{ pound avoirdupois} = 0.453\,592\,37 \text{ kilogram}$$

2.6   TIME AND FREQUENCY STANDARDS

Until 1956, the second was defined as 1/86,400 of the average period of revolution of the earth on its axis. Although this seems to be a relatively simple and straightforward definition, problems remained. There is a gradual slowing of the earth's rotation (about 0.001 second per century) [19], and, in addition, the rotation is irregular.

Therefore, in 1956, an improved standard was agreed on; the second was defined as 1/31,556,925.9747 of the time required by the earth to orbit the sun in the year 1900. This is called the *ephemeris second*. Although the unit is defined with a high degree of exactness, implementation of the definition was dependent on astronomical observation, which was incapable of realizing the implied precision.

In the 1950s, atomic research led to the observation that the frequency of electro-magnetic radiation associated with certain atomic transitions may be measured with great repeatability. One—the hyperfine transition of the cesium atom—was related to the ephemeris second with an estimated accuracy of two parts in 109. On October 13, 1967, in Paris, the Thirteenth General Conference on Weights and Measures officially adopted the following definition of the second as the unit of time in the SI System [7]:

> The second is the duration of 9,192,631,770 periods of the radiation corresponding to the transition between the two hyperfine levels of the ground state of the cesium 133 atom.

Atomic apparatuses, commonly called "atomic clocks," are used to produce the frequency of the transition [20]. In a *fountain clock* [21], a gas of cesium atoms is introduced into a vacuum chamber, where a set of laser beams is used to slow the molecular motion, pushing a group of atoms into a ball and cooling them to a temperature near absolute zero. Another laser is then used to toss the ball of atoms upward into a microwave cavity, where some of the atoms are excited to higher energy levels. When the ball falls again, yet another laser is used to force the emission of radiation. This radiation is detected, yielding the desired frequency. The best cesium standards reproduce the second to an accuracy better than one part in 10^{15}.

2.7   TEMPERATURE STANDARDS

The basic unit of temperature, the kelvin (K), is defined as the fraction 1/273.16 of the thermodynamic temperature of the triple point of water, the temperature at which the solid, liquid, and vapor phases of water coexist in equilibrium. The *degree Celsius* (°C) is defined by the relationship

$$t = T - 273.15$$

where t and T represent temperatures in degrees Celsius and in kelvins, respectively.

In reality, two temperature scales are defined, a *thermodynamic* scale and a *practical* scale. The latter is the usual basis for measurement. The thermodynamic temperature scale is defined in terms of entropy and the properties of heat engines [22]. It can be implemented directly only with specialized thermometers that use media having a precisely known equation of state (a constant volume, ideal gas thermometer, for example). Such thermometers are difficult and time consuming to use if accuracy is desired, and, as a result, a corresponding scale which is more easily realizable is needed [7]. Thus, the thermodynamic scale is normally approximated using a so-called practical scale.

A practical scale has two components. The first is a set of fixed reference temperatures, defined by specific states of matter. The second is a procedure for interpolating between those reference points, for example, by measuring a temperature-dependent electrical resistance. Using the interpolation formulae and fixed points of the practical scale, one can calibrate any other temperature measuring device.

The fixed reference temperatures must correspond to thermodynamic states that are very accurately reproducible. Zero degrees Celsius is the temperature of equilibrium between pure ice and air-saturated pure water at normal atmospheric pressure. However, a more precise datum, independent of both ambient pressure and possible contaminants, is the triple point temperature of water. As noted above, the value 273.16 K (or 0.0100°C) is assigned to this temperature. Relatively simple apparatus can be used to reproduce this temperature fixed point [23].

In 1927, the national laboratories of the United States, Great Britain, and Germany proposed a practical temperature standard that became known as the *International Temperature Scale (ITS-27)*. This standard, adopted by 31 nations, conformed as closely as possible to the thermodynamic scale that had been proposed by Lord Kelvin in 1854. It was based on six fixed-temperature points dependent on physical properties of certain materials, including the ice and steam points of water. Several revisions have since been made, notably in 1948, 1968, and 1990. The practical temperature scale currently in effect is the International Temperature Scale of 1990 (ITS-90), adopted by the International Committee on Weights and Measures and authorized by the Eighteenth General Conference [24].

The ITS-90 defines a number of fixed reference temperatures points, some of which are shown in Fig. 2.1 and Table 2.5. Between these fixed points, elaborate interpolation equations are specified by ITS-90 for use with the various interpolation standards. From 0.65 K to 5.0 K, the standard is based on measurement of the vapor pressure of helium and

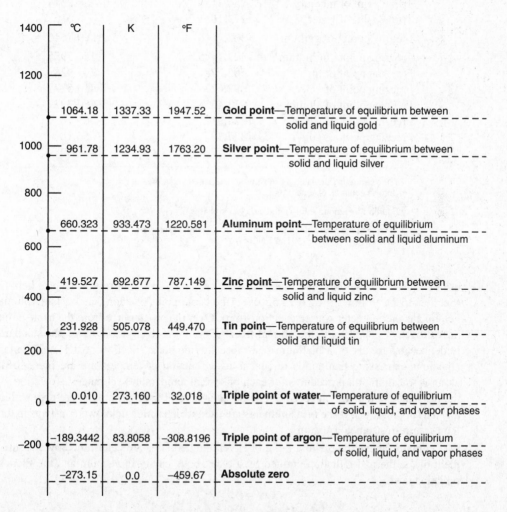

°C	K	°F	
1064.18	1337.33	1947.52	**Gold point**—Temperature of equilibrium between solid and liquid gold
961.78	1234.93	1763.20	**Silver point**—Temperature of equilibrium between solid and liquid silver
660.323	933.473	1220.581	**Aluminum point**—Temperature of equilibrium between solid and liquid aluminum
419.527	692.677	787.149	**Zinc point**—Temperature of equilibrium between solid and liquid zinc
231.928	505.078	449.470	**Tin point**—Temperature of equilibrium between solid and liquid tin
0.010	273.160	32.018	**Triple point of water**—Temperature of equilibrium of solid, liquid, and vapor phases
−189.3442	83.8058	−308.8196	**Triple point of argon**—Temperature of equilibrium of solid, liquid, and vapor phases
−273.15	0.0	−459.67	**Absolute zero**

FIGURE 2.1: Some of the fixed-point temperatures established by the ITS-90.

TABLE 2.5: Defining Fixed Points of the ITS-90

Equilibrium State	Assigned Values of Temperature	
	K	**°C**
Vapor pressure of helium*	3 to 5	−270.15 to −268.15
Triple point[†] of hydrogen	13.8033	−259.3467
Vapor pressure of hydrogen	≈ 17	≈ −256.15
Vapor pressure of hydrogen	≈ 20.3	≈ −252.85
Triple point of neon	24.5561	−248.5939
Triple point of oxygen	54.3584	−218.7916
Triple point of argon	83.8058	−189.3442
Triple point of mercury	234.3156	−38.8344
Triple point of water	273.16	0.01
Melting point[‡] of gallium	302.9146	29.7646
Freezing point[‡] of indium	429.7485	156.5985
Freezing point of tin	505.078	231.928
Freezing point of zinc	692.677	419.527
Freezing point of aluminum	933.473	660.323
Freezing point of silver	1234.93	961.78
Freezing point of gold	1337.33	1064.18
Freezing point of copper	1357.77	1084.62

*Temperature is calculated by substituting measured vapor pressure into an equation of state.

[†]Equilibrium among solid, liquid, and vapor phases.

[‡]Melting and freezing point temperatures correspond to standard atmospheric pressure ($101,325$ N/m^2).

use of equations describing the vapor-pressure-versus-temperature relationship of helium (Section 16.3). From 3.0 K to 24.5561 K (the triple point of neon), a constant-volume helium gas thermometer is used (Section 16.3). Over the broad range from the triple point of hydrogen (13.8033 K) to the normal freezing point of silver (1234.93 K), the standard is defined by means of a platinum resistance thermometer (Section 16.4.1). Complex equations expressing platinum's resistance as a function of temperature are prescribed, along with calibration procedures for each of several temperature subranges. To calibrate, the resistance of a platinum sensor is measured at several fixed-point temperatures within a given subrange, and these measurements are used to determine unknown constants in the temperature-resistance equations.

Finally, above the melting point of silver, temperatures are determined by measurement of the thermal radiation emitted by a black-body cavity in vacuum and the Planck radiation law:

$$\frac{E_\lambda(T)}{E_\lambda(T_{\text{ref}})} = \frac{\exp(C_2/\lambda T_{\text{ref}}) - 1}{\exp(C_2/\lambda T) - 1}$$

where

$E_\lambda(T), E_\lambda(T_{\text{ref}})$ the radiant energy emitted by the black body per unit time, per unit area, and per unit wavelength at a wavelength λ, and at a temperature T or T_{ref}, respectively

T_{ref} The freezing point temperature of either silver (1234.93 K), gold (1337.33 K), or copper (1357.77 K)

C_2 0.014388 m $\cdot$ K

The radiant energy is typically measured by optical pyrometry (Section 16.8). The unknown temperature is then calculated by comparing the emission of a source at the unknown temperature to that from a source at the reference temperature.

The International Temperature Scale of 1990 thus establishes means of determining any temperature from 0.65 K to more than 4000 K. In actual applications, the standardized pyrometer, the standardized resistance thermometer, or the standardized gas thermometers are used as secondary standards for calibration of working instruments (Section 16.12). Apart from any uncertainties introduced in the calibration procedures, the major uncertainties in ITS-90 arise in realizing the fixed points. At a 1σ level (Sections 3.5–3.6), the uncertainties in the fixed-point temperatures are ±0.5 to 1.5 mK for temperatures up to the melting point of gallium, increasing to ±60 mK at the freezing point of copper [25].

The temperature units of the English Engineering System are defined in terms of the kelvin. Absolute temperature takes units of *degrees Rankine* (°R), which differ from the kelvin by a factor of 1.8:

$$T(°\text{R}) = 1.8 \times T(\text{K})$$

The *degree Fahrenheit* (°F) is defined by subtracting 459.67 from the temperature in degrees Rankine:

$$T(°\text{F}) = T(°\text{R}) - 459.67$$

2.8   ELECTRICAL STANDARDS

In the SI System, all electrical units originate from the definition of the ampere. One ampere is defined as the current that produces a magnetic force of 2×10^{-7} N/m on a pair of thin parallel wires carrying that current and separated by one meter. The force on an appropriate pair of conductors can be measured directly, using a so-called *current balance* [26]. The current may be then calculated from the relations of electromagnetic theory. The remaining electrical units, such as volts and ohms, can all be derived from the value of the ampere and the mechanical units of mass, length, and time, again using the results of electromagnetic theory.

The measurement of current from the SI definition of the ampere is cumbersome, just as is the measurement of temperature from the thermodynamic temperature scale. Obtaining the volt and the ohm from the SI definition is also difficult. Consequently, practical standards are normally used in place of the SI definitions in order to obtain the volt and the ohm.

Traditionally, the practical realization of the volt was a so-called *standard cell*, an electrochemical cell of relatively high stability. National standards laboratories maintained such cells as their voltage standards. By the early 1970s, however, the superconducting Josephson junction effect, discovered in 1962, had displaced the standard cell. Josephson junctions create voltages that are repeatable to about four parts in 10^{10}, whereas the voltages of standard cells had shown greater drift [27,28]. The ultimate accuracy of the Josephson junction voltage is limited primarily by the accuracy of the time standard.

The traditional practical standard for the ohm was the *standard resistor*, typically a specially alloyed wire held in an oil bath to stabilize its temperature. These standard resistors have an accuracy of a few parts in 10^7, but were subject to aging and to resistor-to-resistor variations. The quantum Hall effect, discovered in 1980, quickly became an alternative resistance standard. The quantized Hall resistance allows the ohm to be determined directly from fundamental physical constants to a repeatability of about 2 parts in 10^9 [28].

The Eighteenth General Conference on Weights and Measures declared that, from 1990 onward, the Josephson junction and the quantum Hall effect would be the practical standards for the volt and the ohm, respectively. The Conference also standardized the values of the physical constants that characterize the Josephson junction and the quantum Hall effect, so that the same volt and ohm would be obtained in any lab using these devices. The resulting realizations of the volt and the ohm are believed to agree with the formal SI definitions to within a few parts in 10^7 [7].

2.9 CONVERSIONS BETWEEN SYSTEMS OF UNITS

Over the centuries, various systems of units have evolved. Five systems are listed in Table 2.6. To be acceptable, each system of units must be compatible with the physical laws of the universe. If compatible with the laws of nature, the values expressed in one system must be convertible to equally legitimate values in any of the other systems.

Systems of units differ in their use of defined and derived units. This becomes most apparent in the English Engineering System's treatment of weight and mass. To see how this issue arises, let us consider Newton's second law:

> A particle acted upon by an external force will be accelerated in proportion to the force magnitude and in inverse proportion to the mass of the particle; the direction of the acceleration will coincide with the line-of-action of the force.

Algebraically,

$$F = ma \tag{2.1}$$

where

$$F = \text{the magnitude of the applied force,}$$
$$m = \text{the mass of particle,}^4 \text{ and}$$
$$a = \text{the resulting acceleration}$$

---

[4]*Caution:* Particular note should be made of the use throughout this text of the symbols "m" and "w." The symbol *m* is used to represent the magnitude of the dimension mass, and it carries units of *kilogram* (kg) or *pound-mass* (lbm). Weight, *w*, which is a force, carries the units *pounds-force* (lbf) or *newtons* (N). Note should also be made of the use of the symbol "m" to denote the unit of length, meter. Context should always make clear the intent.

TABLE 2.6: Systems of Units. Four dimensions are considered for each system, with derived units underscored. The English Engineering System assigns all four dimensions, resulting in the need for a dimensional constant not equal to unity.

	System				
Quantity	SI (mass, length, time)	CGS (mass, length, time)	English Engineering (force, mass, length, time)	Absolute English (mass, length, time)	Technical English (force, length, time)
Length	meter (m)	centimeter (cm)	foot (ft)	foot (ft)	foot (ft)
Time	second (s)	second (s)	second (s)	second (s)	second (s)
Mass	kilogram (kg)	gram (g)	pound-mass (lbm)	pound-mass (lbm)	slug
Force	newton (N)	dyne	pound-force (lbf)	poundal	pound-force (lbf)
Dimensional constant, g_c	$1 \, \dfrac{\text{kg} \cdot \text{m}}{\text{N} \cdot \text{s}^2}$	$1 \, \dfrac{\text{g} \cdot \text{cm}}{\text{dyne} \cdot \text{s}^2}$	$32.17 \, \dfrac{\text{lbm} \cdot \text{ft}}{\text{lbf} \cdot \text{s}^2}$	$1 \, \dfrac{\text{lbm} \cdot \text{ft}}{\text{poundal} \cdot \text{s}^2}$	$1 \, \dfrac{\text{slug} \cdot \text{ft}}{\text{lbf} \cdot \text{s}^2}$

From experiment [29], we know that near the earth's surface a body acted on solely by gravitational attraction accelerates at a rate of about 32.2 ft/s^2 (9.81 m/s^2).[5] In this situation, the acting force is *weight*, w, which may be expressed in pounds-force (lbf), newtons, dynes, and so on, depending on the particular system of units that is used, and the magnitude of mass may be expressed variously as slugs, pounds-mass (lbm), kilograms, and so on. In any case, whichever system is used, a consistent, compatible balance of units must be maintained. Newton's inertial law is of particular interest in this regard because it demands a careful distinction between the units of force and mass. In the United States, it has long been the habit to use the abbreviation lb as the unit for both mass and force, except when a distinction is absolutely required; then the abbreviations lbm and lbf are used.

This distinction shows how the English Engineering System differs from the SI System in its treatment of the unit of force. The SI System assigns the units of kilograms, meters, and seconds to the dimensions mass, length, and time, respectively. The unit of force, the newton, is a derived unit, equal to 1 kg · m/s^2. In contrast, the English Engineering System assigns the units pounds-mass, feet, seconds, *and* pounds-force, so that force has an assigned, rather than derived, unit.

When the assigned units are applied to Eq. (2.1), we must introduce a factor called the *dimensional constant* g_c, which converts units of mass, length, and time to units of force. Equation (2.1) is modified as follows:

$$F = ma/g_c \quad \text{or} \quad g_c \equiv ma/F \tag{2.2}$$

If we select the English System as an example and assume that 1 lbf acts on 1 lbm, which we know results in an acceleration of 32.2 ft/s^2, then we find that

$$g_c = (1 \text{ lbm})(32.2 \text{ ft/s}^2)/(1 \text{ lbf})$$
$$= 32.2 \text{ (lbm} \cdot \text{ft/lbf} \cdot \text{s}^2)$$

For the SI System, force is a derived unit: 1 N is defined as the force required to accelerate 1 kg at 1 m^2/s. Hence,

$$g_c = (1 \text{ kg})(1 \text{ m/s}^2)/(1 \text{ N})$$
$$= 1 \text{ (kg} \cdot \text{m/N} \cdot \text{s}^2) = 1$$

As a result, the factor g_c has the value unity in the SI System.

The other systems in Table 2.6 use derived units of either force or mass, with the result that g_c is also unity in those systems. Such systems are called "consistent." In the Technical English System, for example, the unit of mass is the slug, which is defined as the mass that a force of 1 lbf accelerates at 1 ft/s^2. In the CGS System (centimeter-gram-second system), the derived unit is that for force, the dyne, which is equal to 1 g · cm/s^2. In the Absolute English System, the derived unit is again that for force, the poundal, equal to 1 lbm · ft/s^2.

In the past, physicists have been partial to the CGS System, whereas engineers have mainly used the English Engineering System and the Technical English System. Throughout this book, we shall use both the SI and the English Engineering Systems. All systems other than SI, however, are regarded as obsolete.

---

[5]The standard gravitational body force ("acceleration due to gravity") is taken as 9.80665 m^2/s (exactly) or 32.174 ft/s^2. Of course, the actual value depends on the specific locality.

EXAMPLE 2.1

Determine the conversion factor between pounds-force and newtons.

Solution   The conversion factors between inches and meters and between pounds-mass and kilograms were given in Sections 2.4 and 2.5.

$$1 \text{ lbf} = 32.174 \text{ lbm} \cdot \text{ft/s}^2$$
$$= 32.174(0.453\,592\,37 \text{ kg})(12 \times 0.0254 \text{ m})/\text{s}^2$$
$$= 4.4482 \text{ kg} \cdot \text{m/s}^2$$
$$= 4.4482 \text{ N}$$

Many more conversion factors are listed in Appendix A.

EXAMPLE 2.2

Water of density ρ and dynamic viscosity μ flows with velocity V through a pipe of diameter D. Calculate the Reynolds number, Re, from the data supplied, using (a) the English Engineering System of units and (b) the SI System. Before making the numerical calculations, check the balance of units.

Referring to Section 15.2, we see that Re $= \rho V D / \mu$ and, as discussed in that section, its value is unitless and hence is independent of the system of units used: thus, we should obtain the same numerical answers for both parts (a) and (b).

Data (see Appendix A for conversion factors):

$$D = 8.00 \text{ in.} = 8/12 \text{ ft} = 0.203 \text{ m}$$
$$\rho = 62.3 \text{ lbm/ft}^3 = 998 \text{ kg/m}^3 \quad \text{(see Appendix D)}$$
$$V = 4.00 \text{ ft/s} = 1.22 \text{ m/s}$$
$$\mu = 2.02 \times 10^{-5} \text{ lbf} \cdot \text{s/ft}^2$$
$$= 9.67 \times 10^{-4} \text{ N} \cdot \text{s/m}^2 \quad \text{(see Appendix D)}$$

Solution

1. If we enter the units for each of the separate quantities appearing in the equation for Re, we have

$$(\text{lbm/ft}^3)(\text{ft/s})(\text{ft})(\text{ft}^2/\text{lbf} \cdot \text{s})(1/g_c)$$

or, entering the units for g_c,

$$(\text{lbm/ft}^3)(\text{ft/s})(\text{ft})(\text{ft}^2/\text{lbf} \cdot \text{s})(\text{lbf} \cdot \text{s}^2/\text{lbm} \cdot \text{ft})$$

We see that the various units cancel, confirming the statement that the Reynolds number is unitless.

In magnitude,

$$\text{Re} = (62.3)(4.00)(8/12)/(2.02 \times 10^{-5})(32.2) \approx 255{,}000$$

2. In terms of SI units, we have

$$(\text{kg/m}^3)(\text{m/s})(\text{m})(\text{m}^2/\text{N} \cdot \text{s})(1/g_c)$$

or, when the units for g_c are entered,

$$(\text{kg/m}^3)(\text{m/s})(\text{m})(\text{m}^2/\text{N} \cdot \text{s})(\text{N} \cdot \text{s}^2/\text{kg} \cdot \text{m})$$

Again, we see that the units cancel. The same result would be obtained by omitting g_c entirely and replacing the newton by its the definition (this is the usual practice in the SI System).

In magnitude, using SI units,

$$\text{Re} = (998)(1.22)(0.203)/(9.67 \times 10^{-4})(1) \approx 256,000$$

Note: The lack of exact numerical agreement in the final numbers results from inexact conversions and rounding of the numbers.

2.10 SUMMARY

All measurements are based on defined standards, most of which have been established by international agreement and U.S. laws. A measurement consists of quantifying the dimensional magnitude of an unknown relative to that established by the standard.

1. The SI System of units employs seven base units (meter, kilogram, second, ampere, kelvin, mole, and candela), a number of derived units, and various multiplying factors (Section 2.3).
2. The standard of length is the meter, defined as the distance travelled by light in 1/299,792,458 of a second (Section 2.4).
3. The standard of mass is the kilogram, defined by the International Prototype Kilogram (Section 2.5).
4. The standard of time is the second, defined as 9,192,631,770 periods of hyperfine-transition radiation from a cesium atom (Section 2.6).
5. The operational (or "practical") temperature standard is the International Temperature Scale of 1990, defined with respect to the thermodynamic temperatures of specific states of matter (Section 2.7).
6. SI electrical units are derived from the ampere. The ampere is defined by the mechanical force present in a particular type of electrical circuit. Practical standards are normally used to obtain the volt and the ohm (Section 2.8).
7. Although SI is the preferred system of measurement, other systems are still commonly used. The most important of these is the English Engineering System (Section 2.9). Conversion factors for changing English Engineering units to SI units are listed in Appendix A.

SUGGESTED READINGS

Many of the following publications may be downloaded from Web pages of the National Institute of Standards and Technology.

Butcher, T., L. Crown, R. Suiter, and J. Williams (eds.). General tables of units of measurement, Appendix C of *NIST Handbook 44: Specifications, Tolerances, and Other Technical Requirements for Weighing and Measuring Devices*. Gaithersburg, Md.: National Institute of Standards and Technology, 2005.

Judson, L. V. *Weights and Measures Standards of the United States: A Brief History,* Gaithersburg, Md.: National Bureau of Standards, Special Publication 447, March 1976.

The United States and the Metric System: A Capsule History. Gaithersburg, Md.: Office of Weights and Measures/Metric Program, National Institute of Standards and Technology, Letter Circular 1136, October 1997.

Preston-Thomas, H. The International Temperature Scale of 1990 (ITS-90). *Metrologia*, 27:3–10 and 107, 1990.

Taylor, B. N. *Guide for the Use of the International System of Units (SI).* Gaithersburg, Md.: National Institute of Standards and Technology, Special Publication 811, 1995.

Taylor, B. N. (ed.) *The International System of Units.* Gaithersburg, Md.: National Institute of Standards and Technology, Special Publication 330, 2001.

PROBLEMS

2.1. Determine the speed of light in vacuum in

(a) miles/hour
(b) feet/s

2.2. Convert the following temperatures to equivalent temperatures in K:

(a) 100°C
(b) 100°F
(c) 595°R

2.3. Determine the following temperatures in °R:

(a) Freezing point of tin
(b) Freezing point of aluminum
(c) Freezing point of copper

2.4. Calculate the force (lbf) necessary to accelerate a weight of 0.5 lbf at 5 ft/s^2.

2.5. Determine the force (N) necessary to accelerate a mass of 200 g at 25 cm/s^2. What is the magnitude of this force in units of lbf?

2.6. Prepare a list of the best secondary standards that are available to you at present as calibration sources for

(a) Length
(b) Time
(c) Mass
(d) Temperature
(e) Pressure

2.7. Determine by calculation the relationship for the following conversions:

(**a**) Pressure in lbf/in.^{2} units to N/m^2 units

(**b**) Viscosity in lbf · s/ft^2 units to kg/m · s units

(**c**) Specific heat in kJ/kg · K units to Btu/lbm · °F

(**d**) Dynamic viscosity in poise (1 dyne · s/cm^2) units to lbm/h · ft

(**e**) Heat flux in W/cm^2 units to Btu/h · ft^2

2.8. Determine the factor for converting volume flow rate in cm^3/s units to gal/min.

2.9. Express the universal gas constant of 1545 ft · lbf/lbm · mol · °R in SI units.

REFERENCES

[1] Judson, L. V. *Weights and Measures Standards of the United States: A Brief History,* Gaithersburg, Md.: National Bureau of Standards, Special Publication 447, March 1976.

[2] *Units of Weights and Measure.* Gaithersburg, Md.: National Bureau of Standards, Misc. Publication 214, July 1955.

[3] Terrien, J. Scientific metrology on the international plane and the Bureau International des Poids et Mesures. *Metrologia* 1(2):15, January 1965.

[4] *U.S. Coast and Geodetic Survey*, Bull. 26, April 1893.

[5] Cochrane, R. D. *Measures for Progress, A History of the National Bureau of Standards.* Washington, D.C.: U.S. Dept. of Commerce, 1966, p. 47.

[6] United States Congress, *Omnibus Trade and Competitiveness Act of 1988*, Public Law 100-418, Section 5164.

[7] Taylor, B. N. (ed.) *The International System of Units.* Gaithersburg, Md.: National Institute of Standards and Technology, Special Publication 330, 2001.

[8] *The United States and the Metric System: A Capsule History.* Gaithersburg, Md.: Office of Weights and Measures/Metric Program, National Institute of Standards and Technology, Letter Circular 1136, October 1997.

[9] Taylor, B. N. *Guide for the Use of the International System of Units (SI).* Gaithersburg, Md.: National Institute of Standards and Technology, Special Publication 811, 1995.

[10] Terrien, J. International agreement on the velocity of light. *Metrologia*, 10:9, 1974.

[11] Svenson, K. M., et al. Speed of light from direct frequency and wavelength measurements of the methane-stabilized laser. *Phys. Rev. Letters*, 29(19):1346–49, 1972.

[12] Documents concerning the new definition of the metre. *Metrologia*, 19:163–177, 1984.

[13] Refinement of values for the yard and the pound, *Federal Register*, Document 59–5442, June 25, 1959.

[14] *Units and Systems of Weights and Measures: Their Origin, Development, and Present Status.* Gaithersburg, Md.: National Bureau of Standards, Letter Circular 1035, November 1985.

[15] Seyfried, P., and P. Becker. The role of N_A in the SI: an atomic path to the kilogram. *Metrologia*, 31:167–172, 1994.

[16] Quinn, T. J. Conclusions of the International Workshop on the Avogadro Constant and the Representation of the Silicon Mole. *Metrologia*, 31:275–276, 1994.

[17] Taylor, B. N. Determining the Avogadro constant from electrical measurements. *Metrologia*, 31:181–194, 1994.

[18] Taylor, B. N., and P. J. Mohr. On the redefinition of the kilogram. *Metrologia*, 36:64–65, 1999.

[19] Clemence, G. M. Time and its measurement. *Am. Scientist*, 40(2):260, April 1952.

[20] Sullivan, D. B., et al. Primary atomic frequency standards at NIST. *J. Res. Natl. Inst. Stand. Technol.*, 106(1):47–63, 2001.

[21] http://tf.nist.gov/cesium/fountain.htm. This National Institute of Standards and Technology Web page provides clear description of the NIST-F1 fountain clock, including animation.

[22] Bejan, A. *Advanced Engineering Thermodynamics*. New York: John Wiley, 1988.

[23] Mangum, B. W., and G. T. Furukawa. *Guidelines for realizing the International Temperature Scale of 1990 (ITS-90)*. Gaithersburg, Md.: National Institute of Standards and Technology, Technical Note 1965, August 1990.

[24] Preston-Thomas, H. The International Temperature Scale of 1990 (ITS-90). *Metrologia*, 27:3–10, 1990 (with corrections in *Metrologia*, 27:107, 1990).

[25] Rusby, R. L., et al. Thermodynamic basis of the ITS-90. *Metrologia*, 28:9–18, 1991.

[26] Driscoll, R. L., and R. D. Cutkoskey. Measurement of current with the National Bureau of Standards current balance. *Natl. Bur. Stand. J. Res.*, 60, April 1958.

[27] Taylor, B. N. New measurements standards for 1990. *Phys. Today*, 23–26, August 1989.

[28] Petley, B. W. Electrical units—the last thirty years. *Metrologia*, 32:495–502, 1994/95.

[29] Cook, A. H. The absolute determination of the acceleration due to gravity. *Metrologia*, 1(3):84, 1965.

CHAPTER 3

Assessing and Presenting Experimental Data

3.1 INTRODUCTION

"How good are the data?" is the first question put to any experimentalist who draws a conclusion from a set of measurements. The data may become the foundation of a new theory or the undoing of an existing one. They may form a critical test of a structural member in an aircraft wing that must never fail during operation. Before a data set can be used in an engineering or scientific application, its quality must be established.

The answer to the question revolves around the meaning we assign to the word *good*. Our first temptation may be to call the data "good" if they agree well with a theoretically derived result. Theory, however, is simply a model intended to mimic the behavior of the real system being studied; there is no guarantee that it actually does represent the physical system well. The accuracy of even the most fundamental theory, such as Newton's laws, is limited both by the accuracy of the data from which the theory was developed and by the accuracy of the data and assumptions used when calculating with it. Thus, measurements should not be compared to a theory in order to assess their quality. What we are really after is the *actual value* of the physical quantity being measured, and that is the standard against which data should be tested. The *error* of a measurement is thus defined as the difference between the measured value and the true physical value of the quantity. The original question could be more clearly phrased as, "What is the error of the data?"

The definition of *error* is helpful, but it suffers from one major flaw: The error cannot be calculated exactly unless we know the true value of the quantity being measured! Obviously, we can never know the true value of a physical quantity without first measuring it, and, because some error is present in every measurement, the true value is something we can never know exactly. Hence, we can never know the error exactly, either.

The definition of *error* is not as circular as it seems, however, because we can usually estimate the likelihood that the error exceeds some specific value. For example, 95% of the readings from one particular flowmeter will have an error of less than 1 L/s. Thus, we can say with 95% confidence (19 times out of 20) that a reading taken from that meter has an error of 1 L/s or less, or, equivalently, that the reading has an *uncertainty* of 1 L/s at confidence level of 95%. A theoretical result that disagrees with the reading by more than 1 L/s shows a measurable inaccuracy; a theory within 1 L/s is supported by the reading at that level of confidence.

Error or uncertainty may be estimated with statistical tools when a large number of measurements are taken. However, the experimentalist must also bring to bear his or her own knowledge of how the instruments perform and of how well they are calibrated in order to establish the possible errors and their probable magnitudes. This chapter describes how to estimate the uncertainty in a measurement and how to present the corresponding experimental data in an easily interpreted way.

3.2   COMMON TYPES OF ERROR

We have defined the error in measuring a quantity x as the difference between the measured value, x_m, and the true value, x_{true}:

$$\text{Error} = \varepsilon \equiv x_m - x_{\text{true}} \qquad (3.1)$$

A primary objective in designing and executing an experiment is to minimize the error. However, after the experiment is completed, we must turn our attention to estimating a bound on ε with some level of confidence. This *bound* is typically of the form

$$-u \le \varepsilon \le +u \quad (n:1) \qquad (3.2)$$

where u is the *uncertainty* estimated at odds of $n:1$. In other words, only one measurement in n will have an error whose magnitude is greater than u. This bound is equivalent to saying that

$$x_m - u \le x_{\text{true}} \le x_m + u \quad (n:1) \qquad (3.3)$$

We would, of course, give higher odds that the true value would lie within a wider interval and lower odds that it would lie within a narrower interval.

The first step in bounding a measurement's error is to identify its possible causes. The specific causes of error will vary from experiment to experiment, and even a single experiment may include a dozen sources of inaccuracy. But, in spite of this diversity, most errors can be placed into one of two general classes [1]: *bias errors* and *precision errors*.

Bias errors, also referred to as *systematic errors*, are those that occur the same way each time a measurement is made. For example, if the scale on an instrument consistently reads 5% high, then the entire set of measurements will be biased by +5% above the true value. Alternatively, the scale may have a fixed offset error, so that the indicated value for every reading of x is higher than the true value by an amount x_{offset}.

Precision errors, also called *random errors*, are different for each successive measurement but have an average value of zero. For example, mechanical friction or vibration may cause the reading of a measuring mechanism to fluctuate about the true value, sometimes reading high and sometimes reading low. This lack of mechanical precision will cause sequential readings of the same quantity to differ slightly, creating a distribution of values surrounding the true value.

If enough readings are taken, the distribution of precision errors will become apparent. The successive readings will generally cluster about a central value and will extend over a limited interval surrounding that central value. In this situation, we may use statistical analysis to estimate the likely size of the error or, equivalently, the likely range of x in which the true value lies.

In contrast, bias errors cannot be treated using statistical techniques, because such errors are fixed and do not show a distribution. However, bias error can be estimated by comparison of the instrument to a more accurate standard, from our knowledge of how the instrument was calibrated, or from our experience with instruments of that particular type.

In practice, bias and precision errors occur simultaneously. The combined effect on repeated measurements of x is shown in Figs. 3.1(a) and (b). In Fig. 3.1(a), the bias error is larger than the typical precision error. In Fig. 3.1(b), the typical precision error exceeds the bias error. In other situations, bias and precision errors may be of the same size. The *total error* in a particular measured x_m is the sum of the bias and precision errors for that measurement.

3.2.1 Classification of Errors

A full classification of all possible errors as either bias or precision error would be convenient but is nearly impossible to make, since categories of error overlap and are at times ambiguous. Some errors behave as bias error in one situation and as precision error in other situations; some errors do not fit neatly into either category. However, for purposes of discussion, typical errors may be roughly sorted as follows:

1. Bias or systematic error

    (a) Calibration errors
    (b) Certain consistently recurring human errors
    (c) Certain errors caused by defective equipment
    (d) Loading errors
    (e) Limitations of system resolution

2. Precision or random error

    (a) Errors caused by disturbances to the equipment
    (b) Errors caused by fluctuating experimental conditions
    (c) Errors derived from insufficient measuring-system sensitivity

3. Illegitimate error

    (a) Blunders and mistakes during an experiment
    (b) Computational errors after an experiment

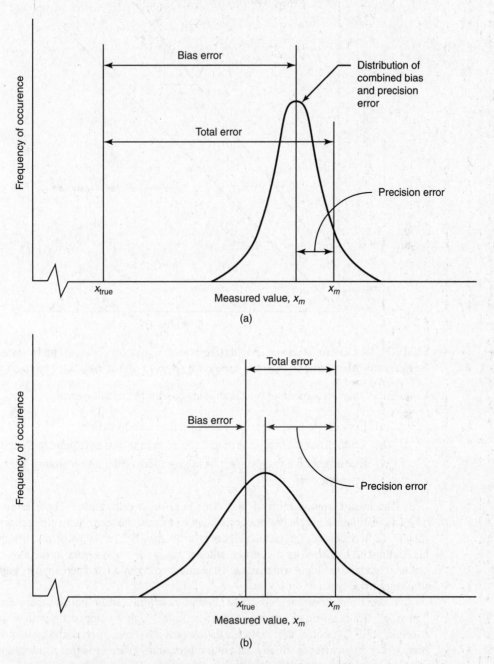

FIGURE 3.1: Bias and precision errors: (a) bias error larger than the typical precision error, (b) typical precision error larger than the bias error.

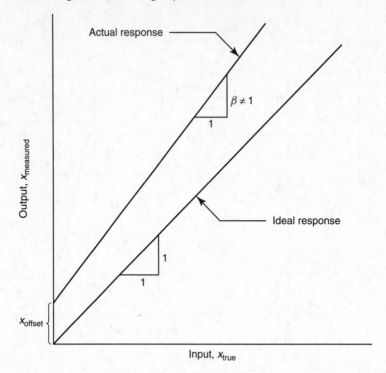

FIGURE 3.2: Calibration errors. For ideal response, $x_{\text{measured}} = x_{\text{true}}$. Actual response may include zero-offset error (x_{offset}) and scale error ($\beta \neq 1$) so that $x_{\text{measured}} = \beta x_{\text{true}} + x_{\text{offset}}$.

4. Errors that are sometimes bias error and sometimes precision error

    (a) Errors from instrument backlash, friction, and hysteresis

    (b) Errors from calibration drift and variation in test or environmental conditions

    (c) Errors resulting from variations in procedure or definition among experimenters

The most common form of bias error is error in calibration. These errors occur when an instrument's scale has not been adjusted to read the measured value properly. As mentioned in Section 1.8, typical calibration errors may be *zero-offset* errors, which cause all readings to be offset by a constant amount x_{offset}, or *scale errors* in the slope of the output relative to the input which cause all readings to err by a fixed percentage. Figure 3.2 illustrates these types of error.

Calibration procedures normally attempt to identify and eliminate these errors by "proving" the measuring system's readout scales through a comparison with a standard (Section 1.7). Of course, the standards themselves also have uncertainties, albeit smaller ones.[1] The impreciseness of any calibration procedure guarantees that some calibration-related bias error is present in all measuring systems.

---

[1] The uniqueness of certain primary standards makes them exceptions to this statement. In particular, the mass standard (the International Prototype Kilogram) has by *definition* a mass of exactly 1 kg. In the sense of practical applications, however, uncertainty will nevertheless occur: Even primary standards require the use of ancillary apparatus, which necessarily introduces some uncertainty.

Human errors may well be systematic, as when an individual experimenter consistently tends to read high or to "jump the gun" when synchronized readings are to be taken.

The equipment itself may introduce built-in errors resulting from incorrect design, fabrication, or maintenance. Such errors result from defective mechanical or electrical components, incorrect scale graduations, and so forth. Errors of this type are often consistent in sign and magnitude, and because of their consistency they may sometimes be corrected by calibration. When the input is time varying, however, introducing a correction is more complicated. For example, distortion caused by poor frequency response cannot be corrected by the usual "static calibration," one based on a signal that is constant in time (see Section 5.20). Such frequency-response errors arise in connection with seismic motion detectors, as discussed in Chapter 17.

Loading error is of particular importance. It refers to the influence of the measurement procedure on the system being tested. *The measuring process inevitably alters the characteristics of both the source of the measured quantity and the measuring system itself; thus, the measured value will always differ by some amount from the quantity whose measurement is sought.* For example, the sound-pressure level sensed by a microphone is not the same as the sound-pressure level that would exist at that location if the microphone were not present. Minimizing the influence of the measuring instrument on a measured variable is a major objective in designing any experiment.

Precision errors are also of several typical forms. The experimenter may be inconsistent in estimating successive readings from his or her instruments. Precision errors in the instrumentation itself may arise from outside disturbances to the measuring system, such as temperature variations or mechanical vibrations. The measuring system may also include poorly controlled processes that lead to random variations in the system output.

Variations in the actual quantity being measured may also appear as precision error in the results. Sometimes these variations are a result of poor experimental design, as when a system designed to run at a constant speed instead has a varying speed. Sometimes the variations are an inherent feature of the process under study, as when manufacturing variations create a distribution in the operating lifetimes of a group of light bulbs. Strictly speaking, variation in the measured quantity is not a measurement "error"; however, it is possible to apply the same statistical techniques to variations in the measured variable and to treat them as if they were errors. In particular, if you wish to find the mean value of the measured quantity, its variations may be averaged (together with the precision errors of the equipment), and the mean value may be calculated along with its uncertainty.

Illegitimate errors are errors that would not be expected. These include outright mistakes (which can be eliminated through exercise of care or repetition of the measurement), such as incorrectly writing down a number, failing to turn on an instrument, or miscalculating during data reduction. Sometimes a statistical analysis will reveal such data as being extremely unlikely to have arisen from precision error.

Backlash and *mechanical friction* are important sources of variation in measuring systems. For example, friction may cause a mechanical element, such as a galvanometer needle, to lag behind advances in its intended position, thus reading low while the measured variable is increased and reading high while the measured variable is decreased. Such *hysteresis* is illustrated in Fig. 3.3. Since this error depends on how a sequence of measurements is taken, it may behave as either a bias error or a precision error. One means of detecting—and often correcting—this type of error is to make measurements while first

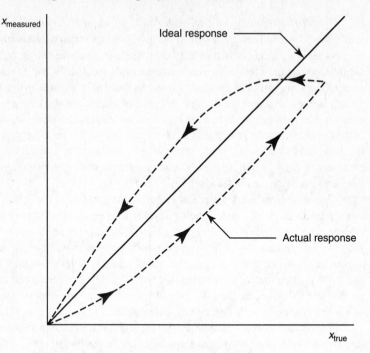

FIGURE 3.3: Hysteresis error.

increasing and then decreasing the measured quantity, an approach sometimes called the *method of symmetry*.

Drift of an instrument's calibration may occur if the response of an instrument varies in time. Often drift results from the sensitivity of an instrument to temperature or humidity fluctuations. If changes in environmental conditions occur between the time an instrument is calibrated and the time it is used, a bias error may appear in the readings. Conversely, if the test duration is long, environmental conditions may fluctuate throughout the test, causing different calibration errors for each successive measurement. In this case, the fluctuations create a precision error.

When an experiment is repeated using different equipment or by different experimenters, the bias errors of successive experiments are unrelated. If enough different experiments are performed, the bias errors are effectively *randomized*, and they become another form of precision error in the set of all experiments. For example, the speed of light in vacuum has been measured by many experimenters, each using different techniques and apparatus to obtain what is supposedly a unique quantity. Each experiment included its own bias and precision errors; but taken together their results show a distribution about a mean value, which may be estimated statistically (Fig. 3.4).

In contrast, the randomness of precision error may sometimes work against itself. For example, computer signal-processing techniques can extract desired information from a very noisy signal, as when photographs from satellites are enhanced to reveal planetary topography. In such cases, the systematic nature of the desired information enables it to be separated from the completely random overlying noise.

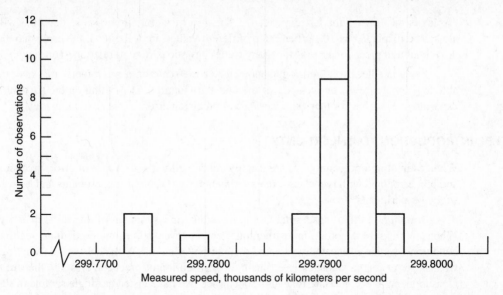

FIGURE 3.4: Measured values of the speed of light, 1947–1967 (Data from Froome and Essen [2]).

3.2.2   Terms Used in Rating Instrument Performance

The following terms are often employed to describe the quality of an instrument's readings. They are related to the expected errors of the instrument.

- *Accuracy.* The difference between the measured and true values. Typically, a manufacturer will specify a *maximum* error as the accuracy; manufacturers often neglect to report the odds that an error will not exceed this maximum value.

- *Precision.* The difference between the instrument's reported values during repeated measurements of the same quantity. Typically, this value is determined by statistical analysis of repeated measurements.

- *Resolution.* The smallest increment of change in the measured value that can be determined from the instrument's readout scale. The resolution is often on the same order as the precision; sometimes it is smaller.

- *Sensitivity.* The change of an instrument or transducer's output per unit change in the measured quantity. A *more sensitive* instrument's reading changes significantly in response to smaller changes in the measured quantity. Typically, an instrument with higher sensitivity will also have finer resolution, better precision, and higher accuracy.

   Reading error refers to error introduced when reading a number from the display scale of an instrument. This type of error may sometimes be a bias error caused by truncation or rounding of the actual value to one within the resolution of the display. Reading error will also include error from inadequate instrument sensitivity if the instrument does not respond to the smallest fluctuations of the measured quantity. For example, a digital display may truncate an actual value of 10.4 to a displayed value of 10. The reading error of the digital display is thus $\pm\frac{1}{2}$ of the last digit read.[2] This is a bias error in the sense that 10.4 will always be displayed as 10. When many different values are to be read, the error may be thought of as a precision error if the many values have no particular relation to one another.

For a needle display on a galvanometer or the scale on a micrometer one may be able to estimate to $\pm\frac{1}{2}$ or even $\pm\frac{1}{5}$ of the finest graduation. Depending on the particular experimenter, such error may be either bias or precision error, as discussed previously.

3.3   INTRODUCTION TO UNCERTAINTY

When estimating uncertainty, we are usually concerned with two types of error, precision and bias error, and with two classes of experiments, *single-sample* experiments and *repeat-sample* experiments.

A *sample*, in this sense, refers to an individual measurement of a specific quantity. When we measure the strain in a structural member several times under identical loading conditions, we have repeatedly sampled that particular strain. With such repeat sampling, we can, for example, statistically estimate the distribution of precision errors in the strain measurement. If we measure that strain only once, we have instead a single sample of the strain, and our result does not reveal the distribution of precision error. In that case, we must resort to other means for estimating the precision error in our result.

Much of the remainder of this chapter is devoted to methods of estimating bias and precision error. Procedures for statistical analysis of precision error in repeat-sampled data are described in Sections 3.4–3.7. The estimation of bias uncertainty is covered in Section 3.9. The estimation of precision uncertainty for single-sample experiments is also considered in Section 3.9. Section 3.10 describes how uncertainty in measured variables leads to uncertainty in results calculated from those variables. Examples of uncertainty analysis are given in Section 3.11.

After determining the individual bias and precision uncertain in a measurement of x, we must combine them to obtain the total uncertainty in our result for x. If the bias uncertainty is B_x and the precision uncertainty is P_x,[3] then the two may be combined in a root-mean-square sense as

$$U_x = \left(B_x^2 + P_x^2 \right)^{1/2} \tag{3.4}$$

to yield the total uncertainty, U_x.

The justification for combining the two uncertainty estimates this way is largely empirical [1]. However, the underlying assumption is that B_x and P_x are associated with independent sources of error, so that the errors are unlikely to have their maximum values simultaneously. When B_x and P_x are each estimates for 95% confidence, then U_x is also a 95% confidence estimate; under the same conditions, it turns out [1] that simply adding B_x and P_x algebraically yields an uncertainty that roughly covers 99% of the data.

---

[2]It may be shown that 95% of the readings (19 out of 20) will differ from the value displayed by less than $\pm 0.5(19/20) = \pm 0.475 \approx \pm 0.5$. The uncertainty is ± 0.5 at 19 : 1 odds.

[3]The precision uncertainty is evaluated in Section 3.6. In terms defined later it has the value $P_x = t S_x / \sqrt{n}$, where t is the t-statistic, S_x is the sample standard deviation, and n is the sample size.

EXAMPLE 3.1

A brass rod is repeatedly loaded to a fixed tensile load and the axial strain in the rod is determined using a strain gage. Thirty results are obtained under fixed test conditions, yielding an average strain of $\varepsilon = 520$ μ-strain (520 ppm). Statistical analysis of the distribution of measurements gives a precision uncertainty of $P_\varepsilon = 21$ μ-strain at a 95% confidence level. The bias uncertainty is estimated to be $B_\varepsilon = 29$ μ-strain with odds of 19 : 1 (95% confidence). What is the total uncertainty of the strain?

Solution   The total uncertainty for 95% coverage is

$$U_\varepsilon = \left(B_\varepsilon^2 + P_\varepsilon^2 \right)^{1/2} = 36 \ \mu\text{-strain} \quad (95\%)$$

In other words, at a confidence level of 95% the true strain lies in the interval 520 ± 36 μ-strain:

$$484 \ \mu\text{-strain} \le \varepsilon \le 556 \ \mu\text{-strain}$$

3.4 ESTIMATION OF PRECISION UNCERTAINTY

Two fundamental concepts form the basis for analyzing precision errors. The first is that of a *distribution* of error. The distribution characterizes the probability that an error of a given size will occur. The second concept is that of a *population* from which *samples* are drawn. Usually, we have only a limited set of observations, our sample, from which to infer the characteristics of the larger population.

Statistical analysis of error usually assumes a model for the distribution of errors in a population, generally the *Gaussian*, or *normal*, distribution. Using this assumed distribution, we may estimate the probable difference between, say, the average value of a small sample and the true mean value of the larger population. This probable difference, or *confidence interval*, provides an estimate of the precision uncertainty associated with our measured sample.

This section and the next four examine some basic probability distributions, the characteristics of a population that satisfies a Gaussian distribution, and the accuracy with which the statistics of samples represent an underlying population. The t-distribution is introduced for treating small samples, and the χ^2-distribution is introduced for other statistical purposes. These sections should provide you with sufficient background to estimate the precision uncertainty in elementary engineering experiments.

3.4.1 Sample versus Population

Manufacturing variations in a production lot of marbles will create a distribution of diameters. To estimate the mean diameter, we may take a handful of marbles, measure them, and average the result (Fig. 3.5). The handful is our *sample*, drawn from the production lot, which is our *population*.

No two samples from the same population will yield precisely the same average value; however, each should approximate the average of the population to some level of

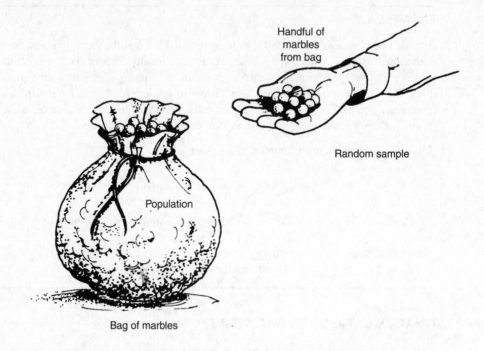

Handful of
marbles
from bag

Random sample

Population

Bag of marbles

FIGURE 3.5: Sample taken from a population.

uncertainty. The difference between the sample characteristics and those of the population will decrease as the sample is made larger.

Because our handful of marbles includes a number of members of the population, it may be regarded as a *repeated* sample. By contrast, if we had drawn only one marble, we would obtain a *single* sample, which would give no direct evidence of the distribution of marble diameters.

Experimental errors can also be viewed in terms of population and sample. If we measure the diameter of a single marble repeatedly, the set of measured diameters gives a sample of the precision error in measuring the diameter. In this case, we could measure the diameter as many times as we liked, and each measurement would include a slightly different precision error. Thus, the population of precision errors is theoretically infinite. (Note, however, that this particular repeat sampling of the precision error is performed on a single sample from the marble population.)

The discussion of this and subsequent sections applies to both of the following two classes of sampling:

1. A sample of size n is drawn from a finite population of size p. The sample is used to estimate properties of the population. Additional data cannot be added to the population; for example, we assume that no more marbles can be added to the particular production lot from the same source. Further, the sample size is assumed to be small compared to the population size: $n \ll p$.

2. A *finite* number of items, n, is randomly drawn from what is assumed to be a population of indefinite size. The properties of the *assumed* population are inferred from the sample.

An important qualification underlies this discussion: The sample must be *randomly* selected from the population. If we select only the largest marbles from the bag, our sample will not accurately represent the whole population of marbles.

3.4.2 Probability Distributions

Probability is an expression of the likelihood of a particular event taking place, measured with reference to *all* possible events. Specifically, suppose that one of n equally likely cases will occur and that m of these cases correspond to an event A. The probability that event A will occur is m/n.

A penny is tossed. The total number of possible outcomes is two—heads and tails. If we choose heads (or tails) as event A, then the probability of A is 1 in 2, or 50%.

A slightly more complex example is that of throwing a pair of dice. One possible outcome yields a sum of 2, six outcomes yield a sum of 7, and the remaining outcomes are as distributed in Fig. 3.6. The bar chart used here is termed a *histogram*. If we divide the ordinate of the chart by the total number of possible outcomes (36), we obtain a graph of the *probability distribution*. For example, the probability of rolling a 7 is 6 in 36, or 16.6%.

Other distributions that will be considered in the following sections are these:

1. *The Gaussian, or normal, probability distribution.* When examining experimental data, this distribution is undoubtedly the first that is considered. The Gaussian distribution describes the population of possible errors in a measurement when many independent sources of error contribute simultaneously to the total precision error in a measurement. These sources of error must be unrelated, random, and of roughly the

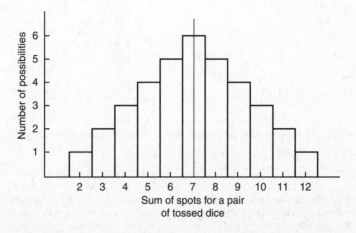

FIGURE 3.6: Distribution of results for a pair of thrown dice.

same size.[4] Although we will emphasize this particular distribution, you must keep in mind that data do not always abide by the normal distribution. For tabulation and calculation, the Gaussian distribution is recast in a standard form, sometimes called the *z-distribution* [see Eq. (3.11) and Table 3.2].

2. *Student's t-distribution.* This distribution is used in predicting the mean value of a Gaussian population when only a small sample of data is available [see Eq. (3.21) and Table 3.5].

3. The χ^2-*distribution.* This distribution helps in predicting the width or scatter of a population's distribution, in comparing the uniformity of samples, and in checking the *goodness of fit* for assumed distributions (see Section 3.7 and Table 3.6).

3.5   THEORY BASED ON THE POPULATION

From a practical standpoint, we are limited to samples from which to extract statistical information. In most cases, it is either impractical or impossible to manipulate the entire population. Nevertheless, some useful and important results can be established at the outset by considering the properties of the entire population.

Consider an infinite population of data, each datum representing a measurement of a single quantity, and assume each datum, x, differs in magnitude from the rest only as a result of precision error. Effectively, each time a different member of the population is randomly selected and measured, a value of x is determined and, a different precision error occurs. Some measurements are larger and some are smaller. The probability of obtaining a specific value of x depends the magnitude of x, and the probability distribution of x-values is described by a *probability density function*, $f(x)$ (Fig. 3.7).

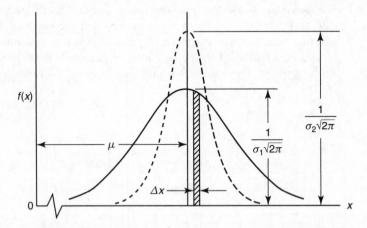

FIGURE 3.7: Normal distribution curve. More precise data are represented by the dashed curve than by the solid curve.

---

[4]Note that the distribution of *precision errors* accompanying *experimental data* and the distribution of *dimensional variations* in *manufacturing operations* are very similar. For example, a drawing may specify tolerances within which the diameter of a shaft is allowed to vary about a nominal value. The variations that actually exist within a production lot are often found to be distributed in normal, or Gaussian, manner. Quality control of machining is based on essentially the same theories as applied to distribution of experimental error.

Since the population is infinite, the probability density function, or PDF, is a continuous curve, unlike the previous bar-graph histogram describing rolled represents dice. Consequently the ordinate of the PDF must be carefully interpreted: It represents the probability of occurrence per unit change of x. The probability of measuring a given x is *not* $f(x)$ itself; instead, the probability of measuring an x in the interval $\Delta x = x_2 - x_1$ is the *area* under the PDF curve between x_1 and x_2

$$\text{Probability}_{(x_1 \to x_2)} = \int_{x_1}^{x_2} f(x)\, dx \qquad (3.5)$$

In any measurement, some value of x will be observed, so the total area under the PDF curve is unity (i.e., the probability of measuring some x value is 100%).

PDFs come in a variety of shapes, which are determined by the nature of the data considered. Precision error in experimental data is often distributed according to the familiar bell-shaped curve given by the *Gaussian*, or *normal, distribution* (Fig. 3.7). Most of the remaining statistical discussion in this chapter is based on that premise. For an infinite population, the mathematical expression for the Gaussian probability density function is

$$f(x) = \frac{1}{\sigma\sqrt{2\pi}} \exp\left[-\frac{(x-\mu)^2}{2\sigma^2}\right] \qquad (3.6)$$

where

$x =$ the magnitude of a particular measurement,

$\mu =$ the mean value of the entire population, and

$\sigma =$ the standard deviation of the entire population

The mean value, μ, is that which would be obtained if every x in the population could be averaged together; for an infinite population, such averaging is clearly impossible, and thus μ remains unknown. Since we assume that the various x's differ as a result of precision error, μ also represents the true value of the quantity we are attempting to measure, and the average sought amounts to averaging out all the precision errors.

If a large number of measurements are taken with equal care, then the arithmetic average of these n measurements,

$$\bar{x} = \frac{x_1 + x_2 + \cdots + x_n}{n}$$

$$= \sum_{i=1}^{n} \frac{x_i}{n} \qquad (3.7)$$

can be shown to be *the most probable single value for the quantity*, μ. Averaging a large sample allows us to estimate the true value.

The amount by which a single measurement is in error is termed the *deviation d*:

$$d = x - \mu \qquad (3.8)$$

TABLE 3.1: Summary of Probability Estimates Based on the Normal Distribution

Common Name for "Error" Level	Error Level in Terms of σ	% Confidence That Deviation of x from Mean is Smaller	Odds That Deviation of x is Greater
Standard deviation	$\pm\sigma$	68.3	abt. 1 in 3
Two-sigma error	$\pm1.96\sigma$	95.0	1 in 20
Three-sigma error	$\pm3\sigma$	99.7	1 in 370

The mean squared deviation, σ^2, is approximated by averaging the squared deviations of a very large sample[5]:

$$\sigma \approx \sqrt{\frac{d_1^2 + d_2^2 + \cdots + d_n^2}{n}} \tag{3.9}$$

The quantity, σ, is called the *standard deviation* of the population; it characterizes both the typical deviation of measurements from the mean value and the width of Gaussian distribution (Fig. 3.7). When σ is smaller, the data are more precise. The standard deviation is a very important parameter in characterizing both populations and samples.

The actual deviation or error for a particular measurement is, of course, never known. However, the likely size of the error, or uncertainty, can be estimated with various levels of confidence by using our knowledge of the distribution of the population. For example, if a population has a Gaussian distribution, then the probability that a single measurement will have a deviation greater than $\pm\sigma$ is 31.7%, or about one chance in three. For a single measurement, then, we can be 68% confident that the deviation is less than $\pm\sigma$. A deviation greater than $\pm1.96\sigma$ has a probability of 5.0% (1 in 20); greater than $\pm3\sigma$, about 0.27% (1 in 370); and greater than $\pm4\sigma$, about 0.0063% (1 in 16,000). The most popular uncertainty estimate is that for 95% confidence, $\pm1.96\sigma$. Table 3.1 summarizes the various levels of probability for the normal distribution.

One common criterion for discarding a data point as illegitimate is that the data point exceeds the 3σ level: since the probability of an error larger than this is 1 in 370, such values are unlikely in modest-sized data sets. Such data are sometimes called *outliers*.

For purposes of tabulation, the Gaussian PDF may be transformed by introducing the variable z:

$$z = \frac{x - \mu}{\sigma} \tag{3.10}$$

Equation (3.6) is now

$$f(z) = \frac{1}{\sigma\sqrt{2\pi}} e^{-z^2/2} \tag{3.11}$$

---

[5]Formally, the mean and the standard deviation may be calculated as integrals of the PDF: $\mu = \int_{-\infty}^{+\infty} xf(x)\,dx$; $\sigma^2 = \int_{-\infty}^{+\infty} (x - \mu)^2 f(x)\,dx$. This amounts to summing the probable contribution to the observed value of each x or $(x - \mu)^2$.

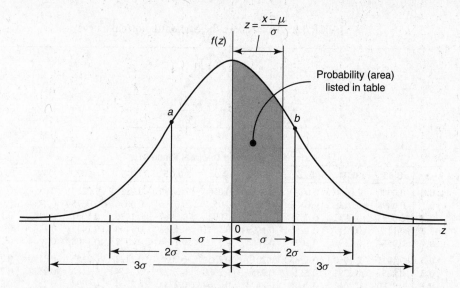

FIGURE 3.8: Standard normal distribution curve. Note that a and b are inflection points.

which is the *standard* curve shown at the top of Table 3.2 and in Fig. 3.8. The table lists the *areas* under the curve between 0 and various values of z. Since the curve is symmetric about zero, the tabulation lists values for only half the curve. Bear in mind that the total area beneath the curve is equal to unity. This tabulation is sometimes called the *z-distribution*.

The following examples illustrate the nature and use of the tabulated data.

EXAMPLE 3.2

(a) What is the area under the curve between $z = -1.43$ and $z = 1.43$?

(b) What is the significance of this area?

Solution

(a) From Table 3.2, read 0.4236. This represents half the area sought. Therefore, the total area is $2 \times 0.4236 = 0.8472$.

(b) The significance is that for data following the normal distribution, 84.72% of the population lies within the range $-1.43 < z < 1.43$.

EXAMPLE 3.3

What range of x will contain 90% of the data?

Solution   We need to find z such that $90\%/2 = 45\%$ of the data lie between zero and $+z$; the other 45% will lie between $-z$ and zero. Reading from Table 3.2, we find $z_{0.45} \approx 1.645$

TABLE 3.2: Areas under the Standard Normal Curve

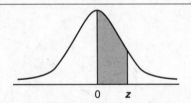

z	0.00	0.01	0.02	0.03	0.04	0.05	0.06	0.07	0.08	0.09
					Second Decimal Place in z					
0.0	0.0000	0.0040	0.0080	0.0120	0.0160	0.0199	0.0239	0.0279	0.0319	0.0359
0.1	0.0398	0.0438	0.0478	0.0517	0.0557	0.0596	0.0636	0.0675	0.0714	0.0753
0.2	0.0793	0.0832	0.0871	0.0910	0.0948	0.0987	0.1026	0.1064	0.1103	0.1141
0.3	0.1179	0.1217	0.1255	0.1293	0.1331	0.1368	0.1406	0.1443	0.1480	0.1517
0.4	0.1554	0.1591	0.1628	0.1664	0.1700	0.1736	0.1772	0.1808	0.1844	0.1879
0.5	0.1915	0.1950	0.1985	0.2019	0.2054	0.2088	0.2123	0.2157	0.2190	0.2224
0.6	0.2257	0.2291	0.2324	0.2357	0.2389	0.2422	0.2454	0.2486	0.2517	0.2549
0.7	0.2580	0.2611	0.2642	0.2673	0.2704	0.2734	0.2764	0.2794	0.2823	0.2852
0.8	0.2881	0.2910	0.2939	0.2967	0.2995	0.3023	0.3051	0.3078	0.3106	0.3133
0.9	0.3159	0.3186	0.3212	0.3238	0.3264	0.3289	0.3315	0.3340	0.3365	0.3389
1.0	0.3413	0.3438	0.3461	0.3485	0.3508	0.3531	0.3154	0.3577	0.3599	0.3621
1.1	0.3643	0.3665	0.3686	0.3708	0.3729	0.3749	0.3770	0.3790	0.3810	0.3830
1.2	0.3849	0.3869	0.3888	0.3907	0.3925	0.3944	0.3962	0.3980	0.3997	0.4015
1.3	0.4032	0.4049	0.4066	0.4082	0.4099	0.4115	0.4131	0.4147	0.4162	0.4177
1.4	0.4192	0.4207	0.4222	0.4236	0.4251	0.4265	0.4279	0.4292	0.4306	0.4319
1.5	0.4332	0.4345	0.4357	0.4370	0.4382	0.4394	0.4406	0.4418	0.4429	0.4441
1.6	0.4452	0.4463	0.4474	0.4484	0.4495	0.4505	0.4515	0.4525	0.4535	0.4545
1.7	0.4554	0.4564	0.4573	0.4582	0.4591	0.4599	0.4608	0.4616	0.4625	0.4633
1.8	0.4641	0.4649	0.4656	0.4664	0.4671	0.4678	0.4686	0.4693	0.4699	0.4706
1.9	0.4713	0.4719	0.4726	0.4732	0.4738	0.4744	0.4750	0.4756	0.4761	0.4767
2.0	0.4772	0.4778	0.4783	0.4788	0.4793	0.4798	0.4803	0.4808	0.4812	0.4817
2.1	0.4821	0.4826	0.4830	0.4834	0.4838	0.4842	0.4846	0.4850	0.4854	0.4857
2.2	0.4861	0.4864	0.4868	0.4871	0.4875	0.4878	0.4881	0.4884	0.4887	0.4890
2.3	0.4893	0.4896	0.4898	0.4901	0.4904	0.4906	0.4909	0.4911	0.4913	0.4916
2.4	0.4918	0.4920	0.4922	0.4925	0.4927	0.4929	0.4931	0.4932	0.4934	0.4936
2.5	0.4938	0.4940	0.4941	0.4943	0.4945	0.4946	0.4948	0.4949	0.4951	0.4952
2.6	0.4953	0.4955	0.4956	0.4957	0.4959	0.4960	0.4961	0.4962	0.4963	0.4964
2.7	0.4965	0.4966	0.4967	0.4968	0.4969	0.4970	0.4971	0.4972	0.4973	0.4974
2.8	0.4974	0.4975	0.4976	0.4977	0.4977	0.4978	0.4979	0.4979	0.4980	0.4981
2.9	0.4981	0.4982	0.4982	0.4983	0.4984	0.4984	0.4985	0.4985	0.4986	0.4986
3.0	0.4987	0.4987	0.4987	0.4988	0.4988	0.4989	0.4989	0.4989	0.4990	0.4990
3.1	0.4990	0.4991	0.4991	0.4991	0.4992	0.4992	0.4992	0.4992	0.4993	0.4993
3.2	0.4993	0.4993	0.4994	0.4994	0.4994	0.4994	0.4994	0.4995	0.4995	0.4995
3.3	0.4995	0.4995	0.4995	0.4996	0.4996	0.4996	0.4996	0.4996	0.4996	0.4997
3.4	0.4997	0.4997	0.4997	0.4997	0.4997	0.4997	0.4997	0.4997	0.4997	0.4998
3.5	0.4998	0.4998	0.4998	0.4998	0.4998	0.4998	0.4998	0.4998	0.4998	0.4998
3.6	0.4998	0.4998	0.4999	0.4999	0.4999	0.4999	0.4999	0.4999	0.4999	0.4999
3.7	0.4999	0.4999	0.4999	0.4999	0.4999	0.4999	0.4999	0.4999	0.4999	0.4999
3.8	0.4999	0.4999	0.4999	0.4999	0.4999	0.4999	0.4999	0.4999	0.4999	0.4999
3.9	0.5000*									

* For $z \geq 3.90$, the areas are 0.5000 to four decimal places.

(by interpolation). Hence, since $z = (x - \mu)/\sigma$, 90% of the population should fall within the range

$$(\mu - z_{0.45}\sigma) < x < (\mu + z_{0.45}\sigma) \tag{3.12}$$

or

$$(\mu - 1.645\sigma) < x < (\mu + 1.645\sigma)$$

3.6   THEORY BASED ON THE SAMPLE

In any real-life situation, we deal with samples from a population and not the population itself. Typically, our objective is to use average values from the sample to estimate the mean or standard deviation of the population. Thus, we would calculate the sample mean

$$\bar{x} = \sum_{i=1}^{n} \frac{x_i}{n} = \frac{x_1 + x_2 + \cdots + x_n}{n} \tag{3.13}$$

as an approximation to the population mean μ and the sample standard deviation

$$S_x = \sqrt{\frac{(x_1 - \bar{x})^2 + (x_2 - \bar{x})^2 + \cdots + (x_n - \bar{x})^2}{n - 1}} = \sqrt{\frac{\left(\sum_{i=1}^{n} x_i^2\right) - n\bar{x}^2}{n - 1}} \tag{3.14}$$

as an approximation to the population standard deviation σ. Here n is the number of data in the sample. The denominator of the standard deviation, $(n - 1)$, is called the number of *degrees of freedom*.[6]

The difference between population and sample values is emphasized by the use of different symbols for each:

	For Population	For Sample
Mean	μ or μ_x	$\bar{x}$
Standard deviation	σ or σ_x	S_x

Naturally, we'd like to have some assurance that the sample mean and standard deviation accurately approximate the corresponding values for the population; more specifically, we'd like to have an estimate of the *uncertainty* in approximating μ and σ by $\bar{x}$ and S_x.

A second objective is often to infer the probability distribution of the population from that of the sample. As it turns out, these two objectives can be accomplished independently for large samples. Conversely, for small samples ($n < 30$), knowledge of the distribution is assumed in estimating the uncertainty of $\bar{x}$.

---

[6]The basis for the number of degrees of freedom is the number of independent discrete data that are being evaluated. In computing the sample average, $\bar{x}$, all n data are independent. However, the standard deviation uses the result of the previous calculation for $\bar{x}$, which is not independent of the remaining data. Thus, the number of degrees of freedom is reduced by one in calculating S_x: We divide by $(n - 1)$ rather than n. In other instances, the data may have been used to calculate two or three necessary quantities (see Section 3.7 and Appendix F); the number of degrees of freedom is then reduced by two or three. The number of degrees of freedom is often denoted as ν.

TABLE 3.3: Results of a 12-Hour Pressure Test

Pressure, p, in MPa	Number of Results, m
3.970	1
3.980	3
3.990	12
4.000	25
4.010	33
4.020	17
4.030	6
4.040	2
4.050	1

3.6.1 An Example of Sampling

During a 12-hour test of a steam generator, the inlet pressure is to be held constant at 4.00 MPa. For proper performance, the pressure should not deviate from this value by more than about 1%. The inlet pressure was measured 100 times during the test. Various factors caused the readings to fluctuate, and the resulting data are listed in Table 3.3. The resolution of the digital pressure gauge used was 0.001 MPa. The number of results, m, is the number of readings falling in an interval of ±0.005 MPa centered about the listed pressure.

A first step in assessing the dispersion of the pressure readings might be to plot a histogram of the readings, as in Fig. 3.9.[7] A clear central tendency is apparent, as is the approximate width of the distribution. To quantify these values, we can compute $\bar{p}$ and S_p (Table 3.4) to obtain

$$\bar{p} = 4.008 \text{ MPa}$$
$$S_p = 0.014 \text{ MPa}$$

Is the distribution of readings Gaussian? A simple test is just to substitute $\bar{p}$ and S_p for μ and σ in Eq. (3.6) and plot the resultant curve over the histogram, as in Fig. 3.9. (The vertical scale for the distribution has been arbitrarily increased, for purpose of comparison.) An eyeball comparison indicates an approximate fit, albeit not a perfect one. How good must the fit be in order that we can claim a Gaussian distribution and apply Gaussian results? Goodness of fit is a legitimate concern, which is addressed further in Section 3.7.1.

Assuming that the population of pressure readings is in fact Gaussian distributed, with $\bar{p} \approx \mu$ and $S_p \approx \sigma$, the results of the previous section can be used to estimate the interval containing 95% of the pressure readings: $\mu \pm 1.96\sigma \approx \bar{p} \pm 1.96 S_p = 4.008 \pm 0.027$ MPa One objective of the pressure test was to verify that the pressure did not deviate from 4.00 MPa by more than 1% = 0.04 MPa. In terms of the standard deviation, 0.04 MPa $\approx 2.86\sigma$. For a Gaussian distribution, the probability of a 2.86σ fluctuation is about one chance in 240.

---

[7]The choice of bin width is, on the one hand, arbitrary, but on the other hand, it can greatly affect the appearance of the resulting graph. An empirical rule for the number of intervals to plot, N, is the *Sturgis rule*: $N = 1 + 3.3 \log n$, for n the total number of points.

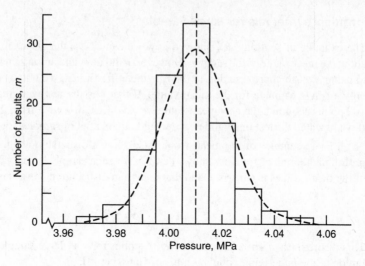

FIGURE 3.9: Histogram of the pressure data.

Two final comments should be made. First, these data do not separate measurement error from actual variations in the pressure; however, the actual pressure fluctuations are unlikely to be larger than the combined variation from measurement error and real fluctuations. Second, the analysis tells us nothing about possible bias errors in the readings.

TABLE 3.4: Calculation of Sample Mean and Standard Deviation

Pressure, p, in MPa	Number of Results, m	Deviation, d	d^2
3.970	1	−0.038	144.4×10^{-5}
3.980	3	−0.028	78.4
3.990	12	−0.018	32.4
4.000	25	−0.008	6.4
4.010	33	0.002	0.4
4.020	17	0.012	14.4
4.030	6	0.022	48.4
4.040	2	0.032	102.4
4.050	1	0.042	176.4

$\sum p = 400.770$ $n = \sum m = 100$ $\sum d^2 = 1858.0 \times 10^{-5}$

$\bar{p} = 400.77/100 = 4.008$ MPa $S_p = \sqrt{1858 \times 10^{-5}/99} = 0.014$ MPa

3.6.2   Confidence Intervals for Large Samples

The example in Section 3.6.1 showed how we can assess the dispersion of sample values about the mean value of the sample. However, it did not yield an estimate for the uncertainty in using $\bar{x}$ as an approximation to the true mean μ. In fact, we obtained from that sample only a single estimate for the mean value. If the pressure test were repeated and another 100 points acquired, the new mean value would differ somewhat from the first mean value. If we repeated the test many times, we would obtain a set of samples for the mean pressure.

These samples of the mean would also show a dispersion about a central value. A profound theorem of statistics shows that if n for each sample is very large the distribution of the mean values is *Gaussian* and that Gaussian distribution has a standard deviation

$$\sigma_{\bar{x}} = \sigma/\sqrt{n} \tag{3.15}$$

This theorem (the *Central Limit Theorem*) applies for very large samples even if the distribution of the underlying population is *not* Gaussian [3].

Armed with this important result, we can attack the uncertainty in our estimate that $\bar{x} \approx \mu$. Since $\bar{x}$ is Gaussian distributed, with standard deviation $\sigma_{\bar{x}}$, it follows that

$$z = \frac{\bar{x} - \mu}{\sigma_{\bar{x}}} = \frac{\bar{x} - \mu}{\sigma/\sqrt{n}} \tag{3.16}$$

where z is the same z-distribution given in Table 3.2. Hence, following Example 3.3, we can assert that $c\%$ of all readings of $\bar{x}$ will lie in the interval

$$\mu \pm z_{c/2}\frac{\sigma}{\sqrt{n}} \tag{3.17}$$

In other words, with $c\%$ confidence, the true mean value, μ, lies in the following interval about any single reading of $\bar{x}$:

$$\bar{x} - z_{c/2}\frac{\sigma}{\sqrt{n}} < \mu < \bar{x} + z_{c/2}\frac{\sigma}{\sqrt{n}} \tag{3.18}$$

This interval is termed a *c% confidence interval*.

The confidence interval suffers from only one limitation: σ is usually unknown. However, a reasonable approximation to σ is S_x when n *is large*. Thus, for large samples, standard practice is to set $\sigma \approx S_x$ in estimating the confidence interval for $\bar{x}$:

$$\bar{x} - z_{c/2}\frac{S_x}{\sqrt{n}} < \mu < \bar{x} + z_{c/2}\frac{S_x}{\sqrt{n}} \tag{3.19}$$

Often, the *standard deviation of the sample mean, $S_{\bar{x}}$*, is introduced in this context:

$$S_{\bar{x}} = \frac{S_x}{\sqrt{n}} \tag{3.20}$$

EXAMPLE 3.4

Determine a 99% confidence interval for the mean pressure calculated in Section 3.6.1.

Solution   First evaluate $z_{c/2} = z_{0.495} = 2.575$ from Table 3.2. Then, with $\sigma \approx S_p$,

$$\bar{p} - z_{0.495} \frac{S_x}{\sqrt{n}} < \mu_p < \bar{p} + z_{0.495} \frac{S_x}{\sqrt{n}}$$

or

$$\mu_p = 4.008 \pm 2.575 \frac{0.014}{\sqrt{100}} \text{ MPa} = 4.008 \pm 0.0036 \text{ MPa} \quad (99\%)$$

Note that this interval is much narrower than the dispersion of the data itself, because we are focusing on the accuracy of the estimate of the population mean.

Equation (3.19) is appropriate when we want the uncertainty in using the sample mean, $\bar{x}$, as an estimate for the population mean, μ. In contrast, the approach of Table 3.1 and Section 3.6.1 is appropriate when we desire an estimate of the width of the population distribution or the likelihood of observing a value that deviates from the mean by some particular amount. For instance, in Section 3.6.1 we used Table 3.1 to estimate the interval containing 95% of all individual readings of p; in the last example, we estimated an interval containing 99% of all measurements of $\bar{p}$. The $c\%$ confidence interval for the mean value is narrower than that of the data by a factor of $1/\sqrt{n}$, because n observations have been used to average out the random deviations of individual measurements.

3.6.3 Confidence Intervals for Small Samples

Equations (3.18) and (3.19) provide confidence intervals for the sample mean, $\bar{x}$, when σ is known or can be approximated by S_x. The condition $\sigma \approx S_x$ will generally apply when the sample is "large," which, from a practical viewpoint, means $n \geq 30$. Unfortunately, in many engineering experiments, n is substantially less than 30, and the preceding intervals are really not much help.

An amateur statistician, writing under the pseudonym Student, addressed this matter by considering the distribution of a quantity t,

$$t = \frac{\bar{x} - \mu}{S_x/\sqrt{n}} \tag{3.21}$$

which replaces σ in Eq. (3.16) by S_x. Student calculated the probability distribution of the t-statistic under the assumption that *the underlying population satisfies the Gaussian distribution*. This PDF, $f(t)$, is shown in Fig. 3.10. Note that the distribution depends on the number of samples taken, through the *degrees of freedom*, $\nu = n - 1$.

The t-distribution is qualitatively similar to the z-distribution. The PDF is symmetric about $t = 0$, and the total area beneath the distribution is again unity. Moreover, since the t-distribution is a PDF, the probability that t will lie in a given interval $t_2 - t_1$ is equal to

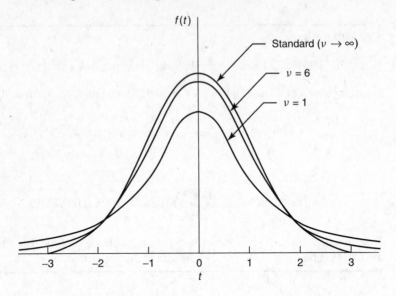

FIGURE 3.10: Probability distribution for the *t*-statistic.

the area beneath the curve between t_2 and t_1. As n (or ν) becomes large, the *t*-distribution approaches the standard Gaussian PDF; for $n > 30$, the two distributions are identical.

By analogy to the previous treatment of the *z*-distribution, the area beneath the *t*-distribution is tabulated in Table 3.5. However, in this case, the area α between t and $t \rightarrow \infty$ is listed; selected areas are given for a range of sample sizes. Thus the area α corresponds to the probability, for a sample size of $= \nu + 1$, that t will have a value *greater* than that given in the table [Fig 3.11(a)]. Since the distribution is symmetric, α is also the probability that t has a value less than the negative of the tabulated value [Fig 3.11(b)]. Conversely, we can assert with a confidence of $c\% = (1 - \alpha)$ that the actual value of t does *not* fall in the shaded area (i.e., if $\alpha = 0.05$, then $c = 0.95 = 95\%$).

Often, we want a two-sided confidence interval for the mean $\bar{x}$ of a small sample, so that both upper and lower bounds on the mean are stated. This interval follows directly from the preceding intervals; with a confidence of $c\%$, the true mean value lies in the interval

$$\bar{x} - t_{\alpha/2,\nu} \frac{S_x}{\sqrt{n}} < \mu < \bar{x} + t_{\alpha/2,\nu} \frac{S_x}{\sqrt{n}} \quad (c\%) \tag{3.22}$$

where $\alpha = 1 - c$ and $\nu = n - 1$ [Fig. 3.11(c)]. This equation is the small-sample analog of Eq (3.19). Sometimes, α is referred to as the *level of significance*.

This confidence interval defines the precision uncertainty in the value $\bar{x}$. From Eq. (3.22), the precision uncertainty in $\bar{x}$ is

$$P_x = t_{\alpha/2,\nu} \frac{S_x}{\sqrt{n}} \quad (c\%) \tag{3.23}$$

at a confidence level of $c\%$. This precision uncertainty is that needed for Eq. (3.4).

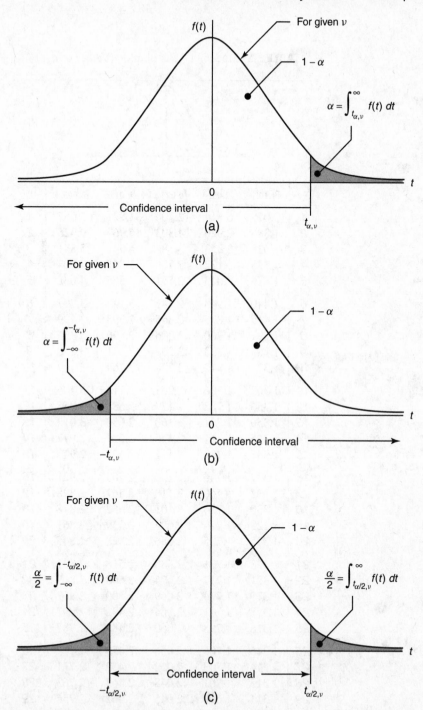

FIGURE 3.11: Confidence intervals for the t-statistic: (a) one-sided, right, (b) one-sided, left, (c) two-sided. With a confidence of $c\% = (1 - \alpha)$, the value of t lies in the unshaded interval.

TABLE 3.5: Student's t-Distribution (Values of $t_{\alpha,\nu}$)

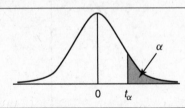

ν	$t_{0.10,\nu}$	$t_{0.05,\nu}$	$t_{0.025,\nu}$	$t_{0.01,\nu}$	$t_{0.005,\nu}$	ν
1	3.078	6.314	12.706	31.821	63.657	1
2	1.886	2.920	4.303	6.965	9.925	2
3	1.638	2.353	3.182	4.541	5.841	3
4	1.533	2.132	2.776	3.747	4.604	4
5	1.476	2.015	2.571	3.365	4.032	5
6	1.440	1.943	2.447	3.143	3.707	6
7	1.415	1.895	2.365	2.998	3.499	7
8	1.397	1.860	2.306	2.896	3.355	8
9	1.383	1.833	2.262	2.821	3.250	9
10	1.372	1.812	2.228	2.764	3.169	10
11	1.363	1.796	2.201	2.718	3.106	11
12	1.356	1.782	2.179	2.681	3.055	12
13	1.350	1.771	2.160	2.650	3.012	13
14	1.345	1.761	2.145	2.624	2.977	14
15	1.341	1.753	2.131	2.602	2.947	15
16	1.337	1.746	2.120	2.583	2.921	16
17	1.333	1.740	2.110	2.567	2.898	17
18	1.330	1.734	2.101	2.552	2.878	18
19	1.328	1.729	2.093	2.539	2.861	19
20	1.325	1.725	2.086	2.528	2.845	20
21	1.323	1.721	2.080	2.518	2.831	21
22	1.321	1.717	2.074	2.508	2.819	22
23	1.319	1.714	2.069	2.500	2.807	23
24	1.318	1.711	2.064	2.492	2.797	24
25	1.316	1.798	2.060	2.485	2.787	25
26	1.315	1.706	2.056	2.479	2.779	26
27	1.314	1.703	2.052	2.473	2.771	27
28	1.313	1.701	2.048	2.467	2.763	28
29	1.311	1.699	2.045	2.462	2.756	29
∞	1.282	1.645	1.960	2.326	2.576	∞

EXAMPLE 3.5

Twelve values in a sample have an average of $\bar{x}$ and a standard deviation of S_x. What is the 95% confidence interval for the true mean value, μ?

Solution   The required level of significance is $\alpha = 1 - 0.95 = 0.05$ and the degree of freedom is $\nu = 11$. The necessary value of t is $t_{\alpha/2,\nu} = t_{0.025,11} = 2.201$ (from Table 3.5). Hence, the 95% confidence interval is

$$\bar{x} - 2.201\frac{S_x}{\sqrt{n}} < \mu < \bar{x} + 2.201\frac{S_x}{\sqrt{n}} \quad (95\%)$$

EXAMPLE 3.6

A simple postal scale of the equal-arm balance type, is supplied with $\frac{1}{2}$-, 1-, 2-, and 4-oz machined brass weights. For a quality check, the manufacturer randomly selects a sample of 14 of the 1-oz weights and weighs them on a precision scale. The results, in ounces, are as follows:

1.08	1.03	0.96	0.95	1.04
1.01	0.98	0.99	1.05	1.08
0.97	1.00	0.98	1.01	

Question: Based on this sample and the assumption that the parent population is normally distributed, what is the 95% confidence interval for the *population mean*?

Solution   Using the t-test, we first calculate the sample mean and standard deviation, with $n = 14$. They are, respectively,

$$\bar{x} = 1.009 \text{ oz} \quad \text{and} \quad S_x = 0.04178$$

From Table 3.5, for $\nu = n - 1 = 13$, we find $t_{0.025,13} = 2.160$.
Calculating the *two-sided* confidence limits, we have

$$\frac{\pm t_{0.025,13}S_x}{n^{1/2}} = \frac{\pm 2.160(0.04178)}{(14)^{1/2}} = \pm 0.02412$$

Hence we may write $\mu = 1.009 \pm 0.024$ oz, with a confidence of 95%.

3.6.4   Hypothesis Testing for a Single Mean for a Small Sample Size $(n \leq 30)$

We often find it necessary to employ statistics in order to make certain decisions regarding a measurand. For example, consider the pressure measurements listed in Table 3.3. These measurements represent a single sample containing 100 pressure measurements obtained during a 12-hour period which produce a mean pressure of 400.77 MPa. If additional samples were taken, and the mean pressure of each sample obtained, the mean pressure can be assumed to be normally distributed based on the Central Limit Theorem. Often we are concerned whether the pressures observed are representative of a set point pressure of 4.00 MPa. One of the most commonly used methods for making such decisions is hypothesis testing. Typically there are two hypotheses in a hypothesis test. One hypothesis is called the *null hypothesis* and the other is often referred to as the *alternate hypothesis*.

The first step in setting up a hypothesis test is too choose the *null hypothesis*. Since the hypothesis testing often refers to a simple mean (although it can be used for any parameter) it takes the form

$$H_0 : \mu = \mu_0 \tag{3.24}$$

where μ_0 is some constant specified value.

The second step in hypothesis testing involves specifying an *alternate hypothesis*. The choice of the alternate hypothesis should reflect on what we are attempting to show. There are three possibilities for the choice of the alternative hypothesis.

1. *Two-tailed test:* If we are primarily concerned with determining whether a population mean, μ, is different from a specific value, μ_0, then the alternate hypothesis should read as

$$H_a : \mu \neq \mu_0 \tag{3.25}$$

2. *Left-tailed test:* If we are primarily concerned with determining whether a population mean, μ, is less than a specific value, μ_0, it should read as

$$H_a : \mu < \mu_0 \tag{3.26}$$

3. *Right-tailed test:* Finally if we want to determine whether a population mean, μ, is greater than μ_0, it should read as

$$H_a : \mu > \mu_0 \tag{3.27}$$

We can construct useful graphical representations of hypothesis testing as shown in Fig. 3.12.

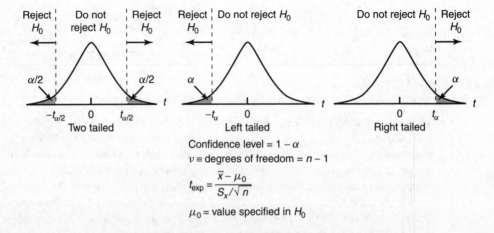

FIGURE 3.12: Criteria for the rejection of the null hypothesis.

For the application of hypothesis testing, we must perform the following steps:

1. Define both the null and alternate hypothesis.
2. Define a level of confidence, $c\%$.
3. Calculate the value of $t_{\exp}$ from the data.
4. From Table 3.5 determine the proper value of $t_{\alpha,\nu}$ using the degrees of freedom ν.
5. If $t_{\exp}$ falls in the reject H_0 region, we reject H_0 and accept the alternate hypothesis H_a.
6. If $t_{\exp}$ falls in the do not reject H_0 region, we conclude that we do not have sufficient data to reject H_0 at the level of confidence specified. (The strongest statistical statement occurs when we can reject H_0 and accept H_a.)

EXAMPLE 3.7

Using the data of Example 3.6, determine if the sample of 14 of the 1-oz weights comes from a population of weights whose true mean weight is greater than 1.00 oz, assuming a confidence level of 99%.

Solution

$$H_0 : \mu = 1.00 \text{ oz}$$
$$H_a : \mu \geq 1.00 \text{ oz}$$
$$t_{\exp} = \frac{\bar{x} - \mu_0}{S_x/\sqrt{n}} = \frac{1.009 - 1.000}{0.04178/\sqrt{14}} = 0.806$$
$$\nu = (14 - 1) = 13$$
$$t_{\alpha,\nu} = t_{0.01,13} = 2.650$$

Since this is a right tailed test and $t_{\exp}$ falls in the do not reject H_0 region, we conclude that the 99% confidence level that the population mean was not significantly different than 1.00 oz.

3.6.5 Hypothesis Testing for a Single Mean for a Large Sample Size ($n \geq 30$)

For a large sample size, the six steps listed in Section 3.6.4 are followed exactly except that $t_{\alpha,\nu}$ is replaced by z_α and $t_{\exp}$ is replaced by

$$\frac{\bar{x} - \mu_0}{S_x/\sqrt{n}} = z_{\exp} \qquad (3.28)$$

Figure 3.12 can now be interpreted as the similar criteria for rejecting or not rejecting the null hypothesis when the $t_{\alpha,\nu}$ limits are replaced by z_α.

EXAMPLE 3.8

From the pressures in Table 3.3, can we conclude that the experimental data obtained indicate a target pressure of 4.00 MPa at a confidence level of 99%?

Solution

$$H_0 : \mu = 4.00 \text{ MPa}$$

$$H_a : \mu \neq 4.00 \text{ MPa}$$

$$z_{\text{exp}} = \frac{\bar{x} - 4.00 \text{ MPa}}{S_x/\sqrt{n}} = \frac{4.008 - 4.000}{0.014/\sqrt{100}} = 5.714$$

Since $n \geq 30$ we use $\pm z_{\alpha/2}$ as our limits. From Table 3.2, $z_{0.005} = 2.575$.

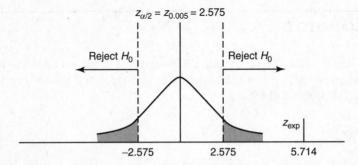

Since z_{exp} is in the reject H_0 region, we reject H_0 and accept H_a. Thus, the data indicate that the target pressure of 4.00 MPa is not being accurately controlled based on a confidence level of 99%.

3.6.6 The *t*-Test Comparison of Sample Means

If we wish to compare two samples solely on the basis of their means, we can use the following form of Eq. (3.21) [4]:

$$t = \frac{\bar{x}_1 - \bar{x}_2}{\sqrt{\left(S_1^2/n_1\right) + \left(S_2^2/n_2\right)}} \tag{3.29}$$

in which $\bar{x}_1$, S_1, n_1 and $\bar{x}_2$, S_2, n_2 are the means, standard deviations and sizes of the two respective samples. This value of t is compared to the interval $\pm t_{\alpha/2,v}$ found in Table 3.5, in which α is for an arbitrarily chosen confidence level $(1 - \alpha)$. The degree of freedom v may be approximated by the following expression:

$$v = \frac{\left[\left(S_1^2/n_1\right) + \left(S_2^2/n_2\right)\right]^2}{\dfrac{\left(S_1^2/n_1\right)^2}{n_1 - 1} + \dfrac{\left(S_2^2/n_2\right)^2}{n_2 - 1}} \tag{3.29a}$$

where v is rounded down to the nearest integer [4]. The hypothesis testing procedures of Section 3.6.4 can be used here to determine specific conditions regarding the two sample means.

EXAMPLE 3.9

An apartment manager wishes to determine if the lifetimes are different, under similar conditions, for two major brands of light bulbs. In the following sample data, the lifetime is in months.

Bulb A	7.2, 7.6, 6.9, 8.2, 7.3, 7.8, 6.6, 6.9, 5.5, 7.4, 5.7, 6.2
Bulb B	7.5, 8.7, 7.7, 7.5, 6.7, 11.2, 7.0, 10.7, 7.0, 8.6, 6.1, 6.3,
	7.8, 8.7, 6.1

Solution   When we calculate the means and standard deviations of each sample, we find

Bulb A	**Bulb B**
$\bar{x}_A = 6.94$ mo	$\bar{x}_B = 7.84$ mo
$S_A = 0.82$ mo	$S_B = 1.53$ mo
$n_A = 12$	$n_B = 15$

The hypothesis to be tested is

$$H_0 : \bar{x}_A = \bar{x}_B$$
$$H_a : \bar{x}_A \neq \bar{x}_B$$

This results in a two-tailed test as illustrated in Figure 3.12. From Eq. (3.29a) we determine the degrees of freedom

$$\nu = \frac{\left[(0.82)^2/12 + (1.53)^2/15\right]^2}{\dfrac{\left[(0.82)^2/12\right]^2}{12-1} + \dfrac{\left[(1.53)^2/15\right]^2}{15-1}}$$
$$\approx 22 \quad \text{(rounded down)}$$

and from Eq. (3.29) we calculate the test statistic

$$t_{\exp} = \frac{6.94 - 7.84}{\sqrt{(0.82)^2/12 + (1.53)^2/15}} = -1.954$$

For a confidence level $(1-\alpha) = 0.95$, we find the critical values of t from Table 3.5 to be

$$\pm t_{0.05/2,22} = \pm t_{0.025,22} = \pm 2.074$$

Since the value of $t_{\exp}$ falls within the do not reject H_0 region, we conclude that there is not a significant difference in the lifetimes of bulbs A and B at a 95% confidence level. [For comparing large sample means ($\nu > 30$), we simply replace t in Eq. (3.29) by z.]

3.7   THE CHI-SQUARE (χ^2) DISTRIBUTION

A variable is considered to have a chi-square distribution if its distribution has the shape of a right-skewed curve as shown in Fig. 3.13. The total area under the curve is 1.0; and as the degrees of freedom, ν, becomes large, the chi-square distribution approaches a symmetric distribution which resembles the normal distribution. Table 3.6 presents the values of χ^2 for various values of α and ν.

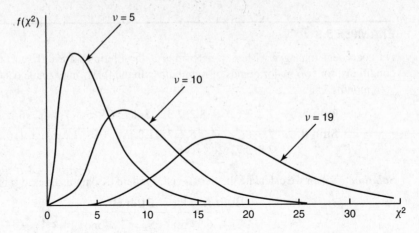

FIGURE 3.13: Probability distributions for the χ^2-statistic.

The chi-square goodness-of-fit-test statistic is defined as

$$\chi^2 = \sum_{i=1}^{k} \frac{(O_i - E_i)^2}{E_i} \qquad (3.30)$$

where

O_i = observed frequency

E_i = expected frequency

k = total number of variables being compared

$v = k - 1$

Since the chi-square statistic, χ^2, is by definition a positive value, a right-tailed test is applied for hypothesis testing (see Fig. 3.14). Note that when the observed frequencies approach the expected frequencies the value of χ^2 approaches zero.

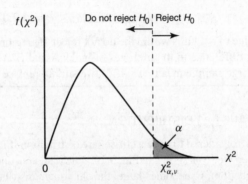

FIGURE 3.14: χ^2 Criteria for rejecting the null hypothesis.

TABLE 3.6: χ^2-Distribution (values $\chi^2_{\alpha,\nu}$)

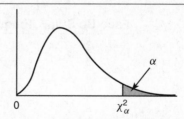

ν	$\chi^2_{0.995}$	$\chi^2_{0.99}$	$\chi^2_{0.975}$	$\chi^2_{0.95}$	$\chi^2_{0.05}$	$\chi^2_{0.025}$	$\chi^2_{0.01}$	$\chi^2_{0.005}$	ν
1	0.000	0.000	0.001	0.004	3.841	5.024	6.635	7.879	1
2	0.010	0.020	0.051	0.103	5.991	7.378	9.210	10.597	2
3	0.072	0.115	0.216	0.352	7.815	9.348	11.345	12.838	3
4	0.207	0.297	0.484	0.711	9.488	11.143	13.277	14.860	4
5	0.412	0.554	0.831	1.145	11.070	12.832	15.086	16.750	5
6	0.676	0.872	1.237	1.635	12.592	14.449	16.812	18.548	6
7	0.989	1.239	1.690	2.167	14.067	16.013	18.475	20.278	7
8	1.344	1.646	2.180	2.733	15.507	17.535	20.090	21.955	8
9	1.735	2.088	2.700	3.325	16.919	19.023	21.666	23.589	9
10	2.156	2.558	3.247	3.940	18.307	20.483	23.209	25.188	10
11	2.603	3.053	3.816	4.575	19.675	21.920	24.725	26.757	11
12	3.074	3.571	4.404	5.226	21.026	23.337	26.217	28.300	12
13	3.565	4.107	5.009	5.892	22.362	24.736	27.688	29.819	13
14	4.075	4.660	5.629	6.571	23.685	26.119	29.141	31.319	14
15	4.601	5.229	6.262	7.261	24.996	27.488	30.578	32.801	15
16	5.142	5.812	6.908	7.962	26.296	28.845	32.000	34.267	16
17	5.697	6.408	7.564	8.672	27.587	30.191	33.409	35.718	17
18	6.265	7.015	8.231	9.380	28.869	31.526	34.805	37.158	18
19	6.844	7.633	8.907	10.117	30.144	32.852	36.191	38.582	19
20	7.434	8.260	8.591	10.851	31.410	34.170	37.566	39.997	20
21	8.034	8.897	10.283	11.591	32.671	35.479	38.932	41.401	21
22	8.643	9.542	10.982	12.338	33.924	36.781	40.289	42.796	22
23	9.260	10.196	11.689	13.091	35.172	38.076	41.638	44.181	23
24	9.886	10.856	12.401	13.848	36.415	39.364	42.980	45.558	24
25	10.520	11.524	13.120	14.611	87.652	40.646	44.314	46.928	25
26	11.160	12.198	13.844	15.379	38.885	41.923	45.642	48.290	26
27	11.808	12.879	14.573	16.151	40.113	43.194	46.963	49.645	27
28	12.461	13.565	15.308	16.928	41.337	44.461	48.278	50.993	28
29	13.121	14.256	16.047	17.708	42.557	45.722	49.588	52.336	29
30	13.787	14.953	16.791	18.493	43.773	46.979	50.892	53.672	30

EXAMPLE 3.10

A pair of dice are tested to determine if they are "true." When the dice are tossed, the face-up sum total is read. After 360 such tosses, the results shown in the following table were obtained:

Face-Up Sum	Frequency
2	8
3	25
4	29
5	42
6	53
7	55
8	46
9	39
10	29
11	22
12	12

Determine if the dice are true at the 99% confidence level.

Solution In this case, the null hypothesis and alternate hypothesis are

$$H_0 : \text{Frequencies are the same}$$
$$H_a : \text{Frequencies are different}$$

We can form the following table using the expected distribution shown in Fig. 3.6:

Face-Up Sum	Observed Frequency	Expected Frequency
2	8	10
3	25	20
4	29	30
5	42	40
6	53	50
7	55	60
8	46	50
9	39	40
10	29	30
11	22	20
12	12	10

The experimental χ^2-statistic is

$$\chi^2_{exp} = \sum_{i=1}^{11} \frac{(O_i - E_i)^2}{E_i} = \frac{(8-10)^2}{10} + \frac{(25-20)^2}{20} + \cdots + \frac{(12-10)^2}{10}$$

$$= 3.358$$

For a 99% confidence level, $\alpha = 1 - 0.99 = 0.01$; and the number of degrees of freedom is $\nu = k - 1 = 10$. Therefore the appropriate χ^2-statistic is

$$\chi^2_{0.01,10} = 23.209$$

Thus, since $\chi^2_{exp} < \chi^2_{0.01,10}$, the dice are true at the 99% confidence level.

3.7.1   Goodness of Fit Based on the Gaussian Distribution

As stated previously, distribution of experimental data often abides by the Gaussian form expressed by Eq. (3.6). One must keep in mind, however, that this approximation is not always justified. For example, fatigue strength data for some metals approximate the so-called Weibull distribution; other distributions are described in the Suggested Readings at the end of the chapter. Since a given set of data may or may not follow the assumed distribution and since, at best, the degree of adherence can be only approximate, some estimate of *goodness of fit* should be made before critical decisions are based on statistical error calculations. In the following paragraphs, we discuss tests of fit that may be applied to the common Gaussian distribution, Eq. (3.6).

At the outset we advise the reader that there is no absolute check in the sense of producing some perfect and indisputable figure of merit. At best, a qualifying *confidence level* must be applied, with the final acceptance or rejection left to the judgment of the experimenter.

The simplest method is simply to plot a histogram and to "eyeball" the result: Yes, the distribution appears to approximate a bell shaped one; or no, it does not. This approach can easily result in misleading conclusions. The appearance of the histogram can sometimes be altered radically, simply by readjusting the number of class intervals.

A second method, which is relatively easy and much more effective is to make a graphical check using a *normal probability plot*. This technique requires a special graph paper[8] available from most bookstores that deal in technical supplies. One axis of the graph represents the cumulative probability (in percent) of the summed data frequencies. The other must be scaled to accommodate the range of data values in the sample. The more nearly the data plots as a straight line and the more nearly the mean corresponds to the 50% point, the better the fit to normal distribution. The final determination is subjective; it depends on the judgment of the experimenter. Considerable deviation from a straight line should raise serious doubts as to the value of any Gaussian-based calculations, particularly the significance of the calculated standard deviation. The following example demonstrates the procedure.

EXAMPLE 3.11

We will illustrate the graphical method by using the data of pressures given in Table 3.3.

Solution   In treating these data we will arbitrarily center our class intervals, or bins, on the mean and will assume eight intervals, each 0.010 MPa in width. Using these ground rules, we prepare Table 3.7.

---

[8]See reference [5], page 25, for directions for preparing one's own normal probability paper.

TABLE 3.7: Pressure Data of Table 3.3 Arranged for a Normal Probability Plot

A	B	C	D
3.965	1	1	1.01
3.975	3	4	4.04
3.983	12	16	16.16
3.995	25	41	41.41
4.005	33	74	74.75
4.015	17	91	91.92
4.025	6	97	97.98
4.035	2	99	99.00
4.045	1*	100	100.00

A = Limits on class intervals, arbitrarily taken as 0.010 MPa.
B = Number of data items falling within respective class intervals.
C = Cumulative number of data items.
D = Cumulative number of data items in percent.
*A rule of thumb often used is to discard arbitrarily any data falling outside a $\pm 3S_x$ limit. Theoretically, discarding out-of-tolerance items could make a readjustment of the mean and the standard deviation necessary. In this particular case, the changes would be so slight as to make the additional work unprofitable.

The ordinate of the graph is in terms of the upper limit of each interval. This quantity correlates with the cumulative values, which are plotted as the abscissa. Data from column A are plotted versus the percentages in column D, yielding Fig. 3.15.

To plot either 0 or 100% is impossible. For this reason, and also because either absence or presence of even one extra data point in the extreme intervals unduly distorts the plot, the two endpoints are generally given little consideration in making a final judgment. On the basis of Fig. 3.15, we can say that the pressure data show a reasonably good Gaussian distribution. Figure 3.16 illustrates the general discrepancies that may be discerned from a non-straight-line normal probability plot and their causes.

Another common test for goodness of fit is based on the χ^2-distribution. Implementation of the method requires considerable data manipulation and, as with other methods, a judgment on the part of the experimenter. In addition, the method does not lend itself well to small samples of data. The usual practice is to divide the test data into a reasonable number of *class intervals*, or *bins*, to determine the number of observations O_i in each interval, and then to compare these numbers with an *expected* number E_i of data items. The expected numbers are based on a "standard," the source of which depends on the objective of the test. If the test is to determine the normalcy of test data then the standard is the normal probability distribution. On the other hand, the standard may simply be another set of data that, in terms of an objective, is considered satisfactory; for example, how well do test data fit a standard norm?

Definite limitations apply:

1. The original, *experimentally determined* values of O_i must be numerical counts. They are *integer frequencies*; fractional events do not occur.

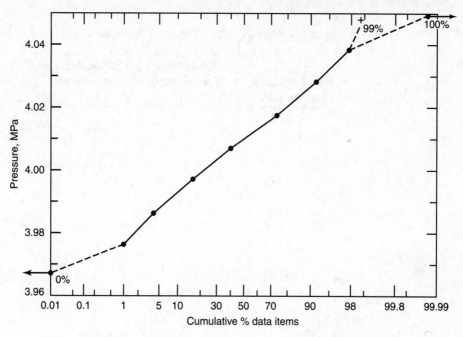

FIGURE 3.15: Normal probability plot of data listed in Table 3.7.

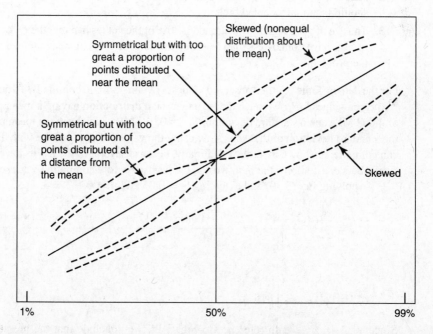

FIGURE 3.16: Graphical effects of data skew and offset as displayed on a normal probability plot.

TABLE 3.8: Pressure Data of Table 3.3 Arranged for a χ^2-Test

Pressure, in MPa	Observed Frequency, O_i	Expected Frequency, E_i
3.965		
	1	0.80
3.975		
	1	4.14
3.985		
	12	12.57
3.995		
	25	20.06
4.005		
	33	31.47
4.015		
	17	19.63
4.025		
	6	8.54
4.035		
	2	2.27
4.045		

2. Frequency values for O_i in each bin should be equal to or greater than to unity. There should be no unoccupied bins.

3. The use of χ^2 is usually questioned if 20% of the values in either the O_i or the E_i cells or bins have counts less than 5. Often the cells or bins can be combined to eliminate the problem.

For the data in Table 3.7 the "observed values," O_i, are those in column B. To determine the "expected values," E_i, we must define a Gaussian distribution having a mean pressure, $\bar{p}$, of 4.008 MPa and a standard deviation, S_p, of 0.014 MPa, which are the mean pressure and the standard deviation of the data. Now using the values of $\bar{p}$, S_p and n with Table 3.2 we can compute E_i shown in Table 3.8. Finally, because several of the 8 bins have fewer than 5 members, we combine the first two and the last two before computing χ^2_{exp}, so that $k = 6$.

Applying Eq. (3.30) and assuming $\alpha = 0.05$ and $\nu = k - 3 = 3$,

$$\chi^2_{\text{exp}} = \sum_{i=1}^{6} \frac{(O_i - E_i)^2}{E_i} = \frac{(2 - 4.94)^2}{4.94} + \frac{(12 - 12.57)^2}{12.57} + \cdots + \frac{(8 - 10.81)^2}{10.81}$$

$$= 4.15$$

$$\chi^2_{\alpha,\nu} = \chi^2_{0.05,3} = 7.815$$

Since $\chi^2_{\text{exp}} < \chi^2_{\alpha,\nu}$, with reference to Fig. 3.14, we conclude that the pressure data are normally distributed at the 95% confidence level.

Of special importance is the fact that $v = k - 3$. This is because the number of bins, the mean, and the standard deviation were chosen from the experimental data.

3.8 STATISTICAL ANALYSIS BY COMPUTER

Statistical analysis can often involve very large sets of data or require the application of a broad range of statistical tests. Consequently, a number of statistical software packages have been developed to assist in data analysis. Commercially marketed versions include MINITAB[9] and SPSS.[10] Packages such as these are available for use on machines ranging from mainframes to personal computers.

The statistical methods of the preceding sections were developed in the last century to reduce the laborious calculations that would otherwise be required in drawing statistical conclusions from samples. These methods are possible largely because the population's probability distribution has been assumed known. However, the digital computer makes detailed statistical computations easier. As a result, some current statistical research is directed toward using computer methods to relax the assumptions associated with classical statistics. For example, can we determine small-sample confidence intervals *without* assuming that the population is Gaussian distributed?

3.9 BIAS AND SINGLE-SAMPLE UNCERTAINTY

Precision error in repeat-sampled data reveals its own distribution, enabling us to bound its magnitude using statistical methods. Bias error, by virtue of its systematic nature, provides no direct evidence of either its magnitude or its presence. The only direct method for uncovering bias error in a measurement is by comparison with measurements made using a separate, and presumably more accurate, apparatus.

Unfortunately, a second set of apparatus is seldom used owing to cost and time constraints. Instead, we rely on knowledge of our own equipment to make estimates for the likely sizes of bias errors. Estimation of bias uncertainty relies heavily on experience and on an understanding of calibration accuracy and dimensional tolerances. Even with such experience and understanding, unexpected sources of bias error can be overlooked. Diligence, persistence, and careful examination of one's results are essential in identifying and eliminating such errors.

Estimates of bias uncertainty should be accompanied by odds or a confidence level [6]. Unlike statistical confidence levels, odds for bias uncertainty cannot be rigorously determined. The level of confidence assigned is a product of our knowledge of the system, reflecting our assessment of the fraction of bias errors likely to land within the uncertainty interval. Sometimes the term *coverage* is used in place of confidence to reflect the empirical nature of these estimates in contrast to those derived using statistical methods.

We have previously discussed general sources of bias error. Let's look at a few in more detail.

Data reduction often requires knowledge of physical properties, system dimensions, or electrical characteristics. For example, a flowmeter measurement may depend on the density of water and a tube diameter; an amplifier gain may depend on the value of a resistor. Differences between the assumed values of these components and their actual values can systematically shift all data taken, creating a bias error.

[9]MINITAB is registered to Minitab, Inc., 3081 Enterprise Drive, State College, PA, 16801.
[10]SPSS is a registered trademark of SPSS, Inc., 444 N. Michigan Avenue, Chicago, IL, 60611.

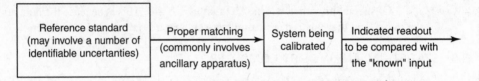

FIGURE 3.17: Block diagram showing calibration procedure.

To find water's density, the temperature is specified and a tabulated handbook value is taken, giving, for example, 998 kg/m³. Inaccuracy or ambiguity in the specified temperature may cause a bias uncertainty in density of 2 kg/m³, for instance. This uncertainty may cover 95% of the temperature range we expect is possible for the system. This inaccuracy will *systematically* affect all data reduced using the particular value of density.

Similarly, the nominal diameter of a pipe may differ from the production diameter by a percent or so; and a manufacturer's rated value for a resistor can vary substantially from the resistor's actual resistance. For carbon resistors, the manufacturing tolerance may be ±5% or ±10%; for higher-quality, metal-film resistors, the tolerance may be ±0.01%. Typically, these tolerances might represent 95% coverage—that is, the variation of 95% of all resistors. Potential sources of error such as these remain unchanged for each measurement made with the system.

If the uncertainty related to a manufacturing tolerance is unacceptably large, taking our own measurement of the specific part can usually reduce the uncertainty substantially. The uncertainty in a resistor's value can be reduced to the accuracy of the ohmmeter measuring it; or a pipe's diameter can be measured to the accuracy of a set of calipers. Physical property data can also be measured, if need be, although it is more common to trust the carefully determined handbook values.

Calibration uncertainty is another very common source of bias uncertainty. Calibration requires a reference or standard against which system response can be compared. The reference may be *fixed* or *one-valued*, such as the triple point of water or the other triple points and melting points used to define the practical temperature scale (Section 2.7). Alternatively, the standard may be capable of supplying a range of inputs comparable to the range of the system, as do various commercially available voltage references. Naturally, the uncertainty of the standard should be considerably less than that of the system being calibrated. A rule of thumb is that the uncertainty of the standard should be no more than one-tenth that of the system being calibrated.

Figure 3.17 shows a typical calibration arrangement. Normally, the indicated readout is compared to the reference standard and a relation between the two is determined. Sometimes the readout scale can be adjusted until agreement with the standard is obtained; sometimes a line fit is used to relate the readout to the standard's value. In either case, additional uncertainty appears in the comparison and adjustment process.

Instrument manufacturers often supply calibration data with their products, which can assist in estimating the uncertainty of the instrument. For example, a particular position transducer is rated at 0.8 V output per millimeter of sensor displacement. The manufacturer has not specified the calibration uncertainty directly. However, we might assume that the uncertainty is roughly 0.05 V/mm, since this is the apparent resolution of the calibration. The coverage is also unknown, but our experience using the device may suggest that 90 to

95% coverage is a reasonable assumption. If necessary, we could reduce the uncertainty by conducting our own calibration.

Examples of estimating bias uncertainty are given in Section 3.11.

3.9.1   Single-Sample Precision Uncertainty

When only one, two, or three repeat observations are made, the confidence intervals calculated statistically can be quite large. In that circumstance, you may determine a narrower range for the mean value by treating precision errors like bias errors and estimating a standard deviation based on your knowledge of the instruments. For example, random variations in test conditions may cause a digital multimeter (DMM) reading to fluctuate; but if the reading is made only once, the random variation simply produces an overall range of uncertainty for the true value of this variable. The uncertainty (at 19 : 1 odds) is twice the standard deviation of the test condition. In other words, $\pm 1.96\sigma \approx \pm 2\sigma$ will cover 95% (or 19 out of 20) of the readings made. If, on the other hand, the same measurement is made several times, the random variations can be averaged out, and statistics can be used to place a narrower bound on the mean value.

We can construct the single-sample estimate a little more formally. We begin by estimating σ as a value (σ_e, say) that is based on our knowledge of the experimental system. Thus, we are assuming that σ of the population is *known*, and, with Eq. (3.18), the precision uncertainty can be estimated as

$$P_x \approx z_{c/2}\frac{\sigma_e}{\sqrt{n}} \tag{3.31}$$

With a single reading, $n = 1$, so that no averaging is performed to reduce the uncertainty. Taking a 95% confidence level, $z_{0.95/2} = 1.96$ and

$$P_x \approx 1.96\sigma_e \quad (95\%) \tag{3.32}$$

The potential precision error underlying a single measurement of a random variable can usually be estimated from your knowledge of how finely an instrument will resolve, of how precisely an instrument may repeat a reading, or of how much the test condition fluctuates. Often, these estimates can be made in advance of performing the experiment, in order to gauge the expected uncertainty in the result.

Section 3.11 includes an example of single-sample uncertainty analysis.

3.10   PROPAGATION OF UNCERTAINTY

Often several quantities are measured, and the results of those measurements are used to calculate a desired result. For example, experimental values of density are usually determined by dividing the measured mass of a sample by the measured volume of that sample. Each measurement includes some uncertainty, and these uncertainties will create an uncertainty in the calculated result. What is that uncertainty?

Finding the uncertainty in a result due to uncertainties in the independent variables is called finding the *propagation of uncertainty*. For uncertainties in the independent variables, the procedure rests on a statistical theorem that is exact for a linear function y of several independent variables x_i with standard deviations σ_i; the theorem states that the standard deviation of y is

$$\sigma_y = \sqrt{\left(\frac{\partial y}{\partial x_1}\sigma_1\right)^2 + \left(\frac{\partial y}{\partial x_2}\sigma_2\right)^2 + \cdots + \left(\frac{\partial y}{\partial x_n}\sigma_n\right)^2} \qquad (3.33)$$

Likewise, a calculated result y is a function of several independent measured variables, $\{x_1, x_2, \ldots, x_n\}$; for example, density is a function of mass and volume. Each measured value has some uncertainty, $\{u_1, u_2, \ldots, u_n\}$ and these uncertainties lead to an uncertainty in y, which we call u_y. To estimate u_y, we assume that each uncertainty is small enough that a first-order Taylor expansion of $y(x_1, x_2, \ldots, x_n)$ provides a reasonable approximation:

$$y(x_1 + u_1, x_2 + u_2, \ldots, x_n + u_n)$$
$$\approx y(x_1, x_2, \ldots, x_n) + \frac{\partial y}{\partial x_1}u_1 + \frac{\partial y}{\partial x_2}u_2 + \cdots + \frac{\partial y}{\partial x_n}u_n \quad (3.34)$$

Under this approximation, y is a linear function of the independent variables. Now we can apply the theorem, assuming that uncertainties will behave much like standard deviations:

$$u_y = \sqrt{\left(\frac{\partial y}{\partial x_1}u_1\right)^2 + \left(\frac{\partial y}{\partial x_2}u_2\right)^2 + \cdots + \left(\frac{\partial y}{\partial x_n}u_n\right)^2} \quad (n:1) \qquad (3.35)$$

Here, all uncertainties must have the same odds and must be independent of each other. This approach was established by S. J. Kline and F. A. McClintock in 1953 [6].

The uncertainties, u_i, may be either bias uncertainties or precision uncertainties. Normally, the bias uncertainties and precision uncertainties in y are propagated separately. The overall uncertainty, U_y, is then calculated by combining B_y and P_y using Eq. (3.4). Section 3.11.2 illustrates this procedure.

Equation (3.35) can be simplified when y has certain common functional forms, as shown in the following examples.

EXAMPLE 3.12

Suppose that y has the form

$$y = Ax_1 + Bx_2$$

and that the uncertainties in x_1 and x_2 are known with odds of $n : 1$. What is the uncertainty in y?

Solution

$$\frac{\partial y}{\partial x_1} = A$$
$$\frac{\partial y}{\partial x_2} = B$$

Using Eq. (3.35),

$$u_y = \sqrt{(Au_1)^2 + (Bu_2)^2} \qquad (n:1) \qquad (3.36)$$

For additive functions, the *absolute* uncertainties in each term are combined in root-mean-square (rms) sense.

EXAMPLE 3.13

Suppose that y has the form

$$y = A \frac{x_1^m x_2^n}{x_3^k}$$

and that the uncertainties in x_1, x_2, and x_3 are known with odds of $n : 1$. What is the uncertainty in y?

Solution

$$\frac{\partial y}{\partial x_1} = mA \frac{x_1^{(m-1)} x_2^n}{x_3^k}$$

$$\frac{\partial y}{\partial x_2} = nA \frac{x_1^m x_2^{(n-1)}}{x_3^k}$$

$$\frac{\partial y}{\partial x_3} = -kA \frac{x_1^m x_2^n}{x_3^{(k+1)}}$$

Using Eq. (3.35),

$$u_y = \sqrt{\left(mA \frac{x_1^{(m-1)} x_2^n}{x_3^k} u_1\right)^2 + \left(nA \frac{x_1^m x_2^{(n-1)}}{x_3^k} u_2\right)^2 + \left(-kA \frac{x_1^m x_2^n}{x_3^{(k+1)}} u_3\right)^2} \quad (n:1)$$

so that, for this case,

$$\frac{u_y}{y} = \sqrt{\left(m \frac{u_1}{x_1}\right)^2 + \left(n \frac{u_2}{x_2}\right)^2 + \left(k \frac{u_3}{x_3}\right)^2} \quad (n:1) \tag{3.37}$$

For multiplicative functions, the *fractional* uncertainties are combined in an rms sense. Note carefully the weighting factors, m, n, and k, in Eq. (3.37) and their sources.

Normally, each source of error is independent of the other sources. The errors will not all be of the same sign, nor will they all take on their maximum values simultaneously. For that reason, Eq. (3.35) combines the uncertainties in a root-mean-square sense. In some situations, however, various sources of uncertainty are not independent. Dependent errors should be added together, before combining them in the root-mean-square sense with other independent sources of error.

3.11 EXAMPLES OF UNCERTAINTY ANALYSIS

3.11.1 Rating Resistors

EXAMPLE 3.14

Carbon resistors are painted with color-coded bands that specify their nominal resistance. The actual resistance of each resistor varies randomly about the nominal value, owing to manufacturing variations. The percentage variation in the resistance of the population of resistors is referred to as the *tolerance* of the resistors. For commercial carbon resistors, this variation is 5, 10, or 20%.

A lab technician has just received a box of 2000 resistors. As a result of a production error, the color-coded bands have not been painted on this lot. To determine the nominal resistance and tolerance, the technician selects 10 resistors and measures their resistances with a digital multimeter. His results are as tabulated.

Number	Resistance (kΩ)
1	18.12
2	17.95
3	18.17
4	18.45
5	16.24
6	17.82
7	16.28
8	16.32
9	17.91
10	15.98

What is the nominal value of the resistors? What is the uncertainty in that value? Consider both precision and bias uncertainty. Can you estimate the tolerance?

Solution   The precision error in the resistors can be averaged to find a 95% confidence interval. From the tabulated data,

$$\bar{R} = 17.32 \text{ k}\Omega$$

$$S_R = 0.982 \text{ k}\Omega$$

The mean resistance, $\bar{R} = 17.32$ kΩ, is clearly the apparent nominal value of the resistors.

To find the uncertainty in this mean value, both the precision and the bias uncertainties must be estimated. Consider the precision uncertainty first. From Table 3.5 with $v = 10 - 1 = 9$ and $\alpha = (1 - 0.95)/2 = 0.025$,

$$t_{\alpha,v} = t_{0.025,9} = 2.262$$

Applying Eq. (3.22), the (unbiased) population mean, μ_R, is in the range

$$\mu_R = \bar{R} \pm t_{\alpha,v} \frac{S_R}{\sqrt{n}}$$

$$= 17.32 \pm 2.262 \frac{0.982}{\sqrt{10}} \text{ k}\Omega$$

$$= 17.32 \pm 0.70 \text{ k}\Omega \quad (95\%)$$

However, this answer accounts for only precision error, specifically, $P_R = 0.70$ kΩ. What is the bias uncertainty in this result?

The manual for the DMM describes its calibration; possible bias uncertainty (from temperature drift, connecting-lead resistances, and other sources) is rated as

$$\pm(0.5\% \text{ of reading} + 0.05\% \text{ of full scale} + 0.2 \text{ } \Omega)$$

The confidence is not given, but we shall assume it to be 95%. The full-scale reading of the meter is 20 kΩ, and after evaluating the terms and summing, the meter's bias uncertainty can be estimated as

$$B_R = \pm 96.8 \ \Omega = \pm 0.10 \ \text{k}\Omega \quad (95\%)$$

Notice that the reading error in the DMM scale is only 0.005 kΩ, which is much lower than the actual uncertainty in the DMM reading! This DMM has relatively high resolution and precision but much lower accuracy.

The total uncertainty in the mean of the population is, from Eq. (3.4),

$$\begin{aligned}
U_R &= \left(B_R^2 + P_R^2 \right)^{1/2} \\
&= \left[(0.10)^2 + (0.70)^2 \right]^{1/2} \ \text{k}\Omega \\
&= 0.71 \ \text{k}\Omega \quad (95\%)
\end{aligned}$$

The uncertainty of the nominal value is $U_R = 0.71$ kΩ (95%), or about $\pm 4\%$.

The precision uncertainty in the mean is the major source of uncertainty. On the other hand, if a sample of 1000 resistors were used, the precision uncertainty would be reduced by a factor of ten (why?), and the bias uncertainty would be dominant.

The tolerance of the resistors remains to be found. What we'd like is an estimate of the percentage deviation from the nominal value which includes, say, 95% of the resistors. One approach is to note that 95% of a Gaussian population lies within $\pm 1.96\sigma$ of the population mean μ (see Table 3.1). On that basis, we could approximate $\sigma_R \approx S_R$ and $\mu_R \approx \bar{R}$, so that

$$\text{Tolerance } \% = \frac{1.96\sigma_R}{\mu_R} \approx \frac{1.96 S_R}{\bar{R}} = \frac{1.96 \cdot 0.982}{17.32} = 0.111$$

that is, a tolerance of 11% (or about 10%, since that's the nearest production tolerance).

In a manufacturing situation, engineers are usually more interested in estimating an interval that is $c\%$ certain to contain at *least* some percentage b of the population. For example, the manufacturer might wish to report, with $c = 95\%$ confidence, that $b = 95\%$ of resistors will have resistances within some specific range of resistances. As it turns out, the approach used in the preceding paragraph is a very poor way to estimate such *tolerance limits*, because it ignores the inaccuracy of S_R and $\bar{R}$ as estimates of the population's σ_R and μ_R. Although 95% of a Gaussian population lies in the interval $\mu_R \pm 1.96\sigma_R$, that is not true of the interval $\bar{R} \pm 1.96 S_R$. For example, our estimate of the mean has a 4% precision uncertainty; this means that the interval likely to contain 95% of the population should be broadened by something like an additional $\pm 4\%$ of $\bar{R}$ beyond $\bar{R} \pm 1.96 S_R$. A proper estimate of tolerance must allow for this uncertainty as well as that in σ_R. More advanced statistical methods [3] show that, at a confidence of 95%, the 95% tolerance interval or the population is almost *twice* as large as that estimated previously (i.e., the interval that is 95% certain to contain at least $\mu \pm 1.96\sigma$ turns out to be $\bar{R} \pm 3.532 S_R$).

After this extended discussion, it may interest you to learn that the resistors actually tested were nominally 18 kΩ with a tolerance of 10%.

3.11.2 Expected Uncertainty for Flowmeter Calibration

EXAMPLE 3.15

Obstruction meters such as venturis and orifice plates are commonly used to measure the steady flow rates of fluids (see Section 15.3). Tables of empirical calibration coefficients, K, are available for use in theoretically based relationships such as Eq. (15.8c):

$$Q = K A_2 \sqrt{2g_c(P_1 - P_2)/\rho} \qquad (3.38)$$

This technique provides a means of measuring flow rate in terms of the pressure drop across the obstruction. The published coefficients will yield approximate flow rates, but accurate measurement requires careful experimental determination of the coefficient for each specific installation.

Figure 3.18 shows a proposed arrangement for calibration of a thin-plate orifice meter. In this case, calibration consists of experimentally determining the coefficient K in Eq. (3.38) by collecting the flowing fluid (water, in this case) in a weigh tank for some period of time. During the calibration period, the flow rate is held as constant as possible, and the pressure difference, $\Delta P = P_1 - P_2$, is recorded. The flow rate, Q, is the measured weight, W, divided by the liquid density, ρ, and the elapsed time, t ($Q = W/\rho t$); the area, A_2, is $\pi D^2/4$, for D the orifice diameter.

Substituting these values into Eq. (3.38), we may solve for the calibration constant, K:

$$K = \frac{4W}{\pi D^2 t} \sqrt{\frac{1}{2\rho\,\Delta P}} \qquad (3.39)$$

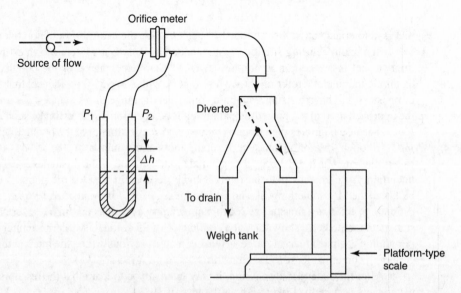

FIGURE 3.18: Setup for calibrating an orifice.

By inserting the observed values of W, t, and ΔP, along with a measured value of D and tabulated data for ρ, the experimental value of K is obtained.

Before undertaking this experiment, we'd like to estimate the expected accuracy of the result. We could, of course, make the following calculations *after* the experiment, but by doing it ahead of time, we can identify those parts of the experiment that contribute most of the uncertainty and, if necessary, improve them.

Solution   Both bias and precision error should be considered. However, since we have no samples for statistical analysis, we can only estimate the expected size of potential precision errors, using estimates for the standard deviations. Thus this analysis is a *single-sample* uncertainty estimate. To proceed, we first estimate the single-sample precision and bias uncertainty in each measured variable and then propagate these uncertainties into K.

The weight, W, is measured with a platform scale. What bias uncertainty exists in the scale measurement? Has the scale's calibration been checked? How recently and against what standard? (See Section 13.1 for further discussion of this point.) Presumably, we have made some sort of check, at a minimum several point calibrations using reliable proof weights. If not, the user's manual should include such data. Let us say that, in our judgment, an uncertainty of $\pm 1\%$ is justified, with a confidence of about 95%. In practice, the bias uncertainty is undoubtedly dependent to some degree on the magnitude of the weight measured relative to the scale range; this effect could also be taken into account if warranted.

Precision uncertainty in the weight measurement will be caused by reading error in the scale, and possibly by hysteresis, friction, or backlash in the scale mechanism. The size, or standard deviation, of the reading error will depend on how finely the scale is graduated; perhaps 0.1% error covers one standard deviation in the scale reading (so that $\pm 0.1\%$ covers 68% of the reading errors). Hysteresis, friction, and backlash should be negligible if the scale is in good condition; however, if these effects are observed, their errors should also be taken into account.

The diameter of the orifice must be determined. Assume that it is a sharp-edged orifice and that we have checked it with an inside micrometer. Did we check the micrometer against gauge blocks, or are we accepting its scale as is? How experienced are we in using a micrometer? We intend to measure the orifice only once, using this value in all subsequent applications of the meter, so the diameter's uncertainty all appears as bias in the results. After these considerations, we estimate a 95% bias uncertainty of ± 0.008 cm in the nominal 4-cm diameter ($\pm 0.2\%$ uncertainty).

How accurate is the determination of the time period, t? If a hand-operated diverter is used, precisely when did the flow start and stop? That is, how well is the time period *defined*? If a stopwatch is used, how good is the synchronization between the diverter and watch actions?

Note that these time uncertainties are essentially *precision* errors, which are likely to vary from run to run. If we carefully make a series of repeated runs of this particular procedure, we could accumulate enough data to perform a precision uncertainty analysis, were such accuracy required. In the present case, we simply estimate the likely standard deviation to be ± 3 s; if the total time period is 5 min, the standard deviation of t is $\pm 1\%$.

Bias error in the time determination may arise if the watch is fast or slow (probably a *very* small error in 5 min) or if we systematically stop or start the watch too soon. Without other information, we'll assume that bias in the time reading is negligible compared to the precision error already discussed.

The density will be read from a handbook table at the temperature of the experiments. Between 0°C and 38°C, the density of water decreases by about 0.7%. Temperature may vary slightly between each experiment, leading to a precision error if only one value of density is used. However, if we use a thermocouple to measure the correct temperature for each experiment, precision error will still arise from the reading error of the temperature measurement. If standard deviation in temperature is ±0.1°C, then the corresponding density variation is only 0.002%.

On the other hand, the bias error in the temperature reading may be higher, perhaps ±1°C, and the resulting bias uncertainty in the density used would then be ±0.02% (again at 95%). If we don't bother to measure the temperature, the bias uncertainty would be larger still, maybe 0.2% if we just assume a standard room temperature of 20°C.

The value of ΔP is measured using a manometer (see Section 14.4). The dominant uncertainty is that in reading the difference in the heights of the manometer columns, essentially a precision uncertainty resulting from reading error. This uncertainty has relatively constant size, independent of the magnitude of ΔP, so that a percentage uncertainty is a bit misleading. At small ΔP, it may be 3%, but at high ΔP it may only be 0.1%. To keep the uncertainty analysis simple, let's take a representative value of ±1% for the standard deviation in the pressure. Bias uncertainty in the manometer turns out to be substantially smaller, about 0.1% at about 95% confidence.

Summarizing our results gives the following:

Variable	Bias Uncertainty, B_x/x (95%)	Standard Deviation, σ_x/x
Weight, W	1%	0.1%
Diameter, D	0.2%	≈ 0
Time, t	≈ 0	1.0%
Density, ρ	0.02%*, 0.2%†	0.002%
Pressure, ΔP	0.1%	1%

* If temperature is measured.

† If temperature is not measured.

Now we can apply Eq. (3.37) to K:

$$\frac{u_K}{K} = \left[\left(\frac{u_w}{W} \right)^2 + \left(2\frac{u_d}{D} \right)^2 + \left(\frac{u_t}{t} \right)^2 + \left(\frac{1}{2}\frac{u_\rho}{\rho} \right)^2 + \left(\frac{1}{2}\frac{u_{\Delta P}}{\Delta P} \right)^2 \right]^{1/2}$$

First, we calculate the bias uncertainty in K:

$$\frac{B_K}{K} = \left[\left(\frac{B_w}{W} \right)^2 + \left(2\frac{B_d}{D} \right)^2 + \left(\frac{B_t}{t} \right)^2 + \left(\frac{1}{2}\frac{B_\rho}{\rho} \right)^2 + \left(\frac{1}{2}\frac{B_{\Delta P}}{\Delta P} \right)^2 \right]^{1/2}$$

$$= \left[(1)^2 + (2 \times 0.2)^2 + (0)^2 + \left(\frac{1}{2} \times 0.02 \right)^2 + \left(\frac{1}{2} \times 0.1 \right)^2 \right]^{1/2}$$

$$= 1.08\% \quad (95\%)$$

Likewise, the standard deviation of K is

$$\frac{\sigma_K}{K} = \left[\left(\frac{\sigma_w}{W}\right)^2 + \left(2\frac{\sigma_d}{D}\right)^2 + \left(\frac{\sigma_t}{t}\right)^2 + \left(\frac{1}{2}\frac{\sigma_\rho}{\rho}\right)^2 + \left(\frac{1}{2}\frac{\sigma_{\Delta P}}{\Delta P}\right)^2 \right]^{1/2}$$

$$= \left[(0.1)^2 + (2 \times 0)^2 + (1)^2 + \left(\frac{1}{2} \times 0.002\right)^2 + \left(\frac{1}{2} \times 1\right)^2 \right]^{1/2}$$

$$= 1.12\%$$

Our estimate for the single-sample precision uncertainty in K is, from Eq. (3.32),

$$\frac{P_K}{K} = \frac{1.96\,\sigma_K}{K} = 1.96(0.0112) = 2.20\% \quad (95\%)$$

From Eq. (3.4), the total uncertainty in K is

$$\frac{U_K}{K} = \left[\left(\frac{B_K}{K}\right)^2 + \left(\frac{P_K}{K}\right)^2 \right]^{1/2}$$

$$= \left[(0.0108)^2 + (0.0220)^2 \right]^{1/2}$$

$$= 2.45\% \quad (95\%)$$

Inspection of these results quickly reveals the parameters having the greatest contribution to the uncertainty. Improvement of the timing and weighing procedures would improve the results the most. Improvement of the pressure and diameter measurements would contribute significantly less improvement.

Most of the total uncertainty is caused by precision uncertainty. We can reduce that uncertainty considerably by repeating the calibration experiment several times and averaging the results. Since P_K will decrease as $\sqrt{n}$, taking $n = 4$ experiments will reduce P_K to about 1.0%.

Note that the density contributes almost nothing to the total uncertainty. Even if we don't bother to measure the temperature, the contribution of density uncertainty remains negligible; that is, $\left(\frac{1}{2} \cdot 0.2\right)^2 \ll (1)^2$. We conclude that careful temperature measurement, in this case, would be a waste of effort.

Are the various uncertainty estimates simply good guesses? To a degree, they are, *but* dismissing them as nothing but guesses would be flippant. The specific considerations leading to each estimate were not arbitrary; when properly made, such "guesses" have a strong foundation in the actual performance of the equipment and the method of taking the data. Even the estimated confidence percentages (usually 95%) are a quantitative assessment of our expectation for the variability of the data, although they *are* essentially just educated guesses. But if we admit to guessing, why not simply guess the overall uncertainty and skip all the intermediate steps? In answer, the detailed analysis provides a means for evaluating the relative effect of each identifiable source of error, thereby separating the more important ones from the less important ones. Furthermore, one can evaluate the uncertainties of each of the individual variables with considerably more assurance than one could judge the total.

3.12 MINIMIZING ERROR IN DESIGNING EXPERIMENTS

The best time to minimize experimental error is in the design stage, when an experimental procedure is being developed. First and foremost, one should perform a single-sample uncertainty analysis of the proposed experimental arrangement *prior* to beginning construction, in order to determine whether the expected uncertainty is acceptably small and to identify the major sources of uncertainty. Some general precautions to observe when designing an experiment are as follows:

1. Avoid approaches that require two large numbers to be measured in order to determine the small difference between them. For example, large uncertainty is likely when measuring $\delta = (x_1 - x_2)$ if $\delta \ll x_1$, unless x_1 can be measured with great accuracy.

2. Design experiments or sensors that amplify the signal strength in order to improve sensitivity. For example, a thermopile uses several thermocouples to resolve a single temperature, and a strain gage uses many loops of wire to measure a single strain.

3. Build "null designs," in which the output is measured as a change from zero rather than as a change in a nonzero value. This reduces both bias and precision error. Such designs often make the output proportional to the difference of two sensors. An excellent example of this approach is the Wheatstone bridge circuit (see Sections 7.8 and 7.9).

4. Avoid experiments in which large "correction factors" must be applied as part of the data-reduction procedure.

5. Attempt to minimize the influence of the measuring system on the measured variable.

6. Calibrate entire systems, rather than individual components, in order to minimize calibration-related bias errors.

3.13 GRAPHICAL PRESENTATION OF DATA

According to the American Standards Association [7],

> When used to present facts, interpretations of facts, or theoretical relationships,
> a graph usually serves to communicate knowledge from the author to his readers,
> and to help them visualize the features that he considers important.

A graph should be used when it will convey information and portray significant features more efficiently than words or tabulations.

A graph is nearly always the most effective format for conveying the interrelation of experimental variables. Graphs are invaluable in constructing acceptable curve fits of experimental data and in identifying outliers or erroneous measurements. Graphs are also useful in testing theoretical calculations against real experimental results and in identifying the conditions under which a theoretical model fails.

The clarity imparted by graphing data can be substantial. Table 3.9 shows the atmospheric pressure measured in Cambridge, Massachusetts, during Hurricane Bob on August 19, 1991. In tabular form the trend is unclear. If, on the other hand, the data are graphed (Fig. 3.19) the progress of the storm is apparent. Atmospheric pressure declined steadily until about 4:00 P.M. and then began to increase. The 11:30 A.M. reading lies well below the trend defined by the other data, and it can probably be assumed to be in error (some checking of the data reductions verified this conclusion). A faired curve has been

TABLE 3.9: Atmospheric Pressure during Hurricane Bob

Time of Day	Pressure (mbar)
10:00 A.M.	1009.0
11:30 A.M.	984.2
1:00 P.M.	999.8
2:15 P.M.	989.0
3:40 P.M.	977.1
4:40 P.M.	981.2
5:40 P.M.	990.0

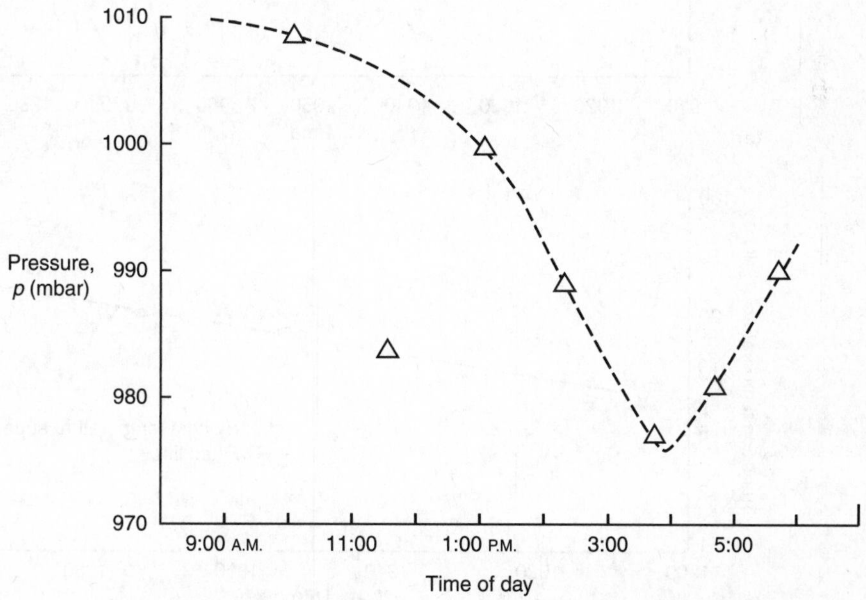

FIGURE 3.19: Atmospheric pressure during Hurricane Bob.

empirically sketched through the remaining data.[11] This curve may be used to estimate the pressure at times other than those measured.

3.13.1 General Rules for Making Graphs

By observing the following guidelines, you can help to ensure that your graph will be easy for your readers to understand. Figures 3.20(a) and (b) illustrate some of these points [7].

1. The graph should be designed to require minimum effort from the reader in understanding and interpreting the information it conveys.

---

[11] *Faired* in this sense means smooth and without irregularity: a draftsman's curve through the data.

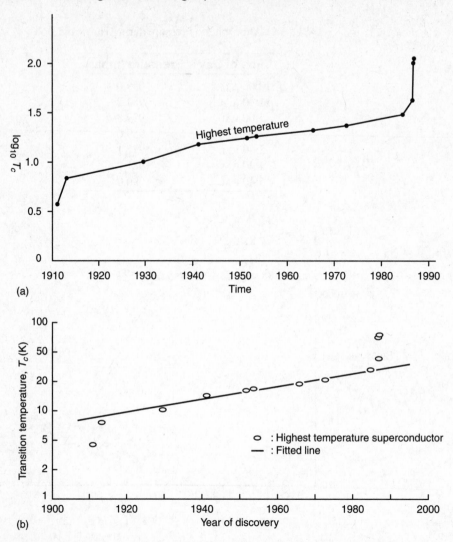

FIGURE 3.20: (a) A poor graph; (b) graph of Fig. 3.20(a) improved by following graphing guideline numbers 2, 3, 5, 8, 10, and 12.

2. The axes should have clear labels that name the quantity plotted, its units, and its symbol if one is in use.

3. Axes should be clearly numbered and should have tick marks for significant numerical divisions. Typically, ticks should appear in increments of 1, 2, or 5 units of measurement multiplied or divided by factors of 10 (1,10,100, ...). Not every tick needs, to be numbered; in fact, using too many numbers will just clutter the axes. Tick marks should be directed toward the *interior* of the figure.

4. Use scientific notation to avoid placing too many digits on the graph. For example, use 50×10^3 rather than 50,000. A particular power of 10 need appear only once

along each axis. Avoid confusing labels such as "Pressure, Pa $\times 10^6$;" use units such as MPa instead.

5. When plotting on semilog or full-log coordinates, use real logarithmic axes; do *not* plot the logarithm itself (e.g., plot 50, not 1.70).[12] Logarithmic scales should have tick marks at powers of 10 and intermediate values, such as 10, 20, 50, 100, 200, ...

6. The axes should usually include zero; if you wish to focus on a smaller range of data, include zero and break the axis, as shown in Fig. 3.4.

7. The choice of scales and proportions should be commensurate with the relative importance of the variations shown in the results. If variations by increments of 10 are significant, the graph should not be scaled to emphasize variations by increments of 1.

8. Use symbols such as $\bigcirc$, $\square$, $\triangle$, and $\Diamond$ for data points. Do not use dots ($\cdot$) for data. Open symbols should be used before filled symbols. You may place a legend defining symbols on the graph (if space permits) or in the figure caption.

9. Place error bars on data points to indicate the estimated uncertainty of the measurement or else use symbols that are the same size as the range of uncertainty.

10. When several curves are plotted on one graph, different lines (solid, dashed, dash-dot, . . .) should be used for each if the curves are closely spaced. The graph should include labels or a legend identifying each curve. Avoid using colors to differentiate curves, since colors are usually lost when the graph is photocopied. Theoretical curves should be plotted as lines, without showing calculated points. Curves fitted to data do not need to pass through every measurement like a dot-to-dot cartoon; however, if a data point lies far from the fitted curve, a discrepancy may be indicated [as for the first and the last three points in Fig. 3.20(b)].

11. Lettering on the graph should be held to the minimum necessary for clarity. Too much text (or too much data) creates crowding and confusion.

12. Labels on the axes and curves should be oriented to be read from the bottom or from the right. Avoid forcing the reader to rotate the figure in order to read it.

13. The graph should have a descriptive but concise title. The title should appear as a caption to the figure rather than on the graph itself.

Good graphing software can help produce graphs that adhere to these guidelines. However, some graphing packages violate even the simplest of these rules. Discretion is advised!

3.13.2 Choosing Coordinates and Producing Straight Lines

The first step in making any graph is to decide which variables to plot and on what scale to plot them. Four basic graphical scales occur frequently in engineering work (Fig. 3.21). *Linear* coordinates have a linear variation of both the x- and y-scales. If a variable changes by several orders of magnitude or is exponentially related to another variable, then a logarithmically scaled axis may be preferable. Graphs having one logarithmically scaled axis and one linearly scaled axis are called *semilogarithmic* (or *semilog*). Those for which both axes are logarithmic are called *full logarithmic* (*full log*, or *log-log*). When a quantity varies with an angle, *polar coordinates* provide a physically suggestive format for the data.

---

[12] An exception is made when the unit *decibels* is plotted.

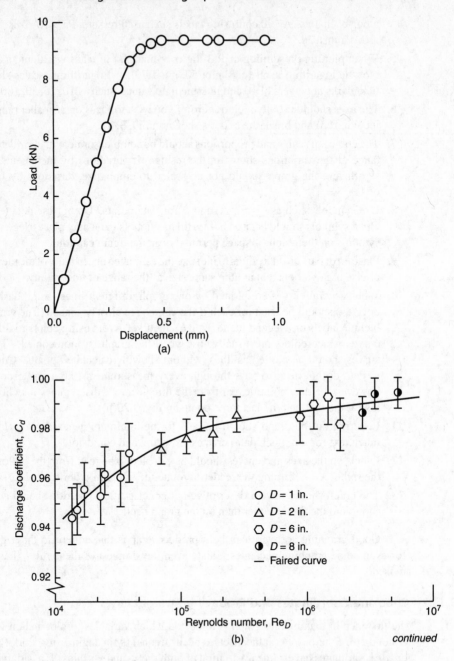

FIGURE 3.21: (a) Linear coordinates, (b) semilogarithmic coordinates.

continued

The choice of which scaling to use is normally guided by your expectations for the physical behavior of the system being studied. You may also attempt to deduce the right scaling by studying a test graph made on linear coordinates. Often, the objective in selecting

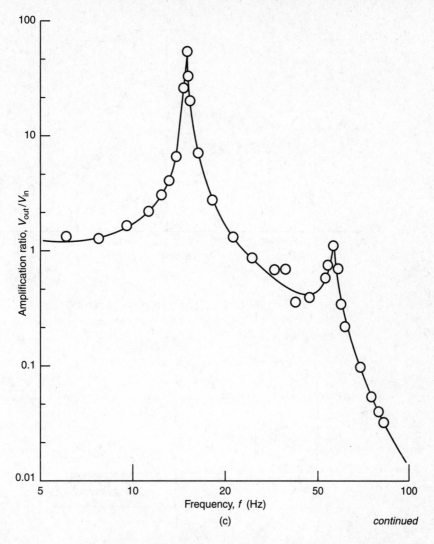

FIGURE 3.21: (c) Full logarithmic coordinates.

a scaling is to find coordinates in which the plotted data fall on a straight line, because straight lines are the easiest curves to fit.

Figure 3.22(a) shows a set of data that represent the cooling of a warm metal slug suddenly immersed in cold liquid. The difference between the slug's temperature and the liquid temperature was recorded at several times after the slug was submerged. The graph has the form of an exponential decay; indeed, heat transfer theory suggests that the cooling curve should have the form

$$\Delta T = \Delta T_0 \exp\left(-\frac{t}{\tau}\right) \qquad (3.40)$$

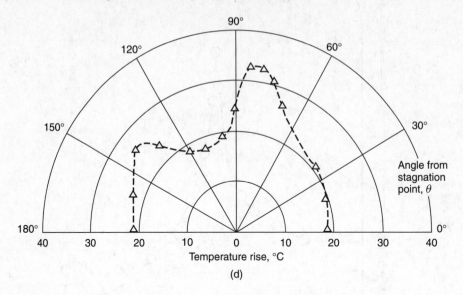

FIGURE 3.21: (d) Polar coordinates.

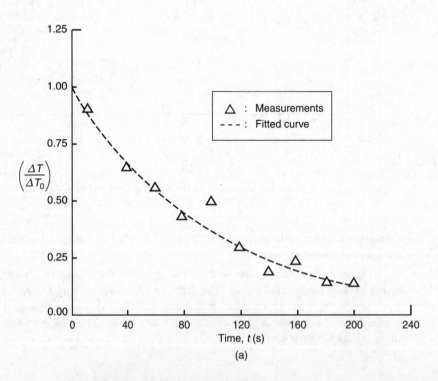

FIGURE 3.22: Cooling data. (a) Linear coordinates.

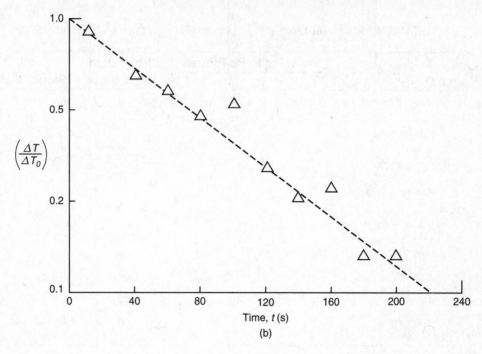

FIGURE 3.22: Cooling data. (b) Semilogarithmic coordinates. (Note logarithmic variation of $\Delta T / \Delta T_0$.)

where ΔT is the measured temperature difference at any time, ΔT_0 is the temperature difference before the slug is immersed, and τ is a time constant for the cooling. You may desire to find an experimental value for τ, so that you can use Eq. (3.40) to estimate ΔT at values of t where you have no measurements. That task is not straightforward using the linear scaling of Fig. 3.22(a).

Instead, you could plot $\log(\Delta T / \Delta T_0)$ as a function of t. Then the relationship between ΔT and t is

$$\log\left(\frac{\Delta T}{\Delta T_0}\right) = -\frac{0.4343\,t}{\tau} \tag{3.41}$$

which is the equation of a line with slope $-0.4343/\tau$ and intercept zero.[13] The graph is most easily made using semilogarithmic coordinates [Figure 3.22(b)], and τ can be calculated from the slope of the line:

$$-0.4343/\tau = \frac{\log\left[\Delta T(t_1)/\Delta T_0\right] - \log\left[\Delta T(t_2)/\Delta T_0\right]}{t_1 - t_2} \tag{3.42}$$

resulting in $\tau = 98$ s. Note that while $\Delta T / \Delta T_0$ is plotted on the logarithmic coordinates, $\log(\Delta T / \Delta T_0)$ must be used in calculating the slope.

---

[13]Base 10 logarithms are standard in graphical work; $\log_{10} e = 0.4343$.

TABLE 3.10: Straight-Line Transformations [8]: $y = f(x) \longrightarrow Y = A + BX$

| $f(x)$ | Variables to Be Plotted | | Straight-Line | |
	Y	X	Intercept, A	Slope, B
$y = a + b/x$	y	$1/x$	a	b
$y = 1/(a + bx)$ or $1/y = a + bx$	$1/y$	x	a	b
$y = x/(a + bx)$ or $x/y = a + bx$	x/y	x	a	b
$y = ab^x$	$\log y$	x	$\log a$	$\log b$
$y = ac^{bx}$	$\log y$	x	$\log a$	$b \log c$
$y = ax^b$	$\log y$	$\log x$	$\log a$	b
$y = a + bx^n$, if n is known	y	x^n	a	b

Semilog paper, full-log paper, and polar graph paper are available from most university bookstores or drafting suppliers. Many computer spreadsheet and plotting programs can also generate these coordinate systems. Moreover, plotting software can expedite experimentation with different scalings and coordinates, so that the most informative ones can be quickly identified.

Logarithmic scaling is only one approach to creating straight-line representations of data. For example, the function

$$y = a + \frac{b}{x} \tag{3.43}$$

does not give a straight-line variation of y with x [Fig. 3.23(a)]. However, it does give a straight-line variation of y with $1/x$. The solution is to plot y as a function of $1/x$ rather than of x itself; then a can be determined as the intercept of resultant line and b as its slope [Fig. 3.23(b)].

Table 3.10 offers a guide to straight-line transformations in which a function $y = f(x)$ is transformed to

$$Y = A + BX \tag{3.44}$$

by plotting an appropriate pair of modified variables, Y and X.

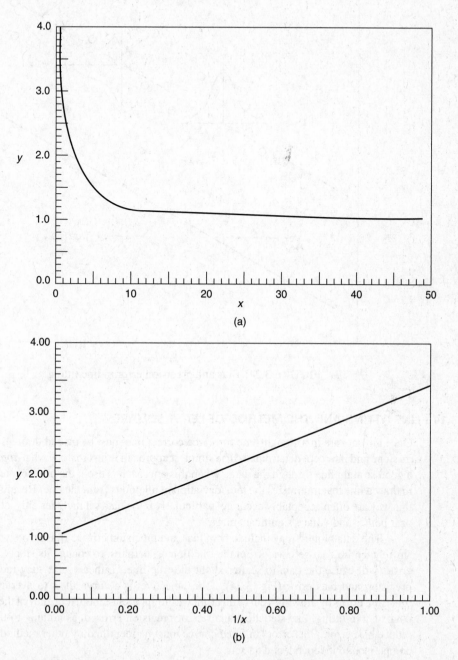

FIGURE 3.23: Plot of $y = 1.0 + (2.5/x)$ as (a) y versus x and (b) y versus $(1/x)$.

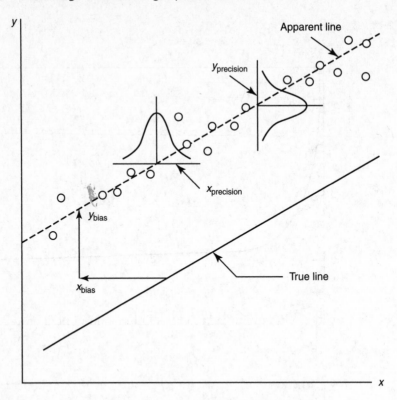

FIGURE 3.24: Bias and precision error in line fitting.

3.14 LINE FITTING AND THE METHOD OF LEAST SQUARES

Once the data are in a straight-line form, the correct line must be passed through them and its slope and intercept determined. The simplest approach is just to draw what *appears* to be a good straight line through the data. When this approach is used, the probable tendency is to draw a line that minimizes the total deviation of all points from the line. Results obtained this way are often acceptably accurate, particularly if the data set includes only a few points or if both x and y have significant error.

The data plotted may include both bias and precision errors. Bias errors will tend to shift the entire data set away from the true line or, perhaps, to change its slope. Precision errors will cause the data to scatter about the true line. Either y or x may include both precision and bias error (Fig. 3.24). The objective of a curve fit is to average out the *precision* errors by drawing a curve that follows the apparent central tendency of the scattered data. Curve fitting, like statistical analysis of precision error, does nothing to uncover or reduce bias error. Bias errors in fitted curves must be identified by other methods, such as comparison to independent data sets.

3.14.1 Least Squares for Line Fits

When the precision error in y is substantially greater than that in x, the *method of least squares*, or linear regression, enables us to calculate a line, $y = a + bx$, through the data [3]. The method identifies the slope b and intercept a that minimize the sum of the squared deviations of the data from the fitted line, S^2:

$$S^2 = \sum_{i=1}^{n} [y_i - y(x_i)]^2 \tag{3.45}$$

Here, for the various measured values of x_i, y_i is the experimentally determined ordinate and $y(x_i) = a + bx_i$ is the corresponding value calculated from the fitted line; n is the number of experimental observations used. The result is

$$a = \frac{\sum y_i \sum x_i^2 - \sum x_i \sum x_i y_i}{n \sum x_i^2 - \left(\sum x_i\right)^2} \tag{3.46}$$

$$b = \frac{n \sum x_i y_i - \sum x_i \sum y_i}{n \sum x_i^2 - \left(\sum x_i\right)^2}$$

Most scientific calculators incorporate programs for calculating least squares lines, a great convenience to the experimentalist. Consequently, least squares has become increasingly popular as *the* method for fitting lines. But to some degree this predominance has also promoted misuse of the method. The least squares method addresses *only* the precision error in y_i; poor results are obtained if x_i also includes large precision error. Least squares assumes, in effect, that the experimental x_i are *error free*.

To indicate the reliability of the fit, most pocket calculators and software packages return the *correlation coefficient, r*, along with the least squares results:

$$r^2 = \frac{\text{explained squared variation about } y_m}{\text{total squared variation about } y_m} \tag{3.47}$$

where y_m is the mean of the measured y_i:

$$y_m = \frac{1}{n} \sum_{i=1}^{n} y_i \tag{3.48}$$

The *explained* squared variation results from the straight-line change of y with x:

$$\sum_{i=1}^{n} [y(x_i) - y_m]^2$$

The *total* squared variation also includes the precision error:

$$\sum_{i=1}^{n} (y_i - y_m)^2 = \cdots = S^2 + \sum_{i=1}^{n} [y(x_i) - y_m]^2$$

Thus,

$$r^2 = \frac{\sum [y(x_i) - y_m]^2}{S^2 + \sum [y(x_i) - y_m]^2} \tag{3.49}$$

94 Chapter 3   Assessing and Presenting Experimental Data

If the sum of squared deviations S^2 is assumed to result only from precision error, then a "perfect fit" occurs when $S^2 \to 0$ and $r^2 \to 1$. Hence, the nearer r is to ± 1, the "better" the fit.

Unfortunately, when the data look basically linear, one usually obtains $|r| > 0.9$; the correlation coefficient is not a very sensitive indicator of the precision of the data. It turns out that $(1 - r^2)^{1/2}$ is a better indicator of the fit's quality (Appendix F, [9]); the closer it is to zero, the lower the precision error in the data. The quantity $(1 - r^2)^{1/2}$ is roughly the ratio of the vertical standard deviation of the data about the line to the total vertical variation of the data.

When both y and x have significant precision error, least squares should not be used; this case is usually identifiable by the highly scattered appearance of the data. In addition to eyeball estimates, various other semiempirical line-fitting procedures are available in that situation, as discussed in reference [10]. Conversely, if both y and x have precision errors, but these errors are *small* relative to the *overall* variation of the data, then least squares results may still be acceptably accurate. In this situation, the data will still appear to fall on a straight line.

3.14.2   Uncertainty in Line Fits

Sometimes the real issue is to estimate an uncertainty for the slope or intercept of a fitted line. Eyeball estimates of the uncertainty are often acceptably accurate; for example, you may be able to vary the slope of a hand-fitted line to determine what range of slopes will still fit the data with 95% confidence. Similar estimates can be applied to finding the intercept. This approach is best either when the sample is small or when both y and x have large precision errors. Statistical confidence intervals may also be derived, as discussed in Appendix F.

Another major concern in line or curve fitting is that of identifying outliers. Points well beyond the trend of the remaining data can often be identified by eye. When a particular point is in doubt, exclude it temporarily, and fit a new line through the remaining data. If that line shows a much better fit, the point can probably be dropped. Again, within the restrictive assumptions of least squares, statistical tests can be applied; essentially, these consist of estimating a 3σ band around the fitted curve (Appendix F). Sometimes, however, data deviate from a fitted line because the actual relationship is *not* a straight line. One must always avoid forcing nonlinear data to fit an assumed straight-line form, since valuable information can be lost in the process.

3.14.3   Software for Curve Fitting

Line fitting is actually a special case of the method of least squares; least squares can be generalized to fit polynomials of any order. Many other methods of curve fitting have been developed, some considerably more sophisticated than least squares. Good discussions of such methods can be found in most texts on numerical analysis. (Reference [11] gives an introduction to these issues.)

One example of higher-order curve fitting can be found in Table 16.6 of this book, where high-order polynomials have been used to fit thermocouple voltages as functions of temperature.

EXAMPLE 3.16

A cantilever beam deflects downward when a mass is attached to its free end. The deflection, δ (m), is a function of the *beam stiffness*, K (N/m), the applied mass, M (kg), and the gravitational body force, $g = 9.807$ m/s:

$$K\delta = Mg$$

To determine the stiffness of a small cantilevered steel beam, a student places various masses on the end of the beam and measures the corresponding deflections. The deflections are measured using a scale (a ruler) marked in 1-mm increments. Each mass is measured in a balance. His results are as follow:

Mass (g)	Deflection (mm)
0	0
50.15	0.6
99.90	1.8
150.05	3.0
200.05	3.6
250.20	4.8
299.95	6.0
350.05	6.2
401.00	7.5

The estimated precision uncertainty in the measured mass, largely from reading error, is ± 0.05 g (95%) and the bias uncertainty, largely from calibration uncertainty, is ± 0.1 g (95%). Thus, the overall uncertainty in the mass is $U_M = 0.11$ g (95%), corresponding to about 0.05% of a typical load.

For the deflection, reading error is the most likely cause of precision uncertainty; the student estimates this uncertainty as ± 0.2 mm (95%). The bias uncertainty in the ruler, from manufacturing error, is estimated at ± 0.1 mm (95%). The overall uncertainty in the deflection is typically 5–10% of the measured value.

What is the stiffness of the beam, and what is its uncertainty?

Solution   The stiffness can be found by taking a least squares fit through the data, using the deflection as the y-variable, since it has a much greater precision uncertainty than does the mass. Setting $y = \delta$ and $x = M$, we calculate the required sums (perhaps using a pocket calculator subroutine that processes the entered data):

$$n = 9$$
$$\sum x = 1801 \text{ g}$$
$$\sum x^2 = 5.109 \times 10^5 \text{ g}^2$$
$$\sum y = 33.50 \text{ mm}$$
$$\sum y^2 = 179.3 \text{ mm}^2$$
$$\sum xy = 9959 \text{ g} \cdot \text{ mm}$$

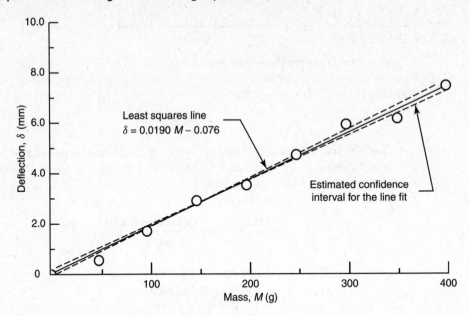

FIGURE 3.25: Beam deflection for various masses.

The least squares results are then

$$y = a + bx \qquad \left(\text{or } \delta = 0 + \frac{g}{K} M\right)$$

$$a = -0.0755 \text{ mm}$$

$$b = \frac{g}{K} = 0.0190 \text{ mm/g}$$

$$r = 0.995886$$

The data and the line fit are shown in Fig. 3.25. The experimental stiffness of the beam is

$$K = \frac{g}{b} = \frac{9.807}{0.0190} = 516 \text{ N/m}$$

From the figure, these data do appear to fall on a straight line. The correlation coefficient, r, is nearly unity, but a better test is to consider $(1 - r^2)^{1/2} = 0.0906$. This value indicates that the vertical standard deviation of the data (from precision error) is only about 9% of the total vertical variation caused by the straight-line relationship between y and x.

What is the uncertainty in the stiffness? The answer is equivalent to the uncertainty in the line's slope. One way to estimate this is just to vary the fitted line by eye to see what range of slopes fit the data with 95% certainty. These bounds are shown in Fig. 3.25 giving a variation in b of about 10%. Thus, the 95% uncertainty in K is also about $\pm 10\%$. A more precise statistical interval is calculated in Appendix F, but the result is essentially the same.

3.15 SUMMARY

Every measurement includes some level of error, and this error can never be known exactly. However, a probable bound on the error can usually be estimated. This bound is called *uncertainty*. Uncertainty should always be accompanied by the odds (or confidence percentage) that a particular error will fall within this bound. When presenting data either graphically or numerically, the uncertainty should also be shown.

1. Errors can usually be classified as either *bias* error or *precision* error. Bias (or systematic) errors occur same way for each measurement made. Precision (or random) errors vary in size and sign with a zero average value. Bias uncertainty must be estimated from our knowledge of the measuring equipment or by comparison to other, more accurate systems. Precision uncertainty can be estimated statistically (Sections 3.1, 3.2).

2. The total uncertainty in a measurement includes both bias and precision error:

$$U_x = \left(B_x^2 + P_x^2 \right)^{1/2}$$

The bias and precision uncertainty should have the same confidence level, typically 95% (Sections 3.3, 3.11).

3. Random variables, such as precision error, may be characterized in terms of a *population* and its *distribution*. The most common distribution for precision error is the *Gaussian* or *normal* distribution (Sections 3.4, 3.5).

4. Properties of a population are estimated by taking a *sample* from it. The sample mean, $\bar{x}$, and sample standard deviation, S_x, are used to estimate the population mean and population standard deviation. The precision uncertainty in $\bar{x}$ is

$$P_x = t_{\alpha/2,\nu} \frac{S_x}{\sqrt{n}} \quad (c\%)$$

with $\alpha = 1 - c$ and $\nu = n - 1$ (Sections 3.6, 3.11).

5. The χ^2-distribution may be used to compare the distribution of a sample to an expected distribution. The accuracy with which a set of data fit the Gaussian distribution can be checked using either the χ^2 "goodness-of-fit" test or a normal probability plot (Section 3.7).

6. Bias uncertainty and single-sample precision uncertainty are estimated from our knowledge of the measuring system (Sections 3.9, 3.11).

7. When experimental data are used to compute a final result, the uncertainty of the data must be *propagated* to determine the uncertainty in the result:

$$u_y = \sqrt{\left(\frac{\partial y}{\partial x_1} u_1 \right)^2 + \cdots + \left(\frac{\partial y}{\partial x_n} u_n \right)^2}$$

Here u_i is either a bias uncertainty, B_i, or a precision uncertainty, P_i (Sections 3.10, 3.11).

8. The accuracy of experiments can be improved before they are conducted by identifying and eliminating major sources of uncertainty (Sections 3.11.2, 3.12).

9. Graphical presentation of data is often the most effective way to convey your results and conclusions. Applying a few simple techniques when making your graphs can help ensure that your readers will fully and easily understand your work (Section 3.13).

10. Line fits enable your results to take an analytical form. Line fits also help to average out precision errors in data. The method of least squares will yield good line fits when (a) the y precision error is much larger than the x precision error or (b) both x and y precision errors are small compared to the overall variation of the data. Line fits do *not* compensate for bias error (Section 3.14).

SUGGESTED READINGS

ANSI/ASME PTC 19.1-1985. ASME Performance Test Codes, Supplement on Instruments and Apparatus, Part 1, Measurement Uncertainty. New York, 1985. This source contains a multitude of valuable references.

Barker, T. B. *Quality by Experimental Design*. 2nd ed. New York: Marcel Dekker, 1994.

Coleman, H. W., and W. G. Steele. *Experimentation and Uncertainty Analysis for Engineers*. 2nd ed. New York: Wiley-Interscience, 1999.

Collection of papers related to engineering measurement uncertainties. *Trans. ASME, J. Fluids Engrg*. 107, June 1985. This source contains a multitude of valuable references.

Johnson, R., I. R. Miller, and J. E. Freund. *Miller and Freund's Probability and Statistics for Engineers*. 7th ed. Englewood Cliffs, N.J.: Prentice Hall, 2004.

Kline, S. J., and F. A. McClintock. Describing uncertainties in single-sample experiments. *Mech. Eng.* 75: 3–8, January 1953.

Moffat, R. J. Contributions to the theory of single-sample uncertainty analysis. *Trans. ASME, J. Fluids Engrg*. 104, June 1982.

Tufte, E. R. *The Visual Display of Quantitative Information*. Chesire, Conn.: Graphics Press, 1983.

Weiss, N. A., and M. J. Hassett. *Introductory Statistics*. 5th ed. Reading, Mass.: Addison-Wesley, 1999.

PROBLEMS

Note: Unless otherwise specified, all uncertainties are assumed to represent 95% coverage (19 : 1 odds).

3.1. For a very large set of data, the measured mean is found to be 200 with a standard deviation of 20. Assuming the data to be normally distributed, determine the range within which 60% of the data are expected to fall.

3.2. From long-term plant-maintenance data, it is observed that pressure downstream from a boiler in normal operation has a mean value of 303 psi with a standard deviation of 33 psi. What is the probability that the pressure will exceed 350 psi during any one measurement in normal operation?

Note: The following information should be helpful in solving Problems 3.3–3.8.

For pure electrical elements in series

- Resistances add directly.
- Reciprocals of capacitances add to yield the reciprocal of the overall capacitance.
- Inductances add directly.

For pure electrical elements in parallel

- Reciprocals of resistances add to yield the reciprocal of the overall resistance.
- Capacitances add directly.
- Reciprocals of inductances add to yield the reciprocal of the overall inductance.

3.3. (a) A 68-kΩ resistor is paralleled with a 12-kΩ resistor. Each resistor has a ±10% tolerance. What will be the nominal resistance and the uncertainty of the combination?

(b) If the values remain the same except that the tolerance on the 68-kΩ resistor is dropped to ±5%, what will be the uncertainty of the combination?

3.4. Five 100-Ω resistors, each having a 5% tolerance, are connected in series. What is the overall nominal resistance and tolerance of the combination?

3.5. Three 1000-Ω resistors are connected in parallel. Each resistor has a ±5% tolerance. What is the overall nominal resistance and what is the best estimate of the tolerance of the combination?

3.6. A 47-Ω resistor is connected in series with a parallel combination of a 100-Ω resistor and a 180-Ω resistor. What is the overall resistance of the array and what is the best estimate of its tolerance?

(a) For individual tolerances equal to 1%

(b) For tolerances of the 47-Ω resistors equal to 10% and for the 180-Ω resistors equal to 5%

3.7. A capacitor of 0.05 μF ± 10% is parallel with a capacitor of 0.1 μF ± 10%.

(a) What are the resulting nominal capacitance and uncertainty?

(b) What would be the nominal capacitance and uncertainty if the two elements were connected in series?

3.8. Two inductances are connected in parallel. Their values are 0.5 mH and 1.0 mH. Each carries a tolerance of ±20%. Assuming no mutual inductance, what is the nominal inductance and uncertainty of the combination?

3.9. Power can be measured as $I^2 R$ or VI. Using the data of Problem 3.17, which method will give the most accurate measurement?

3.10. The volume of a cylinder is to be determined from measurements of the diameter and length. If the length and diameter are measured at four different locations by means of a micrometer with an uncertainty of 0.5% of reading, determine the uncertainty in the measurement.

Diameter	Length (in.)
3.9920	4.4940
3.9892	4.4991
3.9961	4.5110
3.9995	4.5221

3.11. A tube of circular section has a nominal length of 52 cm ± 0.5 cm, an outside diameter of 20 cm ± 0.04 cm, and an inside diameter of 15 cm ± 0.08 cm. Determine the uncertainty in calculated volume.

3.12. A cantilever beam of circular section has a length of 6 ft and a diameter of $2\frac{1}{2}$ in. A concentrated load of 350 lbf is applied at the beam end, perpendicular to the length of the beam. If the uncertainty in the length is ±1.5 in., in the diameter is ±0.08 in., and in the force is ±5 lbf, what is the uncertainty in the calculated maximum bending stress?

3.13. If it is determined that the overall uncertainty in the maximum bending stress for Problem 3.12 may be as great as, but no greater than, 6%, what maximum uncertainty may be tolerated in the diameter measurement if the other uncertainties remain unchanged?

3.14. It is desired to compare the design of two bolts based on their tensile strength capabilities. The following lists the results of the sample testing.

Group	Failure Load	Std. Deviation	Number of Tests
A	30 kN	2 kN	21
B	34 kN	6 kN	9

Is there a difference between the two samples at the 95% confidence level?

3.15. From a sample of 150 marbles having mean diameter of 10 mm and a standard deviation of ±3.4 mm, how many marbles would you expect to find in the range from 10 to 15 mm?

3.16. During laboratory testing of a thin-wall pressure vessel, the cylinder diameter and thickness were measured at 10 different locations; the resulting data were as follows:

$$\overline{D} = 10.25 \text{ in.} \quad S_D = 0.25 \text{ in.}$$
$$\overline{t} = 0.25 \text{ in.} \quad S_t = 0.05 \text{ in.}$$

If the pressure inside the vessel is measured to be 100 psi with an estimated uncertainty of ±10 psi, determine the best estimate of the tangential or hoop stress in the vessel. (*Note:* $\sigma_\theta = PD/2t$.)

3.17. In order to determine the power dissipated across a resistor, the current flow and resistance values are measured separately. If $I = 3.2$ A and $R = 1000$ Ω are measured values, determine the uncertainty if the following instruments are used:

Instrument	Resolution	Uncertainty (% of reading)
Voltmeter	1.0 mV	0.5%
Ohmmeter	1.0 Ω	0.1%
Ammeter	0.1 A	0.5%

3.18. A total of 120 hardness measurements are performed on a large slab of steel. If, using the Rockwell C scale, the mean of the measurements is 39 and the standard deviation is 4.0, how many of the measurements can be expected to fall between the hardness readings of 35 and 45?

3.19. In order to determine whether the use of a rubber backing material between a concrete compression sample and the platen of a testing machine affects the compressive strength, six samples with packing and six without were tested. The strengths are listed in the following table. Determine if the packing material has any effect. Use a 99% confidence level.

	Tensile Strength MN/m^2	
Sample No.	*With Packing*	*Without Packing*
1	2.48	2.18
2	2.76	2.48
3	2.96	2.38
4	2.72	2.00
5	2.62	2.10
6	2.65	2.28

3.20. Results from a chemical analysis for the carbon content of two materials are as follows:

	Carbon Content, %					
Material A	93.52	92.81	94.32	93.77	93.57	93.12
Material B	92.38	93.21	92.55	92.05	92.54	

Determine if there is a significant difference in carbon content at the 99% confidence level.

3.21. Figure 3.9 shows a histogram based on the values listed in Table 3.3. As suggested in Section 3.6.1, prepare histograms representing the data, based on (a) seven bins, (b) eight bins, and (c) ten bins.

3.22. The manufacturer of inexpensive outdoor thermometers checks a sample of ten against a 68°F standard. The following results were obtained:

68.5   67.5   67   69   68   67   67.5   69   69.

Using Student's *t*-test, calculate the range within which the population mean may be expected to lie with a confidence level of 95%.

3.23. Spacer blocks are manufactured in quantity to a nominal dimension of 125 mm. A sample of 12 blocks was selected and the following measurements were made.

1.28   1.32   1.29   1.23
1.26   1.26   1.20   1.29
1.24   1.23   1.26   1.22

Using Student's *t*-test, determine the upper and lower tolerance values within which the population mean may be expected to fall with a significance level of 10%.

3.24. In a laboratory it is suspected that the results from two different viscometers do not agree. Ten fluid samples were tested using apparatus A and corresponding samples were tested using apparatus B. The results are as follows:

Viscosity (Dimensionless)

Sample No.	Using Apparatus A	Using Apparatus B
1	72	73
2	43	45
3	54	56
4	75	75
5	50	53
6	48	50
7	73	72
8	55	54
9	48	48
10	50	52

Determine whether there is a significant difference in the two systems at the 99% confidence level.

3.25. Consider the equation

$$y = 1.0 - 0.2x + 0.01x^2 \qquad (0 \le x \le 3)$$

Determine the maximum uncertainty in y for $\pm 2\%$ uncertainty in the variable x.

3.26. For the following data determine the equation for $y = y(x)$ by graphical analysis.

x	0	0.43	0.76	1.21	2.60	3.5
y	1.00	1.54	3.61	5.25	10.0	13.50

3.27. For the following data, determine the equation $y = y(x)$ by graphical analysis.

x	1.21	1.35	2.75	5.1	8.1
y	12.0	18.2	88.0	325.0	800

3.28. For the following data, determine the equation $y = y(x)$ by graphical analysis.

x	0	0.43	2.6	2.9	4.3
y	94	71	26	19.5	11.5

3.29. The influence of the size of the test specimen on the tensile strength of an epoxy resin was determined by casting seven samples of each size and testing them accordingly. The experimental data are as follows:

Specimen Strengths (kN/m^2)

Sample of Small Specimens	Sample of Large Specimens
3475	1813
4326	3145
2262	4140
7415	6867
3418	3842
4404	3984
3788	3053

Determine whether there is a significant difference between the two samples at the 95% confidence level.

3.30. In 200 tosses of a coin, 116 heads and 84 tails were observed. Determine if the coin is fair using a confidence level of 95% or a significance level of 5%.

3.31. A random number table of 100 digits showed the following distribution of the digits $0, 1, 2, \ldots, 9$. Determine if the distribution of the digits differs significantly from the expected distribution at the 1% significance level.

Digit	0	1	2	3	4	5	6	7	8	9
Observed frequency	7	12	12	7	6	8	14	12	8	14

3.32. A quality control engineer wants to determine if the diameters of ball bearings produced by a machine are normally distributed. From a random sample of 300 bearings, he determines that the sample mean is 10.00 mm with a sample standard deviation of ± 0.10 mm. Moreover, he obtains the following frequency distribution for the diameters.

Diameter, mm	Observed Frequency
Under 9.80	8
9.80–under 9.90	42
9.90–under 10.00	107
10.00–under 10.10	97
10.10–under 10.20	38
10.20 and over	8

Are the bearing diameters normally distributed at the 5% significance level?

3.33. Using the data of Problem 3.32, construct a normal probability plot. What conclusions can you determine from this graphical representation regarding the normalcy of the data?

3.34. A sample of 100 test specimens of a steel alloy provides the following breaking strengths. Determine whether the data are normally distributed at the 1% significance level if the mean breaking strength is 67.45 ksi and the standard deviation is 2.92 ksi.

Breaking Strength, ksi	Observed Frequency
59.5–62.5	5
62.5–65.5	18
65.5–68.5	42
68.5–71.5	27
71.5–74.5	8

3.35. A company subcontracts the mass production of a die casting of fixed design. Four primary types of defects have been identified and records have been kept providing a "standard" against which defect distribution for batches may be judged. For a given batch of 2243 castings the following data apply. Do the batch data vary significantly from the standard?

Defect Identification	Distribution Percent	Results for Batch 2073
Type A	7.2	125
Type B	4.6	60
Type C	1.9	75
Type D	0.9	31
Nondefective	85.4	1952
	100.0	2243

3.36. A system is calibrated statically. The accompanying table lists the results.

Input	Output (Increasing Input)	Output (Decreasing Input)
0.12	1.6	2.2
0.17	2.7	2.3
0.27	3.7	3.2
0.32	3.9	4.2
0.38	4.3	4.8
0.46	5.6	5.2
0.53	6.7	6.5
0.64	7.4	7.4

(a) Plot output versus input.

(b) Calculate the best straight-line fit, first for the increasing output, then for the decreasing output and finally for the combined data.

(c) What is the maximum deviation in each case?

(d) If it is assumed that zero input should yield zero output, what is the zero offset (bias) that should be assigned?

3.37. The following data describe the temperature distribution along a length of heated pipe. Determine the best straight-line fit to the data.

Temperature, °C	Distance from a Datum, cm
100	11.0
200	19.0
300	29.0
400	39.0
500	50.5

3.38. The force-deflection data for a spring are given in the following table. Determine a least squares fit.

Deflection, in.	Force, lbf
0.10	9
0.20	19
0.30	22
0.40	40
0.50	52
0.60	59

3.39. Solve Problem 3.38 adding one more data point —namely, zero deflection under zero load.

3.40. The data in the accompanying tabulation (from several sources) provide the resistivity of platinum at various temperatures.

   (a) Make a linear plot of the data points.

   (b) Determine the constants for a linear least squares fit of the entire data set. Plot the fitted line on your graph of the data.

   (c) Because the resistivity is not a perfectly linear function of temperature, a more accurate fit can be obtained by limiting the range of temperature considered. Obtain the constraints for a linear least fit over the range of 0°C to 1000°C only. Plot the result on your graph.

Temperature, °C	Resistivity, $\Omega \cdot cm$
0	10.96
20	10.72
100	14.1
100	14.85
200	17.9
400	25.4
400	26.0
800	40.3
1000	47.0
1200	52.7
1400	58.0
1600	63.0

3.41. In constructing a spring-mass system, a deflection constant of 50 lbf/in. is required. Four springs are available, two having deflection constants of 25.0 lbf/in. with an uncertainty (tolerance) of ±2.0 lbf/in. and two having deflection constants of 100 lbf/in. with uncertainties of ±4.0 lbf/in. What combinations can be used for a system deflection constant of 50 lbf/in.? What will be the uncertainty in each case?

3.42. Show that $y = a + bx^n$ will plot as a straight line on linear graph paper when y is plotted as the ordinate and x^n is plotted as the abscissa. Show that the intercept is equal to a and the slope is equal to b.

3.43. Show that if $1/y$ versus $1/x$ is plotted on linear paper, the function $x/(ax + b)$ (which may also be written $1/y = a + b/x$) will yield a straight line, with a as the intercept and b as the slope.

3.44. Show that $y = ac^{bx}$ will plot as a straight line on linear paper when $\log y$ is plotted as the ordinate and x is plotted as the abscissa and that the intercept is equal to $\log a$ and the slope is equal to $b \log c$. Note that with the slope known, b and c may be found by simultaneous solution of the slope equation and the original equation written for a selected (x_i, y_i) point.

3.45. Select a range for x and make an x versus y plot of $y = 12x^{2/3}$ on linear graph paper. Now transform the data to $\log y$ and $\log x$ and plot on linear paper. The second set of data should plot as a straight line with a slope of $\frac{2}{3}$ and an intercept of $\log 12$.

3.46. From 1960 to 1983, the standard meter was defined as 1,650,763.73 wavelengths of the light emitted during the transition between the $2p_{10}$ and the $5d_5$ levels of the krypton-86 atom. That emission line is slightly asymmetric, and its wavelength has a total uncertainty of about $\pm 2.4 \times 10^{-5}$ Å (68%).

By the early 1970s, laser technology permitted highly-precise determination of the speed of light, c, by using the relation $c = \lambda f$ and measured values of laser wavelength, λ, and frequency, f. Laser frequency measurements had at that time reached relative uncertainties of $u_f/f = 6 \times 10^{-10}$ (68%).

(a) What was the uncertainty (95%) in the measured speed of light at that time? What factor limited the accuracy of this measurement?

(b) In 1983, the meter was redefined as "the distance traveled by light in vacuum during a time interval of 1/299792458 of a second." How did this affect the uncertainty in the speed of light (in meters per second)?

REFERENCES

[1] ANSI/ASME 19.1-1985. ASME Performance Test Codes. Supplement on Instruments and Apparatus, Part 1, *Measurement Uncertainty*. New York, 1985.

[2] Froome, K. D., and L. Essen. *The Velocity of Light and Radio Waves*. New York: Academic Press, 1969.

[3] Miller, I. R., J. E. Freund, and R. Johnson. *Probability and Statistics for Engineers*. 4th ed. Englewood Cliffs, N.J.: Prentice Hall, 1990.

[4] Weiss, N. A., and M. J. Hassett. *Introductory Statistics*. 5th ed. Reading, Mass.: Addison-Wesley, 1999.

[5] Schenk, J., Jr. *Theories of Engineering Experimentation*. 2nd ed. New York: McGraw-Hill, 1968.

[6] Kline, S. J., and F. A. McClintock. Describing uncertainties in single-sample experiments. *Mech. Engr.* 75: 3–8, January 1953.

[7] *Engineering and Scientific Graphs for Publications*, American Standards Association. New York: American Society of Mechanical Engineers, July 1947.

[8] Natrella, M. G. Experimental Statistics. *National Bureau of Standards Handbook* 91. Washington, D.C.: U.S. Government Printing Office, 1963.

[9] McClintock, F. A. *Statistical Estimation: Linear Regression and the Single Variable*, Research Memo 274, Fatigue and Plasticity Laboratory. Cambridge: Massachusetts Institute of Technology, February 14, 1987.

[10] Rabinowicz, E. *Introduction to Experimentation*. Reading, Mass.: Addison-Wesley, 1970.

[11] Hornbeck, R. W. *Numerical Methods*. New York: Quantum Publishers, 1975.

C H A P T E R 4

The Analog Measurand: Time-Dependent Characteristics

4.1 INTRODUCTION

A parameter common to all of measurement is time: All measurands have time-related characteristics. As time progresses, the magnitude of the measurand either changes or does not change. The time variation of any change is often fully as important as is any particular amplitude.

In this chapter we will discuss those quantities necessary to define and describe various time-related characteristics of measurands. As in Chapter 1, we classify time-related measurands as either

1. Static—constant in time
2. Dynamic—varying in time

    (a) Steady-state periodic

    (b) Nonrepetitive or transient

        i. Single-pulse or aperiodic

        ii. Continuing or random

4.2 SIMPLE HARMONIC RELATIONS

A function is said to be a *simple harmonic* function of a variable when its second derivative is proportional to the function but of opposite sign. More often than not, the independent variable is time t, although any two variables may be related harmonically.

Some of the most common harmonic functions in mechanical engineering relate displacement and time. In electrical engineering, many of the variable quantities in alternating-current (ac) circuitry are harmonic functions of time. The harmonic relation is quite basic to dynamic functions, and most quantities that are time functions may be expressed harmonically.

In its most elementary form, *simple harmonic motion* is defined by the relation

$$s = s_0 \sin \omega t \tag{4.1}$$

where

$s =$ instantaneous displacement from equilibrium position,
$s_0 =$ amplitude, or maximum displacement from equilibrium position,
$\omega =$ circular frequency (rad/s),
$t =$ any time interval measured from the instant when $t = 0$ s

A small-amplitude pendulum, a mass on a beam, a weight suspended by a rubber band—all vibrate with simple harmonic motion, or very nearly so.

By differentiation, the following relations may be derived from Eq. (4.1):

$$v = \frac{ds}{dt} = s_0 \omega \cos \omega t \tag{4.2}$$

and

$$v_0 = s_0 \omega \tag{4.2a}$$

Also,

$$a = \frac{dv}{dt} = -s_0 \omega^2 \sin \omega t \tag{4.3}$$

$$= -s \omega^2 \tag{4.3a}$$

In addition,

$$a_0 = -s_0 \omega^2 \tag{4.3b}$$

In the preceding equations,

$v =$ velocity,
$v_0 =$ maximum velocity or velocity amplitude,
$a =$ acceleration,
$a_0 =$ maximum acceleration or acceleration amplitude

Equation (4.3a) satisfies the description of simple harmonic motion given in the first paragraph of this section: The acceleration a is proportional to the displacement, s, but is of opposite sign. The proportionality factor is ω^2.

4.3   CIRCULAR AND CYCLIC FREQUENCY

The frequency with which a process repeats itself is called *cyclic frequency, f*, and is typically measured in cycles per second, or *hertz*: 1 Hz = 1 cycle/s. However, the idea of *circular frequency*, ω, is also useful in studying cyclic relations. Circular frequency has units of radians per second (rad/s). The connection between the two frequencies is conveniently illustrated by the well-known *Scotch-yoke* mechanism.

Figure 4.1(a) shows the elements of the Scotch yoke, consisting of a crank, *OA*, with a slider block driving the yoke-piston combination. If we measure the piston displacement from its midstroke position, the displacement amplitude will be $\pm OA$. If the crank turns at ω radians per second, then the crank angle θ may be written as ωt. This, of course, is convenient because it introduces time t into the relationship, which is not directly apparent in the term θ. Piston displacement may now be written as

$$s = s_0 \sin \omega t$$

which is the same as Eq. (4.1). One cycle takes place when the crank turns through 2π rad, and, if f is the frequency in hertz, then

$$\omega = 2\pi f \qquad (4.4)$$

Thus the displacement may instead be expressed in terms of cyclic frequency:

$$s = s_0 \sin 2\pi f t \qquad (4.5)$$

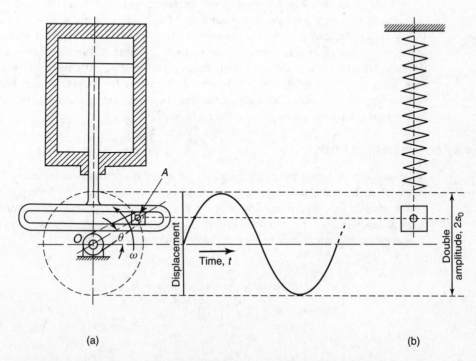

(a)   (b)

FIGURE 4.1: (a) The Scotch-yoke mechanism provides a simple harmonic motion to the piston; (b) a spring-mass system that moves with simple harmonic motion.

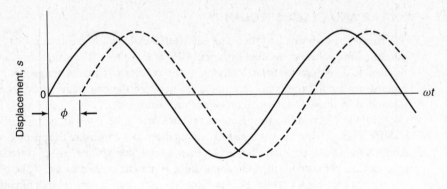

FIGURE 4.2: Motions that are out of phase. The dashed curve lags the solid curve by a phase angle ϕ.

Either displacement equation shows that the yoke-piston combination moves in simple harmonic motion. Many other mechanical and electrical systems display simple harmonic relationships. The spring-mass system shown in Fig. 4.1(b) is such an example. If its amplitude and natural frequency just happened to match the amplitude and frequency of the Scotch-yoke mechanism, then the mass and the piston could be made to move up and down in perfect synchronization.

To put it another way, *for every simple harmonic relationship, an analogous Scotch-yoke mechanism may be devised or imagined.* The crank length *OA* will represent the vector amplitude, and the angular velocity ω of the crank, in radians per second, will correspond to the circular frequency of the harmonic relation. If the mass and piston have the same frequencies and simultaneously reach corresponding extremes of displacement, their motions are said to be *in phase*. When they both have the same frequency but do not oscillate together, the time difference (lag or lead) between their motions may be expressed by an angle referred to as the *phase angle*, ϕ (Fig. 4.2).

4.4 COMPLEX RELATIONS

Most complex dynamic mechanical signals, steady state or transient, whether they are functions such as pressure, displacement, strain, or something else, may be expressed as a combination of simple harmonic components. Each component will have its own amplitude and frequency and will be combined in various phase relations with the other components. A general mathematical statement of this may be written as follows:

$$y(t) = \frac{A_0}{2} + \sum_{n=1}^{\infty}(A_n \cos n\omega t \pm B_n \sin n\omega t) \qquad (4.6)$$

where

$A_0, A_n,$ and B_n = amplitude-determining constants called *harmonic coefficients*,

n = integers from 1 to ∞, called *harmonic orders*

When n is unity, the corresponding sine and cosine terms are said to be *fundamentals*. For $n = 2, 3, 4$, and so on, the corresponding terms are referred to as second, third, fourth *harmonics*, and so on. Equation (4.6) is sometimes called a *Fourier series* for $y(t)$.

The variable part of Eq. (4.6) may be written in terms of either the sine or the cosine alone, by introducing a phase angle. Conversion is made according to the following rules:

Case I: For $y = A \cos x + B \sin x$,

$$y = C \cos(-x + \phi_2) \tag{4.6a}$$

or

$$y = C \sin(x + \phi_1) \tag{4.6b}$$

Case II: For $y = A \cos x - B \sin x$,

$$y = C \cos(x + \phi_2) \tag{4.6c}$$

or

$$y = C \sin(-x + \phi_1) \tag{4.6d}$$

In both cases, $C = \sqrt{A^2 + B^2}$; and ϕ_1 and ϕ_2 are *positive acute angles,* such that

$$\phi_1 = \tan^{-1} \frac{|A|}{|B|}$$

and

$$\phi_2 = \tan^{-1} \frac{|B|}{|A|}$$

Note that in calculating ϕ_1 and ϕ_2, A and B are taken as absolute values.

Although Eq. (4.6) indicates that all harmonics may be present in defining the signal-time relation, such relations usually include only a limited number of harmonics. In fact, all measuring systems have some upper and some lower frequency limits beyond which further harmonics will be attenuated. In other words, no measuring system can respond to an infinite frequency range.

Although it would be utterly impossible to catalog all the many possible harmonic combinations, it is nevertheless useful to consider the effects of a few variables such as relative amplitudes, harmonic orders n, and phase relations ϕ. Therefore, Figs. 4.3 through 4.7 are presented for two-component relations, in each case showing the effect of only one variable on the overall waveform. Figure 4.3 shows the effect of relative amplitudes; Fig. 4.4 shows the effect of relative frequencies; Fig. 4.5 shows the effect of various phase relations; Fig. 4.6 shows the appearance of the waveform for two components having considerably different frequencies; and Fig. 4.7 shows the effect of two frequencies that are very nearly the same.

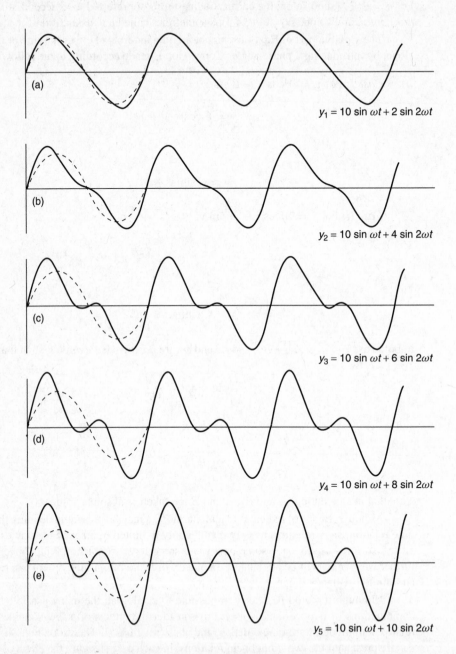

(a)

$$y_1 = 10 \sin \omega t + 2 \sin 2\omega t$$

(b)

$$y_2 = 10 \sin \omega t + 4 \sin 2\omega t$$

(c)

$$y_3 = 10 \sin \omega t + 6 \sin 2\omega t$$

(d)

$$y_4 = 10 \sin \omega t + 8 \sin 2\omega t$$

(e)

$$y_5 = 10 \sin \omega t + 10 \sin 2\omega t$$

FIGURE 4.3: Examples of two-component waveforms with second-harmonic component of various relative amplitudes.

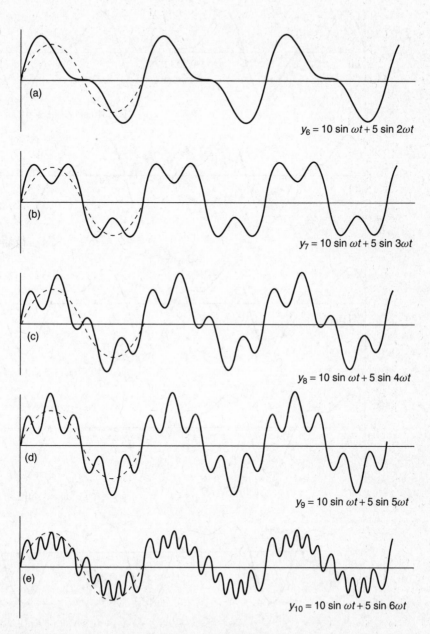

FIGURE 4.4: Examples of two-component waveforms with second term of various relative frequencies.

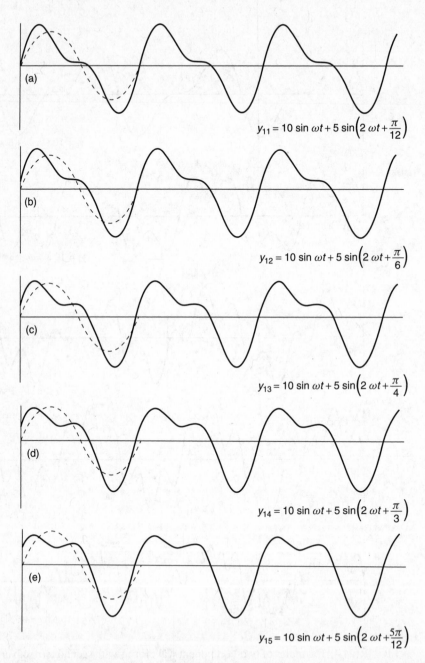

(a)

$$y_{11} = 10 \sin \omega t + 5 \sin\left(2 \omega t + \frac{\pi}{12}\right)$$

(b)

$$y_{12} = 10 \sin \omega t + 5 \sin\left(2 \omega t + \frac{\pi}{6}\right)$$

(c)

$$y_{13} = 10 \sin \omega t + 5 \sin\left(2 \omega t + \frac{\pi}{4}\right)$$

(d)

$$y_{14} = 10 \sin \omega t + 5 \sin\left(2 \omega t + \frac{\pi}{3}\right)$$

(e)

$$y_{15} = 10 \sin \omega t + 5 \sin\left(2 \omega t + \frac{5\pi}{12}\right)$$

FIGURE 4.5: Examples of two-component waveforms with second harmonic having various degrees of phase shift.

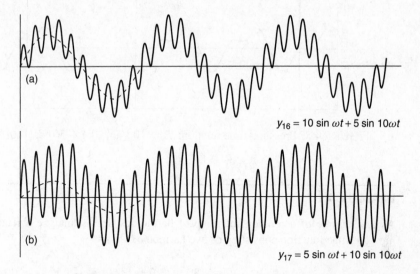

$$y_{16} = 10 \sin \omega t + 5 \sin 10\omega t$$

(a)

$$y_{17} = 5 \sin \omega t + 10 \sin 10\omega t$$

(b)

FIGURE 4.6: Examples of waveforms with the two components having considerably different amplitudes.

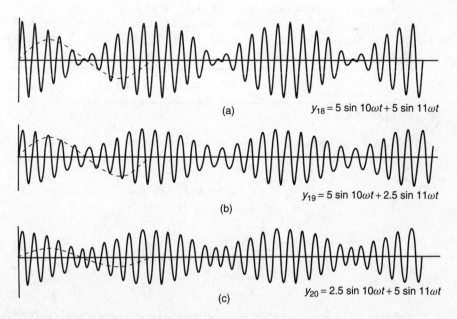

(a)

$$y_{18} = 5 \sin 10\omega t + 5 \sin 11\omega t$$

(b)

$$y_{19} = 5 \sin 10\omega t + 2.5 \sin 11\omega t$$

(c)

$$y_{20} = 2.5 \sin 10\omega t + 5 \sin 11\omega t$$

FIGURE 4.7: Examples of waveforms with the two components having frequencies that are nearly the same.

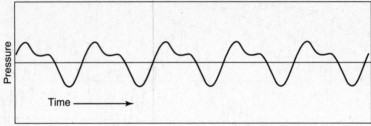

FIGURE 4.8: Pressure–time relation: $P = 100 \sin(80t) + 50 \cos(160t - \pi/4)$.

EXAMPLE 4.1

As an example of a relation made up of harmonics, let us analyze a relatively simple pressure-time function consisting of two harmonic terms:

$$P = 100 \sin 80t + 50 \cos \left(160t - \frac{\pi}{4} \right) \tag{4.7}$$

Solution   Inspection of the equation shows that the circular frequency of the fundamental has a value of 80 rad/s, or $80/2\pi = 12.7$ Hz. The period for the pressure variation is therefore $1/12.7 = 0.0788$ s. The second term has a frequency twice that of the fundamental, as indicated by its circular frequency of 160 rad/s. It also lags the fundamental by one-eighth cycle, or $\pi/4$ rad. In addition, the equation indicates that the amplitude of the fundamental, which is 100, is twice that of the second harmonic, which is 50. A plot of the relation is shown in Fig. 4.8.

EXAMPLE 4.2

As another example, let us analyze an acceleration-time relation that is expressed by the equation

$$a = 3800 \sin 2450t + 1750 \cos \left(7350t - \frac{\pi}{3} \right) + 800 \sin (36,750t) \tag{4.8}$$

where

$$a = \text{angular acceleration (rad/s}^2)$$
$$t = \text{time (s)}$$

Solution   The relation consists of three harmonic components having circular frequencies in the ratio 1 to 3 to 15. Hence the components may be referred to as the fundamental, the third harmonic, and the fifteenth harmonic. Corresponding frequencies are 390, 1170, and 5850 Hz.

4.4.1   Beat Frequency and Heterodyning

The situation shown in Fig. 4.7(a) is the basis of an important method of frequency measurement. Here two waves of equal amplitude and nearly equal frequency have been added. If one wave has a cyclic frequency of f_0 and the second wave has a frequency of $f_1 = f_0 + \Delta f$, then the resultant wave is

$$y = A \sin(2\pi f_0 t) + A \sin[2\pi(f_0 + \Delta f)t]$$

$$= 2A \underbrace{\cos\left(2\pi \frac{\Delta f}{2} t\right)}_{\text{slowly beating amplitude}} \cdot \underbrace{\sin\left(2\pi \frac{f_0 + f_1}{2} t\right)}_{\text{high-frequency wave}} \qquad (4.9)$$

This wave undergoes slow "beats" where the amplitude rises and falls. Although the cosine term in the amplitude has cyclic frequency of $\Delta f/2$, we see that the amplitude itself has two minima per cycle of the cosine. Thus, beats occur at a frequency of Δf.

   This kind of wave addition happens when a tuning fork is used to tune a musical instrument. The tuning fork and the musical instrument produce nearly equal tones, and, when the two sound waves are heard together, a lower beat frequency is also heard. The instrument is adjusted until the beat frequency is zero, so that the instrument's frequency is identical to that of the tuning fork.

   When the difference frequency, Δf, is much smaller than f_0, addition of waves allows us to measure Δf with less uncertainty than if we measured f_0 and f_1 separately and subtracted them. This technique for frequency measurement is called *heterodyning* (Section 10.6). It is very important in radio applications and in laser-doppler velocity measurements (Section 15.10).

EXAMPLE 4.3

Helium-neon laser light has a frequency of 473.8 THz (473.8×10^{12} Hz). A helium-neon laser beam is reflected from a moving target. This creates a doppler shift in the beam, which increases its frequency by 3 MHz (3×10^6 Hz).[1] The reflected beam is "added to" an unshifted beam of equal intensity, by using mirrors to bring the beams together. What is the resulting signal?

Solution   From Eq. (4.9), with $f_0 = 473.8\,\text{THz}$ and $f_1 = f_0 + \Delta f = 473.8\,\text{THz} + 3\,\text{MHz}$,

$$y = 2A \cos\left(2\pi \frac{\Delta f}{2} t\right) \sin\left(2\pi \frac{f_0 + f_1}{2} t\right)$$

$$= 2A \cos\left[2\pi(1.5 \times 10^6)t\right] \sin\left[2\pi(473.8 \times 10^{12})t\right]$$

The amplitude has a cyclic frequency of 1.5×10^6 Hz $= 1.5$ MHz, causing zeros in amplitude twice per cycle for a beat frequency of 3-MHz. This would be manifested

---

[1] A doppler shift is an apparent change in the frequency of a light or sound wave that occurs when the wave source and receiver are in motion relative to one another. One typical example is the change in pitch of a passing train's whistle.

as a 3-MHz variation between bright and dark at the point where the laser beams were added—a variation which could be detected by a sufficiently fast photodetector, such as a photomultiplier tube (Section 6.16). Note that the beat frequency is more than a 100 million times smaller than the frequency of the original light. To find this difference between f_0 and f_1 directly by subtraction, we would have needed to measure each frequency to an accuracy of better than 1 part in 100 million! As a result, the beats provide a much easier way to determine Δf.

4.4.2 Special Waveforms

A number of frequently used special waveforms may be written as infinite trigonometric series. Several of these are shown in Fig. 4.9. Table 4.1 lists the corresponding equations.

Both the square wave and the sawtooth wave are useful in checking the response of dynamic measuring systems. In addition, the skewed sawtooth form, Fig. 4.9(c), gives the voltage–time relation necessary for driving the horizontal sweep of a cathode-ray oscilloscope. All these forms may be obtained as voltage–time relations from electronic signal, or *function*, generators.

For each case shown in Fig. 4.9, all the terms in the infinite series are necessary if the precise waveform indicated is to be obtained. Of course, with increasing harmonic order, their effect on the whole sum becomes smaller and smaller.

As an example, consider the square wave shown in Fig. 4.9(a). The complete series includes all the terms indicated in the relation

$$y = \frac{4A}{\pi}\left(\sin \omega t + \frac{1}{3}\sin 3\omega t + \frac{1}{5}\sin 5\omega t + \cdots\right)$$

By plotting only the first three terms, which include the fifth harmonic, the waveform shown in Fig. 4.10(a) is obtained. Figure 4.10(b) shows the results of plotting terms through and including the ninth harmonic, and Fig. 4.10(c) shows the form for the terms including the fifteenth harmonic. As more and more terms are added, the waveform gradually approaches the square wave, which results from the infinite series.

The analytical calculation of the equations for Figs. 4.9(a) and (d) is performed in Appendix B.

4.4.3 Nonperiodic or Transient Waveforms

In the foregoing examples of special waveforms, various combinations of harmonic components were used. In each case the result was a periodic relation repeating indefinitely in every detail. Many mechanical inputs are not repetitive—for example, consider the acceleration-time relation resulting from an impact test [Fig. 4.11(a)]. Although such a relation is transient, it may be thought of as one cycle of a periodic relation in which all other cycles are fictitious [Fig. 4.11(b)]. On this basis, nonperiodic functions may be analyzed in exactly the same manner as periodic functions. If the nonperiodic waveform is sampled for a time period T, then the fundamental frequency of the fictitious periodic wave is $f = 1/T$ (cyclic) or $\omega = 2\pi/T$ (circular).

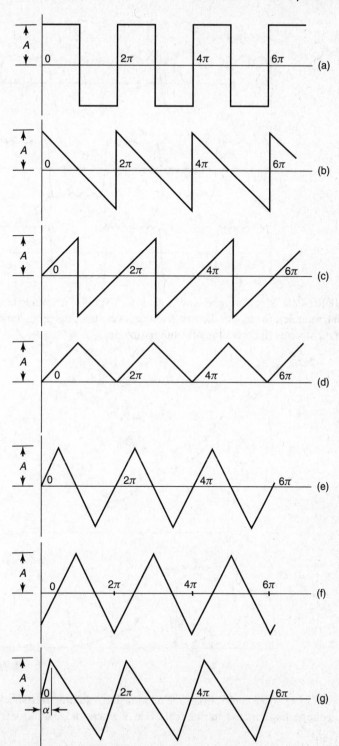

FIGURE 4.9: Various special waveforms of harmonic nature. In each case, the ordinate is y and the abscissa is ωt.

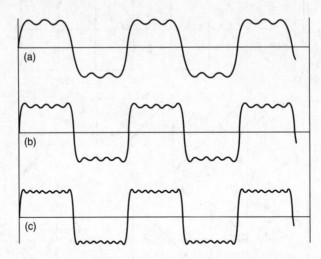

FIGURE 4.10: Plot of square-wave function: (a) plot of first three terms only (includes the fifth harmonic); (b) plot of the first five terms (includes the ninth harmonic); (c) plot of the first eight terms (includes the fifteenth harmonic).

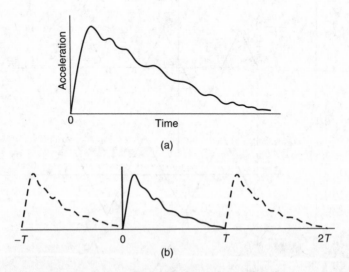

FIGURE 4.11: (a) Acceleration-time relationship resulting from shock test, (b) considering the nonrepeating function as one real cycle of a periodic relationship.

TABLE 4.1: Equations for Special Periodic Waveforms Shown in Fig. 4.9

Figure	Equation*
4.9(a)	$y = \dfrac{4A}{\pi}\left(\sin\omega t + \dfrac{1}{3}\sin 3\omega t + \dfrac{1}{5}\sin 5\omega t + \cdots\right) = \dfrac{4A}{\pi}\displaystyle\sum_{n=1}^{\infty}\left[\dfrac{1}{2n-1}\sin(2n-1)\omega t\right]$
4.9(b)	$y = \dfrac{2A}{\pi}\left(\sin\omega t + \dfrac{1}{2}\sin 2\omega t + \dfrac{1}{3}\sin 3\omega t + \cdots\right) = \dfrac{2A}{\pi}\displaystyle\sum_{n=1}^{\infty}\left[\dfrac{1}{n}\sin n\omega t\right]$
4.9(c)	$y = \dfrac{2A}{\pi}\left(\sin\omega t - \dfrac{1}{2}\sin 2\omega t + \dfrac{1}{3}\sin 3\omega t - \dfrac{1}{4}\sin 4\omega t + \cdots\right) = \dfrac{2A}{\pi}\displaystyle\sum_{n=1}^{\infty}\left[\dfrac{(-1)^{n+1}}{n}\sin n\omega t\right]$
4.9(d)	$y = \dfrac{A}{2} - \dfrac{4A}{(\pi)^2}\left(\cos\omega t + \dfrac{1}{(3)^2}\cos 3\omega t + \dfrac{1}{(5)^2}\cos 5\omega t + \cdots\right) = \dfrac{A}{2} - \dfrac{4A}{\pi^2}\displaystyle\sum_{n=1}^{\infty}\left[\dfrac{1}{(2n-1)^2}\cos(2n-1)\omega t\right]$
4.9(e)	$y = \dfrac{8A}{(\pi)^2}\left(\sin\omega t - \dfrac{1}{(3)^2}\sin 3\omega t + \dfrac{1}{(5)^2}\sin 5\omega t - \cdots\right) = \dfrac{8A}{(\pi)^2}\displaystyle\sum_{n=1}^{\infty}\left[\dfrac{(-1)^{n+1}}{(2n-1)^2}\sin(2n-1)\omega t\right]$
4.9(f)	$y = \dfrac{8A}{(\pi)^2}\left(\cos\omega t + \dfrac{1}{(3)^2}\cos 3\omega t + \dfrac{1}{(5)^2}\cos 5\omega t + \cdots\right) = \dfrac{8A}{(\pi)^2}\displaystyle\sum_{n=1}^{\infty}\left[\dfrac{1}{(2n-1)^2}\cos n\omega t\right]$
4.9(g)	$y = \dfrac{2A}{\alpha(\pi-\alpha)}\left(\sin\alpha\sin\omega t + \dfrac{1}{(2)^2}\sin 2\alpha\sin 2\omega t + \dfrac{1}{(3)^2}\sin 3\alpha\sin 3\omega t + \cdots\right) = \dfrac{2A}{\alpha(\pi-\alpha)}\displaystyle\sum_{n=1}^{\infty}\left[\dfrac{1}{n^2}\sin n\alpha\sin n\omega t\right]$

* n as used in these equations does not necessarily represent the harmonic order.

4.5   AMPLITUDES OF WAVEFORMS

The magnitude of a waveform can be described in several ways. The simplest waveform is a sine or cosine wave:

$$V(t) = V_a \sin 2\pi ft$$

The *amplitude* of this waveform is V_a. The *peak-to-peak amplitude* is $2V_a$. On the other hand, we may want a time-average value of this wave. If we simply average it over one period, however, we obtain an uninformative result:

$$\overline{V} = \frac{1}{T} \int_0^T V_a \sin 2\pi ft \, dt = -\frac{V_a}{2\pi}(\cos 2\pi - \cos 0) = 0$$

The net area beneath one period of a sine wave is zero. Thus it is more useful to work with a root-mean-square (rms) value:

$$V_{\text{rms}} = \sqrt{\frac{1}{T} \int_0^T V^2(t) \, dt} = \sqrt{\frac{1}{T} \int_0^T V_a^2 \sin^2 2\pi ft \, dt} = \cdots = \frac{V_a}{\sqrt{2}} \qquad (4.10)$$

For more complex waveforms, the frequency spectrum (Section 4.6) provides a complete description of the amplitude of each individual frequency component in the signal. However, the spectrum can be cumbersome to use, and a single time-average value is often more convenient to work with. For this reason, the rms amplitude is generally applied to complex waveforms as well. Meters that measure the rms value of a waveform are discussed in Sections 9.3–9.5.

4.6   FREQUENCY SPECTRUM

Figures 4.3 through 4.7 are plotted using time as the independent variable. This is the most common and familiar form. The waveform is displayed as it would appear on the face of an ordinary oscilloscope or on the paper of a strip-chart recorder. A second type of plot is the *frequency spectrum*, in which frequency is the independent variable and the amplitude of each frequency component is displayed as the ordinate. For example, the frequency spectra for the plots of Figs. 4.3(a) and (d) are shown in Fig. 4.12. Spectra corresponding to Figs. 4.4(a) and (c) are shown in Fig. 4.13, and the frequency spectrum for the square wave is shown in Fig. 4.14. Figures 4.12 through 4.14 use circular frequency ω; cyclic frequency f is often used instead.

The frequency spectrum is useful because it allows us to identify at a glance the frequencies present in a signal. For example, if the waveform results from a vibration test of a structure, we could use the frequency spectrum to identify the structure's natural frequencies.

The application of frequency spectrum plots has increased greatly since the development of the *spectrum analyzer* and *fast Fourier transform*. The spectrum analyzer is an electronic device that displays the frequency spectrum on a cathode-ray screen (Sections 9.10 and 18.6). The fast Fourier transform is a computer algorithm that calculates the frequency spectrum from computer-acquired data (see Section 4.7.1).

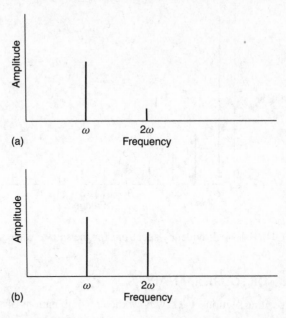

FIGURE 4.12: (a) Frequency spectrum corresponding to Fig. 4.3(a), (b) frequency spectrum corresponding to Fig. 4.3(d).

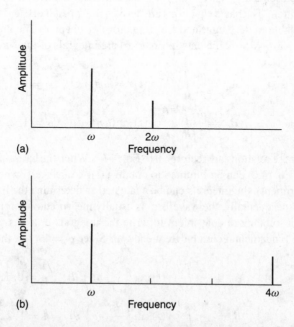

FIGURE 4.13: (a) Frequency spectrum corresponding to Fig. 4.4(a), (b) frequency spectrum corresponding to Fig. 4.4(c).

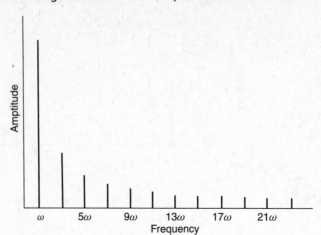

FIGURE 4.14: Frequency spectrum for the square wave shown in Fig. 4.9(a).

4.7 HARMONIC, OR FOURIER, ANALYSIS

In the preceding sections, we saw how known combinations of waves could be summed to produce more complex waveforms. In an experiment, the task is reversed: We measure the complex waveform and seek to determine which frequencies are present in it! The process of determining the frequency spectrum of a known waveform is called harmonic analysis, or Fourier analysis.[2]

Fourier analysis is a branch of classical mathematics on which entire textbooks have been written. The basic equations are derived in Appendix B, and more detailed discussions are available in the Suggested Readings for this chapter. The key to harmonic analysis is that the harmonic coefficients in Eq. (4.6) are integrals of the waveform $y(t)$

$$A_n = \frac{\omega}{\pi} \int_0^{2\pi/\omega} y(t) \cos(\omega n t)\, dt \qquad n = 0, 1, 2, \ldots, \qquad (4.11)$$

$$B_n = \frac{\omega}{\pi} \int_0^{2\pi/\omega} y(t) \sin(\omega n t)\, dt \qquad n = 1, 2, 3, \ldots. \qquad (4.12)$$

These relations are reciprocal to Eq. (4.6). When the harmonic coefficients are already known, Eq. (4.6) can be summed to obtain $y(t)$. Conversely, when $y(t)$ is known (as from an experiment), the integrals can be evaluated to determine the harmonic coefficients.

Experimentally, the waveform is usually measured only for a finite time period T. It turns out to be more convenient to write the integrals in terms of this time period, rather than the fundamental circular frequency, ω. Since $\omega = 2\pi/T$, the integrals are just

$$A_n = \frac{2}{T} \int_0^T y(t) \cos \left(\frac{2\pi}{T} n t \right) dt \qquad n = 0, 1, 2, \ldots, \qquad (4.13)$$

$$B_n = \frac{2}{T} \int_0^T y(t) \sin \left(\frac{2\pi}{T} n t \right) dt \qquad n = 1, 2, 3, \ldots \qquad (4.14)$$

[2]The term *spectral analysis* is also used.

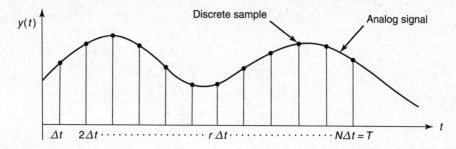

FIGURE 4.15: Discrete sampling of a continuous analog signal. The value of the signal is recorded at intervals Δt apart for a period T.

Practical harmonic analysis usually falls into one of the following four categories:

1. The waveform $y(t)$ is known mathematical function. In this case, the integrals (4.13) and (4.14) can be evaluated analytically. These calculations are illustrated in Appendix B for two cases.

2. The waveform $y(t)$ is an analog signal from a transducer. In this case, the waveform may be processed with an electronic spectrum analyzer to obtain the signal's spectrum (Sections 9.10, 18.5, and 18.6).

3. Alternatively, the analog waveform may be recorded by a digital computer, as discussed in Chapter 8. The computer will store $y(t)$ only at a series of discrete points in time. Integrals (4.13) and (4.14) are replaced by sums and evaluated, as discussed in Section 4.7.1.

4. The waveform is known graphically, for instance, from a strip-chart recorder or the screen of an oscilloscope. In this case, $y(t)$ may be read from the graph at a discrete series of points, and the integrals may again be evaluated as sums.[3]

4.7.1 The Discrete Fourier Transform

The case when $y(t)$ is known only at discrete points in time is very important in practice because of the wide use of computers and microprocessors for recording signals. Normally, a computer will read and store signal input at time intervals of Δt (Fig. 4.15). The computer records a total of N points over the time period $T = N\Delta t$.[4] Therefore, in the computer's memory, the analog signal $y(t)$ has been reduced to a series of points measured at times $t = \Delta t, 2\Delta t, \ldots, N\Delta t$, specifically, $y(\Delta t), y(2\Delta t), \ldots, y(N\Delta t)$. We can write this series more compactly as $y(t_r)$ by setting $t_r = r\Delta t$ for $r = 1, 2, \ldots, N$.

To perform a Fourier analysis of a discrete time signal like this, the integrals in Eqs. (4.13) and (4.14) must be replaced by approximate numerical integration in the form of summations. Likewise, the continuous time t is replaced by the discrete time $t_r = r\Delta t$,

---

[3] An example of this approach is given in [1], Section 4.7.

[4] Assume that the point at $t = 0$ is not recorded and that N is even.

and the period T is replaced by $N\Delta t$. Making these substitutions in Eq. (4.13), we get

$$A_n = \frac{2}{N\,\Delta t} \sum_{r=1}^{N} y(t_r)\cos\left(\frac{2\pi}{N\Delta t}nr\Delta t\right)\,\Delta t$$

$$= \frac{2}{N} \sum_{r=1}^{N} y(r\Delta t)\cos\left(\frac{2\pi rn}{N}\right)$$

Thus, the harmonic coefficients of a discretely sampled waveform are

$$A_n = \frac{2}{N} \sum_{r=1}^{N} y(r\Delta t)\cos\left(\frac{2\pi rn}{N}\right) \qquad n = 0, 1, \ldots, \frac{N}{2} \qquad (4.15)$$

$$B_n = \frac{2}{N} \sum_{r=1}^{N} y(r\Delta t)\sin\left(\frac{2\pi rn}{N}\right) \qquad n = 1, 2, \ldots, \frac{N}{2} - 1 \qquad (4.16)$$

for N an even number. The corresponding expression for the discrete waveform $y(t_r)$ is

$$y(t_r) = \frac{A_0}{2} + \sum_{n=1}^{N/2-1}\left[A_n\cos\left(\frac{2\pi rn}{N}\right) + B_n\sin\left(\frac{2\pi rn}{N}\right)\right] + \frac{A_{N/2}}{2}\cos(\pi r) \qquad (4.17)$$

Equations (4.15) and (4.16) are called the *discrete Fourier transform* (DFT) of $y(t_r)$ [2]. Equation (4.17) is called the *discrete Fourier series*.

In practice, the discrete sample is taken by an *analog-to-digital converter* (Section 8.11) connected to a computer or a microprocessor-driven electronic spectrum analyzer (Section 18.6). The computer or microprocessor evaluates the sums, Eqs. (4.15) and (4.16), often by using the fast Fourier transform algorithm.[5] The result, which approximates the spectrum of the original analog signal, is then displayed.

Like the ordinary Fourier series [Eq. (4.6)], the discrete Fourier series expresses $y(t)$ as a sum of frequency components. In particular,

$$\frac{2\pi rn}{N} = 2\pi\left(\frac{n}{N\,\Delta t}\right)r\,\Delta t = 2\pi(n\,\Delta f)t_r$$

for a fundamental frequency (in hertz) of

$$\Delta f \equiv \frac{1}{N\,\Delta t} = \frac{1}{T} \qquad (4.18)$$

---

[5]The *fast Fourier transform*, or *FFT*, algorithm is special factorization of these sums that applies when N is a power of 2 ($N = 2^m$, for m an integer). The number of calculations normally required to evaluate these sums is proportional to N^2; when the FFT algorithm is used, the number is proportional to $N\log_2 N$. Thus, the FFT requires fewer calculations when N is large.

and harmonic orders $n = 1, \ldots, N/2$. In other words,

$$y(t) = \frac{A_0}{2} + \sum_{n=1}^{N/2-1} [A_n \cos(2\pi n \, \Delta f t) + B_n \sin(2\pi n \, \Delta f t)]$$

$$+ \frac{A_{N/2}}{2} \cos\left(2\pi \frac{N}{2} \Delta f t\right) \quad (4.19)$$

Note that the DFT yields *only* harmonic components up to $n = N/2$, whereas the ordinary Fourier series [Eq. (4.6)] may have an infinite number of frequency components. This very important fact is a consequence of the discrete sampling process itself.

4.7.2   Frequencies in Discretely Sampled Signals: Aliasing and Frequency Resolution

When an analog waveform is recorded by discrete sampling, some care is needed to ensure that the waveform is accurately recorded. The two sampling parameters that we can control are the *sample rate*, $f_s = 1/\Delta t$, which is the frequency with which samples are recorded, and the number of points recorded, N. Typically, the software controlling the data-acquisition computer will request values of f_s and N as input.

Figure 4.16 shows two examples of sampling a particular waveform. In Fig. 4.16 (a), the sample rate is low (Δt is large), and as a result the high frequencies of the original waveform are not well resolved by the discrete samples: The signal seen in the discrete sample (the dashed curve) does not show the sharp peaks of the original waveform. The total time period of sampling is also fairly short (N is small), and thus the low frequencies of the signal are missed as well: It isn't clear how often the signal repeats itself. In Fig. 4.16(b), the sample rate and the number of points are each increased, improving the resolution of both high and low frequencies. The sample rate and total sampling time period clearly determine how well a discrete sample represents the original waveform.

What is the minimum sample rate needed to resolve a particular frequency? Consider the cases shown in Fig. 4.17, where a signal of frequency f is sampled at increasing rates. In (a), when the waveform is sampled at a frequency of $f_s = f$, the discretely sampled signal appears to be constant! No frequency is seen. In (b), the waveform is sampled at a higher rate, between f and $2f$; the discrete signal now appears to be a wave, but it has a frequency *lower* than f. In (c), the waveform is sampled at a rate $f_s = 2f$, and the discrete sample appears to be a wave of the correct frequency, f. Unfortunately, if the sampling begins a quarter-cycle later at this same rate, as in (d), then the signal again appears to be constant. Only when the sample rate is increased *above* $2f$, as in (e), do we always obtain the correct signal frequency with the discrete sample.

The highest frequency resolved at a given sampling frequency is determined by the *Nyquist frequency*,

$$f_{\text{Nyq}} = \frac{f_s}{2} = \frac{1}{2\,\Delta t} \quad (4.20)$$

Signals with frequencies lower than $f_{Nyq} = f_s/2$ are accurately sampled. Signals with frequencies greater than or equal to f_{Nyq} are not accurately sampled and appear as lower frequencies in the discrete sample.

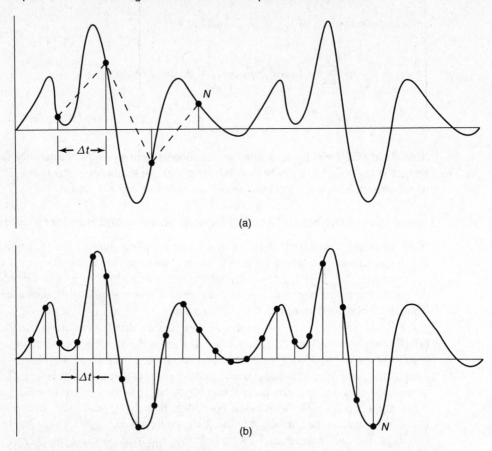

(a)

(b)

FIGURE 4.16: Effect of sample rate and number of samples taken: (a) undersampled, both sample rate and number of points are too low; (b) resolution of waveform is improved by raising the sample rate and number of points.

The phenomenon of a discretely sampled signal taking on a different frequency, as in Fig. 4.17(b), is called *aliasing*.[6] Aliasing occurs whenever the Nyquist frequency falls below the signal frequency. Furthermore, the *phase ambiguity* shown by Figs. 4.17(c) and (d) prohibits sampling at the Nyquist frequency itself. To prevent these problems, the sampling frequency should always be chosen to be *more than twice* a signal's highest frequency, as in Fig. 4.17(e).

The Nyquist frequency tells us how to correctly resolve the highest frequencies of a signal. It also tells us why the DFT contains only a finite number of frequency components: Frequencies higher than the Nyquist are too fast to be resolved at the given sample rate.

In a similar fashion, we can determine the lowest nonzero frequency found by the DFT and explain the discrete spacing of frequencies. If a waveform is to be resolved by discrete sampling, one or more *full periods* of that waveform must be present in the discrete sampling period, as shown for one wave in Fig. 4.18(a). Since the period of sampling is

---

[6]An incorrectly sampled signal takes on a new identity, or *alias*.

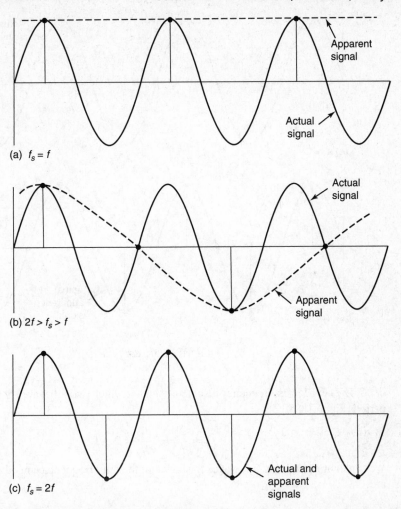

FIGURE 4.17: Effect of varying the sample rate, f_s, on the apparent signal obtained by discrete sampling. (*Continued on page 130*)

$T = N \, \Delta t = N/f_s$, the frequency of this wave is

$$f_{\text{lowest}} = \frac{1}{T} = \frac{1}{N \, \Delta t} = \frac{f_s}{N} \equiv \Delta f \qquad (4.21)$$

Thus the fundamental frequency of the DFT, Δf, is also that of the lowest-frequency full wave that fits within the sampling period. No lower frequency (other than $f = 0$) is resolved.

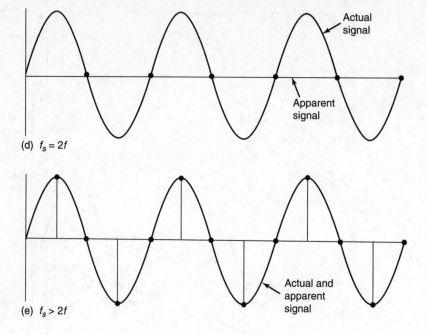

(d) $f_s = 2f$

(e) $f_s > 2f$

FIGURE 4.17: *continued*

The next-lowest frequency resolved is that for which two full waves fit in the sampling period [Fig. 4.18(b)]:

$$f_2 = \frac{2}{T} = 2\frac{f_s}{N} = 2\,\Delta f$$

We can continue adding full waves to show that the only frequencies resolved by the DFT are

$$0, \Delta f, 2\,\Delta f, \ldots, n\,\Delta f, \ldots, \frac{N}{2}\Delta f = f_{\text{Nyq}}$$

Note that although the Nyquist frequency itself is present in the DFT, it may not correctly represent the underlying signal, owing to phase ambiguity.

The frequencies of the DFT are spaced in increments of Δf, and thus Δf is sometimes called the *frequency resolution* of the DFT. If an analog signal contains a frequency f_0 that lies between two resolved frequencies, say $n\,\Delta f < f_0 < (n+1)\,\Delta f$, then this frequency component will "leak" to the adjacent frequencies of the DFT. The adjacent frequencies can each show some contribution from f_0. As a result, each frequency component observed in the DFT has an uncertainty in frequency of approximately $\pm\Delta f/2$ (95%) relative to the frequencies actually present in the original analog signal. We can reduce leakage and sharpen the peaks in the frequency spectrum by decreasing Δf.

This discussion leads us to the following steps for accurate discrete sampling:

1. First, estimate the highest frequency in the signal and choose the Nyquist frequency to be greater than it. In other words, make the sample rate, f_s, *greater than twice* the highest frequency in the signal.

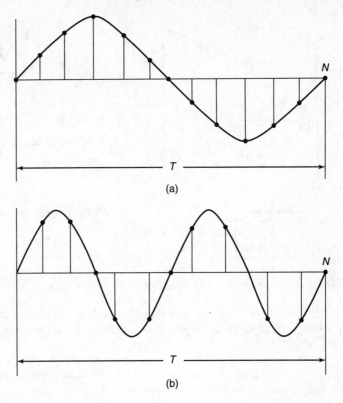

FIGURE 4.18: Resolving low frequencies: (a) one full wave in the sampling period, (b) two full waves in the sampling period.

2. If limitations in the sample rate force you to pick a Nyquist frequency less than the highest frequency in the signal, then use a low-pass filter (Sections 7.16–7.18) to block frequencies greater than the Nyquist frequency so that those higher frequencies will not be aliased into your results.

3. After the sample rate is chosen, estimate the lowest frequency in the signal or estimate the frequency resolution needed to accurately resolve the frequency components in the signal. Then choose the number of points in the sample, N, to yield the desired $\Delta f = f_s/N$ at the previously determined value of the sampling frequency, f_s.

4.7.3   An Example of Discrete Fourier Analysis

Figure 4.19 illustrates a simple experiment. Each of three signal generators was set to produce sine waves and connected to a loudspeaker. Generator A was set to 500 Hz, generator B to 1000 Hz, and generator C to 1500 Hz. A microphone, which converts sound pressure to voltage, was used to measure the sound level. The sound level from each speaker was individually adjusted so that the microphone voltage displayed on the oscilloscope was 50 mV in amplitude. Then all three sources were run simultaneously, so that the three waveforms were mixed, and the voltage signal produced was recorded by the computer.

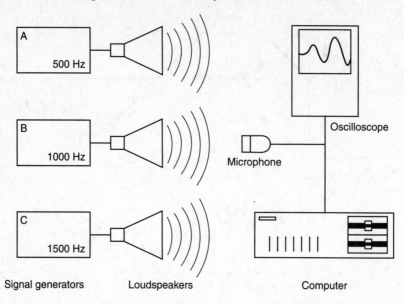

FIGURE 4.19: Experimental setup in which pure tones from three sound sources are mixed, detected by a microphone, displayed on an oscilloscope, and recorded by a computer.

Note that the signal generators were not in phase. The data acquired are shown in Table 4.2.

The computer sampled the microphone voltage at a rate of $f_s = 9000$ Hz, corresponding to a Nyquist frequency of

$$f_{\text{Nyq}} = 4500 \text{ Hz}$$

The lowest frequency in the signal was known to be 500 Hz, so 18 points were used in the DFT ($N = 18$) to yield a frequency resolution of

$$\Delta f = \frac{f_s}{N} = 500 \text{ Hz}$$

Thus, the samples were taken at intervals of 0.11 ms ($\Delta t = 1/f_s = 0.11$ ms), covering a period of $N \, \Delta t = 2.00$ ms.

The test data were analyzed using the computer's fast Fourier transform program. The resulting harmonic coefficients are listed in Table 4.3. Only harmonic components up to $n = N/2 = 9$ can be determined with the DFT (or FFT), and the last of these is the Nyquist frequency component, which suffers from phase ambiguity.

The first, second, and third harmonics were originally set to amplitudes of 50 mV. Since each wave is phase shifted [recall Eqs. (4.6a-d)], we consider the amplitude of the sum of sine and cosine waves at each frequency, $C_n = \sqrt{A_n^2 + B_n^2}$. This frequency spectrum is shown in Fig. 4.20. For the first and third harmonic, the calculated amplitudes are within 10% of what we thought they should be. The second harmonic is 40% higher than we thought. Apart from the lack of precision of the data and any approximation introduced in the DFT calculation, the test condition itself may have contributed to this discrepancy. The test was run in a conventional laboratory environment, and sound reflections from the walls

TABLE 4.2: Experimental Data for Microphone Voltage

Time (ms)	Voltage (mV)
0.11	10
0.22	30
0.33	70
0.44	65
0.56	15
0.67	−40
0.78	−50
0.89	15
1.00	100
1.11	135
1.22	90
1.33	−40
1.44	−130
1.56	−140
1.67	−110
1.78	−50
1.89	−10
2.00	0

TABLE 4.3: Calculated Harmonic Coefficients for the Data of Table 4.2

Harmonic Order, n	Frequency, $n\Delta f$, Hz	A_n, mV	B_n, mV	C_n, mV
0	0	−1.04	0.00	1.04
1	500	−29.11	46.83	55.14
2	1000	49.09	52.70	72.02
3	1500	−21.96	−47.41	52.25
4	2000	2.78	2.23	3.56
5	2500	1.97	−0.47	2.03
6	3000	−0.96	−0.51	1.09
7	3500	0.15	2.93	2.93
8	4000	−0.37	−1.11	1.17
9	4500	−1.08	0.00	1.08

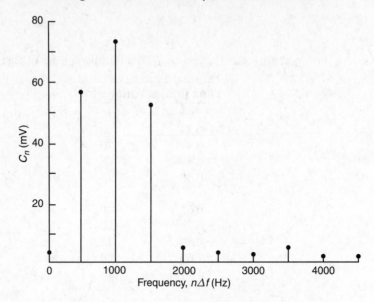

FIGURE 4.20: Frequency spectrum for data of Table 4.2.

and ceiling may have also contributed to some distortion of the original signals. Ideally, the test should have been run in an anechoic chamber.

The sum of the measured harmonic components [Eq. (4.19)] is plotted together with the data in Fig. 4.21. Visually, the computed curve appears to fit the data perfectly. The DFT calculation reconstructs the original signal to within the accuracy of the original data.

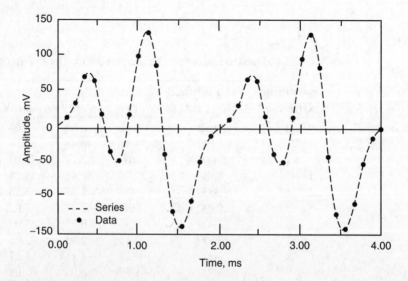

FIGURE 4.21: Comparison of the calculated Fourier series with the measured test data.

4.8 SUMMARY

Mechanical and electrical measuring systems often produce time-varying output signals. We have seen that even fairly complex signals may be broken down and analyzed as a mixture of harmonic components, each having different frequency and amplitude. While reading the following chapters, keep in mind that all dynamic inputs, those whose magnitudes vary with time, are in reality only combinations of simple sinusoidal building blocks.

1. Simple harmonic motion, such as $s = s_0 \sin \omega t$, is the most basic form of time-dependent behavior. The frequency of such motion can be described by either the cyclic frequency, f, or the circular frequency, $\omega = 2\pi f$. The relationship of these two frequencies may be visualized in terms of the Scotch-yoke mechanism (Sections 4.2, 4.3).

2. More complex waveforms can be represented by a sum of simple harmonic components having different amplitudes and frequencies. Even fairly sharp waveforms, such as the square wave, can be described by such sums. These sums are called Fourier series (Sections 4.4, 4.4.2).·

3. When two waves of nearly equal frequency are added, the resulting waveform undergoes periodic beats at a frequency of one-half the frequency difference of the original waves (Section 4.4.1).

4. When a nonperiodic waveform is recorded for a time interval T, the recorded portion may be viewed as a periodic waveform of period T and frequency $f = 1/T$ (Section 4.4.3).

5. The average amplitude of a complex signal is often described using the root-mean-square, or rms, value (Section 4.5).

6. The frequencies present in a complex waveform may be described using the frequency spectrum, which shows the amplitude of each frequency component present (Section 4.6).

7. The process of determining the frequency spectrum of a complex signal is called harmonic analysis or Fourier analysis. Several methods of harmonic analysis are available, depending on the nature of the signal being studied (Section 4.7, Appendix B).

8. When a signal is recorded by a computer, only discrete points are stored. The discrete Fourier transform, or DFT, may be used to find the frequency spectrum of the recorded data (Section 4.7.1, Appendix B).

9. Accurate discrete sampling can be ensured by selecting appropriate values of the sampling frequency, f_s, and the number of sample points recorded, N. The correct values are determined by (a) the Nyquist frequency, $f_{Nyq} = f_s/2$, which must be greater than the highest frequency in the signal; and (b) the frequency resolution, $\Delta f = f_s/N$, which is both the lowest frequency seen in the discrete signal and the spacing of frequencies in the signal's DFT (Section 4.7.2).

SUGGESTED READINGS

Churchill, R. V., and J. W. Brown. *Fourier Series and Boundary Value Problems*. 3rd ed. New York: McGraw-Hill, 1978.

Greenberg, M. D. *Foundations of Applied Mathematics*. Englewood Cliffs, N.J.: Prentice Hall, 1978.

Oppenheim, A. V., A. S. Willsky, and S. H. Nawab. *Signals & Systems*. 2nd ed. Upper Saddle River, N.J.: Prentice Hall, 1997.

PROBLEMS

4.1. The following expression represents the displacement of a point as a function of time:

$$y(t) = 100 + 95 \sin 15t + 55 \cos 15t$$

(a) What is the fundamental frequency in hertz?

(b) Rewrite the equation in terms of cosines only.

4.2. Rewrite each of the following expressions in the form of Eq. (4.6).

(a) $y = 3.2 \cos(0.2t - 0.3) + \sin(0.2t + 0.4)$

(b) $y = 12 \sin(t - 0.4)$

4.3. Construct a frequency spectrum for Fig. 4.9(a).

4.4. Construct a frequency spectrum for Fig. 4.9(c).

4.5. Construct a frequency spectrum for Fig. 4.9(e).

4.6. Construct a frequency spectrum for Fig. 4.9(f).

4.7. Construct a frequency spectrum for Fig. 4.9(g).

4.8. Figure 4.22 represents a trace from an oscilloscope where the ordinate is in volts and the abscissa is in milliseconds. Determine its discrete Fourier transform and sketch its frequency spectrum. Use a sampling frequency of 2000 Hz. Check for periodicity in choosing your data window.

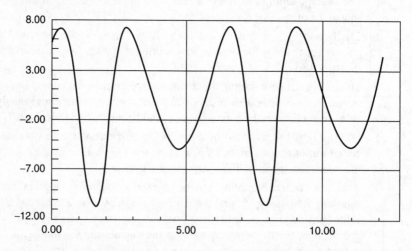

FIGURE 4.22: Oscilloscope trace for Problem 4.8.

4.9. Consider a pressure–time record as shown in Fig. 4.23. Determine its frequency spectrum.

4.10. If the signal of Problem 4.9 were to be sampled digitally for its discrete Fourier transform, what sampling frequency would you recommend?

4.11. Solve Problem 4.8 using a sampling frequency of 1000 Hz and compare your results with those of Problem 4.8.

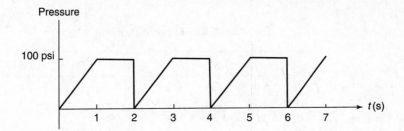

FIGURE 4.23: Pressure–time record for Problem 4.9.

4.12. Using the data files in Table 4.4, if t is in milliseconds and $f(t)$ is in volts, determine the discrete Fourier transform for each set of digital data.

    (**a**) Use $f_1(t)$.
    (**b**) Use $f_2(t)$.
    (**c**) Use $f_3(t)$.
    (**d**) Use $f_4(t)$.
    (**e**) Use $f_5(t)$.
    (**f**) Use $f_6(t)$.

4.13. A 500-Hz sine wave is sampled at a frequency of 4096 Hz. A total of 2048 points are taken.

    (**a**) What is the Nyquist frequency?
    (**b**) What is the frequency resolution?
    (**c**) The student making the measurement suspects that the sampled waveform contains several harmonics of 500 Hz. Which of these can be accurately measured? What happens to the others?

4.14. A 150-Hz cosine wave is sampled at a rate of 200 Hz.

    (**a**) Draw the wave and show the temporal locations at which it is measured.
    (**b**) What apparent frequency is measured?
    (**c**) Describe the relation of the measured frequency to aliasing. Give a numerical justification for your answer.

4.15. (**a**) Suppose that a 500-Hz sinusoidal signal is sampled at 750 Hz. Draw the discrete time signal found and determine the apparent frequency of the signal.
    (**b**) If a 200-Hz component were present in the signal of part (a), would it be detected? Explain.
    (**c**) If a 375-Hz component were present in the signal of part (a), would it be detected? Explain.

TABLE 4.4: Data for Problem 4.12

t	$f_1(t)$	$f_2(t)$	$f_3(t)$	$f_4(t)$	$f_5(t)$	$f_s(t)$
0	5.4	3.76	10.2	4	10.1	−9.4
1	4.74	3.88	10.0	2.58	9.34	−6.8
2	3.01	4.19	9.83	0.99	8.32	−3.5
3	0.8	4.54	9.57	1.54	7.05	−0.2
4	−1.2	4.67	9.26	4.4	5.57	2.52
5	−2.4	4.46	8.92	8.02	3.91	4.11
6	−2.7	4.06	8.56	10.7	2.14	4.58
7	−2.4	3.73	8.21	11.6	0.3	4.17
8	−1.8	3.6	7.89	11.2	−1.6	3.3
9	−1.6	3.55	7.62	10	−3.4	2.47
10	−1.8	3.35	7.43	8.52	−5.1	2.03
11	−2.4	2.89	7.34	7.19	−6.6	2.11
12	−2.7	2.36	7.36	6.66	−7.9	2.57
13	−2.4	2.01	7.5	7.3	−9	3.05
14	−1.2	1.98	7.74	8.45	−9.9	3.11
15	0.8	2.14	8.07	8.46	−10	2.36
16	3.01	2.24	8.48	5.85	−11	0.64
17	4.74	2.16	8.94	0.9	−11	−1.9
18	5.4	2.06	9.41	−4	−10	−4.9
19	4.74	2.16	9.87	−6	−9.3	−7.6
20	3.01	2.57	10.3	−4.5	−8.3	−9.3
21	0.8	3.13	10.6	−1.5	−7.1	−9.5
22	−1.2	3.56	10.9	−0.9	−5.6	−7.9
23	−2.4	3.73	11.1	−4.6	−3.9	−4.5
24	−2.7	3.76	11.2	−11	−2.1	0.06
25	−2.4	3.88	11.3	−15	−0.3	4.93
26	−1.8	4.19	11.2	−15	1.55	9.15
27	−1.6	4.54	11.2	−10	3.35	11.8
28	−1.8	4.67	11.1	−5.1	5.05	12.4
29	−2.4	4.46	10.9	−3.7	6.59	10.7
30	−2.7	4.06	10.8	−6.7	7.94	7.1
31	−2.4	3.73	10.7	−11	9.04	2.29
32	−1.2	3.6	10.6	−12	9.87	−2.7
33	0.8	3.55	10.5	−8.5	10.4	−7
34	3.01	3.35	10.4	−2.4	10.6	−9.7
35	4.74	2.89	10.3	2.57	10.5	−11
36	5.4	2.36	10.2	4	10.1	−9.4

4.16. An engineer is studying the vibrational spectrum of a large diesel engine. Her modeling estimates suggest that a strong resonance is likely at 250 Hz, and that weaker frequencies of up to 2000 Hz may be excited also. She has placed an accelerometer on the machine to measure the vibration spectrum. She samples the accelerometer output voltage using her computer's analog-to-digital converter board.

   (**a**) What is the minimum sample rate she should use?
   (**b**) To reliably test her model of the machine's vibration, she must resolve the peak resonant frequency to ±1 Hz. How can she achieve this level of resolution?

4.17. A temperature measuring circuit responds fully to frequencies below 8.3 kHz; above this frequency, the circuit attenuates the signal. This circuit is to be used to measure a temperature signal with an unknown frequency spectrum. Accuracy of ±1 Hz is desired in the frequency components. If no frequency components above 8.3 kHz are present in the circuit's output, what sample rate and number of samples should be used to sample the output?

4.18. A cantilever beam of stiffness k supports a large mass m on its free end. The vibrational frequency of the beam approximately equal to

$$f = \frac{1}{2\pi}\sqrt{\frac{k}{m}}$$

In an experiment, this frequency was measured by attaching a strain gage to the beam to produce a waveform corresponding to the oscillating motion. The waveform was then discretely sampled using the lab computer, and the frequency of motion was obtained from an FFT of the waveform.

   For one particular case, the mass at the end of the beam was measured to be 60.10 g, to an uncertainty of ±0.11 g (95%). The waveform was sampled at a rate of 128 Hz for 128 points, and a peak frequency of 10 Hz was returned by the FFT calculation.

   (**a**) Calculate the beam stiffness and its uncertainty (95%).
   (**b**) What is the primary source of uncertainty in this result? What is the best way to reduce the uncertainty of the result?
   (**c**) The experimenter's disk is nearly full, so he does not wish to increase the number of points sampled when he repeats the experiment. If the sample rate can be adjusted in increments of 1 Hz, what sample rate would allow the lowest uncertainty with the same number of points? Estimate the uncertainty in stiffness for that sample rate.

REFERENCES

[1] Beckwith, T. G., and R. D. Marangoni. *Mechanical Measurements*. 4th ed. Reading, Mass.: Addison-Wesley, 1990.

[2] Oppenheim, A. V., A. S. Willsky, and S. H. Nawab. *Signals & Systems*. 2nd ed. Upper Saddle River, NJ: Prentice Hall, 1997.

C H A P T E R 5

The Response of Measuring Systems

5.1 INTRODUCTION

Quite simply, *response* is a measure of a system's fidelity to purpose. It may be defined as an evaluation of the system's ability to faithfully sense, transmit, and present all the pertinent information included in the measurand and to exclude all else.

We would like to know if the output information truly represents the input. If the input information is in the form of a sine wave, a square wave, or a sawtooth wave, does the output appear as a sine wave, a square wave, or a sawtooth wave, as the case may be? Is each of the harmonic components in a complex wave treated equally, or are some attenuated, completely ignored, or perhaps shifted timewise relative to the others? These questions are answered by the response characteristics of the particular system—that is, (1) amplitude response, (2) frequency response, (3) phase response, and (4) slew rate.

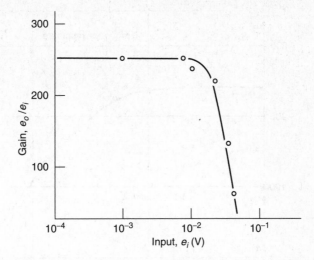

FIGURE 5.1: Gain versus input voltage for amplifier section of a commercially available strain measuring system for a frequency of 1 kHz (gain = output voltage/input voltage).

5.2 AMPLITUDE RESPONSE

Amplitude response is governed by the system's ability to treat all input amplitudes uniformly. If an input of 5 units is fed into a system and an output of 25 indicator divisions is obtained, we can generally expect that an input of 10 units will result in an output of 50 divisions. Although this is the most common case, other special nonlinear responses are also occasionally required. Whatever the arrangement, whether it be linear, exponential, or some other amplitude function, discrepancy between design expectations in this respect and actual performance results in poor amplitude response.

Of course no system exists that is capable of responding faithfully over an unlimited range of amplitudes. All systems can be overdriven. Figure 5.1 shows the amplitude response of a voltage amplifier suitable for connecting a strain-gage bridge to an oscilloscope. The usable range of the amplifier is restricted to the horizontal portion of the curve. The plot shows that for inputs above about 0.01 V the amplifier becomes overloaded and the amplification ceases to be linear.

5.3 FREQUENCY RESPONSE

Good frequency response is obtained when a system reacts to all frequency components in the same way. If a 100-Hz sine wave with an input amplitude of 5 units is fed into a system and a peak-to-peak output of $2\frac{1}{2}$ cm results on an oscilloscope screen, we can expect that a 500-Hz sine-wave input of the same amplitude would also result in a $2\frac{1}{2}$ cm peak-to-peak output. Changing the frequency of the input signal should not alter the system's output magnitude so long as the input amplitude remains unchanged.

Yet here again there must be a limit to the range over which good frequency response may be expected. This is true for any dynamic system, regardless of its quality. Figure 5.2 illustrates the frequency response relations for the same voltage amplifier used in Fig. 5.1. Frequencies above about 10 kHz are attenuated. Only inputs below this frequency limit are amplified in the correct relative proportion.

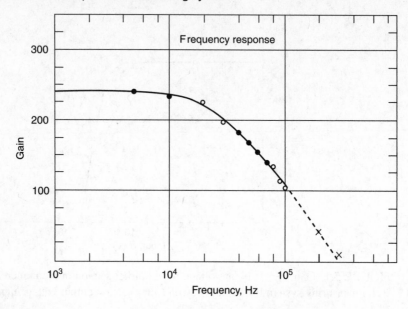

FIGURE 5.2: Frequency response curve for amplifier section of a commercially available strain-measuring system; $e_i = 10$ mV.

5.4 PHASE RESPONSE

Amplitude and frequency responses are important for all types of input waveforms, whether simple or complex. *Phase response*, however, is of primary importance for the complex wave only.

Time is required for the transmission of a signal through any measuring system. Often, when a simple sine-wave voltage is amplified by a single stage of amplification, the output trails the input by approximately 180°, or one-half cycle (see Fig. 5.20). For two stages, the phase shift may be about 360°, and so on. The actual shift will not be an exact multiple of half-wavelengths but will depend on the equipment and the frequency. It is the frequency dependence that defines phase response.

For a single-sine-wave input, any phase shift would normally be unimportant. The output produced on the oscilloscope screen would show the true waveform, and its proper parameters could be determined. The fact that the shape being shown was actually formed on the screen a few microseconds or a few milliseconds after being generated is of no consequence.

Let us consider, however, a complex wave made up of numerous harmonics. If the phase lag is different for each frequency, then each component of the complex wave is delayed by a different amount. The harmonic components then emerge from the system in phase relations different from when they entered. The whole waveform and its amplitudes are changed, a result of poor phase response.

Figure 5.3 illustrates phase response characteristics for a typical voltage amplifier.

5.5 PREDICTING PERFORMANCE FOR COMPLEX WAVEFORMS

Response characteristics of an existing system or a component of a system may be determined experimentally by injecting as input a signal of known form, then determining the

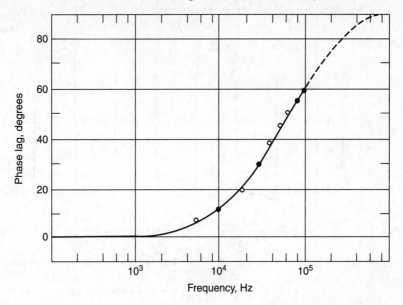

FIGURE 5.3: Phase lag versus frequency (phase response) for the same amplifier used in Fig. 5.2.

output, and finally comparing the results (see Section 10.5). Of course, the most basic test waveform is the sine wave.

   If we know the sine-wave response of a device, can we use this information to predict how it will respond to a complex input, such as a square wave or one of the various sawtooth waveforms? The answer is yes, as we demonstrate in the following example.

EXAMPLE 5.1

Using a computer program and information given in Figs. 5.2 and 5.3, predict the form of amplifier output to be expected if a perfect square wave is the input. Do this for input having fundamental frequencies of 1000 and 2000 Hz and amplitude of 1 mV.

Solution   Recall that in Table 4.1 a square wave is defined by the infinite series

$$y = \frac{4A}{\pi} \sum_{n=1}^{\infty} \frac{1}{(2n-1)} \sin\left[(2n-1)2\pi f t\right] \tag{5.1}$$

We may modify Eq. (5.1) by introducing frequency and phase distortion factors:

$$y = \frac{4A}{\pi} \sum_{n=1}^{\infty} \frac{G_n}{(2n-1)} \sin\left[(2n-1)2\pi f t - \phi_n\right] \tag{5.1a}$$

where

   G_n = an amplitude factor based on frequency,

   ϕ_n = a phase distortion factor

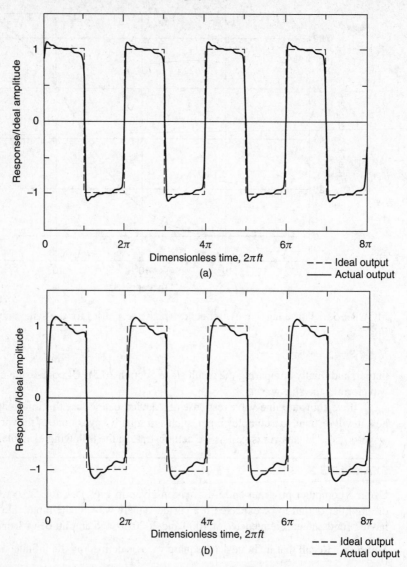

FIGURE 5.4: (a) Computer-determined response to a 1000-Hz square wave by the amplifier whose characteristics are shown in Figs. 5.1 through 5.3, (b) computer-determined response to a 2000-Hz square wave by the amplifier whose characteristics are shown in Figs. 5.1 through 5.3. (Ideal amplitude = 240 mV.)

Magnitudes for G_n and ϕ_n for each harmonic order can be extracted from the response curves, Figs. 5.2 and 5.3. For example, if $n = 15$, then $(2n - 1)f = 29(1000 \text{ Hz}) = 29 \text{ kHz}$, and we read $G_{15} = 200$ from Fig. 5.2 and $\phi_{15} = 30°$ from Fig. 5.3. The computer code can incorporate tables of such frequency-response and phase-response data taken directly from the sine-wave response curves. The results of using these data in Eq. (5.1a) are plotted in Figs. 5.4(a) and (b).

We can make similar calculations for any waveform for which a harmonic series can be written. In particular, to investigate a measuring system's response to a waveform of interest, we can make a Fourier analysis of that waveform and investigate the system's response characteristics using sine-wave test results, as before.

5.6   DELAY, RISE TIME, AND SLEW RATE

Finally, a fourth type of response, which is actually another form of frequency response, is delay, or rise time. When a stepped or relatively instantaneous input is applied to a system, the output may lag, as shown in Fig. 5.5. The time delay after the step is applied, but before proper output magnitude is reached, is known as rise time. It is a measure of the system's ability to handle transients. Sometimes rise time is defined specifically as the time, Δt, required for the system to pass from 10% to 90% of its final response. Alternatively, transient response may be characterized by the *settling time* required for the system response to remain within some small percentage of its final value.

Slew rate is the *maximum rate* of change that the system can handle. In electrical terms, it is de/dt, or volts per unit time (e.g., 25 V/μs). When the voltages changes rapidly, the system can respond no faster than the slew rate.

5.7   RESPONSE OF EXPERIMENTAL SYSTEM ELEMENTS

An experimentalist can usually avoid operating a measuring system under conditions when amplitude response or slew rate are limiting factors. For example, a solid-state amplifier

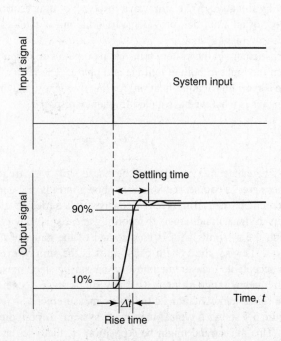

FIGURE 5.5: Response of a typical system to a step input, showing rise time (Δt) and settling time.

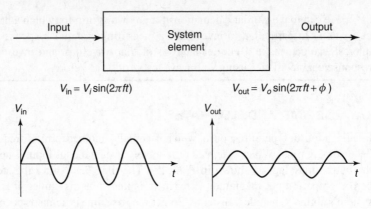

FIGURE 5.6: Response of a system element to a sine-wave input.

that is overloaded typically produces a constant output of ±5 V irrespective of the input; such a condition should be fairly obvious, and its result is clearly useless.

Frequency- and phase-response limitations are not so readily apparent. One would prefer zero phase shift and flat frequency response for all frequencies of interest. However, because experimental inputs often contain a wide range of frequencies, poor response to some subset of frequencies may go unnoticed. An evaluation procedure is generally necessary to establish a particular system's response.

We have seen that a sine-wave test can provide the frequency and phase response frequency by frequency. The sine-wave response of a measuring system element, such as a transducer or an amplifier, provides a foundation for evaluating the performance of the overall measurement system.

Each element in the system transfers its input to an output (Fig. 5.6). The output can differ from the input in both amplitude and phase. The ratio of output amplitude to input amplitude is the gain (or amplification) for that frequency, and the variation of the gain over all frequencies is what we have called frequency response,

$$G(f) = \frac{V_o}{V_i}$$

Similarly, we called the variation of the phase shift with frequency the phase response, $\phi(f)$. These are, of course, the functions shown in Figs. 5.2 and 5.3.[1]

A complete measuring system consists of a series of elements, from sensors and transducers to signal conditioners to recording and display devices. For any given frequency component, the system's gain is the product of the gains of all system elements at that frequency. Likewise, the system phase shift is the sum of every stage's phase shift. To obtain an acceptable measurement, every measuring stage must have acceptable response over the frequency range of interest. If any single stage does not respond properly, it will distort the signal and contaminate the entire measurement.

Figure 5.7 shows a typical measuring system. A periodic measurand is detected by a sensor. This measurand might be a position, x, that oscillates at a frequency f_x. The

---

[1] The reader who has studied system dynamics will recognize that $G(f)$ and $\phi(f)$ are the magnitude and phase, respectively, of the periodic transfer function, $H(j\omega)$, of a linear system [1].

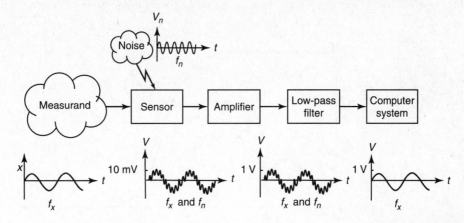

FIGURE 5.7: Measuring system composed of several system elements; the signal is shown as it leaves each element.

sensor produces an output voltage in millivolts, which varies with x. However, the sensor also picks up high-frequency electrical noise from nearby equipment. Thus, the output of the sensor includes both the low frequency of the signal, f_x, and the high frequency of the noise, f_n (see Fig. 4.6). The sensor's signal is received by an amplifier (gain = 100), which raises the signal level to a value convenient for computer recording. Apart from the added noise, both measuring stages respond faithfully to the input signal.

Certain system elements are chosen specifically for the limitations of their frequency-response characteristics. In the case of Fig. 5.7, the measurand varies at a fairly low frequency, whereas the unwanted electrical noise is at a much higher frequency. Since the experimentalist has evaluated the output signal on an oscilloscope, she has a clear idea of which range of frequencies characterizes the measurand and which range characterizes the noise. Thus, she inserted a signal-conditioning *low-pass filter* after the amplifier to eliminate the noise. This filter has flat frequency response at low frequencies (such as f_x) and zero response at high frequencies (such as f_n), as shown in Fig. 5.8. The signal that the filter transmits to the computer recording system excludes the noise component. We discuss filters in more detail when we consider signal conditioning in Chapter 7 (Sections 7.16–7.18).

In contrast to the filter, a well-designed sensor or transducer should respond to all frequencies equally. Unfortunately, most actual sensors and transducers do not. Instead, such devices are characterized by an upper or lower frequency beyond which response is attenuated (much like a filter) or by a high or low frequency at which the sensor resonates with the input, producing an output that is ridiculously large. Determination of such limiting frequencies is extremely important if a dynamic measurand is to be accurately recorded.

The frequency and phase response of a system element (or of the entire chain of elements) can be determined in several ways. For simple elements, we may be able to construct a physical model of the device that accurately predicts its response. For more complex systems, we may wish to test the response experimentally, for example, by using a sine-wave test. In other circumstances, we may rely on response data provided by the manufacturer. In any case, once we have determined the range of frequencies for which the system responds accurately, we will disregard frequencies outside this range, if they can

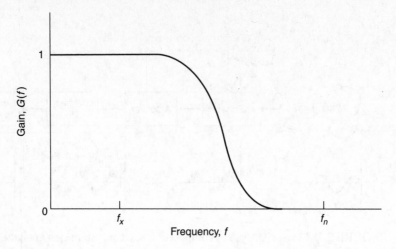

FIGURE 5.8: Frequency response for the low-pass filter of Fig. 5.7.

be identified, and we will take precautions to prevent such frequencies from entering the measuring chain.

The remainder of this chapter is concerned with the identification of system response. Sections 5.8 through 5.19 address the modeling of frequency and phase response for simple physical systems. Section 5.20 returns to the matter of experimental determination and calibration of system response.

5.8 SIMPLIFIED PHYSICAL SYSTEMS

What basic physical factors govern response? In terms of practical systems, we are confronted with two fundamental types of construction: mechanical and electrical. The basic mechanical elements are mass, damping, and some form of equilibrium-restoring element, such as a spring. Corresponding electrical elements are inductance, resistance, and capacitance. Although many, if not most, devices and systems involve both electrical and mechanical elements, for our immediate purposes it is advantageous to consider the two separately. In the next several sections we will discuss some of the mechanical aspects; and beginning with Section 5.17, we will consider the electrical.

5.9 MECHANICAL ELEMENTS

A discussion of the dynamic characteristics of an elementary mechanical system necessitates a short description of the elements composing such a system.

5.9.1 Mass

It is obvious that in all cases mass will be a factor. Under certain conditions, however, the masses making up the device or system will not affect its performance. We will consider such cases in Sections 5.14 and 5.15.

By its very nature, mass must be distributed throughout some volume. In many cases, however, it is not only convenient but also correct, or nearly so, to assume that the mass of a member is concentrated at a point. Depending on the geometry of the member and its application, the point of concentration may or may not be the center of gravity. In certain cases, the center of percussion may be the location of effective concentration.

5.9.2 Spring Force

Many mechanical members deflect in direct proportion to the force exerted on them, that is, $\Delta F / \Delta s = k = $ a constant, where ΔF is an applied force increment and Δs is the resulting deflection increment. Most coil springs, beams, and tension/compression members abide by this relationship. It may be noted that the force is opposed to the deflection; that is, the resulting force always attempts to restore equilibrium.

Torsional members commonly adhere to the relationship $\Delta T / \Delta \theta = k_t = $ a constant, where ΔT is an applied torque increment and $\Delta \theta$ is the resulting torsional deflection increment. The constants k and k_t are called *spring constants*, or *deflection constants*.

Elasticity is not always the source of the restoring force, however. In certain cases, such as for a beam balance (see Section 5.10), the restoring force may be supplied by gravity.

When the motion of a concentrated mass is constrained by an equilibrium-seeking member [Fig. 5.9(a)], simple vibration theory shows that the combination will have a natural frequency

$$\omega_n = \sqrt{\frac{k g_c}{m}} \tag{5.2}$$

where

$$\omega_n = \text{circular frequency in radians per second (see Section 4.3)}$$
$$= 2\pi f,$$
$$f = \text{frequency of vibration in hertz,}$$
$$g_c = \text{the dimensional constant (see Table 2.6)}$$

A system of this sort is said to have a *single degree of freedom*; that is, it is assumed to be constrained in some way to oscillate in a single mode or manner, needing only one coordinate to fully describe its motion.

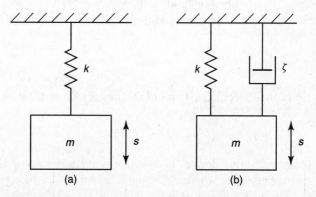

(a)   (b)

FIGURE 5.9: Elementary spring-mass systems: (a) without damping; (b) with viscous damping.

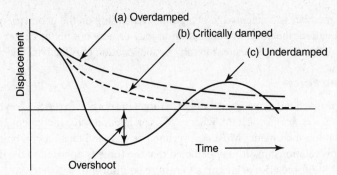

FIGURE 5.10: Time-displacement relations for damped motion: (a) for damping greater than critical; (b) for critical damping; (c) for damping less than critical.

5.9.3 Damping

Another factor important to the usefulness of any system of this type is damping [Fig. 5.9(b)]. Damping in this connection is usually thought of as viscous, rather than Coulomb or frictional, and may be obtained by fluids (including gases) or by electrical means.

Viscous damping is a function of velocity, and the force opposing the motion may be expressed as

$$F = -\zeta \frac{ds}{dt} \tag{5.3}$$

where ζ = the damping coefficient and ds/dt = the velocity. We can see that the damping coefficient is an evaluation of force per unit velocity. The negative sign indicates that the resulting force opposes the velocity. The effect of viscous damping on a freely vibrating single-degree-of-freedom system is to reduce the vibrational amplitudes with respect to time according to an exponential relation.

Damping magnitude is conveniently thought of in terms of *critical damping*, which is the minimum damping that can be used to prevent overshoot when a damped spring-mass system is deflected from equilibrium and then released (see Section 5.16.1 for further discussion). This limiting condition is shown in Fig. 5.10. The value of the critical damping coefficient, ζ_c, for a simple spring-supported mass m is expressed by the relation

$$\zeta_c = 2\sqrt{\frac{mk}{g_c}} \tag{5.4}$$

Damping is often specified in terms of the dimensionless damping ratio,

$$\xi = \frac{\zeta}{\zeta_c}$$

Many measuring devices or system components involve elements constrained by gravity or spring force, whose deflection is analogous to the signal input. The ordinary balance scale is an example, as is the D'Arsonval meter movement (Fig. 9.3). The same is true of most pressure transducers, elastic-force transducers, and many other measuring devices.

If the system is a translational one, a spring-constrained mass may be involved. Our tire gage of Section 1.4 illustrates the case in which the piston and stem and a portion of the spring constitute the mass whose motion is controlled by the interaction of the applied pressure and the spring force. Other examples are the seismic-type instruments discussed in Chapter 17.

These devices depend on equilibrium for correct indication. When equilibrium is disturbed by a change of input, the system requires time to readjust to the new equilibrium, and a number of oscillations may take place before the new output is correctly indicated. The rate at which the amplitude of such oscillations decreases is a function of the system's damping. In addition, the frequency of oscillation is a function of both damping and sensitivity.

5.10   AN EXAMPLE OF A SIMPLE MECHANICAL SYSTEM

Let us consider a symmetrical scale beam *without* damping (Fig. 5.11). For simplification, we will assume that the masses of the scale pans and the weights being compared are concentrated at points A and B and that they are also included in the mass moment of inertia I, which is referred to the main pivot point O. Further, we will assume that a small difference, ΔW, exists between the two weights being compared and that points A, B, and O lie along a straight line. We define sensitivity, η, as the ratio of the displacement of the end of the pointer to the length of the pointer, h, divided by ΔW, or

$$\eta = \frac{1}{\Delta W}\frac{d}{h} = \frac{1}{\Delta W}\tan\theta \qquad (5.5)$$

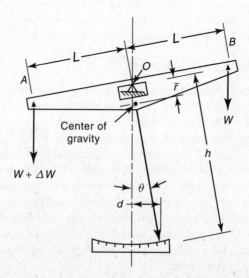

FIGURE 5.11: Schematic diagram of a beam balance.

The system behaves as a compound pendulum, and it can be shown that the period of oscillation will be

$$T = 2\pi \sqrt{\frac{I}{\bar{r} w_b g_c}}$$ (5.6)

where

I = the mass moment of inertia,

$\bar{r}$ = the distance between the center of gravity of the beam alone and pivot point O,

w_b = the weight of the beam

With the weights applied,

$$w_b \bar{r} \sin \theta = \Delta W \, L \cos \theta \quad \text{or} \quad \tan \theta = \frac{L \Delta W}{w_b \bar{r}}$$ (5.7)

Hence, using Eq. (5.5), we find that

$$\eta = \frac{L}{w_b \bar{r}}$$ (5.7a)

Combining Eqs. (5.6) and (5.7a), we have

$$\eta = \frac{L}{I} \left(\frac{T}{2\pi} \right)^2 \cdot g_c = \frac{L}{I} \left(\frac{1}{2\pi f} \right)^2 \cdot g_c$$ (5.7b)

where f = natural frequency.

Equation (5.7b) indicates that the sensitivity is a f nction of T, the *period of oscillation* of the balance scale, with increased sensitivity corresponding to a long beam and low moment of inertia. In other words, the more sensitive instrument oscillates more slowly than the less sensitive instrument. This is an important observation having significant bearing on the dynamic response of most single-degree-of-freedom instruments.

5.11   THE IMPORTANCE OF DAMPING

The importance of proper damping to dynamic measurement may be understood by assuming that our scale beam in the previous example is part of an instrument that is required to come to *different* equilibria as rapidly as possible.

Suppose that our scale beam has very low damping. When a disturbing force is applied, the scale will be caused to oscillate, and the oscillation will continue for a long period of time. A final balance will be obtained only by prolonged waiting, and thus the frequency with which the weighing process may be repeated is limited.

On the other hand, suppose considerable damping is provided—well above critical. An extreme example of this would be to submerge the entire scale in a container of molasses. Balance would be approached at a very slow rate again, but in this case there would be no oscillation. Here, again, excessive time would be required before the next weighing operation could commence.

It appears, therefore, that if we were to design a beam-type scale for quickly determining magnitudes of different masses, the final form would necessarily be a result of compromise. We would like equilibrium to be reached as quickly as possible in order to *get on with the job*. It would seem that there might be an optimum value that should be used. Although this is not exactly the case because of other factors involved, damping of the order of 60% to 75% of critical is provided in many instruments of this type (see Section 5.16.2 for further clarification of this point).

Although damping will tend to decrease the frequency of oscillation, it does not change the inherent sensitivity of the device, which is related to the undamped natural frequency. However, some compromises must still be made in regard to sensitivity. Sensitivity increases in proportion to the undamped natural period, as shown in the previous section. Because a high natural period (or low natural frequency) usually corresponds to a lessened frequency response, sensitivities greater than those required by the application should be avoided in the interest of maintaining adequate response.

5.12   DYNAMIC CHARACTERISTICS OF SIMPLIFIED MECHANICAL SYSTEMS

By making certain simplifying modeling assumptions, we may place the dynamic characteristics of *most* measuring systems in one of several categories. The basic assumptions are that any restoring element (such as a spring) is linear, that damping is viscous, and that the system may be approximated as a single-degree-of-freedom system.

5.13   SINGLE-DEGREE-OF-FREEDOM SPRING-MASS-DAMPER SYSTEMS

Figure 5.12 shows a simple single-degree-of-freedom mechanical system. It is single degree because only one coordinate of motion is necessary to completely define the motion of the system. We will also assume a general form of excitation, $F(t)$, which may or may not be periodic. Forces acting on the mass will result from the spring, damping, and the external force, $F(t)$. Using Newton's second law, we can write

$$F(t) - ks - \zeta \frac{ds}{dt} = \frac{m}{g_c} \left(\frac{d^2 s}{dt^2} \right) \tag{5.8}$$

Note that the spring force will always oppose the displacement and that the damping is proportional to velocity and opposite the velocity direction. This relationship can be rearranged

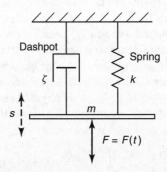

FIGURE 5.12: Mechanical model of a force-excited second-order system.

to read

$$\frac{1}{g_c}\left[m\left(\frac{d^2s}{dt^2}\right)\right] + \zeta\left(\frac{ds}{dt}\right) + ks = F(t) \tag{5.8a}$$

In both equations,

$$s = \text{mass displacement from the equilibrium position}$$

If we assume $F(t)$ to be periodic with time, we can substitute the appropriate Fourier series for $F(t)$ or, in general (see Section 4.4),

$$F(t) = \frac{A_o}{2} + \sum_{n=1}^{\infty} C_n \cos(n\Omega t - \phi_n) \tag{5.9}$$

where Ω = circular forcing frequency and

$$C_n = \sqrt{A_n^2 + B_n^2} \quad \text{and} \quad \tan\phi = \frac{B_n}{A_n} \tag{5.9a}$$

We will consider this general case in Section 5.16.3. First, however, we will consider several special cases.

5.14   THE ZERO-ORDER SYSTEM

A nearly trivial case occurs if we remove the spring and damper. The voltage-dividing potentiometer (see Sections 6.6 and 7.7) is an example. In its simplest form this device is a single slide wire. Aside from the mass of the slider and any member attached to it, there is no appreciable resistance to movement. In particular, an equilibrium-seeking force is not present and the output is independent of time, that is,

$$\text{Output} = \text{constant} \times \text{input}$$

Dynamically, the zero-order system requires no further consideration.

5.15   CHARACTERISTICS OF FIRST-ORDER SYSTEMS

If we assume the mass, m, in Fig. 5.12 and Eq. (5.8a) to be zero, we obtain a *first-order system*. Such systems have a balance between damping forces, $\zeta\ ds/dt$, restoring forces, ks, and externally applied forces, $F(t)$. Among instruments, the most common first-order systems are temperature-measuring devices. A temperature sensor usually responds as a first-order system because the rate of change of its temperature, dT_s/dt, is proportional to its current temperature,[2] T_s. Obviously, a mercury-in-glass thermometer (Section 16.2) or a thermocouple (Section 16.5) bears no physical resemblance to Fig. 5.12. However, the dynamic responses of the three systems are identical.

For the first-order system we can write

$$\zeta\frac{ds}{dt} + ks = F(t) \tag{5.10}$$

---

[2]For further discussion, see Section 16.10.3.

5.15.1 The Step-Forced First-Order System

Let

$$F(t) = 0, \qquad \text{for } t < 0$$

and

$$F(t) = F_0, \qquad \text{for } t \geq 0$$

For force equilibrium on the connecting element (which is assumed to be massless),

$$\zeta \frac{ds}{dt} + ks = F_0 \tag{5.11}$$

where

$$t = \text{time},$$
$$s = \text{displacement},$$
$$\zeta = \text{the damping coefficient},$$
$$k = \text{the deflection constant},$$
$$F_0 = \text{the amplitude of the constant input force}$$

Then

$$\int_0^t dt = \zeta \int_{s_A}^s \frac{ds}{(F_0 - ks)}$$

from which we obtain

$$\frac{F_0 - ks}{F_0 - ks_A} = e^{-kt/\zeta} = e^{-t/\tau} \tag{5.12}$$

The units of $\tau \equiv \zeta/k$ are seconds, and this quantity is known as the *time constant*.
Equation (5.12) can be written

$$s = s_\infty[1 - e^{-t/\tau}] + s_A e^{-t/\tau} = s_\infty + [s_A - s_\infty]e^{-t/\tau} \tag{5.13}$$

where

$$s_\infty = F_0/k = \text{the final displacement of the system as } t \to \infty,$$
$$s_A = \text{the initial displacement at } t = 0$$

We have assumed that the first-order system represents any dynamic condition wherein the elements are essentially massless, the displacement constraint is linear, and a significant viscous rate constraint is present. Generally, Eq. (5.13) can be written

$$P = P_\infty[1 - e^{-t/\tau}] + P_A e^{-t/\tau}$$

or

$$P = P_\infty + [P_A - P_\infty]e^{-t/\tau} \tag{5.14}$$

where

$$P = \text{the magnitude of any first-order process at time } t,$$
$$P_\infty = \text{the limiting magnitude of the process as } t \to \infty,$$
$$P_A = \text{the initial magnitude of the process at } t = 0$$

Although the basic relationship was derived in terms of a spring-dashpot arrangement, other processes that behave in an analogous manner include: (1) a heated (or cooled) bulk mass, such as a temperature sensor subjected to a step temperature change; (2) simple capacitive-resistive or inductive-resistive circuits; and (3) the decay of a radioactive source with time.

Figure 5.13 represents two different process-time conditions for the step-excited first-order system: (a) a *progressive* process, wherein the action is an increasing function of time; and (b) the *decaying* process, wherein the magnitude decreases with time.

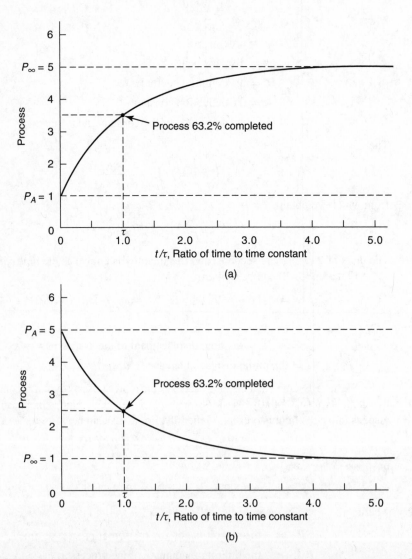

(a)

(b)

FIGURE 5.13: Characteristics of a first-order system subjected to a step input at $t = 0$: (a) for a progressive process; (b) for a decaying process.

Significance of the Time Constant, τ

If we substitute the magnitude of one time constant for t in Eq. (5.14),

$$P = P_\infty + (P_A - P_\infty)(0.368)$$

from which we see that $(1 - 0.368)$, or 63.2%, of the dynamic portion of the process will have been completed. Two time constants yield $(1 - 0.135) = 86.5\%$, three yield 95.0%, four yield 98.2%, five yield 99.3%, and so on. These percentages of completed processes are important because they will always be the same regardless of the process, provided that the process is described by the conditions of the step-excited first-order system.

It is often assumed that a process is completed during a period of five time constants.

___

EXAMPLE 5.2

Assume that a particular temperature probe approximates first-order behavior in a particular application, that it has a time constant of 6 s, and that it is suddenly subjected to a temperature step from 75°F to 300°F. What temperature will be indicated 10 s after the process has been initiated?

Solution   Applying Eq. (5.14), we find that

$$P_\infty = 300°\text{F}, \quad P_A = 75°\text{F}, \quad t = 10 \text{ s},$$
$$P = 300 + (75 - 300)e^{-10/6} = 257°\text{F}$$

___

EXAMPLE 5.3

Assume the same conditions as those of Example 5.2, but with a step from 300°F to 75°F. Find the indicated temperature after 10 s.

Solution

$$P_\infty = 75°\text{F}, \quad P_A = 300°\text{F}, \quad t = 10 \text{ s},$$
$$P = 75 + (300 - 75)e^{-10/6} = 117°\text{F}$$

___

5.15.2   The Harmonically Excited First-Order System

Again referring to Eq. (5.10), let us now consider the case of

$$F(t) = F_0 \cos \Omega t$$

or

$$\zeta \frac{ds}{dt} + ks = F_0 \cos \Omega t \tag{5.15}$$

where

F_0 = the amplitude of the forcing function,

Ω = the circular frequency of the forcing function in radians per second

The solution of Eq. (5.15) yields

$$s = A_1 e^{-t/\tau} + \frac{F_0/k}{\sqrt{1+(\tau\Omega)^2}} \cos(\Omega t - \phi) \qquad (5.16)$$

where

A_1 = a constant whose value depends on the initial conditions,

τ = the time constant = $\dfrac{\zeta}{k}$,

ϕ = the phase lag = $\tan^{-1}\dfrac{\Omega\zeta}{k} = \tan^{-1}\dfrac{2\pi}{T}\tau,$ \qquad (5.17)

$T = \dfrac{2\pi}{\Omega}$ = the period of excitation cycle in seconds

We see that the first term on the right side of Eq. (5.16), the complementary function, is *transient* and after a period of several time constants becomes very small. The second term is the *steady-state* relationship and, except for the short initial period, we can write

$$s = \frac{F_0/k}{\sqrt{1+(\tau\Omega)^2}} \cos(\Omega t - \phi) \qquad (5.18)$$

or

$$\frac{s}{s_s} = \frac{\cos(\Omega t - \phi)}{\sqrt{1+(\tau\Omega)^2}}$$

and

$$\frac{s_d}{s_s} = \frac{1}{\sqrt{1+(\tau\Omega)^2}}$$

$$= \frac{1}{\sqrt{1+(2\pi\tau/T)^2}} \qquad (5.19)$$

where

s_d = the maximum amplitude of the periodic dynamic displacement

and

$$s_s = \frac{F_0}{k}$$

The quantity s_s is the static deflection that would occur should the force amplitude F_0 be applied as a *static* force. The ratio s_d/s_s is often called the amplification ratio. For analogous situations, Eq. (5.19) may be written

$$\frac{P_d}{P_s} = \frac{1}{\sqrt{1+(2\pi\tau/T)^2}} \qquad (5.19a)$$

where P represents the magnitude of the applicable process.

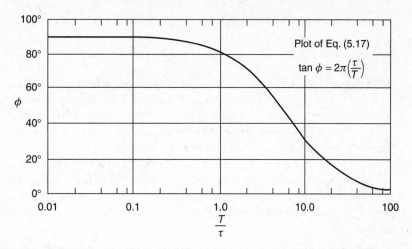

FIGURE 5.14: Phase lag versus ratio of excitation period to time constant for the harmonically excited first-order system.

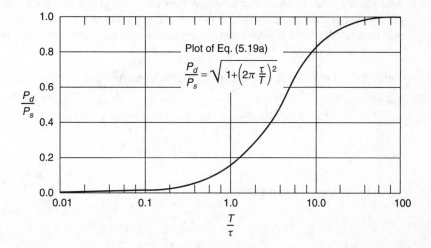

FIGURE 5.15: Amplitude ratio versus ratio of excitation period to time constant for the harmonically excited first-order system.

Figures 5.14 and 5.15 illustrate the relationships of the phase angle and the amplification ratio described by Eqs. (5.17) and (5.19a), respectively.

By calculating the response of the first-order system to a sinusoidal input, we have in fact determined its frequency and phase response. To show this explicitly in the notation of Section 5.7, we can rewrite Eq. (5.19a) in terms of the cyclic forcing frequency, $f = 1/T$:

$$G(f) = \frac{P_d}{P_s} = \frac{1}{\sqrt{1 + (2\pi f \tau)^2}} \tag{5.19b}$$

Likewise, from Eq. (5.17), the phase response is

$$\phi(f) = \tan^{-1}(2\pi f \tau) \tag{5.19c}$$

Note that the ideal response ($G \to 1$ and $\phi \to 0$) is obtained at frequencies small enough that $2\pi f \tau \ll 1$. From Figs. 5.14 and 5.15, we can see that this is equivalent to the statement that *the frequency and phase response of a first-order system are best when the time constant of the system is small compared to the period of forcing, $\tau \ll T$, so that the system responds rapidly in comparison to the variations it is measuring.*

In most mechanical systems, a moving mass exists and cannot be ignored. In such cases, the system is of second order and has characteristics that are discussed in subsequent sections. However, for temperature-sensing systems, first-order response is usually a good model, and we can use this model to examine the response characteristics of temperature probes.

EXAMPLE 5.4

A temperature probe has a time constant of 10 s when used to measure a particular gas flow. The gas temperature varies harmonically between 75°F and 300°F with a period of 20 s (i.e., with a frequency of 0.05 Hz). What is the temperature readout in terms of the input gas temperature? What time constant should the probe have to give 99% of the correct temperature amplitude?

Solution   In this case the temperature input can be expressed as

$$T_{\text{gas}}(t) = \left(\frac{300 + 75}{2}\right) + \left(\frac{300 - 75}{2}\right) \cos\left(\frac{2\pi}{20}t\right)$$

$$= 187 + 112 \cos\left(\frac{2\pi}{20}t\right)$$

From Eq. (5.19a), we find that

$$\frac{T_d}{T_s} = \frac{1}{\sqrt{1 + (2\pi \times 10/20)^2}},$$

$$T_d = \frac{112}{3.3} = 34°\text{F}$$

From Eq. (5.17), we find that the phase lag for a forcing period of $T = 20$ s is

$$\phi = \tan^{-1} 2\pi(\tau/T) = \tan^{-1}\left(\frac{2\pi \times 10}{20}\right) = \tan^{-1}\pi = 72.5°\,(\text{angle})$$

or

$$\text{Time lag} = \frac{72.5}{360} \times 20 = 4 \text{ s}$$

A graphical representation of the situation is shown in Fig. 5.16.

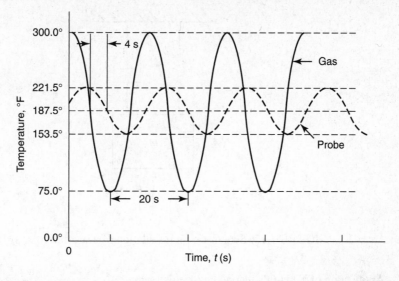

FIGURE 5.16: Response of temperature probe for conditions described in Example 5.4.

To obtain a 99% amplification ratio, we require

$$\frac{T_d}{T_s} = \frac{1}{\sqrt{1 + (2\pi\tau/20)^2}} = 0.99$$

Solving, we find that the probe would need a time constant of $\tau = 0.45$ s. In that case, τ would be very small compared to T.

5.16 CHARACTERISTICS OF SECOND-ORDER SYSTEMS

Figure 5.17 illustrates the essentials of a second-order system. This arrangement approximates many actual mechanical arrangements including simple weighing systems, such as an elastic-type load cell supporting a mass; D'Arsonval meter movements, including the ordinary galvanometers; and many force-excited mechanical-vibration systems such as accelerometers and vibrometers.

As with the first-order system, many excitation modes are possible, ranging from the simple step and simple harmonic functions to complex periodic forms. These modes approximate many actual situations, and, because all periodic inputs can be reduced to combinations of simple harmonic components (Section 4.4), the latter can give us insight into system performance when subject to most forms of dynamic input.

5.16.1 The Step-Excited Second-Order System

Referring to Fig. 5.17, we let

$$F = 0, \text{ when } t < 0$$

and

$$F = F_0, \text{ when } t \geq 0$$

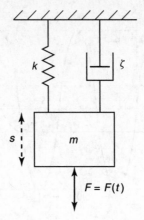

FIGURE 5.17: Schematic representation of a second-order system.

Application of Newton's second law yields

$$\frac{d^2s}{dt^2} + \left(\frac{\zeta g_c}{m}\right)\frac{ds}{dt} + \left(\frac{kg_c}{m}\right)s = F_0\left(\frac{g_c}{m}\right) \qquad (5.20)$$

where

$$s \equiv \text{displacement measured from the equilibrium position}$$

If we assume *underdamping*, that is, $(kg_c/m) > (\zeta g_c/2m)^2$ [see Eq. (5.21a)], the general solution of Eq. (5.20) can be written as

$$s = e^{-(\zeta g_c/2m)t}\,[A\cos\omega_{nd}t + B\sin\omega_{nd}t] + \frac{F_0}{k} \qquad (5.21)$$

where A and B are constants governed by initial conditions and

$$\omega_{nd} = \text{damped natural frequency}$$

$$= \sqrt{\frac{kg_c}{m} - \left(\frac{\zeta g_c}{2m}\right)^2} \qquad (5.21a)$$

Note that the exponential multiplier may be written as $e^{-t/\tau}$, where the time constant $\tau = 2m/\zeta g_c$. If we let $s = 0$ and $ds/dt = 0$ at $t = 0$ and evaluate A and B, then by rearrangement and substitution of terms, we can write Eq. (5.21) as

$$\frac{s}{s_s} = 1 - e^{-\xi\omega_n t}\left[\frac{\xi}{\sqrt{1-\xi^2}}\sin\sqrt{1-\xi^2}\,\omega_n t + \cos\sqrt{1-\xi^2}\,\omega_n t\right] \qquad (5.22)$$

An alternative form is

$$\frac{s}{s_s} = 1 - e^{-\xi\omega_n t}\sqrt{\frac{1}{1-\xi^2}}\cos(\omega_{nd}t - \beta), \qquad (5.22a)$$

$$\beta = \tan^{-1}\left[\frac{\xi}{\sqrt{1-\xi^2}}\right] \qquad (5.22b)$$

where

$\omega_n = \sqrt{kg_c/m}$ = the undamped natural frequency in radians per second,

$\zeta_c = 2\sqrt{mk/g_c}$ = the critical damping coefficient,

$\xi = \zeta/\zeta_c$ = the critical damping ratio,

$s_s = F_0/k$ = the "static" amplitude, or the amplitude that is reached as $t \to \infty$

Here again we may introduce the general idea of an analogous process, P, or

$$\frac{P}{P_s} = \frac{s}{s_s}$$

For the overdamped condition, $\xi = \zeta/\zeta_c > 1$, the solution of Eq. (5.20) can be written as

$$\frac{P}{P_s} = \frac{-\xi - \sqrt{\xi^2 - 1}}{2\sqrt{\xi^2 - 1}} e^{(-\xi + \sqrt{\xi^2 - 1})\omega_n t} + \frac{\xi - \sqrt{\xi^2 - 1}}{2\sqrt{\xi^2 - 1}} e^{(-\xi - \sqrt{\xi^2 - 1})\omega_n t} + 1 \quad (5.23)$$

Figure 5.18 shows the plots for Eqs. (5.22) and (5.23) for various damping ratios.[3] When the system has a nonzero damping, it approaches a static condition with $P/P_s = 1$ as the transient dies out.

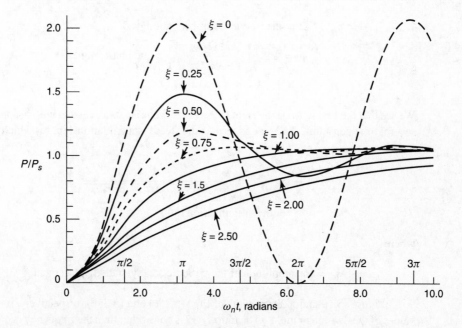

FIGURE 5.18: Response of a second-order system to a step input at $t = 0$.

---

[3] Note that the cases of zero damping and critical damping require special treatment.

5.16.2   The Harmonically Excited Second-Order System

Referring to Fig. 5.17, when

$$F(t) = F_0 \cos \omega t$$

we can write

$$\frac{m}{g_c} \frac{d^2 s}{dt^2} + \zeta \frac{ds}{dt} + ks = F_0 \cos \Omega t \tag{5.24}$$

For underdamped systems the solution becomes

$$s = e^{-(\zeta g_c/2m)t} [A \cos \omega_{nd}t + B \sin \omega_{nd}t] + \frac{(F_0/k)\cos(\omega t - \phi)}{\sqrt{\left[1 - \frac{m\Omega^2}{kg_c}\right]^2 + \left(\frac{\zeta \Omega}{k}\right)^2}} \tag{5.25}$$

$$= e^{-t/\tau} \left[A \cos \sqrt{1 - \xi^2}\, \omega_n t + B \sin \sqrt{1 - \xi^2}\, \omega_n t \right]$$

$$+ \frac{s_s \cos(\Omega t - \phi)}{\sqrt{[1 - (\Omega/\omega_n)^2]^2 + [2\xi(\Omega/\omega_n)]^2}} \tag{5.25a}$$

where A and B are constants that depend on particular initial conditions, and

$$\Omega = \text{the frequency of excitation (rad/s)},$$

$$\phi = \tan^{-1}\left[\frac{2\xi\Omega/\omega_n}{1 - (\Omega/\omega_n)^2}\right] = \text{the phase angle}, \tag{5.25b}$$

$$s_s = F_0/k$$

We see that the first term on the right side of Eq. (5.25a) is transient and will disappear after several time constants. The second term is the steady-state relationship, for which we may write[4]

$$\frac{s_d}{s_s} = \frac{P_d}{P_s} = \frac{1}{\sqrt{[1 - (\Omega/\omega_n)^2]^2 + [2\xi\Omega/\omega_n]^2}}$$

$$= \text{the amplification ratio} \tag{5.25c}$$

where

$$s_d = \text{the amplitude of the periodic steady-state displacement}$$

Figures 5.19 and 5.20 are plots of Eqs. (5.25c) and (5.25b), respectively, for various values of the damping ratio, ξ. The ratio s_s/s_d is none other than the frequency response, G, of the system; we could write it in terms of cyclic forcing frequency, f, by the substitution $\Omega = 2\pi f$.

---

[4]Note that inasmuch as the complementary or homogeneous solution (transient) is not involved, this relationship is valid for under-, over-, and critically damped conditions.

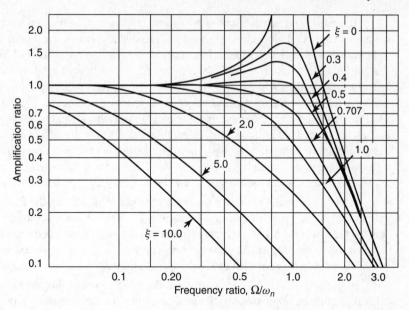

FIGURE 5.19: Plot of Eq. (5.25c) illustrating the frequency response to harmonic excitation of the system shown in Fig. 5.17.

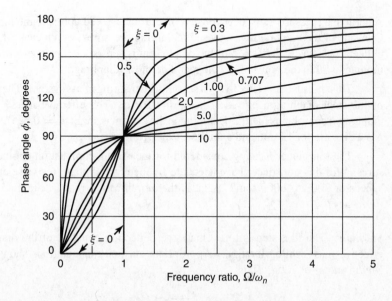

FIGURE 5.20: Plot of Eq. (5.25b) illustrating the phase response of the system shown in Fig. 5.17.

The ratio s_d/s_s is a measure of the system response to the frequency input. Normally, we hope that this relationship is constant with frequency; that is, we would like the system to be insensitive to changes in the frequency of input $F(t)$. Inspection of Fig. 5.19 shows that the amplitude ratio is reasonably constant for only a limited frequency range and then only for certain damping ratios. We see that for a given damping ratio, ideal response ($s_d/s_s = 1$) may sometimes occur at only one or two frequencies. If the system is to be used for general dynamic measurement applications, rather definite damping must be used and an upper frequency limit must be established. Practically, if a damping ratio in the neighborhood of 65%–75% is used, then the amplitude ratio will approximate unity over a range of frequency ratios of about 0%–40%. Even for these conditions, inherent error ($s_d/s_s - 1$) exists, and a usable system can be had only through compromise.

It should be made clear that the basic reason for optimizing the damping ratio is to extend the usable range of exciting frequency Ω. Certain devices, notably piezoelectric sensors (see Section 6.14), commonly possess such high undamped natural frequencies, ω_n, that the range of normal *operating* frequency ratios, Ω/ω_n, may extend from zero to only 10% or even less. In such cases, damping ratio magnitudes are of lesser interest.

Inspection of Fig. 5.20 indicates that damping ratios of the order on 65%–75% of critical provide an approximately linear phase shift for the frequency ratio range of 0%–40%. This approach is desirable if a proper time relationship is to be maintained between the harmonic components of a complex input. (See Sections 17.6.2 and 17.7.1 for further discussion of phase relationships.)

EXAMPLE 5.5

A particular pressure transducer consists of a circular steel diaphragm mounted in a housing. One side of the diaphragm is exposed to varying pressures, which cause the diaphragm to deflect. The elastic deflection of the diaphragm is sensed by a piezoelectric quartz crystal mounted within the housing, on the rear side of the diaphragm.

This transducer is effectively a second-order spring-mass system. The elastic stiffness of the steel diaphragm provides the spring force. The effective mass of the vibrating diaphragm contributes inertia. Damping in the system is slight ($\xi = 0.025$). The diaphragm has a diameter of 12 mm and a thickness of 1.75 mm.

The transducer is being considered for measuring combustion-engine cylinder pressures. Will this transducer have adequate frequency and phase response at typical engine speeds of 3000 to 6000 rpm? What is the amplification ratio at the transducer's natural frequency?

Solution The first step is to estimate the deflection constant of the diaphragm and its effective mass while vibrating. Appropriate elasticity calculations show that $k \approx 131$ MN/m and $m \approx 3.7$ g. Thus

$$\omega_n = \sqrt{\frac{kg_c}{m}} \approx 188 \times 10^3 \text{ rad/s}$$

or 29.9 kHz. The engine's highest operating frequency is

$$\Omega = (6000 \text{ rpm})(2\pi \text{ rad/rev})/(60 \text{ s/min}) = 628 \text{ rad/s}$$

or 100 Hz. Thus

$$\frac{\Omega}{\omega_n} = 0.0033$$

From Figs. 5.19 and 5.20, we see that the transducer will perform with negligible phase shift and an amplification ratio of unity; substitution of ξ and Ω/ω_n into Eqs. (5.25b) and (5.25c) verifies this conclusion. The transducer's response characteristics are well suited to the application.

If the transducer were subjected to pressure variations near its natural frequency, however, the amplification ratio would be enormous; from Eq. (5.25c),

$$\frac{P_d}{P_s} = \frac{1}{\sqrt{0 + [2(0.025)(1)]^2}} = 20$$

This device is undoubtedly not intended to operate at frequencies that high.

5.16.3 General Periodic Forcing

We now return to the general case of periodic forcing as suggested in Section 5.13. For convenience we will restate Eqs. (5.8a), (5.9) and (5.9a)

$$\frac{m}{g_c}\left(\frac{d^2 s}{dt^2}\right) + \zeta\left(\frac{ds}{dt}\right) + ks = F(t) \tag{5.26}$$

where

$$F(t) = \frac{A_0}{2} + \sum_{n=1}^{\infty} C_n \cos(n\Omega t - \phi_n) \tag{5.26a}$$

and

$$C_n = \sqrt{A_n^2 + B_n^2} \quad \text{and} \quad \tan\phi_n = \frac{B_n}{A_n} \tag{5.26b}$$

By substituting Eq. (5.26a) into Eq. (5.26), we obtain

$$\frac{m}{g_c}\left(\frac{d^2 s}{dt^2}\right) + \zeta\left(\frac{ds}{dt}\right) + ks = \frac{A_0}{2} + \sum_{n=1}^{\infty} C_n \cos(n\Omega t - \phi_n) \tag{5.27}$$

Although this expression appears quite formidable, we can easily recognize that it yields a combination of the solutions given by Eqs. (5.21) and (5.25a). Using the reasoning that the cosine terms on the right side of Eq. (5.27) will give results similar to those of a harmonically forced system, we can write a solution for $\xi < 1$ in the form

$$s = e^{-(\zeta g_c/2m)t}\left[A\cos\omega_{nd}t + B\sin\omega_{nd}t\right] + r_0 + \sum_{n=1}^{\infty} r_n \cos(n\Omega t - \phi_n - \psi_n) \quad (5.28)$$

where

$$r_0 = \frac{A_0}{2k},$$

$$r_n = \frac{C_n/k}{\sqrt{\left[1 - \left(\dfrac{n\Omega}{\omega_n}\right)^2\right]^2 + \left[2\xi\dfrac{n\Omega}{\omega_n}\right]^2}},$$

$$\tan\psi_n = \frac{2\xi\left(\dfrac{n\Omega}{\omega_n}\right)}{1 - \left(\dfrac{n\Omega}{\omega_n}\right)^2}$$

It helps to recall that

$$\omega_{nd} = \text{the damped natural frequency} = \sqrt{\frac{kg_c}{m} - \left(\frac{\zeta g_c}{2m}\right)^2},$$

$$\omega_n = \text{the undamped natural frequency} = \sqrt{\frac{kg_c}{m}}$$

As we discussed previously, the first term on the right side of Eq. (5.28) is transient and dies out after several time constants. The remaining terms, then, represent the steady-state response.

EXAMPLE 5.6

(a) Write an expression for the steady-state response of the single-degree-of-freedom system shown in Fig. 5.17 when subjected to the sawtooth forcing function described in Table 4.1 and shown in Fig. 4.9(b).

(b) Let

$$m = 1 \text{ kg, or } 2.2 \text{ lbm,}$$
$$k = 1000 \text{ N/m, or } 5.71 \text{ lbf/in.,}$$
$$\zeta = 31.6 \text{ N} \cdot \text{s/m, or } 0.180 \text{ lbf} \cdot \text{s/in.,}$$
$$A = 10 \text{ N, or } 2.248 \text{ lbf}$$

Using these data (the SI values), obtain computer-plotted waveforms for input frequencies of 10, 30, and 50 rad/s.

Solution

(a) In general, the steady-state response is given by

$$s = \frac{A_0}{2k} + \sum_{n=1}^{\infty} \left(\frac{C_n}{k}\right) \frac{\cos(n\Omega t - \phi_n - \psi_n)}{\sqrt{\left[1 - \left(\frac{n\Omega}{\omega_n}\right)^2\right]^2 + \left[2\xi\frac{n\Omega}{\omega_n}\right]^2}},$$

$$\phi_n = \tan^{-1}\left(\frac{B_n}{A_n}\right),$$

$$C_n = \sqrt{A_n^2 + B_n^2},$$

$$\psi_n = \tan^{-1}\left[\frac{2\xi\left(\frac{n\Omega}{\omega_n}\right)}{1 - \left(\frac{n\Omega}{\omega_n}\right)^2}\right]$$

In this case, the forcing function, $F(t)$, is [see Table 4.1, case 4.9(b)]

$$F(t) = \frac{2A}{\pi} \sum_{n=1}^{\infty} \frac{1}{n} \sin(n\Omega t)$$

so that both $A_0 = 0$ and $A_n = 0$. Thus

$$\phi_n = \tan^{-1} \infty = \frac{\pi}{2},$$

$$C_n = B_n = \frac{2A}{\pi}\frac{1}{n}$$

Therefore,

$$s = \frac{2A}{\pi k} \sum_{n=1}^{\infty} \frac{1}{n} \frac{\cos\left(n\Omega t - \frac{\pi}{2} - \psi_n\right)}{\sqrt{\left[1 - \left(\frac{n\Omega}{\omega_n}\right)^2\right]^2 + \left[2\xi\frac{n\Omega}{\omega_n}\right]^2}}$$

and

$$\omega_n = \sqrt{\frac{kg_c}{m}} = \sqrt{1000} \approx 31.6 \text{ rad/s},$$

$$\zeta_c = 2\sqrt{\frac{mk}{g_c}} = 2\sqrt{1000} \approx 63.3 \text{ N} \cdot \text{s/m},$$

$$\xi = \frac{\zeta}{\zeta_c} = \frac{31.6}{63.3} \approx 0.50,$$

$$\frac{A}{k} = \frac{10}{1000} = 1 \text{ cm},$$

$$\psi_n = \tan^{-1} \left[\frac{2 \times 0.5 \left(\dfrac{n\Omega}{31.6} \right)}{1 - \left(\dfrac{n\Omega}{31.6} \right)^2} \right],$$

$$s = \frac{2}{\pi} \sum_{n=1}^{\infty} \frac{1}{n} \frac{\cos \left(n\Omega t - \dfrac{\pi}{2} - \psi_n \right)}{\sqrt{\left[1 - \left(\dfrac{n\Omega}{31.6} \right)^2 \right]^2 + \left[\dfrac{n\Omega}{31.6} \right]^2}},$$

$$= \frac{2}{\pi} \sum_{n=1}^{\infty} \frac{1}{n} \frac{\cos \left(n\Omega t - \dfrac{\pi}{2} - \psi_n \right)}{\sqrt{\left[1 - \left(\dfrac{n\Omega}{31.6} \right)^2 \right]^2 + \left(\dfrac{n\Omega}{31.6} \right)^2}}$$

Note that s is in centimeters.

The most practical approach to obtaining numerical results for this part of the example is through the use of a computer. Computer-plotted results are shown in Figs. 5.21(a) through 5.21(c).

From the foregoing discussions it is apparent that if simple measurements are to be made as rapidly as possible or, more importantly, if the input signal is continuous and complex, rather definite limitations are imposed by the measuring system. Additional discussion of such limitations can be found in Chapter 17.

5.17   ELECTRICAL ELEMENTS

As we discussed in Section 5.8, most measurement systems are composed of a combination of mechanical and electrical elements. Very often the basic detecting element of the sensor is mechanical and its output is immediately transduced into an electrical signal by a secondary element. The signal conditioning that follows is largely by electrical means; however, termination sometimes requires conversion to something basically mechanical, such as a controller, a galvanometer-type recorder or plotter, and so on. It is clear, then, that overall performance results from a combination of mechanical and electrical responses. In previous sections we have discussed the response of simple, purely mechanical systems. In succeeding sections we will look at the corresponding electrical elements.

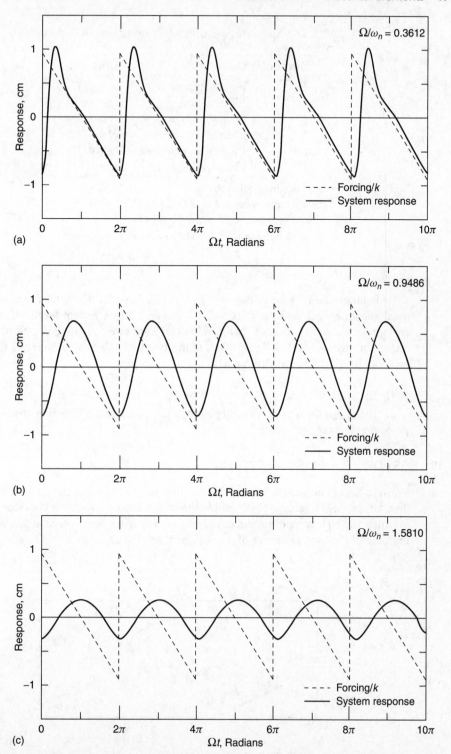

FIGURE 5.21: Computer-plotted sawtooth wave response for the second-order system specified in the text: (a) for $\Omega = 10$ rad/s; (b) for $\Omega = 30$ rad/s; (c) for $\Omega = 50$ rad/s.

TABLE 5.1: Some Basic Electrical Quantities and Relationships

Symbol	Definition	Unit
E	Electric potential	volt (V)
I	Electric current	ampere (A)
Q	Electric charge	coulomb (C)
R	Electrical resistance	ohm (Ω)
L	Electrical inductance	henry (H)
C	Electrical capacitance	farad (F)

Some defining relationships:

For a capacitance: $I = dQ/dt = C(dE/dt)$, $E = Q/C$

For a resistance: $E = IR$ (Ohm's law)

For an inductance: $E = L(dI/dt) = L(d^2Q/dt^2)$

In preparation for the discussion that follows, Table 5.1 lists some fundamental electrical quantities and defining relationships. Table 5.2 lists certain mechanical-electrical analogies. For verification of these, the reader is referred to any basic electric circuits text [2] or physics text [3]. In addition, at this point it is also useful to recall Kirchhoff's two laws for electrical circuits, namely,

1. The algebraic sum of all currents entering a junction point is zero, and
2. The algebraic sum of all voltage drops taken in a given direction around a closed circuit is zero.

5.18   FIRST-ORDER ELECTRICAL SYSTEM

Consider the circuit shown in Fig. 5.22. Assume that the capacitor carries no initial charge; then let the SPDT (single-pole, double-throw) switch be moved to contact A, thereby inserting the battery into the circuit. Now the capacitor begins to charge; what is the response of the circuit in terms of the voltage across the capacitor?

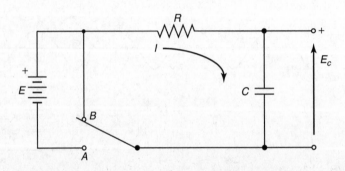

FIGURE 5.22: Series resistance-capacitance circuit.

TABLE 5.2: Dynamically Analogous Mechanical and Electrical System Elements

Symbol	Mechanical Quantity	Symbol	Electrical Quantity
m	Mass, kg (lbm)		
I	Moment of inertia, kg $\cdot$ m^2 (lbm $\cdot$ in.^{2})	L	Inductance, H
k	Deflection constant, N/m (lbf/in.)		
k_t	Torsional deflection constant, N $\cdot$ m/rad (lbf $\cdot$ in./rad)	$1/C$	Reciprocal of capacitance, F^{-1}
ζ	Damping coefficient, N $\cdot$ s/m (lbf $\cdot$ s/in.)		
ζ_t	Torsional damping coefficient, N $\cdot$ m $\cdot$ s/m (lbf $\cdot$ in. $\cdot$ s/in.)	R	Resistance, Ω
F	Force, N (lbf)		
T	Torque, N $\cdot$ m (lbf $\cdot$ in.)	E	Voltage, V
x	Translational displacement, m (in.)		
θ	Angular displacement, rad	Q	Charge, C
dx/dt	Translational velocity, m/s (in./s)		
$d\theta/dt$	Angular velocity, rad/s (rad/s)	dQ/dt	Current, A
Ω	Forcing frequency, rad/s (rad/s)	Ω	Forcing frequency, rad/s
d^2x/dt^2	Translational acceleration, m/s^2 (in./s^2)		
$d^2\theta/dt^2$	Angular acceleration, rad/s^2 (rad/s^2)	d^2Q/dt^2	Rate of change of current, A/s

Note: The uppercase C has long been used as the symbol for capacitance. It has also been assigned as the symbol for the SI unit for electric charge, the coulomb. Likewise, Ω is the SI symbol for resistance, ohms. It is also widely used to represent an exciting frequency in radians per second. In this text we will let the symbols retain each meaning. Should the context not make clear the meanings intended, we include clarifying statements.

By employing Kirchhoff's law of voltages, we may write

$$IR + \frac{Q}{C} - E = 0 \tag{5.29}$$

but

$$I = \frac{dQ}{dt}$$

hence

$$\frac{dQ}{dt} + \frac{Q}{RC} - \frac{E}{R} = 0 \tag{5.29a}$$

Solving, we have

$$Q = CE\left(1 - e^{-t/RC}\right)$$

We define the circuit time constant as

$$\tau = RC$$

The voltage drop across the capacitor is $E_c = Q/C$; hence,

$$E_c = E\left(1 - e^{-t/\tau}\right) \tag{5.30}$$

In a similar manner, if the switch contact is moved from A to B after the capacitor is charged, we may write

$$IR + \frac{Q}{C} = 0 \tag{5.31}$$

for which

$$E_c = E\,e^{-t/\tau} \tag{5.31a}$$

It is apparent, then, that Eqs. (5.14) and (5.10), which apply to a mechanical system, hold equally well for the electrical circuit discussed here. Should the battery in Fig. 5.22 be replaced with a sinusoidal voltage source, analysis would show that equations such as (5.17) and (5.19a,b,c) also apply to electrical systems. In each case, it is necessary to insert the appropriate time constant and to properly interpret the response variable.

EXAMPLE 5.7

Figure 5.23(a) shows a simple circuit consisting of a capacitor, a resistor and a 5-kHz voltage source connected in series. Determine the amplitude and phase shift of the voltage appearing across the capacitor. Compare this to the voltage across the resistor.

Solution From Kirchhoff's law of voltages, we obtain

$$R\frac{dQ}{dt} + \frac{1}{C}Q = E_0\cos\Omega t$$

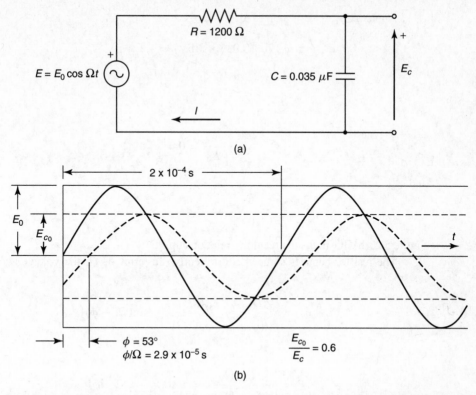

(a)

(b)

FIGURE 5.23: Resistor and capacitor in series: (a) excited by a 5000-Hz voltage source; (b) voltage across the capacitor for the circuit shown in (a). The angle ϕ is a phase *lag*.

By analogy to Section 5.15.2 [Eqs. (5.15), (5.17), and (5.18)], the steady-state solution of this equation is

$$Q = \frac{E_0 C}{\sqrt{1 + (\tau\Omega)^2}} \cos(\Omega t - \phi)$$

with

$$\phi = \tan^{-1}\left(\frac{2\pi}{T}\tau\right)$$

and

$$\tau = RC$$

for T the period of the exciting voltage. The voltage across the capacitor is $E_c = Q/C$. Therefore, the voltage amplitude across the capacitor is

$$\frac{E_{c0}}{E_0} = \frac{1}{\sqrt{1 + (\tau\Omega)^2}} = \frac{1}{\sqrt{1 + (2\pi\tau/T)^2}}$$

Observe that these results are identical to those plotted in Figs. 5.14 and 5.15.

Numerically,

$$\tau = 1200(0.035 \times 10^{-6}) = 4.2 \times 10^{-5}\text{s},$$

$$T = \frac{1}{f} = 2 \times 10^{-4}\text{s}$$

so

$$\frac{E_{c_0}}{E_0} = 0.6$$

and

$$\phi = 53°$$

Figure 5.23(b) illustrates the resulting relationship.

The resistor behaves somewhat differently. In terms of Q, the voltage across the resistor is

$$E_r = R\frac{dQ}{dt} = \frac{E_0 RC\Omega}{\sqrt{1 + (\tau\Omega)^2}}[-\sin(\Omega t - \phi)]$$

or

$$E_r = \frac{E_0 \tau\Omega}{\sqrt{1 + (\tau\Omega)^2}} = \cos(\Omega t + \pi/2 - \phi)$$

Thus, the resistor voltage has an amplitude

$$\frac{E_{r_o}}{E_0} = \frac{\tau\Omega}{\sqrt{1 + (\tau\Omega)^2}} = \frac{2\pi\tau/T}{\sqrt{1 + (2\pi\tau/T)^2}}$$

and a phase *lead* of $\pi/2 - \phi$. Note that the resistor voltage amplitude is small when $\tau/T \ll 1$ (at low excitation frequencies), whereas the capacitor voltage is small when $\tau/T \gg 1$ (at high excitation frequencies). The contrasting behavior of the two elements is the basis for this circuit's application to high-pass and low-pass filters, as we shall see in Chapter 7.

First-order circuits may also be constructed using an inductor and a resistor. Similar results can be developed, the primary difference being that time constant is instead $\tau = L/R$.

5.19   SIMPLE SECOND-ORDER ELECTRICAL SYSTEM

Figure 5.24 illustrates a circuit consisting of R, L, and C elements in series with a voltage source. Referring to Table 5.1 for the voltage drop across each element and then applying Kirchhoff's law of voltages, we can write

$$L\left(\frac{d^2Q}{dt^2}\right) + \left(\frac{dQ}{dt}\right) + \frac{Q}{C} = E_0 \cos \Omega t \qquad (5.32)$$

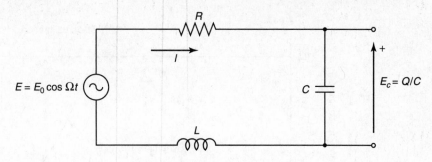

FIGURE 5.24: An *RLC* circuit.

We recognize this expression as having the same form as Eq. (5.24), and thus we can quickly write a solution:

$$Q = e^{-t/\tau}[A\cos\omega_{nd}t + B\sin\omega_{nd}t] + \frac{E_0\cos(\Omega t - \phi)}{C\sqrt{[1/C - L\Omega^2]^2 + (R\Omega)^2}} \qquad (5.32a)$$

If, for example, we consider the steady-state voltage amplitude across the capacitor, we may write (see Table 5.1)

$$E_c = \frac{Q}{C} = \frac{E_0}{C\sqrt{[1/C - L\Omega^2]^2 + (R\Omega)^2}} \qquad (5.32b)$$

Using the analogies in Table 5.2 (see also Section 7.11), we may write

$$\omega_n = \sqrt{\frac{1}{LC}} \qquad (5.32c)$$

and

$$R_c = 2\sqrt{\frac{L}{C}} \qquad (5.32d)$$

where

ω_n = a resonance frequency corresponding to the undamped

      natural frequency of the mechanical system, and

R_c = a critical resistance analogous to critical damping

By algebraic manipulation, Eqs. (5.32a) and (5.32b) may now be written in the dimensionless forms, where E_c is the dynamic amplitude of the voltage across C; that is,

$$\frac{E_c}{E_0} = \frac{1}{\sqrt{[1 - (\Omega/\omega_n)^2]^2 + [2(R/R_c)(\Omega/\omega_n)]^2}} \qquad (5.32e)$$

and

$$\tan\phi = \frac{2(R/R_c)(\Omega/\omega_n)}{1 - (\Omega/\omega_n)^2} \qquad (5.32f)$$

Except for the symbols, we see that Eqs. (5.32e) and (5.32f) are identical to Eqs. (5.25c) and (5.25b), respectively. It follows, then, that Figs. 5.19 and 5.20 apply equally well to the electric circuit that we have just investigated.

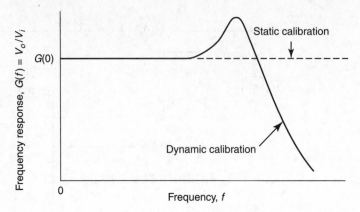

FIGURE 5.25: Comparison of static and dynamic calibrations. In a static calibration, $G(0)$ is measured and used to approximate $G(f)$ at higher frequencies. In a dynamic calibration, $G(f)$ itself is measured.

5.20 CALIBRATION OF SYSTEM RESPONSE

The final and positive proof of a measuring system's performance is direct measurement of the actual system's response to a completely defined and known input. Physical modeling of system response (as in Sections 5.8–5.19) provides important understanding of the device's essential characteristics, such as its time constant or natural frequency. However, if the device is too complex to model well or if the modeling assumptions are tenuous, greater accuracy can be obtained through an experimental test of the system's response. Such testing is always required in high-accuracy experimentation.

Testing a measuring system's response is really a process of *calibration* (Section 1.7). What output is obtained for an input of given amplitude and frequency? Can we experimentally evaluate the behavior of the system when it is confronted with an input that is rapidly changing with time? To do this effectively, we must also have a time-varying input, or *calibration source*, that is precisely known.

Calibration sources having sinusoidal time variation are undoubtedly the easiest to produce and the most used (recall Section 5.7). With electrical input signals, commercial voltage sources are easily applied to this task. Various classical complex waveforms, such as square waves or sawtooth waves, may also be employed. For example, the response of many electrical components can be judged through the use of square-wave inputs. A skilled technician can frequently pinpoint reasons for distortion by observing the tested apparatus's treatment of square-wave input.[5] More importantly, such tests can identify limits beyond which system performance is questionable.

For some measurands, even a sine-wave test is relatively hard to implement: How do you create a well-defined sinusoidal variation in temperature or fluid velocity? For such variables, step inputs or pulse inputs are often easier to produce, and they can also yield useful information about the system's rise time, time constant, or resonant frequencies.

In cases where any time variation of the input is too difficult or too expensive to

---

[5] Square-wave testing is so useful in adjusting the frequency response of hot-wire anemometer bridges (Section 15.9) that many bridges are sold with built-in square-wave generators.

produce, constant inputs are occasionally substituted. For example, we may measure the output of a temperature sensor at several different constant temperatures to obtain $T_{measured}$ as a function of T_{actual}. This tells us how the sensor responds to inputs of different amplitude. We would then assume that the measured amplitude relationship applied for all frequencies of interest.

In this sense, we can distinguish two types of calibrations: *static calibration* and *dynamic calibration*. In Section 5.7, we noted that the frequency response of a system was the ratio of output to input amplitude at a given frequency: $G(f) = V_o/V_i$. In a static calibration, we measure V_o and V_i at zero frequency to obtain $G(0)$. We then assume that $G(f) \approx G(0)$ over the frequency range of interest (Fig. 5.25). In a dynamic calibration, we measure V_o and V_i over a range of frequencies to obtain $G(f)$ itself. Static calibration approximates dynamic calibration at frequencies low enough that the device's frequency response is flat. Static calibrations must always be used with caution and an awareness of their inherent frequency limitations.

Calibration practices for various mechanical measurands are considered in more detail in Part II of this book (see Sections 14.11, 15.6, 16.12, 17.10, and 18.9).

5.21   SUMMARY

Response is a vital feature of a measuring system's ability to accurately resolve time-varying inputs. In this chapter, we have examined various important types of response, procedures for testing response, and simplified physical models of the response of instruments.

1. Amplitude response, frequency response, phase response, and rise time are each important aspects of measuring-system performance. Amplitude response is generally defined by the overload condition of an instrument. Frequency and phase response are considerations in determining the range of frequencies over which an instrument is accurate (Sections 5.2–5.7).

2. Each element in a measuring system must have adequate frequency and phase response for the measurand at hand. Some system elements, like filters, are selected specifically for the limitations of their frequency response (Section 5.7).

3. Often, a physical model of a measuring device can identify the important characteristics of its behavior. Mechanical models rely on the identification of masses, spring forces, and damping (Sections 5.8–5.11).

4. Selection of the correct amount of damping is vital to obtaining optimal frequency response (Sections 5.11, 5.15, 5.16, 5.18).

5. The spring-mass-damper system is a fundamental model of the dynamic response of mechanical systems (Sections 5.12–5.16). Some systems behave as if massless or first order, especially temperature sensors (Section 5.15). Other systems have mass and are second order; examples include accelerometers, elastic diaphragm transducers, and various moving mechanisms (Section 5.16).

6. Electrical-system response is similar to mechanical-system response. Models for electrical systems yield results analogous to those for mechanical systems (Sections 5.17–5.19).

7. For a first-order system, the system's time constant, τ, is of critical importance to its response. The time constant should normally be small compared to the period of forcing, T, in order to obtain good response. First-order response to a step input is

described by Eq. (5.14) and Fig. 5.13; response to harmonic forcing is described by Eqs. (5.17) and (5.19a,b,c) and Figs. 5.14 and 5.15 (Sections 5.15, 5.18).

8. For a second-order system, both the natural frequency, ω_n, and the critical damping ratio, ξ, must be considered. For most systems, good response is obtained when the natural frequency is large compared to the forcing frequency, Ω, and when the damping ratio is 65%–75%. Second-order response to a step input is described by Eqs. (5.22) and (5.23) and Fig. 5.18; response to harmonic input is described by Eqs. (5.25a, b, c) and Figs. 5.19 and 5.20 (Sections 5.16, 5.19).

9. Experimental determination of system response, or calibration, is often required. A sine-wave test is sometimes useful (Section 5.7). Other kinds of test signals, such as square-wave or step input, can also be applied. Static calibrations (for $f = 0$) are occasionally used when dynamic calibrations (for $f > 0$) are too difficult (Section 5.20).

SUGGESTED READINGS

Dorf, R. C., and R. H. Bishop. *Modern Control Systems.* 10th ed. Upper Saddle River, N.J.: Prentice Hall, 2004.

Floyd, T. L. *Principles of Electric Circuits.* 7th ed. Upper Saddle River, N.J.: Prentice Hall, 2003.

Meriam, J. L., L. G. Kraige, and W. J. Palm. *Engineering Mechanics: Vol. 2—Dynamics.* 5th ed. New York: John Wiley, 2002.

Raven, F. *Automatic Control Engineering.* 5th ed. New York: McGraw-Hill, 1994.

Kreyszig, E. *Advanced Engineering Mathematics.* 8th ed. New York: John Wiley, 1998.

PROBLEMS

5.1. A simple U-tube monometer is shown in Fig. 14.5. Show that the period of oscillation may be approximated by

$$T = 2\pi \sqrt{\frac{L}{2g}}$$

where L is the total length of manometer fluid and the mass of the transmitting fluid can be neglected.

5.2. The simple U-tube monometer shown in Fig. 14.5 is subjected to a time-varying pressure with a cyclic circular frequency of $\sqrt{g/8L}$ rad/s. Determine the error in the pressure reading.

5.3. A temperature measuring system (assumed to be a first-order system) is excited by a 0.25-Hz harmonically varying input. If the time constant of the system is 4.0 s and the indicated amplitude is 10°F, what is the true temperature?

5.4. A mercury-in-glass thermometer initially at 25°C is suddenly immersed into a liquid that is maintained at 100°C. After a time interval of 2.0 s, the thermometer reads 76°C. Assuming a first-order system, estimate the time constant of the thermometer.

5.5. A 5-kg mass is statically suspended from a load (force) cell and the load cell deflection caused by this mass is 0.01 mm. Estimate the natural frequency of the load cell.

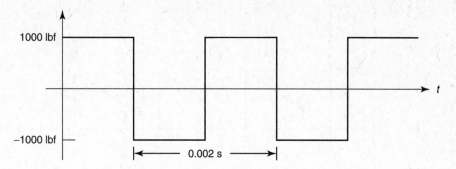

FIGURE 5.26: Dynamic load for Problem 5.8.

5.6. If the load cell of Problem 5.5 is assumed to be a second-order system with negligible damping, determine the practical frequency range over which it can measure dynamic loads with an inherent error of less than 5%.

5.7. If the load cell of Problem 5.5 actually has a damping ratio of 0.707, determine the practical frequency range over which it can measure dynamic loads with an inherent error of less than 5%.

5.8. A manufacturer lists the specifications of a dynamic tension-compression load cell as follows:

> Undamped natural frequency = 1000 Hz
> Damping ratio = 0.707
> Sensitivity = 10 mV/lbf
> Stiffness = 100,000 lbf/in.

If a dynamic force as indicated in Fig. 5.26 is applied to the load cell, determine the steady-state output voltage as a function of time.

5.9. Plot the voltage output of Problem 5.8 as a function of time using at least the first five terms.

5.10. An RC circuit as shown in Fig. 5.22 is required to have a time constant of 1 ms. Determine three different combinations of R and C to accomplish this.

5.11. A tank containing an initial volume of water V_0, discharges water from an opening at the bottom of the tank. If the discharge rate is directly proportional to the volume of water in the tank, determine the time constant for this system.

5.12. For the tank of Problem 5.11, determine the time constant if the initial discharge rate is Q (liters/min).

5.13. A temperature sensor is expected to measure an input having frequency components as high as 50 Hz with an error no greater than 5%. What is the maximum time constant for the temperature sensor that will permit this measurement?

5.14. A thermocouple with a time constant of 0.05 s is considered to behave as a first-order system. Over what frequency range can the thermocouple measure dynamic temperature fluctuations (assumed to be harmonic) with an error less than 5%?

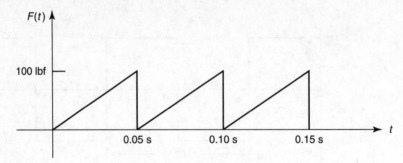

FIGURE 5.27: Forcing for Problem 5.20.

5.15. A 100-μF capacitor is charged to a voltage of 100 V. At time $t = 0$, it is discharged through a 1.0-MΩ resistor. Determine the time for the voltage across the capacitor to reach 10 V.

5.16. A pressure transducer behaves as a second-order system. If the undamped natural frequency is 4000 Hz and the damping is 75% of critical, determine the frequency range(s) over which the measurement error is not greater than 5%.

5.17. What will be the frequency range(s) for Problem 5.16 if the damping ratio is changed to 0.5?

5.18. Consider the pressure transducer of Problem 5.16 to be damaged such that its viscous damping ratio is unknown. When the transducer is subjected to a harmonic input of 2400 Hz, the phase angle between the output and input is measured as 45°. With this in mind, determine the error when the transducer is used to measure a harmonic pressure signal of 1800 Hz. What is the phase angle between the input and output at this frequency?

5.19. A force transducer behaves as a second-order system. If the undamped natural frequency of the transducer is 1800 Hz and its damping is 30% of critical, determine the error in the measured force for a harmonic input of 950 Hz. What is the magnitude of the phase angle?

5.20. Consider a second-order system with a damping ratio of 0.70 and a undamped natural frequency of 50 Hz. If the value of k is 100 lbf/in., determine the steady-state output if the forcing is as shown in Fig. 5.27.

5.21. Refer to Fig. 5.28.

(**a**) What is the time constant if $X = 200 \ \mu F$ and R=10 kΩ?
(**b**) What is the voltage across the resistor 0.5 s after the switch is closed?

5.22. Refer to Fig. 5.28.

(**a**) What is the time constant if $X = 500$ mH and $R = 10 \ \Omega$?
(**b**) What is the voltage across the resistor 0.02 s after the switch is closed?
(**c**) What is the voltage across the resistor 0.05 s after the switch is closed?

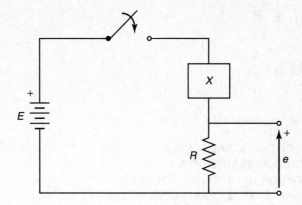

FIGURE 5.28: Circuit for Problems 5.21 and 5.22.

REFERENCES

[1] Dorf, R. C., and R. H. Bishop. *Modern Control Systems*. 10th ed. Upper Saddle River, N.J.: Prentice Hall, 2004.

[2] Floyd, T. L. *Principles of Electric Circuits*. 7th ed. Upper Saddle River, N.J.: Prentice Hall, 2003.

[3] Halliday, D., R. Resnick, and J. Walker. *Fundamentals of Physics*. 7th ed., extended. New York: John Wiley, 2004.

CHAPTER 6

Sensors

6.1   INTRODUCTION

In Section 1.4 we divided the general measuring system into three distinct sections: the sensor-detector stage, the signal-conditioning stage, and the recording-readout stage. In this chapter and in Chapters 7 and 9, we discuss in more detail what goes on in each of these stages.

The first contact that a measuring system has with the measurand is through the input sample accepted by the detecting element of the first stage (see Section 1.4). This act is usually accompanied by the immediate transduction of the input into an analogous form.

The medium handled is information. The detector senses the information input, I_{in}, and then transduces or converts it to a more convenient form, I_{out}. The relationship may be expressed as

$$I_{out} = f(I_{in}) \qquad\qquad (6.1a)$$

further,

$$\text{Transfer efficiency} = \frac{I_{out}}{I_{in}} \qquad\qquad (6.1b)$$

This cannot be more than unity, because the pickup cannot generate information but can only receive and process it. Obviously, as high a transfer efficiency as possible is desirable.

Sensitivity may be expressed as

$$\eta = \frac{d I_{\text{out}}}{d I_{\text{in}}} \tag{6.1c}$$

Very often sensitivity approximates a constant; that is, the output is the linear function of the input.

6.2   LOADING OF THE SIGNAL SOURCE

Energy will always be taken from the signal source by the measuring system, which means that the information source will always be changed by the act of measurement. This is an axiom of measurement. This effect is referred to as *loading*. The smaller the load placed on the signal source by the measuring system, the better.

Of course, the problem of loading occurs not only in the first stage, but throughout the entire chain of elements. While the first-stage detector-transducer loads the input source, the second stage loads the first stage, and finally the third stage loads the second stage. In fact, the loading problem may be carried right down to the basic elements themselves.

In measuring systems made up primarily of electrical elements, the loading of the signal source is almost exclusively a function of the detector. Intermediate modifying devices and output indicators or recorders receive most of the energy necessary for their functioning from sources *other* than the signal source. A measure of the quality of the first stage, therefore, is its ability to provide a usable output without draining an undue amount of energy from the signal.

6.3   THE SECONDARY TRANSDUCER

As an example of a system of mechanical elements only, consider the Bourdon-tube pressure gage, shown in Fig. 6.1. The primary detecting-transducing element consists of a circular tube of approximately elliptical cross section. When pressure is introduced, the section of the flattened tube tends toward a more circular form. This in turn causes the free end *A* to move outward and the resulting motion is transmitted by link *B* to sector gear *C* and hence to pinion *D*, thereby causing the indicator hand to move over the scale.

In this example, the tube serves as the primary detector-transducer, changing pressure into near linear displacement. The linkage-gear arrangement acts as a secondary transducer (linear to rotary motion) and as an amplifier, yielding a magnified output.

A modification of this basic arrangement is to replace the linkage-gear arrangement with either a differential transformer (Section 6.11) or a voltage-dividing potentiometer (Section 6.6). In either case the electrical device serves as a secondary transducer, transforming displacement to voltage.

As another example, let us analyze a simplified compression-type force-measuring *load cell* consisting of a short column or strut, with electrical resistance-type strain gages (see Section 6.7) attached (Fig. 6.2). When an applied force deflects or strains the block, the force effect is transduced to deflection (we are interested in the unit deflection in this case). The load is transduced to strain. In turn, the strain is transformed into an electrical resistance change, with the strain gages serving as secondary transducers.

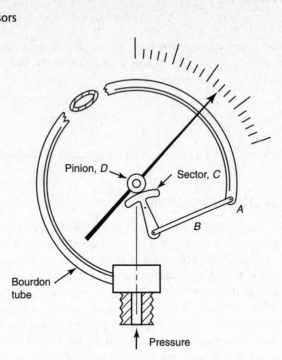

FIGURE 6.1: Essentials of a Bourdon-tube pressure gage.

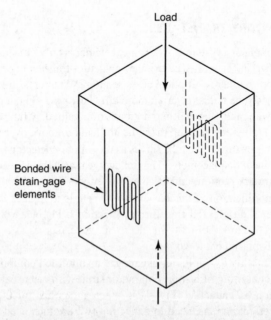

FIGURE 6.2: Schematic representation of a strain-gage load cell. The block forms the primary detector-transducer and the gages are secondary transducers.

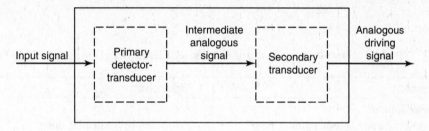

FIGURE 6.3: Block diagram of a first-stage device with primary and secondary transducers.

6.4 CLASSIFICATION OF FIRST-STAGE DEVICES

It appears, therefore, that the stage-one instrumentation may be of varying basic complexity, depending on the number of operations performed. This leads to a classification of first-stage devices as follows:

Class I. First-stage element used as detector only

Class II. First-stage elements used as detector and single transducer

Class III. First-stage elements used as detector with two transducer stages

A generalized first stage may therefore be shown schematically, as in Fig. 6.3.

Stage-one instrumentation may be very simple, consisting of no more than a mechanical spindle or contacting member used to transmit the quantity to be measured to a secondary transducer. Or it may consist of a much more complex assembly of elements. In any event the primary detector-transducer is an *integral* assembly whose function is (1) to sense selectively the quantity of interest, and (2) to process the sensed information into a form acceptable to stage-two operations. It does not present an output in immediately usable form.

More often than not the initial operation performed by the first-stage device is to transduce the input quantity into an analogous displacement. Without attempting to formulate a completely comprehensive list, let us consider Table 6.1 as representing the general area of the primary detector-transducer in mechanical measurements.

We make no attempt now to discuss all the many combinations of elements listed in Table 6.1. In most cases we have referred in the table to sections where thorough discussions can be found. The general nature of many of the elements is self-evident. A few are of minimal importance, included merely to round out the list. However, we can make several pertinent observations at this point.

Close scrutiny of Table 6.1 reveals that, whereas many of the mechanical sensors transduce the input to displacement, many of the electrical sensors change displacement to an electrical output. This is quite fortunate, for it yields practical combinations in which the mechanical sensor serves as the primary transducer and the electrical sensor as the secondary. The two most commonly used electrical means are variable resistance and variable inductance, although others, such as photoelectric and piezoelectric effects, are also of considerable importance.

TABLE 6.1: Some Primary Detector-Transducer Elements and Operations They Perform

Element	Operation
I. Mechanical	
A. Contacting spindle, pin, or finger	Displacement to displacement
B. Elastic member	
1. Load cells (Chapter 13)	
a. Tension/compression	Force to linear displacement
b. Bending	Force to linear displacement
c. Torsion	Torque to angular displacement
2. Proving ring (Chapter 13)	Force to linear displacement
3. Bourdon tube (Chapter 14)	Pressure to displacement
4. Bellows	Pressure to displacement
5. Diaphragm (Chapter 14)	Pressure to displacement
6. Helical spring	Force to linear displacement
7. Torsional spring	Torque to angular displacement
C. Mass	
1. Seismic mass (Chapter 17)	Forcing function to relative displacement
2. Pendulum	Gravitational acceleration to frequency or period
3. Pendulum (Chapter 13)	Force to displacement
4. Liquid column (Chapter 14)	Pressure to displacement
D. Thermal (Chapter 16)	
1. Thermocouple	Temperature to electric potential
2. Bimaterial (includes mercury in glass)	Temperature to displacement
3. Thermistor/RTD	Temperature to resistance change
4. Chemical phase	Temperature to phase change
5. Pressure thermometer	Temperature to pressure
E. Hydropneumatic	
1. Static	
a. Float	Fluid level to displacement
b. Hydrometer	Specific gravity to relative displacement
2. Dynamic (Chapter 15)	
a. Orifice	Fluid velocity to pressure change
b. Venturi	Fluid velocity to pressure change
c. Pitot tube	Fluid velocity to pressure change
d. Vanes	Velocity to force
e. Turbines	Linear to angular velocity

TABLE 6.1: (*continued*)

Element	Operation
II. Electrical	
A. Resistive (Sections 6.5–6.8)	
1. Contacting	Displacement to resistance change
2. Variable-length conductor	Displacement to resistance change
3. Variable-area conductor	Displacement to resistance change
4. Variable dimensions of conductor	Strain to resistance change
5. Variable resistivity of conductor	Temperature to resistance change
B. Inductive (Sections 6.10–6.12)	
1. Variable coil dimensions	Displacement to change in inductance
2. Variable air gap	Displacement to change in inductance
3. Changing core material	Displacement to change in inductance
4. Changing core positions	Displacement to change in inductance
5. Changing coil positions	Displacement to change in inductance
6. Moving coil	Velocity to change in induced voltage
7. Moving permanent magnet	Velocity to change in induced voltage
8. Moving core	Velocity to change in induced voltage
C. Capacitive (Section 6.13)	
1. Changing air gap	Displacement to change in capacitance
2. Changing plate areas	Displacement to change in capacitance
3. Changing dielectric constant	Displacement to change in capacitance
D. Piezoelectric (Section 6.14)	Displacement to voltage and/or voltage to displacement
E. Semiconductor junction (Section 6.15)	
1. Junction threshold voltage	Temperature to voltage change
2. Photodiode current	Light intensity to current
F. Photoelectric (Section 6.16)	
1. Photovoltaic	Light intensity to voltage*
2. Photoconductive	Light intensity to resistance change*
3. Photoemissive	Light intensity to current*
G. Hall Effect (Section 6.17)	Displacement to voltage

* Also sensitive to wavelength of light.

In addition to the inherent compatibility of the mechano-electric transducer combination, electrical elements have several important relative advantages:

1. Amplification or attenuation can be easily obtained.
2. Mass-inertia effects are minimized.
3. The effects of friction are minimized.
4. An output power of almost any magnitude can be provided.
5. Remote indication or recording is feasible.
6. The transducers can often be miniaturized.

Most of the remainder of this chapter is devoted to a discussion of electrical transducers. Modification of their outputs, or signal conditioning, is discussed in Chapter 7.

6.5 VARIABLE-RESISTANCE TRANSDUCER ELEMENTS

Resistance of an electrical conductor varies according to the following relation:

$$R = \frac{\rho L}{A} \tag{6.2}$$

where

$$R = \text{resistance}(\Omega),$$
$$L = \text{the length of the conductor (cm)},$$
$$A = \text{cross-sectional area of the conductor (cm}^2),$$
$$\rho = \text{the resistivity of material } (\Omega \cdot \text{cm})$$

Many sensors are based on changes in the factors determining resistance. Some examples include sliding-contact devices and potentiometers, in which L changes (Section 6.6); resistance strain-gages, in which L, A, and ρ change (Section 6.7 and Chapter 12); and thermistors, photoconductive light detectors, piezoresistive strain gages, and resistance temperature detectors, in which ρ changes (see, respectively, Sections 6.8 and 16.4; 6.15 and 6.16; 6.15; and 16.4).

Probably the simplest mechanical-to-electrical transducer is the ordinary *switch* in which resistance is either zero or infinity. It is a yes–no, conducting–nonconducting device that can be used to operate an indicator. Here a lamp is fully as useful for readout as a meter, since only two values of quantitative information can be obtained. In its simplest form, the switch may be used as a limiting device operated by direct mechanical contact (as for limiting the travel of machine tool carriages) or it may be used as a position indicator. When actuated by a diaphragm or bellows, it becomes a pressure-limit indicator, or if controlled by a bimetal strip, it is a temperature limit indicator. It may also be combined with a proving ring to serve as either an overload warning device or a device actually limiting load carrying, such as a safety device for a crane.

6.6 SLIDING-CONTACT DEVICES

Sliding-contact resistive transducers convert a mechanical displacement input into an electrical output, either voltage or current. This is accomplished by changing the effective

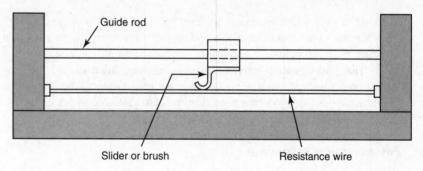

FIGURE 6.4: Variable resistance consisting of a wire and movable contactor or brush. This is often referred to as a slide wire.

length L of the conductor in Eq. (6.2). Some form of electrical resistance element is used, with which a contactor or brush maintains electrical contact as it moves. In its simplest form, the device may consist of a stretched resistance wire and slider, as in Fig. 6.4. The effective resistance existing between either end of the wire and the brush thereby becomes a measure of the mechanical displacement. Devices of this type have been used for sensing relatively large displacements [1].

More commonly, the resistance element is formed by wrapping a resistance wire around a form, or *card*. The turns are spaced to prevent shorting, and the brush slides across the turns from one turn to the next. In actual practice, either the arrangement may be wound for a rectilinear movement or the resistance element may be formed into an arc and angular movement used, as shown in Fig. 6.5(a).

Sliding-contact devices are also made using conductive films as the variable-resistance elements, rather than wires [Fig. 6.5(b)]. Common examples include carbon-composition films, in which graphite or carbon particles are suspended in an epoxy or polyester binder, and ceramic-metal composition films (or *cermet*), in which ceramic and precious metal powders are combined. In each case, the thin film is supported by a ceramic or plastic

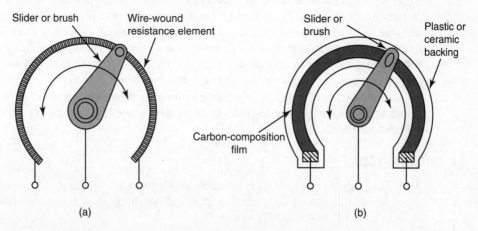

FIGURE 6.5: Angular-motion variable resistance, or potentiometer: (a) wire wound, (b) carbon composition.

backing. Conductive-film devices are less expensive than wire-wound devices. The carbon-film devices, in particular, have outstanding wear characteristics and long life [2], although they are more susceptible to temperature drift and humidity effects.

These devices are commonly called *resistance potentiometers*,[1] or simply *pots*. Variations of the basic angular or rotary form are the multiturn, the low-torque, and various nonlinear types [4]. Multiturn potentiometers are available with various numbers of revolutions, sometimes as many as 40. See also Section 7.7.

6.6.1 Potentiometer Resolution

The resistance variation available from a sliding-contact moving over a wire-wound resistance element is not a continuous function of contact movement. The smallest increment into which the whole may be divided determines the *resolution*. In the case of a wire-wound resistor, the limiting resolution equals the reciprocal of the number of turns. If 1200 turns of wire are used and the winding is linear, the resolution will be 1/1200, or 0.09083%. The meaning of this quantity is apparent: No matter how refined the remainder of the system may be, it will be impossible to divide, or resolve, the input into parts smaller than 1/1200 of the total potentiometer range.

For conductive-film potentiometers, resolution is negligibly small and variation of the slider's contact resistance is a more significant limitation.

6.6.2 Potentiometer Linearity

When used as a measurement transducer, a linear potentiometer is normally required. Use of the term *linear* assumes that the resistance measured between one of the ends of the element and the contactor is a direct linear function of the contactor position in relation to that end. Linearity is never completely achieved, however, and deviation limits are usually supplied by the manufacturer.

6.7 THE RESISTANCE STRAIN GAGE

Experiment has shown that the application of a strain to a resistance element changes its resistance. This is the basis for the resistance strain gage. A resistance element, such as a fine wire, is cemented to the surface of a member to be strained. Assuming a tensile strain, the element lengthens and its cross-sectional area decreases. From Eq. (6.2), we see that both changes increase the wire's resistance. In addition, the wire's resistivity changes when the wire is strained. Thus, every factor in Eq. (6.2) varies simultaneously.

This device is of sufficient importance in mechanical measurements to warrant a more complete discussion than can be given at this point. Chapter 12 is devoted to the theory and use of strain gages.

6.8 THERMISTORS

Thermistors are thermally sensitive variable resistors made of ceramic-like semiconducting materials. Oxides of manganese, nickel, and cobalt are used in formulations having resistances of 100 to 450,000 $\Omega \cdot$ cm.

---

[1]Unfortunately, another entirely different device is also called a potentiometer. It is the voltage-measuring instrument wherein a standard reference voltage is adjusted to counterbalance the unknown voltage (see [3]). The two devices are different and must not be confused.

These devices have two basic applications: (1) as temperature-detecting elements used for the purpose of measurement or control, and (2) as electric-power-sensing devices wherein the thermistor temperature—and hence resistance—are a function of the power being dissipated by the device. The second application is particularly useful for measuring radio frequency power.

Further discussion of thermistors is given in Section 16.4.3.

6.9   THE THERMOCOUPLE

While two dissimilar metals are in contact, an electromotive force exists whose magnitude is a function of several factors, including *temperature*. Junctions of this sort, when used to measure temperature, are called *thermocouples*. Often the junction is formed by twisting and welding together two wires.

Because of its small size, its reliability, and its relatively large range of usefulness, the thermocouple is a very important primary sensing element. Further discussion of its application is reserved for Chapter 16 (see especially Section 16.5).

6.10   VARIABLE-INDUCTANCE TRANSDUCERS

Inductive transducers are based on the voltage output of an inductor (or *coil*) whose inductance changes in response to changes in the measurand. The coil is often driven by an ac excitation, although in dynamic measurements the motion of the coil relative to a permanent magnet may create sufficient voltage. A classification of inductive transducers, based on the fundamental principle used, is as follows.

1. Variable self-inductance

   (a) Single coil (simple variable reluctance)
   (b) Two coil (or single coil with center tap) connected for inductance ratio

2. Variable mutual inductance

   (a) Simple two coil
   (b) Three coil (using series opposition)

3. Variable reluctance with permanent magnet

   (a) Moving iron
   (b) Moving coil
   (c) Moving magnet

The inductance of a coil, L, is influenced by a number of factors, including the number of turns in the coil, the coil size, and especially the permeability of the magnetic flux path that passes through the center of the coil. The magnetic flux path forms a closed loop that extends outside the coil. Often, a magnetic material, such as iron, will be used in the flux path, commonly in conjunction with one or more air gaps. Because the air gaps have a much lower magnetic permeability than the iron, they control the inductance of the coil. Thus, the variation in the thickness of an air gap is often the primary measurand sensed by a variable-inductance device.

Some coils are wound with only air as the core material. They are generally used only with relatively high frequency excitation; however, they will occasionally be found in transducer circuitry. An expression that may be used to estimate the inductance of a straight, cylindrical air-core coil is as follows:

$$L = \frac{d^2 n^2}{18d + 40l} \tag{6.3a}$$

where

$$L = \text{inductance } (\mu\text{H})$$
$$d = \text{coil diameter (in.)}$$
$$l = \text{coil length (in.)}$$
$$n = \text{number of turns}$$

This expression applies for $l > 0.4d$ [5].

When the flux path includes both a magnetic material and an air gap, the inductance may be estimated as follows:

$$L = \frac{n^2}{\dfrac{h_m}{\mu_m A_m} + \dfrac{h_a}{\mu_a A_a}} \tag{6.3b}$$

where

$$L = \text{inductance (H)}$$
$$h_m = \text{length of flux path in magnetic material (m)}$$
$$h_a = \text{length of flux path in air gap (m)}$$
$$\mu_m = \text{permeability of magnetic material (H/m)}$$
$$\mu_a = \text{permeability of air}$$
$$\approx \mu_0, \text{permeability of free space}$$
$$= 4\pi \times 10^{-7} \text{ (H/m)}$$
$$A_m = \text{cross-sectional area of flux path in iron (m}^2)$$
$$A_a = \text{cross-sectional area of flux path in air (m}^2)$$
$$n = \text{number of turns}$$

The quantities $(h/\mu A)$ in the denominator are called *reluctances* and have the same relationship to magnetic flux that resistance has to electric current. If additional gaps or magnetic materials lie in the flux path, the associated values of $(h/\mu A)$ would be added to the sum in the denominator [6].

In many instances, the permeability of the magnetic material is sufficiently high that only the air gaps need to be considered. In such cases, Eq. (6.3b) reduces to

$$L = \frac{\mu_0 n^2 A_a}{h_a} \tag{6.3c}$$

When an ac excitation is used, the inductive reactance and dc resistance of the coil determine the electrical response of the measuring circuit. The inductive reactance is

$$X_L = 2\pi f L \tag{6.4a}$$

where

$$X_L = \text{inductive reactance } (\Omega)$$
$$f = \text{frequency of applied voltage (Hz)}$$
$$L = \text{inductance (H)}$$

The total impedance of the coil is

$$Z = \sqrt{X_L^2 + R^2} \tag{6.4b}$$

in which R is the dc resistance. The higher the inductance of a coil relative to its resistance, the higher is said to be its quality, which is designated by the symbol $Q = X_L/R$. In most cases, high Q is desired.

Inductive transducers may be based on variation of any of the variables indicated in the foregoing equations, and most have been tried at one time or another. The following are representative.

6.10.1 Simple Self-Inductance Arrangements

When a simple single coil is used as a transducer element, the mechanical input usually changes the reluctance of the flux path generated by the coil, thereby changing its inductance. The change in inductance is then measured by suitable circuitry, indicating the value of the input. The flux path may be changed by a change in air gap; however, a change in either the amount or type of core material may also be used [7,8].

The arrangement shown in Fig. 6.6 includes two air gaps whose thickness will vary in response to the movement of an armature. The flux path runs through the upper horseshoe of iron, across one air gap, through the iron armature, and back across the other air gap. Each of the four parts of the path contributes a reluctance, but those from the air gaps will likely dominate the total reluctance, owing to the low permeability of air relative to iron.

Figure 6.7 illustrates a form of *two-coil* self-inductance. (This may also be thought of as a single coil with a center tap.) Movement of the core or armature alters the relative inductance of the two coils. Devices of this type are usually incorporated in some form of inductive bridge circuit (see Section 7.10) in which variation in the inductance ratio between the two coils provides the output. An application of a two-coil self-inductance used as a secondary transducer for pressure measurement is described in Section 14.7.2.

6.10.2 Two-Coil Mutual-Inductance Arrangements

Mutual-inductance arrangements using two coils are shown in Figs. 6.8 and 6.9. Figure 6.8 illustrates the manner in which these devices function. The magnetic flux from a power coil is coupled to a pickup coil, which supplies the output. Input information, in the form of armature displacement, changes the coupling between the coils. In the arrangement shown, the air gaps between the core and the armature govern the degree of coupling. In other

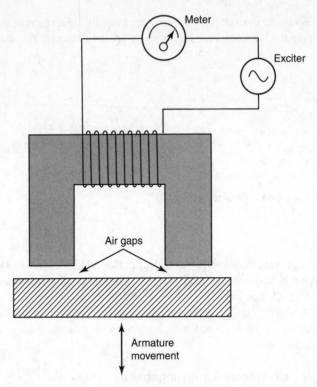

FIGURE 6.6: A simple self-inductance arrangement wherein a change in the air gap changes the pickup output.

arrangements the coupling may be varied by changing the relative positions of the coils and armature, either linearly or angularly.

Figure 6.9 shows the detector portion of an *electronic micrometer* [9]. Inductive coupling between the coils, which depends on the reluctance of the magnetic-flux path, is

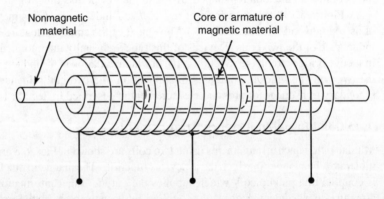

FIGURE 6.7: Two-coil (or center-tapped single coil) inductance-ratio transducer.

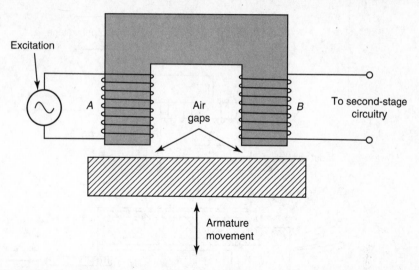

FIGURE 6.8: A mutual-inductance transducer. Coil A is the energizing coil and B is the pickup coil. As the armature is moved, thereby altering the air gap, the output from coil B is changed, and this change may be used as a measure of armature movement.

changed by the relative proximity[2] of a permeable material. A variation of this has been used in a transducer for measuring small inside diameters [10]. In that case the coupling is varied by relative movement between the two coils.

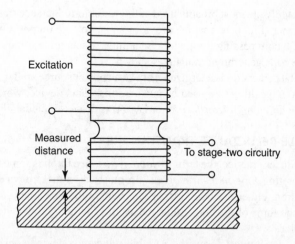

FIGURE 6.9: Two-coil inductive pickup for "an electronic micrometer."

----

[2]Sensors that detect the presence or position of a nearby object are often called *proximity sensors.*

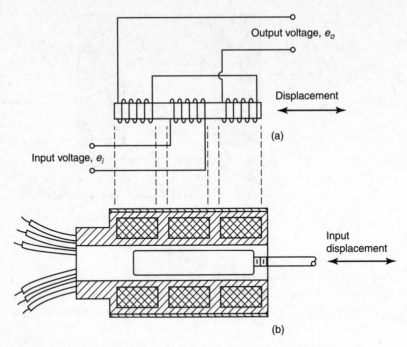

FIGURE 6.10: The differential transformer: (a) schematic arrangement (b) section through typical transformer.

6.11 THE DIFFERENTIAL TRANSFORMER

Undoubtedly the most broadly used of the variable-inductance transducers is the differential transformer (Fig. 6.10), which provides an ac voltage output proportional to the displacement of a core passing through the windings. It is a mutual-inductance device making use of three coils generally arranged as shown.

The center coil is energized from an ac power source, and the two end coils, connected in phase opposition, are used as pickup coils. This device, which is often called a *linear variable differential transformer* or LVDT, is discussed in detail in Section 11.13.

6.12 VARIABLE-RELUCTANCE TRANSDUCERS

In transducer practice, the term *variable reluctance* implies some form of inductance device incorporating a *permanent magnet*. In most cases these devices are limited to dynamic application, either periodic or transient, where the flux lines supplied by the magnet are cut by the turns of the coil. Some means of providing relative motion is incorporated into the device.

In its simplest form, the variable-reluctance device consists of a coil wound on a permanent magnet core (Fig. 6.11). Any variation of the reluctance of the magnetic flux path causes a change in the flux. As the flux field varies, a voltage is induced in the coil, according to Faraday's law:

$$V = -n\frac{d}{dt}\Phi \qquad (6.5)$$

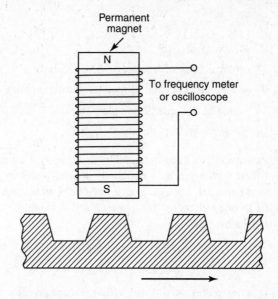

Permanent
magnet

N

To frequency meter
or oscilloscope

S

FIGURE 6.11: A simple variable-reluctance pickup.

where

$$V = \text{induced voltage (V)}$$
$$n = \text{number of turns in coil}$$
$$\Phi = \text{magnetic flux through coil (Wb)}$$

Since the rate of change of the flux depends directly on the speed at which the teeth move past the magnet in Fig. 6.11, the variable-reluctance transducer is sensitive to velocity, rather than displacement. Practical applications of this device are discussed in Sections 10.7 and 15.5.1.

Whereas the preceding arrangement depends upon changes of the reluctance of the magnetic flux path, other devices separate the magnet from the coil and depend upon relative movement between the coil and the flux field (Section 17.7.2).

6.13 CAPACITIVE TRANSDUCERS

When a capacitor is formed from a pair of parallel flat plates, its capacitance is given by the following equation:

$$C = \frac{\varepsilon_0 K A}{d} \tag{6.6a}$$

where

C = capacitance (pF),

ε_0 = permittivity of free space, 8.8542 pF/m,

K = dielectric constant of medium between plates (= 1 for air),

A = area of one side of one plate (m^2),

d = separation of plate surfaces (m)

Greater sensitivity can be obtained by using several capacitors in parallel. This may be accomplished with a stack of n equally spaced plates in which alternate plates are connected to one another. For example, if five plates were stacked, plates 1, 3, and 5 would be connected to one voltage, while plates 2 and 4 would be connected to another. The capacitance of such a stack is

$$C = \frac{\varepsilon_0 K A (n - 1)}{d} \tag{6.6b}$$

All the terms represented in Eqs. (6.6), except possibly the number of plates, have been used in transducer applications [7,8]. The following are examples of each.

Changing Dielectric Constant

Figure 6.12 shows a device developed for the measurement of level in a container of liquid hydrogen [11]. The capacitance between the central rod and the surrounding tube varies with changing dielectric constant brought about by changing liquid level. The device readily detects liquid level even though the difference in dielectric constant between the liquid and vapor states may be as low as 0.05.

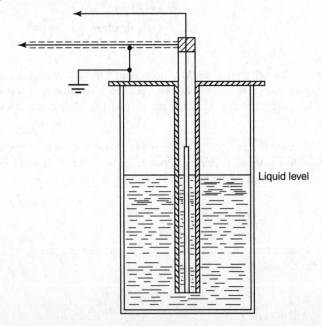

FIGURE 6.12: Capacitance pickup for determining level of liquid hydrogen.

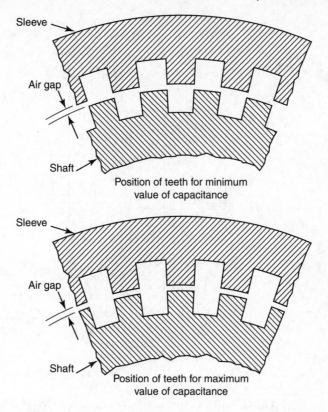

FIGURE 6.13:   Section showing relative arrangement of teeth in capacitance-type torque meter.

Changing Area

Capacitance change depending on changing effective area has been used for the secondary transducing element of a torque meter [12]. The device uses a sleeve with teeth or serrations cut axially, and a matching internal member or shaft with similar axially cut teeth. Figure 6.13 illustrates the arrangement. A clearance is provided between the tips of the teeth, as shown. Torque carried by an elastic member causes a shift in the relative positions of the teeth, thereby changing the effective area. The resulting capacitance change is calibrated in terms of torque.

Changing Distance

Varying the distance between the plates of a capacitor is undoubtedly the most common method for using capacitance in a pickup.

Figure 6.14 illustrates a capacitive-type pressure transducer, wherein the capacitance between the diaphragm to which the pressure is applied and the electrode foot is used as a measure of the diaphragm's relative position [13–15]. Flexing of the diaphragm under pressure alters the distance between it and the electrode.

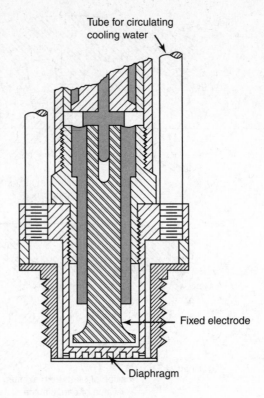

Tube for circulating
cooling water

Fixed electrode

Diaphragm

FIGURE 6.14: Section through capacitance-type pressure pickup.

6.14 PIEZOELECTRIC SENSORS

Certain materials can generate an electrical charge when subjected to mechanical strain or, conversely, can change dimensions when subjected to voltage [Fig. 6.15(a)]. This is known as the *piezoelectric*[3] *effect*. Pierre and Jacques Curie are credited with its discovery in 1880. Notable among these materials are quartz, Rochelle salt (potassium sodium tartarate), properly polarized barium titanate, ammonium dihydrogen phosphate, certain organic polymers, and even ordinary sugar.

Of all the materials that exhibit the effect, none possesses all the desirable properties, such as stability, high output, insensitivity to temperature extremes and humidity, and the ability to be formed into any desired shape. Rochelle salt provides a very high output, but it requires protection from moisture in the air and cannot be used above about 45°C (115°F). Quartz is undoubtedly the most stable, but its output is low. Because of its stability, quartz is commonly used for regulating electronic oscillators (Section 10.4). Often the quartz is shaped into a thin disk with each face silvered for the attachment of electrodes. The thickness of the plate is ground to the dimension that provides a mechanical resonance frequency corresponding to the desired electrical frequency. This crystal may then be incorporated in an appropriate electronic circuit whose frequency it controls.

---

[3]The prefix *piezo* is derived from the Greek *piezein*, meaning "to press" or "to squeeze."

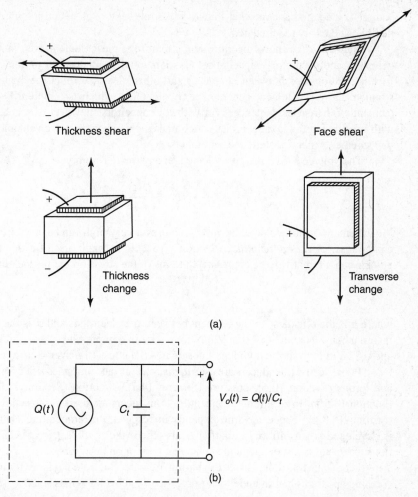

FIGURE 6.15: (a) Basic deformation modes for piezoelectric plates; electrodes ($+$ and $-$) shown; (b) equivalent circuit for a piezoelectric element.

Rather than existing as a single crystal, as are many piezoelectric materials, barium titanate is polycrystalline; thus it may be formed into a variety of sizes and shapes. The piezoelectric effect is not present until the element is subjected to polarizing treatment. Although exact polarizing procedure varies with the manufacturer, the following procedure has been used [16]. The element is heated to a temperature above the Curie point of 120°C, and a high dc potential is applied across the faces of the element. The magnitude of this voltage depends on the thickness of the element and is on the order of 10,000 V/cm. The element is then cooled with the voltage applied, which results in an element that exhibits the piezoelectric effect.

Piezoelectric polymers, such as polyvinylidene fluoride [17], provide low-cost piezo-electric transducers with relatively high voltage outputs. These semicrystalline polymers are formed into thin films, perhaps 30 μm thick, with silvered electrodes on either side and are coated onto a somewhat-thicker Mylar backing. The resulting transducer is light, flexible, and easily manipulated.

Figure 6.15(b) shows an equivalent circuit for a piezoelement, consisting of a charge generator and a shunting capacitance, C_t. (The reader should refer to Tables 5.1 and 5.2 for the relationships between charge Q and other electrical units.) When mechanically strained, the piezoelement generates a charge $Q(t)$, which is temporarily stored in the element's inherent capacitance. As with all capacitors, however, the charge dissipates with time owing to leakage, a fact which makes piezodevices more valuable for dynamic measurements than for static measurements.

The voltage across the piezoelement at any time is simply

$$V_o(t) = \frac{Q(t)}{C_t} \tag{6.7a}$$

Measurement of this voltage, however, requires a very high impedance circuit to prevent charge loss. A *charge amplifier* (Section 7.15.2) is normally used for this purpose. The voltage can alternatively be expressed in terms of the stress on the piezoelement,

$$V_o = Gh\sigma \tag{6.7b}$$

where h is the thickness of the element between the electrodes and σ is the stress. G is a material constant equal to 0.055 Vm/N for quartz in compressive stress (thickness change) and 0.22 Vm/N for polyvinylidene fluoride in axial stress (transverse change).

Piezoelectric transducers are used to measure surface roughness (Section 11.14), force and torque (Section 13.6), pressure (Section 14.7.3), motion (Section 17.7), and sound (Section 18.5.1). In addition, the piezofilm transducers are used to sense thermal radiation (Section 16.8.4): Since the film expands in response to temperature change, a charge is developed when infrared radiation is absorbed (the *pyroelectric* effect). Pyroelectric transducers are common elements in household motion detectors.

Ultrasonic transducers may use barium titanate. Such elements are found in industrial cleaning apparatus and in underwater detection systems known as *sonar*.

6.15 SEMICONDUCTOR SENSORS

The semiconductor revolution has profoundly influenced measurement technology. In addition to digital voltmeters, computer data-acquisition systems, and other readout and data-processing systems, semiconductor technology has produced compact and inexpensive sensors. A principal strength of semiconductor sensors is that they take advantage of microelectronic fabrication techniques. Thus, the sensors can be quite small, mechanical structures (such as diaphragms and beams) can be etched into the device, and other electronic components (resistors, transistors, etc.) can be directly implanted with the sensor to form a transducer having onboard signal conditioning.

6.15.1 Electrical Behavior of Semiconductors

Semiconducting materials include elements, such as silicon and germanium, and compounds, such as gallium arsenide and cadmium sulphide. Semiconductors differ from metals in that relatively few free electrons are available to carry current. Instead, when a bound electron is separated from a particular atom in the material, a positively charged *hole* is formed and will move in the direction opposite the electron. Both negatively charged electrons and positively charged holes contribute to the flow of current in a semiconductor.

The number of charge carriers (electrons or holes) in a semiconductor, n_c, is a strong function of temperature, T. Typically,

$$n_c = \text{number per unit volume} \propto T^{3/2} \exp\left(\frac{-\text{constant}}{T}\right) \qquad (6.8)$$

Since the resistivity of a material is proportional to $1/n_c$, a semiconductor's resistance decreases rapidly with increasing temperature. For silicon near room temperature, the resistivity decreases by about 8%/°C [18].

Greater control over a semiconductor's electrical behavior is obtained by *doping* it with impurity atoms. These atoms may be either *electron donors* or *electron acceptors*. Electron donor atoms (such as phosphorus or arsenic) raise the number of free electrons in the material. Electron acceptor atoms (such as gallium and aluminum) hold electrons, thus raising the number of holes. Since the number of doping atoms is usually large relative to the number of free electrons in the undoped material, the dopant sets the *majority current carrier* of the material. Specifically, doping with donor atoms creates a predominance of *negative* charge carriers (electrons), giving an *n-type* semiconductor. Doping with acceptor atoms creates a predominance of *positive* charge carriers (holes), giving a *p-type* semiconductor.

Semiconductors, either doped or undoped, are useful as temperature sensors. For undoped semiconductors, the number of carriers increases rapidly with temperature [Eq. (6.8)], so that the resistance is a strongly decreasing function of temperature. Thermistors (Sections 6.8, 16.4.3) are based on this effect. Because such sensors have a *negative temperature coefficient* of resistance, they are sometimes called *NTC* sensors. When semiconductors are heavily doped, the mobility of the carriers *decreases* with increasing temperature, so that the resistance increases with temperature; these *positive-temperature-coefficient* devices are called *PTC* sensors.

Semiconductors also respond to strain. For example, a *p*-type region diffused into an *n*-type base functions as a resistor whose resistance increases strongly when it is strained (this behavior is called *piezoresistivity*). Such resistors are the basis of semiconductor strain gages, semiconductor diaphragm pressure sensors (Fig. 6.16, Section 14.6.3), and semiconductor accelerometers [19,20].

6.15.2 *pn*-Junctions

Most semiconductor devices involve a *junction*, at which *n*-type and *p*-type doping meet (Fig. 6.17). Current flows easily from the *p*-type to the *n*-type material, since holes (+charge) easily enter the *n*-type material and electrons (−charge) easily enter the *p*-type material. Current flow in the opposite direction meets much greater resistance. Thus, this junction behaves like a diode.

When a voltage is applied to the junction, the current through it varies as shown in Fig. 6.18. When V is positive, we say that the junction is *forward biased*. The current becomes very large once the voltage reaches a threshold level. If the voltage is instead negative, the junction is *reverse biased*, and only a very small current flows. As the reverse-bias voltage is raised, the current quickly reaches a value $-I_0$, which is nearly independent of V. The current I_0 is called the *reverse saturation current*. Typical values of I_0 are on the order of nanoamperes for silicon and microamperes for germanium. The reverse saturation

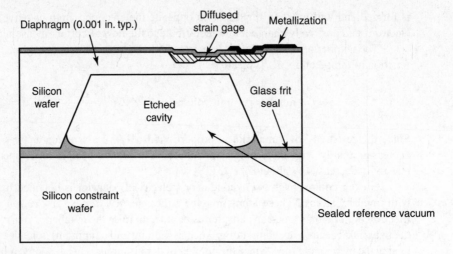

FIGURE 6.16: Cross section of a semiconductor-diaphragm absolute-pressure sensor (Motorola, not to scale). External pressure change causes diaphragm to deflect, straining the gage.

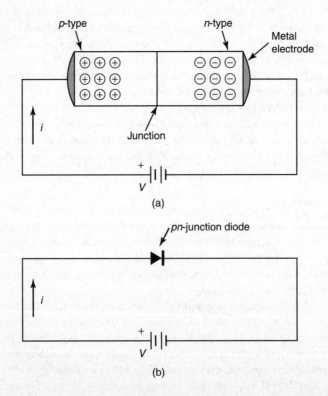

FIGURE 6.17: (a) *pn*-junction with applied voltage (b) circuit representation as a diode.

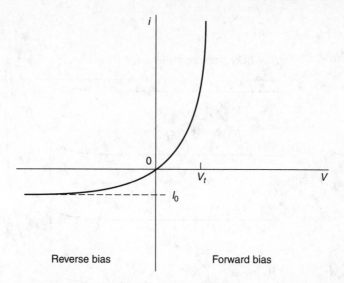

FIGURE 6.18: Voltage–current curve for a *pn*-junction. I_0 is the reverse saturation current and V_t is the forward threshold voltage.

current is more nearly voltage-independent for germanium than for silicon, as a result of various secondary effects [21].

The voltage current curve in Fig. 6.18 is described by the equation

$$i = I_0 \left[\exp\left(\frac{q}{kT} V \right) - 1 \right] \tag{6.9}$$

which shows a strong dependence of current on temperature (q is the charge of an electron and k is Boltzmann's constant). The saturation current is also a function of temperature, roughly doubling with every 10°C increase in temperature for silicon and germanium diodes near room temperature [18].

When forward biased, the voltage drop (or threshold voltage), V_t, across a silicon *pn*-junction limits to about 0.7 V at room temperature. This voltage drop decreases by about 2 mV/°C as temperature increases, and the changing voltage forms the basis of some semiconductor-junction temperature sensors. Through integrated-circuit techniques, junction-temperature dependencies have been applied to make linear-response, temperature-sensing chips (Section 16.6).

One disadvantage of many semiconductor-junction sensors is that the operating temperature must remain below about 150°C to prevent degradation of the junction's electrical characteristics.

6.15.3 Photodiodes

Semiconductor junctions are sensitive to light as well as heat. If a junction is formed near the surface of a semiconductor, photons reaching the junction can create new pairs of electrons and holes, which then separate and flow in opposite directions. Thus, the irradiating light

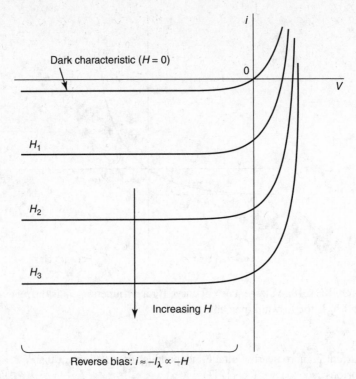

FIGURE 6.19: Photodiode *i*-*V* characteristic for various incident light intensities, *H*.

produces an additional current, I_λ, at the junction:

$$i = I_0 \left[\exp\left(\frac{q}{kT} V\right) - 1 \right] - I_\lambda \qquad (6.10)$$

The photocurrent, I_λ, is directly proportional to the intensity, H, of the incoming light (in W/m^2):

$$I_\lambda = \text{constant} \times H$$

where the constant is proportional to the area of the diode exposed to light. The voltage-current characteristics of a photodiode are shown in Fig. 6.19 [22].

The photocurrent is typically on the order of milliamperes and thus is much larger than the reverse saturation current ($I_\lambda \gg I_0$). By operating a photodiode with a reverse-bias voltage, the output current is made directly proportional to the incident light intensity ($i \approx -I_\lambda \propto -H$).

Semiconductor junctions are most responsive to infrared wavelengths, but sensitivity can extend to visible wavelengths and near-ultraviolet wavelengths as well. Common photodiodes are usually made from inexpensive silicon junctions, although several other semiconductors, such as germanium, are also in use.

The sensitivity of a photodiode is limited by its "dark current," which is the usual junction current with no incident light [i.e., Eq. (6.9)]. Since this current decreases with temperature, sensitivity can be improved by cooling the diode to very low temperatures. For example, in high-performance infrared sensing, photodiodes may be operated at liquid-nitrogen temperatures or even liquid-helium temperatures ($-198°C$ or $-269°C$, respectively) [23].

Some bulk semiconductors, without a *pn*-junction, also respond to light. Photons create additional electron-hole pairs in the material, thereby reducing its resistance. Common examples of such *photoconductive* materials include cadmium sulphide (CdS) and cadmium selenide (CdSe). For high-performance infrared sensing, low-temperature doped germanium may be used.

Photodetectors are discussed further in the next section.

6.16   LIGHT-DETECTING TRANSDUCERS

Light-sensitive transducers, or photosensors, are used to detect light of all types: thermal radiation from warm objects, laser light, light emitted by diodes, or even sunshine. These transducers may be categorized as either *thermal detectors* or *photon detectors*. The thermal detectors use a temperature-sensitive element which is heated by incident light. The photon detectors respond directly to absorbed photons, either by emitting an electron from a surface (the *photoelectric effect*) or by creating additional electron-hole pairs in a semiconductor (as discussed in Section 6.15.3).

Among the issues to be considered in selecting a photodetector are the wavelength to be sensed, the speed of response needed, and the sensitivity required. In general, thermal detectors are much slower than photon detectors but respond to a broader range of wavelengths. For any detector, the speed of response will also depend upon the supporting circuitry. For visible and near-infrared light, semiconductor detectors are commonly used. Photoemissive detectors can be sensitive well into the ultraviolet range. To detect long-wave infrared light, which is the heat emitted by objects near room temperature, either thermal detectors or cryogenically cooled semiconductors may be used.

6.16.1   Thermal Detectors

Thermal detectors create a temperature change in the detecting element when light heats the detector. Some other property of the detecting element changes in response to the temperature change, and that property is measured to determine the light intensity [24]. Some examples follow.

Thermopile detecting elements use several thermocouples in series to produce a voltage output proportional to the detector temperature (see Section 16.5.6). *Pneumatic* detecting elements use a chamber containing a gas. When the gas is heated, its pressure rises, and the pressure is measured to determine the light intensity (see Section 16.3). *Bolometers* use temperature-dependent electrical resistance as a sensor (see Section 16.4). *Pyroelectric* detectors respond to temperature changes by generating an electric charge (see Section 6.14); they are widely used as infrared motion sensors for automotic light switches and intruder alarms [24,25].

The response times of typical thermal detectors range from a few milliseconds to several seconds, depending mainly upon the size and configuration of the detecting element. Specialized pyroelectric sensors can achieve response times below 1 ns. Thermal detectors are discussed further in Section 16.8.

6.16.2 Photon Detectors

Photoemissive detectors (Type A, Table 6.2) consist of a cathode anode combination in an evacuated envelope made of glass or synthetic quartz. In the proper circuit (commonly requiring a dc source of several hundred volts), light impingement on the cathode causes electrons to be emitted. The electrons travel to the anode, thereby providing a small current. By adding several successively higher-voltage electrodes (or *dynodes*) to the envelope, substantial current amplification is obtained, producing a *photomultiplier tube*, or *PMT*. PMTs can be extraordinarily sensitive, to the point of detecting *single* photons, and they have rise times as short as 1 ns. Since the invention of smaller, cheaper semiconductor photosensors, these devices are used only in rather specialized applications that require extremely high sensitivity [26].

Semiconductor sensors are of several types. In general, they perform best at near-infrared wavelengths. The major classes of semiconductor photodetectors are as follow.

Photoconductive sensors (Type B, Table 6.2) consist of a layer of semiconducting material between two electrodes. When the layer is exposed to light, absorbed photons create additional conductors in the semiconductor (electrons and holes), lowering its resistance. In conjunction with resistance-sensitive circuitry (Sections 7.6–7.9), an output may be obtained that is a function of the intensity of the light source. Typical photoconducting detectors use cadmium sulfide, cadmium selenide, lead sulfide, indium antimonide, or mercury cadmium telluride. Among these, inexpensive CdS and CdSe are probably most commonly used, although they are limited to visible and near-infrared wavelengths and respond rather slowly (with typical rise times of 50 ms). Devices using InSb or HgCdTe can operate at far-infrared wavelengths and have rise times on the order of 100 ns. For best performance, these devices require cooling to liquid nitrogen temperatures [27].

Photodiodes (Types C1, C2, and C3, Table 6.2) utilize a *pn*-junction and are similar to photoconductive detectors (see Section 6.15.3). When operated with a reverse-bias voltage (Fig. 6.19), a photodiode behaves as a light-sensitive current source. Rise times may be as low as 1 ns. The *pin*-photodiode differs from the common variety in that a layer of undoped (or intrinsic) semiconductor is sandwiched between the *p* and *n* layers. This increases both the speed of response and the sensitivity. Some *pin*-photodiodes have rise times below 0.1 ns [28].

Avalanche photodiodes are obtained by operating photodiodes with a reverse bias voltage that approaches the diode's breakdown voltage (typically several hundred volts). The result is that the current generated at a given light level is 10 to 100 times greater than that produced by a *pn* or *pin*-photodiode. APDs have the advantages of high gain and fast response (rise times down to 0.1 ns), with the disadvantages of requiring a stable voltage supply and a stable operating temperature.

Phototransistors and *photodarlingtons* (Types D1 and D2, Table 6.2) are basically photodiodes followed one or two stages of amplification incorporated into the same package to enhance the sensitivity. Phototransistors have gains similar to APDs, but much slower rise times (tens of μs). Photodarlingtons have higher gain, but with rise times of hundreds of μs [29].

6.16.3 Applications

Applications of photodetectors in mechanical measurements are wide ranging. They include simple counting, where the interruption of a beam of light is used (Fig. 10.1), strain

TABLE 6.2: Photon Detectors

Type	Symbol and Typical Circuit	Form of Output	Relative Frequency Response	Comments
A. Photoemissive or photomultiplier		Current	Very fast	Cathode–anode in evacuated glass or quartz envelope. PMT gain can be 10^3 to 10^8. Bulky; requires high voltage. Highest sensitivity.
B. Photoconductive (or photoresistive)		Resistance change	Extremely slow	Light-sensitive resistor. Increased light intensity causes reduced resistance. Expensive types can be fast.
C1. Photodiode (pn-junction)		Current	Very fast	Primary disadvantage is low output current. "Dark current" very low (nanoampere range), but not zero.
C2. pin-photodiode		Current	Extremely fast	pin diode has "intrinsic" layer between p and n layers, which creates faster response.
C3. Avalanche photodiode		Current	Extremely fast	APD is operated near breakdown voltage. Has much higher internal gain and is much more sensitive than other photodiodes.
D1. Phototransistor		Current	Slow	Produces much higher current for given input than photodiode does because of its amplifying ability. Base lead, if accessible, is seldom used.
D2. Photodarlington		Current	Very slow	Much more sensitive than phototransistor.

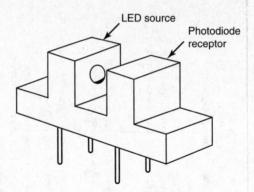

FIGURE 6.20: A photointerrupter consisting of an LED light source (often infrared) and a photodetector. Mechanical interruption of the light path can be used for various purposes, such as counting, triggering, and synchronization.

measurements, temperature measurements (Section 16.8), dewpoint controls, and a wide range of process monitors. In conjunction with lasers, they form the basis of fluid velocity measurement systems (Section 15.10) and vibration measurement systems (Section 17.12).

Photodetectors are available in one- and two-dimensional arrays, as well. These configurations allow the contruction of digital cameras, which reduce a visual image to a discrete set of photodetector voltages (Section 8.12). Of particular importance are *charge-coupled diode* arrays (CCD arrays), which use silicon photodiodes to capture digital images. Arrays of photodetectors enable whole-field temperature detection, machine vision systems, fluid flow visualization, and many other measurements.

Other special packages include optointerrupters and optoisolators, in which a photodiode and a light-emitting diode (LED) are packaged so that the light from the LED impinges on the photodiode (see Figs. 6.20 and 6.21). The interrupter is configured so that some form of mechanical mask may be used to break the light beam between the LED and the detector, thereby providing on off switching for counting or a variety of other purposes.

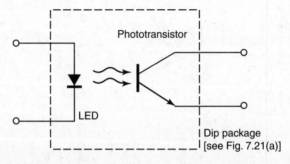

FIGURE 6.21: The essentials of a photoisolator, used for connecting low-impedance current circuits to high-impedance voltage circuits. The isolator is also useful for providing complete electrical isolation between cirucits, sometimes imperative in health-related electronics.

The optoisolator is used to match low-impedance current circuits to high-impedance voltage circuits, or vice versa. It also provides a high-impedance isolation between circuits, which is an important feature in some forms of health-related electronics [29].

6.17   HALL-EFFECT SENSORS

The *Hall effect* is the appearance of a transverse voltage difference on a conductor carrying a current perpendicular to a magnetic field [30]. This voltage is directly proportional to the magnetic field strength. If the magnetic field is made to vary with the position of a nearby object, the Hall effect can be the basis of a proximity sensor.

In Fig. 6.22(a), a conductor carries current in the x-direction, so that electrons flow in the x-direction with a velocity $\mathbf{v}_d$. The magnetic field runs in the y-direction. Because the electrons carry a charge $-q$, they experience a magnetic force $\mathbf{F}_B$ in the z-direction:

$$\mathbf{F}_B = -q\mathbf{v}_d \times \mathbf{B}$$

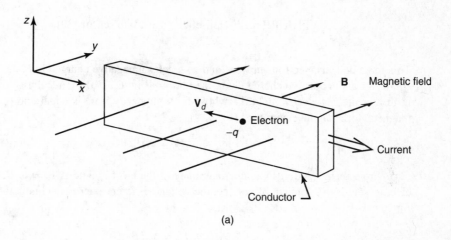

(a)

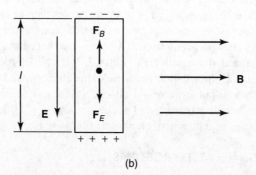

(b)

FIGURE 6.22: The Hall effect: (a) A conductor carries current in a perpendicular magnetic field; (b) electrons are driven upward by magnetic force, creating an opposing electric field.

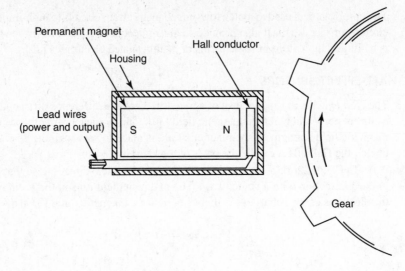

FIGURE 6.23: Hall-effect gear-tooth sensor [30].

This force deflects electrons upward and so creates a negative charge along the top of the conductor and a positive charge along the bottom [Fig. 6.22(b)]. This charge distribution in turn creates an electric field, **E**, whose force in steady state is equal and opposite the magnetic force on the electrons:

$$\mathbf{F}_E = -q\mathbf{E} = -\mathbf{F}_B$$

From these relationships, the magnitude of the electric field is $E = v_d B$. Since the electric field is the gradient of voltage, the voltage difference across a conductor of height l is

$$V_{\text{Hall}} = lE = v_d l B \tag{6.11}$$

This is the Hall-effect voltage.

The Hall beffect is present in any conductor carrying current in a magnetic field, but it is much more pronounced in semiconductors than in metals. Thus most Hall-effect transducers use a semiconducting material as the conductor, often in conjunction with an integrated-circuit signal conditioner. A permanent magnet may be built into the transducer to provide the needed magnetic field (Fig. 6.23). If a passing object, such as another magnet or a ferrous metal, alters the magnetic field, the change in the Hall-effect voltage is seen at the transducer's output terminals. Hall-effect transducers are used as position sensors, as solid-state keyboards actuators, and as current sensors [31]. Low-cost, ruggedly packaged versions are used as automotive crankshaft-timing sensors [32].

6.18 SOME DESIGN-RELATED PROBLEMS

Accuracy, sensitivity, dynamic response, repeatability, and the ability to reject unwanted inputs are all qualities highly desired in each component of a measuring system. Many of the parameters that combine to provide these qualities present conflicting problems and must be compromised in the final design. Uncertainty considerations in experimental design were discussed in Sections 3.11.2 and 3.12. Response requirements were presented in Chapter 5. At this point we will discuss some additional design problems.

6.18.1 Manufacturing Tolerances

The conception of a component or system on paper is a necessary and important beginning; to be useful, however, the apparatus must be produced, and no manufacturing process can reproduce *exact* length or angular dimensions. Dimensions must always be assigned with some specified or implied tolerances. How can one predict the effect of such variations on performance? The following example describes one approach to the problem.

EXAMPLE 6.1

Suppose a spring scale such as that shown in Fig. 6.24 is to be designed. We will assume a force capacity of 50 N and a maximum deflection of 10 cm. This gives us a deflection constant of 5 N/cm.

Solution   Using conventional coil-spring design relations [33], we find that a spring made with a mean coil diameter D_m of 2 cm and a steel wire diameter of 2 mm will meet stress requirements provided we ensure against overload by including appropriate deflection limits or stops. The deflection equation commonly used for coil springs is

$$K = \frac{F}{y} = \frac{E_s D_w^4}{8 D_m^3 n}$$

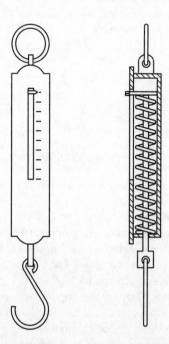

FIGURE 6.24: Common spring-type "fish" scale.

where

K = the deflection constant (N/m),

F = the design load (N),

y = the corresponding deflection (m),

E_s = the torsional elastic modulus (about 80×10^9 Pa for steel),

n = the number of coils

Using this relation and the preceeding design values, we find that 40 coils are needed to provide the required deflection constant. If we apply reasonable tolerances, our specifications become

$$D_w = 2 \pm 0.01 \text{ mm},$$

$$D_m = 2 \pm 0.05 \text{ cm},$$

$$n = 40 \pm \frac{1}{3} \text{ coils, and}$$

$$E_s = 80 \times 10^6 \pm 3.5 \times 10^6 \text{ kPa}$$

Let us now consider how various uncertainties (manufacturing tolerances) may affect the deflection constant, K. (We direct attention to reference [34] at this point.) We assume that 99% of the coils will not exceed *any* of these tolerances; in other words, we take the tolerances to represent 99% uncertainty limits for each variable. Then, using the approach described in Section 3.10, we have

$$\frac{U_K}{K} = \sqrt{\left(\frac{4 \times 0.01}{2}\right)^2 + \left(\frac{3 \times 0.05}{2}\right)^2 + \left(\frac{0.33}{40}\right)^2 + \left(\frac{3.5}{80}\right)^2}$$

$$= 0.0895 \approx 9\% \qquad (99\% \text{ confidence})$$

or

$$K = 500 \pm 45 \text{ N/m} \quad (99\%)$$

Note that mm, cm, and m have been used in the preceding example; hence care must be used in placing decimal points.

We see then that if we lay out the graduations corresponding to nominal values, a force of 50 N may actually be indicated as anything in the range from 45.5 to 54.5 N, depending on how the manufacturing tolerances may fall. Should this result not be satisfactory, our only recourse is (1) to provide better control of the manufacturing tolerances or (2) to provide some means for adjusting calibration.

Various methods of calibration can be used, depending on the intended "quality" of the device. In this instance, two or three faceplates can be provided, each with graduations scaled to cover a portion of the calibration range. At the time of assembly, a simple calibration would determine the most appropriate plate to use. Section 13.4.1 presents another approach that could be applied to our particular example.

At this point, it is appropriate to make an additional observation. *Weight* is basically a *force*; hence we should express the calibration in newtons rather than in kilograms. Should we wish a scale calibrated in kilograms, then, to be completely correct we should include the assumed value of gravitational acceleration on the faceplate. The standard acceleration due to gravity is 9.80665 m/s^2, and the kilogram range corresponding to our 50 N range becomes 0 to 50/9.80665 $\approx$ 5.1 kg. (Use of the non-SI symbol kgf is discouraged.)

It should be clear that the procedures used in the preceding example are applicable to most elastic transducer configurations, as well as to many other tolerance problems.

6.18.2   Some Temperature-Related Problems

An ideal measuring system will react to the design signal only and ignore all else. Of course, this ideal is never completely fulfilled. One of the more insidious adverse stimuli affecting instrument operation is temperature. It is insidious in that it is almost impossible to maintain a constant-temperature environment for a general-purpose measuring system. The usual solution is to accept the temperature variation and to devise methods to compensate for it.

Temperature variations cause dimensional changes and changes in physical properties, both elastic and electrical, resulting in deviations (bias error) referred to as *zero shift* and *scale error* (see Fig. 3.2) [35]. Zero shift, as the name implies, results in a change in the no-input reading. Factors other than temperature may cause zero shift; however, temperature is probably the most common cause. In most applications the zero indication on the output scale would be made to correspond to the no-input condition. For example, the indicator or the spring scales referred to earlier should be set at zero when there is no weight in the pan. If the temperature changes after the scale has been set to zero, there may be a differential dimensional change between spring and scale, altering the no-load reading. This change would be referred to as *zero shift*. Zero shift is primarily a function of linear dimensional change caused by expansion or contraction with changing temperature.

Dimensional changes are expressed in terms of the coefficient of expansion by the following familiar relations:

$$\alpha = \frac{1}{\Delta T} \frac{\Delta L}{L_0} \tag{6.12}$$

and

$$L_1 = L_0(1 + \alpha \, \Delta T) \tag{6.13}$$

where

$$\alpha = \text{the coefficient of linear expansion (ppm/deg temp.} \times 10^{-6}),$$
$$L/L_0 = \text{the unit change in length,}$$
$$\Delta T = \text{the change in temperature, } T_1 - T_0,$$
$$L_0 = \text{the length dimension at the reference temperature } T_0,$$
$$L_1 = \text{the length dimension at any other temperature } T_1$$

In addition to causing zero shift, temperature changes usually affect scale calibration when resilient load-carrying members are involved. The coil and wire diameters of our spring would be altered with temperature change, and so too would the modulus of elasticity of the spring material. These variations would cause a changed spring constant, and hence changed load-deflection calibration, resulting in what is referred to as *scale error*.

The thermoelastic coefficient is defined by the relations

$$c = \frac{1}{\Delta T} \frac{\Delta E}{E_0} \tag{6.14}$$

and

$$E_1 = E_0(1 + c\,\Delta T) \tag{6.15}$$

where

c = the temperature coefficient for the tensile modulus of elasticity (ppm/deg temp. $\times 10^{-6}$),

$\Delta E / E_0$ = the unit change in the tensile modulus of elasticity,

E_0 = the tensile modulus of elasticity at temperature T_0,

E_1 = the tensile modulus of elasticity at temperature T_1

Similarly, the coefficient for torsional modulus may be written

$$m = \frac{1}{\Delta T} \frac{\Delta E_s}{E_{s0}} \tag{6.16}$$

and

$$E_{s_1} = E_{s0}(1 + m\,\Delta T) \tag{6.17}$$

where

m = the temperature coefficient for the torsional modulus of elasticity (ppm/deg temp. $\times 10^{-6}$),

$\Delta E_s / E_{s0}$ = the unit change in the torsional modulus of elasticity,

E_{s0} = the torsional modulus of elasticity at temperature T_0,

E_{s_1} = the torsional modulus of elasticity at temperature T_1

Representative values of these quantities are given in Table 6.3.

The manner in which temperature changes in elastic properties affect instrument performance can be demonstrated by the following example. Assume that a restoring element in an instrument is essentially a single-leaf cantilever spring of rectangular section, for which the deflection equation at reference temperature T_0 is

$$K_0 = \frac{F}{y} = \frac{3E_0 I_0}{L_0^3} = \frac{E_0 w_0 t_0^3}{4L_0^3} \tag{6.18}$$

TABLE 6.3: Temperature Characteristics for Some Materials

Material	Tensile Modulus of Elasticity, E Pa × 10⁻¹⁰ (psi × 10⁻⁶)	Torsional Modulus of Elasticity, E Pa × 10⁻¹⁰ (psi × 10⁻⁶)	Coefficient of Linear Expansion, α ppm/°C ppm/°F	Coefficient of Tensile Modulus of Elasticity, c* ppm/°C ppm/°F
High-carbon spring steel	20.7 (30)	7.93 (11.5)	11.6 (6.5)	−220 (−122)
Chrome-vanadium steel	20.7 (30)	7.93 (11.5)	12.2 (6.8)	−260 (−145)
Stainless steel, type 302	19.3 (28)	6.9 (10)	16.7 (9.3)	−439 (−244)
Spring brass	10.3 (15)	3.8 (5.5)	20.2 (11.2)	−391 (−217)
Phosphor bronze	10.3 (15)	4.3 (6.3)	17.8 (9.9)	−380 (−211)
Invar†	14.8 (21.4)	5.6 (8.1)	1.1 (0.6)	+48.1 (+27)
Isoelastic†	18.0 (26)	6.3 (9.2)	7.2 (4)	−36 to +13 (−20 to +7.3)
Aluminum	6.9 (10)	2.6 (3.8)	23 (13)	−270 to −400 (−150 to −220)

*c may be used for torsional modulus also.
†Trade names.

where

$$K_0 = \text{the deflection constant,}$$
$$I_0 = \text{the moment of inertia,}$$
$$w_0 = \text{the width of the section at reference temperature,}$$
$$t_0 = \text{the thickness of the section at reference temperature,}$$
$$L_0 = \text{the length of the beam at the reference temperature}$$

A second equation may be written for any other temperature, T_1, as follows:

$$K_1 = \frac{[E_0(1 + c\ \Delta T)][w_0(1 + \alpha\ \Delta T)][t_0(1 + \alpha\ \Delta T)]^3}{4[L_0(1 + \alpha\ \Delta T)]^3} \tag{6.19}$$

Thus we have

$$\text{Percent error in deflection scale} = \left(\frac{K_0 - K_1}{K_0}\right) \times 100$$
$$= [1 - (1 + c\ \Delta T)(1 + \alpha\ \Delta T)] \times 100$$

which we may simplify, by expanding and neglecting the second-order term, to read

$$\text{Percent scale error} = -(c + \alpha)\ \Delta T \times 100 \tag{6.20}$$

If our spring is made of spring brass,

$$\text{Percent scale error/}^\circ\text{F} = -(-217 + 11.2) \times 10^{-6} \times 100 = 0.021\%$$

Hence a temperature change of $+50^\circ$F would result in a scale error of about $+1\%$. (This means that the reading is too high; our spring is too flexible, and a given load deflects the spring more than it should.)

It is interesting to note that for our example the scale error is a function of material or materials. It should be clear that we are speaking of the load-deflection relation for resilient members in this connection and that this would not include members whose duty it is simply to transmit motion, such as the linkage in a Bourdon-tube pressure gage.

Although not a mechanical quantity, another item affected by temperature change is electrical resistance. The basic resistance equation may be written in the form

$$R = \rho \frac{L}{A} \tag{6.21}$$

where

$$R = \text{the electrical resistance } (\Omega),$$
$$\rho = \text{the resistivity } (\Omega \cdot \text{cm}),$$
$$L = \text{the length of the conductor (cm)},$$
$$A = \text{the cross-sectional area of the conductor (cm}^2)$$

As temperature changes, a change in the resistance of an electrical conductor will be noted. This will be caused by two different factors: dimensional changes due to expansion or contraction and changes in the current-opposing properties of the material itself. For an unconstrained conductor, the latter is much more significant than the former, causing more than 99% of the total change for copper [36]. Therefore, in most cases it is not very important whether the dimensional effect is accounted for or not. If dimensional changes caused by temperature are ignored, the change in resistivity with temperature may be expressed as

$$b = \frac{1}{\Delta T} \frac{\Delta \rho}{\rho_0} \tag{6.22}$$

or

$$\rho_1 = \rho_0 (1 + b \, \Delta T) \tag{6.23}$$

where

b = the temperature coefficient of resistivity $[(\Omega \cdot cm)/(\Omega \cdot cm \cdot deg)]$,

ΔT = the temperature change(deg),

$\Delta \rho / \rho_0$ = the unit change in resistivity,

ρ_0 = the resistivity at the reference temperature $T_0 (\Omega \cdot cm)$,

ρ_1 = the resistivity at any temperature $T_1 (\Omega \cdot cm)$

If we account for temperature-dimensional changes, the equation reads

$$\rho_1 = \frac{R_0 A_0}{L_0} (1 + b \, \Delta T)(1 + \alpha \, \Delta T)$$
$$= \rho_0 (1 + b \, \Delta T)(1 + \alpha \, \Delta T) \tag{6.24}$$

Table 6.4 lists values of the coefficients of resistivity for selected materials.

6.18.3 Methods for Limiting Temperature Errors

Three approaches to a solution of the temperature problem in instrumentation are as follows: (1) *minimization* through careful selection of materials and operating temperature ranges, (2) *compensation* through balancing of inversely reacting elements or effects, and (3) *elimination* through temperature control. Although each situation is a problem unto itself, thereby making specific recommendations difficult, a few general remarks with regard to these possibilities may be made.

Minimization   As we pointed out earlier, temperature errors may be caused by thermal expansion in the case of simple motion-transmitting elements, by thermal expansion and modulus change in the case of calibrated resilient transducer elements, and by thermal expansion and resistivity change in the case of electrical resistance transducers. All these effects may be minimized by selecting materials with low-temperature coefficients in each of the respective categories. Of course, minimum temperature coefficients are not always combined with other desirable features such as high strength, low cost, corrosion resistance, and so on.

TABLE 6.4: Resistivity and Temperature Coefficients of Resistivity for Selected Materials

Material	Composition (for alloys)	Resistivity at 20°C (68°F) $\Omega \cdot cm \times 10^6$	Coefficient of Resistivity, b $\Omega/\Omega \cdot deg \times 10^6$ Per°C	Per°F
Aluminum	—	2.8	3900	2170
Constantan*	60% Cu, 40% Ni	44	11	6
Copper (annealed)	—	1.72	3900	2180
Iron	99.9% pure	10	5000	2800
Isoelastic*	36% Ni, 8% Cr, 4% Mn, Si, and Mo, remainder Fe	48	470	260
Manganin*	9–18% Mn, $1\frac{1}{2}$–4% Ni; remainder Cu	44	11	6
Monel*	33% Cu, 67% Ni	42	2000	1100
Nichrome*	75% Ni, 12% Fe, 11% Cr, 2% Mn	100	400	220
Nickel	—	7	6400	3550
Silver	—	1.6	4000	2250

Note: Values should be considered as quite approximate. Actual values depend on exact composition and, in certain cases, degree of cold work.
*Trade names.

Compensation   Compensation may take a number of different forms, depending on the basic characteristics of the system. If a mechanical system is being used, a form of compensation making use of a composite construction may be employed. If the system is electrical, compensation is generally possible in the electrical circuitry.

An example of composite construction is the balance wheel in a watch or clock. As the temperature rises, the modulus of the spring material reduces and, in addition, the moment of inertia of the wheel (if of simple form) increases because of thermal expansion, both of which cause the watch to *slow down*. If we incorporate a bimetal element of appropriate characteristics in the rim of the wheel, the moment of inertia decreases with temperature enough to compensate for both expansion of the wheel spokes and change in spring modulus. (See also Section 11.6 for a discussion of temperature effects on linear measuring devices.)

Electrical circuitry may use various means of compensating for temperature effects. The thermistor, discussed in some detail in Sections 6.8 and 16.4.3, is quite useful for this purpose. Most circuit elements possess the characteristic of increasing dc resistance with rising temperature. The thermistor has an opposite temperature-resistance property, along with reasonably good stability, both of which make it ideal for simple temperature-resistance compensation.

Resistance strain gages are particularly susceptible to temperature variations. The actual situation is quite complex, involving thermal-expansion characteristics of both the base material and all the gage materials (support, cement, and grid) and temperature-resistivity properties of the grid material, combined with the fact that heat is dissipated by the grid since it is a resistance device. Temperature compensation is very nicely handled, however, by pitting the temperature effect output from like gages against one another while subjecting them differentially to strain. This outcome is accomplished by use of a resistance bridge circuit arrangement, which is used extensively in strain-gage work (see Section 12.10). In addition through careful selection of grid materials, so-called self-compensating gages have been developed. (See also Section 13.5.)

Elimination   The third method—eliminating the temperature problem by temperature control—really requires no discussion. Many methods are possible, extending from the careful control of large environments to the maintenance of constant temperature in small instrument enclosures. An example of the latter is the "crystal oven," often used to stabilize a frequency-determining quartz crystal.

6.19   SUMMARY

We have in no sense exhausted the list of possible devices or principles suitable for sensing mechanical inputs. In certain instances, we discuss others elsewhere in this book, and in Table 6.1 have attempted to reference some of these. For further information on basic sensing devices, we refer you to the Suggested Readings.

1. Sensors often include both a first, detecting stage and a second, transducing stage. Each stage may convert the sensed information into a different form, often resulting a final electrical signal. Specific sensor design and selection is normally guided by the requirements of sufficient sensitivity and minimal source loading (Sections 6.1–6.4).

2. A wide range of transducers are based on changes in the resistance of a sensing element (Sections 6.5–6.8). Variations in inductance (Sections 6.10–6.12) and capacitance (Section 6.13) are also used frequently.

3. Some sensing elements are self-powering. These include thermocouples (Section 6.9), which generate an electromotive force dependent upon temperature, and piezoelectric sensors (Section 6.14), which generate a charge when loaded.

4. Semiconductors devices are increasingly common among sensing elements (Section 6.15). These sensors can take advantage of microelectronic fabrication technology.

5. Light-detecting transducers may be divided into thermal and photon devices. A variety of photon devices are described in Section 6.16.

6. Hall-effect sensors are often used in position sensing and related applications (Section 6.17).

7. Manufacturing tolerance and temperature errors must be considered when designing a sensing system. Often, the magnitude of these problems can be estimated in advance, using the methods of uncertainty analysis (Section 6.18; also see Sections 3.10, 3.11.2, and 3.12). Such estimates can provide guidelines for improving transducer performance.

SUGGESTED READINGS

ISA Directory of Instrumentation. Research Triangle Park, N.C.: Instrument Society of America, yearly editions.

Gautschi, G. H. *Piezoelectric Sensorics.* Berlin: Springer-Verlag, 2002.

Juds, S. M. *Photoelectric Sensors and Controls.* New York: Marcel Dekker, 1988.

Khazan, A. D. *Transducers and Their Elements.* Englewood Cliffs, N.J.: Prentice Hall, 1994.

Kovacs, G. T. A. *Micromachined Transducers Sourcebook.* New York: McGraw-Hill, 1998.

Norton, H. N. *The Handbook of Transducers.* Englewood Cliffs, N.J.: Prentice Hall, 1989.

Nunley, W., and J. S. Bechtel. *Infrared Optoelectronics: Devices and Applications.* New York: Marcel Dekker, 1987.

Pallàs-Areny, R., and J. G. Webster. *Sensors and Signal Conditioning.* 2nd ed. New York: John Wiley, 2001.

Sydenham, P. H. (ed.). *Handbook of Measurement Science.* New York: John Wiley, 1992.

Todd, C. D. *The Potentiometer Handbook.* New York: McGraw-Hill, 1975.

Trietley, H. L. *Transducers in Mechanical and Electronic Design.* New York: Marcel Dekker, 1986.

Wilson, J., and J. F. B. Hawkes. *Optoelectronics: An Introduction.* 3rd ed. Harlow, UK: Prentice Hall Europe, 1998.

PROBLEMS

6.1. Consider an inductive displacement probe having a diameter of 0.25 in. If the probe is set at a "stand-off" distance of 0.050 in. relative to a shaft, determine the probe sensitivity (mV/0.001 in. displacement) when the probe is used as shown in the circuit shown in Figure 6.25. Assume Eq. (6.3c) is valid here with $n = 100$ and that the excitation frequency is 1000 Hz.

6.2. It is desired to construct a dynamic compression force cell capable of measuring forces in the range of ± 1000 N. If a quartz disk 1.0 mm thick and 10 mm in diameter is used as the sensing element, determine the force cell sensitivity (mV/N).

6.3. For a capacitive displacement transducer whose behavior can be represented by Eq. (6.6a), determine an expression for the sensitivity $de_o/d(d)$ for an excitation frequency f if the transducer is used as shown in Fig. 6.26.

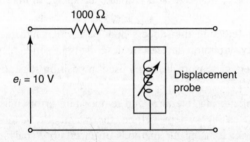

FIGURE 6.25: Circuit for Problem 6.1.

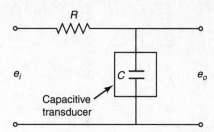

FIGURE 6.26: Circuit for Problems 6.3, 6.6, and 6.8.

6.4. A proving-ring-type force transducer is a very reliable device for checking the calibration of material-testing machines. An equation for estimating the deflection constant of the elemental ring, loaded in compression, is given Table 13.1. If $D = 10$ in. (25.4 cm) $\pm$ 0.010 in. (0.25 mm), $t = $ the radial thickness of the section $= 0.6$ in. (15.24 mm) $\pm$ 0.005 in. (0.127 mm), $w = $ the axial width of the section $= 2$ in. (5.08 cm) ± 0.015 in. (0.381 mm), and $E = 30 \times 10^6$ lbf/in.^{2} (20.68×10^{10} N/m^2) $\pm 0.5 \times 10^6$ lbf/in.^{2} (0.34×10^{10} N/m^2), calculate the value of K and its uncertainty, using English units.

6.5. Solve Problem 6.4 using SI units.

6.6. Consider the capacitive displacement transducer in Fig. 6.26 to be governed by the following relationship:

$$C = \frac{0.225A}{d}$$

where

$C = $ capacitance (pF),

$A = $ cross-sectional area of transducer tip (in.^{2}),

$d = $ air-gap distance (in.)

Determine the change in e_o when the air gap changes from 0.010 in. to 0.015 in.

6.7. A capacitive displacement transducer as shown in Fig. 6.27 is constructed of two plates with area 2.0 in.^{2} separated by a distance of 0.006 in. If air is the separating medium, determine the sensitivity of the transducer in picofarads per 0.001 in. change in x.

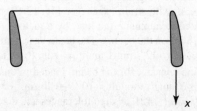

FIGURE 6.27: Transducer arrangement for Problem 6.7.

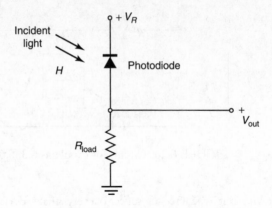

FIGURE 6.28: Circuit for Problem 6.10.

6.8. If the transducer of Problem 6.7 is inserted into the circuit of Fig. 6.26, determine the change in output voltage when x changes from 0.010 in. to 0.015 in. Is the sensitivity constant in this range?

6.9. (a) A commercial force sensor uses a piezoelectric quartz crystal as the sensing element. The quartz element is about 0.2 in. thick and has a cross section of about 0.3 in. by 0.3 in. The sensing element is compressed in the thickness direction when a load is applied over its cross section. The output voltage is measured across the thickness. What is the output of the sensor in volts per newton?

   (b) A polyvinylidene fluoride film is used as a piezoelectric load sensor. The film is 25 μm thick, 1 cm wide, and 2 cm in the axial direction. It is stretched in the axial direction by the load. The output voltage is measured across the thickness. What is the output in volts per newton?

6.10. The circuit of Figure 6.28 may be used to operate a photodiode. The voltage V_r is a reverse-bias voltage large enough to make diode current, i, proportional to the incident light intensity, H. Under this condition, $i/H = 1\mu\text{A}/(\text{W/m}^2)$.

   (a) Show that the output voltage, V_{out}, varies linearly with H.

   (b) If $H = 1000$ W/m^2, $V_r = 5$ V, and an output voltage of 1 V is desired, determine an appropriate value of R_{load}.

6.11. A digital readout weighing scale with a 300 lbf (1334 N) capacity uses an aluminum cantilever beam with strain gages (see Fig. 12.19) as the force-sensing element. A lever system is used to attenuate the load by a factor of 18 (i.e., a 300 lbf load on the scale exerts 300/18 lbf on the beam). The beam has an effective length $L = 7$ in., a rectangular section $w = \frac{1}{2}$ in. (12.7 mm), and $t = \frac{1}{4}$ in. (6.35 mm), oriented as shown in Fig. 12.19. The deflection constant for the beam, loaded in this manner, is $3EI/L^3$ (see Table 13.1, Item C). E is Young's modulus (10×10^6lbf/in.^{2} or 6.89×10^{10} Pa), and I is the moment of inertia ($I = wt^3/12$). Assign tolerances and determine the uncertainty in K.

6.12. Solve Problem 6.11 using SI units.

6.13. Assuming that the specifications for Example 6.1 in Section 6.18.1 are for a nominal temperature of 20°C, calculate the nominal value for the deflection constant K for temperatures of (a) 40°C and (b) −20°C. (Use values for high-carbon spring steel.)

6.14. Calculate the deflection constant for the force transducer specified in Problem 6.4 for temperatures of (a) 40°C and (b) −20°C.

6.15. Calculate the deflection constant for the force transducer specified in Problem 6.11 for temperatures of (a) 40°C and (b) −20°C.

REFERENCES

[1] Kneen, W. A. A review of the electric displacement gages used in railroad car testing. *ISA Proc.*, 6:74, 1951.

[2] Michael, P. C., N. Saka, and E. Rabinowicz. Burnishing and adhesive wear of an electrically conductive polyester-carbon film. *Wear*, 132:265–285, 1989.

[3] Beckwith, T. G., and R. D. Marangoni. *Mechanical Measurements*. 4th ed. Reading, Mass: Addison-Wesley, 1990, Section 7.8.

[4] Todd, C. D. *The Potentiometer Handbook*. New York: McGraw-Hill, 1975.

[5] Hutchinson, C. (ed). *The ARRL Handbook for Radio Amateurs*. 78th ed. Newington, Conn.: American Radio Relay League, 2000, p. 6.22. Revised annually.

[6] Carlson, A. B., and D. G. Gisser. *Electrical Engineering*. Reading, Mass.: Addison-Wesley, 1981, Chapter 17.

[7] Khazan, A. D. *Transducers and Their Elements*. Englewood Cliffs, N.J.: Prentice Hall, 1994, Chapter 3.

[8] Pallàs-Areny, R., and J. G. Webster. *Sensor and Signal Conditioning*. 2nd ed. New York: John Wiley, 2001, Chapter 4.

[9] Electronic micrometer uses dual coils. *Prod. Engr.*, 19:134, January 1948.

[10] Brenner, A., and E. Kellogg. An electric gage for measuring the inside diameter of tubes. *NBS J. Res.*, 42:461, May 1949.

[11] Low-temperature liquid-level indicator for condensed gases. *NBS Tech. News Bull.*, 38:1, January 1954.

[12] Heteny, M. *Handbook of Experimental Stress Analysis*. New York: John Wiley, 1950, p. 287.

[13] Sihvonen, Y. T., G. M. Rassweiler, A. F. Welch, and J. W. Bergstrom. Recent improvements in a capacitor-type pressure transducer. *ISA J.*, 2, November 1955.

[14] Leggat, J. W., G. M. Rassweiler, A. F. Welch, and Y. T. Sihvonen. Engine pressure indicators, application of a capacitor type. *ISA J.*, 2, August 1955.

[15] Welch, Weller, Hanysz, and Bergstrom. Auxiliary equipment for the capacitor-type transducer. *ISA J.*, 2, December 1955.

[16] Fleming, L. T. A ceramic accelerometer of wide frequency range. *ISA Proc.*, 5:62, 1950.

[17] *Kynar Piezo Film Technical Manual*. Valley Forge, Pa.: Penwalt Corporation, 1987.

[18] Carlson, A. B., and D. G. Gisser. *Electrical Engineering*. Reading, Mass.: Addison-Wesley, 1981, pp. 287–294.

[19] *Pressure Sensors*, Catalog BR121/D. Phoenix, Ariz.: Motorola, Inc., 1991.

[20] *Solid-State Sensor Handbook*. Sunnyvale, Calif.: SenSym, Inc., 1989.

[21] Alley, C. L., and K. W. Atwood. *Microelectronics*. Englewood Cliffs, N.J.: Prentice Hall, 1986, pp. 56–57.

[22] Wilson, J., and J. F. B. Hawkes. *Optoelectronics, an Introduction*. 2nd ed. Hemel Hempstead, UK: Prentice Hall International, 1989, pp. 280–286.

[23] *Handbook of Infrared Radiation Measurements*. Stamford, Conn.: Barnes Engineering Company, 1983, pp. 51–56.

[24] Dennis, P. N. J. *Photodetectors: An Introduction to Current Technology*. New York: Plenum Press, 1986.

[25] Pallàs-Areny, R., and J. G. Webster. *Sensors and Signal Conditioning*, 2nd ed. New York: John Wiley, 2001, Chapter 6.

[26] *Photomultiplier Handbook*. Lancaster, Pa.: Burle Technologies, Inc., 1980.

[27] Wilson, J., and J. F. B. Hawkes. *Optoelectronics: An Introduction*. 3rd ed. Harlow, UK: Prentice Hall Europe, 1998, Section 7.3.5.

[28] Kasap, S. O. *Optoelectronics and Photonics: Principles and Practices*. Upper Saddle River, N.J.: Prentice Hall, 2001.

[29] Horowitz, P., and W. Hill. *The Art of Electronics*. 2nd ed. New York: Cambridge University Press, 1989.

[30] Tipler, P. A. *Physics*. New York: Worth Publishers, Inc., 1976, pp. 848–850.

[31] *Hall Effect Transducers: How to Apply Them as Sensors*. Freeport, Ill.: Micro Switch, A Honeywell Division, 1982.

[32] Shuller, J., and A. Lee. Personal Communication to J. H. Lienhard, Chrysler Motors Corporation, February 1992.

[33] Shigley, J. E., and C. R. Mischke. *Mechanical Engineering Design*. 5th ed. New York: McGraw-Hill, 1989, Ch. 10.

[34] Haugen, E. B. *Probabilistic Approaches to Design*. New York: John Wiley, 1968.

[35] Gitlin, R. How temperature affects instrument accuracy. *Control Eng.*, 2, May 1955.

[36] Laws, F. A. *Electrical Measurements*. 2nd ed. New York: McGraw-Hill, 1938, p. 217.

CHAPTER 7

Signal Conditioning

7.1 INTRODUCTION

Once a mechanical quantity has been detected and possibly transduced, it is usually necessary to modify the stage-one output further before it is in satisfactory form for driving an indicator or becoming the input to an electronic control or display. We will now consider some of the methods used in this intermediate, signal-conditioning step.

Measurement of dynamic mechanical quantities places special requirements on the elements in the signal-conditioning stage. Large amplifications, as well as good transient response, are often desired, both of which are difficult to obtain by mechanical, hydraulic, or pneumatic methods. As a result, electrical or electronic elements are usually required.

An input signal is often converted by the detector-transducer to a mechanical displacement (see Table 6.1). It is then commonly fed to a secondary transducer, which converts it into a form, usually electrical, that is more easily processed by the intermediate stage. In some cases, however, such a displacement is fed to mechanical intermediate elements, such as linkages, gearing, or cams (see Fig. 6.1, for example); these mechanical elements present design problems of considerable magnitude, particularly if dynamic inputs are to be handled.

In the field of dynamic measurements, strictly mechanical systems are much more uncommon than they were in years past, largely because of several inherent disadvantages, which we will discuss only briefly.[1]

Mechanical amplification by these elements is quite limited. When amplification is required frictional forces are also amplified, resulting in considerable undesirable signal loading. These effects, coupled with backlash and elastic deformations, result in poor response. Inertial loading results in reduced frequency response and in certain cases, depending on the particular configuration of the system, phase response is also a problem.

7.2 ADVANTAGES OF ELECTRICAL SIGNAL CONDITIONING

As we have already seen, many detector–transducer combinations provide an output in electrical form. In these cases, of course, it is convenient to perform further signal conditioning electrically. Such conditioning may typically include converting resistance changes to voltage changes, subtracting offset voltages, increasing signal voltages, or removing unwanted frequency components. In addition, in order to minimize friction, inertia, and structural flexibility requirements, we also prefer electrical methods for their ease of *power amplification*. Additional power may be fed into the system to provide a greater output power than input by the use of power amplifiers, which have no important mechanical counterpart in most instrumentation.[2] Electronic signal conditioning is, obviously, always needed when the output is to be recorded or processed by a computer, electronic control, or digital display.

7.3 MODULATED AND UNMODULATED SIGNALS

Measurands may be "pure" in the sense that the analog electrical signal contains nothing other than the real-time variation of the measurand information itself. On the other hand, the signal may be "mixed" with a *carrier*, which consists of a voltage oscillation at some frequency higher than that of the signal. A common rule of thumb is that the frequency ratio should be at least 10 to 1. The signal is said to *modulate* the carrier. The measurand affects the carrier by varying either its amplitude or its frequency. In the former case the carrier frequency is held constant and its amplitude is varied by the measurand. This process is known as amplitude modulation, or AM [Fig. 7.1(a)]. In the latter case the carrier amplitude is held constant and its frequency is varied by the measurand. This is known as frequency modulation, or FM [Fig. 7.1(b)]. The most familiar use of AM and FM transfer of signals is in AM and FM radio broadcasting.

When modulation is used in instrumentation, amplitude modulation is the more common form. Nearly any mechanical signal from a passive pickup can be transduced into an analogous AM form. Sensors based on either inductance or capacitance *require* an ac excitation. The differential transformer (Section 6.11) is an example of the former, whereas

---

[1]The first and second editions of this book contain a more thorough discussion of strictly mechanical signal-conditioning methods and problems.

[2]It is true that hydraulic and pneumatic systems may be set up to increase signal power; however, their use is limited to relatively slow-acting control applications, primarily in the fields of chemical processing and electric power generation. As in the case of mechanical systems, friction and inertia severely limit transient response of the type required for measurement of dynamic inputs.

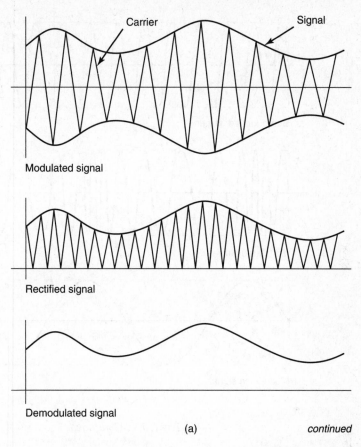

Modulated signal

Rectified signal

Demodulated signal

(a)   *continued*

FIGURE 7.1: (a) Amplitude modulation, whereby the envelope of the carrier contains the signal information; (b) frequency modulation, whereby the signal information is contained in the frequency variation of the carrier.

the capacitive pickup for liquid level (Fig. 6.12) is an example of the latter. In addition, some resistive-type sensors use ac excitation.

Extracting the signal information from the modulated carrier is required. When AM is used, this operation may take several forms. The simplest is merely to display the entire signal using an oscilloscope or oscillograph, and then to "read" the result from the envelope of the carrier. More commonly, the mixed signal and carrier are *demodulated* by rectification and filtering, as shown in Fig. 7.1(a). FM demodulation is a more complex operation and may be accomplished through the use of frequency discrimination, ratio detection, or IC phase-locked loops. Further discussion is beyond the scope of this text.

7.4 INPUT CIRCUITRY

Electrical detector–transducers are of two general types: (1) *passive*, those requiring an auxiliary source of energy in order to produce a signal; and (2) *active*, those that are self-powering. The simple bonded strain gage is an example of the former, whereas the piezoelectric accelerometer is an example of the latter.

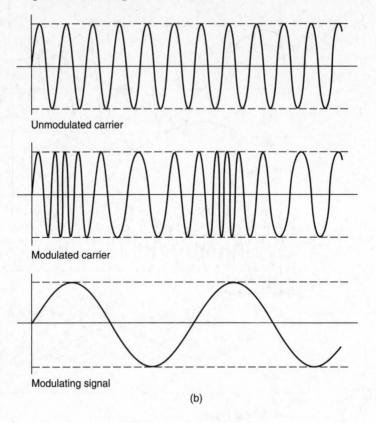

Unmodulated carrier

Modulated carrier

Modulating signal

(b)

FIGURE 7.1: *Continued*

Whereas it may be possible to use an active, or self-powering, detector–transducer directly with a minimum of circuitry, the passive type, in general, requires special arrangements to introduce the auxiliary energy. The particular arrangement required will depend on the operating principle involved. For example, resistive-type pickups may be powered by either an ac or a dc source, whereas capacitive and inductive types, with an exception or two, require an ac source.

Although not all inclusive, the following list classifies the most common forms of input circuits used in transducer work: (1) simple current-sensitive circuits, (2) ballast circuits, (3) voltage-dividing circuits, (4) bridge circuits, (5) resonant circuits, and (6) amplifier input circuits. Often, the input circuits will be followed by some type of filter circuit. These circuits are discussed in the following sections.

7.5   THE SIMPLE CURRENT-SENSITIVE CIRCUIT

Figure 7.2(a) illustrates a simple current-sensitive circuit in which the transducer may use any one of the various forms of variable-resistance elements. We will let the transducer resistance be kR_t, where R_t represents the maximum value of transducer resistance and k represents a percentage factor that may vary between 0.0 and 1.0 (0% and 100%), depending on the magnitude of the input signal. Should the transducer element be in the form of a sliding contact resistor, the value of k could vary through the complete range of 0% to

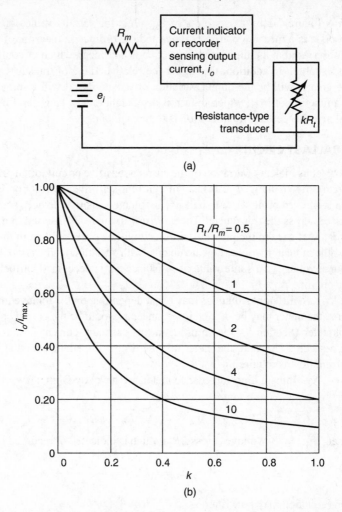

FIGURE 7.2: (a) Simple current-sensitive circuit; (b) plot of Eq. (7.2), showing variation of current in terms of input signal k for a simple current-sensitive circuit.

100%. On the other hand, if R_t represents, say, a thermistor, then k would fall within some limiting range not including 0.0%. We will let R_m represent the remaining circuit resistance, including both the meter resistance and the internal resistance of the voltage source.

If i_o is the current flowing through the circuit and hence the current indicated by the readout device, we have, using Ohm's law (Section 5.17),

$$i_o = \frac{e_i}{kR_t + R_m} \tag{7.1}$$

Note that maximum current flows when $k = 0$, at which point the current is $i_{max} = e_i / R_m$. Equation (7.1) may thus be rewritten as

$$\frac{i_o}{i_{max}} = \frac{i_o R_m}{e_i} = \frac{1}{1 + \left(\dfrac{R_t}{R_m}\right)k} \tag{7.2}$$

Figure 7.2(b) shows plots of Eq. (7.2) for various values of resistance ratio. The abscissa is a measure of *signal input* and the ordinate a measure of *output*. First of all, it is observed that the input–output relation is nonlinear, which of course would generally be undesirable. In addition, the higher the relative value of transducer resistance R_t to R_m, the greater will be the output variation or sensitivity. It will also be noted that the output is a function of i_{max}, which in turn is dependent on e_i. Thus careful control of the driving voltage is necessary if calibration is to be maintained.

7.6 THE BALLAST CIRCUIT

Now let us look at a variation of the current-sensitive circuit, often referred to as the *ballast circuit*, shown in Fig. 7.3. Instead of a current-sensitive indicator or recorder through which the total current flows, we shall use a voltage-sensitive device (some form of voltmeter) placed across the transducer. The *ballast resistor* R_b is inserted in much the same manner as R_m was used in the previous circuit. It will be observed that in this case, were it not for R_b, the indicator would show no change with variation in R_t; it would always indicate full source voltage. So some value of resistance R_b is necessary for the proper functioning of the circuit.

Two different situations may exist, depending on the relative impedance of the meter. First, the meter may be of high impedance, as would be the case if some form of electronic voltmeter (Section 9.4) were used; in this case any current flow through the meter may be neglected. Second, the meter may be of low impedance, so that consideration of such current flow is required.

Assuming a high-impedance meter, we have, by Ohm's law,

$$i = \frac{e_i}{R_b + kR_t} \tag{7.3}$$

Then, if e_o = the voltage across kR_t (which is indicated or recorded by the readout device),

$$e_o = i(kR_t) = \frac{e_i kR_t}{R_b + kR_t} \tag{7.4}$$

This equation may be written as

$$\frac{e_o}{e_i} = \frac{kR_t/R_b}{1 + (kR_t/R_b)} \tag{7.5}$$

For a given circuit, e_o/e_i is a measure of the output, and kR_t/R_b is a measure of the input.

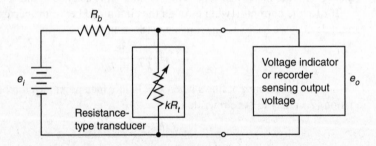

FIGURE 7.3: Schematic diagram of a ballast circuit.

Defining η as the sensitivity, or the ratio of change in output to change in input, we have

$$\eta = \frac{de_o}{dk} = \frac{e_i R_b R_t}{(R_b + kR_t)^2} \tag{7.6}$$

We may change R_b by inserting different values of resistance. In that case the sensitivity would be altered, which would mean that there may be some optimum value of R_b so far as sensitivity is concerned. By differentiation with respect to R_b, we should be able to determine this value:

$$\frac{d\eta}{dR_b} = \frac{e_i R_t(kR_t - R_b)}{(R_b + kR_t)^3} \tag{7.7}$$

The derivative will be zero under two conditions: (1) for $R_b = \infty$, which results in minimum sensitivity, and (2) for $R_b = kR_t$, for which maximum sensitivity is obtained.

The second relation indicates that for full-range usefulness, the value R_b must be based on compromise because R_b, a constant, cannot always have the value of kR_t, a variable. However, R_b may be selected to give maximum sensitivity for a certain point in the range by setting its value to correspond to that value of kR_t.

This circuit is occasionally used for dynamic applications of resistance-type strain gages [1, 2]. In this case the change in resistance is quite small compared with the total gage resistance, and the relations above indicate that a ballast resistance equal to gage resistance is optimal.

Figure 7.4 shows the relation between input and output for a circuit of this type as given by Eq. (7.5).

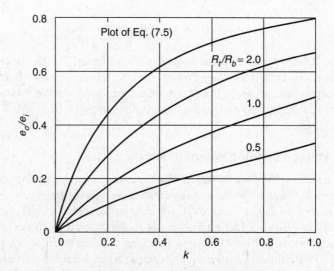

FIGURE 7.4: Curves showing relation between input and output for a ballast circuit.

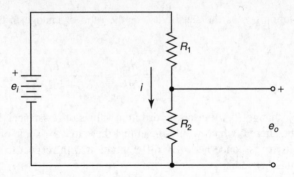

FIGURE 7.5: The voltage-divider circuit.

It will be noted that the same disadvantages apply to this circuit as to the current-sensitive circuit discussed previously—namely, (1) a percentage variation in the supply voltage, e_i, results in a greater change in output than does a similar percentage change in k, so that very careful voltage regulation must be used; and (2) the relation between output and input is not linear.

7.7 VOLTAGE-DIVIDING CIRCUITS

The *voltage divider* (Fig. 7.5) is a ubiquitous element of instrumentation circuits. Very simply, this circuit uses a pair of resistors to divide an input voltage, e_i, into a smaller output voltage, e_o. If a negligible current is drawn from the output terminals, the current through the resistors follows from Ohm's law:

$$i = \frac{e_i}{(R_1 + R_2)}$$

The output voltage measured across R_2 is then

$$e_o = i R_2 = \frac{R_2}{R_1 + R_2} e_i \tag{7.8}$$

Voltage dividers appear throughout this chapter. The ballast circuit of the preceding section is essentially a voltage divider [see Eq. (7.4)] in which the fraction of input voltage at the output depends on the transducer resistance; bridge circuits (Section 7.9) are essentially pairs of voltage dividers; and the noninverting amplifier (Example 7.5) also incorporates a voltage divider.

7.7.1 The Voltage-Dividing Potentiometer

Figure 7.6 is a very useful voltage-divider arrangement for sliding contact resistance transducers. It is known as the *voltage-dividing potentiometer circuit*. Note that the circuit is connected not to the slider, as it would be in the ballast circuit, but across the complete resistance element. The terminating, or readout, device is connected to sense the voltage drop across the portion of resistance element R_p determined by k.

Two different situations may occur with this arrangement, depending on the relative impedance of the resistance element and the indicator-recorder. If the terminating instrument is of sufficiently high relative impedance, no appreciable current will flow through it.

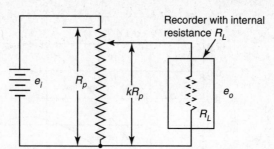

FIGURE 7.6: Simple voltage-dividing potentiometer circuit.

The circuit then becomes a true voltage divider, and the indicated output voltage e_o may be determined from Eq. (7.8),

$$e_o = \frac{kR_p}{R_p}e_i = ke_i$$

or

$$k = \frac{e_o}{e_i} \tag{7.8a}$$

On the other hand, if the readout device draws appreciable current, a *loading error* (see Section 6.2) will result.

7.7.2 Loading Error

The loading error may be analyzed as follows. Referring to Fig. 7.6, we find that the total resistance *seen* by the source of e_i will be

$$R = R_p(1 - k) + \frac{kR_pR_L}{kR_p + R_L}$$

and

$$i = \frac{e_i}{R} = \frac{e_i(kR_p + R_L)}{kR_p^2(1 - k) + R_pR_L}$$

The output voltage will then be

$$e_o = e_i - iR_p(1 - k)$$

or

$$\frac{e_o}{e_i} = \frac{k}{1 + (R_p/R_L)k - (R_p/R_L)k^2} \tag{7.9}$$

If we assume the simpler relation given by Eq. (7.8a) to hold, an error in e_o will be introduced according to the following relation:

$$\text{Error} = e_i\left[k - \frac{k}{k(1 - k)(R_p/R_L) + 1}\right]$$

$$= e_i\left[\frac{k^2(1 - k)}{k(1 - k) + (R_L/R_p)}\right] \tag{7.10}$$

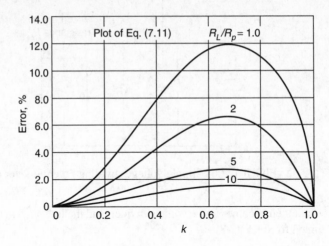

FIGURE 7.7: Curves showing error caused by loading a voltage-dividing potentiometer circuit.

By comparing to the *full-scale output*, e_i, this relation may be written as

$$\text{Percent error} = \left[\frac{k^2(1-k)}{k(1-k) + (R_L/R_p)} \right] \times 100 \qquad (7.11)$$

Except for the endpoints ($k = 0.0$ or 1.0), where the error is zero, the error will always be on the negative side; that is, the measured value of voltage will be lower than would be the case if the system performed as a linear voltage divider. Figure 7.7 shows a plot of the variation in error with slider position for various ratios of load to potentiometer resistance. Obviously, the higher the value of load resistance compared with potentiometer resistance, the lower will be the error; thus high-input resistance is a desirable feature in voltage-reading devices.

7.7.3 Use of End Resistors

It will be observed that the nonlinearity in the relation between the potentiometer output and the input displacement k may be reduced if only a portion of the available potentiometer range is used. For example, a 1000-Ω potentiometer may be selected, but the input could be limited to only a 500-Ω portion of the total range. This limitation would reduce the potentiometer resolution and would be generally impractical; however, it would result in a reduction in the deviation from linearity. A similar result may be obtained through use of what are known as *end resistors* (Fig. 7.8). When either an upper- or lower-end resistor, or both, is used, it is often possible to compensate for the reduced potentiometer output caused by the increased resistance by increasing the voltage input e_i by a proportional amount.

7.8 SMALL CHANGES IN TRANSDUCER RESISTANCE

Some resistance transducers show only very small changes in their resistance. For example, the resistance of a foil strain gage may vary by only about 0.0001% during use! The

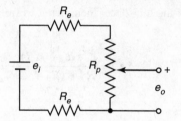

FIGURE 7.8: Method for improving linearity of potentiometer circuits when low-impedance indicating devices are used. Resistors R_e are termed end resistors.

smallness of the resistance change has important ramifications for the choice of signal-conditioning circuit.

Suppose that a voltage-divider (or ballast) circuit is formed from a transducer of resistance R_2 and a second resistor R_1 [Fig. 7.9(a)]. The resistances are initially made equal, $R_1 = R_2 = R_0$, so that the initial output voltage is

$$e_o = \frac{R_2}{R_1 + R_2}\, e_i = \frac{R_0}{R_0 + R_0}\, e_i = \frac{e_i}{2}$$

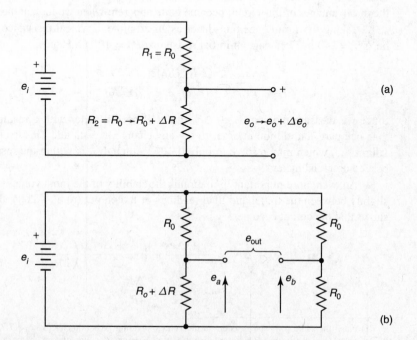

FIGURE 7.9: The use of voltage dividers in measuring small resistance changes.

If the resistance of the transducer then increases from $R_2 = R_0$ to $R_2 = (R_0 + \Delta R)$, the output changes to

$$
\begin{aligned}
e_o + \Delta e_o &= \frac{(R_0 + \Delta R)}{R_0 + (R_0 + \Delta R)} e_i \\
&= \frac{R_0 + \Delta R}{2R_0 + \Delta R} e_i \\
&= \frac{R_0}{2R_0}\left(\frac{1 + \Delta R/R_0}{1 + \Delta R/2R_0}\right) e_i \\
&= \frac{e_i}{2}\left(1 + \frac{\Delta R/2R_0}{1 + \Delta R/2R_0}\right) \\
&= \frac{e_i}{2} + \frac{e_i}{2}\left(\frac{\Delta R}{2R_0}\right)\left(\frac{1}{1 + \Delta R/2R_0}\right)
\end{aligned}
$$

Assuming that the resistance change is small, so that $\Delta R/2R_0 \ll 1$, we can approximate the last factor on the right-hand side as unity; hence,

$$
e_o + \Delta e_o \approx \frac{e_i}{2} + \left(\frac{e_i}{2}\right)\left(\frac{\Delta R}{2R_0}\right) \tag{7.12}
$$

$$
= e_o + \frac{\Delta R}{4R_0} e_i \tag{7.12a}
$$

Thus, for small resistance changes, the output voltage shows straight-line variation with ΔR.[3] Such variation is advantageous because it simplifies data reduction. Unfortunately, the disadvantages of this circuit become quite apparent when we look at the numbers.

Taking the strain gage transducer as an example, a typical resistance change might be $\Delta R = 240\ \mu\Omega$ in a gage of initial resistance $R_0 = 120\ \Omega$. Hence,

$$
\frac{\Delta e_o}{e_o} = \frac{(\Delta R/4R_0)e_i}{e_i/2} = \frac{\Delta R}{2R_0} = 10^{-6}
$$

Since we measure the sum $e_o + \Delta e_o$, we will need a meter with a resolution of better than one part in a million in order to see any change in e_o at all. This excludes common voltmeters, which may resolve to only 0.01%, although it is within the reach of the very best commercial meters.

An even more important limitation is the stability of the input voltage, e_i. If e_i drifts slightly between the initial and final readings of the output (to $e_i + \Delta e_i$), then Eq. (7.12) shows that the output becomes

$$
e_o + \Delta e_o \approx \frac{e_i + \Delta e_i}{2} + \left(\frac{e_i + \Delta e_i}{2}\right)\left(\frac{\Delta R}{2R_0}\right)
$$

$$
\approx e_o + \frac{\Delta e_i}{2} + \frac{\Delta R}{4R_0} e_i
$$

---

[3]In much the same way, end resistors in the voltage-dividing potentiometer (Section 7.7.3) serve to make the transducer resistance change small relative to the other resistances, creating an approximately straight-line variation of the output.

Thus if e_i drifts by even 0.1% ($\Delta e_i = 0.001 e_i$), the change in Δe_o caused by voltage drift will be $0.001 (e_i/2)$—a thousand times larger than the strain-induced voltage change $(\Delta R/4R_0)e_i = 10^{-6}(e_i/2)$!

The difficulty, of course, is that we are trying to resolve a voltage change which is a tiny fraction of the total output voltage, and it illustrates an important principle in measurement: *Avoid measurements based on a small difference between large numbers.* Such measurements are limited by the accuracy with which the *large* numbers can be measured.

In this case, the solution is to design a circuit having output voltage proportional to Δe_o itself, without the large offset voltage, e_o. We can do this by introducing another voltage divider with fixed resistors R_0 [Fig. 7.9(b)], which has a midpoint voltage of e_o. We now measure the difference between the midpoint voltages of the two dividers as the output voltage of the circuit:

$$e_{\text{out}} = e_a - e_b = (e_o + \Delta e_o) - e_o = \Delta e_o = \frac{\Delta R}{4R_0}e_i \qquad (7.13)$$

The problems caused by the offset voltage, e_o, are thus eliminated.

This arrangement of two voltage dividers is, in fact, identical to the Wheatstone bridge circuit discussed in the next section; however, the Wheatstone bridge is not always restricted to small resistance changes.

7.9 RESISTANCE BRIDGES

Bridge circuits are the most common method of connecting passive transducers to measuring systems. Of all the possible configurations, the Wheatstone resistance bridge devised by S. H. Christie in 1833 [3, 4] is undoubtedly used to the greatest extent. Figure 7.10 shows a dc Wheatstone bridge circuit consisting of four resistor *arms* with a voltage source (battery) and a detector (meter). In applications, one or more of the arms is a resistance transducer whose resistance is to be determined. Typical resistance transducers used with a circuit of

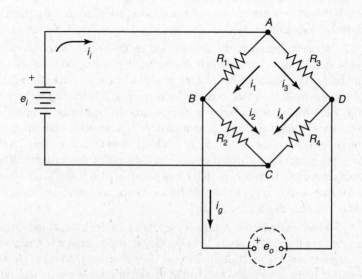

FIGURE 7.10: Simple Wheatstone bridge circuit.

this kind include resistance strain gages, resistance thermometers, or thermistors.

Bridge circuits enable high-accuracy resistance measurements. These measurements are accomplished either by *balancing* the bridge—making known adjustments in one or more of the bridge arms until the voltage across the meter is zero—or by determining the magnitude of *unbalance* from the meter reading. If the circuit appears complicated to you, it may help to recognize that, when negligible current flows through the meter, the bridge is simply a pair of voltage-divider circuits (*ABC* and *ADC*) with the output taken between the midpoints of the two dividers (*B* to *D*). The great advantage of the bridge circuit is that the offset voltages of the two dividers cancel, so that the bridge output voltage can be directly related to changes in transducer resistance (see Section 7.8).

Using Fig. 7.10, we may analyze the requirements for balance. At balance, the voltage across the meter is zero and no current flows through it; hence, $i_g = 0$. In that case, we also know that $i_1 = i_2$ and $i_3 = i_4$. Since the potential across the meter is zero, $i_1 R_1 = i_3 R_3$ and $i_2 R_2 = i_4 R_4$. By eliminating i_1 and i_3 from these relations, we obtain the condition for balance, namely,

$$\frac{R_1}{R_2} = \frac{R_3}{R_4} \tag{7.14}$$

or

$$\frac{R_1}{R_3} = \frac{R_2}{R_4} \tag{7.14a}$$

From these two equations we may formulate a statement to assist us in remembering the necessary balance relation. *In order for the Wheatstone resistance bridge to balance, the ratio of resistances of any two adjacent arms must equal the ratio of resistances of the remaining two arms, taken in the same sense.* (*Note*: "Taken in the same sense" means that if the first resistance ratio is formed from two adjacent resistances reading from left to right, the balancing ratio must also be formed by reading from left to right, etc.)

Basic bridge types are summarized in Table 7.1. When a *null* bridge is used, the resistance of one unknown arm is determined by finding values of the other arms for which the bridge is balanced. Thus, some provision must be made for adjusting the resistance of one or two arms so as to reach balance. Some balancing arrangements are shown in Fig. 7.11. An important factor in determining the type to use is bridge sensitivity. If large resistance changes are to be accommodated, large resistance adjustments must be provided; thus one of the series arrangements would be most useful and could well be the type to use for sliding-contact variable-resistance transducers or thermistors. When small resistance changes are to take place, as in the case of resistance strain gages, then the shunt balance would be used. In order to provide for a range of resistances, a bridge with both series and shunt balances might be utilized.

When the *deflection* bridge is used, bridge unbalance, as indicated by the meter reading, is the measure of input. Usually, the deflection bridge is balanced initially and later changes in transducer resistance cause the unbalance. Manufacturing variations in real resistors make it virtually impossible to obtain three resistors that will match the initial transducer resistance well enough to satisfy Eq. (7.14a) precisely; hence, provision is gen-

TABLE 7.1: Types of Electrical Bridge Circuits

Bridge Type	Bridge Features
Null balance bridge vs. Deflection bridge	Adjustment is required to maintain balance. This becomes source of readout (e.g., a manually adjusted strain indicator). Readout is deviation of bridge output from initial balance (e.g., as required by a computer's analog-to-digital converter [A/D]).
Voltage-sensitive bridge vs. Current-sensitive bridge	Readout instrument does not "load" bridge; that is, it requires no current (e.g., electronic voltmeter or analog-to-digital converter). Readout requires current (e.g., a low-impedance indicator such as a simple galvanometer is used).
ac bridge vs. dc bridge	Alternating-current voltage excitation is used. Direct-current voltage excitation is used.
Constant voltage vs. Constant current	Voltage input to bridge remains constant (e.g., battery or voltage-regulated power supply is used). Current input to bridge remains constant regardless of bridge unbalance (e.g., current-regulated power supply is used).
Resistance bridge vs. Impedance bridge	Bridge arms made up of "pure" resistance elements. Bridge arms may include reactance elements.

erally made for initial balancing by adjusting one or more arms, again using an arrangement from Fig. 7.11. For static inputs, an ordinary voltmeter may be used to display the output; for dynamic signals, however, the output may be displayed by an oscilloscope (Section 9.6) or the output may be fed to an analog-to-digital converter and a computer for display, recording, or immediate application (Section 8.11).

The output from a deflection bridge may be connected to either a high- or a low-impedance device. If the bridge is connected to a simple D'Arsonval meter or most galvanometers, the output circuit will be of low impedance, and an appreciable current (i_g) is drawn from the bridge. In most cases in which amplification or digital processing is necessary, the bridge output will be connected to a high-impedance device and the bridge would supply essentially no current. Such is the case when either an an electronic voltmeter or an analog-to-digital converter is used. In the former instance the bridge is *current sensitive*; in the latter it is *voltage sensitive*.

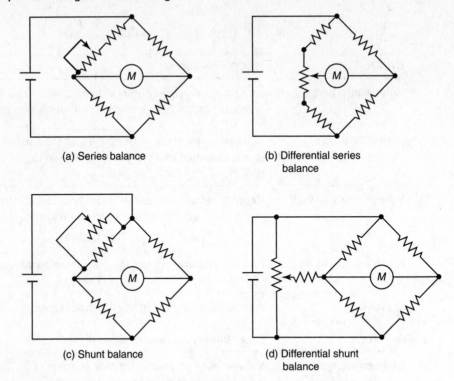

(a) Series balance

(b) Differential series balance

(c) Shunt balance

(d) Differential shunt balance

FIGURE 7.11: Arrangements used to balance dc resistance bridges.

7.9.1 The Voltage-Sensitive Wheatstone Bridge

Let us consider the simplest case first, in which the bridge output is connected directly to a high-impedance device, say an oscilloscope. Referring to Fig. 7.10, we see that the output voltage is the difference between the voltages at B and D

$$e_o = e_B - e_D$$

and, making use of the voltage-divider relation [Eq. (7.8)], we may write

$$e_o = e_i \left(\frac{R_2}{R_1 + R_2} - \frac{R_4}{R_3 + R_4} \right) \tag{7.15}$$

$$= e_i \left(\frac{R_2 R_3 - R_4 R_1}{(R_1 + R_2)(R_3 + R_4)} \right) \tag{7.15a}$$

We will now assume that the resistance R_2 changes by an amount ΔR_2, or

$$e_o + \Delta e_o = e_i \left[\frac{(R_2 + \Delta R_2)R_3 - R_4 R_1}{(R_1 + R_2 + \Delta R_2)(R_3 + R_4)} \right]$$

$$= e_i \left\{ \frac{1 + (\Delta R_2/R_2) - (R_4 R_1/R_2 R_3)}{[(1 + (R_1/R_2) + (\Delta R_2/R_2)][1 + (R_4/R_3)]} \right\} \tag{7.16}$$

The relation may be simplified by assuming all resistances to be initially equal to R (in which case $e_o = 0$). Then

$$\frac{\Delta e_o}{e_i} = \frac{\Delta R_2/R}{4 + 2(\Delta R_2/R)} \qquad (7.17)$$

Figure 7.12(a), plotted from Eq. (7.17), shows the relation for the output of a voltage-sensitive deflection bridge whose resistance arms are initially equal. Inspection of the curve indicates that this type of resistance bridge is inherently nonlinear. In many cases, however,

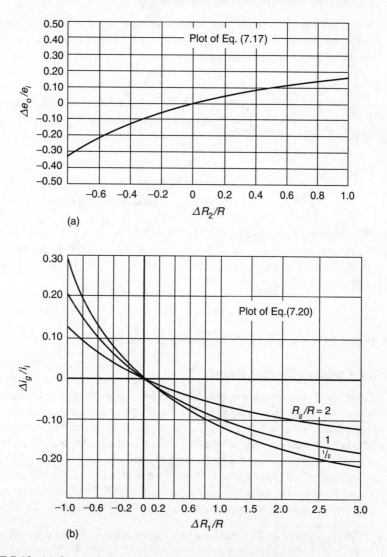

(a)

(b)

FIGURE 7.12: (a) Output from a voltage-sensitive deflection bridge whose resistance arms are initially equal; (b) output from a current-sensitive deflection bridge whose resistance arms are initially equal, plotted for different relative galvanometer resistances.

the actual resistance change is so small that the arrangement may be assumed linear. This assumption applies to most resistance strain-gage circuits. In those cases, $\Delta R_2/2R \ll 1$ and the linearized output is

$$\frac{\Delta e_o}{e_i} = \frac{\Delta R_2}{4R} \qquad (7.18)$$

which is identical to Eq. (7.13).

7.9.2 The Current-Sensitive Wheatstone Bridge

When the deflection-bridge output is connected to a low-impedance device such as a galvanometer, appreciable current flows and the galvanometer resistance must be considered in the bridge equation. Galvanometer current may be expressed by the following relation [5]:

$$i_g = \frac{i_i(R_2 R_3 - R_1 R_4)}{R_g(R_1 + R_2 + R_3 + R_4) + (R_2 + R_4)(R_1 + R_3)} \qquad (7.19)$$

where

$$i_g = \text{the galvanometer current,}$$
$$i_i = \text{the input current,}$$
$$R_g = \text{the galvanometer resistance}$$

The remaining symbols are as defined in Fig. 7.10.

If we assume that an initial bridge balance is upset by an incremental change in resistance ΔR_1 in arm R_1 and all arms are of equal initial resistance R, we may write

$$\frac{\Delta i_g}{i_i} = \frac{-\Delta R_1/R}{4[1 + (R_g/R)] + [2 + (R_g/R)](\Delta R_1/R)} \qquad (7.20)$$

Figure 7.12(b) shows Eq. (7.20) plotted for various values of R_g/R.

7.9.3 The Constant-Current Bridge

To this point our discussion of bridge circuits has assumed a constant-voltage energizing source (a battery, for example). As the bridge resistance is changed, the total current through the bridge will, therefore, also change. In certain instances (see Section 12.9), use of a *constant-current* bridge[4] may be desirable [6, 7]. Such a circuit is usually obtained through the application of a commercially available *current-regulated* dc power supply,[5] whereby the total current flow i_i through the bridge (Fig. 7.10) is maintained at a constant value. It should be noted that such a bridge may still be either voltage sensitive or current sensitive, depending on the relative impedance of the readout device.

---

[4]The term *Wheatstone*, as applied to bridge circuits, is commonly limited to the *constant-voltage resistance bridge*. We shall abide by this convention and avoid referring to the constant-current bridge as a Wheatstone bridge.

[5]Constant current is obtained by using the voltage drop across a series resistor in the supply-output line to provide a regulating feedback voltage. It may also be approximated by placing a large ballast resistor between the bridge and the voltage source; the resistor is made large enough that variations in the bridge resistors have a negligible effect on i_i.

Relationships for the *voltage-sensitive constant-current* bridge may be developed as follows. Referring to Fig. 7.10, we may write

$$i_i = \frac{e_i}{R_1 + R_2} + \frac{e_i}{R_3 + R_4} \tag{7.21}$$

or

$$e_i = i_i \left[\frac{(R_1 + R_2)(R_3 + R_4)}{R_1 + R_2 + R_3 + R_4} \right]$$

Substituting in Eq. (7.15a), we have

$$e_o = i_i \left[\frac{R_2 R_3 - R_4 R_1}{R_1 + R_2 + R_3 + R_4} \right] \tag{7.22}$$

which is the basic equation for the voltage-sensitive constant-current bridge, provided that i_i is maintained at a constant value. If the resistance of one arm, say R_2, is changed by an amount ΔR, then

$$e_o + \Delta e_o = i_i \left[\frac{(R_2 + \Delta R)R_3 - R_4 R_1}{R_1 + (R_2 + \Delta R) + R_3 + R_4} \right]$$

and

$$\Delta e_o = i_i \left[\frac{(R_2 + \Delta R)R_3 - R_4 R_1}{R_1 + (R_2 + \Delta R) + R_3 + R_4} - \frac{R_2 R_3 - R_4 R_1}{R_1 + R_2 + R_3 + R_4} \right] \tag{7.23}$$

For equal initial resistances ($R_1 = R_2 = R_3 = R_4 = R$),

$$\Delta e_o = i_i \left[\frac{\Delta R}{4 + (\Delta R / R)} \right] \tag{7.24}$$

The constant-current bridge has better linearity than the constant-voltage bridge, as is apparent upon comparing Eqs. (7.17) and (7.24).

7.9.4   The AC Resistance Bridge

Resistance bridges powered by ac sources may also be used. An additional problem, however, is the necessity for providing reactance balance. In spite of the fact that the Wheatstone bridge, strictly speaking, is a resistance bridge, it is impossible to completely eliminate stray capacitances and inductances resulting from such factors as closely placed lead wires in cables to and from the transducer, and wiring and component placement in associated equipment. In any system of reasonable sensitivity, such unintentional reactive components must be accounted for before satisfactory bridge balance can be achieved.

Reactive balance can usually be accomplished by introducing an additional balance adjustment in the circuit. Figure 7.13 shows how this may be provided. Balance is accomplished by alternately adjusting the resistance and reactance balance controls, each time reducing bridge output, until proper balance is finally achieved.

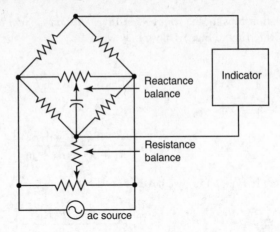

FIGURE 7.13: Circuit arrangement for balancing an ac bridge.

7.9.5 Compensation for Leads

Frequently a sensor and a bridge-type instrument must be separated by an appreciable distance. Wires, or leads, are used to connect the two as illustrated in Fig. 7.14(a), which shows the sensor as some type of resistance element such as a resistance thermometer or strain gage. In addition to the extra resistance introduced by the leads, temperature gradients along the wires may, in certain cases, cause error (see Sections 12.8.3 and 16.4.2). We can

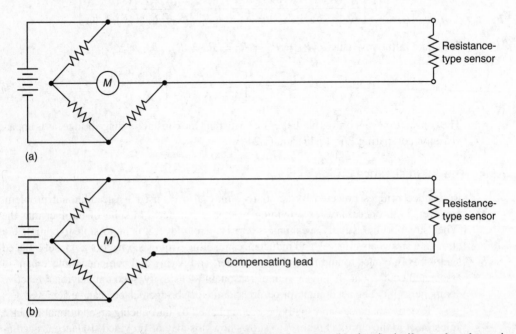

FIGURE 7.14: (a) Simple bridge with remotely located sensor; (b) circuit similar to that shown in (a), but with a compensating wire.

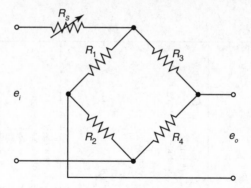

FIGURE 7.15: Method for adjusting bridge sensitivity through use of variable series resistance, R_s.

compensate for this type of error by using a three-wire circuit as illustrated in Fig. 7.14(b). Inspection shows that the additional lead serves to balance the total lead-wire lengths in the two adjacent arms, thereby eliminating any unbalance from this source.

7.9.6 Adjusting Bridge Sensitivity

Adjustable bridge sensitivity may be desired for several reasons: (1) Such adjustment may be used to attenuate inputs that are larger than desired; (2) it may be used to provide a convenient relation between system calibration and the scale of the readout instrument; (3) it may be used to provide adjustment for adapting individual transducer characteristics to precalibrated systems (this method is used to insert the gage factor for resistance strain gages in some commercial circuits); (4) it provides a means for controlling certain extraneous inputs such as temperature effects (see Section 13.5).

A very simple method of adjusting bridge output is to insert a variable series resistor in one or both of the input leads, as shown in Fig. 7.15. If we assume equal initial resistance R in all bridge arms, the resistance seen by the voltage source will also be R. If a series resistance is inserted as shown, then, thinking in terms of a voltage-dividing circuit, we see that the input to the bridge will be reduced by the factor

$$ n = \frac{R}{R + R_s} = \frac{1}{1 + (R_s/R)} \tag{7.25} $$

We call n the *bridge factor*. The bridge output will be reduced by a proportional amount, which makes this method very useful for controlling bridge sensitivity.

7.10 REACTANCE OR IMPEDANCE BRIDGES

Reactance or impedance bridge configurations are of the same general form as the Wheatstone bridge, except that reactive elements (capacitors and inductors) are involved in one or more of the arms. Because such elements are inherently frequency sensitive, impedance bridges are ac excited. Obviously the multitude of variations that are possible preclude more than a general discussion in a work of this nature; thus the reader is referred to more specialized works for detailed coverage [8, 9].

Figure 7.16 shows several of the more common ac bridges, along with the type of element usually measured and the balance requirements.

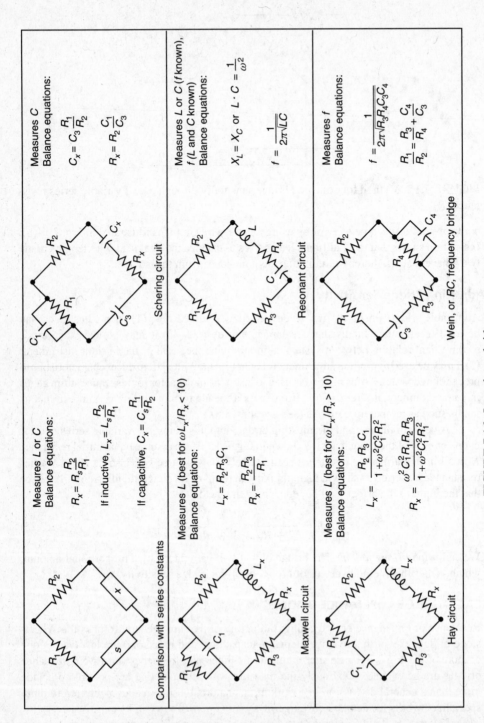

FIGURE 7.16: Impedance bridge arrangements.

Comparison with series constants

Measures L or C
Balance equations:

$$R_x = R_s \frac{R_2}{R_1}$$

If inductive, $L_x = L_s \dfrac{R_2}{R_1}$

If capacitive, $C_x = C_s \dfrac{R_1}{R_2}$

Schering circuit

Measures C
Balance equations:

$$C_x = C_3 \frac{R_1}{R_2}$$

$$R_x = R_2 \frac{C_1}{C_3}$$

Maxwell circuit

Measures L (best for $\omega L_x / R_x < 10$)
Balance equations:

$$L_x = R_2 R_3 C_1$$

$$R_x = \frac{R_2 R_3}{R_1}$$

Resonant circuit

Measures L or C (f known),
f (L and C known)
Balance equations:

$$X_L = X_C \text{ or } L \cdot C = \frac{1}{\omega^2}$$

$$f = \frac{1}{2\pi\sqrt{LC}}$$

Hay circuit

Measures L (best for $\omega L_x / R_x > 10$)
Balance equations:

$$L_x = \frac{R_2 R_3 C_1}{1 + \omega^2 C_1^2 R_1^2}$$

$$R_x = \frac{\omega^2 C_1^2 R_1 R_2 R_3}{1 + \omega^2 C_1^2 R_1^2}$$

Wein, or RC, frequency bridge

Measures f
Balance equations:

$$f = \frac{1}{2\pi\sqrt{R_3 R_4 C_3 C_4}}$$

$$\frac{R_1}{R_2} = \frac{R_3}{R_4} + \frac{C_4}{C_3}$$

250

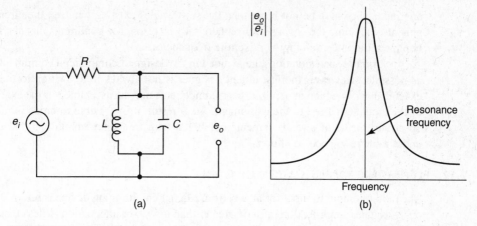

FIGURE 7.17: Parallel LC circuit with curve showing output versus frequency characteristics.

7.11 RESONANT CIRCUITS

Capacitance-inductance combinations present varying impedance, depending on their relative values and the frequency of the applied voltage. When connected in parallel, as in Fig. 7.17(a), the inductance offers small opposition to current flow at low frequencies, whereas the capacitive reactance is low at high frequencies. At some intermediate frequency, the opposition to current flow, or impedance, of the combination is a maximum [Fig. 7.17(b)]. A similar but opposite variation in impedance is obtained in the series-connected combination.

The frequency corresponding to maximum effect, known as the *resonance frequency*, may be determined by the relation

$$f = \frac{1}{2\pi\sqrt{LC}} \qquad (7.26)$$

where

$$f = \text{the frequency (Hz)},$$
$$L = \text{the inductance (H)},$$
$$C = \text{the capacitance (F)}$$

It is evident that should, say, a capacitive transducer element be used, it could be in combination with an inductive element to form a resonant combination. Variation in capacitance caused by variation in an input signal (e.g., mechanical pressure) would then alter the resonance frequency, which could then be used as a measure of input.

7.11.1 Undesirable Resonance Conditions

On occasion, resonance conditions that occur may introduce spurious outputs. Most circuits are susceptible because they use some combination of inductance and capacitance and most are called on to handle dynamic signal inputs. In certain cases the capacitance and inductance may be not more than the stray values existing between the circuit components, including the wiring. Hence there may be resonance conditions that can result in nonlinearities at certain input or exciting frequencies.

Normally such situations are avoided in the design of commercial equipment insofar as possible. However, the instrument designer is not always in a position to predict the exact manner in which general-purpose components may be assembled or the exact nature of the input signal fed to the equipment. As a result, it is possible to unintentionally set up arrangements of circuit elements combined with frequency conditions that result in undesirable resonance conditions.

7.12 ELECTRONIC AMPLIFICATION OR GAIN

The ratio of output to input for an electronic signal-conditioning device is referred to variously as gain, amplification ratio (if greater than unity), or attenuation (if less than unity). It may be defined in terms of voltages, currents, or powers, that is,

Voltage gain = voltage output/voltage input,

Current gain = current output/current input,

Power gain = power output/power input

Another way of expressing *power gain* is through use of the *decibel*. A decibel (dB) is one-tenth of a *bel* and is based on a ratio of powers:

$$\text{Decibel (dB)} = 10\log_{10}(P_o/P_i) \tag{7.27}$$

where P_o = the output power and P_i = the input power, both expressed in the same units.

The average human ear can just detect a loudness change from an audio amplifier when a power ratio change of one decibel is made. It has also been observed that this is nearly true regardless of the power level.

Solving Eq. (7.27) for the ratio (P_o/P_i) corresponding to one decibel yields a ratio of 1.26. In other words, for the average human ear to just detect an increase in sound output from an amplifier (feeding some form of earphone or loudspeaker), an increase of approximately 26% in power is required. Some other useful power ratios, as expressed in decibels, are

$$P_o = 2 \times P_i : \quad 3 \text{ dB}, \qquad\qquad P_o = \frac{1}{2} \times P_i: \quad -3 \text{ dB}.$$

$$P_o = 10 \times P_i : \quad 10 \text{ dB}, \qquad\qquad P_o = \frac{1}{10} \times P_i: \quad -10 \text{ dB},$$

$$P_o = 100 \times P_i : \quad 20 \text{ dB}, \qquad\qquad P_o = \frac{1}{100} \times P_i: \quad -20 \text{ dB}$$

The half-power point, -3 dB, is often used in characterizing the frequency response of amplifiers and, especially, of filters.

For a pure resistance, electric power may be expressed as

$$\text{Power} = ei = e^2/R = i^2 R$$

where e = the voltage, i = the current, and R is a pure resistance. Substituting either of the last two forms into Eq. (7.27) yields

$$dB = 20 \log_{10} \left(\frac{e_o}{e_i} \right) - 10 \log_{10} \left(\frac{R_o}{R_i} \right) \tag{7.28}$$

or

$$dB = 20 \log_{10} \left(\frac{i_o}{i_i} \right) + 10 \log_{10} \left(\frac{R_o}{R_i} \right) \tag{7.28a}$$

Should $R_i = R_o$, then the last term in each case reduces to zero. The relationship of the decibel to power and voltage ratios is illustrated in Fig. 7.18 for $R_i = R_o$.

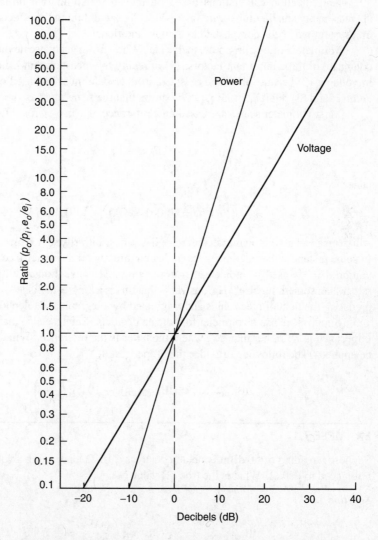

FIGURE 7.18: Relationship of power and voltage to the decibel.

One should remember that the decibel is fundamentally a *power ratio* and that "forgetting" the R's in the preceding equations is strictly legitimate only if the two loads, with and without amplification, are equal. Nevertheless, output/input ratios are often described using the decibel even when no load is directly involved, and one frequently sees voltage ratios expressed in decibels as

$$dB = 20 \log_{10} \left(\frac{e_o}{e_i} \right) \tag{7.28b}$$

Another common use of the decibel is in constructing a *Bode plot* of frequency response. In such a graph, the gain in decibels is plotted against a logarithmic frequency axis, rather than showing e_o/e_i versus f on linear coordinates (compare Figs. 7.26 and 7.27).

Amplification calculations based on the decibel offer two important advantages: (1) reasonably small numbers are involved, and (2) combining the effects of various stages of a system may be accomplished by simple addition.

Voltmeters often carry a decibel scale. When using such a scale one must always be cognizant of three important factors: (1) in reality the measurement is not in decibels, but in voltage; (2) because the decibel is a ratio, the scale must be based on some *reference voltage*; and (3) reference to Eq. (7.28) shows that the scale must assume a *reference load*. Most voltmeter scales are based on a reference of 1 mW across 600 Ω, or

$$P = \frac{e^2}{R}$$

hence,

$$e = (PR)^{1/2} = (0.001 \times 600)^{1/2} = 0.7746 \text{ V}$$

which means that zero on the decibel scale has been arbitrarily set to correspond to 0.7746 V. In some instances the references are indicated directly on the meter face. Often the abbreviation dBm is used to indicate the aforementioned conventions. Why the 600-Ω load rather than something else? The answer is that this is a long-established industrial standard, predating the field of electronics and originated by telegraph and telephone practices.

Suppose we use a voltmeter to indicate decibels. Suppose also that the signal source impedance is R_s rather than R_r, where the latter is the reference. What correction should be applied? The following provides the proper result:

$$dB_{(corrected)} = dB_{(indicated)} + 10 \log \left(\frac{R_r}{R_s} \right) \tag{7.28c}$$

EXAMPLE 7.1

Suppose a reading of 50 dBm is obtained across a 16-Ω load, using a voltmeter with scale referenced to 600 Ω. What is the true dB value?

Solution

$$dB_{(corrected)} = 50 + 10 \log \left(\frac{600}{16} \right) = 65.7 \text{ dB}$$

As we discussed before, corrections must be made to obtain *true* dB values when load and reference conditions differ. Very conveniently, however, if we require only *differences* or changes in decibels, then we may not need corrections in individual readings. This situation exists if the loads remain unchanged during the actual measurements.

7.13 ELECTRONIC AMPLIFIERS

Some form of amplification is almost always used in circuitry intended for mechanical measurement. It is not the purpose of this section to be concerned with electronics or electronic theory beyond the barest minimum required to make intelligent use of such equipment for the purposes of mechanical measurement. The following discussion, therefore, is brief and is directed primarily to applications rather than to specific theory of operation.

Electronic amplification originated with the invention of the triode vacuum tube. Thomas Alva Edison discovered that electrons could flow from a heated cathode to an anode in an evacuated space; hence the term *Edison effect.* Lee de Forest is credited with showing that the flow could be *controlled* by inserting a third element, the grid, between the cathode and the anode. This resulted in the triode electron tube and, in various configurations, many with additional elements, provided the basis for electronic amplification.

Of course, vacuum tubes are little used in instrumentation today, having been almost entirely superseded by less fragile, less costly, and increasingly sophisticated semiconductor devices. Historically, the term *electronic*, as opposed to the word *electrical*, meant that in some part of the circuit electrons are caused to flow through space in the absence of a physical conductor, thus implying the use of vacuum tubes. Today, the word *electronics* has taken on a broader meaning, encompassing the *solid-state* devices most often used in instrumentation: diodes, transistors, integrated circuits, and the like. Throughout the remainder of the book it will be understood that, unless more specific reference is made, the word *electronics* is being used in its broadest sense.

Electronic amplifiers are used in mechanical measurements to provide one or a combination of the following basic services: (a) voltage gain; (b) current gain, or power, gain; and (c) impedance transformations. In most cases in which mechanical or electrical transduction is used, voltage is the electrical output that is the analogous signal. Often the voltage level available from the transducer is very low; thus a voltage amplifier is used to increase the level for subsequent processing. Occasionally, the input signal must finally be used to drive another device, such as a control mechanical or mechanical indicator. In this case, voltage gain may not be sufficient in itself because power must be increased; hence a current or power amplifier is needed.

In certain instances a transducer produces sufficient signal level but is accompanied by an unacceptably high output impedance level. This is true of most piezoelectric-type transducers, for example. High-impedance lines have the disadvantage of susceptibility to noise. If the signal is to be transmitted any appreciable distance (even a few inches in some cases), the noise pickup from the environment may be unacceptable. Low-impedance lines are much less prone to this problem. Hence it may well be desirable to insert an impedance transformation in the form of an amplifier that will accept a high-impedance input but produce a low-impedance output. This type of amplifier is often called a *buffer*.

There are several generalities that can be listed for the ideal (but nonexistent) electronic amplifier:

1. Infinite input impedance: no input current, hence no load on the previous stage or device

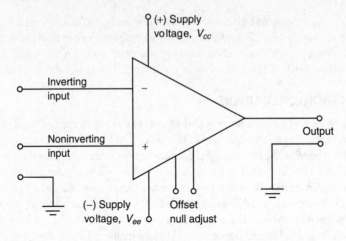

FIGURE 7.19: Diagram showing typical operational amplifier connections.

2. Infinite gain (lower gain can be obtained by adding attenuation circuits)
3. Zero output impedance (low noise)
4. Instant response (wide frequency bandwidth)
5. Zero output for zero input
6. Ability to ignore or reject extraneous inputs

Although none of these aims can be completely realized, it is often possible to approach them, and their assumption simplifies circuit analysis.

Today's amplifiers are most often constructed as *integrated circuits*. As the name implies, integrated circuits are groups of circuit elements combined into a single device. For the most part the elements consist of transistors, diodes, resistors, and, to a lesser extent, capacitors, all connected and packaged in convenient plug-in or surface-mount units. They form the building blocks used to construct more complex circuits: amplifiers, mixers (for combining signals), timers, filters, audio preamps, audio power amplifiers, voltage references, regulators and comparators, and many of the digital devices discussed in Chapter 8. Of particular importance to mechanical measurements is the operational amplifier, or op amp. In the following paragraphs we will discuss this device in more detail.

7.14 OPERATIONAL AMPLIFIERS

The op amp is an integrated circuit that functions as a dc differential voltage amplifier. By dc we mean that it will process input signals over a frequency range extending down to and *including* a dc voltage. As a differential amplifier it accepts two inputs and responds to the *difference* in the voltages applied to the input terminals. One of these inputs, called *noninverting*, is conventionally identified with the (+) symbol (Fig. 7.19). The other, called the *inverting* input, carries the (−) symbol. The voltage at the output terminal, e_o, is the product of the amplifier gain, G, and the voltage difference:

$$e_o = G(e_+ - e_-) \tag{7.29}$$

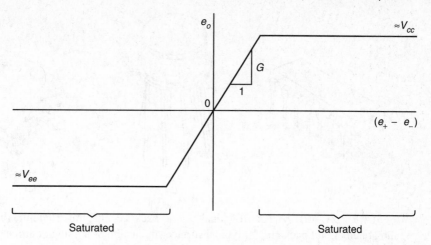

FIGURE 7.20: Op-amp output response.

The output voltage is roughly limited to the power supply voltages, V_{cc} and V_{ee}, as the voltage difference increases; if the voltage difference becomes too large, the output *saturates* near one of these values and remains constant if the voltage differential increases further. Op-amp response is illustrated in Fig. 7.20.

The op amp's differential characteristic has great importance in instrumentation because it eliminates offset voltages and noise signals common to both input terminals. For example, nearby power lines may induce 60-cycle noise in the exterior circuitry leading to the amplifier. Such line noise is often present in identical form at both input terminals, and it is thus canceled by the differential amplification. This behavior is known as *common-mode rejection*. If, instead, an op amp receives the output of a voltage-sensitive Wheatstone bridge, the common offset voltages of the two voltage dividers are canceled, and only the desired difference voltage is amplified.

Figure 7.19 shows the configuration of the exterior circuitry of the op amp. Two power sources of equal magnitude but opposite polarity are generally required ($-V_{cc} = V_{ee}$). These voltages usually fall somewhere in the range of 5 to 30 V dc. Quite often, common 9-V dc batteries may be used. Op amps are usually packaged in dual-in-line package (DIP) form, one of the standard "TO" cans, or in surface-mount packages [Figs. 7.21(a), (b), and (c)].

The op amp very nearly satisfies the ideal voltage amplifier characteristics of Section 7.13 for the following reasons:

1. It has very high input impedance (megaohms to teraohms).
2. It is capable of very high gain (10^5–10^6 or 100 dB–120 dB).
3. It has very low output impedance (down to a fraction of an ohm with feedback).
4. It has very fast response or high slew rate (output can change several volts per μs.)
5. It is quite effective in rejecting common-mode inputs.

One nonideal characteristic of most op amps is that they do not completely satisfy the differential amplifying property: With both inputs grounded, a residual output voltage

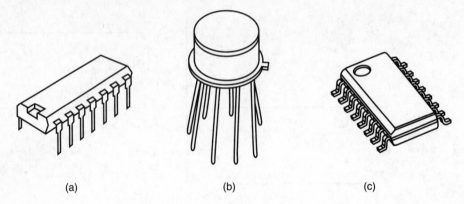

(a) (b) (c)

FIGURE 7.21: (a) Typical DIP (dual in-line package) integrated circuit; (b) typical TO integrated circuit package; (c) typical surface-mountable SOP (small outline package).

remains. The multitude of transistors, resistors, and other elements within the op amp are never perfectly matched, so the amp output actually reaches zero at some small *nonzero* input voltage. To accommodate this *input offset voltage*, the common op amp is provided with pins marked "offset null" or "balance," which provide a means for adjusting the unwanted offset voltage toward zero (see Example 7.7).

A second limitation is that the actual common-mode rejection is finite. If the two input signals each include a common-mode voltage, e_{cm}, the op amp's actual output will be

$$e_o = G(e_+ - e_-) + G_{cm}e_{cm}$$

The finite common-mode rejection is characterized by the *common-mode rejection ratio* (CMRR) in decibels:

$$\text{CMRR} = 20\log_{10}\left(\frac{G}{G_{cm}}\right) \tag{7.30}$$

Typical op amps have a CMRR of 60 to 120 dB; thus, the common-mode gain is typically 10^3 to 10^6 times smaller than the differential gain. Obviously, a high CMRR is desirable.

In addition, thermal drift can limit op-amp performance. Both internal and external circuit elements may be temperature sensitive, and the design of each circuit usually includes compensating features. A wide variety of op amps are available, and their differences largely represent attempts to improve thermal stability, CMRR, offset voltages, or frequency response. Understandably, such refinements are reflected in cost.

7.14.1 Typical Op-Amp Specifications

Op amps are often designed to optimize those aspects of performance needed for a specific application. One common general-purpose amplifier is the LF411. In comparison to more sophisticated op amps, the LF411 is quite simple; yet in a package the size of a finger-nail, it incorporates 23 transistors, 11 resistors, 3 diodes, and 1 capacitor. Typical LF411 specifications are as follows:

Open-loop gain	2×10^5 (depends on frequency)
Input impedance	$10^{12}\ \Omega$
Input offset voltage	0.8 mV
Input offset voltage drift	$7\ \mu V/^\circ C$
Input offset current	25 pA
Input bias current	50 pA
CMRR	100 dB
Maximum output current	25 mA
Slew rate	$15\ V/\mu s$
Maximum power supply voltages	± 18 V
Power supply current	1.8 mA
Maximum input voltage range	± 15 V
Maximum differential input	± 30 V
Short-circuit output time	Indefinite

7.14.2   Applications of the Op Amp

Operational amplifiers may be used as the basic components of linear voltage amplifiers, differential amplifiers, integrators and differentiators, voltage comparators, function generators, filters, impedance transformers, and many other devices. They are *not* power amplifiers, nor do they have exceptionally wide bandwidth capabilities. Undistorted frequency response is typically limited to about 1 MHz when the circuit gain is low, and it decreases as the gain is raised. In general, an op amp's maximum voltage output is limited by the supply voltage.

Since the number of applications of the op amp to mechanical measurements is almost limitless, we can describe here only a few. Yet this will give the reader some idea of the tremendous versatility of the device and will suggest additional uses (see also the Suggested Readings at the end of the chapter).

One feature common to most op-amp circuits is a *negative feedback loop*. Because op-amp gain is so high, even a slight input-voltage difference will drive the amplifier to saturation. To prevent this, a connection is made between the output terminal and the inverting $(-)$ terminal. With this connection in place, an increase in e_o will be fed back to e_-, reducing the input voltage-difference. The net effect is to produce a circuit that holds $e_- \approx e_+$, preventing saturation. Example 7.2 describes a circuit with no feedback, and several subsequent examples treat circuits having feedback loops.

EXAMPLE 7.2

The *open-loop configuration*[6] has the following characteristics:

1. No feedback loop. R_L is the load resistance powered by e_o. The circuit may be free floating or grounded.

2. Amplifier is run wide open: Any input other than zero will drive the amplifier to saturation (i.e., a very small input will drive the output to the limit permitted by the power supply).

---

[6]It is conventional in op-amp circuit diagrams to show only those terminals that are used in the particular configuration. Power supply inputs are always required, whether shown or not. Null adjustment is often not shown, although it may be required for optimal performance (see Example 7.7).

3. It is seldom used; however, it may be employed as a voltage comparator. With different voltages applied to (+) and (−), open-loop output polarity (positive or negative saturation) will be controlled by the larger input. For sinusoidal input, a square-wave output would result.

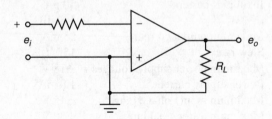

EXAMPLE 7.3

The *voltage follower*, or *impedance transformer*, has a feedback loop connecting the full output to the inverting input.

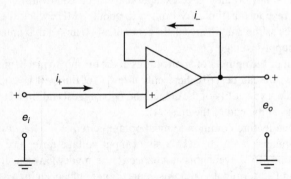

The feedback loop prevents saturation by holding $e_- \approx e_+$. Since $e_i = e_+$ and $e_o = e_- \approx e_+$, the output voltage is equal to (follows) the input voltage: $e_o = e_i$. The circuit gain is $G = 1$.

This circuit capitalizes on the high input impedance of the op amp: Since the input impedance is so large, the input current i_+ is in nanoamps (nA) or even picoamps (pA). Source loading is minimized and can often be entirely neglected ($i_+ \approx 0$). In contrast, the output terminal can deliver up to the maximum current of the op amp. This circuit acts as an impedance transformer in that the input impedance is in gigaohms, whereas the output impedance is a fraction of an ohm.

This example demonstrates two important rules of thumb that can be applied any op-amp circuit having negative feedback:

1. The input currents, i_+ and i_-, are essentially zero: $i_+, i_- \approx 0$.
2. The input voltages, e_+ and e_-, are held equal by the negative feedback: $e_+ \approx e_-$.

EXAMPLE 7.4

The *inverting amplifier* is one of the most used op-amp circuits. Feedback is provided through resistor R_2.

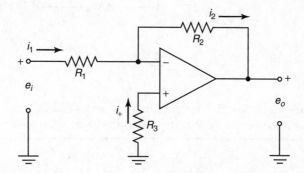

Since $i_+ \approx 0$, Ohm's law shows $e_+ = 0$. Because negative feedback is present, $e_- = e_+ = 0$. The inverting input also draws no current, so that $i_1 = i_2$. Thus we can apply Ohm's law to resistors R_1 and R_2 to find the relation between e_i and e_o:

$$i_1 = \frac{e_i - 0}{R_1} = \frac{e_i}{R_1}$$

$$i_2 = \frac{0 - e_o}{R_2} = -\frac{e_o}{R_2}$$

or

$$e_o = -\frac{R_2}{R_1} e_i$$

The output is opposite in sign from the input (inverted, or 180° out of phase), and the gain of the circuit is $G = -R_2/R_1$.

The resistor R_3 is commonly made approximately equal to the parallel value of R_1 and R_2, i.e., $R_3 \approx R_1 R_2/(R_1 + R_2)$. This choice provides nearly equal input impedances at the $(-)$ and $(+)$ terminals.

EXAMPLE 7.5

The *noninverting amplifier* is as shown at the top of the next page.

The input voltage is applied to the $(+)$ terminal $(e_i = e_+)$; because negative feedback is present, $e_- = e_+ = e_i$. The output voltage is related to the voltage at the inverting terminal by the voltage-divider relation:

$$e_- = \left(\frac{R_1}{R_1 + R_2} \right) e_o$$

Rearranging,

$$e_o = \left(\frac{R_1 + R_2}{R_1} \right) e_i$$

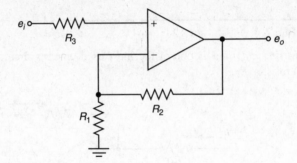

Thus, the output and the input are in phase and the circuit gain is $G = (R_1 + R_2)/R_1$. Resistor R_3 serves the same purpose as in the inverting amplifier.

EXAMPLE 7.6

In the *differential*, or *difference, amplifier*:

1. If $R_1 = R_2$ and $R_3 = R_4$, then $e_o = -(R_3/R_1)(e_{i_1} - e_{i_2})$ (see Problem 7.23).
2. The need for offset null adjustment (see Example 7.7) is minimized by making input resistances at $(-)$ and $(+)$ equal.
3. Precise resistor matching is necessary to achieve high CMRR.

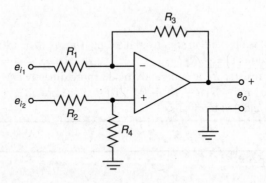

EXAMPLE 7.7

An amplifier with offset null adjustment is exemplified by the accompanying diagram.

1. The circuit allows trimming to zero output with zero input.
2. Specific example shown illustrates pin numbering.

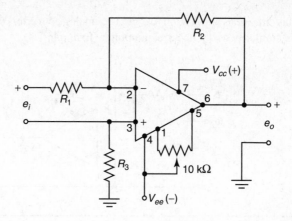

EXAMPLE 7.8

The *voltage comparator* has the following features:

1. A small voltage difference between e_i and e_{ref} swings output to limit permitted by power supplies; e_{ref} is set to desired reference voltage. No feedback is used.
2. When $e_i > e_{ref}$, output is positively saturated; when $e_i < e_{ref}$, output is negatively saturated. This provides output indication for the size of e_i relative to e_{ref}. For example, should e_i be gradually rising, when its value reaches e_{ref} the output polarity would reverse. This could be used to trigger external action. (See Section 8.11.2 for application to analog-to-digital conversion.)
3. Diodes serve to limit differential input.

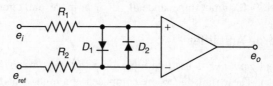

EXAMPLE 7.9

The *summing amplifier* shown at the top of the next page has the following characteristics:

1. $e_o = -[e_1(R_4/R_1) + e_2(R_4/R_2) + e_3(R_4/R_3)]$ (see Problem 7.24).
2. If $R_1 = R_2 = R_3 = R$, then $e_o = -(R_4/R)(e_1 + e_2 + e_3)$.

3. This circuit has application to digital-to-analog converters (Section 8.11.1). Also note the similarity to the inverting amplifier (Example 7.4).

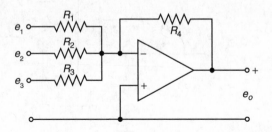

7.15   SPECIAL AMPLIFIER CIRCUITS

7.15.1   Instrumentation Amplifiers

In practice, transducer signals are often small voltage differences that must be accurately amplified in the presence of large common-mode signals. Simultaneously, the current drawn from the transducer must remain small to avoid loading the transducer and degrading its signal. Standard op-amp circuits, such as the differential amplifier (Example 7.6), may not provide adequate input impedance or CMRR when high-accuracy measurements are needed.

The *instrumentation amplifier* uses three op amps to remedy these problems (Fig. 7.22). The instrumentation amp is essentially a differential amplifier with a voltage follower placed at each input (this is easily seen if R_1 is temporarily removed). The voltage followers increase the (+) and (−) input impedances to the op-amp impedances. The addition of R_1 between the two followers has the effect of raising CMRR. Resistor matching is less critical for this circuit than for a differential op-amp circuit alone.

Instrumentation amplifiers may be built from discrete components, or they may be purchased as single integrated circuits. The typical instrumentation amp may have CMRR reaching 130 dB, input impedance of 10^9 Ω or more, and circuit gain of up to 1000.

7.15.2   The Charge Amplifier

The *charge amplifier* is used with piezoelectric transducers (Sections 6.14, 13.6, 14.7.3, and 18.5.1). These transducers are composed of a high-impedance material that generates electric charge $Q(t)$ in response to a varying load. The charge amp produces an output proportional to the charge while avoiding the potential noise difficulties of a high-impedance source. The complete circuit is shown in Fig. 7.23.

The transducer, cable, and feedback capacitances are C_t, C_c, and C_f, respectively (see Sections 5.17–5.19 for a brief review of charge and capacitance). If the large feedback resistor R_f is ignored, the output of the circuit can be expressed as

$$e_o = \frac{-Q(t)}{C_f + (C_t + C_c + C_f)/G}$$

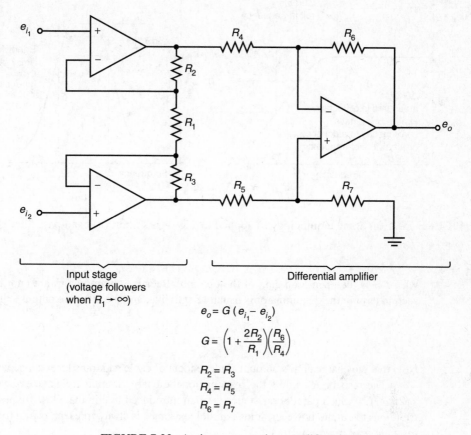

$$e_o = G\,(e_{i_1} - e_{i_2})$$

$$G = \left(1 + \frac{2R_2}{R_1}\right)\left(\frac{R_6}{R_4}\right)$$

$$R_2 = R_3$$
$$R_4 = R_5$$
$$R_6 = R_7$$

FIGURE 7.22: An instrumentation amplifier circuit.

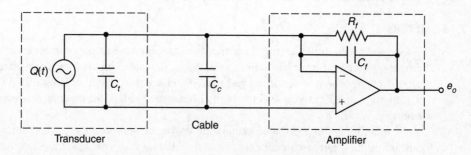

FIGURE 7.23: A charge amplifier circuit.

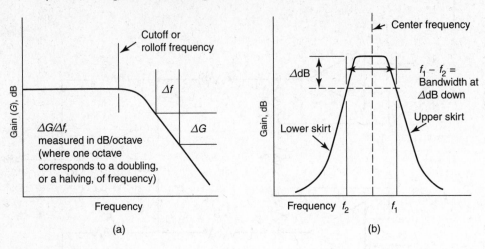

FIGURE 7.24: (a) Some terminology as applied to a low-pass filter; (b) band-pass filter characteristics.

where G is the open-loop gain of the op amp. Because op-amp gains are enormous, the second term in the denominator is usually negligible, and the effective output is just

$$e_o = -\frac{Q(t)}{C_f}$$

Note that the charge amp's output is independent of cable and transducer capacitance [10].

The resistor R_f limits the response of the charge amp at frequencies below $f = 1/2\pi R_f C_f$. Such parallel resistance is often introduced to eliminate low-frequency contributions to output; however, some parallel resistance is always present, owing to the finite resistances of real capacitors.

Although the piezoelectric effect was known in the nineteenth century, it did not become technologically important until very-high-input-impedance amplifiers were developed in the 1950s and 1960s. The charge amp itself was patented by W. P. Kistler in 1950 and gained wide use following the development of MOSFET circuits and high-grade electrical insulators such as Teflon and Kapton [11].

7.16 FILTERS

As we have seen, time-varying measurands commonly consist of a combination of many frequency components or harmonics. In addition, unwanted inputs (noise) are often picked up, thereby resulting in distortion and masking of the true signal. It is usually possible to use appropriate circuitry to selectively filter out some or all of the unwanted noise (see, for example, Section 5.7).

Filtering is the process of attenuating unwanted components of a measurand while permitting the desired components to pass. Filters are of two basic classes, *active* and *passive*. An active filter uses powered components, commonly configurations of op amps, whereas a passive filter is made up of some form of *RLC* arrangement. In addition, filters may be classified by the descriptive terms *high-pass, low-pass, band-pass,* and *notch* or *band-reject*. In each case, reference is to the signal frequency; for example, the high-pass filter permits components above a certain cutoff frequency to pass through. The notch filter attenuates a selected band of frequency components, whereas the band-pass filter permits only a range of components about its center frequency to pass. Figures 7.24(a) and (b) illustrate certain terms applied in filter design and use. Similar terms are applicable to the high-pass and notch filters, respectively.

7.17   SOME FILTER THEORY

The simplest low-pass and high-pass filters are made from a single resistor and capacitor. Electrically, these passive RC filters are first-order systems (Section 5.18). The RC low-pass and high-pass filters are shown in Figs. 7.25(a) and (b), respectively.

Consider first the RC low-pass filter. Since a capacitor tends to block low-frequency currents and pass high-frequency currents, the basic effect of the capacitor in this filter is to short-circuit the high-frequency components of the input signal. To determine the frequency characteristics, we must find the filter output, e_o, for a harmonic input voltage, e_i:

$$e_i = V_i \sin(2\pi f t)$$

If negligible current is drawn at the output, the currents through the resistor and the capacitor are equal, so that

$$i = \frac{e_i - e_o}{R} = C \frac{d}{dt} e_o$$

or

$$\frac{d}{dt} e_o + \frac{1}{RC} e_o = \frac{1}{RC} e_i = \frac{V_i}{RC} \sin(2\pi f t) \tag{7.31}$$

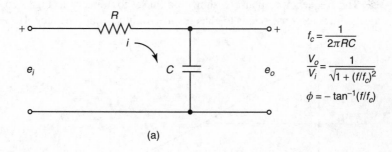

$$f_c = \frac{1}{2\pi RC}$$

$$\frac{V_o}{V_i} = \frac{1}{\sqrt{1 + (f/f_c)^2}}$$

$$\phi = -\tan^{-1}(f/f_c)$$

(a)

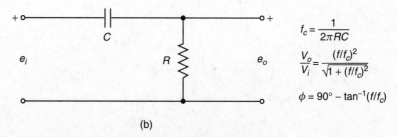

$$f_c = \frac{1}{2\pi RC}$$

$$\frac{V_o}{V_i} = \frac{(f/f_c)^2}{\sqrt{1 + (f/f_c)^2}}$$

$$\phi = 90° - \tan^{-1}(f/f_c)$$

(b)

FIGURE 7.25: First-order RC filters: (a) low pass, (b) high pass.

Solution of this equation gives (cf. Example 5.7)

$$e_o = V_o \sin(2\pi f t + \phi) = \frac{V_i}{\sqrt{1 + (2\pi RCf)^2}} \sin(2\pi f t + \phi) \tag{7.32}$$

where the phase lag, ϕ, is

$$\phi = -\tan^{-1}(2\pi RCf) \tag{7.32a}$$

Filter performance is normally characterized by defining a *cutoff frequency*; f_c:

$$f_c \equiv \frac{1}{2\pi RC} \tag{7.33}$$

In terms of f_c, the frequency response (or gain), from Eq. (7.32), is

$$\frac{V_o}{V_i} = \frac{1}{\sqrt{1 + (f/f_c)^2}} \tag{7.33a}$$

and the phase response, from Eq. (7.32a), is

$$\phi = -\tan^{-1}\left(\frac{f}{f_c}\right) \tag{7.33b}$$

At the cutoff frequency,

$$\frac{V_o}{V_i} = \frac{1}{\sqrt{2}} \quad \text{and} \quad \frac{P_o}{P_i} = \frac{1}{2} \tag{7.33c}$$

either of which indicates a -3 dB change in the signal strength (see Section 7.12).

The frequency response is shown on linear coordinates in Fig. 7.26. Graphed this way, the filter response seems to change only slightly with frequency. However, the graph

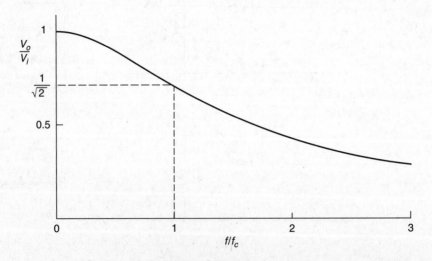

FIGURE 7.26: Frequency response of the RC low-pass filter (linear coordinates).

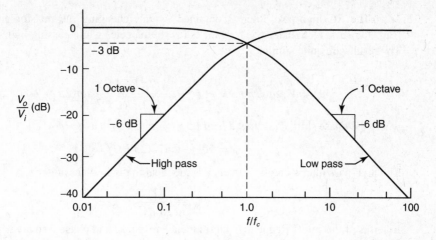

FIGURE 7.27: Frequency response of *RC* low-pass and high-pass filters (Bode plot).

shows only a factor-of-three increase in frequency, while, in practice, filters are used to separate frequencies that may differ by orders of magnitude. A logarithmic graph, such as a Bode plot (Section 7.12), is needed to illustrate such variation.

A Bode plot of the low-pass filter's response is given in Fig. 7.27, illustrating the -3 dB reduction in signal at the cutoff frequency. The frequency range plotted spans four orders of magnitude, and the amplitude attenuation runs from 0 to -40 dB. For frequencies well below f_c, the filter's response is flat and shows no signal reduction. The transition from the passband to rejection band occurs gradually with increasing frequency. In the rejection band itself, at frequencies well above f_c, the amplitude rolloff is -6 dB/octave (an octave being a factor-of-two change in frequency) or -20 dB/decade (a decade being a factor of ten).

In addition to reducing amplitude, this filter also produces an increasing phase shift as signal frequency rises (Fig. 7.28). At the -3 dB point, the output lags the input by $45°$.

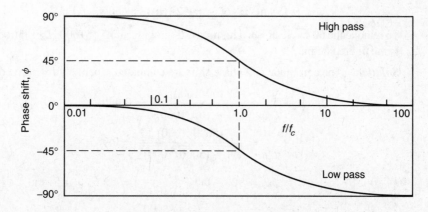

FIGURE 7.28: Phase response of *RC* high-pass and low-pass filters.

The RC high-pass filter is obtained by interchanging the resistor and capacitor [Fig. 7.25(b)]. Now the capacitor blocks low frequencies while passing high frequencies. The results are quite similar:

$$e_o = V_o \sin(2\pi ft + \phi) = \frac{V_i (2\pi RCf)}{\sqrt{1 + (2\pi RCf)^2}} \sin(2\pi ft + \phi)$$

where the phase shift, ϕ, is now a *lead* ($\phi > 0$) rather than a lag ($\phi < 0$)

$$\phi = 90° - \tan^{-1}(2\pi RCf)$$

The high-pass filter's cutoff frequency is identical to the low-pass filter's:

$$f_c \equiv \frac{1}{2\pi RC} \tag{7.34}$$

In terms of the cutoff frequency, the frequency response and phase lead are

$$\frac{V_o}{V_i} = \frac{(f/f_c)}{\sqrt{1 + (f/f_c)^2}}, \tag{7.34a}$$

$$\phi = 90° - \tan^{-1}\left(\frac{f}{f_c}\right) \tag{7.34b}$$

The -3 dB point is again f_c, and the rolloff in the rejection band is again -6 dB/octave or -20 dB/decade. The high-pass frequency and phase response are shown in Figs. 7.27 and 7.28. One common use of this filter is to remove dc ($f = 0$) offsets.

First-order RC filters have a fairly slow rolloff above the cutoff frequency (not many decibels per octave), but their simplicity still gains them wide use in situations where the desired and undesired frequencies are widely separated. Similar high-pass and low-pass filters can be made using a single resistor–inductor pair. However, first-order RL filters are seldom used.

EXAMPLE 7.10

A transducer responding to a 5000-Hz signal also picks up 60 Hz noise. The resulting output is

$$\{5 \sin(2\pi \cdot 60 \cdot t) + 25 \cos(2\pi \cdot 5000 \cdot t)\} \text{ mV}$$

To remove the 60-cycle noise, a high-pass filter with cutoff of 1000 Hz is introduced. What is the filtered output?

Solution   The amplitude and phase shift are computed separately for each component:

$$\left(\frac{V_o}{V_i}\right)_{60} = \frac{(60/1000)}{\sqrt{1 + (60/1000)^2}} = 0.060,$$

$$\left(\frac{V_o}{V_i}\right)_{5000} = \frac{(5000/1000)}{\sqrt{1 + (5000/1000)^2}} = 0.98,$$

$$\phi_{60} = 90° - \tan^{-1}\left(\frac{60}{1000}\right) = 86.6° = 1.51 \text{ rad},$$

$$\phi_{5000} = 90° - \tan^{-1}\left(\frac{5000}{1000}\right) = 11.3° = 0.197 \text{ rad}$$

Then

$$e_0 = \{0.3\sin(2\pi \cdot 60 \cdot t + 1.51) + 24.5\cos(2\pi \cdot 5000 \cdot t + 0.197)\}\ \text{mV}$$

The noise amplitude is reduced from 20% of the signal amplitude to only 1.2%. Note that the signal itself undergoes a slight amplitude reduction as well as a small phase shift. Such changes in the signal are undesirable, and they often motivate the use of more complex filters.

Three desirable elements of filter performance are as follows:

1. Nearly flat response over the pass and rejection bands;
2. High values of rolloff for low-and high-pass filters, as measured in decibels per octave;
3. Steep skirts for band-pass and band-rejection filters.

Significant improvements in performance are obtained by using combinations of several capacitors, inductors, or resistors to produce second-order (or higher-order) electrical response. Such filters can have steeper rolloff and sharper transition from pass to rejection bands. In addition, such compound *RLC* arrangements can produce band-pass and notch filters. For example, Figs. 7.29(a) and (b) show an *RC* band-pass filter and its response:

$$\frac{V_o}{V_i} = \frac{1}{\sqrt{[1 + R_1/R_2 + C_2/C_1]^2 + [2\pi R_1 C_2 f - (1/2\pi R_2 C_1 f)]^2}} \tag{7.35}$$

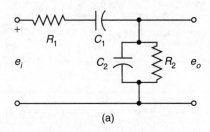

(a)

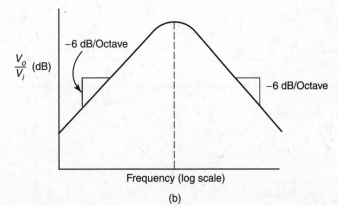

(b)

FIGURE 7.29: (a) A circuit for a simple band-pass filter; (b) performance characteristics of band-pass filter shown in (a).

Inductors and capacitors used together allow resonant behavior, which can produce steeper filter skirts than are possible with first-order *RC* circuits. In fact, the resonant circuit of Section 7.11 is sometimes used to build very narrow band-pass filters known as *tuned filters*. Some additional *LC* designs are shown in Fig. 7.30.

Two practical issues influence the design and use of passive filters. First, the filters considered here are all designed as if negligible current is drawn from the output terminals. If several filters are placed in series, to steepen rolloff, then the current drawn by one filter

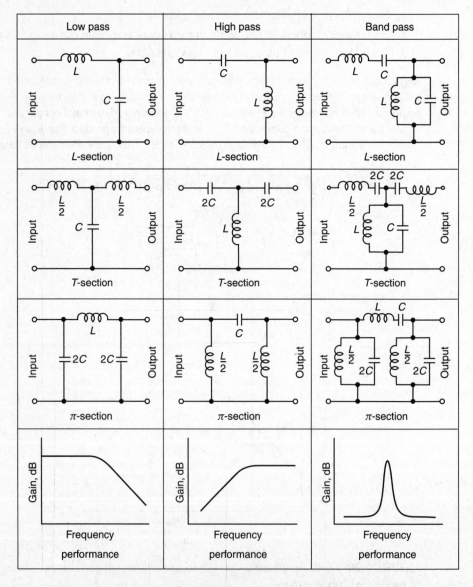

FIGURE 7.30: Examples of *LC* filter arrangements and their output characteristics.

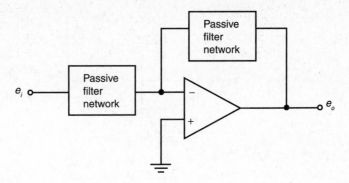

FIGURE 7.31: Basic active filter circuit.

can alter the performance of the filter that precedes it. To avoid this output loading, a voltage follower (Example 7.3) should be introduced as a buffer between each successive filter.

The inductors themselves are the second problem. At the frequencies encountered in mechanical measurements, which rarely exceed 100 kHz, the required inductors may be quite large and bulky. In addition, the optimal inductor values are not always easily obtained, and the inductors may have substantial internal resistances as well. As often happens in engineering, inductors can be much less satisfactory in practice than they seem on paper. The usual way of avoiding these problems is to employ an active filter, as described next.

7.18   ACTIVE FILTERS

Op amps can be used to construct filter circuits without inductors and without the problems of output loading. These active filters can also have very steep rolloff, arbitrarily flat passbands, and even adjustable cutoff frequencies. Active filters are a rich subject, and entire textbooks have been devoted to their design.

The basic active filter is shown in Fig. 7.31. Passive filter networks are linked to an op amp, which provides power and improves impedance characteristics. The passive network is built from resistors and capacitors only: Inductive characteristics are simply simulated by the circuit. Since the output impedance is generally low, these filters can deliver an output current without reduced performance. Some typical active filters are shown in Fig. 7.32.

Active filters are available with rolloffs of 80 dB/octave and more than 60 dB attenuation in the rejection band. High-order active filters are even sold as integrated circuits contained in a single DIP package. For further study, see the Suggested Readings at the end of this chapter.

7.19   DIFFERENTIATORS AND INTEGRATORS

A final op-amp application is in circuits that respond to the rate of change or the time history of an input signal, called *differentiators* and *integrators*, respectively [Figs. 7.33(a) and (b)].

In the differentiator, the currents through the resistor and capacitor are equal, and $e_- = e_+ = 0$. Thus

$$C \frac{d}{dt} e_i = -\frac{e_o}{R}$$

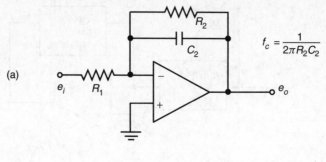

$$f_c = \frac{1}{2\pi R_2 C_2}$$

(a)

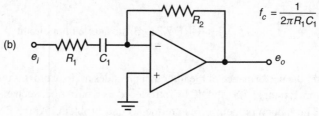

$$f_c = \frac{1}{2\pi R_1 C_1}$$

(b)

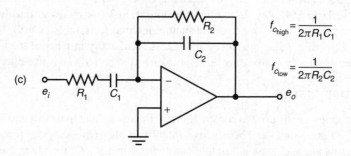

$$f_{c_{high}} = \frac{1}{2\pi R_1 C_1}$$

$$f_{c_{low}} = \frac{1}{2\pi R_2 C_2}$$

(c)

FIGURE 7.32: First-order active filters: (a) low pass; (b) high pass, (c) band pass.

or

$$e_o = -RC \frac{d}{dt} e_i \qquad (7.36)$$

In the integrator, the capacitor charges in proportion to the time summation of e_i. Again, the resistor and capacitor currents are equal:

$$\frac{e_i}{R} = C \frac{d}{dt}(-e_o)$$

or

$$e_o = -\frac{1}{RC} \int e_i \, dt + \text{constant} \qquad (7.36a)$$

To prevent drift in the capacitor's charge over long time intervals, a large resistor, R_f, may be placed in parallel with it. In that case, the integrator circuit is restricted to signal frequencies high enough that $f \gg 2\pi R_f C$.

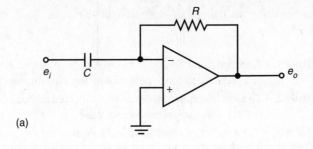

(a)

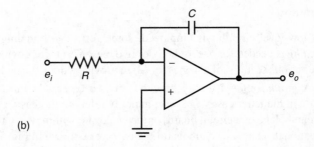

(b)

FIGURE 7.33: (a) Op-amp differentiator; (b) op-amp integrator.

7.20 SHIELDING AND GROUNDING

7.20.1 Shielding

Shielding applied to electrical or electronic circuitry is used for either or both of two related but different purposes:

1. To isolate or retain electrical energy within an apparatus
2. To isolate or protect the apparatus from outside sources of energy

An example of the former is the shielding required by the Federal Communications Commission to minimize radio frequency radiation from computers. In the second case, shielding may be required to protect low-level circuitry from the entry of unwanted outside signals. A very common source of outside energy is the ubiquitous 60-Hz power line.

Shielding is of two basic types: (a) electrostatic and (b) electromagnetic. In each case the shielding normally consists of some form of metallic enclosure; for example, metallic braid may be used to shield signal-carrying wiring, or circuitry may be partially or entirely enclosed in metal boxes.

Only nonmagnetic metals may be used for electromagnetic shielding, whereas almost any conducting metal, such as steel, aluminum, or copper, may be used for electrostatic shielding. Circuits within a device often must be shielded from each other; however, connections must still be made between the subcircuits through use of special amplifiers or transformers. For example, power transformers are often provided with copper shielding between primary and secondary windings. The copper provides electrostatic shielding without hindering the transfer of electromagnetic power. Some rules for shielding are as follows [12, 13]:

Rule 1. An electrostatic shield enclosure, to be effective, should be connected to the zero-signal reference potential of any circuitry contained within the shield.

Rule 2. The shield conductor should be connected to the zero-signal reference potential at the signal-to-earth connection.

Rule 3. The number of separate shields required in a system is equal to the number of independent signals being processed, plus one for each power entrance.

7.20.2 Grounding

When low-level circuitry is employed, some form of grounding is inevitably required. *Grounding* is needed for one, or both, of two reasons: (1) to provide an electrical reference for the various sections of a device, or (2) to provide a drainage path for unwanted currents.

A *ground reference* may be either of two types, (1) earth ground, or (2) chassis ground. In the latter case, *chassis* commonly refers to the basic mounting structure (e.g., the ground plane of a circuit board) or the enclosure within which the circuitry is mounted. Conventional schematic symbols for the two are as shown in Fig. 7.34.

In a text such as this, only superficial coverage of this complicated topic is possible. However, certain "rules" and observations may be listed as follows:

1. An entire system can be "grounded" and need not involve earth at all. For example, circuitry in aircraft and spacecraft are referenced to some common datum.

2. The word *circuit* need not imply wires or components. Each of the various elements in a device may, unless effectively shielded, possess capacitive paths to one or more of the others.

3. Shielding can be at any potential and still provide shielding.

4. The assumption that two nearby points are at the same potential is often invalid—not only earth points, but also points in any ground plane.

5. Potential characteristics of an element are not the same at "radio" or high frequencies as they are at "power" or low frequencies. For example, a capacitance exists between a bonded strain gage element and the structure on which it is mounted. At dc or low frequency, such capacitance may be unimportant, but at radio frequencies the capacitance may provide a ready electrical path.

6. A ground bus is protection against effects of equipment faulting, but it is not the source of zero potential for the solution of instrumentation processes.

7. All metal enclosures and housings should be earthed and bonded together, but no current should be permitted to flow in these connections.

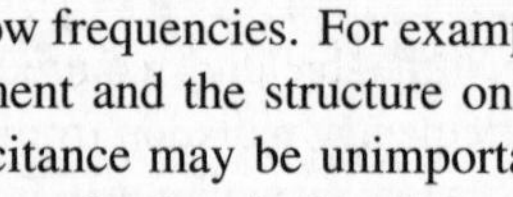

Earth   Chassis

FIGURE 7.34: Conventional symbols for ground references.

8. Good practice suggests that it is wise to insulate an equipment rack from the obvious ties, such as building earths and conduit connections, so that the rack can be ohmically connected to a potential most favorable to the instrumentation processes.

9. Rules that are applicable at one frequency range may be inadequate at another.

10. Safety practices demanded by various civil codes can seem to be in direct conflict with good instrumentation practice.

11. Electrostatic shields are simply metallic enclosures that surround signal processes. To be effective, these shields should be tied to a zero signal potential where the signal makes its external, or ground connection.

To reiterate, shielding and grounding are very complex subjects, and often some degree of trial and error, coupled with experience, is required to find a solution.

7.21   COMPONENT COUPLING METHODS

When electrical circuit elements are connected, special attention must often be given to the coupling methods used. In certain cases, transducer–amplifier, amplifier–recorder, or other component combinations are inherently incompatible, making direct coupling impossible or, at best, causing nonoptimal operation. Coupling problems include obtaining proper impedance matching and maintaining circuit requirements such as damping. These problems are usually caused by the desire for maximum energy transfer and optimum fidelity of response.

The importance of impedance matching, however, varies considerably from application to application. For example, the input impedances of most cathode-ray oscilloscopes and electronic voltmeters are relatively high, but satisfactory operation may be obtained from directly connected low-impedance transducers. In this case, voltage is the measured quantity and power transfer is incidental. *In most cases, driving a high-impedance circuit component with a low-impedance source presents fewer problems than does the reverse.*

As a simple example of transfer, consider Fig. 7.35. Shown is a *source* of energy E_S and a *sink* or *load* having impedance Z_L. Z_S is the source impedance. To simplify our example further, let the impedances be simple electrical resistances, R_S and R_L, respectively. Then the voltage across R_L will be

$$E_L = E_S \left(\frac{R_L}{R_L + R_S} \right)$$

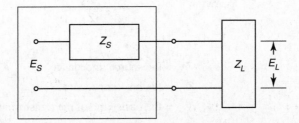

FIGURE 7.35: Simple circuit for demonstrating transfer concepts.

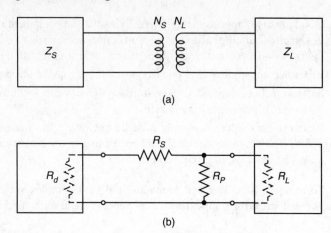

FIGURE 7.36: Impedance matching (a) by means of a coupling transformer and (b) by means of a resistance pad.

and the power delivered to R_L is

$$P = \frac{E_L^2}{R_L} = \frac{E_S^2}{R_L} \left(\frac{R_L}{R_L + R_S} \right)^2 \qquad (7.37)$$

To determine the values of R_L for maximum power transfer, set dP/dR_L to zero and solve. We find that maximum power is transferred if $R_L = R_S$. Although this is not really a proof, in general terms, maximum power is transferred when $Z_L = Z_S$.

In addition to, or instead of, optimizing *power transfer*, proper coupling may be important in providing adequate dynamic response. Three special methods of coupling, depending on the circuit elements, are common; they utilize (1) matching transformers (discussed shortly), (2) impedance transforming (see Example 7.3), and (3) coupling networks (discussed shortly).

An example of transformer coupling is the use of electronic power amplifiers to drive vibration exciters such as those discussed in Section 17.14.1. The problem is similar to that of connecting a speech amplifier to a loudspeaker. In both cases the output impedance of the driving power source in the amplifier is often higher than the load impedance to which it must be connected. Transformer coupling is generally used as shown in Fig. 7.36(a). Matching requirements may be expressed by the relation

$$\frac{N_S}{N_L} = \frac{Z_S}{Z_L} \qquad (7.38)$$

where

$$Z_S = \text{the source impedance,}$$
$$Z_L = \text{the load impedance,}$$
$$N_S/N_L = \text{the turns ratio of the transformer}$$

General-purpose devices such as simple voltage amplifiers and oscillators often incorporate a final amplifier stage, called a buffer, to supply a low-impedance output. By reducing the impedance source, one can minimize losses in the connecting lines and the possibility of extraneous signal pickup.

Proper coupling may also be accomplished through use of matching resistance pads. Figure 7.36(b) illustrates one simple form.

If we assume that the driver output and load impedances are resistive, then the matching requirements may be put in simple form as follows: The driving device, which may be a voltage amplifier, *looks* into the resistance network and *sees* the resistance R_s in series with the paralleled combination of R_L and R_p. Hence, for proper matching,

$$R_d = R_s + \frac{R_p R_L}{R_p + R_L} \tag{7.39}$$

where

R_d = the output impedance of the driver (Ω),

R_L = the load resistance (Ω),

R_p = the paralleling resistance (Ω),

R_s = the series resistance (Ω)

The driven device *sees* two parallel resistances, made up of R_p and the series-connected resistances R_s and R_d. Hence, for matching,

$$R_L = \frac{R_p(R_s + R_d)}{R_p + (R_s + R_d)} \tag{7.39a}$$

Solving for R_s and R_p, we have

$$R_s = [R_d(R_d - R_L)]^{1/2} \tag{7.40}$$

and

$$R_p = \left[R_L \left(\frac{R_d R_L}{R_d - R_L} \right) \right]^{1/2} \tag{7.40a}$$

Now if R_d and R_L are known, values of R_s and R_p may be determined to satisfy the matching requirements by use of Eqs. (7.40) and (7.40a). In using resistive elements, a loss in signal energy is unavoidable. Such losses are often referred to as *insertion losses*. In general, however, by providing proper match, the network will provide optimum gain.

7.22 SUMMARY

Electrical signal conditioning can serve many purposes: to convert sensor outputs to more easily used forms (e.g., resistance change into voltage); to separate small signals from large offsets or noise; to increase signal voltage; to remove unwanted frequency components from the signal; or to enable the signal to drive output devices.

1. For resistance-type transducers, several simple signal-conditioning circuits are useful if the resistance change is relatively large. These include current-sensitive circuits, ballast circuits, and voltage-dividing circuits (Sections 7.5, 7.6, 7.7).

2. For small resistance changes, bridge circuits provide a more sensitive method of detection, in which offsets are eliminated. Output can vary linearly with resistance change, but when the resistance changes are too large, bridge circuits become non-linear (Sections 7.8, 7.9).

3. AC excitation of signal-conditioning circuits is necessary with inductive and capacitive sensors. AC-excited circuits include reactance bridges (Section 7.10) and resonant circuits (Section 7.11).

4. Decibels (dB) provide a convenient method of quantifying gain or attenuation. The decibel is a logarithmic measure of signal power (Section 7.12).

5. Electronic amplifiers are ubiquitous elements of signal-conditioning circuits. Amplifiers serve to increase voltage, to increase power or current, or to change impedance (Section 7.13).

6. Operational amplifiers are among the most common elements in instrumentation circuits. Many different op-amp configurations are possible, with the choice depending on the characteristics of the specific sensor involved and the type of output response desired (Sections 7.14, 7.15, 7.18, 7.19).

7. Some signal-conditioning techniques apply primarily to time-varying measurands. These include carrier modulation (Section 7.3), filtering (Sections 7.16–7.18), and differentiating and integrating (7.19).

8. Filters allow unwanted frequency components to be removed from a signal. Filters are characterized by a cutoff frequency (or -3 dB point) separating the pass and rejection bands and by the steepness of rolloff from the passband to the rejection band. *RC* filters are the simplest type of passive filter. Common applications of filters include noise removal, dc-offset removal, and carrier demodulation (Sections 7.16–7.18).

9. Shielding and grounding are essential considerations in building and using measurement circuits. Shielding for noise prevention is especially important when precise measurements are to be made (Section 7.20).

10. Component coupling is often designed to maximize power transfer or dynamic response. If voltage detection is of greater importance than power transfer, however, it is often sufficient that a device's input impedance be large compared to the output impedance of the device driving it (Section 7.21).

SUGGESTED READINGS

Carr, J. J. *Designer's Handbook of Instrumentation and Control Circuits*. San Diego: Academic Press, 1991.

Clayton, G. B., and S, Winder. *Operational Amplifiers*. 4th ed. Boston: Newnes, 2000.

Coughlin, R. F., and F. F. Driscoll. *Operational Amplifiers and Linear Integrated Circuits*. 5th ed. Upper Saddle River, N.J.: Prentice Hall, 1998.

Deliyannis, T., Y. Sun, and J .K Fidler. *Continuous-Time Active Filter Design*. Boca Raton, Fla.: CRC Press, 1999.

Floyd, T. L. *Electronic Devices: Electron Flow Version*. 5th ed. Upper Saddle River, N.J.: Prentice Hall, 2005.

Floyd, T. L. *Principles of Electric Circuits*. 7th ed. Upper Saddle River, N.J.: Prentice Hall, 2003.

Hague, B., and T. R Foord. *Alternating Current Bridge Methods*. 6th ed. London: Pitman, 1971.

Horowitz, P., and W. Hill. *The Art of Electronics*. 2nd ed. New York: Cambridge University Press, 1989.

Kibble, B. P., and G. H Rayner. *Coaxial AC Bridges*. Bristol, UK: Adam Hilger, 1984.

Morrison, R. *Grounding and Shielding Techniques*. 4th ed. New York: Wiley-Interscience, 1998.

Newby, B. W. G. *Electronic Signal Conditioning*. Oxford: Butterworth-Heinemann, 1994.

Ott, H. W. *Noise Reduction Techniques in Electronic Systems*. 2nd ed. New York: John Wiley, 1988.

Pallàs-Areny, R., and J. G. Webster. *Sensors and Signal Conditioning*. 2nd ed. New York: John Wiley, 2001.

Schaumann, R., and M. E Van Valkenburg. *Design of Analog Filters*. New York: Oxford University Press, 2001.

Stanley, W. D. *Operational Amplifiers with Linear Integrated Circuits*. 4th ed. Upper Saddle River, N.J.: Prentice Hall, 2002.

Zverev, A. I. *Handbook of Filter Synthesis*. New York: John Wiley, 1967.

PROBLEMS

7.1. A force cell uses a resistance element as the sensing element. It is connected in a simple current-sensitive circuit in which the series resistance R_m is 100 Ω, which is one-half the nominal resistance of the force cell. Determine the current for force inputs of (a) 25%, (b) 50%, and (c) 75% of full range if the input voltage is 10 V.

7.2. If the force cell of Problem 7.1 is placed into a ballast circuit ($R_b = R_m$), determine the output voltage for the conditions of Problem 7.1.

7.3. For the ballast circuit of Problem 7.1, determine the sensitivity, η, for the three percentages of full range.

7.4. Equations (7.5) and (7.6) are derived on the basis of a high-impedance indicator. Analyze the circuit assuming that the indicator resistance R_m is comparable in magnitude to R_t.

7.5. For $E_s = 10$ V and $R_s = 75$ Ω, use Eq. (7.37) to plot P versus R_L over the range $0 < R_L < 200$ Ω.

7.6. The circuit shown in Figure 7.37 is used to determine the value of the unknown resistance R_2. If the voltmeter resistance, R_L, is 10 MΩ and the voltmeter reads $e_o = 4.65$ V, what is the value of R_2?

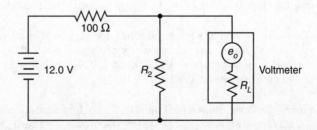

FIGURE 7.37: Circuit for Problem 7.6.

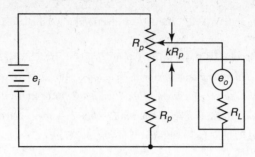

FIGURE 7.38: Circuit for Problem 7.7.

7.7. The voltage-dividing potentiometer shown in Fig. 7.5 is modified as shown in Fig. 7.38. Determine the relationship for e_o/e_i as a function of k. Compare the results with Eq. (7.9). What advantages or disadvantages does this circuit have over the general voltage-dividing potentiometer?

7.8. Write a spreadsheet template to solve Eq. (7.15a), permitting each term to be varied by a delta amount. [*Suggestion*: Rewrite the equation, multiplying each term by $(1 + k)$, where k is the delta plus/minus term—for example, $R_i(1 + k_i)$.]

7.9. A simple Wheatstone bridge as shown in Fig. 7.10 is used to determine accurately the value of an unknown resistance R_1 located in leg 1. If upon initial null balance R_3 is 127.5 Ω and if, when R_2 and R_4 are interchanged, null balance is achieved when R_3 is 157.9 Ω, what is the value of the unknown resistance R_1?

7.10. Consider the voltage-sensitive bridge shown in Fig. 7.10. If a thermistor whose resistance is governed by Eq. (16.3) is placed in leg 1 of the bridge while $R_2 = R_3 = R_4 = R_0$, determine the bridge output when $T = 400°C$ if $R_0 = 1000\ \Omega$ at $T_0 = 27°C$ and $\beta = 3500$. Plot the bridge output from $T = 27°C$ to $T = 500°C$ and determine the maximum deviation from linearity in this temperature range.

7.11. Referring to Fig. 7.10, show that if initially $R_1 = R_2 = R_3 = R_4 = R$ and if $\Delta R_1 = -\Delta R_2$, the bridge output will be linear. (*Note*: This bridge configuration is very commonly used when strain gages are applied to a beam in bending situations; see Table 12.4.)

7.12. Referring to Fig. 7.10, initially let $R_1 = R_2 = R_3 = R_4 = R$. In addition, assume that $\Delta R_4/R = -\Delta R_1/R$. Demonstrate the nonlinearity of the bridge output by plotting e_o/e_i over the range $0 < \Delta R_i/R < 0.1$. (*Suggestion*: Use a spreadsheet program.)

7.13. A resistive element of a force cell forms one leg of a Wheatstone bridge. If the no-load resistance is 500 Ω and the sensitivity of the cell is 0.5 Ω/N, what will be the bridge outputs for applied loads of 100, 200, and 350 N if the bridge excitation is 10 V and each arm of the bridge is initially 500 Ω?

7.14. Figure 7.39 shows a differential shunt bridge configuration. One or more of the resistances, R_i, may be resistance-type transducers (thermistor, resistance thermometer, strain gage, etc.), with the remaining resistances fixed. Resistance R_6 is a conventional voltage-dividing potentiometer, usually of the multiturn variety. It may be used either for initial

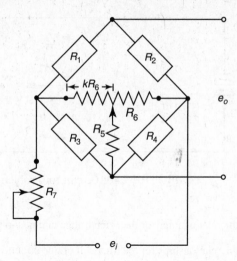

FIGURE 7.39: Circuit for Problem 7.14.

nulling of the bridge output or as a readout means. The variable k is a proportional term varying from 0 to 1 (or 0% to 100%); see Section 7.9. R_5 is sometimes called a *scaling resistor*. Its value largely determines the range of effectiveness of R_6. R_7 is used to adjust bridge sensitivity. Devise a spreadsheet template to be used for designing a bridge of this type.

7.15. Using the spreadsheet created in answer to the preceding problem, determine the null-balance range of ΔR_1 that the bridge can accommodate if

$$R_1 = 1000 \ \Omega \ (\text{nominal}),$$
$$R_2 = R_3 = R_4 = 1000 \ \Omega \ (\text{fixed}),$$
$$R_5 = 10,000 \ \Omega,$$
$$R_6 = 12,000 \ \Omega,$$
$$R_7 = 0 \ \Omega$$

7.16. Using the spreadsheet from Problem 7.14 and the nominal resistance values listed in Problem 7.15, investigate changes in measurement range of the bridge as affected by (a) changes in R_5 and (b) changes in R_6. Investigate the linearity of the circuit when used in the null-balance mode, using k as the calibrated readout.

7.17. Referring to Problem 7.14, assume that the tolerances for resistances R_2, R_3, and R_4 are $\pm5\%$. What will now be the effective null-balance range, ΔR_1, that can be accommodated?

7.18. Figure 7.40 shows a shunt balance arrangement for nulling a Wheatstone bridge. Suppose that $R_1 = R_3 = 120 \ \Omega$, $R_{\text{trim}} = 127 \ \Omega$, and $R_{\text{pot}} = 10 \ \text{k}\Omega$. What is the maximum value of R_2 for which the bridge can be brought into balance by adjusting R_{pot}? What would be the maximum value if $R_1 = 119 \ \Omega$ and $R_3 = 121 \ \Omega$?

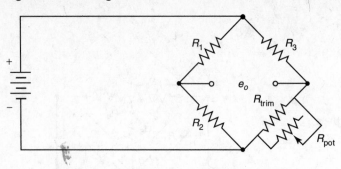

FIGURE 7.40: Circuit for Problem 7.18.

7.19. Derive the relationships for the Wien bridge circuit shown in Fig. 7.16.

7.20. Derive the relationships for the Maxwell bridge circuit shown in Fig. 7.16.

7.21. Show that an increase of 1 dB corresponds to a power increase of about 26%. Also show that an increase of n dB corresponds to a power increase to approximately $(1.26)^n$.

7.22. Equation (7.28c) may be written as

$$K = \frac{dB_{(corrected)}}{dB_{(indicated)}} = 1 + 10\frac{\log(R_r/R_s)}{dB_{(indicated)}}$$

where

$$K = \text{correction factor,}$$
$$dB_{(corrected)} = \text{corrected decibels,}$$
$$dB_{(indicated)} = \text{indicated decibels,}$$
$$R_r = \text{reference impedance,}$$
$$R_s = \text{source impedance.}$$

Plot K versus R_r/R_s for $dB_{(indicated)} = 50, 100, 150$, and 200. Make separate families of plots (a) for $R_r/R_s \leq 1$ and (b) for $R_r/R_s \geq 1$.

7.23. Show that the input–output relationship for the differential amplifier described in Example 7.6 is $e_o = -(R_3/R_1)(e_{i_1} - e_{i_2})$.

7.24. Show that the input–output relationship for the summing amplifier described in Example 7.9 is $e_o = -[e_1(R_4/R_1) + e_2(R_4/R_2) + e_3(R_4/R_3)]$.

7.25. The circuit shown in Fig. 7.41 is a voltage-to-current converter. Determine the value of R_3 for which i_f in milliamps equals e_i in volts and find e_o.

7.26. The circuit shown in Fig. 7.42 accepts two input voltages, e_1 and e_2. Determine its output voltage, e_o.

7.27. The circuit shown in Fig. 7.43 uses a potentiometer to achieve variable gain. By adjusting the potentiometer, the gain can be raised or lowered. The total resistance of the pot is R_{pot} and the resistance of the pot between its movable wiper and ground is kR_{pot} for $0 \leq k \leq 1$. Find e_+ in terms of e_i and k, and then determine e_o and the gain.

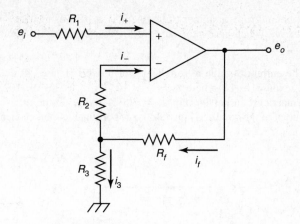

FIGURE 7.41: Circuit for Problem 7.25.

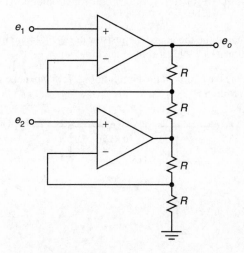

FIGURE 7.42: Circuit for Problem 7.26.

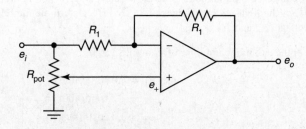

FIGURE 7.43: Circuit for Problem 7.27.

7.28. A noninverting amplifier (Example 7.5) is built using 1-kΩ precision resistors R_1, R_2, and R_3, which have a tolerance of $\pm 1\%$. Determine the circuit's gain and the tolerance in the gain.

7.29. The current through a semiconductor diode is related to the voltage across it by $i = I_o(\exp(\lambda e) - 1)$, where $I_o = 1 \times 10^{-9}$A, $\lambda = 1.17 \times 10^4/T$, and T is the absolute temperature in kelvin. For $T = 300$ K and e_i on the order of one volt, show that the output of the circuit in Fig. 7.44 is proportional to the exponential of the input voltage.

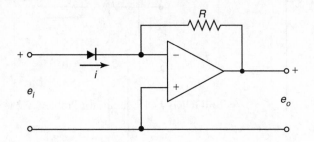

FIGURE 7.44: Circuit for Problems 7.29 and 7.30.

7.30. Show that, if the diode and the resistor in Fig. 7.44 are interchanged, the circuit output is proportional to the logarithm of the input voltage.

7.31. Determine the output of the circuit in Fig. 7.45 when the switch is in position A and when it is in position B.

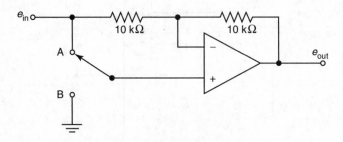

FIGURE 7.45: Circuit for Problem 7.31.

7.32. In the circuit of Fig. 7.46, a photodiode is reverse biased with a voltage e_b. In this condition, the photodiode generates a current, i, which is proportional to the intensity, H, of the light which irradiates it: $i = kH$. Determine the relation between the output voltage, e_o, and H.

7.33. For the summing amplifier of Example 7.9, if

$$e_1 = 8.0 \sin(400t) \text{ V},$$
$$e_2 = -2.0 \sin(400t) \text{ V},$$
$$e_3 = 3.0 \cos(400t) \text{ V}$$

and $R_1 = R_2 = R_4 = 5$ kΩ, and $R_3 = 2.5$ kΩ, what is the rms output voltage?

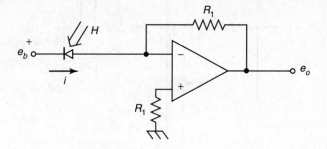

FIGURE 7.46: Circuit for Problem 7.32.

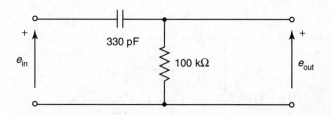

FIGURE 7.47: Circuit for Problem 7.34.

7.34. Consider the filter circuit shown in Fig. 7.47.

   (**a**) What type of filter is this? Calculate its cutoff frequency in Hz.
   (**b**) The following input signal is applied to the circuit:

$$e_{in} = \{5\sin(2\pi\,200t) + 2.5\cos(2\pi\,1000t) + 1.5\sin(2\pi\,10000t)\}\ \text{mV}$$

Determine e_{out}.

7.35. Consider the following data from an RC filter. From experimental testing, these data were obtained for a 1.0 V sine-wave input.

Frequency (Hz)	e_{out} (volts)
10.0	1.00
20.0	1.00
50.0	0.97
100.0	0.92
200.0	0.71
500.0	0.37
1,000.0	0.21
2,000.0	0.10
5,000.0	0.04
10,000.0	0.02

If the capacitor $C = 0.033\ \mu\text{F}$, determine the value of the resistor R.

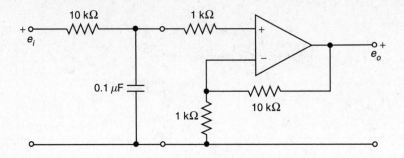

FIGURE 7.48: Circuit for Problem 7.37.

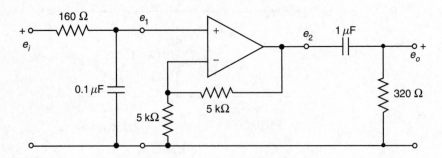

FIGURE 7.49: Circuit for Problem 7.38.

7.36. A set of RC low-pass filters are built from 100-kΩ resistors and 0.01-μF capacitors. The resistors' tolerance is $\pm 10\%$ and the capacitors' tolerance is $\pm 20\%$. Determine the cutoff frequency of these filters and the tolerance of the cutoff frequency.

7.37. Consider the amplifier circuit shown in Fig. 7.48. Determine the amplitude e_o if $e_i = 5\cos(40\pi t)$ if e_i is in millivolts and t is in seconds. What is the amplitude e_o if $e_i = 3.0\sin(4000\pi t)$?

7.38. Consider the circuit shown in Fig. 7.49.

  (**a**) Qualitatively speaking, what does this circuit do to an ac signal?
  (**b**) Make a Bode plot of the amplitude ratio $|e_1/e_i|$.
  (**c**) Add a Bode plot of the amplitude ratio $|e_2/e_i|$ to the graph of part (b).
  (**d**) Add a Bode plot of the amplitude ratio $|e_o/e_i|$ to the graph of parts (b) and (c).

7.39. Referring to Fig. 7.35 and Eq. (7.37), show that the maximum power is transferred when $R_L = R_s$.

REFERENCES

[1] Geldmacher, R. C. Ballast circuit design. *SESA Proc.* 12(1): 27, 1954.

[2] Meier, J. H. Discussion of Ref. [1], in same source, p. 33.

[3] Wheatstone, C. An account of several new instruments and processes for determining the constants of a voltaic circuit. *Phil. Trans. Roy. Soc. (London)* 133:303, 1843.

[4] Wheatstone's bridge. In *Encyclopedia Britannica*, vol. 23. Chicago: William Benton, Publisher, 1957, p. 566.

[5] Laws, F. A. *Electrical Measurements*. 2nd ed. New York: McGraw-Hill, 1938, p. 217.

[6] Bowes, C. A. Variable resistance sensors work better with constant current excitation. *Instrument Technol.*, March 1967.

[7] Sion, N. Bridge networks in transducers. *Instrument Control Systems*, August 1968.

[8] Hague, B., and T. R. Foord. *Alternating Current Bridge Methods*. 6th ed. London: Pitman, 1971.

[9] Kibble, B. P., and G. H Rayner. *Coaxial AC Bridges*. Bristol, UK: Adam Hilger, 1984.

[10] Gautschi, G. H. *Piezoelectric Sensorics*. Berlin: Springer-Verlag, 2002, Chap. 11.

[11] Kistler Instrument Corporation. *Kistler Piezo-Instrumentation General Catalog*. 1st ed. Amherst, N.Y.: Kistler Instrument Corp., 1989.

[12] Morrison, R. *Grounding and Shielding Techniques*. 4th ed. New York: Wiley-Interscience, 1998.

[13] Ott, H. W. *Noise Reduction Techniques in Electronic Systems*. 2nd ed. New York: John Wiley, 1988.

CHAPTER 8

Digital Techniques in Mechanical Measurements

8.1   INTRODUCTION

In this chapter we discuss some basic uses of digital logic and circuitry as they apply to mechanical measurements, but it must be understood at the outset that our purpose is not to cover digital electronics in depth: The subject is too large for such a treatment. Rather, our intent is to survey the subject sufficiently so that those in fields of engineering other than electrical will gain some appreciation for the advantages, disadvantages, and general workings of digital circuitry in the context of measurements.

Most measurands originate in analog form. An analog variable or signal is one that varies smoothly in time, without discontinuity. In many cases the amplitude is the basic variable; in others, the frequency or phase might be. A common example of a quantity in analog form is the ordinary 120-V ac, 60-Hz power-line voltage [Fig. 8.1(a)]. An analog signal, however, need not be simple sinusoidal or periodic in form. The stress–time relationship accompanying a mechanical shock [Fig. 8.1(b)] is considered analog in form. The pressure variations associated with the transmission of the human voice through the air are also analog. The readout from the pointer-over-scale D'Arsonval meter (Fig. 9.3) is also considered analog because of the uninterrupted movement of the pointer over the scale. An analog scale can be compared to the range of brightness between black and white, including all the variations of gray in between. Digital information, on the other hand, would permit only the time variation between the two extreme levels of brightness—black or white.

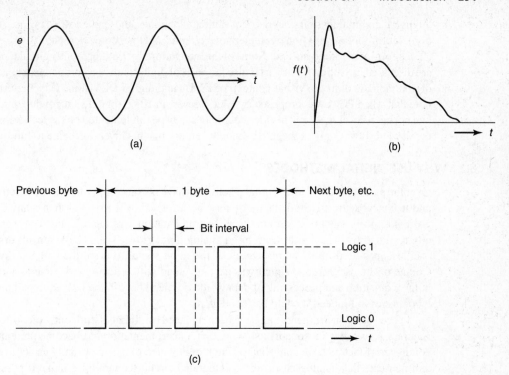

FIGURE 8.1: Examples of voltage–time relationships for analog signals (a) and (b), and a digital signal (c).

Digital information is transmitted and processed in the form of *bits* [Fig. 8.1(c)], each bit being defined by (a) one or the other of two predefined "logic levels" and (b) the time interval assigned to it, called a *bit interval*. The most common basis for the two logic states is predetermined voltage levels, say 0 and 5 V dc. Current or shifts in carrier frequency are also used. The time rate of the bits is closely controlled, commonly by a crystal-controlled oscillator (or *clock*). The information is then carried by specific bit groupings, coded in predetermined sequences; for example, alphanumeric data may be handled by sequences of three, four, or more bits sent in the various possible combinations, with each combination or group forming a *word* of information (see Section 8.6). The term *byte* is applied to an 8-bit word, whereas the term *word* may be applied to *any* unit of digital information. A 16-bit word is two bytes in length. A 4-bit word is sometimes referred to as a *nibble*.

Figure 8.1(c) shows one possible combination of bits grouped to form one byte. In this case, the sequence of bit values is 1010 0110. From this we see that a bit need not be a pulse in the sense that it must be a *completed* off/on/off sequence. Indeed a byte of information could well be 0000 0000 or 1111 1111, in which case no bit-to-bit changes occur throughout the byte. One bit corresponds to either of two different logic states held constant during the one-bit interval.

Because most measurement inputs originate in analog form, some type of analog-to-digital (A/D) converter (or ADC) is usually required (Section 8.11). In certain instances it may also be desirable or necessary to use a digital-to-analog (D/A) converter (or DAC) somewhere in the measurement chain. A sophisticated example of digital information handling is *pulse-code modulation* (PCM) of the human voice, which is used in essentially all digital communication systems. For digital telephones, the spoken word is converted to digital form for transmission over telephone circuits and then converted back to its original analog form at the receiving end. Some of the motivations for this apparently circular process are discussed in Section 8.2. PCM is also the basis of digital audio on computers, music CDs, digital radio, and, in the visual context, of DVDs and digital television. It is interesting to note that when PCM was proposed by Alec Reeves, in 1937, digital semiconductor circuits had not been developed, and his idea could not easily be implemented on the technologies of the day [1]. It was only in the late twentieth century that his invention came to full fruition.

8.2   WHY USE DIGITAL METHODS?

Over the past several decades, digital electronics have become ubiquitous in both household and technical settings. Students today may be surprised to learn that when many of their professors were students, essentially all instrumentation was analog and data were usually recorded by hand. Voltmeters used analog pointer-over-scale (D'Arsonval) displays, oscilloscopes used direct amplification of the input signal to drive the cathode ray tube, storage of the oscilloscope signal was accomplished with a film-based camera, and computer acquisition and processing of data would be found only in the best equipped research laboratories—if indeed it could be found at all!

The analog instruments that have been replaced by digital instruments were often very accurate, so we may reasonably ask why digital instrumentation has become predominant. A number of factors have contributed. First, digital electronics are usually easier to design and fabricate than analog electronics. Integrated circuit technologies make it possible at low cost to mass produce relatively sophisticated instruments. These devices can be quite small, and they generally operate in the range of 5 to 12 V dc. This may be compared to the relative bulk of some analog devices (for example, those requiring large, high-quality capacitors) and to the high operating voltages of some analog instruments (perhaps ranging to several hundred volts).

A second factor, of course, is the ease of data recording, storage, and display. A digital voltmeter provides a direct numerical display of the voltage, often automatically scaled into the proper range. The analog voltmeter had to be visually interpolated if the pointer lay between adjacent graduations of the scale. A digital voltmeter may also be directly coupled to a computer for data recording and processing. Thus, good-quality graphs may be produced and printed with ease—your professor may very well have used graph paper and a French curve to plot data when he was a student. Of equal importance is that many steps necessary in data reduction can be handled automatically. For example, temperatures and pressures acquired from a fluid-flow process may be combined immediately and reduced to provide an on-the-spot overall result, the mass flow rate perhaps. Further, data stored on a computer can easily be rescaled and replotted so as to test various different theories for its interpretation or to investigate the cause of some anomaly.

Digital instrumentation has another tremendous advantage: Digital signals are inherently noise resistant. The informational content of the digital signal is *not* amplitude dependent. Rather, it is dependent on the particular sequence of on/off pulses that apply. Therefore, so long as the sequence is identifiable, the *true and complete* form of the input remains unimpaired. Maintenance of accuracy, lack of distortion induced by signal processing and noise pickup, and greater stability are all enhanced in comparison to analog methods. The nature of the digital signal and circuitry permits the signal to be regenerated or reconstituted from point to point throughout the processing chain. The voltage amplitude of the informational pulses is commonly 5 V dc. Unless noise pulses approach this magnitude—a highly unusual condition—they are ignored. This is of particular value in the central control of a large processing system, such as a refinery or power plant, where low-frequency signals might be relayed over relatively great distances, perhaps a mile or more. This advantage is even more obvious when radio links are used, as in the ground recording of signals originating from a space vehicle. On the level of consumer electronics, one reason that the digital compact disc format for music displaced the earlier analog phonograph record is that digital sound reproduction is substainally less susceptible to the effects of dust, vibration, and wear than is the magnetic phonograph needle; and CDs are smaller and contain more music than did a vinyl LP.

8.3   DIGITIZING MECHANICAL INPUTS

To be processed digitally, analog measurands must be

1. Converted to yes/no pulses;
2. Coded in a form meaningful to the remainder of the system; and
3. Synchronized so as to mesh properly with other inputs or control or command signals.

Meeting these three requirements is collectively referred to as *interfacing*.

When some form of computer is a part of the system, not only must the input be converted to digital pulses, but also the pulses must be converted to the language used by the computer, that is, *binary words*. In addition, of course, the computer is unable to give undivided attention to any one signal source: It will also be receiving inputs from other sources, processing the inputted data, and outputting data and control commands. Input from any one source must wait its turn for attention. In other words, all the inputs and outputs must be synchronized through proper interfacing. Before attempting further coverage of computer data acquisition, however, we will discuss some of the fundamentals and a few of the simpler types of digital instrumentation.

Single digital-type instruments whose end purpose is simply to *display* the magnitude of an input in digital form (as opposed to an input to be interfaced into a system) often require only that the input be transduced to a frequency. Conventional transducers may be used to sense the magnitude of the measurand and to convert it into an analogous voltage. The voltage can then be amplified and, by a circuit called a *voltage-controlled oscillator* (VCO) [2], transduced to a proportional frequency. A frequency-measuring circuit (Section 8.7.3) might then be calibrated to display the magnitude of the input. Many transducers in the field of mechanical measurements produce voltage outputs (e.g., strain-gage bridges, thermistor bridges, differential transformers, thermocouples, etc.). In addition, mechanical motion, both rotational and translational, may often be quickly, easily, and completely converted to digital voltage pulses by proximity transducers (Sections 6.12 and 6.17), photodetectors (Section 6.16), optical interrupters (Section 6.16), and so on.

8.4   FUNDAMENTAL DIGITAL CIRCUIT ELEMENTS

8.4.1   Basic Logic Elements

An ordinary single-pole single-throw (SPST) switch [Fig. 8.2(a)] is a digital element in its simplest form. When actuated, it is capable of producing and controlling an on/off sequence. It this sense, its output is binary. The ordinary electromechanical relay [Fig. 8.2(b)] is a

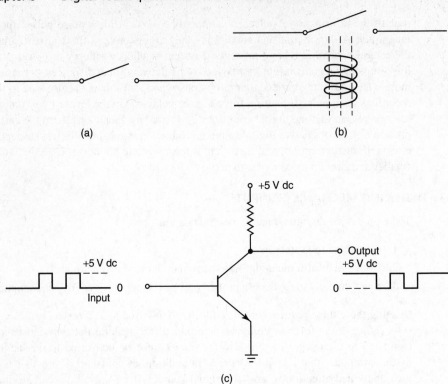

FIGURE 8.2: Digital switching devices: (a) a simple mechanically operated switch, (b) an electrically controlled relay, (c) a transistor-type switch. For the latter, when the input is "high" the transistor conducts, thereby effectively shorting the output to ground, and hence providing near-zero or "low" output. When the input is low the transistor does not conduct, thereby providing near +5 V dc at the output. Additional arrangements are possible if signal inversion is not desired.

slightly more advanced digital device in which an electrical input may be used to change the output condition. A more sophisticated electronic switch is provided by the transistor [Fig. 8.2(c)]. When properly biased, a transistor can be made to conduct or not to conduct, depending on the input signal. It is a near-ideal switching device. It can function at relatively low control voltages, is capable of switching at rates of hundreds of GHz, can be made extremely small and rugged, and can be designed to consume relatively little power. Originally, discrete transistors were hardwired into the various circuits, usually as a replacement for vacuum tubes. (Thus, one still hears the term *transistor radio* even though vacuum tube radios have virtually vanished.) Today, transistors are almost always fabricated together with other simple elements such as resistors, capacitors, and diodes as a single special-purpose device on a single semiconductor chip—a so-called *integrated circuit*, or IC. Integrated circuits are typically housed in the DIP or surface-mount packages shown in Fig. 7.21, and they may be used as building blocks from which more complicated devices are built.

A multitude of special-purpose IC chips are available to the electronic design engineer. Their variations and complexity have grown at an astonishing rate over the past decades.

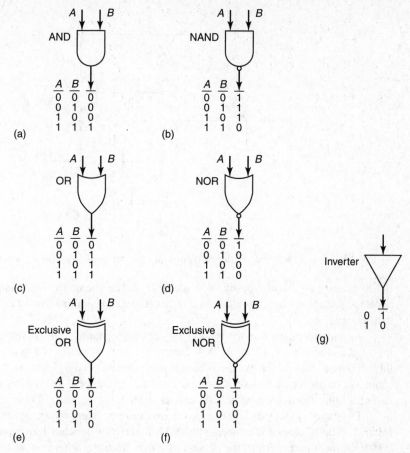

FIGURE 8.3: Symbols for some common digital logic units. Also shown are the respective truth tables for the units.

These range from very simple components such as op amps (Section 7.14) through a myriad of digital-circuit building blocks, such as logic gates and flip-flops (which we discuss shortly), through instrumenation circuits, such as voltmeters or frequency counters, up to highly sophisticated microprocessors.

At the heart of most digital devices are *logic gates*. These are simple circuits that involve just a handful of transistors and provide an output that depends upon two or more inputs. For convenience, these circuits are represented by special shorthand symbols. Figure 8.3 illustrates symbols for common logic elements used in various combinations to form many of the IC chips. Elements 8.3(a) through 8.3(f) are *gates*. Figure 8.3(g) represents a simple inverter. Also shown are *logic*, or *truth, tables*, which list all the possible combinations of inputs and their corresponding outputs. Recall that basic digital operations are based on simple yes/no, on/off, 1/0, states. For example, the truth table for the AND gate shows that the output is high *only* if the inputs to *both* A and B are high, hence, the *AND rule*:

> For the AND gate, *any* low input will cause a low output; that is, *all* inputs must be high to yield a high output.

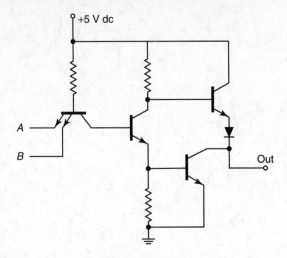

FIGURE 8.4: A schematic diagram showing the internal structure of a NAND gate.

In like manner, we can also easily state rules for the other elements. Truth tables are of particular importance to the circuit designer, especially when combinations of circuits increase their complexity.

As a matter of interest, the NAND gate actually contains four transistors, three resistors, and a diode, as shown in Fig. 8.4. Suppose a chain of pulses is applied to input *B* of the AND gate. We can see that their passage may be controlled by input *A*. If *A* is high, the chain will be permitted to pass; if low, the chain will be stopped. From this simple example the origin and significance of the term *gate* is clear.

The various gates may be expanded to provide more than two inputs; see, for example, Fig. 8.5. The IC shown is a three-input NAND gate, for which a zero input level at any one of the input ports permits the passage of pulses from any other port; in other words, all inputs must be high in order for the output to be low. With this arrangement, a combination of several control conditions must be met simultaneously to block passage of a signal.

Integrated circuits are often characterized by the effective number of gates they incorporate. The so-called MSI (medium-scale integration) chips incorporate 12 to 100 individ-

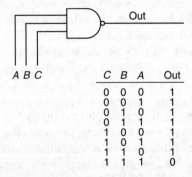

C	B	A	Out
0	0	0	1
0	0	1	1
0	1	0	1
0	1	1	1
1	0	0	1
1	0	1	1
1	1	0	1
1	1	1	0

FIGURE 8.5: NAND gate with three inputs and $2^3 = 8$ possible input combinations. A NAND gate with eight inputs has $2^8 = 256$ input combinations of which only one causes an output of logic 0.

ual gates per chip. LSI (large-scale integration) chips contain 100 to 10,000, VLSI (very large-scale integration) chips contain 10^4 to 10^5 gates, and so on.

8.4.2   Combination of Logic Elements: The Flip-Flop

Figure 8.6(a) shows two NAND gates connected to form a very useful circuit. The circuit has two inputs, S and R, and two outputs, Q and $\overline{Q}$, that are always meant to be opposite one another (one at logic 1 and the other at logic 0). As a simple flip-flop, element A is called the SET gate and element B the RESET gate. Consideration of the individual truth tables, along with their particular interconnections, shows that the following are the only workable conditions:

	S	R	Q	$\overline{Q}$
Condition I	1	1	1	0
Condition II	1	1	0	1
Condition III	0	1	1	0
Condition IV	1	0	0	1

Now suppose that both S and R are initially at logic 1; then either of Conditions I or II may exist, depending on random or programmed preconditions. If either of two outputs can correspond to a given input, the device is referred to as having *bistable logic*. In some contexts, the circuit is called a *latch*.

Let us momentarily ground input S—that is, impose logic 0. Condition III, called the SET condition, will result. This is true regardless of whether the circuit is initially in Condition I or II. Return of S to the high state will cause no change in Q or $\overline{Q}$: It is *latched*.

Now, if R is momentarily grounded, Condition IV will be instituted, and this state will continue even when R is returned to logic 1. This is called RESET, and we see that the outputs are caused to flip and flop between SET and RESET.

Grounding both inputs simultaneously—setting both S and R to logic 0—would amount to attempting SET and RESET simultaneously. This state is normally avoided. Because the changes in state occur when S or R is at low logic, this circuit is referred to as an *active-low S-R latch*.

This ciruit has various important uses. For example, as a latch it may be used to hold (latch) a count in an electronic events counter, then await a RESET input for initiating the next count. The flip-flop is used as a memory cell, capable of holding one bit of information for later use. It also provides the basis for the *switch debouncer*. When an ordinary electric switch depending on mechanical contacts is closed, numerous contacts are actually made and lost before solid contact is finalized. In a counting circuit, for example, this sort of switch hash cannot be tolerated. By placing a flip-flop or latch in the switch circuit [Fig. 8.6(b)] we cause the latch to respond to that first momentary contact and then to ignore all that follows until a RESET signal reinitializes the circuit. These are only several of the uses to which the circuit may be applied; and the particular circuit discussed is only one of a number of different circuits referred to as flip-flops [3].

8.4.3   IC Families

Several *families* of integrated-circuit chips are available, each of which has special characteristics but, in general, all of which perform essentially the same basic tasks. The

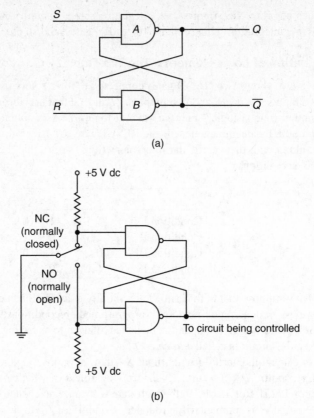

(a)

(b)

FIGURE 8.6: (a) Two NAND gates configured to form a flip-flop, or latch; (b) a switch-debouncer circuit.

most commonly used groups are the newer *complementary metal oxide semiconductor*, or CMOS, group, and the older *transistor-transistor-logic*, or TTL, group. In each family the various logic units are combined to perform special functions. For example, the TTL family consists of more than 150 different types of chips. Table 8.1 is a partial list. All have more or less the same outward appearance (see Fig. 7.21), but each is designed to perform a different function. Schematics of several of the TTL family are illustrated in Fig. 8.7. The circuits in Figs. 8.7(a), (b), and (c) are simple enough that the functional performance symbols may be shown. The circuitry of Fig. 8.7(d), however, is so complex (because of the number of elements used) that we have made no attempt to indicate the internal architecture. Application of most of the chips selected for listing in Table 8.1 is covered in later sections of this chapter. All of these basic devices are also available as CMOS IC chips.

Within each family of circuits, various subcategories are available, each with some-what different current, voltage, and speed characteristics. The subtype is usually indicated by a set of letters that is placed between the 74 and the following digits. For example, an "advanced high speed" CMOS (AHC subtype) quad two-input AND gate would be designated 74AHC08. Here, the word *quad* indicates that four separate gates are on a single chip.

CMOS circuits are based on a different type of transistor than the TTL family (field-effect transistors rather than bipolar junction transistors), and for most purposes CMOS

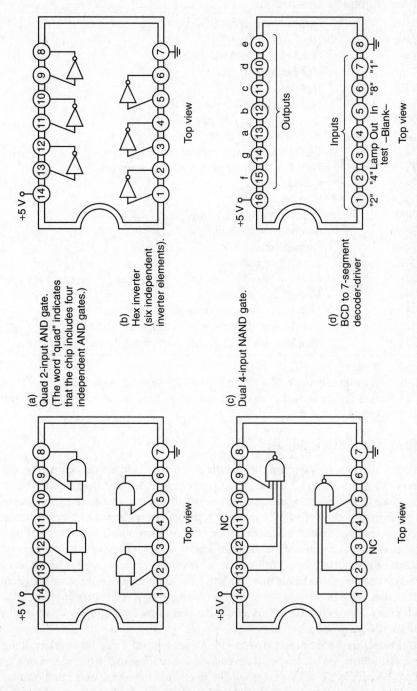

FIGURE 8.7: Diagram showing four typical IC chips: (a) quad 2-input AND gate, (b) hex inverter (six independent inverter elements), (c) dual 4-input NAND gate, (d) BCD (binary-coded decimal) to 7-segment decoder-driver.

299

TABLE 8.1: A Partial Listing of TTL IC Chips

Type Number	Description
7400	Quad two-input NAND gate
7404	Hex inverter
7408	Quad two-input AND gate
7414	Hex Schmidt trigger
7416	Hex driver, inverting
7430	Eight-input NAND gate
7432	Quad two-input OR gate
7445	BCD to decimal (1-of-10) decoder driver
7447	BCD to seven-segment decoder driver
7474	Dual D-type flip-flop, positive-edge trigger
7475	Quad latch
7483	Four-bit full adder
7485	Four-bit magnitude comparator
7489	64-bit (16×4) memory
7490	Decade counter
7492	Base-12 counter
74109	Dual J-K flip-flop, positive-edge trigger
74121	Monostable multivibrator
74150	1-of-16 data selector/multiplexer
74154	1-of-16 data distributor/demultiplexer
74160	Synchronous BCD decade counter with asynchronous reset

circuits have substantially more desirable electrical characteristics than TTL circuits. As a result, the CMOS family is used in most new circuit designs, and the entire TTL family is gradually becoming obsolete.

8.4.4 IC Oscillators and Clock Signals

Figure 8.8 illustrates an interesting IC oscillator—the 555 chip. This chip contains 23 transistors, 15 resistors, and 2 diodes. The package is about half the size of a common postage stamp, has eight terminals, and costs less than a cup of coffee and a doughnut. When configured with two or three external resistors and a capacitor, it can produce an oscillating output that switches back and forth between two states. In this type of *astable oscillation*, the 555 is capable of yielding a square-wave output covering a frequency range from 0.1 Hz to over 100 kHz. With additional outboard circuit elements, triangular and linear ramp waveforms can be obtained. In still other configurations, this chip can produce a single output pulse in response to an input (so-called *monostable operation*), and many other uses have been devised [4]. This IC oscillator is an example of the wide versatility of a single special-purpose integrated circuit.

Chips such as the 555 are used to provide a time-base reference for a variety of digital systems, and so they are sometimes referred to as *timers*. The oscillating signal of a timer circuit can be used as a *clock signal* to synchronize the actions of several components in a system. In these *synchronous* systems, the time at which any subcircuit's output can change

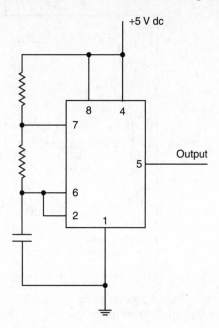

FIGURE 8.8: A 555 astable/monostable oscillator or multivibrator.

is controlled by the clock signal. For example, flip-flop can be designed so that when an input is changed, the change in output is delayed until the clock signal changes; the D-type and J-K flip-flops in Table 8.1 (7474 and 74109, respectively) both have this feature. Synchronous operation usually simplifies circuit design. Circuits that are not synchronous are called *asynchronous*.

The output of a 555-based clock circuit may be stable to about 1% or so but it is susceptible to drift. When higher accuracy is required, a quartz-crystal oscillator is used instead. The heart of such a circuit is a small piece of quartz cut to precise dimensions, which resonates at a highly stable frequency. Owing to the piezoelectric properties of quartz (see Section 6.14), the resonating crystal is easily coupled to an electrical circuit. Quartz-crystal oscillators are available as off-the-shelf ICs at low cost with frequencies from roughly 10 kHz to 100 MHz. Frequency stability to a few parts per million is easily obtained. These crystals are used in all microprocessors and microcomputer systems. Crystals resonating at 32.768 kHz are at the heart of digital wristwatches: When additional digital circuits are used to divide this frequency by 2^{15}, the result is a 1-Hz signal for counting seconds [5]. Note that an error of 10 parts per million in the crystal frequency would correspond to about one second per day of error in a watch.

8.4.5 Digital Displays

In some situations we may require only simple, single lights for an indication or readout, perhaps to show that a device is powered or ready for use. In other cases we may use combinations of discrete lights or indicators to produce an alphanumeric display.

Alphanumeric readout elements are usually of either the liquid-crystal diode (LCD) type or the light-emitting diode (LED) type. The liquid crystal has a decided advantage in some cases, as, for example, in a digital watch requiring very low power consumption.

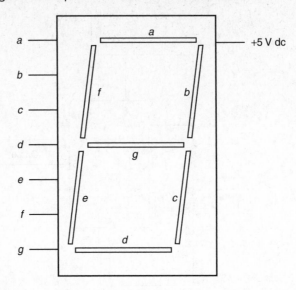

FIGURE 8.9: Typical solid-state numeric display. The +5 V dc anode is common to all segments. Grounding pin a, for example, through a suitable voltage-dropping resistor lights segment a.

LCDs are the usual choice for battery-powered instruments. A disadvantage is the need for either proper external illumination or a separate lamp for backlighting the display. The LED provides a bright display, usually in red or green, but it consumes much more power. It is therefore more commonly used in instrumentation that connects to an external power source.

Figure 8.9 shows an arrangement of a typical LED seven-segment digital display. Each segment consists of one or more LED elements and with proper switching supplied by an appropriate IC driver, each of the decimal digits, 0 through 9, can be formed. Commonly, the input signal and a 5-V dc power source are all that is required for power.

The operation of a digital display requires a process of converting information between binary and decimal forms. This process is known as *encoding* when information is converted to a binary form, and it is called *decoding* when information is converted out of a binary form. These operations clearly require some understanding of the binary and decimal number systems. In the next two sections, we look at number systems and the common binary codes.

8.5 NUMBER SYSTEMS

Whether it is in digital or analog form, a measurement signal conveys a magnitude. Magnitudes are expressed in numbers and numbers imply some sort of numbering system or structure. Digital devices with their high/low, yes/no, on/off sequencing suggest the use of a base 2, or binary, system for counting. In this section we will present the essentials of number systems other than decimal that are pertinent to digital operations.

—— We know that the *position* of each of the digits in a decimal number is important. For example, consider the decimal number 347.25. The 3 is the most significant digit and the 5 is the least significant digit. We know that the number can be expanded to read

$$347.25 = 3 \times 10^2 + 4 \times 10^1 + 7 \times 10^0 + 2 \times 10^{-1} + 5 \times 10^{-2}$$

It is clear that the various positions of the digits determine the power to which the base, 10, is raised.

For the binary number system, only two different digits are required, a 1 and a 0. The digit 1 may correspond to a *high* condition, say +5 V dc, and the digit 0, to a *low* condition, say 0 V dc. In the binary system, as in the decimal system, position has meaning. Consider the binary number 11010.01_2. (We add the subscript 2 to make it clear that we are using the binary system.) The first digit, 1, is the most significant digit and the last digit, 1, the least significant. Each digit is called a *bit* in the sense that it supplies an elemental "bit" of information. Hence, the terms *most significant bit*, or MSB, and *least significant bit*, or LSB, are used. What corresponds to the decimal point in the decimal system is called the *binary point* in the binary system.

As we have observed, in a decimal number each position corresponds to an integral power of 10. By the same token, in a binary number, each position corresponds to an integral power of 2. Whereas the coefficients for the decimal number could be anything between 0 and 9, for the binary system we are limited to 0 and 1.

Let's convert the binary number written above to the equivalent decimal number. *Equivalent*, of course, means that the two numbers signify the same actual magnitude or quantity (it may be convenient to refer to Table 8.2).

$$11010.01_2 = 1 \times 2^4 + 1 \times 2^3 + 0 \times 2^2 + 1 \times 2^1 + 0 \times 2^0 + 0 \times 2^{-1} + 1 \times 2^{-2}$$

$$= 16 + 8 + 0 + 2 + 0 + 0 + \frac{1}{4}$$

$$= 26.25_{10}$$

TABLE 8.2: Decimal Values of Bases Raised to Various Powers*

n	2^n	4^n	8^n	16^n
⋮	⋮	⋮	⋮	⋮
−3	1/8	1/64	1/512	1/4096
−2	1/4	1/16	1/64	1/256
−1	1/2	1/4	1/8	1/16
0	1	1	1	1
1	2	4	8	16
2	4	16	64	256
3	8	64	512	4,096
4	16	256	4,096	65,536
5	32	1,024	32,768	1,048,576
6	64	4,096	262,144	16,777,216
⋮	⋮	⋮	⋮	⋮

*n is both the *power to which the base is raised* and the *positional weight*.

We can see that the positional significance of the 0s and 1s lies in the integral powers to which the base (also called the *radix*) is raised. This is true for both the binary and the decimal systems.

Two other systems are commonly used in digital manipulations: the *octal*, or base 8, system; and the *hexadecimal*, or base 16, system. These two systems have the marked advantage over the decimal system in the ease and convenience of their conversion, by either machine or human, to binary.

For further discussion of number systems, see Appendix C.

8.6   BINARY CODES

In addition to the binary number, which is, of course, limited in magnitude only by the number of positional bits that may be arbitrarily permitted, there are various binary *codes*. At this point we will consider several of them.

8.6.1   Binary-Coded Decimal

Suppose we limit the number of positions to four, i.e., we provide only four on/off switches or their equivalents. We are then limited to the decimal range of 0 to 15 inclusive. The result is what is known as the four-bit *word*, also referred to as a *four-level code*.

A modification of this code is known as binary-coded decimal, or *BCD*. Although BCD is also a 4-bit word, it arbitrarily makes illegal all words greater than 1001_2, or 9_{10}. We see that the BCD code is useful because of its convenient relationship to the decimal digits 0 through 9. Table 8.3 lists the equivalencies. When the BCD code is used, each digit in a decimal number is processed separately as a 4-bit sequence. For example, the decimal number 875_{10} translates to $1000\ 0111\ 0101_2$ in BCD.

TABLE 8.3: Four-Bit Binary Numbers and Binary-Coded Decimal

Decimal Digit	Binary	BCD
0	0000	0000
1	0001	0001
2	0010	0010
3	0011	0011
4	0100	0100
5	0101	0101
6	0110	0110
7	0111	0111
8	1000	1000
9	1001	1001
10	1010	*Illegal*
11	1011	*Illegal*
12	1100	*Illegal*
13	1101	*Illegal*
14	1110	*Illegal*
15	1111	*Illegal*

8.6.2   Position Encoders and Gray Code

Figure 8.10 shows a schematic diagram of one type of binary displacement encoder based upon a 5-bit binary code. The "card" shown consists of five active tracks plus a reference track, which may or may not be needed. Pickups (not shown) sense the relative displacement of the card. One sensor is used per track. Optical sensors are most commonly used to detect the on/off status of of each track at a given position. The cards may be transparent or opaque for use with transmitted or reflected light, respectively. Printed-circuit methods may also be used. The output from the sensors is easily processed by onboard or remote digital circuitry.

A direct binary encoding such as shown in Figure 8.10 has an inherent disadvantage. When several bits are different between adjacent levels, the associated position sensors may not all change at the exact same instant, leading to a drastic sensing error. For example, in going from the eighth to the ninth position, the binary encoder shown would change from

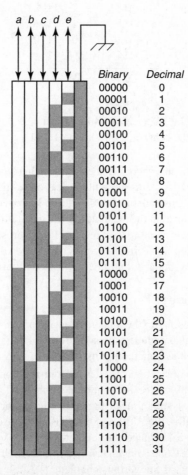

Binary	Decimal
00000	0
00001	1
00010	2
00011	3
00100	4
00101	5
00110	6
00111	7
01000	8
01001	9
01010	10
01011	11
01100	12
01101	13
01110	14
01111	15
10000	16
10001	17
10010	18
10011	19
10100	20
10101	21
10110	22
10111	23
11000	24
11001	25
11010	26
11011	27
11100	28
11101	29
11110	30
11111	31

FIGURE 8.10: Binary displacement encoder.

TABLE 8.4: Four-Bit Binary Numbers and Gray Code

Decimal Digit	Binary	Gray Code
0	0000	0000
1	0001	0001
2	0010	0011
3	0011	0010
4	0100	0110
5	0101	0111
6	0110	0101
7	0111	0100
8	1000	1100
9	1001	1101
10	1010	1111
11	1011	1110
12	1100	1010
13	1101	1011
14	1110	1001
15	1111	1000

00111_2 to 01000_2 (i.e., from 7_{10} to 8_{10}). If just one bit were to change too soon or too late, the position could be sensed as 00000_2 or 01111_2, corresponding to either the first or the sixteenth level!

Gray code is a binary code that eliminates this type of error, and it is thus preferred for use in position encoders. It is not a numerical code, in the sense that there is no positional weight to the digits. Instead, the digits of the code are sequenced so that only one bit changes in going from one level in the sequence to the next. If one track's sensor changes too early or too late, the error is no more than one level—the drastic errors possible with direct binary encoding are avoided. A 4-bit Gray code code is shown in Table 8.4. Figure 8.11 illustrates a circular card for a rotary shaft position encoder based upon four-bit Gray code.

8.6.3 Alphanumeric Codes

The preceding codes are for the transmission and processing of numeric data. Alphanumeric information must also include provisions for the letters of the alphabet and perhaps certain other symbols, such as punctuation marks.

One of the simplest binary codes for transmission of general information (as opposed to numeric data only) is the International Morse Code, which uses *pulse-duration modulation*, or PDM (also sometimes called pulse-length or pulse-width modulation). The two different pulse widths, the "dot" and the "dash," are used in various combinations to transmit the alphabet, the decimal digits, and certain other special-purpose telegraphic symbols. International Morse Code is an *uneven-length* code in that various time intervals are required for the transmission of the various characters. A related code is the *Baudot*, or common teletypewriter, code, which is a five-unit, *even-length code*. All characters are formed by a

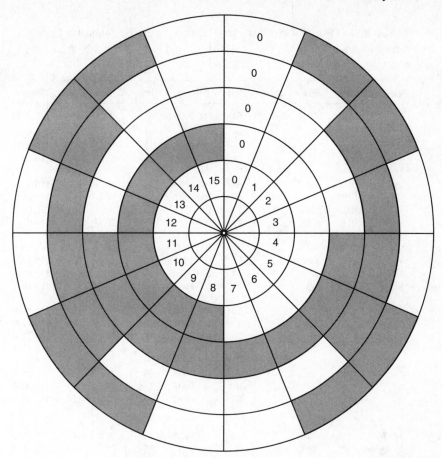

FIGURE 8.11: Circular encoder using Gray code.

combination of five possible on/off states, each of *uniform* duration. These two codes are not especially important in mechanical measurements.

The most well-known code for alphanumeric data is the *American Standard Code for Information Interchange* or *ASCII* [6]. This code is of 7-bit binary form, and it thus has 2^7 or 128 characters. Both the uppercase and lowercase letters of the English alphabet, plus the decimal digits zero through 9 and certain other control symbols, are included (Table 8.5). ASCII originated as a 7-bit bit teleprinter code, essentially as a more sophisticated successor to the 5-bit Baudot code. Thus, the first 32 characters and the final character are control codes that might govern such functions as carriage returns, line feeds, and form feeds—all handy for driving a printer. These codes are not associated with displayed characters, and so they are omitted from the table.

When writing the binary equivalents of the ASCII code, one generally divides the binary number into two groups of 3 and 4 bits each; for example, for the letter a, the binary equivalent is written 110 0001. The lefthand binary digit is the MSB (most significant bit) and the righthand, the LSB. In processing, the ASCII number for a would appear as

TABLE 8.5: Displayable Characters of the American Standard Code for Information Interchange (ASCII) with Decimal, Binary, and Hexadecimal Numbering

Dec	Binary	Hex	Char	Dec	Binary	Hex	Char	Dec	Binary	Hex	Char
32	010 0000	20	Space	64	100 0000	40	@	96	110 0000	60	`
33	010 0001	21	!	65	100 0001	41	A	97	110 0001	61	a
34	010 0010	22	"	66	100 0010	42	B	98	110 0010	62	b
35	010 0011	23	#	67	100 0011	43	C	99	110 0011	63	c
36	010 0100	24	$	68	100 0100	44	D	100	110 0100	64	d
37	010 0101	25	%	69	100 0101	45	E	101	110 0101	65	e
38	010 0110	26	&	70	100 0110	46	F	102	110 0110	66	f
39	010 0111	27	'	71	100 0111	47	G	103	110 0111	67	g
40	010 1000	28	(	72	100 1000	48	H	104	110 1000	68	h
41	010 1001	29	)	73	100 1001	49	I	105	110 1001	69	i
42	010 1010	2A	*	74	100 1010	4A	J	106	110 1010	6A	j
43	010 1011	2B	+	75	100 1011	4B	K	107	110 1011	6B	k
44	010 1100	2C	,	76	100 1100	4C	L	108	110 1100	6C	l
45	010 1101	2D	–	77	100 1101	4D	M	109	110 1101	6D	m
46	010 1110	2E	.	78	100 1110	4E	N	110	110 1110	6E	n
47	010 1111	2F	/	79	100 1111	4F	O	111	110 1111	6F	o
48	011 0000	30	0	80	101 0000	50	P	112	111 0000	70	p
49	011 0001	31	1	81	101 0001	51	Q	113	111 0001	71	q
50	011 0010	32	2	82	101 0010	52	R	114	111 0010	72	r
51	011 0011	33	3	83	101 0011	53	S	115	111 0011	73	s
52	011 0100	34	4	84	101 0100	54	T	116	111 0100	74	t
53	011 0101	35	5	85	101 0101	55	U	117	111 0101	75	u
54	011 0110	36	6	86	101 0110	56	V	118	111 0110	76	v
55	011 0111	37	7	87	101 0111	57	W	119	111 0111	77	w
56	011 1000	38	8	88	101 1000	58	X	120	111 1000	78	x
57	011 1001	39	9	89	101 1001	59	Y	121	111 1001	79	y
58	011 1010	3A	:	90	101 1010	5A	Z	122	111 1010	7A	z
59	011 1011	3B	;	91	101 1011	5B	[	123	111 1011	7B	{
60	011 1100	3C	<	92	101 1100	5C	\	124	111 1100	7C		
61	011 1101	3D	=	93	101 1101	5D	]	125	111 1101	7D	}
62	011 1110	3E	>	94	101 1110	5E	^	126	111 1110	7E	~
63	011 1111	3F	?	95	101 1111	5F	_				

110 0001. As a function of time, it would appear as shown in Fig. 8.12.

ASCII does not include very many of the characters found for world's written languages. By itself, it is not even sufficient for the languages of Western Europe as it lacks accented characters such as ç, ñ, and ö. This defect was partially addressed by adding one more bit to ASCII, creating an 8-bit *extended ASCII* having 256 characters. Unfortunately, several versions of extended ASCII were promulgated by different manufacturers, and no single standard exists under that name.

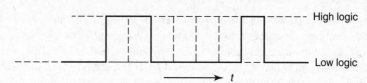

FIGURE 8.12: The ASCII logic sequence for the letter a.

Subsequent work did lead to standardized 8-bit character sets, which are very commonly used in coding Web pages. For Western European languages, including English, the most important of these is ISO 8859-1, the so-called *Latin 1 character set* [7]. In the Latin 1 encoding, the printable characters below 127_{10} are the same as those in ASCII. Additional standard encodings are have been defined for other families of languages.

Internationalization of commerce and information exchange has grown dramatically in recent decades, and so it is desirable to have character encoding systems that need not be changed whenever the language of application changes, or, more particularly, which can support the use of several different languages within a single document. Further, the encoding system should ideally be the same for every piece of software. These needs have driven the development of the Unicode standard [8], which aims to assign a unique numerical code to *every* character in *every* language. When one considers that tens of thousands of characters are found in some Asian languages, such as Han Chinese, it becomes clear that far more than 8 bits are required. The early versions of Unicode ran to 32 bits (facilitating up to 65,535 characters), but that barrier was removed in subsequent versions. Unicode 4.0 assigns unique numbers to more than 90,000 characters.

Serial and Parallel Transmission

The International Morse and the Baudot codes are necessarily of a *serial* type, i.e., the various on/off states occur in sequence, one following another in "bucket-brigade" fashion. If they were transferred by an electrical conductor, only a single transmission circuit would be required. In some cases, an alternative to serial transmission, called *parallel* transmission, is used. This means that the bits in a single word are transmitted simultaneously. In its simplest form, this requires as many transmission circuits as there are code levels, but with the obvious benefit of higher speed. Parallel circuitry is often used within a single instrument or device, whereas serial circuitry is used when long cables are involved, as, for example, when connecting a computer to the internet through a telephone modem.

Parity and Error Detection

We have previously noted that digital signal transmission is less susceptible to transmission errors than analog signal transmission. Nevertheless, errors still occur, albeit with a low probability. Digital transmission systems are capable of sending many millions of bits per second, so even with a low probablility of occurrence, errors will arise often enough that a means of detecting them is needed.

One approach to this problem is the *parity method* of detection. In this scheme, an extra bit is attached to each word of information. For example, a 7-bit ASCII code would have an eighth *parity bit* attached to it, so that the transmitted word would have 8 bits. In an *even parity* method, the parity bit is selected so that the total number of ones in the word is even. If 1 bit in the received word is incorrect, the number of ones would be odd, thus enabling the receiver to determine that a transmission error has occurred. In an *odd parity* scheme, the total number of ones would be odd. The parity method clearly cannot detect errors in which 2 bits of one word are incorrect; however, in situations where even a single bit error is improbable, a 2-bit error will be *extremely* improbable and is thus of little concern.

8.6.4 Bar Codes

Everyone is familiar with at least some of the applications of bar codes. The familiar black/white lines are arranged to codify data of various sorts—addresses, inventory, and virtually any other information that can be put into a binary representation. A low-intensity laser may be reflected off of the bar pattern, with the signal received by an appropriate photodetector. The output is electronically processed and sent on to the cash register or inventory tracking system.

A number of standards have been used to create bar codes, the most common being the *Universal Product Code* (UPC) and the *European Article Numbering* (EAN) system. In a typical implementation, a black line corresponds to 1 and a white line corresponds to 0. Seven lines are used for each character; however, the first and last lines are always opposite each other (black and white or white and black, respectively) in order that adjacent characters can be distinguished. The five remaining bits offer 32 possible combinations, but not all are used. Instead, only combinations yielding just two black bars and two white bars are employed, where a bar's width may be one, two, three, or four lines. There are only 20 such combinations. These combinations are assigned to the ten decimal digits from 0 to 9 in two different sequences, one "left handed" and the other "right handed."

Depending upon the particular type of bar code, as many as 14 digits may be encoded. In addition to the bars for the digits, the codes usually include guard bars at left, center, and right. The numerical code itself is assigned to particular companies and products by a standardization agency.

In addition to the basic one-dimensional bar code, various two-dimensional *matrix style* coding systems are also in use. These systems can encode greater amounts of alphanumeric data. They are sometimes used on shipping labels from commercial delivery services.

8.7 SOME SIMPLE DIGITAL CIRCUITRY

8.7.1 Events Counter

Figure 8.13 shows the outward simplicity of one form of digital events counter. Two consecutive stages, or *decades*, are shown. Each decade consists of a seven-segment LED display, two IC chips, and a few current-limiting resistors. The 7490 chip is called a *decade counter* (see Table 8.1). It accepts input pulses in serial form through pin 14 and sends as output parallel BCD pulses through pins *d, c, b*, and *a*, where *d* corresponds to the most significant bit and *a* to the least significant. In addition, this chip provides an output pulse at the *end* of the ninth input pulse. This pulse is available from pin 11 and can be used as the input to the next higher counting decade. We can see that additional decades may be cascaded easily. As we will note later, this arrangement obviously also provides a "divide-by-10" capability.

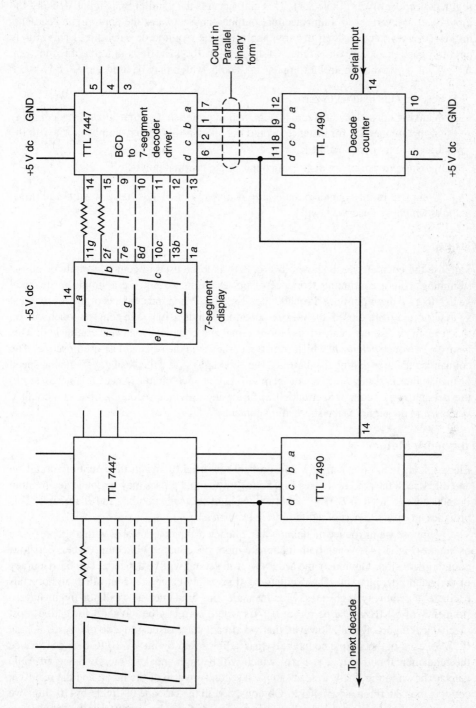

FIGURE 8.13: Schematic circuit for a simple digital events counter.

The BCD outputs from the 7490 chip are fed to a 7447 chip, called a BCD-to-seven-segment decoder driver (Table 8.1). This chip accepts the parallel BCD input, decodes the input—i.e., converts it to a unit decimal output—and activates the appropriate segments in the seven-segment readout, thereby displaying the required decimal digit. This simple circuit is capable of counting at rates of up to about 100 kHz. In addition to this function, IC 7447 also provides control terminals 3, 4, and 5, which may be used for

1. resetting the readout to zero;
2. blanking unused leading zeros—e.g., making provision so that a five-decade display of the number 25, for instance, would not show as 00025 but simply as 25, with the unnecessary zeros blanked;
3. making provisions for displaying appropriate decimal points.

Complete event counter circuits are available as LSI ICs, usually with additional features which we describe next.

8.7.2 Gating

Suppose the counter circuit shown in Fig. 8.13 is to be used to count a sample of pulses stemming from a continuing sequence of pulses. Can we not use a simple mechanical switch to *gate* the input line? Probably not directly! When the contacts of the mechanical switch (or relay) are closed, there exists a period of indecision. The contacts touch, break, touch again, and so on, many times before a final and complete contact is established. This is the source of *switch hash*, which sometimes shows on the screen of an oscilloscope. The counter is unable to distinguish between the "good guys" and the "bad guys" and the speed of the counter is sufficient to count them all! Under such circumstances it is necessary to use debouncing circuitry (Section 8.4.2) in the input, which recognizes the first contact, latches, and then ignores the succeeding contacts.

8.7.3 Frequency Meter

Electronic switching or gating would probably be used to switch the counting circuit on and off. Recall the AND gate [see Fig. 8.3(a)]. The input pulses may be connected to input A and a control input to B. The output will be high only when both A and B are high. This provides an essential part of a digital *frequency meter*.

Suppose we introduce an unknown frequency at A and control B with a square-wave oscillator (Fig. 8.14). When both inputs are high, the counter will count. When B is low, counting will stop. Obviously the accuracy of the count will be dependent on the accuracy of the gated time interval. The oscillator will probably be crystal controlled, and crystals oscillate at relatively high rates. For instance, the fundamental oscillator might have a frequency of 5 MHz. The period of 0.1 μs would certainly be too short for gating most mechanical inputs. Recall, however, that we already have discussed a divide-by-10 IC, the IC 7490 used in the simple counter. By cascading seven divide-by-10 ICs we can reduce the fundamental period to 2 s, 1 s of which will be high and 1 s low. By using an AND gate as shown in Fig. 8.14, we can sample a second's worth of the input, and the resulting count is thus the frequency in hertz. Obviously, with the cascade of divide-by-10 chips we can use panel-controlled switches for different gating times by simply tapping the cascade at other points in the chain. Other features can be added as required.

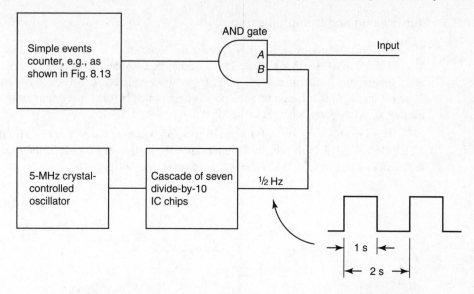

FIGURE 8.14: Schematic diagram for a simple frequency meter circuit.

8.7.4 Wave Shaping

Digital logic circuits prefer instant toggling from low to high and back again. Suppose we wish to feed the counter or frequency meter described previously with a sine wave or some other varying waveform. If the signal level is sufficient, our counter may work; but, if the peaks cannot easily be discriminated from the rest of the signal, the counter may miss or double count them. We can convert the sine wave to a square wave signal, or an irregularly varying wave into a sequence of high and low voltages, by using a *Schmitt trigger* (see TTL 7414 in Table 8.1, for instance).

The Schmitt trigger accepts a gradually rising input but stays at low-output voltage until its "rising trigger level" is reached. It then switches to a high output voltage, and it will remain there until the signal drops below its "falling trigger level." These two trigger levels are made slightly different so that a signal that lingers near one of the trigger levels will not cause multiple transitions. We are therefore able to reshape the input into a series of on/off pulses, much preferred by the common IC.

8.7.5 Integrated-Circuit Counter and Frequency Meter

The close relationship of event counters, frequency meters, and period or time-interval meters allows all these functions to be incorporated on a single LSI chip. An example popular in the 1990s was the Intersil ICM7226A, an eight-digit, multifunction, frequency counter/timer chip. It incorporated a high-frequency oscillator, a decade time-base counter, triggers, an eight-decade data counter, and seven-segment display decoder/drivers in a single package. It could function as any of the meters mentioned previously and could run an eight-digit LED display. It measured periods from 0.1 μs to 10 s and frequencies from dc to 10 MHz.

8.7.6 Multiplexing and Demultiplexing

Figure 8.15(a) illustrates an IC chip called a *multiplexer*, or a 1-of-16 data selector (see TTL 74150, Table 8.1). In mechanical terms it may be considered a selectable commutator; quite simply, it is similar to a single-pole, 16-position switch [Fig. 8.15(b)]. The particular input (of up to 16) that is permitted to pass is determined by the binary number inserted at the d, c, b, and a ports, where d is the MSB and a is the LSB.

The demultiplexer, or 1-of-16 data distributor, is similar but with reversed action (see TTL 74154, Table 8.1). It accepts a digital signal through its one input and then routes it to the particular output selected by the binary value at the control ports.

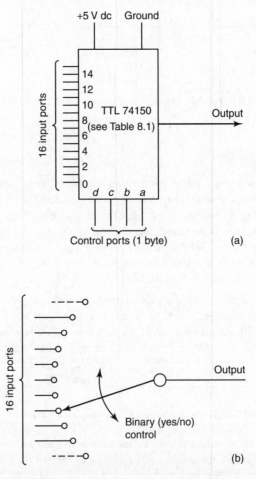

FIGURE 8.15: (a) The TTL 74150 multiplexer or 1-of-16 selector. Each of the 16 possible logic combinations of the *dcba* control ports is employed to select a corresponding signal input for connection to the output. (b) Illustration of a nearly equivalent mechanical switching arrangement. A difference, however, lies in the fact that the mechanical device must switch through the sequence in order, whereas the multiplexer need not.

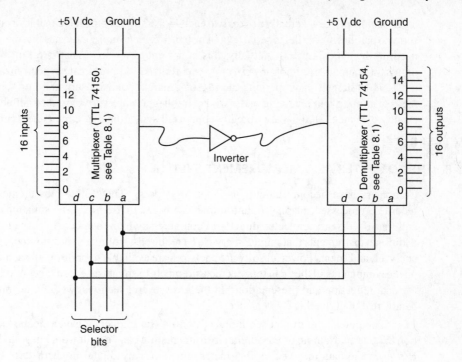

FIGURE 8.16: A multiplexer–demultiplexer circuit.

Why concern ourselves with multiplexers and demultiplexers? In many cases continuous monitoring or recording of a given data source is not necessary: Periodic sampling will suffice. We can see that by sequencing the binary control through 0 to 15, we can consecutively connect 16 different inputs to a given readout/recording/computing system. This approach is economical in many cases.

Multiplexing may also be used, in certain instances, within the circuitry of a single instrument. Recall the events counter we discussed in Section 8.7.1. Each decade required a seven-segment decoder driver plus seven current-limiting resistors. More sophisticated counter circuitry uses a multiplexer–demultiplexer combination arranged so that each readout element is sequentially connected to a single driver–resistor combination. By time sharing in this manner we can reduce the number of circuit elements and realize quite a saving in cost. The sequencing rate is sufficiently high that the readout appears to be illuminated continuously.

Data processing may take place at quite some distance from the data source, as in a large industrial complex such as a refinery or power plant. By using multiplexer–demultiplexer combinations, we may also use single, rather than separate, circuits to connect the two positions (Fig. 8.16). We hasten to add that this may not be precisely the case because it would also be necessary to synchronize our binary control circuits at the two locations. A simple solution is to run four additional wires connecting ports d, c, b, and a. Thus we would have 5 circuits instead of 16 (for the particular combination cited).

There is, however, another possibility through use of more sophisticated ICs. The universal asynchronous receiver/transmitter (UART) contains what amounts to a multiplexer and a demultiplexer—actually a set of two each—on a single chip. It is capable of converting serial to parallel or parallel to serial data. The word *asynchronous* indicates that the operation need not be synchronized. Actually, this is not entirely true. What is meant is that the inputs and outputs need not be *perfectly* synchronized: Approximate synchronization, say $\pm 5\%$, is sufficient. This permits the use of separate control oscillators, or "clocks," at the transmitting and receiving ends, whose frequencies are very close but not necessarily exactly the same. In this case, directly connected synchronizing circuits between the two locations are not necessary.

8.8 THE COMPUTER AS A MEASUREMENT SYSTEM

Computers have become ubiquitous laboratory tools. They allow measurements to be recorded, processed, displayed, and printed with great ease. In some situations, sensors and transducers are connected directly to computers, which then take over the role of any additional instruments that might otherwise be required—voltmeter, oscilloscope, spectrum analyzer, and so on. Powerful software packages are available to perform all such functions on the computer, and they can transfer measurements to spreadsheet packages or other codes for data reduction and presentation. In the next several sections, we look at some of the details that underlie this convenience.

Computers are, of course, digital systems. The frequency meter mentioned in Section 8.7.5 is an example of how digital instrumentation can be built from integrated circuits. However, the data received by that meter are essentially digital in character: All that is required is to determine whether the input voltages are high or low. More often, we may need to deal with transducer signals that are entirely analog, as when a thermocouple voltage is used to measure temperature. In this case, we must convert the voltage to digital form if a digital device such as a computer is to process it. This transformation is accomplished by an *analog-to-digital converter* or ADC (Section 8.11). A computer system used for recording analog signals will incorporate an ADC, either as an integral part of the computer or within the transducer that is connected to the computer. In every case, the sample rate and voltage range of an ADC must be chosen to avoid aliasing and other forms of signal distortion (Section 8.11.3).

Data from the ADC will be transferred to the computer's storage or central processing units along one of the computer's *buses*, which are internal circuits for passing data and control signals. For most computer users, only the *external bus* (Section 8.10.4) will be of concern. The external bus governs the operation of the computer's input and output *ports*, which connnect printers, remotely operated instruments or controls, external data storage devices, modems, and other peripherals. Buses may use either serial or parallel transmission of data. Common types of buses include the IEEE 1394 (*FireWire*), the RS-232C, the IEEE 488 (GPIB), and the *Universal Serial Bus* or USB.

At the heart of most current computers is a VLSI integrated circuit called a *microprocessor* (Section 8.9). The microprocessor is a collection of digital circuits that form the *central processing unit*, or *CPU*, of the computer. The microprocessor executes the instructions of computer programs and controls the transfer of information between the input ports, memory, and the output ports. Computers using microprocessor CPUs are sometimes called *microcomputers*, even though today's microcomputers are so powerful that the term *micro* may be a bit misleading! We describe the internal architecture of the microcomputer in Section 8.10.

8.9   THE MICROPROCESSOR

A microcomputer is built is around a microprocessor chip that acts as its central processing unit (CPU). The CPU serves as the control center for directing the flow of digitized information. It is more than a traffic controller because it not only provides the organizational plan for the flow of elemental bits of information but also assigns the pathways, temporary "parking" spaces, and stop/go gating and can perform a limited manipulation of the traffic. It accepts inputs in digital form, either data or command instructions, and routes them to predetermined (programmed) destinations over buses (pathways) to displays, memories, controllable devices, and so on. Sources of the data or commands may be external memories, keyboards, transducers, or other devices.

   A wide range of CPUs are available, with many levels of complexity—4-bit, 8-bit, 16-bit, 32-bit, 64-bit, and so on. The CPU selected for a given purpose will depend on the application. For some simple low-end dedicated uses a 4-bit or 8-bit CPU may suffice, whereas 32-bit or 64-bit CPUs may be selected for general-purpose or high-end microcomputers. For purposes of illustration, the 8-bit Motorola 6800 microprocessor serves very well. This chip is obsolete for most purposes, but the general principles of its design are characteristic of most CPUs.

   Figure 8.17 is a highly simplified schematic diagram of the external connections and some of the internal features of the Motorola 6800. The diagram shows the primary buses into and out of the processor plus some essential internal devices. To be functionally useful, the system requires additional supporting circuitry external to the CPU, including interfacing devices sometimes referred to as buffers, input/output (I/O) facilities, synchronizing clocks, memory, and so on.

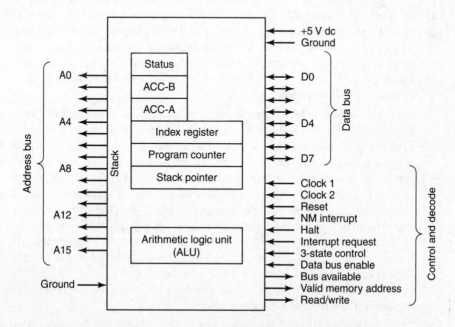

FIGURE 8.17: Simplified schematic diagram of the Motorola 6800 microprocessor.

The Motorola 6800 microprocessor is a single integrated circuit chip housed in a dual-in-line package (DIP) similar to, but larger than, the one shown in Fig. 7.21. The dimensions, exclusive of the pins that provide electrical connections, are about $6 \times 20 \times 50$ mm. The power requirement for this CPU by itself is 0.6 to 1.2 W at 5 V dc—less than 1% of the wattage of a large CPU today.

The figure shows the following:

1. An *address bus* consisting of 16 parallel lines for accessing (connecting to) $2^{16} = \text{FFFF}_{16} = 65,536_{10}$ different memory locations. The actual number available is, of course, dependent on what may be provided by supporting hardware.

2. A *data bus* consisting of eight bidirectional lines for simultaneously handling 8 bits (1 byte) of data. This bus is a two-way street, and the direction of flow is controlled by gating (Section 8.7.2). Eight bits provide for $2^8 = \text{FF}_{16} = 256_{10}$ combinations.

3. Various *control and decode lines*. Bus synchronization, closely akin to multiplexing–demultiplexing (Section 8.7.6), is controlled through use of two external clock signals. Additional lines are shown, and in many cases their titles provide a clue to their uses.

Figure 8.17 also shows some of the internal structure of the 6800 chip. Various *registers* are used. These may be considered as *temporary* storage bins or momentary "parking" locations for data, instructions, or addresses as they are being shunted from one location to another. Essentially, all data and instructions must pass through at least one of the accumulators, *A* or *B*, each having a 1-byte (8-bit) handling capability. Primarily for providing manipulation of 2-byte (16-bit) addresses, the *index register* (IR), *program counter* (PC), and *stack pointer* (SP) are added.

As illustrated in Fig. 8.17, the *stack* consists of the registers as shown. Their contents are stored in contiguous addresses, and the purpose of the stack pointer is to keep track of where the stack information is stored in the external random access memory (RAM). The *program counter* controls the sequencing of any program steps, including the starting point, and the *index register*, in addition to other functions, provides a channel through which 2-byte addresses may be handled in a program. The contents of the registers are always available on command.

The arithmetic logic unit (ALU) has a relatively limited capability of manipulating numbers. It can add one byte to another or determine their difference, or it can perform several logic functions, such as AND, OR, and EOR (Section 8.4.1). To handle data in magnitudes requiring several bytes, the add and subtract functions of the ALU may involve carryovers (for addition) or borrows (for subtraction). It is the responsibility of the status register (SR) (also called the condition code register) to monitor such requirements for possible further program use. *Flags* (primary data bits) in the status register are either set (made equal to 1) or not set (made equal to 0), depending on predetermined conditions. If an addition results in a carryover from one byte to the next byte, a bit momentarily indicating that fact, stored in the status register, will be added to the byte of the next higher order. Or, if the operation results in a zero, a zero flag in the SR may be used to trigger the decision to branch (or not to branch) to some other point in a program. These are only two of a number of powerful functions of the status register.

For a more detailed discussion of microprocessors and for information on current designs, refer to the Suggested Readings at the end of the chapter.

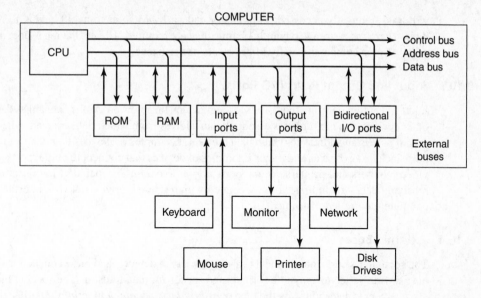

FIGURE 8.18: Typical microcomputer system.

8.10 THE MICROCOMPUTER

As noted previously, a microprocessor must be surrounded by a number of servants before it can claim to be a microcomputer. Figure 8.18 is a schematic diagram of a microcomputer system showing some of the essential components. The various peripheral devices communicate with the CPU through the three microprocessor buses mentioned in the previous section.

8.10.1 Read-Only Memory (ROM)

ROM is used to store programs that direct the computer's operation at start-up and that guide basic video and keyboard communications. Collectively, these instructions are usually referred to as the system BIOS, or *basic input–output system*. When the computer is turned on, the BIOS will direct the computer to read and load the operating system software from the computer's hard disk drive. Data stored in ROM are usually semipermanent, in the sense that they are preserved when the computer is powered off but can be updated by the user from time to time.

8.10.2 Random Access Memory (RAM)

RAM is the "bank" that is used for temporary deposits and withdrawals of data or information required for the operation of programs. RAM may be both read and written, as program instructions are loaded, sent to the CPU, or replaced by other program instructions. All addresses in RAM may be accessed in any order—thus the term *random*. Random access memory is usually *volatile*—if the power is turned off or if the system is rebooted, all stored data are lost.

Data can usually be accessed much faster when they are in RAM than when they are on a nonvolatile mass storage device such as a hard disk drive or CD-ROM. Thus, computers will hold frequently accessed programs or data in RAM so as to avoid having to recover them from a mass storage device each time they are required. Hence, having a large amount of RAM on a computer usually improves its performance.

8.10.3 Input and Output Ports (I/O Ports)

Input and output ports are the portals through which the microprocessor exchanges data with other devices. An *input port* is essentially an address from which data are read, whereas an *output port* is an address to which data are sent. Examples of devices that connect to input ports include keyboards and mice. Examples of devices that connect to output ports include video monitors and printers. Some ports are *bidirectional*—capable of both sending and receiving data—including those associated with hard disk drives, network interfaces, and computer-controlled instruments.

8.10.4 External Buses

The computer's ports may connect to a wide variety of devices, and these connections in turn have a wide variety of possible formats and speeds for data transfer. The specific interface arrangement is determined by the type of *external bus* or *communications interface* through which the port reaches a particular device. Some common types are listed in Table 8.6.

Either serial or parallel data transfer (Section 8.6.3) may be used to connect measuring instruments or sensors to a computer. The speed of a serial connection is measured in bits of data transferred per second. The speed of a parallel connection is measured in bytes transferred per second; parallel connections are usually 8 or 16 bits wide; that is, 1 or 2 bytes. Parallel cables generally have a large number of wires, ranging from 24 to 68 in most cases. Network and wireless connections are also used by some measuring devices (data transfer on these buses is serial). An instrument with a network interface has a network address that computers and devices on the network may use to obtain readings or send instructions.

Like most other areas of digital technology, communications arrangements are constantly evolving and the available speed of data transfer is constantly increasing. Recently, the IEEE 1394 and USB 2 buses have become the most common serial standards for connecting devices to computers. Older serial buses, based on RS-232C or USB 1, are being phased out quickly. Parallel connections were standard for printers until the late 1990s, but they are also being replaced by USB 2. In the past, much effort might be required to establish proper connections of devices to computers, possibly involving a great deal of time soldering wires to RS-232 connectors and trial-and-error adjustment of software settings. A particular problem was the synchronization of two-way communication over bidirection lines, a process called *handshaking*. Today's buses are better standardized and better integrated with computer operating systems, so that establishing two-way communication is usually as simple as plugging in a connector.

8.11 ANALOG-TO-DIGITAL AND DIGITAL-TO-ANALOG CONVERSION

As we discussed in Section 8.3, some measurands originate in digital form. Most mechanical inputs, however, exist in analog form. Hence, before digital data processing can be accomplished, an analog-to-digital conversion is necessary. In a like manner, if a computer's digital output is used to drive an analog device, a digital-to-analog conversion must be performed.

TABLE 8.6: External Buses

Standard	Maximum Speed*	Comments
Serial Interfaces		
RS-232C	~10 kbps	Cables limited to 15 m. 9- or 25-pin connectors. Was once very widely used.
RS-422A	10 Mbps	Up to 1200 m cables at lower speeds.
USB 1	12 Mbps	Universal Serial Bus. Also has 1.5 Mbps speed. Up to 127 devices may be daisy-chained on a single port. Four-pin connector. Superceded by USB 2.
IEEE 1394	400 Mbps	Trade-named FireWire. Commonly used for digital video. Up to 63 devices may be daisy-chained on a single port. Six-pin connector.
USB 2	480 Mbps	Fast enough for digital video.
Parallel Interfaces		
Centronics	150 kBps	Orignal PC parallel port connector. Output only. Speed often much slower than maximum. Up to 10 m cable. 8 bit. 25- and 36-pin connectors. Obsolete.
IEEE 488	1 MBps	Developed by Hewlett Packard Corp. in 1960s as the HP-IB bus. Up to 14 instruments controlled by one device. Up to 20 m cable length. 8 bit. 24-pin connectors. Also called General Purpose Interface Bus, or GPIB.
IEEE 1284	2 MBps	Bidirectional successor to Centronics bus. Released in 1994. Up to 10 m cable. 8 bit. 25- and 36-pin connectors.
SCSI	320 MBps	Small Computer System Interface. Many different versions exist, including 8 and 16 bit. 50- and 68-pin connectors. Up to 12 m cable. Up to 16 devices. Commonly used as hard disk drive bus.
Network and Wireless Interfaces		
Dial-up modem	56 kbps	Early modems were just 300 bps!
ADSL modem	varies	Maximum speeds of several Mbps; actual speeds often a few hundred kbps.
Bluetooth	3 Mbps	Wireless connection in 2.4-GHz band. 10 m range.
IEEE 802.11b	11 Mbps	Wireless ethernet in the 2.4-GHz band.
IEEE 802.11g	54 Mbps	Wireless ethernet in the 2.4-GHz band.
IEEE 802.11a	54 Mbps	Wireless ethernet in the 5-GHz band.
IEEE 802.3	10 Gbps	Ethernet. 10-Mbps and 100-Mbps networks are common. Higher-speed networks sometimes use optical fiber.

*bps = bits per second; Bps = bytes per second.

Analog-to-digital (A/D) and digital-to-analog (D/A) conversions can be executed using a variety of different circuits. A/D and D/A converters are often manufactured as integrated-circuit chips for incorporation into other devices, but they may also be obtained as cards that plug into computers or as stand-alone data recording systems. The basic measures of performance are the number of bits used and the speed of processing. A wide range of designs exist for both A/D and D/A converters. Here we describe only typical examples of each, leaving further development to the Suggested Readings.

8.11.1   A Digital-to-Analog Converter

A simple D/A converter is based on the summing amplifier (Example 7.9, Section 7.14.2). Referring to Fig. 8.19, we see that the currents summed are controlled by a set of digital switches. Four basic elements are involved:

1. A stable reference voltage (for instance, $E_{ref} = 1$ V dc).

2. A ladder arrangement of summing resistors. For this 8-bit converter, eight ladder resistors are used. The resistance values increase in a sequence of powers of 2 from $2^0 R$ to $2^7 R$.

3. A series of switches. These are not mechanical switches; rather, they are solid-state gates [e.g., simple AND ICs (Section 8.4.1)]. The eight switches can be activated by digital inputs, so their operation may be controlled by the respective bits contained in a single byte of data.

4. Op-amp output circuitry. The op-amp output voltage is equal to $-R_G$ times the sum of the currents from each ladder branch. The gain-controlling resistor, R_G, is selected to *scale* the output voltage, perhaps to a maximum magnitude of 10 V dc.

When a switch is closed, a current is delivered to the op amp in proportion to the power of 2 for that circuit's resistor: Switch 0 contributes $i_0 = E_{ref}/128R = 2^{-7}E_{ref}/R$, switch 1 contributes $i_1 = E_{ref}/64R = 2^{-6}E_{ref}/R$, and so on. Hence, switch 0 corresponds

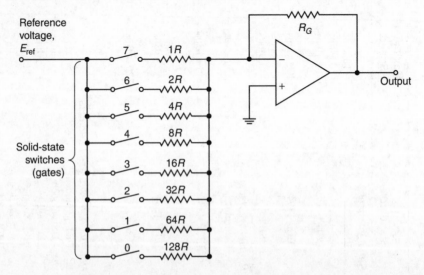

FIGURE 8.19: A simple 8-bit DAC (digital-to-analog converter).

to the least significant bit, b_0; switch 7 corresponds to the most significant bit, b_7; and so on. By closing selected switches, the output voltage can be made proportional to any particular 8-bit number. For example, if switches 0, 5, and 6 are closed, the output voltage is proportional to $0110\,0001_2$ ($= 61_{16} = 97_{10}$). Since R_G sets the constant of proportionality, the output gain can be scaled to provide an appropriate range of analog values. Additional output circuitry could be used to invert the signal or to add offset voltages.

8.11.2   An Analog-to-Digital Converter

One typical A/D converter is the *parallel encoder* (Fig. 8.20). This circuit uses a set of voltage comparators (Example 7.8, Section 7.14.2) and a series of resistors to compare

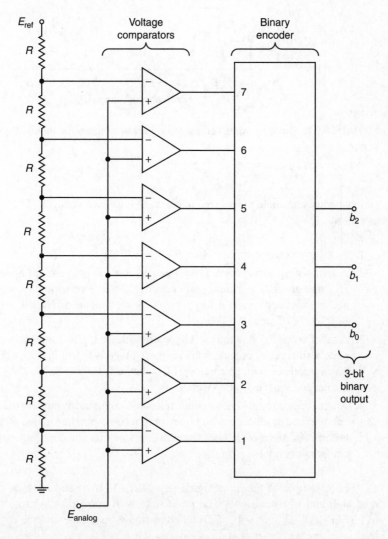

FIGURE 8.20: A 3-bit parallel A/D converter.

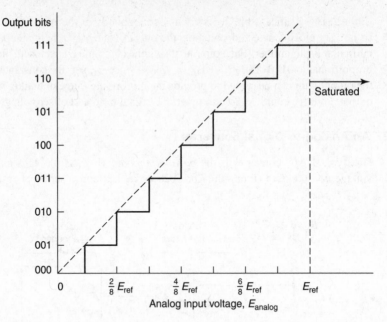

FIGURE 8.21: Binary output versus analog input voltage for the 3-bit parallel A/D converter.

simultaneously an analog input signal to a set of reference voltages. The basic elements are as follows.

1. A stable reference voltage (such as $E_{ref} = 1$ V dc).
2. A series of equal resistors. These resistors form a voltage-dividing sequence between E_{ref} and ground. For this 3-bit converter, $2^3 = 8$ resistors are used. The voltages at the nodes between resistors increase in increments of $(R/8R) \cdot E_{ref}$, specifically, $(\frac{1}{8})E_{ref}, (\frac{2}{8})E_{ref}, \ldots, (\frac{7}{8})E_{ref}$.
3. A set of voltage comparators. The analog voltage is simultaneously compared to each node's voltage. A comparator's output voltage is high (on) when E_{analog} is above a given reference voltage and low (off) when it is below. Seven $(2^3 - 1)$ comparators are needed for the 3-bit converter.
4. An encoder circuit. The encoder reads the comparator outputs (each a high or low voltage) and produces a 3-bit binary output corresponding to one of the eight possible on/off conditions of inputs 1 through 7. (Note that one condition is to have all seven comparators off.)

For example, if the input signal is between $(\frac{3}{8})E_{ref}$ and $(\frac{4}{8})E_{ref}$, comparators 1–3 read high and 4–7 read low. The input state is 3, corresponding to a binary output of $011_2 (=3_{10})$. If $E_{analog} > (\frac{7}{8})E_{ref}$, all comparators read high and the output is 111_2; and if $E_{analog} < (\frac{1}{8})E_{ref}$, all are low and the output is 000_2. The digital output is shown as a function of E_{analog} in Fig. 8.21.

Parallel A/D conversion is particularly fast, since all bits are set simultaneously, for which reason it is sometimes called *flash encoding*. Other A/D converters may use a successive approximation technique that sets one bit at a time. Yet another type of A/D converter uses a voltage-controlled oscillator (VCO) to convert voltage to frequency and a digital frequency meter to digitize the resulting frequency.

If the output of an A/D converter is sent to a binary-to-BCD encoder and from there to a BCD-to-seven-segment decoder/driver, and then to an LED display, the result is a *digital voltmeter*. Integrated-circuit technology allows all these functions, and others, to be placed on a single (and inexpensive) chip, which might form the heart of a handheld instrument. We discuss digital meters further in Chapter 9.

8.11.3   Analog-to-Digital Conversion Considerations

High-quality analog-to-digital converters are readily available to the experimentalist, so it may be tempting treat them as black boxes. They cannot, however, be used carelessly if accurate results are to be obtained. In the following, we outline some of the most important considerations for the end user of an ADC.

Saturation Error

The most obvious limitation of an A/D converter is that it has definite upper and lower limits of voltage response. Typical full-scale input ranges are 0 to 10 V and -10 to $+10$ V. If the input signal exceeds the upper or lower limits of response, the converter saturates and the recorded signal does not vary with the input. This situation can be prevented by appropriate signal conditioning, such as amplitude attenuation or dc offset removal.

Resolution and Quantization Error

As we have seen, an A/D converter responds to discrete changes in the input voltage. For example, the 3-bit parallel encoder's output steps correspond to changes in E_{analog} of $E_{ref}/2^3$ (Fig. 8.21). Thus there is a smallest increment of voltage change that can be resolved by an A/D converter. In general, the voltage resolution per bit, ε_V, depends on the full-scale voltage range and the number of bits of the converter:

$$\varepsilon_V = \frac{\Delta V_{fs}}{2^n} \tag{8.1}$$

where

$$\Delta V_{fs} = \text{the full-scale voltage range, and}$$
$$n = \text{the number of bits of the A/D converter}$$

Typical A/D converters have 8, 12, or 16 bits, corresponding to division of ΔV_{fs} into a total of $2^8 = 256$, $2^{12} = 4096$, and $2^{16} = 65,536$ increments. A 16-bit converter with a -10 to $+10$ V range has a voltage resolution of 0.3 mV. The voltage resolution is a known value for a given A/D converter, and it is needed to process the digitized data.

The finite resolution of the A/D converter introduces error in the recorded values, since the actual analog voltage usually lies between the available bit levels. This is called *quantization* error (since the bit levels are "quantized"), and it is entirely analogous to the reading error of a digital display (Section 3.2.2). An estimate for the *quantization uncertainty* is $u_q = \varepsilon_V/2$ (95%). Quantization errors may be reduced by using an A/D converter with more bits.

Conversion Errors

A/D converters may also suffer from slight nonlinearity, zero-offset errors, scale errors, or hysteresis. Such errors are a direct by-product of the particular method of input quantization. For example, the conversion illustrated in Fig. 8.21 is, on the average, low by one-half of the least significant bit (the amount by which the solid curve lies beneath the dashed line). Normally, the manufacturer will provide specifications for the potential size of such conversion errors.

Sample Rate

The rate at which an A/D converter records successive values of a time-varying input is called the sample rate. Each A/D converter has a maximum possible sample rate, of which typical values range from about 1000 Hz to more than 100 MHz. Software often allows the user to specify any sample rate up to this maximum value. The influence of sampling rate on the accuracy of the recorded signal is discussed in Sections 4.7.1 and 4.7.2: *Aliasing* and *frequency resolution* are of particular concern.

Signal Conditioning for A/D Conversion

To make the best use of an A/D converter, conditioning of the analog signal is often required. The most important considerations are the prevention of aliasing, the minimization of quantization errors, and the prevention of saturation errors. Aliasing can be prevented by using a low-pass, or *antialiasing*, filter to remove frequencies of $f_{sample}/2$ or more from the analog signal. Quantization error can be minimized by amplifying the signal to span as much of the full-scale range as possible. However, this approach sometimes conflicts with the need to avoid saturation errors.

As an example, a hot-wire anemometer signal (Section 15.9) may include both a 3-V dc component and important ac components of only 5 mV rms. Suppose that a ±10-V, 16-bit A/D converter is to be used for sampling. The ac components are near the A/D converter's resolution ($20\ \text{V}/2^{16} = 0.3$ mV), but large amplification of the sum of the ac and dc components will saturate the converter. To resolve the ac components accurately, an offset-and-gain (or "buck-and-gain") amplifier may be used. This amplifier subtracts (or "bucks") a precisely specified dc voltage from the analog signal and amplifies the remaining ac signal by a gain of several hundred. The dc offset voltage is recorded with one A/D converter channel and the amplified ac component is recorded with another. In this way, both components can be digitized accurately.

8.11.4   Digital Signal Processing

Digital techniques make it possible to filter or enhance analog signals such as those carrying sound or images in telecommunications systems, audio systems, and video systems. A *digital signal processor*, or DSP, is an integrated circuit that performs functions such as filtering using digital algorithms. Upon combining an analog-to-digital converter, an digital signal processor, and a digital-to-analog converter, one obtains a circuit that can, say, filter noise out of a telephone signal while the users are speaking.

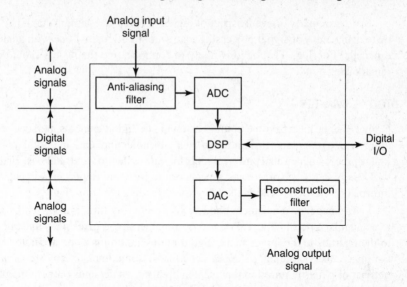

FIGURE 8.22: A digital signal processing system.

How does this work? As discussed in Section 4.7.1, digitized data can be processed numerically, for example, by using the Fast Fourier Transform (FFT). The FFT allows identification of the frequencies present in a signal, and with that information it becomes possible to filter the signal digitally: Numerical algorithms can be used to eliminate unwanted frequencies or to enhance frequencies of interest [9]. With sufficient computational power, these digital filtering techniques can produce sharp frequency cutoffs that would be very difficult to obtain from the analog filters described in Sections 7.16–7.18. Furthermore, digital filters are not subject to drift when the temperature shifts, as analog filters are, and they may be programmed to achieve different effects as required without changing any components in the circuit.

Figure 8.22 shows a schematic arrangement for a digital signal processing system, all of which might reside on a single IC chip. An analog signal is passed through an antialiasing filter to an A/D converter. The DSP then processes the signal and sends the output to a D/A converter. The reconstructed analog signal is smoothed by a final low pass (or *reconstruction*) filter as it leaves the chip. The DSP itself may be controlled by through an external I/O interface. For an audio DSP, this might enable varying levels of artificial reverberation to be added to the original signal, so as to simulate a concert hall sound for part of a music track.

A DSP is essentially a specialized microprocessor, designed for high-speed numerical calculations. Like other microprocessors, as DSPs have become cheaper, smaller, and more powerful, their applications have multiplied. They are now found in cell phones, video recorders, CD players, computer sound cards, and digital television sets. DSP is used in medical image processing, as for magnetic resonance imaging (MRI) and computed tomography (CT) scanning. DSP is essential to modern telecommunications sytems, where it provides echo reduction, signal compression, and signal multiplexing. DSP has found application in speech recognition systems, radar, sonar, seismology, high-speed modems, MP3 players, and many other areas.

A particularly relevant aspect of digital signal processing is that it has come to dominate many areas of signal processing that were traditionally done with analog techniques, especially filtering. This trend is likely to continue, and the results should continue to be remarkable.

8.12 DIGITAL IMAGES

Digital images have become ubiquitous owing to digital cameras, personal computers, and digital video recorders. In addition to their consumer application, digital images and digital image processing are also powerful tools for data collection and analysis. In this section, we will briefly describe some basic aspects of this vast and important subject. More detailed information may be found in the Suggested Readings for this chapter.

An image, be it a camera photograph or an x-ray film, is *digitized* by dividing it into a rectangular grid of cells, called *pixels*. In general, the pixel sizes must be small relative to the features in the image if the digital representation is to be accurate. When it is not, the image takes on a jagged appearance that is sometimes referred to as *pixelation*. The number of pixels required to represent an image thus depends on both the physical size of the image and the amount of fine detail to be recorded. For an image digitized to a grid of 640 pixels by 480 pixels, a total of $640 \times 480 = 307,200$ pixels are required. If 1600 by 1200 pixels are used, the image has 19.2 million pixels!

Each pixel carries information about the intensity and color of the image at that point. For a purely black-and-white image, only one bit is needed per pixel: 0 = black and 1 = white, say. For a grayscale image, 8 bits (or 1 byte) might be used to carry the intensity, allowing $2^8 = 256$ possible shades of gray to be applied to a single pixel. For color images, a common approach is to use three colors per pixel: red, green, and blue, or RGB. Each color has a different intensity in each pixel. If 8 bits are assigned to each of the three colors (for a total of 24 bits per pixel), then each pixel has $2^8 \times 2^8 \times 2^8 = 16.8$ million possible mixtures of color.

The total number of bits needed per image can thus be quite large. A black-and-white image at 640 by 480 pixels needs only 307,000 bits (or 38,400 bytes). The same image in a 24-bit color format would be $24 \times 640 \times 480 = 7.4$ million bits (or 921,600 bytes), and a 24-bit color image at 1600 by 1200 pixels would require a total of 46 million bits (or about 5.8 million bytes). For comparison, the full text James Joyce's novel *Ulysses* can be stored in ASCII format in about 12 million bits (or 1.5 million bytes) [10]. Perhaps this gives added meaning to the saying "A picture is worth 1000 words."

A digital video is simply a succession of digital images, or frames, shown at high speed. A typical *framing rate* for normal viewing is 30 frames/second. For purposes of measurement, much higher framing rates may be employed, ranging up to 100,000 frames/second or more, as when it is desired to record processes happening on millisecond or microsecond timescales. At 30 frames/second and a video image size of 640 by 480 pixels in 24-bit color, the rate of data transfer for recording or display is $30 \times 640 \times 480 \times 24 = 221$ million bits/second. Comparison to Table 8.6 shows that connection of a video camera to a computer will require either a USB 2 or IEEE 1394 bus.

We may compare video data rates to those for high-quality audio, as on a music CD. The CD must reproduce the full spectrum of human hearing, which contains frequencies up to 20 kHz, so it is sampled at 44.1 kHz (giving a Nyquist frequency of 22 kHz). Each sample records the volume in a 16-bit format. Two-channel audio (stereo) therefore has a data rate of 1.4 million bits/second. Interfacing audio signal to computers is clearly much less demanding of speed than is video. Similarly, audio files usually require less storage space than video files.

Once an image has been digitized, computer software may be used to manipulate it in various ways, perhaps to enhance dim features or to smooth out pixelation. One of the most common digital processes applied to images is *image compression*. Because much of data in an image will be similar in adjacent pixels, it is possible to use numerical algorithms to reduce the amount of information that must be stored. A very common compression method for single images is JPEG compression; for video streams, MPEG compression is a typical format.[1] Many other compression algorithms are in use. With appropriate compression, digital image or video files can be reduced in size by an order of magnitude with little apparent loss of clarity.

Digital techniques may also be used to identify features and to extract measurements from images. The possibilities are almost limitless. To name just two, digital video may be used to record the motion of particles suspended in a liquid, and software may be used to locate the particles in the image and to compute the velocity field from the sequence of particle positions (see Section. 15.11). A second example is infrared temperature measurement. An appropriate detector array can be used to photograph the infrared radiation from a surface, and software can then be used to calculate the temperature at the location of each pixel giving the temperature distribution on the surface (see Section 16.8.5).

Digital images can be produced by nonphotographic methods, such as computer graphics programs, but for photographic recording the essential detecting element is often a CCD array. The CCD array, or *charge-coupled diode* array, is a semiconductor device. It consists of an array of photodetecting cells that generate charge in response to photons they absorb [11]. A CCD array is situated behind the focusing lenses of a camera (in the place where one would have once found film). After the camera shutter is opened and closed, the amount of charge accumulated is proportional to the intensity of light received. The charge is converter to voltage using a charge amplifier (Section 7.15.2), and the voltages for each pixel in the array are digitized by an A/D converter and recorded. The digital number associated with each pixel (or numbers, for color photography) define the intensity for that cell of the digital image.

8.13 GETTING IT ALL TOGETHER

The possibilities for putting together a sophisticated digital measurements system for a given project are entirely open ended. The degree of completeness is usually limited by resources. In many situations the design and assembly of a complex system may be more costly in time and money than the project warrants. On the other hand, when great masses of data are to be collected, requiring extensive computational time to digest, or when continuous monitoring and control is needed, funds and time expended in putting together an "automatic" system may very well be cost-conservative.

Figure 8.23 shows some of the tremendous possibilities for advanced data gathering and processing. It should be noted that the transducers and A/D subsystem may be partially or entirely incorporated into some type of digital measuring instrument which connects

---

[1] JPEG is an acronym for Joint Photographic Experts Group. MPEG is an acronym for Moving Picture Experts Group. These groups work in coordination with the International Organization for Standardization (ISO) to develop international standards in their respective areas.

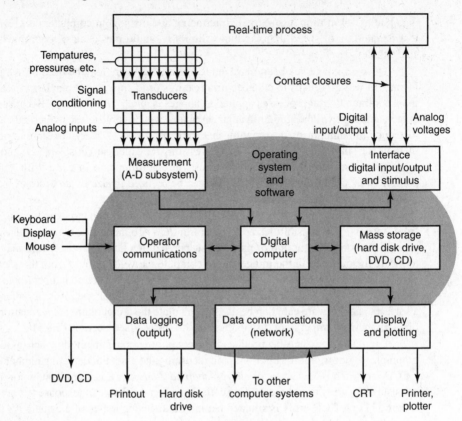

FIGURE 8.23: A block diagram illustrating the potential of an integrated measurement/control system.

to the computer via an appropriate external bus. We consider such instruments further in Chapter 9.

8.14 SUMMARY

In bringing this chapter to a close, we reiterate that it is presumptuous to attempt to summarize digital techniques in so few pages. We hope, however, that the material presented will serve as an introduction to further study. There is no doubt as to the value of the topic as it relates to mechanical measurements, and further developments are continually increasing this importance.

1. A digital signal has only two values—on or off, 0 or 1, yes or no, black or white, low voltage or high. Digital information may be composed of sets of bits, each of which takes one of the two possible values (Section 8.1).

2. Digital instrumentation has the advantage of direct computer interfacing, inherent noise resistance, low system voltages, and direct numerical display of readings (Section 8.2). Analog signals must be converted to an appropriate digital form to be processed by a digital measuring system (Section 8.3).

3. The basic elements of digital circuits are switches (transistors) and logic gates. These elements are the building blocks for more complex devices, such as flip-flops, innumerable integrated circuits, clock or timer circuits, and digital displays (Section 8.4).

4. Digital information, when expressed as a string of bits, may be interpreted using the binary number system by assigning the values 0 and 1 to the bit levels (Section 8.5). Binary numbers, in turn, allow information to be expressed in digital codes, such as binary-coded decimal (BCD), the American Standard Code for Information Interchange (ASCII), and the Latin 1 character set (ISO 8859-1). Digital patterns (strips of black and white) may be used to encode positional information or product labels (Section 8.6).

5. Simple digital circuits include events counters, frequency meters, waveform-shaping devices, and multichannel switches or multiplexers (Section 8.7).

6. The digital computer is a natural component of measuring systems, both for recording and processing of data. The microcomputer is composed of a microprocessor, permanent and temporary memory (ROM and RAM), input and output ports, and external buses or communications interfaces (Sections 8.8–8.10).

7. Conversion of signals between analog and digital form is essential when analog transducers or devices are coupled to a digital computer or microprocessor. These operations are achieved using either analog-to-digital (A/D) or digital-to-analog (D/A) converters. By combining an A/D and a D/A converter with a digital signal processor (DSP), analog signals can be efficiently filtered, enhanced, and processed (Section 8.11).

8. Images may be digitized by breaking them into a two-dimensional grid of pixels. Each pixel is assigned a digital number or numbers that describe the intensity and/or color at that point in the grid. Digital video is a succession of digital images played at high speed. Digital image and video files may be manipulated by computer to extract measurements of various sorts. The associated files and data transfer rates can be relatively large (Section 8.12).

SUGGESTED READINGS

Balch, M. *Complete Digital Design*. New York: McGraw-Hill, 2003.

Brey, B. B. *The Intel Microprocessors: 8086/8088, 80186/80188, 80286, 80386, 80486, Pentium, Pentium Pro Processor, Pentium II, Pentium III, and Pentium 4: Architecture, Programming, and Interfacing*. Upper Saddle River, N.J.: Prentice Hall, 2003.

Floyd, T.L. *Digital Fundamentals*. 8th ed. Upper Saddle River, N.J.: Prentice Hall, 2003.

Gonzalez, R. C., and R. E. Woods. *Digital Image Processing*. 2nd ed. Upper Saddle River, N.J.: Prentice Hall, 2002.

Horowitz, P., and W. Hill. *The Art of Electronics*. 2nd ed. Cambridge, U.K.: Cambridge University Press, 1989.

Korneev, V. V., and A. Kiselev. *Modern Microprocessors*. 3rd ed. Hingham, Mass.: Charles River Media, 2004.

Kularatna, N. *Digital and Analogue Instrumentation*. London: The Institution of Electrical Engineers, 2003.

Rathore, T. S. *Digital Measurement Techniques*. 2nd ed. Pangborne, U.K.: Alpha Science International, 2003.

Smith, S. W. *Digital Signal Processing: A Practical Guide for Engineers and Scientists*. Boston: Newes, 2002.

Tocci, R. J., N. S. Widmer, and R. L. Moss. *Digital Systems: Principles and Applications*. 9th ed. Upper Saddle River, N.J.: Prentice Hall, 2004.

PROBLEMS

8.1. Write the following binary numbers as base 10 numbers.
   (**a**) 1010111
   (**b**) 1010
   (**c**) 1111
   (**d**) 10111011

8.2. Write the following decimal numbers as binary numbers.
   (**a**) 16
   (**b**) 87
   (**c**) 419
   (**d**) 40177

8.3. Use a truth table to show that the output from the circuit in Fig. 8.24 is high except when the two inputs to either one or both of the AND gates are simultaneously high.

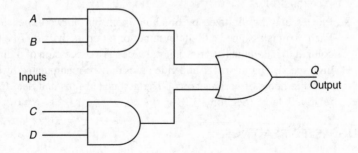

FIGURE 8.24: AND/OR circuit for Problem 8.3.

8.4. Use a truth table to show that the output from the circuit in Fig. 8.25 will be high only when both inputs of either of the OR gates are simultaneously low.

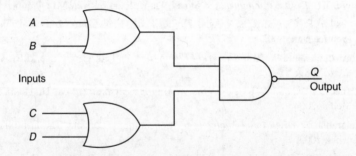

FIGURE 8.25: OR/NAND circuit for Problem 8.4.

8.5. Find the percent resolution of each.
  (**a**) An 8-bit A/D converter
  (**b**) A 12-bit A/D converter

8.6. The output from a temperature sensor is expected to vary from 2.500 mV to 3.500 mV. If the signal is fed to a 12-bit A/D converter having a ±5.0 V range, estimate the voltage increment represented by LSB. By what gain should the signal be amplified?

8.7. Each step of an 8-bit D/A converter represents 0.10 V. What will the output of the D/A converter be for the following digital inputs?
  (**a**) 01000111
  (**b**) 10101101

8.8. Use a truth table to determine under what set of conditions the output from the three-input NAND gate shown in Fig. 8.26 will be low.

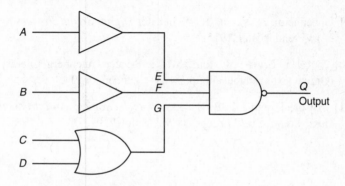

FIGURE 8.26: Circuit for Problem 8.8.

8.9. Compact disc digital audio tracks are usually recorded using 16 bits to digitize the volume of each sample on each of the two stereo tracks. Samples are taken at a frequency of 44.1 kHz. Some recordings are made using 24-bit samples and sample rates of 96 kHz.

  (**a**) What file size (in bits) does each format require to record a three minute song?
  (**b**) What is the Nyquist frequency for each format and how does it relate to the range of human hearing, which ends at about 20 kHz?
  (**c**) If the dynamic range compares the highest voltage amplitude recordable to the lowest nonzero amplitude recordable, what is the dynamic range of each format in dB? Assume that the DA converter driving the speakers is noise free. How does this compare to human hearing, which spans roughly 120 dB?

REFERENCES

[1] Robertson, D. *Alec Reeves 1902–1971*. Sacramento, Calif.: Privateline.com, 2002.

[2] Horowitz, P., and W. Hill. *The Art of Electronics*. 2nd ed. Cambridge, U.K.: Cambridge University Press, 1989, p. 240.

[3] Floyd, T. L. *Digital Fundamentals*. 8th ed. Upper Saddle River, N.J.: Prentice Hall, 2003, Chap. 8.

[4] Berlin, H. M. *The 555 Timer Application Source Book with Experiments*. Derby, Conn.: E & L Instruments, Inc., 1976.

[5] Horowitz, P., and W. Hill. *The Art of Electronics*. 2nd ed. Cambridge, U.K.: Cambridge University Press, 1989, Section 5.19.

[6] American National Standards Institute. *Coded Character Set—7-Bit American National Standard Code for Information Interchange (7-Bit ASCII)*, ANSI X3.4-1986. New York: ANSI, 1986.

[7] ISO 8859-1:1998. *Information technology—8-Bit Single-Byte Coded Graphic Character Sets—Part 1: Latin Alphabet No. 1*. Geneva, Switzerland: International Organization for Standardization, 1998.

[8] Aliprand, J., et al. (eds.) *The Unicode Standard, Version 4.0*. Boston: Addison-Wesley, 2003.

[9] Oppenheim, A. V., and R. W. Schafer *Digital Signal Proceesing*. Englewood Cliffs, N.J.: Prentice Hall, 1975.

[10] Joyce, J. *Ulysses*. Oxford, Miss.: Project Gutenberg Literary Foundation, 2002. (Online text available at http://www.gutenberg.org/)

[11] Wilson, J., and J. F. B. Hawkes. *Optoelectronics: An Introduction*. 3rd ed. Harlow, U.K.: Prentice Hall Europe, 1998, Section 7.3.7.

CHAPTER 9

Readout and Data Processing

9.1   INTRODUCTION

Final usefulness of any measuring system depends on its ability to present the measured output in a form that is comprehensible to the human operator or the controlling device. The primary function of the terminating device is to accept the analogous driving signal presented to it and to either provide the information in a form for immediate reading or record it for later interpretation.

For direct human interpretation, except for simple yes-or-no indication, the terminating device presents the readout (1) as a relative displacement, or (2) in digital form. Examples of the first are a pointer moving over a scale, a scale moving past an index, a light beam and scale, and a liquid column and scale. Examples of digital output are an odometer in an automobile speedometer, an electronic decade counter, and a rotating drum mechanical counter.

Examples of exceptions to the two forms just mentioned are any form of yes-or-no limiting-type indicator, such as the red oil-pressure lights in some automobiles, pilot lamps on equipment, and—an unusual kind—litmus paper, which also provides a crude measure of magnitude in addition to a yes-or-no answer. Perhaps the reader can think of other examples.

As we have stated on numerous occasions in previous chapters, measurement of *dynamic* mechanical quantities practically presupposes use of electrical equipment for stages one and two. In many cases the electronic components used consist of rather elaborate systems within themselves. This is true, for example, of the cathode-ray oscilloscope. Sweep circuitry is involved, providing a time basis for the measurement. In addition, the input is carried through further stages of amplification before final presentation. The primary purpose of the complete system, however, is to present the input analogous signal in a form acceptable for interpretation. Such a self-contained system will therefore be classified as an integral part of the terminating device itself.

For the most part, dynamic mechanical measurement requires some form of voltage-sensitive terminating device. Rapidly changing inputs preclude strictly mechanical, hydraulic-pneumatic, and optical systems, either because of their extremely poor response characteristics or because the output cannot be interpreted. Therefore, the major portion of this chapter will be concerned with electric indicators and recorders.

The most basic readout device is undoubtedly the simple counter of items or events. Mechanically constructed counting devices are quite familiar. Old-style automobile odometers were of this type, simply counting the turns of a drive shaft through a gear reduction, which scales the readout to miles or kilometers. Modern laboratory-type counters, however, are electronic, such as those discussed in Section 8.7 and in the following section.

9.2 THE ELECTRONIC COUNTER

The electronic counter is a multipurpose digital counting device. A basic understanding of how the electronic counter operates suggests uses for this instrument and also makes the user aware of its limitations.

To determine the principles of operation, let us first consider the astable oscillator or timer (see Section 8.4.4). The output from the device is a square wave, which may be used as a timing signal [see Fig. 9.1(a)]. A second IC of importance is the J-K flip-flop, which is a variation of the S-R flip-flop discussed in Section 8.4.2. This circuit may be used as a divide-by-2 device. The J-K output is triggered by the negative side of the input square wave and the flip-flops may be cascaded with the output from one providing the input for the next. For example, for each *period* of the clock input, the value of the 4-bit binary output, A3, A2, A1, A0, is as shown in Fig. 9.1(b). A3 is the most significant bit (MSB) and A0 is the least significant bit (LSB).

This circuitry is called a binary counter and is the governing circuit for the universal counter. As shown, the counter will count only to 1111_2, or 15_{10}. By cascading *eight* J-K flip-flops, the count may be increased to $1111\ 1111_2$, or 255_{10}. See Section 8.7 for further discussion of electronic counters.

The schematic diagram for the electronic counter is shown in Fig. 9.1(c). The binary numbers are converted to decimals and displayed by means of light-emitting diodes (LEDs) or via a liquid-crystal display (LCD).

9.2.1 Event Counter

The event count is the simplest measurement to perform using an electronic counter. If the clock is disconnected and replaced by a device that produces a square pulse (presumably due to a single physical event), the result is an instrument that totalizes events, or an event counter, as shown in Fig. 9.2(a).

9.2.2 Time-Interval Meter

Figure 9.2(b) illustrates a binary counter with the addition of an AND gate (see Fig. 8.3) and an electronic circuit called an S-R latch [see Fig. 8.6(a)]. If the counter is armed (by pulsing the reset), $Q = 0$ on latch A and $Q = 1$ on latch B. At instant of time t_1, the voltage to latch A causes Q for latch A to go high. This, in turn, switches the AND gate on, and the clock signal passes through the AND gate, causing the counter to start counting. As soon as the voltage to latch B reaches time t_2, latch B output (Q) goes low, which stops

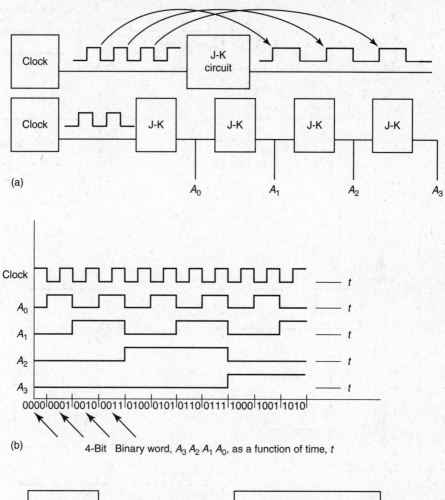

(a)

(b)   4-Bit   Binary word, $A_3 A_2 A_1 A_0$, as a function of time, t

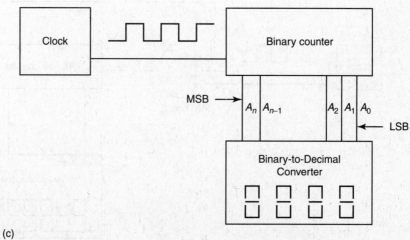

(c)

FIGURE 9.1: (a) Frequency-dividing circuit, (b) 4-bit binary counter, (c) schematic diagram for the electronic counter.

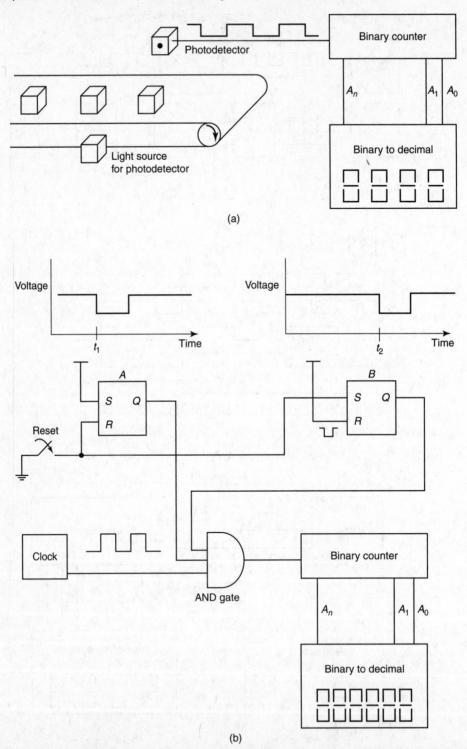

FIGURE 9.2: (a) The electronic counter used as an event counter, (b) time-interval meter, (c) EPUT meter.

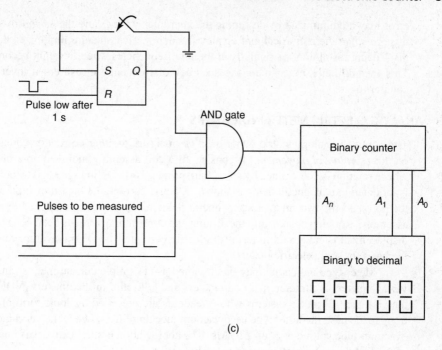

Pulse low after
1 s

AND gate

Pulses to be measured

Binary counter

A_n A_1 A_0

Binary to decimal

(c)

FIGURE 9.2: *continued*

the clock signal from passing through the AND gate. The final count is held in the counter, which represents the time $(t_2 - t_1)$. Obviously, the higher the clock frequency, the greater the accuracy of the measurement.

9.2.3 Events per Unit Time (EPUT) Meter

By rearranging the basic elements in an electronic counter, an EPUT meter [see Fig. 9.2(c)] may be obtained. In this mode the electronic counter is configured to count pulses in a specified length of time, called the *gate time*. The gate time is the interval between closing the switch on S and receiving the low pulse at R. If the gate time is 1 s, then the number of pulses on the counter is in pulses per second.

9.2.4 Count Error

Counting accuracy may involve time base error, trigger error, and a "± 1 count ambiguity." Time-base error is concerned with any deviation of the time-base oscillator frequency from intended frequency. This may result from a lack of both short-and long-term frequency stability. For most mechanical measurements, this error may be negligible (e.g., after proper warm-up a drift or aging rate of less than 3 parts in 10^7 per month may be attained). Trigger error is concerned with the preciseness with which the gating action is known or controlled. Uncertainties may be reduced by averaging over a longer period.

Finally, a ± 1 count error in electronic counting often exists because of the normal lack of synchronization between the gating and the measured pulses (whether from internal or external sources). It results from the possibility that the gate closing (or opening) may occur so as to barely miss the count of a passing cycle, but still, in fact, include (or exclude) the greater part of that particular cycle's period. For frequency measurement this source of error may be minimized by designing the instrument to measure the *period* of a cycle and then *compute* the reciprocal and display frequency. This feature minimizes the effect of the 1 count resolution, particularly at the low frequencies where it might become serious. This method helps to maintain the accuracy of frequency measurement over the entire counter range.

9.3 ANALOG ELECTRIC METER INDICATORS

The common analog electric meter used for measuring either current or voltage is based on the *D'Arsonval movement*. It consists of a coil assembly mounted on a pivoted shaft whose rotation is constrained by spiral hairsprings, as shown in Fig. 9.3. The coil assembly is mounted in a magnetic field, as shown. Electric current, the measurand, passes through the coil, and the two interacting magnetic fields result in a torque applied to the pivoted assembly. Rotation occurs until the driving and constraining torques balance. The resulting displacement is calibrated in terms of electric current. The D'Arsonval movement forms the basis for some electric meters.

Meter-type indicators may be classified as (1) simple current meters (ammeters) or voltage meters (voltmeters); (2) ohmmeters and volt-ohm-milliammeters (VOM or multi-meters); and (3) meter systems whose readouts are preceded by some form of amplification. In the past, the latter type used vacuum tube amplifiers; hence they became known as "vacuum-tube voltmeters," or VTVMs. The abbreviation is still heard occasionally in spite of the fact that solid-state amplifiers are now used.

The simple D'Arsonval meter movement is often used as the final indicating device. However, moving-iron meters may be used for measuring alternating current. In the more versatile types, such as the volt-ohm-milliammeter or multimeter, internal shunts or multipliers are provided with switching arrangements for increasing the usefulness of the instrument.

Basically, the D'Arsonval movement is current sensitive; hence, regardless of the

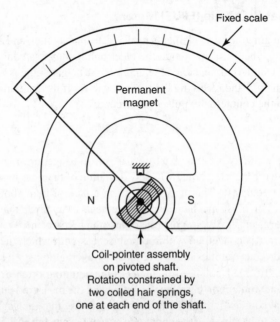

Fixed scale

Permanent
magnet

N   S

Coil-pointer assembly
on pivoted shaft.
Rotation constrained by
two coiled hair springs,
one at each end of the shaft.

FIGURE 9.3: The D'Arsonval meter movement.

application, whether it be as a current meter or as a voltmeter, current must flow. Naturally, in most applications, the smaller the current flow, the lower will be the *loading* on the circuit being measured. The meter movement itself possesses internal resistance varying from a few ohms for the less sensitive milliammeter to roughly 2000 Ω for the more sensitive microammeter. Actual meter range, however, is primarily governed by associated range resistors.

Figure 9.4 shows schematically the basic dc voltmeter and dc ammeter circuits. Either multiplier or shunt resistors are used in conjunction with the same basic meter movement. To minimize circuit loading, it is desirable that total *voltmeter* resistance be much greater than the resistance of the circuit under test. For the same reason, the *ammeter* resistance should be as low as possible. In both cases, meter movements providing large deflections for given current flow through the dc meter are required for high sensitivity.

9.3.1   Voltmeter Sensitivity

Voltmeter resistance is determined primarily by the series multiplier resistance. High multiplier resistance means that the current available to actuate the meter movement is low and that a sensitive basic movement is required. Because sensitivity may differ from meter to meter even though the meters may be of the same range, it is insufficient to rate voltmeters

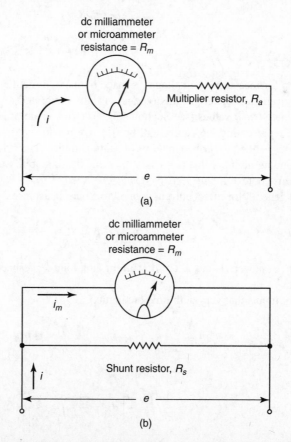

FIGURE 9.4: (a) dc voltmeter circuit, (b) dc ammeter circuit.

simply by stating total resistance. Rating is commonly stated in terms of *ohms per volt.*
*This value may be thought of as the total voltmeter resistance that a given movement must
possess in order for the application of* 1 V *to provide full-scale deflection.* This value
combines both resistance and movement sensitivity, and the higher the value, the lower will
be the loading effect for a given meter indication.

Simple pocket multimeters generally use a meter of 1 mA and 1000 Ω/V rating,
whereas more expensive multimeters may use movements with a rating of 50 μA and
20,000 Ω/V.

The value of the series multiplying resistor, R_a, as shown in Fig. 9.4(a), may be
determined from the relation

$$R_a = \frac{e}{i} - R_m \tag{9.1}$$

9.3.2 The Current Meter

Since current meters are connected in series with the test circuit, the voltage drop across
the meter must be kept as low as possible. This means that the combination of meter and
shunt must have as low a combined resistance as practical. Referring to Fig. 9.4(b), we may
write the following relation, based on equal voltage drops across meter and shunt:

$$R_s = \frac{i_m R_m}{i - i_m} \tag{9.2}$$

9.3.3 AC Meters

Provision for measuring ac voltages is made by using a rectifier in conjunction with a dc
meter movement. Meters of this type are usually calibrated to read in terms of the root-
mean-square (rms) values (see Section 4.5). The rms value of ac current or voltage is the
dc value representing the equivalent, or effective *power,* content of the corresponding ac
value. This is often described in terms of heating ability. The rms current (or voltage) is the
corresponding dc input that possesses the same heating ability as does the ac input. Note
that, in this context, a pure resistive load is assumed.

In general, for direct current applied to a pure resistance,

$$P = I^2 R = \frac{E^2}{R} \tag{9.3}$$

where P = power; I and E are dc current and voltage, respectively; and R is the resistive load.

For inputs that vary with time (ac inputs),

$$P = \frac{1}{T} \int_0^T i^2 R \, dt = \frac{1}{2\pi} \int_0^{2\pi} i^2 R \, d(\omega t) \tag{9.4}$$

or

$$P = \frac{1}{T} \int_0^T \frac{e^2}{R} \, dt = \frac{1}{2\pi} \int_0^{2\pi} \frac{e^2}{R} d(\omega t) \tag{9.5}$$

where i and e are current and voltage as functions of time and T is the period of the cycle.

If we equate the dc powers expressed by Eq. (9.3) to corresponding powers expressed by Eq. (9.4), for effective current I_{eff} we obtain

$$I_{\text{eff}} = I_{\text{rms}}$$

$$= \sqrt{\frac{1}{T} \int_0^T i^2 \, dt} \tag{9.6}$$

$$= \sqrt{\frac{1}{2\pi} \int_0^{2\pi} i^2 \, d(\omega t)}, \tag{9.6a}$$

or, in terms of voltage,

$$E_{\text{eff}} = E_{\text{rms}} = \sqrt{\frac{1}{T} \int_0^T e^2 \, dt} \tag{9.7}$$

$$= \sqrt{\frac{1}{2\pi} \int_0^{2\pi} e^2 \, d(\omega t)} \tag{9.7a}$$

For the most common case (i.e., sinusoidal variations),

$$i = I_0 \cos \frac{2\pi t}{T} = I_0 \cos \omega t, \tag{9.8}$$

or

$$e = E_0 \cos \frac{2\pi t}{T} = E_0 \cos \omega t \tag{9.8a}$$

where I_0 and E_0 are current and voltage amplitudes, respectively. Substituting in Eqs. (9.6a) and (9.7a) and evaluating, we obtain

$$I_{\text{rms}} = \frac{I_0}{\sqrt{2}} \approx 0.707 I_0 \tag{9.9}$$

and

$$E_{\text{rms}} = \frac{E_0}{\sqrt{2}} \approx 0.707 E_0 \tag{9.9a}$$

The results indicate that a sinusoidal source delivers about 71% of the power that a dc source of like amplitude delivers. Considering an ordinary line voltage of 120 V ac, we see that the voltage amplitude is

$$E_0 = \frac{120}{0.707} \approx 170 \text{ V}$$

This is the voltage that must be considered when required insulation is specified. The peak-to-peak magnitude in this case is about 340 V.

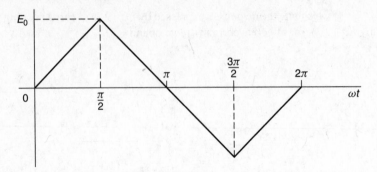

FIGURE 9.5: Waveform for Example 9.1.

The following two cases are examples of waveforms that are not sinusoidal.

EXAMPLE 9.1

Consider the waveform shown in Fig. 9.5. Inspection suggests that because of symmetry we need only deal with the shape over the range $0 < \omega t < \pi/2$.

Solution   For this range,

$$e = \frac{E_0}{\pi/2}\omega t$$

Substituting in Eq. (9.7a) and evaluating, we have

$$E_{\text{rms}} = \left[\frac{1}{\pi/2} \int_0^{\pi/2} \left(\frac{E_0}{\pi/2} \right)^2 (\omega t)^2 \, d(\omega t) \right]^{1/2}$$

$$= \frac{E}{\sqrt{3}} \approx 0.577 E_0$$

For equal amplitudes the waveform shown in Fig. 9.5 is only about 58% as effective as dc of the same amplitude.

EXAMPLE 9.2

Figure 9.6 shows a pulsed square wave for which we wish to determine the rms voltage.

Solution   There are three distinct intervals in each cycle, which will be treated separately.

$$0 < \omega t < \frac{\pi}{2}, \qquad e = 0,$$

$$\frac{\pi}{2} < \omega t < \frac{3\pi}{4}, \qquad e = E_0,$$

$$\frac{3\pi}{4} < \omega t < 2\pi, \qquad e = 0$$

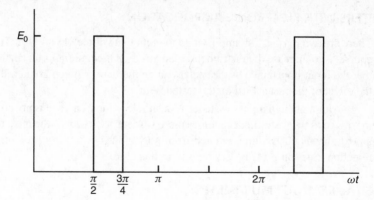

FIGURE 9.6: Waveform for Example 9.2.

Applying Eq. (9.7a),

$$
\begin{aligned}
E_{\text{rms}} &= \left\{ \frac{1}{2\pi} \left[\int_0^{\pi/2} (0)\, d(\omega t) + \int_{\pi/2}^{3\pi/4} E_0^2\, d(\omega t) + \int_{3\pi/4}^{2\pi} (0)\, d(\omega t) \right] \right\}^{1/2} \\
&= E_0 \left\{ \frac{1}{2\pi} \left(\frac{3\pi}{4} - \frac{\pi}{2} \right) \right\}^{1/2} \\
&= \frac{E_0}{\sqrt{8}} \approx 0.35 E_0
\end{aligned}
$$

It should be noted that the integration circuitry of the ac meter yields rms readouts independent of the sine waveform. Based on a voltage amplitude of E_0, the reader may wish to confirm this statement by using the setup suggested in Problem 9.7 and showing that, conforming with theory,

1. For a square waveform, $E_{\text{rms}} = E_0$;
2. For a sine waveform, $E_{\text{rms}} = 0.707 E_0$;
3. For a triangular waveform, $E_{\text{rms}} = 0.577 E_0$.

9.3.4 The Multimeter and Resistance Measurement

A versatile tool around any laboratory is the basic volt-ohm-milliammeter (VOM), which uses switching arrangements for connecting multiplier and shunt resistors and a rectifier into or out of a circuit in order to cover ranges of dc and ac voltages. In addition, the meter is arranged to measure resistances, using an internal source of current. The VOM applies a known voltage to the resistor and senses the resulting current flow. By switching to the ohmmeter function and connecting the leads to the unknown resistance, one can determine from the meter movement the current flowing through the resistor. In ohmmeter mode, the current flow indication is calibrated in terms of resistance, thereby providing a direct means for measurement. Generally, the resistor must be removed from its circuit before it is measured.

9.4 METERS WITH ELECTRONIC AMPLIFICATION

There are two reasons for amplifying the input to a voltmeter or VOM. The obvious one, of course, is to increase the instrument's sensitivity. Of equal importance, however, is the fact that the input impedance of the meter can be made very much greater, thereby decreasing the effect of the meter load on the tested item.

Although high input resistance is desirable in most cases, it is not an unmixed blessing inasmuch as the instrument becomes more susceptible to *noise*, the most troublesome being an extraneous 60-Hz hum radiated from power lines. Circuitry providing common-mode rejection (Section 7.14) is very helpful in this case.

9.5 DIGITAL-READOUT MULTIMETERS

The advent of the digital counter brought about its application to numerous measurement problems. Basically the counter is simply that—a counter of events. In Section 8.7 we showed how a simple counter circuit can be arranged to display a frequency. A simple way to make use of this capability for measuring a voltage is to combine the meter with a voltage-controlled oscillator (VCO). The frequency output of a VCO (e.g., the National Semiconductor LM566) is determined by the magnitude of the applied voltage. It is easy to visualize that through use of the VCO, a frequency counter, and proper scaling circuitry, a digitally reading voltmeter can be devised. Although this is a simple approach, it possesses certain disadvantages and is not commonly used.

Dual-slope integration is a much more common method. The essential building blocks for this method are an op-amp integrator (Fig. 7.33b, Section 7.19), a clock, and a frequency counter, combined with the necessary scaling and control circuitry. The clock is simply a fixed-frequency oscillator, usually crystal controlled, that supplies timing pulses.

Through use of IC gating the integrator's capacitor is charged for a predetermined length of time (referenced to the clock frequency). It is then discharged at a constant current rate, and the clock pulses occurring during the discharge period are counted. This count becomes the measure of the input voltage. With proper scaling, the count is equated to the input magnitude. Other circuitry can be incorporated within the meter, making it a general-purpose multimeter for measuring dc or ac voltages or currents, or resistances.

Advantages of the double-slope circuit over others is that aging of either the clock or the integrator causes little or no error. We can see that charge and discharge of the integrator capacitor are each dependent on capacitance value and time interval and that changes in either will be self-compensating. Should the clock slow down with age, the charging time will be reduced, but the discharge time will be increased in like proportion. A similar effect results from small changes in capacitor value.

Other approaches include

1. single-slope conversion,
2. charge balance,
3. linear ramp conversion, and
4. successive approximation.

These approaches are not discussed here; however, the reader is referred to the Suggested Readings at the end of the chapter for further information.

FIGURE 9.7: (a) Digital-readout volt-ohm-milliammeter (VOM). (Courtesy of Agilent Technologies).

Many digital multimeters also include automatic polarity indication and self-ranging ability (automatic placement of the decimal point). Many meters of this type are said to display one-half digits (e.g., a "$3\frac{1}{2}$-digit display"). This means that the most significant digit can be only a 0 or a 1, excluding all others. A $3\frac{1}{2}$-digit meter, for instance, is not capable of displaying a number greater than 1999.

Figure 9.7(a) illustrates a compact, digital-readout VOM. Typicial specifications for such a meter are:

dc voltmeter ranges to 1200 V dc with accuracies equal to or better than
  0.1% of reading +2 digits

ac voltmeter ranges to 1200 V rms with accuracies equal to or better than
  1.5% of reading: 10 digits

dc input resistance = 10 MΩ

dc and ac ammeter ranges to 2.0 A

ohmmeter ranges to 20 MΩ

A/D conversion: dual slope

Figure 9.7(b) illustrates how a typical digital voltmeter operates.

9.6 THE CATHODE-RAY OSCILLOSCOPE (CRO)

Probably the most versatile readout device used for mechanical measurements is the cathode-ray oscilloscope (CRO). This is a voltage-sensitive instrument, much the same as the electronic voltmeter, but with an inertialess (at mechanical frequencies) beam of electrons sub-

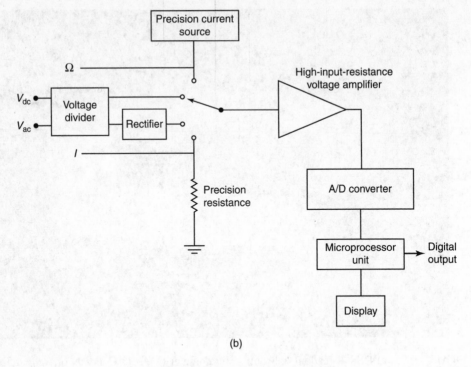

(b)

FIGURE 9.7: (b) Typical digital multimeter.

stituted for the meter pointer and a fluorescent screen replacing the meter scale. Figure 9.8 shows a typical general-purpose CRO.

The heart of the instrument is the cathode-ray tube (CRT), shown schematically in Fig. 9.9. A stream of electrons emitted from the cathode is focused sharply on the fluorescent screen, which glows at the point of impingement, forming a bright spot of light. Deflection plates control the direction of the electron stream and hence the position of the bright spot on the screen. If an electrical potential is applied across the plates, the effect is to bend the pencil of electrons, as shown in Fig. 9.10. With the use of two sets of deflection plates arranged to bend the electron stream both vertically and horizontally, an instantaneous relation between two separate deflection voltages may be obtained.

Figure 9.11 is a block diagram of a typical general-purpose cathode-ray oscilloscope. The nature of the CRO is such that it may appear with many different variations in the form of special controls and input and test terminals. The diagram shown is not for any particular commercial instrument. Certain oscilloscopes will have features not shown here, and others may not use certain ones that are shown.

9.6.1   Oscilloscope Amplifiers

The sensitivity of the typical electrostatic cathode-ray *tube* is relatively low, varying from about 0.010 to 0.15 cm deflection per volt dc, or from about 6 to 100 V/cm of deflection. This means that in order to be widely useful for measurement work, the CRO should provide means for signal amplification before the signal is applied to the deflection plates. All

FIGURE 9.8: Model DSO81004A digitizing oscilloscope. (Courtesy of Agilent Technologies).

general-purpose oscilloscopes provide such amplification. Some are equipped for both dc and ac amplification on both the vertical and horizontal plates. During ac amplification, or *ac coupling*, a capacitor is used to block the dc component of the signal, so that only the ac component is amplified.

Some means for varying gain is provided in order to control the amplitude of the trace on the screen. This is often accomplished through use of fixed-gain amplifiers, preceded by variable attenuators.

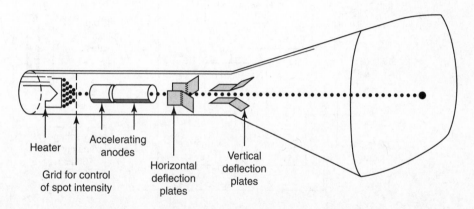

FIGURE 9.9: Elements of the basic cathode-ray tube (CRT).

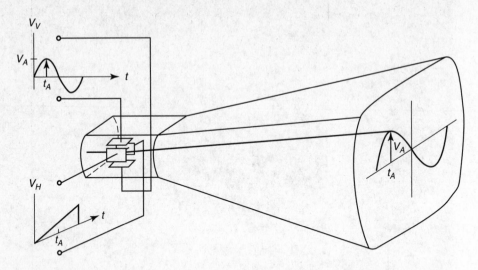

FIGURE 9.10: Time-varying voltage V_V applied to the vertical plates and sawtooth waveform voltage V_H applied to the horizontal plates cause the electron beam to display patterns on the screen.

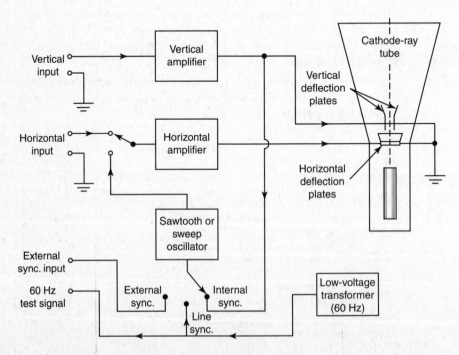

FIGURE 9.11: Block diagram of a typical general-purpose oscilloscope.

9.6.2   Sawtooth Oscillator or Time-Base Generator

Except for special-purpose applications, the usual cathode-ray oscilloscope is equipped with an integral *sawtooth*, or *sweep*, oscillator. This variable-frequency oscillator produces an output voltage–time relation in the form shown in Fig. 4.9(c). Ideally, the voltage increases uniformly with time until a maximum is reached, at which point it collapses almost instantaneously.

When the output from the sawtooth oscillator is applied to the horizontal deflection plates of the cathode-ray tube, the bright spot of light will traverse the screen face at a uniform velocity. As the voltage reaches a maximum and collapses to zero, the spot is whipped back across the screen to its starting point, from which it repeats the cycle. The length of the path will then be a measure of the period of the oscillator frequency (called sweep frequency) in seconds, and each point along the path will represent a proportional time interval measured from the beginning of the trace. (By convention, increasing time is measured to the right.) In this manner a very useful *time base* is obtained along the *x*-axis of the tube face.

As a simple example, let us suppose that ordinary 60-Hz line voltage is applied to the *y*-deflection plates of the tube and the output from a variable-frequency sweep oscillator is applied to the *x*-deflection plates. (Usually the sawtooth oscillator is within the case of the CRO and a switch is simply set to "Internal Sweep.") With the two voltages applied, the frequency of the sawtooth oscillator would be adjusted by means of the sweep range and sweep vernier controls on the control panel. If the sawtooth frequency is adjusted *exactly* to 60 Hz, then one complete cycle of the vertical input waveform will appear stationary on the screen, as shown in Fig. 9.12(a). If the sweep frequency is slightly greater or slightly less than 60 Hz, then the waveform will appear to creep backward or forward across the screen. *The reciprocal of the time in seconds required for the waveform to creep exactly one complete wavelength on the screen will be the discrepancy in hertz between the sweep frequency and the input frequency.* In certain cases this relationship may be used in making precise measurement of frequency or period.

If the sweep frequency is changed to *exactly* 30 Hz, then two complete cycles of the input signal will appear and remain stationary on the screen, as shown in Fig. 9.12(b).

9.6.3   Synchronization or Triggering

In the example just mentioned, one cycle of the 60-Hz waveform will appear stationary on the screen only if the sweep frequency is exactly 60 Hz. Frequencies from all types of electronic signal generators tend to shift or drift with time. This is caused by a change in component characteristics brought about by temperature changes due to the warm-up of the instrument. Therefore, to hold a pattern on the screen without creep, one must continuously monitor the trace, making adjustments in sweep frequency as required.

When a steady-state signal is applied to the vertical terminals, however, it is possible to lock the sweep oscillator frequency to that of the input frequency, provided the sweep frequency is *first* adjusted to approximately the input frequency or some multiple thereof. This is controlled through use of "Trigger Source" and "Trigger Level" controls.

In our example, we would wish to use the vertical input as our synchronizing signal source, so we would set the trigger source to "Internal" [see Fig. 9.12(c)]. Voltage pulses from the input signal would then be applied to the sweep oscillator and would be used to control the oscillator frequency over a small range. If the frequency is initially adjusted

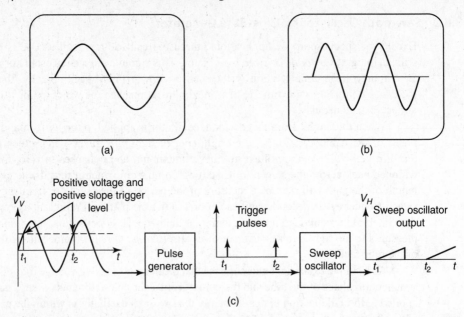

FIGURE 9.12: Example of internal triggering: (a) 60-Hz voltage applied to vertical plates and horizontal sweep adjusted to 60 Hz, (b) trace obtained when horizontal sweep is changed to 30 Hz, (c) general internal trigger synchronization.

to some integral multiple of the input frequency, the sweep oscillator would then lock in step with the input signal and the trace would be held stationary on the screen. The trigger level voltage would need to be adjusted to lie in the range of the periodic signal in order for trigger pulses to be generated.

Electrical engineers are often concerned with measurements at 60 Hz; hence oscilloscopes are usually equipped to provide a direct synchronizing signal from the power line. This setting on the trigger source selector is often simply marked "Line."

Finally, it is often desirable to trigger a CRO trace from an external source closely associated with the input signal. As an example, suppose some form of electrical pressure pickup is being used for measuring the cylinder pressures in a reciprocating-type air compressor. Although the pressure signal from the pickup may be steady state, making internal triggering a possibility, changing load, erratic valve action, or the like may make this signal an undesirable source for synchronization. An external circuit may be used in a case of this sort. A simple make-and-break contactor could be attached to the compressor shaft, and a voltage pulse could be provided for synchronization through use of a simple battery. Such a circuit would be connected between the external trigger input and the ground, and the trigger source would be set at "External." In this case the horizontal sweeps would take place only when initiated by the external contactor.

This arrangement is also useful when a *single sweep* only is desired. In such a case, the synchronizing contactor or switch may be simply hand operated, or it may be incorporated in the test cycle. As an example, a photocell circuit could be arranged so that a beam of light intercepted by a projectile or the like would provide the initiating pulse in synchronization with the test signal of interest.

When the driven sweep is initiated as just outlined, through use of an external source of triggering, the sweep occurs once for each triggering pulse. The sweep rate in this case is still controlled by the sweep range and sweep vernier. Of course, the sweep cannot be pulsed at a rate greater than that provided by the sweep control settings. That is, the electron beam must have returned from the previous excursion before it can be triggered again.

9.7   ADDITIONAL CRO FEATURES

9.7.1   Multiple Trace

In many cases it is desirable to make an accurate time comparison between two continuing inputs, and very often CRO multiple-trace capability is the solution to this problem. Although oscilloscopes are available that permit simultaneous writing of more than two traces, the dual-trace type is the most common.

There are two different basic methods for accomplishing double traces: (1) through use of two separate electron "guns" within a single tube envelope, and (2) by high-speed gating (switching) two inputs to the vertical plates of a conventional one-gun cathode-ray tube. In either case, duplicate circuitry (terminals, amplifiers, positioning controls, etc.) is required. The second approach is by far the more common.

When the gating method is used, the oscilloscope design engineer has a choice of either or both of two different schemes. The first, called the *chopped-trace* method, successively switches from input A to input B and back again many times during a single sweep across the CRT screen. A switching rate of 200 kHz is typical. For many sweep rates the gating is so fast that the two traces appear to be continuous. In addition to being dependent on the relative rates of gating to sweep, the illusion of continuity also depends somewhat on the persistence of the phosphor. However, as the sweep rate is increased, depending on the demands of the measurement, the actual discontinuity of the traces may become a problem.

Alternate gating, the second method, alternately displays the entire traces, first for input A, then B, back to A, and so on. Screen persistence will permit simultaneous viewing of near-simultaneous traces. Many dual-trace scopes provide switch selection of either method.

9.7.2   Magnification and Delayed Sweep

Amplification may be used to stretch out or magnify the horizontal sweep to a number of times the size of the CRO screen. This means, coupled with adjustment of the horizontal position control, allows us to magnify a portion of the normal sweep for closer inspection. Oscilloscopes with more advanced circuitry (and higher cost) may use what is called *delayed sweep* to accomplish similar results. That is, the operator may select any small portion of the normal display, which may then be shown at a selected higher sweep rate. The effect of the increased rate over the selected portion of the normal sweep is to expand or magnify the portion that has been pinpointed.

9.7.3   Digital Storage Oscilloscopes

The digital storage oscilloscope differs from its analog counterpart in that it converts the analog input waveform into a digital signal that is stored in memory and then converted back into analog form for display on a conventional CRT (Fig. 9.13). The input signal is supplied to an A/D converter (see Section 8.11) that digitizes the signal before storage. The

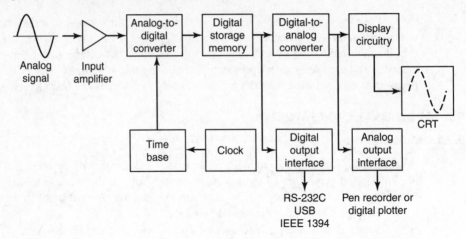

FIGURE 9.13: Block diagram of a digital storage oscilloscope.

contents of the memory are outputted to a D/A converter (see Section 8.11) and then to the vertical and horizontal (y and x) deflection sections of the CRT circuitry. The data are displayed most frequently in the form of individual dots that collectively make up the CRT trace. The vertical screen position of each dot is given by the binary number stored in each memory location and the horizontal screen position is derived from the binary address of that memory location. The number of dots displayed depends on three factors: the frequency of the input signal with respect to the digitizing rate, the memory size, and the rate at which the memory contents are read out. The greater the frequency of the input signal with respect to the digitizing rate, the fewer the data points captured in the oscilloscope memory in a single pass, and the fewer the dots available in the reconstructed waveform.

Digital storage oscilloscopes and analog oscilloscopes each have distinct advantages, but digital oscilloscopes have created the most recent excitement because of their dramatic improvements in performance. In addition to merely capturing and displaying waveforms, the digital oscilloscopes can perform the following tasks: indefinite storage of waveform data for comparison, transferring stored data to other digital instruments, and in most cases providing onboard processing of the data: Waveform parameters such as maximum, minimum, peak-to-peak, mean, rms, rise time, fall time, waveform frequency, and pulse delay can be computed and made available for presentation in decimal form on the oscilloscope screen. Digital oscilloscopes are also well suited for capturing transient signals. If the oscilloscope is set to the single-sweep mode, data can be captured automatically and stored on the occurrence of the trigger event. As a result, the problem of synchronization inherent in analog oscilloscopes is eliminated.

The analog-to-digital converter of the digital oscilloscope determines some of its most important operating characteristics. The voltage resolution is dictated by the bit resolution of the A/D converter, and the storage speed by the maximum speed of the converter. For example, a 0 to 10-V converter using 8, 10, or 12 bits is able to resolve to 0.0391 V, 0.0098 V, and 0.0025 V, respectively. The time resolution is selectable in that the user can define how much memory space is needed for each waveform stored.

The output of digital oscilloscopes is available in other forms than the trace on the CRT. Digital output in USB, IEEE-1394, or RS-232C format (Section 8.10.4) is generally available. For more flexible output recording, some oscilloscopes are fitted with disk drives so that any captured signals can be immediately stored on CD-ROM drives for subsequent computer data analysis. Analog output may also be provided for driving pen recorders or other indicators.

9.7.4 Single-Ended and Differential Inputs

There are two types of inputs through which a signal can be connected to the oscilloscope: the single-ended input and the differential input.

Single-ended inputs have only one input terminal besides the ground terminal at each amplifier channel. Only voltages relative to ground can be measured with a single-ended input. (Note that the most common input connector to oscilloscopes is the BNC coaxial connector. The external conductor of the BNC connector is the ground terminal of the input.)

A differential input has three terminals (two input terminals besides the ground terminal at each amplifier channel). With a differential input, the voltage between two non-grounded points in a circuit can be measured. The amplifier electronically subtracts the voltage levels applied at the two terminals and displays the difference on the screen. In addition, differential amplifiers are able to reduce unwanted common-mode interference problems. This feature is especially important when it is necessary to measure small signals in the presence of much larger, undesired common-mode signals. The differential input is sometimes available as a plug-in unit on those oscilloscopes that have interchangeable plug-in capabilities.

9.8 *XY*-PLOTTERS

The term *xy-plotter* is nearly self-explanatory. It refers to an instrument used to produce a Cartesian graph originated by two dc inputs, one plotted along the x-axis and the other along the y-axis. Of course, the great advantage in its use is that the graph is plotted automatically, thereby sidestepping the laborious point-by-point plotting by hand. In addition, families of curves may be plotted easily by varying a third parameter in step fashion from plot to plot. Figure 9.14 shows a typical *xy*-plotter. Basic components consist of a platen to which the graph paper is either mechanically attached or held by vacuum or by electrostatic means and one or more servo-driven styluses. In addition, amplification of the input signals is normally required.

Performance variables include input ranges (amplitude and frequency), sensitivity, stylus slewing rate and acceleration limit, resolution, resetability, and provision for common-mode rejection. Slewing rate in cm/s is the maximum velocity with which the stylus can be driven. This becomes a limiting response characteristic, especially when large amplitudes are to be plotted. Limits on the maximum acceleration of the stylus are more often a factor when low-amplitude, high-frequency inputs are plotted.

Common chart sizes are 22×28 cm ($8\frac{1}{2} \times 11$ in.) and 28×44 cm (11×17 in.). Two-stylus (*xyy*) models are available for the simultaneous plotting of two curves. These models, of course, require three separate drive systems, unless the x-axis is time as provided by an internal drive system.

9.9 DIGITAL WAVEFORM RECORDERS

Most analog *xy*-recorders are limited by a relatively slow stylus slewing rate and acceleration limit. The digital plotter has generally replaced the older strip-chart recorders because of its improved performance characteristics. The inked pen or hot stylus strip-chart recorder

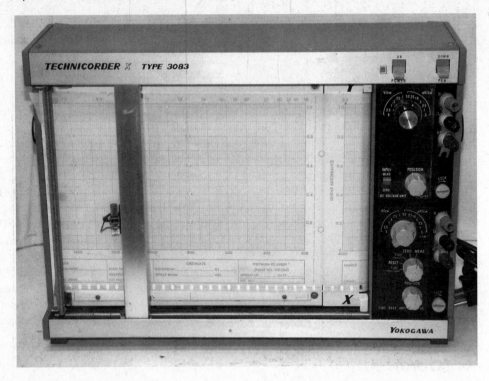

FIGURE 9.14: An *xy*-recorder. (Courtesy of Labequip.com.)

plotted analog time-dependent input voltages in real time on long rolls of chart paper. Because of the pen or stylus inertia, the top recording frequency is below 50 Hz in order to maintain the full-amplitude swing of the pen or stylus arm. These features are minimized by the use of a digital plotter, which combines the features of an *xy*-recorder, a waveform recorder or storage oscilloscope, and a digital voltmeter. It samples analog signals, digitizes them, stores them in digital memory, and plots them in analog form.

Figure 9.15 illustrates the versatility of an HP 7090A low-frequency measurement-plotting system, which can be used for recording most signals from sensors used in mechanical measurements. The figure illustrates just one of three channels which may be input simultaneously. Each channel has a buffer, which can capture and store 1000 data points. During a buffered recording the input buffers are filled simultaneously. These buffered data can in turn be recalled and plotted or viewed on an oscilloscope.

A direct recording mode is also available to produce real-time recordings as a hard-copy plot. The input signal is converted to digital data at a fixed rate of 250 samples/s (but not stored in memory), which are used to drive the plotter stepper motors. The data of any two channels can be plotted versus time or versus a third channel.

9.10 THE SPECTRUM ANALYZER

Spectrum analyzers, which measure the frequency spectrum of a signal, are of two forms: the swept type and the digital type. Figure 9.16 shows a simplified block diagram of the workings of a swept-type spectrum analyzer. Pertinent items are as follows:

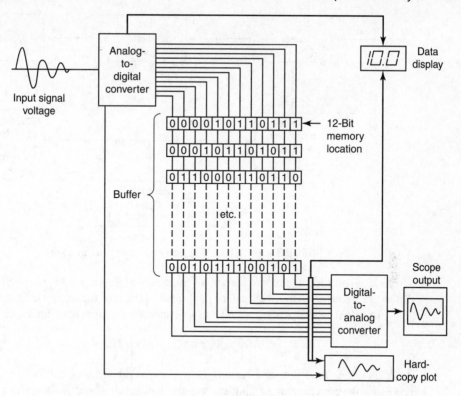

FIGURE 9.15: HP 7090A plotter system.

1. A *sawtooth waveform generator* running at a fixed frequency but whose voltage output varies linearly in ramp fashion.

2. A *voltage-controlled oscillator* (VCO), whose output frequency, f_{VCO}, sweeps linearly across a given frequency range. Its voltage amplitude is constant.

3. A *mixer* that combines (mixes) the input signal with the VCO output. This produces sum-and-difference frequency components. For an input frequency, f_{in}, two side frequencies, $(f_{VCO} - f_{in})$ and $(f_{VCO} + f_{in})$, are generated (see Section 10.6).

4. An *IF* (intermediate-frequency) *band-pass amplifier*, whose passband is designed to accommodate a single (ideally) value of $(f_{VCO} - f_{in})$ to the exclusion of other frequencies.

5. A *detector*, which is basically a voltage rectifier, passing a voltage of one polarity (say, positive).

6. A *cathode-ray tube* (CRT), used for display.

In operation, as the output of the sawtooth generator linearly rises from zero, it drives the CRT electron beam along the x-axis, across the face of the tube. At the same time, the output frequency from the VCO sweeps linearly upward. When the VCO frequency and the frequency component of the input produce a difference frequency, $(f_{VCO} - f_{in})$, matching the passband frequency of the IF amplifier, a signal component passes whose amplitude is proportional to that of the input component. This is then rectified and displayed on the CRT

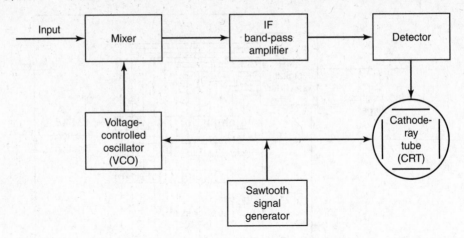

FIGURE 9.16: Block diagram of a spectrum analyzer.

screen as a spike located on the horizontal frequency axis at a point corresponding to the difference frequency. Both axes can be calibrated in terms of the input parameters.

To help us visualize the result, let us consider a two-component input:

$$A(t) = 10 \cos \, \Omega t + 5 \cos 2\Omega t$$

As an amplitude-time plot, the function would appear as shown in Fig. 9.17(a). Figure 9.17(b) represents the corresponding amplitude-frequency plot as it would be shown by a spectrum analyzer. For an ideal input and an ideal response, the two spikes would be indicated by perfect vertical lines. However, as with all instrumentation, there are limitations and as input frequencies are increased, a broadening of the spikes becomes apparent.

Various adjustments are provided on even the simplest analyzers to accommodate ranges of frequency and amplitude. In addition, the IF band-pass analyzer can often be varied to help in isolating frequency components in the input signal. Analyzers are selected on the basis of application: low-frequency analyzers for vibration and sound work, radio frequency analyzers for RF work, UHF for TV, gigahertz for microwaves, and so on.

Material on digital fast fourier transform (FFT) analysis and digital spectrum analyzers is given in Sections 4.7 and 18.6.

9.11   LabVIEW

LabVIEW is a programming language and environment that was developed by National Instruments primarily to create "virtual instruments" for measurements with a personal computer.[1] A virtual instrument can be any instrument that you would find in a laboratory (e.g., a multimeter, oscilloscope, filter, etc.); but in LabVIEW, the computer collects the data through a data acquisition card, and a program is used to analyze and display the data on the computer monitor in a format that one would normally see on the instrument itself. The benefit of LabVIEW is that it enables a user to create essentially an unlimited number of instruments using only a personal computer equipped with a data acquisition card. Note that there are limitations to this approach, namely the limited sampling speed and resolution

---

[1] LabVIEW is a commercial software package which is available from National Instruments, Inc.

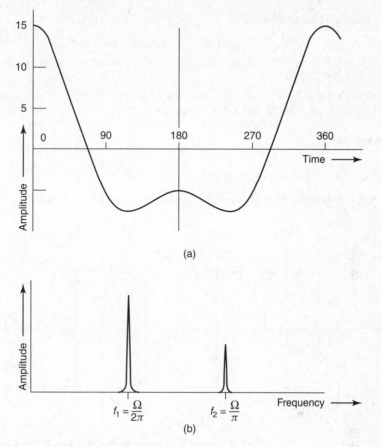

FIGURE 9.17: (a) Time domain plot of $A(t) = 10\cos \Omega t + 5\cos 2\Omega t$, (b) frequency domain plot of $A(t) = 10\cos \Omega t + 5\cos 2\Omega t$.

provided by the analog-to digital converter on the data acquisition card. Nonetheless, such a system can be extremely useful as a multipurpose data collection device. In addition to being able to mimic many instruments, the concept of using the computer to collect data (so-called automated data collection) can be invaluable in many situations such as in remote sites, at regular intervals, or at speeds that are too high for human operators.

The heart of LabVIEW is its programming language, called *graphical programming*. Programming graphically is quite different from programming in conventional languages, in which the code is written with text and symbols in a sequential fashion. In graphical programming the "code" is drawn on the computer screen with icons and connection wires. Data "flow" through the program from icon to icon along the wires. The program itself is called a *virtual instrument*.

Data are displayed in two ways in LabVIEW: with *charts* and *graphs*. Charts are used for real-time data display. Their method of handling data is faster than that for graphs. Graphs are used for off-line analysis and display of data.

SUGGESTED READINGS

Hickman, I. *Oscilloscopes.* 5th ed. Boston: Newnes, 2000.

Ibrahim, K. F. *Instruments and Automatic Test Equipment: An Introductory Textbook.* New York: John Wiley, 1986.

Wolf, S., and R. F. M. Smith. *Student Reference Manual for Electronic Instrumentation Laboratories.* 2nd ed. Upper Saddle River, N.J.: Prentice Hall, 2003.

PROBLEMS

9.1. Verify Eqs. (9.6a) and (9.7a).

9.2. Determine the rms voltage for the waveform shown in Fig. 9.18.

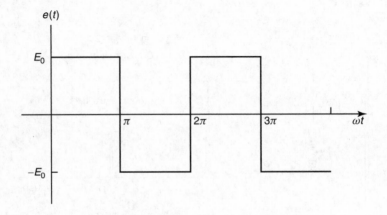

FIGURE 9.18: Waveform for Problem 9.2.

9.3. Determine the rms voltage for the waveform shown in Fig. 9.19.

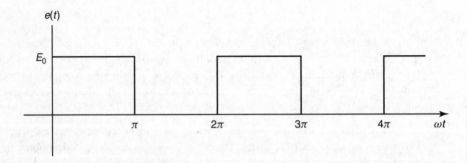

FIGURE 9.19: Waveform for Problem 9.3.

9.4. Figure 9.20 illustrates a clipped sine wave. Determine the rms voltage for this waveform.

9.5. What is the rms value for the waveform defined by the following?

$$y = \sin \omega t \cos 2\omega t$$

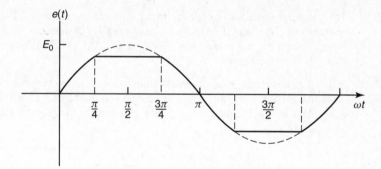

FIGURE 9.20: Waveform for Problem 9.4.

9.6. What is the rms value for the waveform defined by by the following?

$$y = 10 \sin \omega t + 10 \sin 2\omega t$$

9.7. Assemble the apparatus shown in Fig. 9.21.

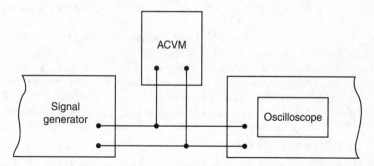

FIGURE 9.21: Circuit for Problem 9.7 (signal generator has sine, square, and triangular waveform capabilities).

 (**a**) Input selected waveforms and record the rms values indicated by the ac voltmeter (ACVM) and E_0 as determined by the oscilloscope.

 (**b**) Use Eq. (9.7a) and calculate the rms value for the selected waveform and compare with the meter readout.

9.8. Collect the following equipment: a general-purpose oscilloscope (preferably a single-channel basic oscilloscope) and two variable-frequency signal generators, along with an assortment of appropriate leads. Connect one signal source to the vertical input terminals of the oscilloscope and the other to the horizontal terminals. Proceed to "turn the knobs." Experiment until you are familiar with the purpose and action of each of the controls. (Feel free to make any front-panel adjustments that are available, except for possible screwdriver balance adjustments. In addition, avoid holding an intense, concentrated, fixed spot on the screen. To do so could cause local burning of the phosphor.) Refer to Figs. 9.12(b), 10.6, and 10.9. Can you reproduce these patterns?

9.9. The cathode-ray oscilloscope and electronic voltmeters are considered "high-input-impedance" devices. The simple D'Arsonval meter is usually considered to be of "low impedance." What is meant by "high" and "low" in this sense? Discuss the relative merits and disadvantages of each category.

9.10. Figure 9.22 illustrates an experimental setup for determining the time constant for various combinations of R and C (or L). Insert a range of components and compare experimentally determined values for the resulting time constants with theoretical values. (Refer to Section 5.18.)

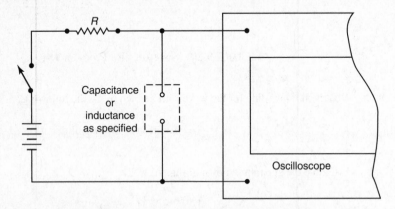

FIGURE 9.22: Experimental setup for Problem 9.10.

Problems 9.11 through 9.16 specify experimental exercises to be performed. Although the circuits are prescribed, specific component values are not specified since a variety of values as may be available will usually be quite satisfactory. So-called decade boxes of resistances, capacitances, and inductances are particularly useful because they permit a wide range of values. A decade box is simply an assembly of components, in switch-selectable decade steps of value.

9.11. Insert the circuit shown in Fig. 9.23 in the circuit of Fig. 9.24. Using components of known value, find the LC resonance frequency experimentally. Check against Eq. (7.26).

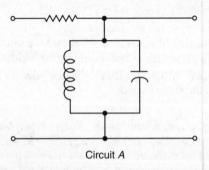

Circuit *A*

FIGURE 9.23: Parallel LC circuit to be used in Problems 9.11 and 9.12.

9.12. Duplicate the experiment given in Problem 9.11; however, use the experimentally determined resonance frequency and a known value of capacitance to determine an unknown inductance.

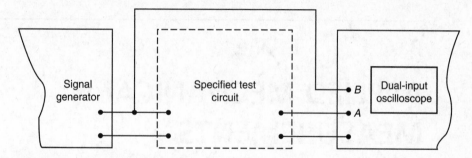

FIGURE 9.24: Circuit for Problems 9.11 through 9.16 (signal generator has sine, square, and triangular waveform capabilities).

9.13. Using the circuit in Fig. 9.24, experimentally determine the characteristics of the circuit shown in Fig. 7.25(a).

9.14. Using the circuit in Fig. 9.24, experimentally determine the characteristics of the circuit shown in Fig. 7.25(b).

9.15. Using the circuit in Fig. 9.24, experimentally determine the characteristics of the circuit shown in Fig. 7.29(a).

9.16. Using the circuit in Fig. 9.24, experimentally determine the characteristics of the selected circuits shown in Fig. 7.30.

9.17. A pressure pickup is used for measuring the pressure time relationship in the cylinder of an internal combustion engine. The output is amplified and then applied to the vertical plates of an oscilloscope. Describe arrangements for synchronizing the scope trace with the engine speed (a) using internal sweep and (b) using external sweep. If the pickup is applied to the determination of the pressure time history resulting from detonations of explosive charges, how should the oscilloscope be configured to obtain satisfactory traces? Assume that the charges are "one shot" but that they may be repeated as desired.

PART TWO

APPLIED MECHANICAL MEASUREMENTS

Measurement of Count, Events per Unit Time, Time Interval, and Frequency

10.1   INTRODUCTION

To be able to count items or events is fundamental to engineering. Items or events to be counted may be pounds of steam, cycles of displacement, number of lightning flashes, or anything divisible into discrete units. Also, time is often introduced, and the number of items or *events per unit of time* (EPUT) must be measured. The expressions *EPUT* and *frequency* usually have slightly different connotations. Frequency is thought of as being the events per unit of time for phenomena under steady-state oscillations, such as mechanical vibrations or ac voltage or current. EPUT, however, is not dependent on a steady rate, and the term includes the counting of events that take place intermittently or sporadically. An example of this is the counting of any of the various particles radiated from a radioactive source.

Time interval is often desired, and this becomes *period* if it is the duration of a cycle of a periodic event. Or the time interval desired may be that which occurs between events in an erratic phenomenon, or perhaps the duration of a "one-shot" event such as an impulsive pressure or force.

Problems in counting or timing emerge primarily when the events are too rapid to determine by direct observation, or the time intervals are of very short duration, or unusual accuracy is desired. In general, counting and timing-measurement problems may be classified as follows:

1. *Basic counting*, either to determine a total or to indicate the attainment of a predetermined count

2. *Number of events or items per unit of time, or EPUT*, independent of rate of occurrence

3. *Frequency*, or the number of cycles of uniformly recurring events per unit of time

4. *Time interval* between two predetermined conditions or events

5. *Phase relation*, or percentage of period between predetermined recurring conditions or events

General laboratory equipment such as oscilloscopes and computer A/D converters, perhaps used in conjunction with frequency sources or standards, may often be applied to EPUT, time-interval, frequency, and phase measurements. The possibilities are limited only by the ingenuity of the user. In some cases, a single lab instrument provides all that is needed. For example, some handheld or benchtop multimeters can measure the frequency of ac input signals. Likewise, a digital storage oscilloscope may have cursors and software menus that allow signal period or frequency to be displayed onscreen, or, for a dual-trace oscilloscope, which allow the direct determination of the time delay between two signals. In the latter case, if the signals are periodic, use of the signal period or frequency would allow the time delay to be converted to a phase lead or lag. No particular explanation is required for such devices and methods.

10.2 USE OF COUNTERS

Various characteristics and uses of electronic counters were discussed in Sections 8.7 and 9.2. In this section, we look further at their application.

10.2.1 Electronic Counters

Electronic counters used as either basic counting devices or EPUT meters require that the counted input be converted to simple voltage pulses, a count being recorded for each pulse. It should be clear that input functions used to trigger the counter need not be analogous to any quantity other than the count; hence even a simple switch may be used, actuated by the function to be counted. In addition, photodetectors, variable resistance, inductance, or capacitance devices, Geiger tubes, and the like may be employed. Simple amplifiers may be used, if necessary, to raise the voltage level to that required by the counter—and, because most electronic counters have a high-impedance input, no particular power requirement is imposed. Signal inputs may include almost any mechanical quantity, such as displacement, velocity, acceleration, strain, pressure, and load, so long as distinct cycles or pulses of the input are provided. The starting or stopping of the counting cycle may be controlled by direct manual-switch operation on the panel or by remote switching. One must not overlook, however, the ± 1 count ambiguity referred to in Section 9.2.4.

A variation of the simple electronic counter is the *count-control* instrument. Provision is made for setting a predetermined count, and when the count is reached, the instrument supplies an electrical output that may be used as a control signal. Figure 10.1 shows how such a device could be used to prepare predetermined batches or lots for packaging.

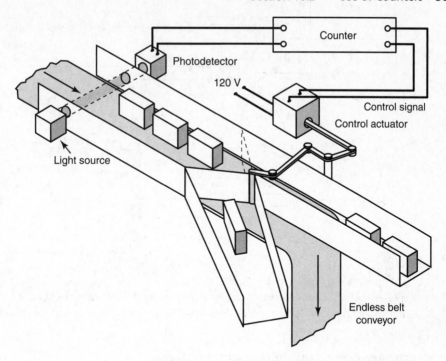

FIGURE 10.1: Counter arrangement to provide a control of predetermined count.

10.2.2 EPUT Meters

EPUT meters combine the simple electronic counter and an internal time base with a means for limiting the counting process to preset time intervals. This permits direct measurement of frequency and is useful for accurate determinations of rotational speeds (see Sections 8.6.2 and 10.7). The instrument is not limited, however, to an input varying at a regular rate; intermittent or sporadic events per unit of time may also be counted. Other applications include its use as a readout device for frequency-sensitive pickups, such as turbine-type flowmeters (see Section 15.5.1).

10.2.3 Time-Interval Meter

By modifying the arrangement of circuitry of an electronic counter, one can obtain a *time-interval meter*. In this case input pulses start and stop the counting process, and the pulses from an internal oscillator make up the counted information. In this manner the time interval taking place between starting and stopping may be determined, provided the frequency of the internal oscillator is known.

Figure 10.2 illustrates a simple application of the time-interval meter. Photodetectors are arranged so that the interruption of the beams of light provide pulses—first to start the counting process and second to stop it. The counter records the number of cycles from the oscillator, which has an accurately known stable output. In the example shown, the count would represent the number of hundred-thousandths of a second required for the projectile to traverse the distance between the light beams. Refer to Fig. 9.2(b) for a more detailed picture of the electronic counter's operation in this application.

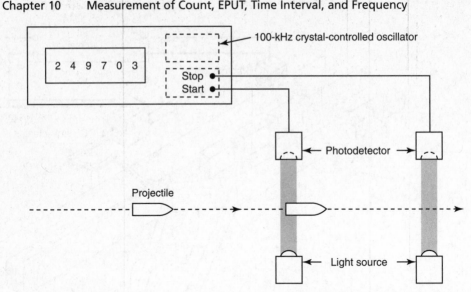

FIGURE 10.2: Time-interval meter arranged to count the number of hundred-thousandths of a second required for the projectile to traverse a known distance between photodetectors.

10.3 STROBOSCOPY AND HIGH-SPEED IMAGING

10.3.1 The Stroboscope

The term *stroboscope* is derived from two Greek words meaning "whirling" and "to watch." Early stroboscopes used a whirling disk as shown in Fig. 10.3. During the intervals when openings in the disk and the stationary mask coincided, the observer would catch fleeting glimpses of an object behind the disk. If the disk speed was synchronized with the motion of the object, the object could be made to appear to be motionless. In some ways the action is the inverse of the illusion produced by the motion picture projector. Also, if the disk were made to rotate with a period slightly less than, or greater than, the period of the observed object, the object could be made to apparently creep either forward or backward. This made possible direct observation of such things as rotating gears, shaft whip, helical spring surge, and the like while the devices were in operation.

Modern stroboscopes operate on a somewhat different principle. Instead of the whirling disk, a controllable, intense flashing light source is used. Repeated short-duration (10 to 40 μs) light flashes of adjustable frequency are supplied by the light source. The frequency, controlled by an internal oscillator, is varied to correspond to the cyclic motion being studied. The readout is the flashing rate required for synchronization. These devices are often called *strobe lights*.

Two different cautions require mention. The first involves a minor problem concerned with the geometry of the item being studied. Suppose, for example, that the gear illustrated in Fig. 10.3 is the study subject, and suppose that the spokes are used as the target for synchronization. A moment's thought makes it clear that each of the six spokes in this example will, in succession, occupy a given position. One must use care in making certain that one, and only one, spoke is identified. The usual practice is to place a distinctive mark on one of the spokes and to use that spoke alone in searching for one-on-one synchronization.

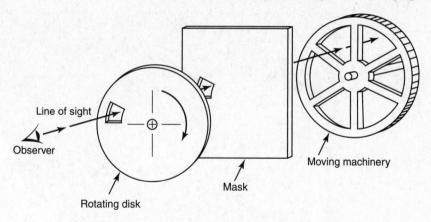

FIGURE 10.3: Essential parts of early disk-type stroboscope.

The second caution concerns multiple ratios of flashing rate to the object's true cycling rate. As an example, consider the rotation of a crank arm. Suppose that the arm is rotating at 1200 revolutions per minute (rpm) and the flashing rate is 600 cycles per minute (cpm). Then the arm would be in the same position at each flash and would appear to be stationary, but at only half the actual frequency. Further, if the flash rate were 400 cpm or 300 cpm, the arm would again occupy the same position for successive flashes and would again appear to be stationary. This is another example of *aliasing* (see Section 4.7.2).

An obvious approach is to "stop" the motion, note the flashing rate, then double the rate and check again. Another approach is to use the following convenient procedure:

1. Determine a flashing rate f_1 that freezes the motion.
2. Slowly reduce the rate until the motion is frozen once more. Note this rate, f_2.

Then

$$f_0 = \frac{f_1 f_2}{f_1 - f_2} \qquad (10.1)$$

where

$$f_0 = \text{actual cycling rate of the object}$$

It should be noted that it is not always necessary to obtain a one-to-one synchronization. Note also that the procedure described makes it possible to extend the upper frequency limit beyond the stroboscope's normal range.

In addition, as shown in Fig. 10.4, stroboscopic lighting can be used to study non-repeating action. By using a still camera with the shutter locked in the open position, stroboscopic lighting can be used to track the position of a moving object. For example, if the flashing rate is known, the position or displacement of the object can be determined at various instants of time. These data can be numerically differentiated to determine the instantaneous velocity or acceleration.

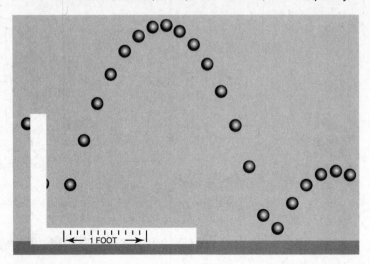

FIGURE 10.4: Photo obtained by "open-shutter" camera technique and using a Strobotac® set at a flash rate of 20 Hz (Courtesy: GenRad, Inc., Concord, Massachusetts).

10.3.2 High-Speed Imaging

Stroboscopy is closely related to high-speed photography or movie making. By using short exposures taken in quick succession, it is possible to capture the dynamics of motion that might otherwise pass in the blink of an eye. For example, Fig. 10.5 shows a sequence of images of a water-filled balloon as it drops onto a surgical scapel and bursts. These images were recorded by a high-speed digital video camera at rate of 10,000 frames per second. The entire process shown took just 0.070 s.

The basic requirements of high-speed imaging are that the exposure time be short enough to capture an unblurred image of the moving object and that images be taken frequently enough to track the changing features of the process. The faster the object moves, the shorter must be the exposure and the higher the frame rate. A rule of thumb is that if an object requires a time T to move its own length, the exposure time should be $0.001\,T$ in order to obtain adequate focus [7]. Framing rates of up to 1000 frames per second (fps) with exposure times down to microseconds are adequate for a wide range of relatively low-speed applications, including studies of moving machinery and crash testing. In studies of ballistics and explosives, the velocities are much higher, and exposures of just 10 to 100 ns with frame rates of up to 10^6 fps may be required.

While film cameras were long used for high-speed movie making, most high speed imaging today is done using digital video cameras. The advantages of digital systems over film systems are considerable. These include immediate availability of the image, the ability to use computers to process the images and extract data from them, ease of operation, longer recording times, and the elimination of chemical film processing. Film retains some advantages, including lower equipment cost, potentially larger image format or resolution, and, in some cases, significantly higher framing rates.

High-speed digital video cameras with frame rates from 1000 fps to more than 100,000 fps are available. Not suprisingly, the cost tends to rise as the speed rises. These high-speed cameras are one to two orders of magnitude more expensive than the common digital video camera used for taking home movies, which runs at just 30 fps. By using

(a) $t = 0.0$ ms

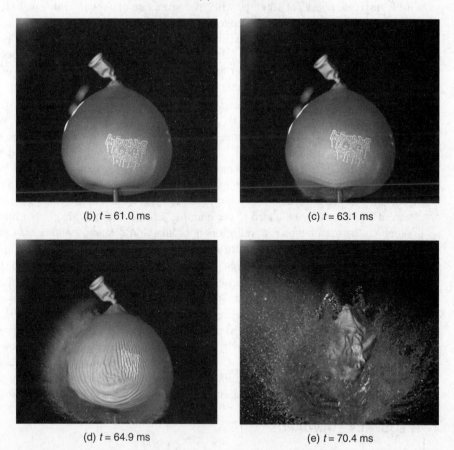

(b) $t = 61.0$ ms

(c) $t = 63.1$ ms

(d) $t = 64.9$ ms

(e) $t = 70.4$ ms

FIGURE 10.5: A water-filled balloon dropping onto a surgical scapel. (Courtesy: Professor Ian W. Hunter, Massachusetts Institute of Technology)

rotating mirrors, multiple sensor arrays, and other tricks, frame rates beyond 10^6 may be achieved, although only a small number of frames may be shot at such speeds. Certain film cameras may reach framing rates above 2×10^6 fps, but again only for short intervals.

The basic components of a digital video camera are a lens system, an electronic image sensor, and a shutter. Unlike film cameras, which use mechanical shutters, many digital video cameras control the exposure electronically (a so-called *electronic shutter*). The shutter speed can be varied, with some cameras capable of exposures ranging from 50 ns to 5 s. Electronic image sensors are increasingly often made with CMOS arrays [1], which offer a variety of advantages over the earlier CCD arrays, particular in terms of fabrication and system integration. Digital data from either type of array are transferred to memory within the camera as each image is shot. Data from the camera's memory may then be transferred to a computer through an ethernet or USB interface. The computer itself may have complete control of the camera's operation.

The resolution of the digital camera, as measured by the number of pixels in the image, may decrease as the frame rate rises. This is because the camera may not be able to exceed a particular data rate, in bit/s. For example, a camera taking 1024 by 1024 pixel 8-bit grayscale images at 1000 fps will have a data rate of about 8 Gbit/s or 1 Gbyte/s (see Section 8.12). This may be the upper limit of whatever data bus (see Section 8.10.4) is used to transfer the images from the image sensor to memory. If the frame rate is increased by a factor of 10, the number of pixels per image must be correspondingly reduced. Color digital images are three times larger than monochrome images, and so the use of color can require lowering either the frame rate or the image size, or both.

We have already noted that exposure times are governed by the speed of the process being recorded. In order to obtain an image, a sufficient amount of light must be collected by the sensor array or film during the exposure time. Thus, the shorter the exposure, the more intense must be the lighting. This effect can be partially offset by increasing the f-number of the camera (having a more open aperture), but, for the most part, high-speed imaging requires close and intense lighting. Heat from the lamps can become a significant problem when large areas must be illuminated, as in an automobile crash test where many tens of kilowatts may be added to the building's air conditioning load. For more modest situations, halogen lamps with reflectors or ordinary photostudio lighting may suffice.

Finally, there is the question of when to start and stop the photography. If the entire event lasts just a fraction of a second, human reflexes will be too slow to push a start button at just the right moment. Instead, some form of electronic triggering is required [2]. Here, the ingenuity of the experimentalist is paramount. For example, the recording of a crash test might be triggered by a sensitive accelerometer (Chapter 17). As shown in Fig. 10.2, the interruption of a light beam could be used to capture the passage of a moving projectile. A supersonic event might be captured by using an appropriately located microphone to detect the initial passage of a shock wave. Capacitive or magnetic sensors may also be configured to detect the passage of an object, and a thin wire may be placed so that a passing object breaks it, changing the electrical resistance of a circuit and starting the recording.

10.4   FREQUENCY STANDARDS

Frequency is the number of recurrences of a phenomenon or series of events during a given time interval, and the reciprocal of frequency is *period*. A frequency standard chops time into discrete bits that may be used as time standards and, through comparative means, for timing events. Alternatively, a frequency standard may be used in it own right as a source of frequency for driving instruments, such as function generators, or for calibrating other frequency sources.

Frequency standards are often designed to operate at a single frequency, which is typically in the range of tens of kHz to GHz. The signal may be reduced to other desired frequencies, 1 Hz say, by using frequency-dividing circuits, such as the "divide-by-X" IC chips discussed in Section 8.7. Standard frequency signals may be compared to other signals by various methods. One method, which is easily implemented with an oscilloscope, is the Lissajous figure technique described in Section 10.5. Another method, suitable for radio frequency signals, is the heterodyne technique described in Section 10.6.

Frequency sources may be either electrical or mechanical, or some combination of both. Pendulums were long used to provide mechanical frequency standards for clocks. They are also the basis of the mechanical metronome that a musician might use to keep tempo during practice. To take another musical example, tuning forks are mechanical resonators that vibrate at a given pitch—perhaps at a 440-Hz A tone. Mechanical resonators may be combined with electrical circuitry that drives the motion and outputs the resonant frequency. Quartz-crystal oscillators are an extremely important example of this type of source, as discussed in Section 10.4.3.

Quantum mechanical resonances—related to the energy levels of atoms—are the basis of "atomic clocks," such as the cesium clock that forms the international time standard (see Section 2.6). Cesium clocks are extremely accurate, but cost tens of thousands of dollars. The rubidium atom can be used to provide a slightly less accurate atomic standard at about one-tenth the cost. Both types of atomic clock would be used only when the highest precision and stability are required.

Purely electrical oscillators are also employed as frequency sources. These encompass a wide variety of analog and digital circuits, including IC oscillators (Section 8.4.4) and LC resonant circuits (Section 7.11). The accuracy and stability for many of these types of circuits is not even as high as the cheapest quartz-crystal oscillators [3], but they are nonetheless commonly applied as low-precision frequency sources in all kinds of electrical devices. Another readily available electrical frequency source is the ordinary 60-Hz line voltage (with a suitable voltage transformer, of course), which provides a frequency signal stable to about ±0.05 Hz.

Electronic oscillators are sometimes classified by frequency as follows:

1. Audio frequency (0 to 20 kHz)
2. Supersonic (20 to 50 kHz, roughly)
3. Radio frequency (50 kHz to 10 GHz)

In the rest of this section, we discuss some of the high-precision frequency sources that are readily available as laboratory standards.

10.4.1 Global Positioning System Signals

The *global positioning system*, or GPS, is a network of 24 or more satellites which provide navigation signals across the globe. The network was originally launched for military use, but civilian access was granted during the 1980s. Small GPS positioning receivers have become a common consumer product, allowing one to find his or her position to within 10 to 20 m at essentially any point on Earth.

The proper measurement of time is essential to navigation, and GPS accomplishes this by the presence of either cesium or rubidium atomic clocks within the satellites themselves. As a result, GPS transmissions contain not only positioning data, but also high-accuracy time and frequency signals. A GPS *timing receiver* can obtain time to an accuracy of a few nanoseconds. Frequency can be determined to a fractional error better than 4×10^{-14} [5]. This type of GPS receiver usually provides both a timing output of one pulse per second and various standard frequency outputs in the MHz range. (It should be noted that GPS timing receivers are different instruments than the more commonly encountered GPS positioning receivers.)

10.4.2 Radio Time and Frequency Transmission

Precise frequency and time standards are also available through reception of transmissions from the National Institute of Standards and Technology radio stations. Stations WWV and WWVB are located at Fort Collins, Colorado. Station WWVH is in Hawaii. WWV and WWVH are classified as *high-frequency* (HF or *shortwave*) stations, and they broadcast on several bands between 2.5 MHz and 20 MHz. WWVB transmits at 60 kHz and is classified as *low frequency* (LF). Low frequencies have the advantage of providing more stable reception at a distance because of reduced variations in the transmission paths peculiar to those frequencies. The signals of these radio stations are directly traceable to the NIST's primary frequency standard, the cesium fountain clock (Section 2.6). The 60-kHz carrier signal of WWVB, for example, can be used to calibrate electronic equipment or secondary frequency standards.

The radio broadcasts include various time and frequency signals. WWVB sends its time code in binary-coded-decimal format (Section 8.6) are a rate of 1 bit per second. WWV and WWVH include short tones at frequencies of 500 or 600 Hz and announcements of the *Coordinated Universal Time*, or UTC, as well a BCD coded time signal on a sub-carrier frequency. In particular, it is WWVB that provides the signal used to synchronize radio-controlled clocks and wristwatchs in the continental United States.[1] These timepieces contain miniature antennas, not unlike those found in small AM radios [6]. The radio signal is used to reset the clock periodically. The clock itself may operate on an inexpensive quartz-crystal oscillator that accumulates errors of up to seconds per day.

The uncertainty in WWVB's frequency signal is one part in 10^{10} to 10^{12}. The uncertainty in its time signal is 0.1 to 15 ms [8].

10.4.3 Quartz-Crystal Oscillators

Quartz-crystal oscillators are electromechanical resonators that produce an oscillating voltage through the piezoelectric effect (see Section 6.14). We have already mentioned the importance of these oscillators in providing timing signals in digital circuits, such as microprocessors and computers (see Section 8.4.4). Crystal oscillators are probably the most widely used type of frequency standard.

Quartz-crystal oscillators are made from synthetic or natural quartz that is cut to precise dimensions. The dimensions chosen determine the resonant frequency for mechanical vibration of the crystal. The crystal is integrated into an electrical circuit that drives the

---

[1] In addition to the United States, many other nations broadcast timing signals of various types, such as, ATA in India (10 MHz), BPC in China (68.5 kHz), CHU in Canada (3.330, 7.335, 14.670 MHz), DFC77 in Germany (77.5 kHz), HLA in South Korea (5 MHz), JJY in Japan (40, 60 kHz), and MSF in the United Kingdom (60 kHz). Radio-controlled timepieces are thus able to remain calibrated in most parts of the world.

oscillation and produces the output signal. Various grades of oscillator are available. The relatively inexpensive types, such as used in digital wristwatches, are available as IC chips that contain all the driving circuitry. They have frequency accuracies in the range of 10 parts per million and cost about a dollar.

Higher grades of crystal oscillators compensate for the sensitivity of the crystal to temperature changes and other environmental factors. The simplest approach is to incorporate a temperature sensor, such as a thermistor (see Section 16.4.3) into the driving circuit. The sensor's output is used to correct for temperature-induced shifts in the crystal's resonant frequency. Oscillators of this type may have accuracies of 1 to 100 parts per billion, and they are sometimes used as the built-in frequency standards for instruments such as counters and function generators.

The best quartz oscillators eliminate temperature drift entirely by placing the oscillator in a fixed temperature enclosure, called an *oven*. Oven-controlled crystal oscillators may have long-term frequency stabilities significantly better than one part per billion, ranging to better than one part per trillion accuracy over short (1-second) intervals. Oscillators of this type may cost thousands of dollars [4].

10.4.4   Complex-Wave Oscillators and Function Generators

Complex waveforms (Section 4.4) are used in a variety of special applications. Ramp or sawtooth waveforms are used for sweep generators (Section 9.6.2); square waves may be used for evaluating signal conditioner responses (Section 5.20) or for providing synchronization and coding in digital computers. IC chips that provide most or all of these functions are available.

A common lab instrument, the *function generator*, also provides complex waveforms. Users may specify frequency, shape, offsets, modulation, and other characteristics of the output voltage signal using controls on the front panel. Inputs for external triggering may be provided. The sophistication and quality of function generators varies widely among models, as does the cost. High-quality function generators often contain an internal frequency standard from which the user-specified outputs are derived, through the application of appropriate wave shaping circuitry; others may rely on 60-Hz line voltage to provide the frequency standard. In some cases, the waveforms are digitally generated, and the user can program any output signal desired. Programming commands may be sent to the instrument by a computer through an external bus (see Section 8.10.4), such as USB, GPIB (IEEE 488), or Ethernet.

10.5   LISSAJOUS FIGURES FOR FINDING FREQUENCY AND PHASE RELATIONS

Lissajous (Liss-a-ju) figures provide a straightforward approach to determining the relative characteristics of two different frequency sources, primarily their frequency and phase relations. The technique is usually implemented an oscilloscope using a calibrated variable-frequency source to which an unknown signal is compared. However, the original study of Lissajous figures was done by Nathaniel Bowditch in 1815, who produced them using narrow stream of sand issuing from the base of a compound pendulum [9]. The technique is best described through examples.

Suppose that two 60-Hz sinusoidal voltages from different sources are connected to an oscilloscope, one to the vertical and the other to the horizontal input channels. Any of the following several patterns may result.

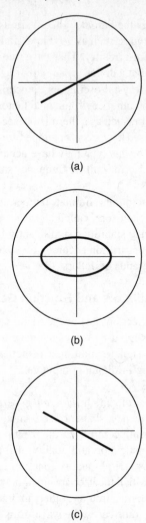

FIGURE 10.6: (a) In-phase Lissajous figure, (b) Lissajous figure for sinusoidal inputs ±90° out of phase, (c) Lissajous figure for inputs 180° out of phase.

In-Phase Relations

If the two voltages are in phase, then as the x-voltage increases, so also does the y-voltage. The x-voltage will deflect the beam along the horizontal axis, and the y-voltage will deflect it in the vertical direction. The resulting trace, then, will be a line diagonally placed across the face of the tube, as shown in Fig. 10.6(a). The angle that the line makes with the horizontal will depend on the relative voltage magnitudes and the oscilloscope gain settings.

90° Phase Relations

Suppose the two 60-Hz sinusoidal voltages are 90° out of phase. Then as one voltage passes through zero, the other will be at a maximum and vice versa. The resulting trace will be that shown in Fig. 10.6(b). In general, it will be an ellipse with axes placed horizontal and vertical.

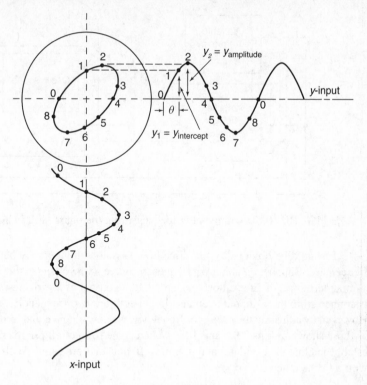

FIGURE 10.7: Lissajous figure for sinusoidal inputs of the same frequency, but with a phase relation of θ degrees.

180° Phase Relations

Figure 10.6(c) shows the pattern that results when the two voltages are 180° out of phase.

Other Forms of Lissajous Figures

Intermediate forms are ellipses with axes inclined to the horizontal. A study of Fig. 10.7 shows that when the horizontal input is at midsweep, the vertical precedes it by θ degrees, corresponding to a vertical input of y_1.

From the sine-wave plot of the curve we see that

$$\sin \theta = \frac{y_1}{y_2} = \frac{y\text{-intercept}}{y\text{-amplitude}}$$

Therefore, by determining the values of y_1 and y_2 from the ellipse, we may determine the phase relation between the two inputs.

An example of the application of this method is the determination of phase shift through an amplifier. A sampling of the amplifier input signal would be applied to the x-input terminals of an oscilloscope, and the amplifier output would be connected to the y-input terminals as shown in Fig. 10.8. By scanning the frequency range for which the amplifier is intended, one could detect any shift in phase relation. Of course, it would be necessary to know that no shift occurs with frequency in the oscilloscope circuitry or in any of the circuitry external to the amplifier.

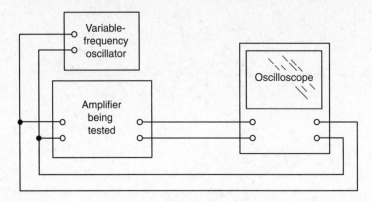

FIGURE 10.8: Arrangement for measuring the phase shift in an amplifier.

It should be obvious by this time how Lissajous figures may be used to determine frequencies. Suppose an unknown frequency source with voltage output is connected to the y-input terminals of an oscilloscope, and that the output of a variable-frequency oscillator is connected to the x-input terminals. In general, the two frequencies would be different. However, by adjusting the oscillator frequency, we may obtain equal frequency figures such as those shown in Figs. 10.6 and 10.7. When some form of ellipse results, proof would be established that the oscillator and unknown frequencies are equal. With one known, so too would be the other.

Fortunately the method is not limited to equal frequencies. Figure 10.9 shows Lissajous figures for several other frequency ratios. By studying these figures, we see that a basic relation may be written as follows:

$$\frac{\text{vertical input frequency}}{\text{horizontal input frequency}} = \frac{\text{number of vertical maxima on Lissajous figure}}{\text{number of horizontal maxima on Lissajous figure}}$$

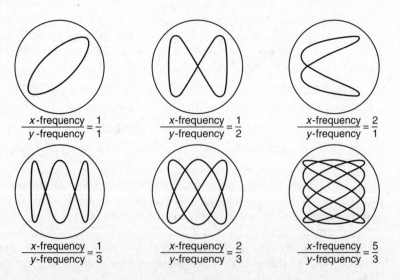

FIGURE 10.9: Lissajous figures for sinusoidal inputs at various frequency ratios.

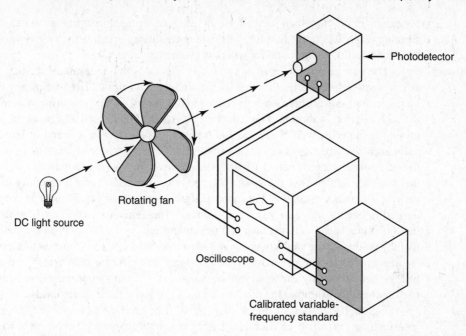

FIGURE 10.10: A method for determining fan speed employing a photoelectric sensor and a frequency standard.

We also see that for the figure to remain fixed on the screen, either the two input frequencies must each be fixed or they must be changing at proportional rates. In addition, the symmetry of the figure will depend on the phase relation between the two inputs.

If the frequencies are reasonably stable, ratios as high as 10 to 1 may be determined without undue difficulty. This suggest a means whereby a variable-frequency source, such as an oscillator or signal generator, may be calibrated. By use of a fixed-frequency source, such as the 60-Hz line voltage (through a small step-down transformer), a variable-frequency source may be calibrated for a number of points using the 10-to-1 relation. For example, 60-Hz line voltage may be used to spot-calibrate from 6 to 600 Hz.

Finally, we note that pure sine-wave inputs are not always necessary. Figure 10.10 shows a simple arrangement for determining the speed of a fan. In this case the resulting "one-to-one" Lissajous diagram approximates a distorted parallelogram rather than an ellipse.

10.6   HETERODYNE AND PHASE-LOCK MEASUREMENTS OF FREQUENCY

Suppose we have two sources of pure audio tones, the two tones having nearly the same frequency. When mixed, a third or beat note is produced (see Fig. 4.7). The frequency of the beat is a function of the *difference* in the two original notes. If the frequency of one of the sources is known and is adjusted to produce *zero beat*, then the frequency of the other source is also known by comparison. Conversely, the unknown source frequency may be adjusted until it matches that of the known source, effectively calibrating it. Musicians, for example, do this when they adjust the pitch of stringed instruments to match that of a tuning fork. This procedure for determining frequency is called the *heterodyne* method, the mathematical background for which is given in Section 4.4.1.

The general advantage of heterodyne methods is that measurements may be made at much lower frequencies that than of the signals in question. High-frequency measurements are more susceptible to errors related to time delays in the measuring system because the signal period is shorter. Thus, the heterodyne method is particularly useful for determining frequencies well above the audio range. One example is found in laser-doppler velocimetry, where measurement of a small change in the frequency scattered light is used to find fluid velocities (Section 15.10). Radio frequencies are also commonly measured by this method. In that case the standardizing signal originates from a carefully calibrated, variable-frequency oscillator. For radiated signals, an ordinary radio receiver covering the desired frequency range may serve as a mixer. The generator frequency is adjusted until the difference between the known and unknown frequencies falls within the audio range, thereby producing the well-known amplitude-modulated squeal so familiar when two radio stations interfere. The generator is then fine-tuned to produce zero beat. True zero beat may be determined by ear to within 20 or 30 Hz. By using appropriate instruments (an oscilloscope, for example) the resolution may be reduced to far lower levels.

Figure 10.11 shows a block diagram of a heterodyne arrangement for frequency measurement, in which an oscillator having a known frequency is compared to a signal with an unknown frequency. The two signals are received by a circuit called a *mixer*. The mixer is a radio-frequency component that actually multiplies, rather than adds, the signals producing an output at the sum and difference frequencies. If sine waves at frequencies f_1 and f_2 are the inputs, the output is

$$\sin(2\pi f_1 t)\sin(2\pi f_2 t) = \frac{1}{2}\cos[2\pi(f_1 - f_2)t] - \frac{1}{2}\cos[2\pi(f_1 + f_2)t] \qquad (10.2)$$

The mixed signal is low-pass filtered to obtain a signal at the beat frequency, $(f_1 - f_2)$. The filtered signal may then be sent to a frequency counter, allowing the unknown frequency

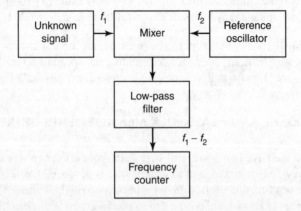

FIGURE 10.11: Heterodyne method of frequency measurement.

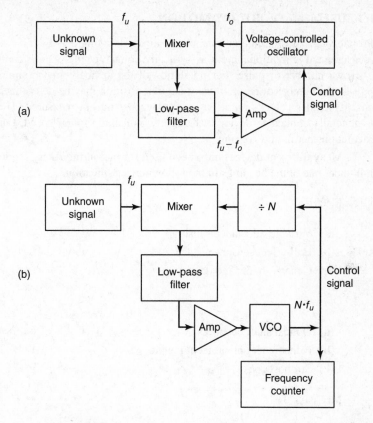

FIGURE 10.12: Phase-lock loop method of frequency measurement: (a) Feedback loop adjusts oscillator frequency f_o to unknown frequency f_u; (b) resolution multiplication arrangement for low-frequency counting.

to be determined. It is also possible to use the beat signal as feedback to an adjustable reference oscillator (a voltage-controlled oscillator), enabling it to "tune-in" on an unknown or fluctuating frequency. The latter technique creates what is called a *phase-locked loop*, since the feedback loop brings the reference oscillator into phase with the unknown oscillator [Fig. 10.12(a)]. The VCO frequency and the beat signal may be monitored in order to measure fluctuations in the unknown signal with great precision. Both of these methods have application to the calibration and characterization of frequency sources [10, 11].

The phase-lock loop also provides a tool for reducing the ± 1 count error when applying frequency counters to low-frequency signals (Section 9.2.4). As shown in Fig. 10.12(b), a divide-by-N circuit may be placed in the feedback loop, so that the VCO is driven to an output frequency of $N \cdot f_u$, that is, a frequency N times greater than the low-frequency input f_u. The output of the VCO is sent to a frequency counter, improving count resolution by a factor of N. For example, if a signal is counted for one second and $N = 1000$, the number of counts is 1000 greater than if f_u itself were counted, so that an error of ± 1 count (corresponding to ± 1 Hz) in $N \cdot f_u$ amounts to $\pm 1/1000$ count (or ± 0.001 Hz) in f_u. The number obtained by the counter is, obviously, divided by N to obtain the measurement.

10.7   MEASUREMENT OF ROTARY MOTION

Probably the most common example of direct counting and EPUT determination is the measurement of rotational motion. Typically, some type of sensor is configured to produce a known number of pulses per revolution of the device being monitored. By counting pulses, the number of revolutions or angular position may be determined. If the counting circuit includes a clock or time base, angular speed may be measured. The term *tachometer* is generally applied to device that directly indicates angular speed, typically in units of revolutions per minute, or rpm.

Many different devices have been used for measuring rotary motion, most of which fall under one of the headings in the following classification:

1. Electrical

   (a) Variable reluctance sensors
   (b) Hall-effect sensors
   (c) Generators (ac and dc)

2. Optical

   (a) Optical shaft encoders
   (b) Noncontact optical tachometers
   (c) Stroboscopes

3. Mechanical

   (a) Direct counters
   (b) Centrifugal speed indicators

Electrical tachometers often make use of a variable-reluctance pickup (Section 6.12) or a Hall-effect device (such as the sensor illustrated in Fig. 6.23). These permit the measurement of angular speed or position by noncontacting means. If the pickup is placed near the teeth of a rotating gear, for example, extremely accurate speed measurements may be made by either an electronic counter or a frequency meter. It is also possible to measure rotary speed using a small permanent magnet dc or ac generator in conjunction with appropriate voltage measuring circuitry.

Optical devices are very widely used for angular speed and position measurement. Photodetectors or optical interruptors (Fig. 6.20) can provide voltage pulses originating by the interruption of a light beam from rotation or movement of a machine member. These pulses may be treated in a manner similar to those from a reluctance or Hall-effect sensor. Optical shaft encoders, based on circular binary encoders (Fig. 8.11), may be similarly adapted to angular speed measurement. Noncontact optical tachometers are available as handheld units. A light beam from the tachometer is reflected from a moving pattern on the rotating surface, which may be something as simple as a piece of reflective tape. By counting the reflected light pulses, rotary speed may be determined to high accuracy.

Any of the various stroboscopic methods discussed in Section 10.3 may also be used for optical speed measurements.

Mechanical counters may be of the *direct-counting* digital type or may be counters with a gear reducer. In the latter case, angular motion available at a shaft end is reduced by a worm and gear, and the output is indicated by rotating scales. In both examples, rpm is measured by simply counting the revolutions for a length of time as measured with a stopwatch, and calculating the turns per minute from the resulting data. Modifications incorporate the timing mechanism in the counter. The timer is used to actuate an internal clutch that controls the time interval during which the count is made. *Centrifugal rpm indicators* use the familiar *flyball-governor* principles, which balance centrifugal force against a mechanical spring. An appropriate mechanism transmits the resulting displacement to a pointer, which indicates the speed on a calibrated scale.

Of course, the considerable advantages of optical or electrical sensors over mechanical types include continuous readings, noncontact operation, and digital output. Mechanical devices for rotational measurement are generally obsolescent.

SUGGESTED READINGS

Diddams, S. A., J. C. Bergquist, S. R. Jefferts, and C. W. Oats. Standards of time and frequency at the outset of the 21st century. *Science*, 306:1318–1324, 2004.

Horowitz, P., and W. Hill. *The Art of Electronics*. 2nd ed. New York: Cambridge University Press, 1989.

Lombardi, M. A. *NIST Time and Frequency Services*. National Institute of Standards and Technology, Special Publication 432. Washington, D.C.: U.S. Government Printing Office, Jan. 2002.

Rathore, T. S. *Digital Measurement Techniques*. 2nd ed. Pangborne, U.K.: Alpha Science International, 2003.

Ray, S. F. (ed.). *High Speed Photography and Photonics*. Oxford, U.K.: Focal Press, 1997.

PROBLEMS

10.1. Prove the validity of Eq. (10.1).

10.2. Using an electronic counter, monitor the local power-line frequency. Use a step-down transformer to avoid the danger of the line voltage. What variations are noted over the period of the test? Compare results using (a) the setting for frequency readout and (b) the setting for period readout.

10.3. Use a flashing-light-type stroboscope and determine the time–speed relationship for a shop-bench-type grinder as it *decelerates*. A suggested technique is as follows: Set the stroboscope flashing rate to a predetermined frequency; then turn off the grinder power, simultaneously starting a timer (e.g., a stopwatch). Determine the time required for the first synchronization between wheel speed and flashing rate. Repeat for a range of flashing rates until sufficient data are obtained to plot a deceleration versus time curve. This approach is a very simple one that requires a minimum of test equipment. Why would this approach not be viable to determine the acceleration–time plot for the typical bench grinder? Other more sophisticated schemes may be devised; propose other approaches to measuring the decelerating speed–time characteristics.

10.4. Using a radio receiver and an oscilloscope, compare the local power-line frequency with the 600-Hz audio tone transmitted by WWV the radio station of the National Institute for Standards and Technology. *Be sure to use an isolation transformer between the power line and the oscilloscope.*

10.5. Use the power-line frequency to calibrate a signal generator over the range 10 to 600 Hz. *Be sure to use a step-down transformer to isolate the line and to obtain a reasonably low voltage (say, no more than 6 V ac).*

10.6. Suggest a means of using an oscilloscope for measuring the actuating time of a simple double-throw electric relay. [*Hint:* Use the relay actuation voltage to trigger the sweep and a circuit using the relay contacts for the timing-limit switches. Would it be practical to use a debouncing element in the circuit? (See Section 8.4.2.)]

10.7. Show that an ellipse is generated on an oscilloscope display when the y-input is $A_0 \sin(\omega t - \phi)$ and the x-input is $A_i \sin \omega t$.

10.8. Make a test arrangement similar to that shown in Fig. 10.10 (or with modifications as desired) and determine the time–speed relationship for a common office fan as it accelerates from rest to full speed. Observe the oscilloscope screen, determining the time for the Lissajous diagram to transit the one-to-one ratio. Adapt the procedural suggestions given in Problem 10.3 to this setup. Use the curve-fitting methods given in Sections 3.13 and 3.14 to find a relationship, $\omega = f(t)$. Check to see whether, by chance, the relationship approximates first-order characteristics. Run a similar set of tests to determine the speed–time curve for deceleration.

10.9. Devise an experimental method for determining the *accelerating* speed-time characteristics of a shop-bench-type tool grinder (see Problem 10.3).

10.10. Reference [12] describes a simple method for determining the fundamental resonance frequency of a turbine blade. In essence, the method consists of striking the blade with a soft mallet and using a microphone to pick up the sound. The microphone output is fed to an oscilloscope and its frequency is compared with that of a signal generator using the Lissajous technique. Any harmonics of significance would also be displayed. Locate a complex elastic member and use the method to study its characteristics of frequency of vibration.

10.11. Using the basic approach and the "complex elastic member" suggested in the final sentence of Problem 10.10, make a harmonic analysis of acoustic signal captured by the microphone. Use available equipment as appropriate, such as a storage oscilloscope or a computer-operated A/D converter with appropriate software.

REFERENCES

[1] Fuller, P. W. W. Project design and planning. In S. F. Ray (ed.), *High Speed Photography and Photonics*. Oxford, U.K.: Focal Press, 1997, Chap. 14.

[2] Brandt, T. High-speed CMOS imagers are flexible. *Laser Focus World*, 40(2), Feb. 2004.

[3] Fuller, P. W. W. Synchronization and triggering. In S. F. Ray (ed.) *High Speed Photography and Photonics*. Oxford, U.K.: Focal Press, 1997, Chap. 3.

[4] Horowitz, P., and W. Hill. *The Art of Electronics*. 2nd ed. New York: Cambridge University Press, 1989, Sects. 5.12–5.19.

[5] Parker, T. E., and D. Matsakis. Time and frequency dissemination: Advances in GPS transfer techniques. *GPS World*, November 2004, pp. 32–38.

[6] Lombardi, M. A. *Radio Controlled Clocks*. Proceedings of the 2003 NCSL International Workshop and Symposium, Tampa. Boulder, Colo.: NCSL Internation, 2003.

[7] Lombardi, M. A. *NIST Time and Frequency Services*. National Institute of Standards and Technology, Special Publication 432. Washington, D.C.: U.S. Government Printing Office, Jan. 2002.

[8] Lombardi, M. A. *NIST Frequency Measurement and Analysis System: Operator's Manual*. NISTIR 6610. App. A. Boulder, Colo.: Time and Frequency Division, National Institute of Standards and Technology, 2001.

[9] Lissajous Figures. In *The New Encyclopædia Britannica*. 15th ed., Vol. 7. Chicago: Encyclopædia Britannica Inc., 1998, p. 393.

[10] Levine, J. Introduction to time and frequency metrology. *Rev. Sci. Instr.* 70(6):2567–2596, 1999.

[11] Howe, D. A., D. W. Allan, and J. A. Barnes. *Properties of Oscillator Signals and Measurement Methods*. Boulder, Colo.: Time and Frequency Division, National Institute of Standards and Technology, 2000.

[12] Rosard, D. D. *Natural Frequencies of Twisted Cantilever Beams*. ASME Paper 52-A-15, 1952.

CHAPTER 11

Displacement and Dimensional Measurement

11.1   INTRODUCTION

The determination of linear displacement is one of the most fundamental of all measurements. The displacement may determine the extent of a physical part, or it may establish the extent of a movement. It is characterized by the determination of a component of space. In *unit* form it may be a measure of either strain (Chapter 12) or angular displacement.

To a greater extent than any other quantity, displacement lends itself to the simplest process of measurement: *direct comparison*. Certainly the most common form of displacement measurement is by direct comparison with a secondary standard. Measurements to least counts of the order of 0.5 mm (about 0.02 in.) may be accomplished without undue difficulty with use of nothing more than a steel rule for a standard. For greater resolutions or sensitivities, measuring systems of varying degrees of complexity are required. Measurements to least counts on the order of 0.005 mm (about 0.0002 in.) may be achieved with precision mechanical mechanisms, which typically use a system of gears to amplify a small displacement. For example, a micrometer converts the linear displacement of its shaft to a much larger rotation along the scale on the shaft barrel. Higher-resolution devices (to about 0.1 μm or a several microinches) may provide direct comparison to a set of calibrated gage blocks; very high resolution may be obtained with an optical interferometer, which enables displacement measurements to a fraction of the wavelength of light (about 0.005 μm or 0.2 microinch).

For purposes of discussion, we classify various measuring devices based on direct comparison in Table 11.1. With a few exceptions, these systems are used for measuring

TABLE 11.1:   Classification of Displacement Measuring Devices That Use Direct Comparison to a Secondary Standard

---

Low-Resolution Devices (to 1/100 in. or 0.25 mm)

1. Steel rule used directly or with assistance of
   a. Calipers
   b. Dividers
   c. Surface gage
2. Thickness gages

Medium-Resolution Devices (to 1/10,000 in. or 2.5×10^{-3} mm)

1. Micrometers (in various forms, such as ordinary, inside, depth, screw thread, etc.) used directly or with assistance of accessories such as
   a. Telescoping gages
   b. Expandable ball gages
2. Vernier instruments (various forms, such as outside, inside, depth, height, etc.)
3. Specific-purpose gages (variously named, such as plug, ring, snap, taper, etc.)
4. Dial indicators
5. Measuring microscopes

High-Resolution Devices (to a Few Microinches or about 2.5×10^{-5} mm)

Gage blocks used directly or with assistance of some form of comparator, such as
a. Mechanical comparators
b. Electronic comparators
c. Optical flats and monochromatic light sources

Super-Resolution Devices

Various forms of interferometers used with special light sources

---

fixed physical dimensions. Before we discuss any of them in detail, let us consider the following measurement problem.

11.2   A PROBLEM IN DIMENSIONAL MEASUREMENT

Suppose a hole is to be bored to the dimensions shown in Fig. 11.1, and that the part is to be produced in quantity. Such a dimension would probably be checked with some form of plug gage, illustrated in Fig. 11.2. One end of the plug gage is the *go* end, and the other the *no-go* end. If the *go* end of the gage fits the hole, we know that the hole has been bored large enough; if the *no-go* end cannot be inserted, the hole has not been bored too large.

Now, the plug gage itself would have to be manufactured, and no doubt drawings of it would be made. A rule of thumb is to dimension the plug gage with tolerances on the order of 10% of the tolerance of the part to be measured. If this rule is followed, the ends of the plug gage may be dimensioned as shown in Fig. 11.2.

It will be noted that the gage tolerance of 0.0004 in. (10% of the part tolerance, 0.004 in.) is applied symmetrically to the *no-go* end, corresponding to the upper limiting dimension of 1.504 in. On the other hand, the gage tolerance as applied to the *go* end penalizes the machinist somewhat because, in effect, an extra ten-thousandth of an inch is taken away from what is the machinist's. This is often done to increase the life of the gage

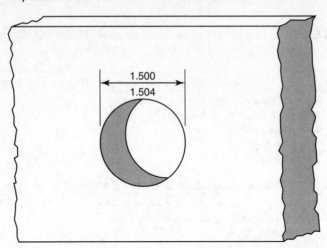

FIGURE 11.1: Typical dimensioning specifications for an internal diameter (English units).

by letting the gage wear toward the specified limit. Ideally, the *go* end will be inserted every time the gage is used and hence will wear, whereas the *no-go* end will never be inserted and will therefore experience no wear.

Provision has now been made for satisfactorily gaging the bored hole, provided the gages themselves are accurately made. How will we know if the gages are within tolerance? We can find out only by measuring them. This leads directly to the *gage block* listed under "High-Resolution Devices" in Table 11.1. A gage-block set is the basic "company" standard for any small dimension (0.01 to 10 in.).

By use of one of the comparison methods (to be described in a later section), the plug gage would be checked dimensionally. But, of course, we must not overlook the fact that to be useful, the gage blocks themselves must be measured, and so on ad infinitum, or at least back to the basic length standard (Section 2.4).

An example may be used to illustrate the extreme importance of measurement standards. Suppose that the 1.500-in. hole described above is in a part to be used by an automobile manufacturer. Very probably the gage would be made by some other company, one that specializes in making gages. Both the gage maker and the automobile manufacturer would undoubtedly "standardize" their measurements by using gage blocks. It is clear that unless the different gage-block sets are accurately derived from the same basic standard, the dimension specified by the automobile manufacturer will not be reproduced by the gage maker.

11.3 GAGE BLOCKS

Gage-block sets are industry's dimensional standards. They are the *known* quantities used for calibration of dimensional measuring devices, for setting special-purpose gages, and for direct use with accessories as gaging devices. They are simply small blocks of steel having parallel faces and dimensions accurate within the tolerances specified by their class. Blocks are normally available in the following classes.

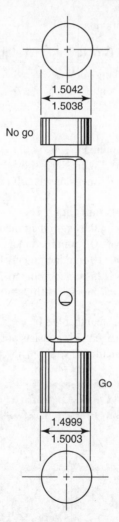

FIGURE 11.2: A go/no-go plug-type gage.

For the SI system,

Grade of block	Tolerance, μm (micrometers)[1]
0	±0.10 to ±0.25
1	±0.15 to ±0.40
2	±0.25 to ±0.70
3	±0.50 to ±1.30

---

[1] For very small displacement or tolerances, the mechanical engineer will commonly use the μm (the micrometer) or the microinch (1×10^{-6} in.). Physicists sometimes use the angstrom (1×10^{-7} mm). Equivalents are $1\ \mu$m = 39.37 μin. and 1Å = 0.003937 μin. *Unit displacement* (e.g., strain and certain coefficients) is unitless. In this case parts per million, or ppm, is conveniently employed.

For the English system of units,

Grade of Block		Tolerance, μin. (microinches)[2]
Class B	"Working" blocks	±8
Class A	"Reference" blocks	±4
Class AA	"Master" blocks	±2 for all blocks up to 1 in., and ±2 μ in./in. for larger blocks

Gage blocks are supplied in sets, with those sets having the largest number of blocks being the most versatile. Figure 11.3 shows a set made up of 76 blocks (plus 2 wear-prevention blocks) having dimensions as follows.

        1 block 1.005 mm thick
        49 blocks with 0.01-mm increments from 1.01 to 1.49 inclusive
        19 blocks with 0.5 mm increments from 0.5 to 9.5 inclusive
        4 blocks with 10-mm increments from 10 to 40 inclusive
        3 blocks with 25-mm increments from 50 to 75 inclusive
        2 carbide wear blocks, each 2 mm thick

Blocks are made of steel that has been given a stabilizing heat treatment to minimize dimensional change with age. This consists of alternate heating and cooling until the metal is substantially without "built-in" strain. They are hardened to about 62 Rockwell C.

FIGURE 11.3: A set of 76 gage blocks. (Courtesy: Mitutoyo America Corporation.)

---

[2]A range of tolerances is listed. Precise values depend on the size of the block.

Distribution of sizes within a set is carefully worked out beforehand, and for the set in Fig. 11.3 accurate combinations are possible in steps of 0.005 mm in over 95,000 dimensional variations.

11.4   ASSEMBLING GAGE-BLOCK STACKS

Blocks may be assembled by *wringing* two or more together to make up a given dimension. Suppose that a dimensional standard of 14.715 mm is desired. The procedure for arriving at a suitable combination might be determined by successive subtraction, as indicated immeadiately below.

		Blocks Used
Desired dimension	= 14.715	
Ten-thousandths place	= <u>1.005</u>	1.005
Remainder	= 13.71	
Hundredths place	= <u>1.21</u>	1.21
Remainder	= 12.5	
Tenths place	= <u>2.5</u>	2.5
Remainder	= 10.0	
Two wear blocks	= <u>4.0</u> (2.0 each)	4.0
Remainder	= 6.0	
Units place	= <u>6.0</u>	6.0
Remainder	= 0.0 Check ...	14.715

Blocks are not stacked by simply resting them one on top of another. They must be wrung together in such a way as to eliminate all but the thinnest oil film between them. This oil film, incidentally, is an integral part of the block itself; it cannot be completely eliminated, since it was present even at manufacture. The thickness of the oil film is always of the order of 0.2 μin. [1].

Properly wrung blocks markedly resist separation because the adhesion between the surfaces is about 30 times that due to atmospheric pressure. Unless the assembled blocks exhibit this characteristic, they have not been properly combined. The resulting assembly of blocks may be used for *direct* comparison in various ways. Two simple ways are shown in Fig. 11.4.

Gage blocks are sometimes used with special accessories, including clamping devices for holding the blocks. When so used, height gages, snap gages, dividers, pin gages, and the like may be assembled, using the basic gage blocks for establishing the essential dimensions. Use of devices of this type eliminates the necessity for transferring the dimension from the gage-block stack to the measuring device.

11.5   SURFACE PLATES

When blocks are used as shown in Fig. 11.4, some accurate reference plane is required. Such a flat surface, known as a *surface plate*, must be made with an accuracy comparable to that of the blocks themselves. In years past, carefully aged cast-iron plates with adequate ribbing

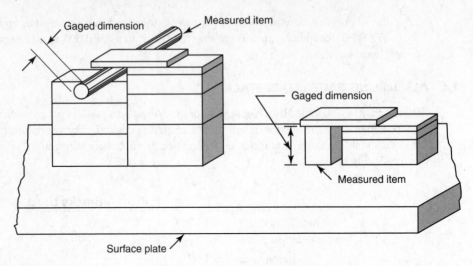

FIGURE 11.4: Two methods for using gage blocks for direct comparison of the length dimension.

on the reverse side were used. Such plates were prepared in sets of three, carefully ground and lapped together. When combinations of two are successively worked together, the three surfaces gradually approach the only possible surface common to all three—the true flat.

Machine-lapped and polished granite surface plates have largely replaced the hand-produced cast-iron type. Granite has several advantages. First, it is probably more nearly free from built-in residual stresses than any other material because it has had the advantage of a long period of time for relaxing. Hence, there is less tendency for it to warp when the plates are prepared. Second, should a tool or work piece be accidentally dropped on its surface, residual stresses are not induced, as they are in metals, causing warpage; the granite simply powders somewhat at the point of impact. Third, granite does not corrode.

11.6 TEMPERATURE PROBLEMS

Temperature differences or changes are major problems in accurate dimensional gaging. The coefficient of expansion of gage-block steels is about 11.2 ppm/°C (6.4 ppm/°F). Hence even a shift of one degree in temperature would cause dimensional changes of the same order or magnitude as the gage tolerances. The standard gaging temperature has been established as 20°C (68°F).

Several solutions to the temperature problem are possible. First, the most obvious solution is to use air-conditioned gaging rooms, with temperature maintained at 20°C. This procedure generally is followed when the volume of work warrants it; however, it is not a complete solution, for mere handling of the blocks causes thermal changes requiring up to 20 minutes to correct. For this reason, use of insulating gloves and tweezers is recommended. In addition, care must be exercised to minimize radiated heat from light bulbs, and so on [2].

A constant-temperature bath of kerosene or some other noncorrosive liquid may be used to bring the blocks and work to the same temperature. They may be removed from the bath for comparison; or, in extreme cases, measurement may be made with the items submerged.

On the other hand, if temperature control is not feasible, corrections may be used, based on existing conditions. A moment's thought will indicate that *if the gage blocks and work piece are of like materials, there will be no temperature error so long as the two parts are at the same temperature.* In by far the greatest number of applications, steel parts are gaged with steel gage blocks, and although there will probably be a slight difference in coefficients of expansion and the gage and parts may be at slightly different temperatures, appreciable compensation exists and the problem is not always as great as suggested in the preceding several paragraphs.

If both the part being gaged and the blocks are at temperature T_r (room temperature), corrections may be made by application of the following:

$$L = L_b \left[1 - (\Delta\alpha)(\Delta T)(10^{-6}) \right] \qquad (11.1)$$

where

$\Delta\alpha = (\alpha_p - \alpha_b)$,

$\Delta T = (T_r - \text{standard reference temperature})$,

L = the true length of the dimension being gaged (at reference temperature),

L_b = the nominal length of gage blocks determined by summation of dimensions etched thereon,

α_p = the temperature coefficient of expansion of the part being gaged (ppm/$°$),

α_b = the temperature coefficient of expansion of the gage-block material (ppm/$°$),

T_r = the ambient temperature

In using these relations, it is necessary that proper signs be applied to $\Delta\alpha$ and ΔT.

EXAMPLE 11.1

Let $L_b = 10$ cm, $\alpha_p = 13$ ppm/$°$C, $\alpha_b = 11.2$ ppm/$°$C, and $T_r = 24°$C. Find L.

Solution   Substituting, we have

$$L = 10 \left[1 - (1.8)(4)(10^{-6}) \right]$$
$$= 9.999928 \text{ cm}$$

EXAMPLE 11.2

Let $L_b = 9.7153$ in., $\alpha_p = 5.9$ ppm/$°$F, $\alpha_b = 6.4$ ppm/$°$F, and $T_r = 62°$F. Find L.

Solution   Substituting, we have

$$L = 9.7153 \left[1 - (-0.5)(-6)(10^{-6}) \right]$$
$$= 9.715271 \text{ in.}$$

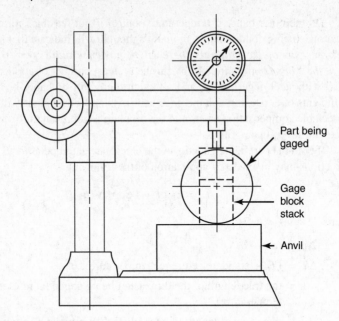

FIGURE 11.5: Comparator employed to measure the difference between a known and an unknown linear dimension.

11.7 USE OF COMPARATORS

One of the primary applications of gage blocks is that of calibrating a device called a *comparator*. As the name suggests, a comparator is used to compare known and unknown dimensions. One form of mechanical comparator is shown in Fig. 11.5. Some form of displacement sensor is used to indicate any dimensional differences as described in the next paragraph. The figure shows an ordinary dial indicator, driven by mechanical displacement. A variety of different sensing devices are found on commercial comparators, including strain-gage types, purely mechanical devices, variable inductance, optical, and so on. Readout for electronic sensors is usually in the form of a digital display. The resolution and accuracy of the sensing device would add to the gage tolerance in establishing the minimum uncertainty. We see that to use a sensor with the ability to sense discrepancies less than block tolerances would be unnecessary.

As an example of a comparator's use, suppose that the diameter of a plug gage is required. The nominal dimension may first be determined by use of an ordinary micrometer. Gage blocks would be stacked to the indicated rough dimension and placed on the comparator anvil, and the indicator would be adjusted on its support post until a zero reading is obtained. The gage blocks would then be removed and the part to be measured substituted. A change in the indicator reading would show the difference between the unknown dimension and the height of the stack of blocks and would thereby establish the value of the dimension in question. Inasmuch as most gage-block sets contain series having very small changes in base dimensions (e.g., in steps of ten-thousandths of an inch for the English system), the gage set may be used conveniently for calibrating the comparator.

11.8 MONOCHROMATIC LIGHT

The method of interferometry, used for accurate measurement of small linear dimension, is described in subsequent sections. A required tool is a source of monochromatic (one color or wavelength) light.

There are various sources of monochromatic light. Optical filters may be used singly or in combination to isolate narrow bands of approximately single-wavelength light. Or a prism may be employed to "break down" white light into its components and, in conjunction with a slit, to isolate a desired wavelength; however, both these methods are inefficient. Most practical sources rely on the electrical excitation of atoms of certain elements that radiate light at discrete wavelengths. The excitation system may be either a laser or, less often, a lamp. Table 11.2 lists the approximate wavelengths for the various primary colors, and the wavelengths of several specific sources are given in Table 11.3.

TABLE 11.2: Approximate Wavelengths of Light of the Various Primary Colors

	Range of Wavelengths	
Color	micrometers	microinches
Violet	0.399 to 0.424	15.7 to 16.7
Blue	0.424 to 0.490	16.7 to 19.3
Green	0.490 to 0.574	19.3 to 22.6
Yellow	0.574 to 0.599	22.6 to 23.6
Orange	0.599 to 0.645	23.6 to 25.4
Red	0.645 to 0.699	25.4 to 27.5

TABLE 11.3: Wavelengths from Specific Sources

Source	Wavelengths		Fringe Interval	
	μin.	μm	μin./fringe	μm/fringe
Lasers				
Argon ion	18.03	0.4579	9.01	0.2290
	19.21	0.4880	9.606	0.2440
	20.26	0.5145	10.13	0.2573
Krypton ion	16.26	0.4131	8.13	0.2066
	25.48	0.6471	12.74	0.3236
	26.63	0.6764	13.32	0.3382
Helium-neon	24.91	0.6328	12.46	0.3164
Lamps				
Mercury 198	21.5	0.546	10.75	0.273
Helium	23.2	0.589	11.6	0.295
Sodium	23.56	0.598	11.78	0.299
Krypton 86	23.85	0.606	11.92	0.303
Cadmium red	25.38	0.644	12.69	0.322

Possible elements for lamps include mercury, mercury 198, cadmium, krypton, krypton 86, thallium, sodium, helium, and neon [3]. Means are provided for vaporizing the element, if not already gaseous, and to produce a visible light through the application of electric potential, often at a high voltage and/or frequency.

The helium lamp, for example, is obtained by means not unlike those used in the familiar neon signs. A tube is charged with helium and connected to a high-voltage source, which causes it to glow. The resulting light has a narrow range of wavelengths and is intense enough for practical use. The wavelength of this source is 0.589 μm (23.2 μin.).

Laser light sources are of various forms, among which gas lasers are probably the type most often used for dimensional measurements. Common examples include the red helium-neon laser, with a wavelength of 0.6328 μm, and the blue-green argon-ion laser, which can provide various wavelengths near 0.5 μm. Gas lasers have several enormous advantages over lamps as sources of monochromatic light: (1) They can be extremely monochromatic, containing only a very narrow wavelength band; (2) their light is highly collimated, forming a narrow and directional beam; (3) they are usually polarized; and (4) their light is highly coherent.

The property of coherence is particularly valuable for interference measurements. A light beam is *coherent* when the light wave maintains the same phase relation over the length of the beam. Thus, one speaks of the *coherence length* of a light beam as the distance over which the beam stays in phase with itself; beyond this length, the beam shows some phase shift. Interference measurements, which relate the phase difference of two coherent beams to a difference in their path lengths, are possible only for path differences less than the coherence length. Because lasers have much greater coherence lengths than lamps, they vastly increase the distances which can be measured by interference. For example, a monochromatic lamp has a coherence length no more than several tens of centimeters and is useful for interference measurements of similar distances at most. A frequency-stabilized HeNe laser, on the other hand, can have a coherence length of more than a kilometer; in principle, laser interference allows distances of hundreds of meters to be measured to a fraction of the wavelength of light.

11.9   THE INTERFEROMETER

By this time the reader is undoubtedly aware that some accurate and reliable means must be available to establish the absolute length of gage blocks. In other words, how are gage blocks calibrated? We have already discussed methods whereby they may be compared with other gage blocks, but somehow a comparison must be made with a more fundamental standard, such as the wavelength of a specific light source, as discussed in Section 2.4. Interferometers make possible this very basic measurement.

An optical interferometer of great historical significance was devised and used by Albert A. Michelson (1852–1931). Until 1960 the length standard was the meter as defined by two finely scribed lines on the platinum-iridium prototype meter bar. Gages of this sort are called *line standards*, whereas the ordinary gage block is an *end standard*. Michelson's primary objective was to determine the wavelengths of light derived from certain sources by comparison to the meter bar. Of course, the tables are now turned. Light now provides the standard, and the problem is to use it to measure lengths, such as gage-block dimensions.

The fundamental elements of a Michelson-type interferometer are shown in Fig. 11.6. A laser beam passes through a beam splitter, which directs 50% of its light to each of two

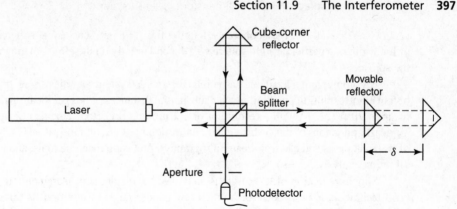

FIGURE 11.6: Michelson-type interferometer.

mirrors or cube-corner reflectors (shown). The light beams reflect back to the beam splitter, and a portion of each again passes through to an aperture and photodetector. At the detector, the two beams will interfere *constructively*, adding to produce brighter light, or *destructively*, cancelling to produce darkness. Whether interference is constructive or destructive depends on the number of wavelengths by which the paths of the two beams differ.

One reflector can be traversed along the length of an unknown dimension. If it moves a distance δ, the path of its light beam increases by 2δ. The number of successive occurrences of destructive interference, or dark *fringes*, at the photodetector during this motion is equal to the number of wavelengths, N, in the path change:

$$2\delta = N\lambda$$

By counting the passing fringes, N is obtained and the distance δ is measured. With care, changes of only 1/100 of a fringe can be resolved.

The distance traversed by the moving reflector between successive fringes is $\lambda/2$, or one half the wavelength of the light source. Thus, $\lambda/2$ is called the *fringe interval* (see Table 11.3).

Many variations on this basic arrangement have been implemented. For example, Michelson reduced the problem of counting large numbers of fringes by devising stepped gages, each representing a known number of fringes, which he termed *etalons*. His procedure was analogous to using a yard-or meter-stick to step off longer distances.

Gage-block interferometers are available from manufacturers of optical apparatus. Although differing in design details, all use essentially the same operating principles. For more information on the construction of gage-block interferometers, see References [3, 4, 5].

As discussed in Section 11.8, the high coherence length of a gas laser makes it possible to measure distances two or three orders of magnitude greater than are possible with a monochromatic-lamp interferometer. The improved coherence length is closely related to the very narrow linewidth of lasers as compared to lamps. For example, a Hg-198 source has a wavelength spread of approximately 0.0005 nm about a center wavelength of 546.1 nm ($\Delta\lambda/\lambda \approx 10^{-6}$). Because of this lack of preciseness, the fringes become more poorly defined as the path difference increases. In fact, the coherence length is somewhat less than $L_c \approx \lambda^2/\Delta\lambda = 60$ cm. In contrast, a moderately priced frequency-stabilized helium-neon laser has a linewidth of about 5×10^{-6} nm about a center wavelength of 632.8 nm

$(\Delta\lambda/\lambda \approx 7.5 \times 10^{-9})$, for a coherence length of $L_c \approx 85$ m. The actual coherence length or linewidth of a particular light source will depend greatly on the efforts made to minimize the spread.

A commercially available laser interferometer system[3] applies these principles to obtain high-resolution positioning over substantial distances. The specifications list a range of up to 70 ft (21 m) with a resolution of 1.2 nm and reflector velocities of up to 1 m/s. Typical applications of this instrument include integrated-circuit fabrication (such as control of wafer steppers and electron-beam lithography), precision machine tools, and mechanical vibration analysis.

Simple reflection of laser light is also used for distance measurement. In the *pulse-echo* technique, a short pulse of light (a few nanoseconds in duration) is directed at the object whose distance is to be found. The time Δt elapsing before the reflected pulse returns is measured, and the distance is calculated as $L = c\Delta t/2$ for c the speed of light. This approach has, for example, been used to measure the distance between the earth and the moon (to ±15 cm) [6]. Changes in the angle or position at which reflected light is received have also been used to measure the displacement of a reflective surface.

11.10   MEASURING MICROSCOPES

Figure 11.7 shows a section through a general-purpose low-power microscope. Basically the instrument consists of an objective cell containing the objective lens, an ocular cell containing the eye and field lenses, and a reticle mounting arrangement, all assembled in optical and body tubes. The ocular cell is adjustable in the optical tube, thereby allowing the eyepiece to be focused sharply on the reticle. The complete optical tube is adjustable in the body tube, by means of a rack and pinion, for focusing the microscope on the work.

Aside from the necessary optical excellence required in the lens system, the heart of the measuring microscope lies in the reticle arrangement. The reticle itself may involve almost any type of plane outline, including scales, grids, and lines. Figure 11.8 illustrates several common forms. In use, the images of the reticle and the work are superimposed, making direct comparison possible. If a scale such as Fig. 11.8(a) is used and if the relation between scale and work is known, the dimension may be determined by direct comparison.

Microscopes used for mechanical measurement are of relatively low power, usually less than $100\times$ and often about $40\times$. They may be classified as follows: (1) fixed-scale, (2) filar, (3) traveling, (4) traveling-stage, (5) draw-tube, and (6) digital. The first two, fixed-scale and filar, are intended for measurement of relatively small dimensional magnitudes, from 0.050 to 0.200 in. (1 to 5 mm) in most cases.

11.10.1   Fixed-Scale Microscopes

The fixed-scale measuring microscope uses reticles of the type shown in Fig. 11.8(a). After proper focusing has been accomplished, the scale is simply compared with the work dimension, and the number of scale units is thereby determined. The scale units, of course, must be translated into full-scale dimensions; that is, the instrument must be calibrated. This is accomplished by focusing on a calibration scale, which is generally made of glass with an etched scale. Typical calibration scales are: 100 divisions with each division 0.1 mm long, and 100 divisions with each division 0.004 in. long. Comparison of the calibration

---

[3]Agilent 108978 Laser Interferometry System.

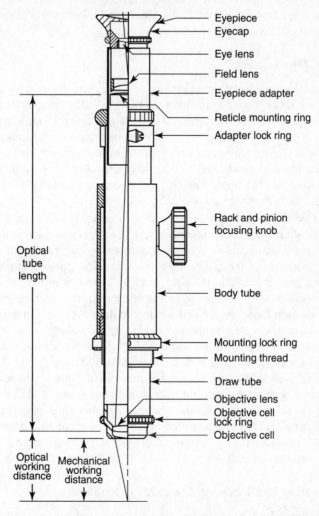

Eyepiece
Eyecap
Eye lens
Field lens
Eyepiece adapter
Reticle mounting ring
Adapter lock ring
Optical tube length
Rack and pinion focusing knob
Body tube
Mounting lock ring
Mounting thread
Draw tube
Objective lens
Objective cell lock ring
Objective cell
Optical working distance
Mechanical working distance

FIGURE 11.7: Section through a simple low-power microscope. (Courtesy: The Gaertner Scientific Corp., Chicago, Illinois)

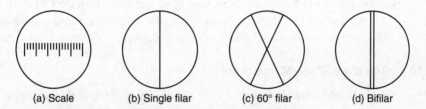

(a) Scale   (b) Single filar   (c) 60° filar   (d) Bifilar

FIGURE 11.8: Examples of measuring microscope reticles. (Courtesy: The Gaertner Scientific Corp., Chicago, Illinois)

and reticle scales provides a positive calibration. Some microscopes are precalibrated and expected to maintain their calibration indefinitely. However, if the objective is changed or tampered with, recalibration will be necessary.

11.10.2   Filar Microscopes

Filar microscopes make use of moving reticles. Actually in most cases a single or double hairline is moved by a fine-pitch screw thread, with the micrometer drum normally divided into 100 parts for subdividing the turns. A total range of about 0.25 in. (6 mm) is common. The kind using the double hairline, called a bifilar type, is more common. In use, the double hairline is aligned with one extreme of the dimension, then moved to the other extreme, with the movement indicated by the micrometer drum. In general, the bifilar type is more easily used than the single-hairline type. A comparison of the views obtained by both the filar and the bifilar microscopes is shown in Fig. 11.8.

One of the problems in using a filar-measuring microscope is to keep track of the number of turns of the micrometer screw. Two methods for accomplishing this are used. In the first case, a simple counter is attached to the microscope barrel, thus providing direct indication of the number of turns. The more common method is to use a built-in notched bar, or *comb*, in the field of view, as shown in Fig. 11.9. Each notch on the comb corresponds to one complete turn of the micrometer wheel. Further minor and major divisions corresponding to 5 and 10 turns of the wheel are also provided. When the comb is used, a mental scale is applied. As an example, referring to Fig. 11.9, assume that for the initial position the micrometer wheel reads 85. The user might mentally designate the major divisions, reading to the right, as 0, 1000, 2000, 3000, and so on. The initial reading would therefore be 085. The hairlines are then moved to the other extreme of the dimension being measured. Suppose now that the micrometer scale reads 27. Reference to the mental scale applied to the comb supplies the hundreds and thousands places, and the reading should therefore be 3127. The dimension then is 3127 less 85, or 3042 micrometer divisions. From previous calibration, each micrometer division has been determined to be, say, 0.000032 in. Therefore, the actual dimension is 3042 × 0.000032, or 0.09734 in.

11.10.3   Traveling and Traveling-Stage Microscopes

A traveling microscope is moved relative to the work by means of a fine-pitch lead screw, and the movement is measured in a manner similar to that used for the ordinary micrometer. In this case the microscope is used merely to provide a magnified index. The traveling-stage type is similar, except that the work is moved relative to the microscope. In both cases the microscope simply serves as an index, and the micrometer arrangement is the measuring means. About 4 or 5 in. is the usual limit of movement, with a least count of 0.0001 in.

An instrument called a *toolmaker's* microscope is an elaborate version of the moving-stage microscope. Special illuminators are used, along with a protractor-type eyepiece.

11.10.4   The Draw-Tube Microscope

The draw-tube microscope uses a scale on the side of the optical tube to give a measure of the focusing position. The microscope is used to determine displacements in a direction along the optical axis. For example, the height of a step could be measured. The instrument would be focused on the first level, or elevation, and a reading made; then it would be moved

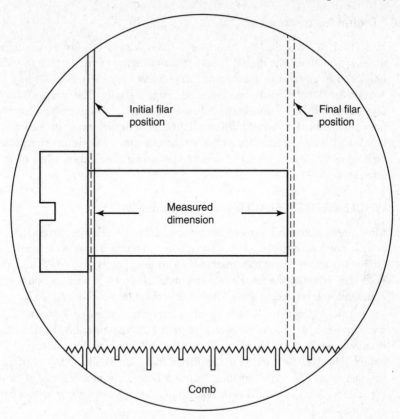

FIGURE 11.9: View through eyepiece of a filar-type microscope showing reference comb for indicating turns of the microscope drum. (Courtesy: The Gaertner Scientific Corp., Chicago, Illinois)

to the second elevation and a second reading made. The difference in readings would be the height of the step.

Vernier scales are normally used, along with microscopes having very shallow depth of focus. A typical range of measurement is $1\frac{1}{2}$ in., with a least count of 0.005 in.

11.10.5   Focusing

Proper focusing of any measuring microscope is essential. First, the eyepiece is carefully adjusted on the reticle without regard to the work image. This adjustment is accomplished by sliding the ocular relative to the optical tube, up or down, until maximum sharpness is achieved. Next, the complete optical system is adjusted by means of the rack and pinion until the work is in sharp focus.

Positive check on proper focus may be obtained by checking for parallax. *When the eye is moved slightly from side to side, the relative positions of the reticle and work images should remain unchanged.* If the reticle image appears to move with respect to the work when this check is made, the focusing has not been done properly.

11.10.6 Digital Microscopes

In addition to the optical or "human-eye" microscopes just discussed, digital microscopes are also available. The digital variety places a digital camera where the human eye would be placed in an optical microscope. The camera typically consists of a CCD array (see Section 8.12) that records the magnified image digitally. The images from a digital microscope can be directly recorded on a disk drive, displayed on screen, or transmitted through a network to other locations. Further, all manner of digital image processing techniques may be applied. Lens systems for such microscopes may be designed somewhat differently, as well, to facilitate adjustment of field of view and magnification while observing the image on screen.

11.11 WHOLE-FIELD DISPLACEMENT MEASUREMENT

Most often mechanical displacements are measured as relative movements between discrete points: Point *A* is displaced by some amount in relation to some reference or datum point. On occasion the relative movements of an array of points may be desired, including whole-field map, which provides the movements of all points within its bounds. Commonly the end purpose is related to some form of experimental stress analysis [7].

Laser holography is the basis of an important technique for whole-field displacement measurement. A three-dimensional image of the initial, undisplaced surface of an object is stored on a holographic plate. The surface is then displaced (perhaps by application of a straining load) and the image of the displaced surface is allowed to interfere with the stored original image. The resulting interference pattern is essentially a map of the displacement of the surface. More information on *holographic interferometry* can be found in the Suggested Readings for this chapter.

11.12 DISPLACEMENT TRANSDUCERS

In Chapter 6, we listed a number of devices that are basically displacement-sensitive. These include

1. Resistance potentiometers (Section 6.6);
2. Resistance strain gages (Section 6.7 and the subject of Chapter 12);
3. Variable-inductance devices (Section 6.10);
4. Differential transformers (Section 6.11 and the subject of further discussion in the Section 11.13);
5. Capacitive transducers (Section 6.13);
6. Piezoelectric transducers (Section 6.14); and
7. Hall-effect transducers (Section 6.17).

Most of the other transducers listed in Chapter 6 can be configured to sense displacement. For example, the variable-reluctance transducer (Section 6.12) is basically velocity sensitive. Combined with an integrating circuit, the output can be made displacement related, and so on.

Usually variable-inductance, capacitance, piezoelectric, and strain-sensitive transducers are suitable only for small displacements (from about 100 nm to perhaps 5 mm). These devices are commonly used as secondary transducers in diaphragm-based pressure sensors (Section 14.6). The differential transformer may be used over intermediate ranges, say 100 nm up to 300 mm. Although wire-wound resistance potentiometers are not as sensitive to small displacements as most of the others, conductive-film potentiometers are sensitive to about 100 nm; and there is practically no limit on the maximum displacement for which potentiometers may be used [8, 9]. With the exception of the piezoelectric type, all may be used for both static and dynamic displacements.

Another important class of displacement transducers are the *binary optical encoders* described in Section 8.6.2. Encoders detecting either linear or angular displacement are available. Rotary encoders have resolutions from 20 to 100,000 pulses per rotation. Linear encoders have accuracies reaching a fraction of a micrometer.

11.13   THE DIFFERENTIAL TRANSFORMER OR LVDT

Because of its singular importance and because it is fundamentally a mechanical displacement transducer, detailed discussion of the differential transformer was reserved for this chapter.

The device, often referred to as a linear-variable differential transformer, or LVDT, provides an ac voltage output proportional to the relative displacement between the transformer core and the windings. Figure 11.10 illustrates the simplicity of its construction. It is a mutual-inductance device using three coils and a core, as shown.

The center coil is energized from an external ac power source, and the two end coils, connected together in phase opposition, are used as pickup coils. Output amplitude and phase depend on the relative coupling between the two pickup coils and the power coil. Relative coupling is, in turn, dependent on the position of the core. Theoretically, there should be a core position for which the voltage induced in each of the pickup coils will be of the same magnitude, and the resulting output should be zero. As we will see later, this condition is difficult to attain perfectly.

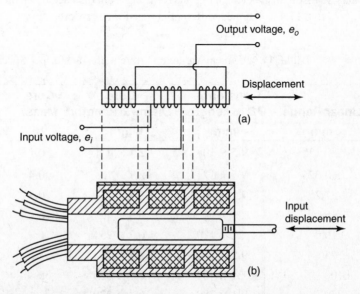

FIGURE 11.10: The differential transformer: (a) schematic arrangement, (b) section through a typical transformer.

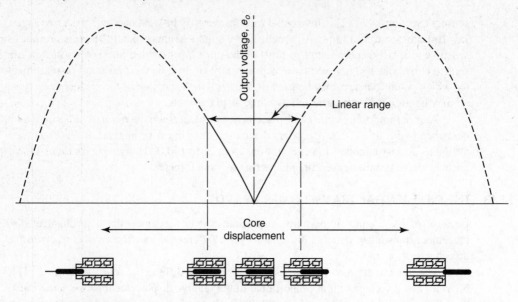

FIGURE 11.11: Typical differential transformer performance characteristics.

Typical differential transformer characteristics are illustrated in Fig. 11.11, which shows output versus core movement. Within limits, on either side of the null position, core displacement results in proportional output. In general, the linear range is primarily dependent on the length of the secondary coils. Although the output voltage magnitudes are ideally the same for equal core displacements on either side of null balance, the phase relation existing between power source and output changes 180° through null. It is therefore possible, through phase determination or the use of phase-sensitive circuitry (discussed later), to distinguish between outputs resulting from displacements on either side of null.

Table 11.4 lists typical differential transformer specifications.

TABLE 11.4: Typical Variable Differential Transformer Specifications

Linear Range, mm	Transformer OD × Length, mm	Core Diam. × Length, mm	Core Mass, g	Sensitivity, mV/(mm · V_{in}) 2.5 kHz	10 kHz
±0.125	9.5 × 10	2.7 × 4.6	0.1	102	335
±0.625	9.5 × 17	2.7 × 10	0.4	180	310
±1.25	9.5 × 20	2.7 × 13	0.4	102	130
±5.0	21 × 64	6.3 × 42	8	100	
±10.0	21 × 110	6.3 × 75	16	32	
±25.	9.5 × 140	2.7 × 76	2.5	28	30
±100.	21 × 400	6.3 × 180	36	8.0	
±250.	21 × 780	6.3 × 220	43	3.0	

11.13.1   Input Power

Input voltage is limited by the current-carrying ability of the primary coil. In most applications, LVDT sensitivities are great enough that very conservative ratings can be applied. In general, a given LVDT is designed for a specific input frequency. Many commonly used commercial transformers are made to operate on 2.5 kHz at 3 V rms. Most of the 2.5-kHz differential transformers draw less than 0.1 watt of excitation power. Separate signal-conditioning units are available which provide both the driving power and a digital readout of the core displacement.

Exciting frequency, sometimes referred to as *carrier* frequency, limits the dynamic response of a transformer. The desired information is superimposed on the exciting frequency, and a minimum ratio of 10 to 1 between carrier and signal frequencies is usually considered to be the limit. For ratios less than 10 to 1, signal definition tends to become lost; and, therefore, when increased dynamic response is required, a higher excitation frequency must be applied.

Transformer sensitivity is usually stated as either "mV output per V input per mm core displacement" or "mV output per V input per 0.001-in. core displacement." It is directly proportional to exciting voltage and varies with frequency [10]. Of course, the output also depends on LVDT design, and in general the sensitivity will increase with increased number of turns on the coils. There is a limit, however, determined by the solenoid effect on the core. In many applications this effect must be minimized; hence design of the general-purpose LVDT is the result of compromise [11].

Solenoid or axial force exerted by the core is zero when the core is centered and increases linearly with displacement. Increasing the excitation frequency reduces this force. For example, a particular LVDT having a linear range of ± 0.03 in. (± 0.76 mm) exerts an axial force of about 1.80×10^{-4} lbf (0.8 mN) at 60 Hz and about 1.80×10^{-5} lbf (0.08 mN) at 1000 Hz, both for a driving voltage of 7 V rms.

When utmost sensitivity is required, attainment of a sharp null balance may be difficult without the addition of external components. First, in addition to a reactive balance, resistive balance may also be required. This balance can be accomplished through use of a paralleled potentiometer, inserted as shown in Fig. 11.12(a), whose total resistance is high enough to minimize output loading: 20,000 Ω or higher may be used. In addition to resistive balance, small reactive unbalances may remain at null. They may be caused by unavoidable differences in physical characteristics of the two pickup coils or from external sources present in a particular installation. These unbalances can usually be nulled through use of small capacitances whose values and locations are determined by trial. One or more of the capacitors shown dotted in Fig. 11.12(a) may be required. The figure also shows two diodes, D_1 and D_2. These may or may not be added, as desired, to effectively demodulate the output signal, providing plus and minus voltages on either side of null, as shown in Fig. 11.12(b).

Integrated-circuit techniques make it possible to incorporate a sine-wave oscillator, a demodulator, and amplifier circuits within the LVDT housing. In this form, the LVDT is powered by a regulated dc power supply (± 15 V, typically) and provides a dc output voltage linearly proportional to the core position (typically 0.04 to 8 V/mm displacement). The onboard circuitry incorporates phase discrimination, so that the sign of the output voltage changes when the null position is passed.

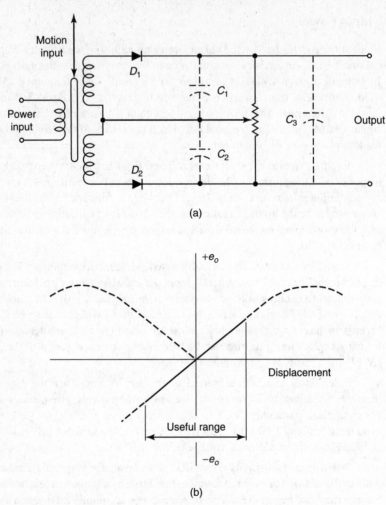

(a)

(b)

FIGURE 11.12: (a) Arrangement for improving the sharpness of null balance, (b) demodulated output.

11.13.2 Advantages of the LVDT

The LVDT offers several distinct advantages over many competitive transducers. First, serving as a primary detector–transducer, it converts mechanical displacement into a proportional electric voltage. As we have found, this is a fundamental conversion. In contrast, the electrical strain gage requires the assistance of some form of elastic member. In addition, the LVDT cannot be overloaded mechanically, since the core is completely separable from the remainder of the device. It provides high linearity (typically ±0.25% of full scale) and potentially high resolution and sensitivity. It is also relatively insensitive to high or low temperatures or to temperature changes. It is reusable, of reasonable cost, and generally of high lifetime and reliability.

Probably its greatest disadvantages lie in the area of dynamic measurement. Its core is of appreciable mass, particularly compared with the mass of the bonded strain gage.

11.14   SURFACE ROUGHNESS

Surface finish may be measured by many different methods, using several different units of measurement. The more commonly used methods include the following:

1. *Visual comparison* with a *standard* surface. This method is based on appearance, which involves more than the surface roughness.

2. The *stylus* or *tracer method*, which uses a stylus that is dragged across the surface, tracing its profile. This method is the most common for obtaining quantitative results.

3. *Optical measurements*. These include reflection of light from the surface (measured by a photodetector), optical followers (which track changes in optical pathlength as the surface height varies); scanning confocal microscopes; and various other techniques using reflection or diffraction of light by a surface.

4. *Interferometry*, using techniques related to those discussed earlier in this chapter.

5. The *capacitance method*, which assumes that the electrical capacitance is a function of the actual surface area, the rough surface providing a greater capacitance than a smooth surface.

6. *Scanning probe microscopes*. These include the *atomic force microscope* (AFM) and the *scanning-tunneling microscope* (STM). The latter device detects quantum-mechanical effects (tunneling) on a tiny stylus passing over a surface to determine the surface structure on an atomic scale [12, 13].

Suppose Fig. 11.13 represents a sample contour of a machined surface. The values listed thereon may be thought of as actual deviations of the surface from the reference plane *x-x*, which is located such that the sectional areas above and below the line are equal. These values are also listed in Table 11.5 as absolute values, and their average is calculated to be 12.89 μ in. In addition, the root-mean-square (rms) height is calculated as 14.58 μ in. We also see that the peak-to-peak height is 27 + 22, or 49 μ in. Each of the values—the peak-to-peak height, the arithmetical average deviation, or the root-mean-square—may be used as measures of roughness.

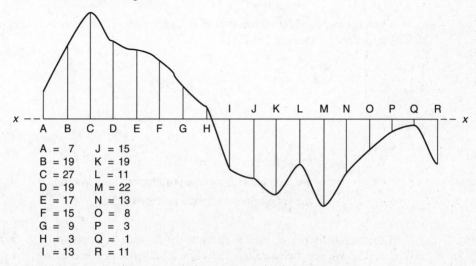

A = 7	J = 15
B = 19	K = 19
C = 27	L = 11
D = 19	M = 22
E = 17	N = 13
F = 15	O = 8
G = 9	P = 3
H = 3	Q = 1
I = 13	R = 11

FIGURE 11.13: Assumed contour of a finished metal surface.

TABLE 11.5: Calculation of Mean Absolute Height and Root-Mean-Square Average

Position	Absolute Elevation from x-x, μin.	Square of Elevation
A	7	49
B	19	361
C	27	729
D	19	361
E	17	289
F	15	225
G	9	81
H	3	9
I	13	169
J	15	225
K	19	361
L	11	121
M	22	484
N	13	169
O	8	64
P	3	9
Q	1	1
R	11	121
	Total = 232	Total = 3828

$$\text{Average absolute height} = \frac{232}{18} = 12.89 \ \mu\text{in.}$$

$$\text{Root-mean-square height} = \sqrt{\frac{3828}{18}} = 14.58 \ \mu\text{in.}$$

The most commonly used measure of surface roughness is the *arithmetical average deviation*, defined by the equation [14]

$$R_a = \frac{1}{l} \int_0^l |z| \, dx \tag{11.2}$$

where

R_a = the arithmetical average deviation,

z = the ordinate of the curve profile from the centerline,

l = the length over which the average is taken

The term *centerline*, as used in defining the distance y, corresponds to the x-x line in Fig. 11.13. We see therefore that the value 12.89 μin. in our example is a practical evaluation of Eq. (11.2).

TABLE 11.6: Typical Range of Surface Roughness, R_a, for Various Production Processes

Process	μm	μin.
Flame cutting	12.5–25	500–1000
Sawing	1.8–25	63–1000
Milling, electron beam	0.8–6.3	32–250
Barrel finishing	0.2–0.8	8–32
Grinding	0.1–1.8	4–63
Honing	0.1–0.8	4–32
Lapping	0.05–0.4	2–16
Sand casting, hot rolling	12.5–25	500–1000
Forging	3.2–12.5	125–500
Cold rolling, drawing, extruding	0.8–3.2	32–125

Many roughness measuring systems using the tracer method yield the rms height. On a typical surface, this value will be about 11% greater than the arithmetical average deviation. The difference between the two values, however, is less than the normal variations from one piece to another and is commonly ignored. An idea of the relative value for practical surface finishes may be obtained from Table 11.6 [15].

Although the tracer method for measuring surface roughness is used more than any of the others, it does present several important problems. First, in order for the scriber to follow the contour of the surface, it should have as sharp a point as possible. Some form of conical point with a spherical end is most common. A point radius of 2 to 5 μm (80 to 160 μin.) is often used. Therefore, the stylus will not always follow the true contour. If the surface irregularities are primarily what might be referred to as *wavy* or *rolling* hills and valleys of appreciable vertical radius, the stylus may indeed follow the actual contour. If the surface is rugged, on the other hand, the stylus will not extend fully into the valleys.

Second, the stylus will probably actually round off the peaks as it is dragged over the surface. This problem increases as the radius of the tip is decreased. It would seem, therefore, that there are two conflicting requirements with regard to stylus-tip radius. The marring of the surface will in part be a function of the material constants, which of course should have no bearing on the measure of surface roughness.

A third problem lies in the fact that the stylus can never inspect more than a very small percentage of the overall surface.

In spite of these problems, the tracer method is undoubtedly the most commonly used. Various forms of secondary transducers have been employed, including piezoelectric elements (Section 6.14), variable inductance (Section 6.10), and variable reluctance (Section 6.12). In each case the stylus motion is transferred to the transducer element, which converts it to an analogous electric signal. Modern systems digitize this signal and use a microprocessor or computer to compute and display appropriate measures of the surface finish.

SUGGESTED READINGS

Doiron, T., and J. Beers. *The Gauge Block Handbook*. NIST Monograph 180. Gaithersburg, Md.: National Institute of Standards and Technology, 2005.

Farago, F. T., and M. A. Curtis. *Handbook of Dimensional Measurement*. 3rd ed. New York: Industrial Press, 1994.

Khazan, A. D. *Transducers and Their Elements*. Englewood Cliffs, N.J.: Prentice Hall, 1994.

Pallàs-Areny, R., and J. G. Webster. *Sensors and Signal Conditioning*. 2nd ed. New York: John Wiley, 2001.

Slocum, A. H. *Precision Machine Design*. Englewood Cliffs, N.J.: Prentice Hall, 1992.

Smith, W. J. *Modern Optical Engineering*. 3rd ed. New York: McGraw-Hill, 2000.

Vest, C. M. *Holographic Interferometry*. New York: John Wiley, 1979.

Whitehouse, D. *Surfaces and Their Measurement*. New York: Taylor & Francis, 2002.

PROBLEMS

11.1. Using the dimensioning "rules" given in Section 11.2 for go/no-go gages, sketch and dimension a plug gage for gaging a hole that carries the dimension 1.125/1.132 in.

11.2. Assuming that the dimensions specified for the plug gage in Fig. 11.2 correspond to a temperature of 68°F, calculate the limiting dimensions corresponding to (a) 90°F and (b) 40°F. Use data for chrome-vanadium steel listed in Table 6.3.

11.3. Determine the temperature change (°F) that will cause a change in the length of a 1-in. gage block equal to the block's nominal tolerance for (a) working blocks, (b) reference blocks, and (c) master blocks.

11.4. Using the set of blocks listed in Section 11.3, specify a combination to provide a dimension of 2.7816 in.: (a) including wear blocks, and (b) without wear blocks.

11.5. Equation (11.1) is written assuming that both the blocks and the gaged dimension are at the same nonstandard temperature. Modify the relation to care for a situation in which the blocks and the gaged part are at different temperatures, neither of which is the standard temperature.

11.6. A Michelson-type interferometer is shown in Fig. 11.14. The laser beam passes through a beam splitter, travels to each of two mirrors, and is reflected back to a photodetector. Mirror A is movable; as its position changes, the interference at the photodetector changes. This interference produces either light or dark fringes, corresponding to high and low voltages at the photodetector output.

(a) If the laser wavelength is 0.5145 μm, how many dark fringes will occur as mirror A is moved a distance of $\delta = 100$ μm?

(b) When the mirror is moved at a constant speed, the photodetector voltage seen on an ac-coupled oscilloscope trace is shown in Fig. 11.14(b). At what speed is the mirror moving?

(c) Mirror A, initially at rest, moves a short distance and stops. The dc-coupled oscilloscope output voltage trace is shown in Fig. 11.14(c). How far did mirror A move?

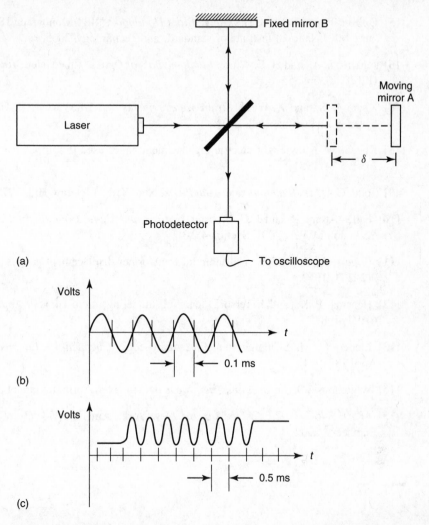

FIGURE 11.14: (a) Michelson interferometer, (b) ac-coupled oscilloscope trace when mirror A moves at constant speed, (c) dc-coupled oscilloscope trace when mirror A, initially at rest, moves a distance δ and stops.

REFERENCES

[1] Peters, C. G., and W. B. Emerson. Interference methods for producing and calibrating end standards. *NBS J. Res.* 44:427, April 1950.

[2] Metrology of gage blocks. *NBS Circular* 581:67, April 1, 1957. Washington, D.C.: U.S. Government Printing Office.

[3] American Society of Tool and Manufacturing Engineers. *Handbook of Industrial Metrology.* Englewood Cliffs, N.J.: Prentice Hall, 1967.

[4] Scarr, A. J. T. *Metrology and Precision Engineering.* New York: McGraw-Hill, 1967.

[5] Doiron, T., and J. Beers. *The Gauge Block Handbook.* NIST Monograph 180. Gaithersburg, Md.: National Institute of Standards and Technology, 2005.

[6] Jenkins, F. A., and H. E. White. *Fundamentals of Optics.* 4th ed. New York, McGraw-Hill, 1976, pp. 655–656.

[7] Khan, A. S., and X. Wang. *Strain Measurements and Stress Analysis.* Upper Saddle River, N.J.: Prentice Hall, 2001.

[8] Kneen, W. A review of electric displacement gages used in railroad car testing. *ISA Proc.* 6:74, 1951.

[9] Todd, C. D. *The Potentiometer Handbook.* New York: McGraw-Hill, 1975.

[10] Pallàs-Areny, R. and J. G. Webster. *Sensors and Signal Conditioning.* 2nd ed. New York: John Wiley, 2001, Section 4.2.3.

[11] Boggis, A. G. Design of differential transformer displacement gauges. *SESA Proc.* 9(2):171, 1952.

[12] Hansma, P. K., and J. Tersoff. Scanning tunneling microscopy. *J. Applied Physics* 61:R1, 1987.

[13] Lieber, C. M. Scanning tunneling microscopy. *Chemical & Engineering News,* April 18, 1994.

[14] Whitehouse, D. *Surfaces and Their Measurement.* New York: Taylor & Francis, 2002.

[15] *ASME B46.1-2002: Surface Texture.* New York: American Society of Mechanical Engineers, 2002.

CHAPTER 12

Strain and Stress: Measurement and Analysis

12.1 INTRODUCTION

All machine or structural members deform to some extent when subjected to external loads or forces. The deformations result in relative displacements that may be normalized as percentage displacement, or strain. For simple axial loading (Fig. 12.1),

$$\varepsilon_a = \frac{dL}{L} \approx \frac{L_2 - L_1}{L_1} = \frac{\Delta L}{L_1} \tag{12.1}$$

where

ε_a = axial strain,

L_1 = initial linear dimension or gage length,

L_2 = final strained linear dimension

More correctly, the term *unit strain* should be used for the preceding quantity and is generally intended when the word *strain* is used alone. Throughout the following discussion, when

413

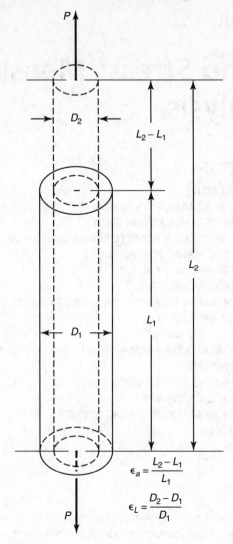

FIGURE 12.1: Defining relations for axial and lateral strain.

the word *strain* is used, we mean the quantity defined by Eq. (12.1). If the net change in a dimension is required, the term *total strain* will be used.

Because the quantity strain, as applied to most engineering materials, is a very small number, it is commonly multiplied by one million; the resulting number is then called *microstrain*,[1] or parts per million (ppm).

The stress strain relation for a uniaxial condition, such as exists in a simple tension test specimen or at the outer fiber of a beam in bending, is expressed by

$$E = \frac{\sigma_a}{\varepsilon_a} \tag{12.2}$$

---

[1]Considerable use of the term *microstrain* will be made throughout this chapter and elsewhere in the book. For convenience, the abbreviation μ-strain will often be used.

where

$$E = \text{Young's modulus,}$$
$$\sigma_a = \text{uniaxial stress,}$$
$$\varepsilon_a = \text{the strain in the direction of the stress}$$

This relation is linear: that is, E is a constant for most materials so long as the stress is kept below the proportional limit.

When a member is subjected to simple uniaxial stress in the elastic range (Fig. 12.1), lateral strain results in accordance with the following relation:

$$\nu = \frac{-\varepsilon_L}{\varepsilon_a} \qquad (12.2a)$$

where

$$\nu = \text{Poisson's ratio,}$$
$$\varepsilon_L = \text{lateral strain}$$

A more general condition commonly exists on the *free surface* of a stressed member. Let us consider an element subject to orthogonal stresses σ_x and σ_y as shown in Fig. 12.2. Suppose

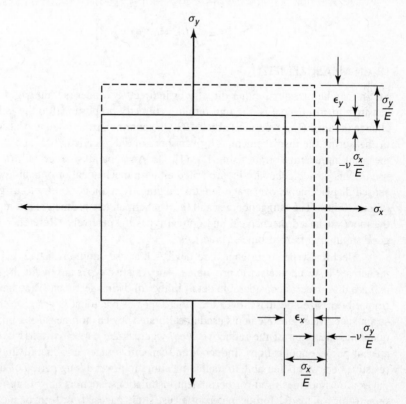

FIGURE 12.2: An element taken from a biaxially stressed condition with normal stresses known.

that the stresses σ_x and σ_y are applied one at a time. If σ_x is applied first, there will be a strain in the x-direction equal to σ_x/E. At the same time, because of Poisson's ratio, there will be a strain in the y-direction equal to $-\nu\sigma_x/E$.

Now suppose that the stress in the y-direction, σ_y, is applied. This stress will result in a y-strain of σ_y/E and an x-strain equal to $-\nu\sigma_y/E$. The net strains are then expressed by the relations

$$\varepsilon_x = \frac{\sigma_x - \nu\sigma_y}{E} \quad \text{and} \quad \varepsilon_y = \frac{\sigma_y - \nu\sigma_x}{E} \tag{12.3}$$

If these relations are solved simultaneously for σ_x and σ_y, we obtain the equations

$$\sigma_x = \frac{E(\varepsilon_x + \nu\varepsilon_y)}{1 - \nu^2} \quad \text{and} \quad \sigma_y = \frac{E(\varepsilon_y + \nu\varepsilon_x)}{1 - \nu^2} \tag{12.4}$$

When a stress σ_z exists, acting in the third orthogonal direction, the more general three-dimensional relations are

$$\varepsilon_x = \frac{1}{E}[\sigma_x - \nu(\sigma_y + \sigma_z)],$$

$$\varepsilon_y = \frac{1}{E}[\sigma_y - \nu(\sigma_z + \sigma_x)], \tag{12.5}$$

$$\varepsilon_z = \frac{1}{E}[\sigma_z - \nu(\sigma_x + \sigma_y)]$$

12.2 STRAIN MEASUREMENT

Strain may be measured either directly or indirectly. Modern strain gages are inherently sensitive to *strain*; that is, the unit output is directly proportional to the unit dimensional change (strain). However, until about 1930 the common experimental procedure consisted of measuring the displacement ΔL over some initial gage length L and then calculating the resulting average strain using Eq. (12.1). An apparatus called an *extensometer* was used. This device generally incorporated either a mechanical or optical lever system and sensed displacements over gage lengths ranging from about 50 mm to as great as 25 cm (about 10 in.). The Huggenberger and the Tuckerman extensometers are representative of the more advanced mechanical and optical types, respectively. Reference [1] provides a good summary of extensometer practices.

Electrical-type strain gages are devices that use simple resistive, capacitive [2, 3], inductive [4], or photoelectric principles. The resistive types are by far the most common and are discussed in considerable detail in the following sections. They have advantages, primarily of size and mass, over the other types of electrical gages. On the other hand, strain-sensitive gaging elements used in calibrated devices for measuring other mechanical quantities are often of the inductive type, whereas the capacitive kind is used more for special-purpose applications. Inductive and capacitive gages are generally more rugged than resistive ones and better able to maintain calibration over a long period of time. Inductive gages are sometimes used for permanent installations, such as on rolling-mill frames for monitoring roll loads. Torque meters often use strain gages in one form or another, including inductive [5] and capacitive [2].

Other strain measuring techniques include optical methods [6], such as photoelasticity, the Moire technique, and holographic interferometry (Section 11.11).

12.3   THE ELECTRICAL RESISTANCE STRAIN GAGE

In 1856 Lord Kelvin demonstrated that the resistances of copper wire and iron wire change when the wires are subject to mechanical strain. He used a Wheatstone bridge circuit with a galvanometer as the indicator [7]. Probably the first wire resistance strain gage was that made by Carlson in 1931 [8]. It was of the unbounded type: Pillars were mounted, separated by the gage length, with wires stretched between them. What was probably the first bonded strain gage was used by Bloach [9]. It consisted of a carbon film resistance element applied directly to the surface of the strained member.

In 1938 Edward Simmons made use of a bonded wire gage in a study of stress–strain relations under tension impact [10]. His basic idea is covered in U.S. Patent No. 2,292,549. At about the same time, Ruge of the Massachusetts Institute of Technology (MIT) conceived the idea of making a preassembly by mounting wire between thin pieces of paper. Figure 12.3 shows the general construction.

During the 1950s advances in materials and fabricating methods produced the foil-type gage, which soon replaced the wire gage. The common form consists of a metal foil element on a thin epoxy support and is manufactured using printed-circuit techniques. An important advantage of this type is that almost unlimited plane configurations are possible; a few examples are shown in Fig. 12.4.

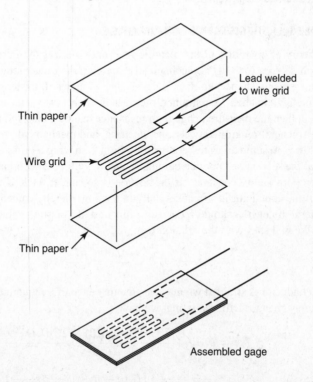

FIGURE 12.3: Construction of bonded-wire-type strain gage.

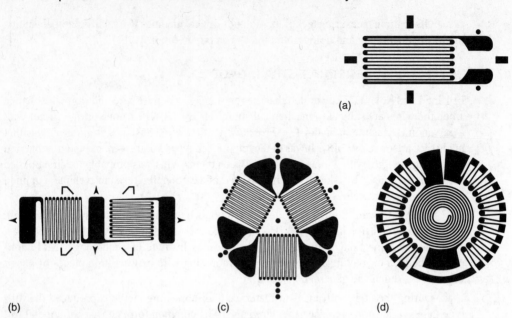

FIGURE 12.4: Typical foil gages illustrating the following types: (a) single element, (b) two-element rosette, (c) three-element rosette, (d) one example of many different special-purpose gages. The latter is for use on pressurized diaphragms.

12.4 THE METALLIC RESISTANCE STRAIN GAGE

The theory of operation of the metallic resistance strain gage is relatively simple. When a length of wire (or foil) is mechanically stretched, a *longer* length of *smaller* sectioned conductor results; hence the electrical resistance changes. If the length of resistance element is intimately attached to a strained member in such a way that the element will also be strained, then the measured change in resistance can be calibrated in terms of strain.

A general relation between the electrical and mechanical properties may be derived as follows: Assume an initial conductor length L, having a cross-sectional area CD^2. (In general the section need not be circular; hence D will be a sectional dimension and C will be a proportionality constant. If the section is square, $C = 1$; if it is circular, $C = \pi/4$, etc.) If the conductor is strained axially in tension, thereby causing an increase in length, the lateral dimension should reduce as a function of Poisson's ratio.

We will start with the relation [Eq. (6.2)]

$$R = \frac{\rho L}{A} = \frac{\rho L}{CD^2} \tag{12.6}$$

If the conductor is strained we may assume that each of the quantities in Eq. (12.6) except for C may change. Differentiating, we have

$$dR = \frac{CD^2(L\,d\rho + \rho\,dL) - 2C\rho LD\,dD}{(CD^2)^2}$$

$$= \frac{1}{CD^2}\left((L\,d\rho + \rho\,dL) - 2\rho L\frac{dD}{D}\right) \tag{12.7}$$

Dividing Eq. (12.7) by Eq. (12.6) yields

$$\frac{dR}{R} = \frac{dL}{L} - 2\frac{dD}{D} + \frac{d\rho}{\rho}$$
(12.8)

which may be written

$$\frac{dR/R}{dL/L} = 1 - 2\frac{dD/D}{dL/L} + \frac{d\rho/\rho}{dL/L}$$
(12.8a)

Now

$$\frac{dL}{L} = \varepsilon_a = \text{axial strain},$$

$$\frac{dD}{D} = \varepsilon_L = \text{lateral strain}$$

and

$$\nu = \text{Poisson's ratio} = -\frac{dD/D}{dL/L}$$

Making these substitutions gives us the basic relation for what is known as the gage factor, for which we shall use the symbol F:

$$F = \frac{dR/R}{dL/L} = \frac{dR/R}{\varepsilon_a} = 1 + 2\nu + \frac{d\rho/\rho}{dL/L}$$
(12.9)

This relation is basic for the resistance-type strain gage.

Assuming for the moment that resistivity should remain constant with strain, then according to Eq. (12.9) the gage factor should be a function of Poisson's ratio alone, and in the elastic range should not vary much from $1 + (2)(0.3) = 1.6$. Table 12.1 lists typical values for various materials. Obviously, more than Poisson's ratio must be involved, and if resistivity is the only other variable, apparently its effect is not consistent for all materials. Note the value of the gage factor for nickel. The negative value indicates that a stretched element with increased length and decreased diameter (assuming elastic conditions) actually exhibits a reduced resistance.

In spite of our incomplete knowledge of the physical mechanism involved, the factor F for metallic gages is essentially a constant in the usual range of required strains, and its value, determined experimentally, is reasonably consistent for a given material.

By rewriting Eq. (12.9) and replacing the differential by an incremental resistance change, we obtain the following equation:

$$\varepsilon = \frac{1}{F}\frac{\Delta R}{R}$$
(12.10)

In practical application, values of F and R are supplied by the gage manufacturer, and the user determines ΔR corresponding to the input situation being measured. This procedure is the fundamental one for using resistance strain gages.

TABLE 12.1: Representative Properties of Various Grid Materials

Grid Material	Composition	Approx. Gage Factor, F	Approximate Resistivity		Approximate Temperature Coefficient of Resistance, ppm/°C	Maximum Operating Temp., °C (approx.)
			Micro-ohm · cm	Ohms per mil · foot		
Nichrome V*	80% Ni; 20% Cr	2.0	108	650	400	1100
Constantan*, Copel*, Advance*	45% Ni; 55% Cu	2.0	49	290	11	480
Isoelastic*	36% Ni; 8% Cr; 0.5% Mo; Fe remainder	3.5	112	680	470	—
Karma*	74% Ni; 20% Cr; 3% Al; 3% Fe	2.4	130	800	18	815
Manganin*	4% Ni; 12% Mn; 84% Cu	0.47	48	260	11	—
Platinum-Iridium	95% Pt; 5% Ir	5.1	24	137	1250	1100
Monel*	67% Ni; 33% Cu	1.9	42	240	2000	—
Nickel		-12†	7.8	45	6000	—
Platinum		4.8	10	60	3000	—

*Trade names.
† Varies widely with cold work.

12.5 SELECTION AND INSTALLATION FACTORS FOR BONDED METALLIC STRAIN GAGES

Performance of bonded metallic strain gages is governed by five gage parameters: (a) grid material and configuration; (b) backing material; (c) bonding material and method; (d) gage protection; and (e) associated electrical circuitry.

Desirable properties of grid material include (a) high gage factor, F; (b) high resistivity, ρ; (c) low temperature sensitivity; (d) high electrical stability; (e) high yield strength; (f) high endurance limit; (g) good workability; (h) good solderability or weldability; (i) low hysteresis; (j) low thermal emf when joined to other materials; and (k) good corrosion resistance.

Temperature sensitivity is one of the most worrisome factors in the use of resistance strain gages. In many applications, compensation is provided in the electrical circuitry; however, this technique does not always eliminate the problem. Two factors are involved: (1) the differential expansion existing between the grid support and the grid proper, resulting in a strain that the gage is unable to distinguish from load strain; and (2) the change in resistivity ρ with temperature change.

Thermal emf superimposed on gage output obviously must be avoided if dc circuitry is used. For ac circuitry this factor would be of little importance. Corrosion at a junction between grid and lead could conceivably result in a miniature rectifier, which would be more serious in an ac than in a dc circuit.

Table 12.1 lists several possible grid materials and some of the properties influencing their use for strain gages. Commercial gages are usually of constantan or isoelastic. The former provides a relatively low temperature coefficient along with reasonable gage factors. Isoelastic gages are some 40 times more sensitive to temperature than are Constantan gages. However, they have appreciably higher output, along with generally good characteristics otherwise. They are therefore made available primarily for dynamic applications where the short time of strain variation minimizes the temperature problem.

The gage factor listed for nickel is of particular interest, not only because of its relatively high value, but also because of its negative sign. It should be noted, however, that the value of F for nickel varies over a relatively wide range, depending on how it is processed. Cold working has a rather marked effect on the strain- and temperature-related characteristics of nickel and its alloys, and this feature is used advantageously to produce special temperature self-compensating gages (see Section 12.10.2).

Common backing materials include phenolic-impregnated paper, epoxy-type plastic films, and epoxy-impregnated fiberglass. Most foil gages intended for a moderate range of temperatures ($-75°C$ to $100°C$) use an epoxy film backing. Table 12.2 lists commonly recommended temperature ranges.

No particular difficulty should be experienced in mounting strain gages if the manufacturer's recommended techniques are followed carefully. However, we may make one observation that is universally applicable: *Cleanliness* is an absolute requirement if consistently satisfactory results are to be expected. The mounting area must be cleaned of all corrosion, paint, and so on, and bare base material must be exposed. All traces of greasy film must be removed. Several of the gage suppliers offer kits of cleaning materials along with instructions for their use. These materials are very satisfactory.

Most gage installations are not complete until provision is made to protect the gage from ambient conditions. The latter may include mechanical abuse, moisture, oil, dust

TABLE 12.2: General Recommendations for Strain-Gage Backing Materials and Adhesives

Grid and Backing Materials	Recommended Adhesive	Permissible Temperature Range, °C
Foil on epoxy	Cyanoacrylate	−75 to 95
Foil on phenol-impregnated fiberglass	Phenolic	−240 to 200
Strippable foil or wire	Ceramic	−240 to 400 (to 1000 for short-time dynamic tests)
Free filament wire	Ceramic	−240 to 650 (to 1100 for short-time dynamic tests)

and dirt, and the like. Once again, gage suppliers provide recommended materials for this purpose, including petroleum waxes, silicone resins, epoxy preparations, and rubberized brushing compounds. A variety of materials is necessary because of the many types of protection required—such things as hot oil, immersion in water, liquefied gases, and so on. An extreme requirement for gage protection is found in the case of gages mounted on the exterior of a ship or submarine hull for the purpose of sea trials [11]. Special methods of protection are used, including vulcanized rubber boots over the gages.

Gages may have one or more elements (see Fig. 12.4). When used for stress analysis, the single-element gage is applied to the uniaxial stress condition; the two-element rosette is applied to the biaxial condition when either the principal axes or the axes of interest are known; and the three-element rosette is applied when a biaxial stress condition is completely unknown (see Section 12.15.2 for a more complete discussion).

12.6 CIRCUITRY FOR THE METALLIC STRAIN GAGE

When the sensitivity of a metallic resistance gage is considered, its versatility and reliability are truly amazing. The basic relation as expressed by Eq. (12.10) is

$$\varepsilon = \frac{1}{F}\frac{\Delta R_g}{R_g} \tag{12.11}$$

Typical gage constants are

$$F = 2.0 \quad \text{and} \quad R_g = 120\ \Omega$$

Strains of 1 ppm (1 microstrain) are detectable with commercial equipment; hence, the corresponding resistance change that must be measured in the gage will be

$$\Delta R_g = F R_g \varepsilon = (2)(120)(0.000001) = 0.00024\ \Omega$$

which amounts to a resistance change of 0.0002%. Obviously, to measure changes as small as this, instrumentation more sensitive than the ordinary ohmmeter will be required.

Three circuit arrangements are used for this purpose: the simple voltage-dividing potentiometer or ballast circuit (Section 7.6), the Wheatstone bridge (Section 7.9), and the constant-current circuit. Some form of bridge arrangement is the most widely used.

12.7 THE STRAIN-GAGE BALLAST CIRCUIT

Figure 12.5 illustrates a simple strain-gage ballast arrangement. Using Eq. (7.4) and substituting R_g for kR_t, we may write

$$e_o = e_i \frac{R_g}{R_b + R_g}$$

and

$$de_o = \frac{e_i R_b \, dR_g}{(R_b + R_g)^2} = \frac{e_i R_b R_g}{(R_b + R_g)^2} \frac{dR_g}{R_g}$$

From Eq. (12.11),

$$de_o = \frac{e_i R_b R_g}{(R_b + R_g)^2} F \varepsilon \qquad (12.12)$$

where

$$e_i = \text{the exciting voltage,}$$
$$e_o = \text{the voltage output,}$$
$$R_b = \text{the ballast resistance } (\Omega),$$
$$R_g = \text{the strain-gage resistance } (\Omega),$$
$$F = \text{the gage factor,}$$
$$\varepsilon = \text{strain}$$

Some of the limitations inherent in this circuit may be demonstrated by the following example. Let

$$R_b = R_g = 120 \ \Omega$$

A resistance of 120 Ω is common in a strain gage, and it will be recalled that in Section 7.6 we showed that equal ballast and transducer resistances provide maximum sensitivity. Also, let

$$e_i = 8 \text{ V}$$

and let

$$F = 2.0$$

which is a common value. Then

$$e_o = 8 \left(\frac{120}{120 + 120} \right) = 4 \text{ V}, \quad \text{and}$$

$$de_o = \frac{8 \times 120 \times 120 \times 2 \times \varepsilon}{(120 + 120)^2} = 4\varepsilon$$

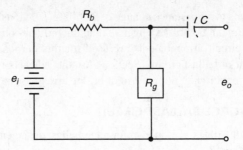

FIGURE 12.5: Ballast circuit for use with strain gages.

If our indicator is to provide an indication for strain of, say, 1 μ-strain, it must sense a 4-μV variation in 4 V, or 0.00010%. This severe requirement practically eliminates the ballast circuit for *static* strain work. We may use it, however, in certain cases for dynamic strain measurement when any static strain component may be ignored. If a capacitor is inserted into an output lead, the dc exciting voltage is blocked and only the variable component is allowed to pass (Fig. 12.5). Temperature compensation is not provided; however, when only transient strains are of interest, this type of compensation is often of no importance.

12.8 THE STRAIN-GAGE BRIDGE CIRCUIT

A resistance-bridge arrangement is particularly convenient for use with strain gages because it may be easily adjusted to a null for zero strain, and it provides means for effectively reducing or eliminating the temperature effects previously discussed (Section 12.5). Figure 12.6 shows a minimum bridge arrangement, where arm 1 consists of the strain-sensitive gage mounted on the test item. Arm 2 is formed by a similar gage mounted on a piece of unstrained material as nearly like the test material as possible and placed near the test location so that the temperature will be the same. Arms 3 and 4 may simply be fixed resistors selected for good stability, plus portions of slide-wire resistance, D, required for balancing the bridge.

If we assume a voltage-sensitive deflection bridge with all initial resistances nominally equal, using Eq. (7.17) we have

$$\frac{\Delta e_o}{e_i} = \frac{\Delta R_1 / R}{4 + 2(\Delta R_1 / R)}$$

In addition,

$$\varepsilon = \frac{1}{F} \frac{\Delta R_1}{R} \quad \text{or} \quad \Delta R = F R \varepsilon$$

Then

$$\Delta e_o = \frac{e_i F \varepsilon}{4 + 2 F \varepsilon}$$

For $e_i = 8$ V and $F = 2$,

$$\Delta e_o = \frac{8 \times 2 \times \varepsilon}{4 + (2)(2)\varepsilon}$$

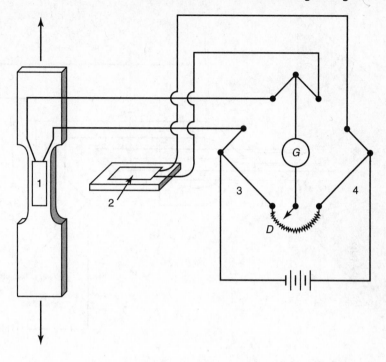

FIGURE 12.6: Simple resistance-bridge arrangement for strain measurement.

If we neglect the second term in the denominator, which is normally negligible, then

$$\Delta e_o = e_o = 4\varepsilon \text{ V}$$

or for $\varepsilon = 1$ μ-strain, $e_o = 4$ μV.

We see that under similar conditions the output increment for the bridge and ballast arrangements is the same. The tremendous advantage that the bridge possesses, however, is that the incremental output is not superimposed on a large fixed-voltage component. Another important advantage, which is discussed in Section 12.10, is that temperature compensation is easily attained through the use of a bridge circuit incorporating a "dummy," or compensating, gage.

12.8.1 Bridges with Two and Four Arms Sensitive to Strain

In many cases bridge configuration permits the use of more than one arm for measurement. This is particularly true if a known relation exists between two strains, notably the case of bending. For a beam section symmetrical about the neutral axis, we know that the tensile and compressive strains are equal except for sign. In this case, both gages 1 and 2 may be used for strain measurement. This is done by mounting gage 1 on the tensile side of the beam and mounting gage 2 on the compressive side, as shown in Fig. 12.7 (see also case F in Table 12.4). The resistance changes will be alike but of opposite sign, and a doubled bridge output will be realized.

This may be carried further, and all four arms of the bridge made strain sensitive, thereby *quadrupling* the output that would be obtained if only a single gage were used. In

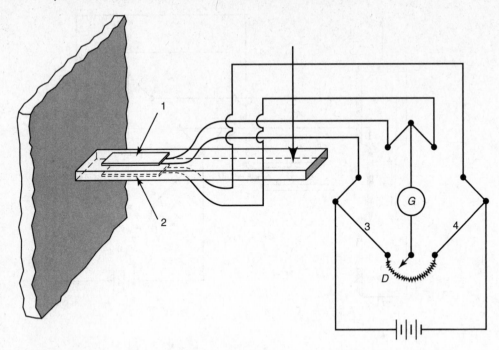

FIGURE 12.7: Bridge arrangement with two gages sensitive to strain.

this case gages 1 and 4 would be mounted to record like strain (say tension) and 2 and 3 to record the opposite type (case G in Table 12.4).

Bridge circuits of these kinds may be used either as null-balance bridges or as deflection bridges (Section 7.9). In the former the slide-wire movement becomes the indicated measure of strain. This bridge is most valuable for strain-indicating devices used for static measurement. Most dynamic strain-measuring systems, however, use a voltage- or current-sensitive deflection bridge. After initial balance is accomplished, the output, amplified as necessary, is used to deflect an indicator, such as an oscilloscope beam or input to a digital recorder. In addition, the constant-current bridge may offer certain advantages (Section 7.9.3).

12.8.2 The Bridge Constant

At this point we introduce the term bridge constant, which we shall define by the following equation:

$$k = \frac{A}{B} \tag{12.13}$$

where

$k =$ the bridge constant,

$A =$ the actual bridge output,

$B =$ the output from the bridge if only a single gage,

sensing maximum strain, were effective

In the example illustrated in Fig. 12.7, the bridge constant would be 2. This is true because the bridge provides an output double of that which would be obtained if only gage 1 were strain sensitive. If all four gages were used, quadrupling the output, the bridge constant would be 4. In certain other cases (Section 12.16), gages may be mounted sensitive to lateral strains that are functions of Poisson's ratio. In such cases bridge constants of 1.3 and 2.6 (for Poisson's ratio = 0.3) are common.

12.8.3 Lead-Wire Error

When it is necessary to use unusually long leads between a strain gage and other instrumentation, *lead-wire error* may be introduced. The reader is referred to Sections 7.9.5 and 16.4.2, where solutions to this problem are discussed.

12.9 THE SIMPLE CONSTANT-CURRENT STRAIN-GAGE CIRCUIT

Measurement of dynamic strains may be accomplished by the simple circuit shown in Fig. 12.8. It is assumed that the power source is a true constant-current supply and that the indicator (an oscilloscope is shown) possesses near-infinite input impedance compared to the gage resistance. As the gage resistance changes as a result of strain, the voltage across the gage, hence the input to the oscilloscope, will be

$$e_i = i_i R \tag{12.14}$$

and

$$\Delta e_i = i_i \, \Delta R \tag{12.14a}$$

Dividing Eq. (12.14a) by Eq. (12.14), we have

$$\frac{\Delta e_i}{e_i} = \frac{\Delta R}{R}$$

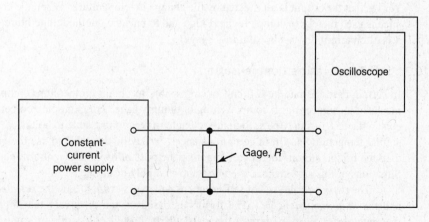

FIGURE 12.8: Single-gage constant-current circuit.

Inserting $\Delta R/R$ in Eq. (12.10) gives us

$$\varepsilon = \frac{1}{F}\frac{\Delta e_i}{e_i} \tag{12.14b}$$

The oscilloscope should be set in the ac mode to cancel the direct dc component, and, of course, the oscilloscope amplification capability must be sufficient to provide an adequate readout.

12.10 TEMPERATURE COMPENSATION

As already implied, resistive-type strain gages are normally quite sensitive to temperature. Both the differential expansion between the grid and the tested material and the temperature coefficient of the resistivity of the grid material contribute to the problem. It has been shown (Table 12.1) that the temperature effect may be large enough to require careful consideration. Temperature effects may be handled by (1) cancellation or compensation or (2) evaluation as a part of the data reduction problem.

Compensation may be provided (1) through use of adjacent-arm balancing or compensating gage or gages or (2) by means of self-compensation.

12.10.1 The Adjacent-Arm Compensating Gage

Consider bridge configurations such as those shown in Figs. 12.6 and 12.7. Initial electrical balance is obtained when

$$\frac{R_1}{R_2} = \frac{R_3}{R_4}$$

If the gages in arms 1 and 2 are *alike* and *mounted on similar materials* and if both gages experience the same resistance shift, ΔR_t, caused by temperature change, then from Eq. (12.19) in Section 12.16 with $\varepsilon_1 = \varepsilon + \varepsilon_T$ and $\varepsilon_2 = \varepsilon_T$

$$\frac{de_o}{e_i} = \frac{F}{4}[\varepsilon_1 - \varepsilon_2] = \frac{F}{4}\varepsilon$$

We see that the output is unaffected by the change in temperature. When the compensating gage is used merely to complete the bridge and to balance out the temperature component, it is often referred to as the "dummy" gage.

12.10.2 Self-Temperature Compensation

In certain cases it may be difficult or impossible to obtain temperature compensation by means of an adjacent-arm compensating or dummy gage. For example, temperature gradients in the test part may be sufficiently great to make it impossible to hold any two gages at similar temperatures. Or, in certain instances, it may be desirable to use the ballast rather than the bridge circuit, thereby eliminating the possibility of adjacent-arm compensation. Situations of this sort make *self-compensation* highly desirable.

The two general types of self-compensated gages available are the *selected-melt* gage and the *dual-element* gage. The former is based on the discovery that through proper manipulation of alloy and processing, particularly through cold working, some control over the temperature sensitivity of the grid material may be exercised. Through this approach grid

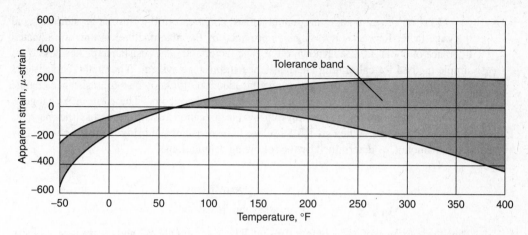

FIGURE 12.9: Approximate range of apparent strain versus temperature for a typical selected-melt gage mounted on the appropriate material (e.g., steel at about 11 ppm/°C).

materials may be prepared that show very low apparent strain versus temperature change over certain temperature ranges when the gage is mounted on a particular test material. Figure 12.9 shows typical characteristics of selected-melt gages compensated for use with a material having a coefficient of expansion of 6 ppm/°F, which corresponds to the coefficient of expansion of most carbon steels. In this case, practical compensation is accomplished over a temperature range of approximately 50°F–250°F. Other gages may be compensated for different thermal expansions and temperature ranges. These curves give some idea of the degree of control that may be obtained through manipulation of the grid material.

The second approach to self-compensation makes use of two grid elements connected in series in one gage assembly. The two elements have different temperature characteristics and are selected so that the net temperature-induced strain is minimized when the gage is mounted on the specified test material. In general, the performance of this type of gage is similar to that of the selected-melt gage shown in Fig. 12.9.

Neither the selected-melt nor the dual-element gage has a distinctive outward appearance. One company uses color-coded backings to assist in identifying gages of different specifications.

12.11 CALIBRATION

Ideally, calibration of any measuring system consists of introducing an accurately known sample of the variable that is to be measured and then observing the system's response. This ideal cannot often be realized in bonded resistance strain-gage work because of the nature of the transducer. Normally, the gage is bonded to a test item for the simple reason that the strains (or stresses) are unknown. Once bonded, the gage can hardly be transferred to a *known* strain situation for calibration. Of course, this is not necessarily the case if the gage or gages are used as secondary transducers applied to an appropriate elastic member for the purpose of measuring force, pressure, torque, and so on. In cases of this sort, it may be perfectly feasible to introduce known inputs and carry out satisfactory calibrations. When the gage is used for the purpose of experimentally determining strains, however, some other approach to the calibration problem is required.

Resistance strain gages are manufactured under carefully controlled conditions, and the gage factor for each lot of gages is provided by the manufacturer within an indicated tolerance of about $\pm 0.2\%$. Knowing the gage factor and gage resistance makes possible a simple method for calibrating any resistance strain-gage system. The method consists of determining the system's response to the introduction of a known small resistance change at the gage and of calculating an equivalent strain therefrom. The resistance change is introduced by shunting a relatively high-value precision resistance across the gage, as shown in Fig. 12.10. When switch S is closed, the resistance of bridge arm 1 is changed by a small amount, as determined by the following calculations.

Let

$$R_g = \text{the gage resistance,}$$

$$R_s = \text{the shunt resistance}$$

Then the resistance of arm 1 before the switch is closed equals R_g, and the resistance of arm 1 after the switch is closed equals $(R_g R_s)/(R_g + R_s)$, as determined for parallel resistances. Therefore, the change in resistance is

$$\Delta R = \frac{R_g R_s}{(R_g + R_s)^2} - R_g = -\frac{R_g^2}{R_g + R_s}$$

Now to determine the equivalent strain, we may use the relation given by Eq. (12.11):

$$\varepsilon = +\frac{1}{F}\frac{\Delta R_g}{R_g}$$

By substituting ΔR for ΔR_g, the equivalent strain is found to be

$$\varepsilon_e = -\frac{1}{F}\left(\frac{R_g}{R_g + R_s}\right) \tag{12.15}$$

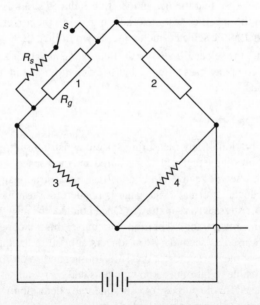

FIGURE 12.10: Bridge employing a shunt resistance for calibration.

EXAMPLE 12.1

Suppose that

$$R_g = 120 \ \Omega,$$
$$F = 2.1,$$
$$R_s = 100 \ k\Omega \ (\text{i.e., } 100{,}000 \ \Omega)$$

What equivalent strain will be indicated when the shunt resistance is connected across the gage?

Solution   From Eq. (12.15),

$$\varepsilon_e = -\frac{1}{2.1}\left[\frac{120}{100{,}0000 + 120}\right] = -0.00057$$
$$= -570 \ \mu\text{-strain}$$

Dynamic calibration is sometimes provided by replacing the manual calibration switch with an electrically driven switch, often referred to as a chopper, which makes and breaks the contact 60 or 100 times per second. When displayed on a oscilloscope screen or recorded, the trace obtained is found to be a square wave. The *step* in the trace represents the equivalent strain calculated from Eq. (12.15).

There are other methods of electrical calibration. One system replaces the strain-gage bridge with a substitute load, initially adjusted to equal the bridge load [12]. A series resistance is then used for calibration. Another method injects an accurately known voltage into the bridge network.

12.12 COMMERCIALLY AVAILABLE STRAIN-MEASURING SYSTEMS

Commercially available systems intended for use with metallic-type gages fall within three general categories:

1. The basic strain indicator, useful for static, single-channel readings
2. The single-channel system either external to or an integral part of a oscilloscope or a computer data acquisition system.
3. Data acquisition systems (e.g., see Fig. 8.23), whereby the strain data may be

   (a) displayed (digitally and/or by a video terminal)
   (b) recorded (on hard disk or hard-copy printout)
   (c) fed back into the system for control purposes

The wide range of availability and divergence of such systems makes it impractical to attempt any but a superficial coverage in this text. Better sources of state-of-the-art details are the brochures and technical "aids" provided by many of commercial suppliers.

12.12.1   The Basic Indicator

Typically, the basic indicator consists of a manually or self-balancing Wheatstone bridge with meter-type or digital readout, an amplifier, and adjustments to accommodate a range of gage factors. Provision is also common for handling bridges with a single active-gage, two-gage, and four-gage configurations. For fewer than four gages, the bridge loop is completed within the instrument. The measurement process consists of zeroing the bridge under initial conditions, then, after applying test conditions, rebalancing the bridge. The difference between initial and final readings provides the strain increment. Such instruments are generally precalibrated to provide direct strain readout, often in digital form. Most strain indicators now have analog output terminals to display dynamic strain data to oscilloscopes or computer data acquisition systems.

12.13   STRAIN-GAGE SWITCHING

Mechanical development problems often require the use of many gages mounted throughout the test item, and simultaneous or nearly simultaneous readings are often necessary. Of course, if the data must be recorded at precisely the same instant, it will be necessary to provide separate channels for each gage involved. However, frequently steady-state conditions may be maintained or the test cycle repeated, and readings may be made in succession until all the data have been recorded. In other cases, the budget may prohibit duplication of the required instrumentation for simultaneous multiple readings or recordings, and it becomes desirable to switch from gage to gage, taking data in sequence.

For high-speed multiple strain measurements, the digital techniques described in Chapter 8 are recommended, possibly including multiplexing methods (Section 8.7.6) or multichannel A/D converters (Section 8.11).

12.14   USE OF STRAIN GAGES ON ROTATING SHAFTS

Strain-gage information may be conducted from rotating shafts in at least three different ways: (1) by direct connection, (2) by wireless telemetery, and (3) by use of slip rings. When a shaft rotates slowly enough and when only a sampling of data is required, direct connections may be made between the gages and the remainder of the measuring system. Sufficient lead length is provided, and the cable is permitted to wrap itself onto the shaft. In fact, the available time may be doubled with a given length of cable if it is first wrapped on the shaft so that the shaft rotation causes it to unwrap and then to wrap up again in the opposite direction. If the machine cannot be stopped quickly enough as the end of the cable is approached, a fast or automatic disconnecting arrangement may be provided. This actually need be no more than soldered connections that can be quickly peeled off. Shielded cable should be used to minimize reactive effects resulting from the coil of cable on the shaft. This technique is somewhat limited, of course, but should not be overlooked, because it is quite workable at slow speeds and avoids many of the problems inherent in the other methods.

A second method is that of actually transmitting the strain-gage information through a radio frequency transmitter mounted on the shaft and with the signal picked up by a receiver placed nearby. This method has been used successfully for a long time [13], and a

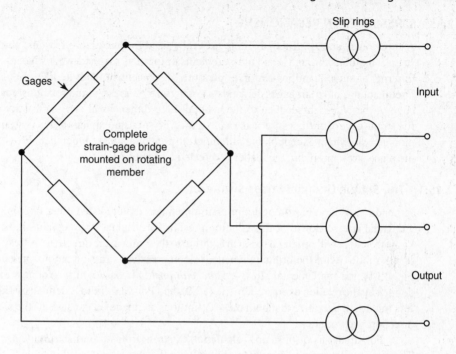

FIGURE 12.11: Slip rings external to the bridge.

diverse array of wireless instrumentation is commercially available. For strain applications in particular, digital telemetry systems are available with onboard multiplexing that allows many sensors to be routed through a single transmitter. Such a system is practical when the added cost can be justified.

Undoubtedly the most common method for obtaining strain-gage information from rotating shafts is through the use of slip rings. Slip-ring problems are similar to switching problems, as discussed in the preceding section, except that additional variables make the problem more difficult. Factors such as ring and brush wear and changing contact temperatures make it imperative that the full bridge be used at the test point and that the slip rings be introduced externally to the bridge as shown in Fig. 12.11.

Commercial slip-ring assemblies are available whose performances are quite satisfactory. Their use, however, presents a problem that is often difficult to solve. The assembly is normally self-contained, consisting of brush supports and a shaft with rings mounted between two bearings. The construction requires that the rings be used at a free end of a shaft, which more often than not is separated from the test point by some form of bearing. This arrangement presents the problem of getting the leads from the gage located on one side of a bearing to the slip rings located on the opposite side. It is necessary to feed the leads through the shaft in some manner, which is not always convenient. Where this presents no particular problem, the commercially available slip-ring assemblies are practical and also probably the most inexpensive solution to the problem.

12.15 STRESS–STRAIN RELATIONSHIPS

As previously stated, strain gages are generally used for one of two reasons: to determine stress conditions through strain measurements or to act as secondary transducers calibrated in terms of such quantities as force, pressure, displacement, and the like. In either case, intelligent use of strain gages demands a good grasp of stress–strain relationships. Knowledge of the *plane*, rather than of the general three-dimensional case, is usually sufficient for strain-gage work because it is only in the very unusual situation that a strain gage is mounted anywhere except on the *unloaded surface* of a stressed member. For a review of the plane stress problem, the reader is directed to Appendix E.

12.15.1 The Simple Uniaxial Stress Situation

In bending, or in a tension or compression member, the unloaded outer fiber is subject to a uniaxial stress. However, this condition results in a triaxial strain condition, because we know that there will be lateral strain in addition to the strain in the direction of stress. Because of the simplicity of the ordinary tensile (or compressive) situation and its prevalence (see Fig. 12.1), the fundamental *stress–strain relationship* is based on it. Young's modulus is defined by the relation expressed by Eq. (12.2), and Poisson's ratio is defined by Eq. (12.2a). It is important to realize that both these definitions are made on the basis of the simple one-direction stress system.

For situations of this sort, calculation of stress from strain measurements is simple. The stress is determined merely by multiplying the strain, measured in the axial direction in microstrains, by the modulus of elasticity for the test material.

EXAMPLE 12.2

Suppose the tensile member in Fig. 12.6 is of aluminum having a modulus of elasticity equal to 6.9×10^{10} Pa (10×10^6 lbf/in.^{2}) and the strain measured by the gage is 326 μ-strain. What axial stress exists at the gage?

Solution

$$\sigma_a = E\varepsilon_a = (6.9 \times 10^{10}) \times (325 \times 10^{-6})$$
$$= 22.4 \times 10^6 \text{ Pa } (3250 \text{ lbf/in.}^2)$$

EXAMPLE 12.3

Strain gages are mounted on a beam as shown in Fig. 12.7. The beam is of steel having an estimated modulus of elasticity of 20.3×10^{10} Pa (29.5×10^6 lbf/in.^{2}). If the total readout from the two gages is 390 μ-strain, what stress exists at the longitudinal center of the gage? Note that the bridge constant is 2.

Solution

$$\sigma_b = E\varepsilon_b = (20.3 \times 10^{10}) \times (390 \times 10^{-6}/2)$$
$$= 3958 \times 10^4 \text{ Pa } (5700 \text{ lbf/in.}^2)$$

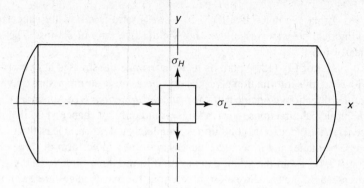

FIGURE 12.12: Element located on the shell of a cylindrical pressure vessel.

12.15.2 The Biaxial Stress Situation

Often gages are used at locations subject to stresses in more than one direction. If the test point is on a free surface, as is usually the case, the condition is termed *biaxial*. A good example of this condition exists on the outer surface, or shell, of a cylindrical pressure vessel. In this case, we know that there are *hoop* stresses, acting circumferentially, tending to open up a longitudinal seam. There are also longitudinal stresses tending to blow the heads off. The situation may be represented as shown in Fig. 12.12.

The stress–strain condition on the outer surface corresponds to that shown in Fig. 12.2. The two stresses σ_L and σ_H are principal stresses (no shear in the longitudinal and hoop directions), and the corresponding stresses may be calculated using Eq. (12.4), if we know (or can estimate) Young's modulus and Poisson's ratio.

EXAMPLE 12.4

Suppose we wish to determine, by strain measurement, the stress in the circumferential or hoop direction on the outer surface of a cylindrical pressure vessel. The modulus of elasticity of the material is 10.3×10^{10} Pa, and Poisson's ratio is 0.28. By strain measurement the hoop and longitudinal strains (in microstrain) are determined to be

$$\varepsilon_H = 425 \quad \text{and} \quad \varepsilon_L = 115$$

Solution   Using Eq. (12.3), we have

$$\sigma_H = \frac{E(\varepsilon_H + v\varepsilon_L)}{1 - v^2} = \frac{10.3 \times 10^{10}(425 + 0.28 \times 115) \times 10^{-6}}{1 - (0.28)^2}$$
$$= 5.11 \times 10^7 \text{ Pa } (7.42 \times 10^3 \text{ lbf/in.}^2)$$

Although we may not be directly interested, we have the necessary information to determine the longitudinal stress also, as follows:

$$\sigma_L = \frac{E(\varepsilon_L + v\varepsilon_H)}{1 - v^2} = \frac{10.3 \times 10^{10}(115 + 0.28 \times 425) \times 10^{-6}}{1 - (0.28)^2}$$
$$= 2.61 \times 10^7 \text{ Pa } (3.8 \times 10^3 \text{ lbf/in.}^2)$$

It may be noted that the 2-to-1 stress ratio traditionally expected for the thin-wall cylindrical pressure vessel does not yield a like ratio of strains. The strain ratio is more nearly 4 to 1.

Use of Eq. (12.4) permits us to determine the stresses in two orthogonal directions. However, this information gives the *complete* stress–strain picture only when the two right-angled directions coincide with the *principal directions* (see Appendix E). If we do not know the principal directions, our readings would only by chance yield the maximum stress. In general, if a plane stress condition is completely unknown, at least three strain measurements must be made, and it becomes necessary to use some form of three-element *rosette* (see Fig. 12.4). From the strain data secured in the three directions, we obtain the complete stress–strain picture. Stress-strain relations for rosette gages are given in Table 12.3.

Although only three strain measurements are necessary to define a stress situation completely, the T-delta rosette, which includes a fourth gage element, is sometimes used to advantage for the following reasons:

1. The fourth gage may be used as a check on the results obtained from the other three elements.
2. If the principal directions are approximately known, gage d may be aligned with the estimated direction. Then, if the readings from gages b and c are of about the same magnitude, it is known that the estimate is reasonably correct, and the principal stresses may be calculated directly from Eqs. (12.4), greatly simplifying the arithmetic. If the estimate of direction turns out to be incorrect, complete data are still available for use in the equations from Table 12.3.
3. If the four readings are used in the T-delta equations in Table 12.3, an averaging effect results in better accuracy than if only three readings are used.

In spite of the advantages of the T-delta rosette, the rectangular one is probably the most popular, with the equiangular (delta) kind receiving second greatest use.

EXAMPLE 12.5

Figure 12.13 illustrates a rectangular rosette used to determine the stress situation near a pressure vessel nozzle. For thin-walled vessels, the assumption that principal directions correspond to the hoop and longitudinal directions is valid for the shell areas removed from discontinuities. Near an opening, however, the stress condition is completely unknown, and a rosette with at least three elements must be used.

Let us assume that the rosette provides the following data (in microstrain):

$$\varepsilon_a = 72, \qquad \varepsilon_b = 120, \qquad \varepsilon_c = 248$$

In addition, we shall say that

$$\nu = 0.3, \qquad E = 20.7 \times 10^{10} \text{ Pa}$$

Solution   A study of the equation forms in Table 12.3 shows that for each case, the principal strain, the principal stress, and maximum shear relations involve similar radical terms. Therefore, in evaluating rosette data, it is convenient to calculate the value of the radical as the first step. It will also be noted that the second term in the principal stress relations is

TABLE 12.3: Stress–Strain Relations for Rosette Gages*†

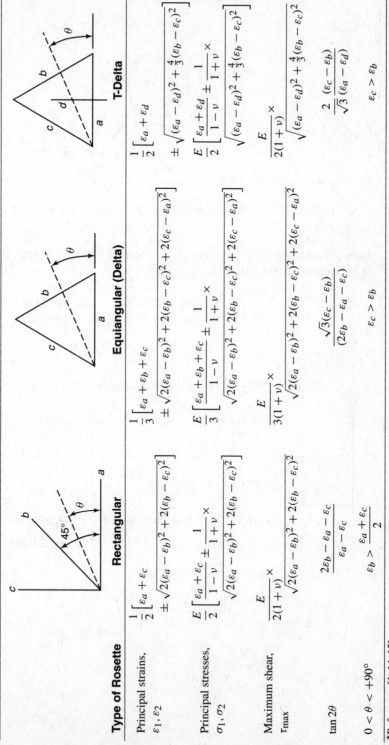

Type of Rosette	Rectangular	Equiangular (Delta)	T-Delta
Principal strains, $\varepsilon_1, \varepsilon_2$	$\dfrac{1}{2}\left[\varepsilon_a + \varepsilon_c \pm \sqrt{2(\varepsilon_a - \varepsilon_b)^2 + 2(\varepsilon_b - \varepsilon_c)^2}\right]$	$\dfrac{1}{3}\left[\varepsilon_a + \varepsilon_b + \varepsilon_c \pm \sqrt{2(\varepsilon_a - \varepsilon_b)^2 + 2(\varepsilon_b - \varepsilon_c)^2 + 2(\varepsilon_c - \varepsilon_a)^2}\right]$	$\dfrac{1}{2}\left[\varepsilon_a + \varepsilon_d \pm \sqrt{(\varepsilon_a - \varepsilon_d)^2 + \dfrac{4}{3}(\varepsilon_b - \varepsilon_c)^2}\right]$
Principal stresses, σ_1, σ_2	$\dfrac{E}{2}\left[\dfrac{\varepsilon_a + \varepsilon_c}{1-\nu} \pm \dfrac{1}{1+\nu}\times \sqrt{2(\varepsilon_a - \varepsilon_b)^2 + 2(\varepsilon_b - \varepsilon_c)^2}\right]$	$\dfrac{E}{3}\left[\dfrac{\varepsilon_a + \varepsilon_b + \varepsilon_c}{1-\nu} \pm \dfrac{1}{1+\nu}\times \sqrt{2(\varepsilon_a - \varepsilon_b)^2 + 2(\varepsilon_b - \varepsilon_c)^2 + 2(\varepsilon_c - \varepsilon_a)^2}\right]$	$\dfrac{E}{2}\left[\dfrac{\varepsilon_a + \varepsilon_d}{1-\nu} \pm \dfrac{1}{1+\nu}\times \sqrt{(\varepsilon_a - \varepsilon_d)^2 + \dfrac{4}{3}(\varepsilon_b - \varepsilon_c)^2}\right]$
Maximum shear, τ_{max}	$\dfrac{E}{2(1+\nu)}\times \sqrt{2(\varepsilon_a - \varepsilon_b)^2 + 2(\varepsilon_b - \varepsilon_c)^2}$	$\dfrac{E}{3(1+\nu)}\times \sqrt{2(\varepsilon_a - \varepsilon_b)^2 + 2(\varepsilon_b - \varepsilon_c)^2 + 2(\varepsilon_c - \varepsilon_a)^2}$	$\dfrac{E}{2(1+\nu)}\times \sqrt{(\varepsilon_a - \varepsilon_d)^2 + \dfrac{4}{3}(\varepsilon_b - \varepsilon_c)^2}$
$\tan 2\theta$	$\dfrac{2\varepsilon_b - \varepsilon_a - \varepsilon_c}{\varepsilon_a - \varepsilon_c}$	$\dfrac{\sqrt{3}(\varepsilon_c - \varepsilon_b)}{(2\varepsilon_b - \varepsilon_a - \varepsilon_c)}$	$\dfrac{2}{\sqrt{3}}\dfrac{(\varepsilon_c - \varepsilon_b)}{(\varepsilon_a - \varepsilon_d)}$
$0 < \theta < +90°$	$\varepsilon_b > \dfrac{\varepsilon_a + \varepsilon_c}{2}$	$\varepsilon_c > \varepsilon_b$	$\varepsilon_c > \varepsilon_b$

*References [1, 14, 15].

†Note: $\theta =$ the angle of reference, measured positive in the counterclockwise direction from the a-axis of the rosette to the axis of the algebraically larger stress.

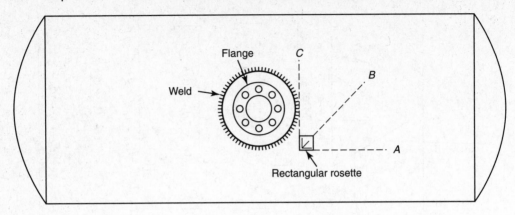

FIGURE 12.13: Rosette installation near a pressure vessel nozzle.

equal to the shear stress; thus arithmetical manipulations may be kept to a minimum if the shear stress is calculated before the principal stresses are determined. Hence,

$$\sqrt{2(\varepsilon_a - \varepsilon_b) + 2(\varepsilon_b - \varepsilon_c)^2} = \sqrt{2(72 - 120)^2 + 2(120 - 248)^2}$$
$$= 193 \ \mu\text{-strain}$$

and

$$\varepsilon_1 = \frac{1}{2}[72 + 248 + 193] = 256 \ \mu\text{-strain},$$

$$\varepsilon_2 = \frac{1}{2}[72 + 248 - 193] = 63 \ \mu\text{-strain},$$

$$\tau_{\max} = \frac{20.7 \times 10^{10}}{2(1 + 0.3)}(193) \times 10^{-6} = 1537 \times 10^4 \text{ Pa (2230 lbf/in.}^2),$$

$$\sigma_1 = \frac{20.7 \times 10^{10}}{2} \times \frac{72 + 248}{0.7} \times 10^{-6} + 1537 \times 10^4$$
$$= (4731 + 1537) \times 10^4 = 6268 \times 10^4 \text{ Pa (9091 lbf/in.}^2),$$

$$\sigma_2 = (4731 - 1537) \times 10^4 = 3194 \times 10^4 \text{ Pa (4632 lbf/in.}^2)$$

To determine the principal planes, we have

$$\tan 2\theta = \frac{(2\varepsilon_b - \varepsilon_a - \varepsilon_c)}{(\varepsilon_a - \varepsilon_c)}$$
$$= \frac{(2 \times 120) - 72 - 250}{(72 - 150)} = 0.46,$$
$$2\theta = 25° \quad \text{or} \quad 205°,$$

or

$$\theta = 12.5° \quad \text{or} \quad 102.5°$$

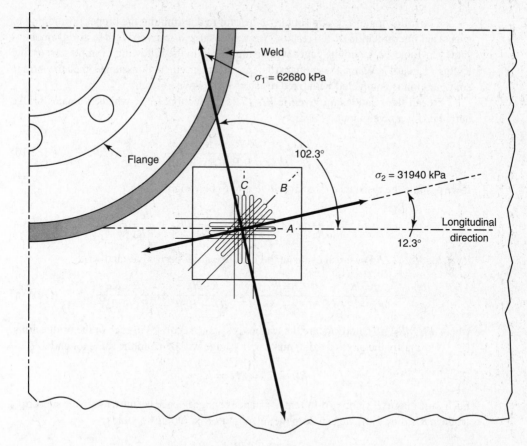

FIGURE 12.14: Stress conditions determined from data obtained by the rosette shown in Fig. 12.13.

measured counterclockwise from the axis of element A. We must test for the proper quadrant as follows (see the last line in Table 12.3):

$$\frac{\varepsilon_a + \varepsilon_c}{2} = \frac{72 + 248}{2} = 160$$

which is greater than ε_b. Therefore, the axis of maximum principal stress does *not* fall between $0°$ and $90°$. Hence, $\theta = 102°$. Figure 12.14 illustrates this condition.

12.16   GAGE ORIENTATION AND INTERPRETATION OF RESULTS

In a given situation it is often possible to place gages in several different arrangements to obtain the desired data. Often there is a best way, however, and in certain instances unwanted strain components may be canceled by proper gage orientation. For example, it is often desirable to eliminate unintentional bending when only direct axial loading is of primary interest; or perhaps only the bending component in a shaft is desired, to the exclusion of torsional strains.

The following discussion should be helpful in determining the proper positioning of gages and interpretation of the results. We will assume a *standard* bridge arrangement as shown in Table 12.4, and the gages will be numbered in the following examples according to this standard. When fewer than four gages are used, it is assumed that the bridge configuration is completed with fixed resistors insensitive to strain.

Recall the relationship given in Eq. (7.15a), Section 7.9.1, which evaluates bridge output e_o for a given input e_i, namely,

$$e_o = e_i \frac{R_1 R_4 - R_2 R_3}{(R_1 + R_2)(R_3 + R_4)} \tag{12.16}$$

If we assume the resistance of each bridge arm to be variable, then

$$de_o = \frac{\partial e_o}{\partial R_1} dR_1 + \frac{\partial e_o}{\partial R_2} dR_2 + \frac{\partial e_o}{\partial R_3} dR_3 + \frac{\partial e_o}{\partial R_4} dR_4 \tag{12.17}$$

By using Eq. (12.16) we can evaluate the various partial derivatives and write

$$\frac{de_o}{e_i} = \frac{R_2 dR_1}{(R_1 + R_2)^2} - \frac{R_1 dR_2}{(R_1 + R_2)^2} - \frac{R_4 dR_3}{(R_3 + R_4)^2} + \frac{R_3 dR_4}{(R_3 + R_4)^2} \tag{12.17a}$$

where dR_1, dR_2, dR_3, and dR_4 are the various resistance changes in each of the bridge arms.

Ordinarily the gages used to make up a bridge will be from the same lot and

$$R_1 = R_2 = R_3 = R_4 = R$$

Each gage may experience a different resistance change: hence we must retain the subscripts on the dR's; however, we can drop them from the R's. Doing so yields

$$\frac{de_o}{e_i} = \frac{dR_1 - dR_2 - dR_3 + dR_4}{4R} \tag{12.17b}$$

From Eq. (12.10), we have

$$\frac{dR_n}{R_n} = F\varepsilon_n \tag{12.18}$$

where

$$F = \text{the gage factor,}$$

$$\varepsilon_n = \text{the strain sensed by gage } n$$

Combining Eq. (12.18) and (12.17b) gives us

$$\frac{de_o}{e_i} = \frac{F}{4}[\varepsilon_1 - \varepsilon_2 - \varepsilon_3 + \varepsilon_4] \tag{12.19}$$

where the ε's are the strains sensed by the respective gages.

Equation (12.19) aids in the proper interpretation of the strain results obtained from the standard four-arm bridge in addition to assisting the stress analyst in the proper placement and orientation of gages for experimental measurements.

TABLE 12.4: Strain-Gage Orientation

Standard Bridge Configuration	

Requirement for null: $\dfrac{R_1}{R_2} = \dfrac{R_3}{R_4}$

$k = $ Bridge constant $= \dfrac{\text{Output of bridge}}{\text{Output of primary gage}}$

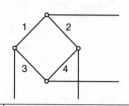

A	$k = 1$ 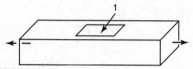	Compensates for temperature if "dummy" gage is used in arm 2 or arm 3. Does not compensate for bending.
B	$k = 2$ 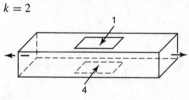	Compensates for bending. Two-arm bridge does not provide temperature compensation. Four-arm bridge ("dummy" gages in arms 2 and 3) provides temperature compensation.
C	$k = 1 + \nu$ 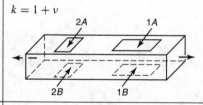	Two-arm bridge compensates for temperature and bending.
D	$k = 2(1 + \nu)$ 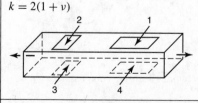	Four-arm bridge compensates for temperature and bending.
E	$k = 1$ 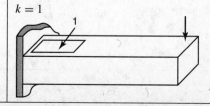	Temperature compensation accomplished when "dummy" gage is used in arm 2 or arm 3. Bridge is also sensitive to axial and torsional components of loading.

TABLE 12.4: *Continued*

Standard Bridge Configuration	Requirement for null: $\dfrac{R_1}{R_2} = \dfrac{R_3}{R_4}$

k = Bridge constant = $\dfrac{\text{Output of bridge}}{\text{Output of primary gage}}$

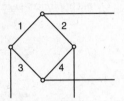

F	$k = 2$ 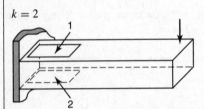	Temperature effects and axial and torsional components are compensated.
G	$k = 4$ 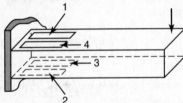	Four-arm bridge. Temperature effects and axial and torsional components are compensated.
H	$k = \dfrac{a+b}{a}$ 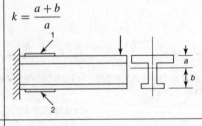	Temperature effects and axial and torsional components are compensated.
I	$k = 1 + \left(\dfrac{b}{a}\right)v$ 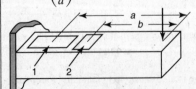	Temperature effects are compensated. Axial and torsional load components are not compensated.

TABLE 12.4: *Continued*

Standard Bridge Configuration	Requirement for null: $\dfrac{R_1}{R_2} = \dfrac{R_3}{R_4}$

$$k = \text{Bridge constant} = \frac{\text{Output of bridge}}{\text{Output of primary gage}}$$

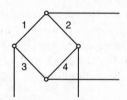

Torsion

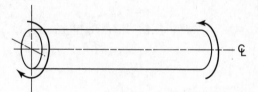

J	$k = 2$	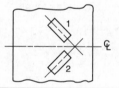	Two-arm bridge.
			Temperature and axial load components are compensated.
			Bending components are accentuated.
K	$k = 2$	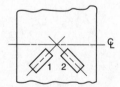	Two-arm bridge.
			Temperature effects and axial load components are compensated.
			Relatively insensitive to bending.
L	$k = 4$	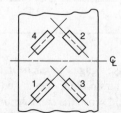	Four-arm bridge.
			Sensitive to torsion only.
			(Gages 1 and 3 are on opposite sides of the shaft from gages 2 and 4.)

For example, when only one active gage is used (gage 1, say), Eq. (12.19) reduces to

$$\frac{de_o}{e_i} = \frac{F\varepsilon_1}{4}$$

In further discussions we will assume that the term *bridge constant* abides by the definition given in Section 12.8.2.

EXAMPLE 12.6

The simplest application uses a single measuring gage with an external compensating gage as shown in Fig. 12.6 (also Case A, Table 12.4). This arrangement is primarily sensitive to axial strain; however, it will also sense any unintentional bending strain. The compensating gage, mounted on a sample of unstrained material identical to the test material, is located so that its temperature and that of the specimen will be the same. In this case,

$$\varepsilon_1 = \varepsilon_a + \varepsilon_b + \varepsilon_T,$$
$$\varepsilon_2 = \varepsilon_T,$$
$$\varepsilon_3 = 0 \text{ (a fixed resistor)},$$
$$\varepsilon_4 = 0 \text{ (a fixed resistor)}$$

where

$$\varepsilon_a = \text{the strain caused by axial loading},$$
$$\varepsilon_b = \text{the strain caused by any bending component},$$
$$\varepsilon_T = \text{the strain caused by temperature changes}$$

Solution Substituting in Eq. (12.19) gives us

$$\frac{de_o}{e_i} = \frac{F}{4}[\varepsilon_a + \varepsilon_b]$$

If the bending strain is negligible,

$$\frac{de_o}{e_i} = \frac{F\varepsilon_a}{4}$$

and the bridge constant is unity. Note that any strains caused by temperature effects cancel.

EXAMPLE 12.7

The arrangement shown in Case G, Table 12.4, uses gages in each of the four bridge arms. Gages 1 and 4 experience positive-bending strain components and gages 2 and 3 sense negative-bending components. All gages would sense the same strains derived from axial load and/or temperature should these be present. In addition, should the member be subjected to an axial torque, gages 1 and 2 would sense like strains from this source, as would gages 3 and 4. All gages would sense strain components of like magnitude from any torque acting about the longitudinal axis of the member; however, strains sensed by gages 1 and 2 would be of opposite sign to those sensed by 3 and 4.

Solution   Substitution of all these effects into Eq. (12.19) yields

$$\frac{de_o}{e_i} = \left(\frac{F}{4}\right)(4\varepsilon_b) = F\varepsilon_b$$

We see that only bending strains will be sensed and that the bridge constant is 4.

12.16.1   Gages Connected in Series

Figure 12.15 shows a load cell element using six gages. three connected in series in each of the bridge arms 1 and 2. At first glance it might be thought that the three gages in series would provide an output three times as great as that from a single gage under like conditions. Such is not the case, for it will be recalled that it is the percentage change in resistance, or dR/R, that counts, not dR alone. It is true that the resistance change for one arm, in this case, is three times what it would be for a single gage, but so also is the total resistance three times as great. Therefore, the only advantage gained is that of *averaging* to eliminate incorrect readings resulting from eccentric loading. The remaining two arms (not shown in the figure) may be made up of either inactive strain gages or fixed resistors. The bridge constant is $1 + \nu$.

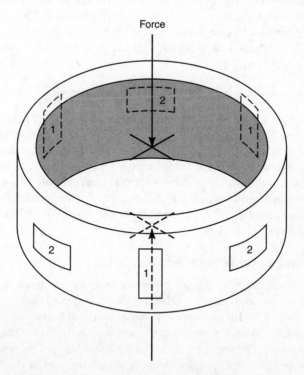

FIGURE 12.15: Load cell employing three series-connected axial gages and three series-connected Poisson-ratio gages.

12.17 SPECIAL PROBLEMS

12.17.1 Cross Sensitivity

Strain gages are arranged with most of the strain-sensitive filament aligned with the sensitive axis of the gage. However, unavoidably, a part of the grid is aligned transversely. The transverse portion of the grid senses the strain in that direction and its effect is superimposed on the longitudinal output. This is known as *cross sensitivity*. The error is small, seldom exceeding 2 or 3%, and the overall accuracy of many applications does not warrant accounting for it. For more detailed consideration the reader is referred to [15], [16], and [17].

12.17.2 Plastic Strains and the Postyield Gage

The average commercial strain gage will behave elastically to strain magnitudes as high as 3%. This represents a surprising performance when it is realized that the corresponding umaxial elastic stress in steel would be almost 1,000,000 lbf/in.^{2} (if elastic conditions in the steel were maintained). It is not very great, however, when viewed by the engineer seeking strain information beyond the yield point. When mild steel is the strained material, strains as great as 15% may occur immediately following attainment of the elastic limit before the stress again begins to climb above the yield stress. Hence, the usable strain range of the common resistance gage is quickly exceeded.

Gages known as *postyield* gages have been developed, extending the usable range to approximately 10% to 20%. Grid material in very ductile condition is used, which is literally caused to flow with the strain in the test material. The primary problem, of course, in developing an "elastic-plastic" grid is to obtain a gage factor that is the same under both conditions. Data reduction presents special problems, and for coverage of this aspect see references [18] and [19].

12.17.3 Fatigue Applications of Resistance Strain Gages

Strain gages are subject to fatigue failure in the same manner as are other engineering structures. The same factors are involved in determining their fatigue endurance. In general, the vulnerable point is the discontinuity formed at the juncture of the grid proper and the lead wire to which the user makes connection. Of course, as with any fatigue problem, strain level is the most important factor in determining life.

Isoelastic grid material performs better under fatigue conditions than does Constantan; the carrier material is also an important factor. Figure 12.16 illustrates the effects of most of the factors just discussed.

12.17.4 Cryogenic Temperature Applications

Strain measurement at extreme cryogenic temperatures must be done with an awareness of both thermal and materials-related problems. As previously discussed, resistance strain gages are sensitive to temperature changes, and in this situation temperature compensation is essential. Dummy gages, as discussed in Section 12.10, may be added to the bridge circuit to cancel temperature effects on gage resistance. The dummy gage must be placed into the same thermal environment as the measuring gage, and it must experience the same temperature and noise level as the measuring gage. On the materials side of the problem, adhesives and backings can become glass-hard and brittle at these temperatures. Whereas

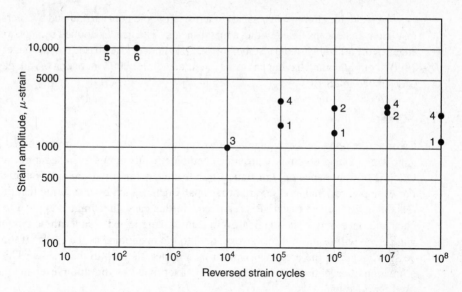

FIGURE 12.16: Relationship of endurance limit to strain level for gages of various materials and constructions (data from various sources including manufacturers' literature).

the mechanical properties of certain grid materials are drastically curtailed, those of others remain only slightly affected. Metal foil gages remain useful in cryogenic work. The bonding agent, typically an epoxy, should ideally have a thermal expansion coefficient similar to the gage back so as to avoid thermally induced strains in the measurement [20].

12.17.5   High-Temperature Applications

Maximum continuous-use temperatures for polyimide or Kapton backed gages range from about 100°C to 250°C. Primary limiting factors are decomposition of cement and carrier materials. At these temperatures grid materials present no particular problems. For applications at higher temperatures (to 1000°C) some form of ceramic-base insulation must be used. The grid may be of the strippable support, free-element type with the bonding as described next, or the gage may be of the "weldable" type.

Use of the free-element-type gage involves "constructing" the gage on the spot. Either brushable or flame-sprayed ceramic bonding materials are used. Application of the former consists of laying down an insulating coating upon which the free-element grid is secured with more cement. The process demands considerable skill and carefully controlled baking or curing-temperature cycling.

Flame spraying involves the use of a plasma-type oxyacetylene gun [21]. Molten particles of ceramic are propelled onto the test surface and used as both the cementing and insulating material for bonding the grid element to the test item. In both cases, leads must be attached by spot-welding to provide the necessary high-temperature properties to the connections. Lead-wire temperature-resistance variations may also present problems. It is obvious that considerable technique must be developed to use either of these types satisfactorily.

A weldable strain gage consists of a resistance element surrounded by a ceramic-type insulation and encapsulated within a metal sheath. The gage is applied by spot-welding the edges of the assembly to the test member [22]. (A novel laser-based extensometer usable at 1900°C or higher and having an overall accuracy of ±0.0002 in. over a 0.3-in. gage length is described in [23].)

12.17.6 Creep

Creep in the bond between gage and test surface is a factor sometimes ignored in strain-gage work. This problem is approximately diametrically opposite to the fatigue problem in that it is of importance only in static strain testing, primarily of the long-duration variety. For example, residual stresses are occasionally determined by measuring the dimensional relaxation as stressed material is removed. In this case, the strain is applied to the gage once and once only. The loading cycle cannot be repeated. Under these circumstances, gage creep will result in direct errors equal to the magnitude of the creep. If the load can be slowly cycled, the creep will appear as a hysteresis loop in the results. This effect is a function of several things but is primarily determined by the strain level and the cement used for bonding.

12.17.7 Residual Stress Determination

Occasionally it is necessary to determine the residual stresses existing in a structure or machine element. These stresses are generally developed during mechanical forming processes, such as casting or heat treatment. These stresses can be determined by using the strain-measuring techniques previously described, although they generally destroy the structure being analyzed.

Consider the pressure vessel of Example 12.5. If it is desired to estimate the residual stresses near a pressure-vessel nozzle due to welding, the rectangular rosette may be applied to the unpressurized vessel as shown. After the various strain-gage lead wires are attached to strain readout equipment, the region of pressure vessel containing the rosette is removed (cut away) from the rest of the material, and the resulting change in strains from the gages is recorded. Using these data (note the change in sign), the residual stresses existing in the unpressurized vessel at this location may be estimated. (See Problem 12.30 for an example of this process.)

Most strain-gage manufacturers provide a special strain rosette, whereby the strain-gage elements are arranged in such a fashion that a single hole may be drilled, relieving the stresses in the region and thus eliminating the need to completely cut away the material.

12.18 FINAL REMARKS

In addition to being the key to experimental stress analysis, strain can be made an analog for essentially any of the various mechanical inputs of interest to the engineer: force, torque, displacement, pressure, temperature, motion, and so on. For this reason strain gages are very widely and successfully used as secondary transducers in measuring systems of all types. Their response characteristics are excellent, and they are reliable, relatively linear, and inexpensive. It is important, therefore, that the engineer concerned with experimental work be well versed in the techniques of their use and application.

SUGGESTED READINGS

Dally, J. W., and W. F. Riley. *Experimental Stress Analysis.* 3rd ed. New York: McGraw-Hill, 1991.

Doyle, J. F. *Modern Experimental Stress Analysis.* Chichester, England: John Wiley, 2004.

Hannah, R. L., and S. E. Reed. *Strain Gage Users' Handbook.* Bethel, Conn.: Society for Experimental Mechanics, 1992.

Khan, A. S., and X. Wang. *Strain Measurements and Stress Analysis.* Upper Saddle River, N.J.: Prentice Hall, 2001.

Murray, W. M., and W. R. Miller. *The Bonded Electrical Resistance Strain Gage: An Introduction.* New York: Oxford University Press, 1992.

Doyle, J. F., and J. W. Phillips. *Manual on Experimental Stress Analysis.* 5th ed. Bethel, Conn.: Society for Experimental Mechanics, 1989.

PROBLEMS

12.1. A simple tension member with a diameter of 0.505 in. is subjected to an axial force of 7215 lbf. Strains of 1640 and −485 μ-strain are measured in the axial and transverse directions, respectively. Assuming elastic conditions, determine the values of Young's modulus and Poisson's ratio for the material. (*Note:* The diameter that is specified is commonly considered a "standard" for circularly sectioned metal specimens. Do you know why?)

12.2. A single strain gage is mounted on a tensile member, as shown in Fig. 12.6. If the readout is 425 μ-strain, what is the axial stress (a) if the member is of steel, and (b) if the member is of aluminum? See Appendix D for values of E.

12.3. A strain gage is centered along the length of a simply supported beam carrying a centrally positioned, concentrated load. The beam is four gage lengths long. What correction factor should be applied to care for the strain gradients if the purpose of the measurement is to determine the maximum strain?

12.4. Referring to the previous problem, what correction should be applied if the beam carries a uniformly distributed load?

12.5. Referring to Problem 12.3, what correction factor should be applied if, instead of being simply supported, the beam has built-in ends?

12.6. A resistance-type strain gage having a factor of 2.00 ± 0.05 and a resistance of $121 \pm 2\ \Omega$ is used in conjunction with an indicator having an uncertainty of $\pm 2\%$. What maximum uncertainty may be introduced by these tolerances? What probable uncertainty?

12.7. The sensing element of a weighing scale is described in Problem 6.11. Four strain gages are located as shown in Fig. 12.17. The gages are connected in a full bridge (see Case G, Table 12.4). Their nominal resistances are 300 Ω and their gage factors are 3.5. If the bridge is powered with a regulated 5.6-V dc source, what will be the voltage output corresponding to the maximum design load of 300/18 = 16.67 lbf?

12.8. Two strain gages are mounted on a cantilever beam as shown in Fig. 12.7. If the total strain readout is 620 μ-strain, what are the outer fiber stresses (a) if the member is of steel, and (b) if the member is of aluminum?

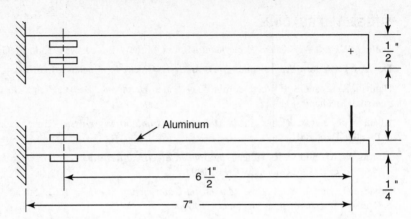

FIGURE 12.17: Strain-gage/beam configuration described in Problem 12.7.

12.9. Assume a system configured as shown in Fig. 12.10, using a conventional oscilloscope for readout.

 (a) Make a list of variables that you feel will have a measurable effect on the overall uncertainty of the system. Indicate those that you would expect to change with input magnitude and those that will be relatively constant; see Eq. (12.11). Include the uncertainty due to limits of resolution of the readout method and note that some form of system calibration must be used, with its attendant uncertainty.

 (b) Assign what you believe to be reasonable uncertainties to each factor in your list and determine the overall uncertainty in the final readout. Finally, divide the uncertainties into two categories: those having a major effect on the overall uncertainty and those of minor importance.

12.10. A plastic specimen is subjected to a biaxial stress condition for which $\sigma_x = 1380$ and $\sigma_y = 605$ lbf/in.^{2}. Measured strains (in microstrain) are $\varepsilon_x = 1780$ and $\varepsilon_y = 139$. Calculate Poisson's ratio and Young's modulus.

12.11. Two identical strain gages are mounted on a constant-moment beam as shown in Fig. 12.18. They are connected into a Wheatstone bridge as shown in Fig. 12.18(b). With no load on the beam the bridge is nulled with all arms having equal resistances. When the loads are applied, a bridge output of 650 μV is measured. Determine the gage factor for the gages on the basis of the following additional data: $E = 29.7 \times 10^6$ lbf/in.^{2}, Poisson's ratio = 0.3, and the gage nominal resistance is 120 Ω.

12.12. A two-element strain rosette is mounted on a simple tensile specimen of steel. One gage is aligned in an axial direction and the other in a transverse direction. The gages are connected in adjacent arms of the bridge. If the total bridge readout (based on single-gage calibration) is 900 μ-strain, what is the axial stress in pascal? $E = 20 \times 10^{10}$ Pa (29×10^6 lbf/in.^{2}) and $\nu = 0.29$.

12.13. Each line in Table 12.5 represents a set of data corresponding to a given plane stress condition. The first three columns are strains (in microstrain) obtained using a three-element *rectangular* rosette. The final two items are material properties. For a selected set of data, determine the following:

 (a) The principal strains

 (b) The principal stresses

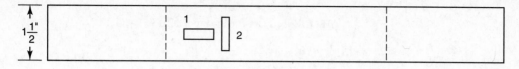

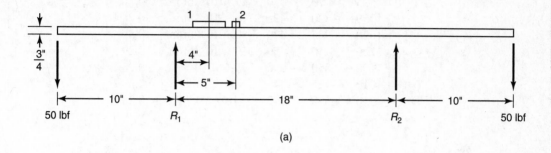

(a)

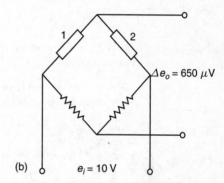

(b)

FIGURE 12.18: Detail of arrangement described in Problem 12.11.

   (c) The maximum shear stress
   (d) Principal directions referred to the axis of gage a

Also, sketch the following:

   (e) Mohr's circles for stress
   (f) Mohr's circles for strain
   (g) An element similar to that shown in Fig. 12.14

Use units corresponding to those given for E.

12.14. Repeat Problem 12.13(a) through (g), assuming an equiangular rosette.

12.15. If a rectangular-type rosette happens to be aligned such that elements a and c coincide with the principal directions, then measured values of ε_a and ε_c will be ε_1 and ε_2 (or vice versa). Show that under these circumstances the strain sensed by b will be $(\varepsilon_1 + \varepsilon_2)/2$.

12.16. Devise a spreadsheet template and/or a computer program to evaluate the rectangular strain rosette relationships in Table 12.3.

TABLE 12.5: Data for Problem 12.13

ε_a	ε_b	ε_c	E	ν
−320	210	680	10×10^6 lbf/in.^{2}	0.29
1585	470	0	29×10^6 lbf/in.^{2}	0.3
1250	−820	425	15×10^6 lbf/in.^{2}	0.28
−1020	985	−420	30×10^6 lbf/in.^{2}	0.3
2220	0	0	30×10^6 lbf/in.^{2}	0.29
0	850	−990	7.5×10^{10} Pa	0.28
−1010	−125	1440	20×10^{10} Pa	0.3
−210	−510	−212	10×10^{10} Pa	0.28
ε	ε	ε	E	ν
ε	ε	$-\varepsilon$	E	ν
ε	$-\varepsilon$	$-\varepsilon$	E	ν

12.17. Devise a spreadsheet template and/or a computer program to evaluate the T-delta strain rosette relationships in Table 12.3.

12.18. Devise a spreadsheet template and/or a computer program to evaluate the equiangular strain rosette relationships in Table 12.3.

12.19. Values of ν and E must be used in equations for converting strains to stresses. For steel, $\nu = 0.3$ and $E = 30 \times 10^6$ lbf/in.^{2} are often assumed. In fact, however, for steel ν may vary over the approximate range 0.27 to 0.32 and E may vary over a range of about 28×10^6 to 32×10^6 lbf/in.^{2}. Using the rectangular rosette strain values given in Example 12.5 in Section 12.15.2, analyze the effects of variations in assumed values ν and E on the calculated principal strains, stresses, and directions. It is suggested that the spreadsheet template (or program) written for Problem 12.16 be used to minimize the drudgery of number crunching.

12.20. Show how strain gages may be mounted on a simple beam to sense temperature change while being insensitive to variations in beam loading.

12.21. Two strain gages are mounted on a steel shaft ($E = 20 \times 10^{10}$ Pa and Poisson's ratio $= 0.29$), as shown in Case J, Table 12.4. The gage resistance is 119 Ω and $F = 1.23$. When a 250,000-Ω resistor is shunted across gage 1, a 3.4-cm upward shift is recorded on the face of the oscilloscope. When the shaft is torqued, a 5.7-cm shift is measured. For these conditions, and assuming bending and axial loading may be neglected,

(a) Calculate the maximum torsional stress.
(b) What are the three principal stresses on the shaft surface?
(c) Plot Mohr's circles for stress.
(d) If a bending moment and/or axial load is present, how would the results be affected?

12.22. Strain gages A, B, C, and D are mounted on a plate subjected to a simple bending moment M and an axial load F, as shown in Fig. 12.19. Gages are to be inserted into a standard bridge (see Table 12.4) in order to accomplish the following:

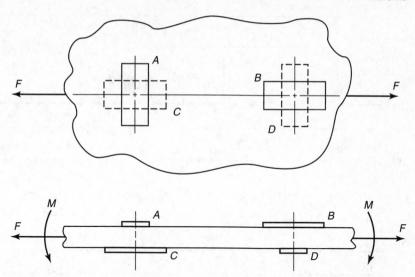

FIGURE 12.19: Arrangement of strain gages described in Problem 12.22.

(a) To sense bending only and, under this requirement, provide maximum bridge output

(b) To sense axial stress only (eliminating bending stress)

In each case, what will be the bridge constant, and will adjacent-arm temperature compensation be accomplished?

12.23. To determine the power transmitted by a 10 cm (3.94 in.) shaft, four strain gages are mounted as shown in Case L, Table 12.4. They are connected as a four-arm bridge and the output is fed to a chart recorder. Gage resistances are 118 Ω with a gage factor of 2.1. A 210,000 Ω calibration resistor may be shunted across one of the gages. Figures 12.20(a) and (b) show the calibration and strain records, respectively. The chart speed is 100 mm/s. The shaft is of steel with $E = 20 \times 10^{10}$ Pa and Poisson's ratio = 0.3. Determine the extreme and mean values of transmitted power in watts.

12.24. Four axially aligned, identical strain gages are equally spaced around a $1\frac{1}{4}$-in. (31.75-mm)-diameter bar, as shown in Fig. 12.21. The basic load on the bar is tensile; however, because of a small load ecentricity a bending moment also exists. If the strain readings shown on the sketch are determined for the individual gages, what axial load and bending moment must exist? Also determine the position of the neutral axis of bending.

12.25. Four gages are mounted on a thin-walled cylindrical pressure vessel. Two of the gages are aligned circumferentially (these are gages 1 and 4 in the standard bridge, Table 12.4), and the remaining gages 2 and 3 are aligned in the axial direction. (Note that this is not necessarily an optimal configuration.) If the bridge output is 27.8 units when a 300,000-Ω resistor is shunted across gage 1, and an output from the bridge of 47 units is recorded when the vessel is pressurized, what is the circumferential stress? Use $F = 3.5$, $R_g = 180 \ \Omega$, $E = 7 \times 10^{10}$ Pa, and Poisson's ratio = 0.3. Assume that the conventional 2-to-1, circumferential-to-longitudinal stress ratio applies. (See Example E.2, Appendix E.)

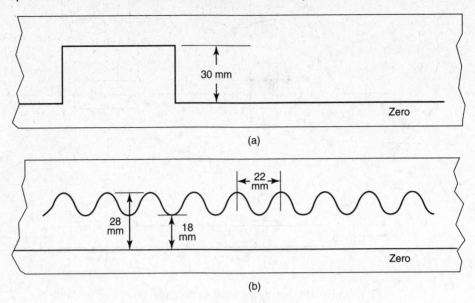

(a)

(b)

FIGURE 12.20: Recorder output for conditions of Problem 12.23.

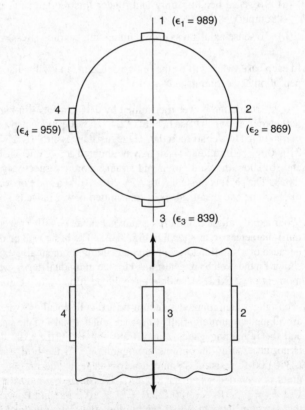

FIGURE 12.21: Configuration of strain gages described in Problem 12.24. Values of ϵ are in microstrain.

12.26. Strain readouts from a rectangular strain rosette are $\varepsilon_a = 620$, $\varepsilon_b = -200$, and $\varepsilon_c = 410$ μ-strain. Assume that under the same conditions an equiangular rosette is mounted and that its a element is aligned with the direction of the a element of the original rectangular rosette. What readouts should be expected from the delta gage? Assume the same gage factor and resistances for both rosettes. (*Hint:* See Appendix E for Mohr's circles for strain.)

12.27. A strain gage having a resistance $R_g = 120$ Ω and a gage factor $F = 2.0$ is used in an optimum ballast circuit. What is the maximum error over a range of $0 < \varepsilon < 1500$ μ-strain relative to a "best" straight line referenced to $\varepsilon = 0$?

12.28. Analyze the effect of lead wire length and wire gage on the sensitivity of the following strain gage circuits:

(**a**) Ballast circuit
(**b**) Circuit shown in Fig. 12.6
(**c**) Circuit shown in Fig. 12.7
(**d**) A four-arm bridge such as shown in Fig. 12.19

The following data may be useful if a quantitative analysis is being made.

Wire Size A.W.G.*	Ohms per 1000 ft at 25°C
12	1.62
15	3.25
18	6.51
20	10.35
24	26.17

* American Wire Gage.

12.29. Analyze the uncertainty inherent in shunt calibration of strain-gage circuits.

12.30. A mechanical engineering student wishes to determine the internal pressure existing in a diet soda can. She proceeds by carefully mounting a single-element strain gage aligned in circumferential direction on the center of the soda can, as shown in Fig. 12.22. After wiring the gage properly to a commercial strain indicator, she "pops" the flip-top lid, which relieves the internal pressure. She notes that the strain indicator reads -400 μ-strain. If the can body is made of aluminum with a thickness of 0.010 in. and a diameter of 2.25 in., what was the original internal pressure of the sealed can?

12.31. Another student also performed the experiment described in Problem 12.30. Unfortunately, he did not have access to the commercial strain indicator, and instead he had to construct his own Wheatstone bridge circuit. His strain gage had an initial resistance of 120 Ω and a gage factor of 2.05. He used the single gage as one leg of the bridge, which he powered with a 6-V battery. The bridge output was fed to an amplifier (gain = 1000), and the amp's output was read by a voltmeter. The student balanced the bridge circuit before he opened the can. After the can was opened, the voltmeter indicated a voltage of -1.57 V. What was the measured strain for his can?

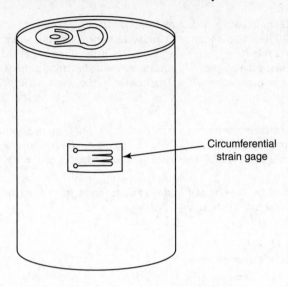

Circumferential strain gage

FIGURE 12.22: Instrumented soda can.

REFERENCES

[1] Hetenyi, M. *Handbook of Experimental Stress Analysis*. New York: John Wiley, 1950.

[2] Brookes-Smith, C. H. W., and J. A. Colls. Measurement of pressure, movement, acceleration and other mechanical quantities by electrostatic systems. *J. Sci. Inst. (London)* 14:361, 1939.

[3] Carter, B. C., J. F. Shannon, and J. R. Forshaw. Measurement of displacement and strain by capacity methods. *Proc. Inst. Mech. Eng.* 152:215, 1945.

[4] Langer, B. F. Design and application of a magnetic strain gage. *SESA Proc.* 1(2):82, 1943.

[5] Langer, B. F. Measurement of torque transmitted by rotating shafts. *J. Appl. Mech.* 67(3):A.39, March 1945.

[6] Khan, A. S., and X. Wang. *Strain Measurements and Stress Analysis*. Upper Saddle River, N.J.: Prentice Hall, 2001.

[7] Thompson, K. On the electro-dynamic qualities of metals. *Phil. Trans. Roy. Soc. (London)* 146:649–751, 1856.

[8] Eaton, E. C. Resistance strain gage measures stresses in concrete. *Eng. News Rec.* 107:615–616, Oct. 1931.

[9] Bloach, A. New methods for measuring mechanical stresses at higher frequencies. *Nature* 136:223–224, Aug. 19, 1935.

[10] Clark, D. S., and G. Datwyler. Stress-strain relations under tension impact loading. *Proc. ASM* 38:98–111, 1938.

[11] Mills, D., III. Strain gage waterproofing methods and installation of gages on propeller strut of USS Saratoga. *SESA Proc.* 16(1):137, 1958.

[12] Frank, E. Series versus shunt bridge calibration. *Instr. Automation* 31:648, 1958.

[13] Campbell, W. R., and R. F. Suit, Jr. A transistorized AM-FM radio-link torque telemeter for large rotating shafts. *SESA Proc.* 14(2):55, 1957.

[14] Baumberger, R., and F. Hines. Practical reduction formulas for use on bonded wire strain gages in two-dimensional stress fields. *SESA Proc.* 2(1):133, 1944.

[15] Perry, C. C., and H. R. Lissner. *The Strain Gage Primer.* 2nd ed. New York: McGraw-Hill, 1962, p. 157.

[16] Meier, J. H. On the transverse sensitivity of foil gages. *Exp. Mech.* 1: July 1961.

[17] Wu, C. T. Transverse sensitivity of bonded strain gages. *Exp. Mech.* 2:338, Nov. 1962.

[18] Plan, T. H. H. Reduction of strain rosettes in the plastic range. *J. Aerospace Sci.* 26:842, December 1959.

[19] Ades, C. S. Reduction of strain rosettes in the plastic range. *Exp. Mech.* 2:345, November 1962.

[20] Timmerhaus, K. D., and T. M. Flynn. *Cryogenic Process Engineering.* New York: Plenum, 1989.

[21] Leszynski, S. W. The development of flame sprayed sensors. *ISA J.* 9:35, July 1962.

[22] Rastogi, V., K. D. Ives, and W. A. Crawford. High-temperature strain gages for use n sodium environments. *Exp. Mech.* 7:525, December 1967.

[23] Karnie, A. J., and E. E. Day. A laser extensometer for measuring strain at incandescent temperatures. *Exp. Mech.* 7:485, November 1967.

CHAPTER 13

Measurement of Force and Torque

13.1 INTRODUCTION

Mass, time, and displacement are fundamental measurement dimensions. *Mass is the measure of quantity of matter. Force* is a derived unit and *weight* is a force having distinctive characteristics.[1]

Mass is one of the fundamental parameters determining the gravitational attraction (force) exerted between two bodies. Newton's law of universal gravitation is expressed by the relation

$$F = \frac{Cm_1m_2}{r^2} \tag{13.1}$$

or

$$C = \frac{Fr^2}{m_1m_2} \tag{13.1a}$$

The units of C are $\text{N} \cdot \text{m}^2 \, \text{kg}^2$ or $\text{lbf} \cdot \text{ft}^2 \, \text{lbm}^2$, where

$$m_1 \text{ and } m_2 \;=\; \text{the masses of bodies 1 and 2, respectively,}$$
$$r \;=\; \text{the distance separating them,}$$
$$F \;=\; \text{the mutual gravitational force exerted, one on the other, and}$$
$$C \;=\; \text{the gravitational constant}$$

---

[1] At this point it may be well for the reader to review Section 2.9, "Conversion between Systems of Units."

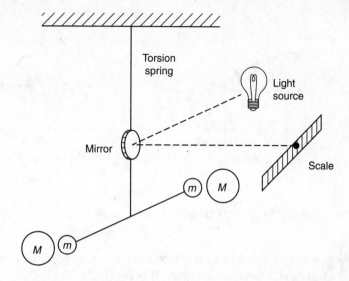

FIGURE 13.1: Balance used by Cavendish to measure gravitational constant.

Henry Cavendish (1731–1810), an English scientist, used a sensitive torsional balance (see Fig. 13.1) to determine the value of C. In SI units, $C = 6.67 \times 10^{-11}$ N · m^2/kg^2.

When one of the attracting masses is the earth and the second is that of some object on the surface of the earth, the resulting force of mutual attraction is called weight. Mass and weight are related through Newton's laws of motion. The gist of his first law is contained in the following statement: *If the resultant of all forces applied to a particle is other than zero, the motion of the particle will be changed.*

Newton's second law may be stated as follows: *The acceleration of a particle is directly proportional to and in the same direction as the resultant applied force.* This may be expressed as

$$\frac{F_1}{a_1} = \frac{F_2}{a_2} = \frac{m}{g_c} = \frac{w}{g} \tag{13.2}$$

To help establish the correctness of Eq. (13.2), let us look at the units (refer to Table 2.6). For the SI system of units we have

$$\frac{F}{a} \left(\frac{\text{N} \cdot \text{s}^2}{\text{m}} \right) = \frac{m}{g_c} \left(\frac{\text{kg} \cdot \text{N} \cdot \text{s}^2}{1 \text{ kg} \cdot \text{m}} \right)$$

Using the English Engineering System, we have

$$\frac{F}{a} \left(\frac{\text{lbf} \cdot \text{s}^2}{\text{ft}} \right) = \frac{m}{g_c} \left(\frac{\text{lbm} \cdot \text{lbf} \cdot \text{s}^2}{32.2 \text{ lbm} \cdot \text{ft}} \right)$$

A most convenient force to apply is the earth's gravitational attraction for the body or particle, which is the weight. If this is the *only* force, then the resulting acceleration is that of the falling body *in vacuo* at the particular location. Both the weight and the gravitational

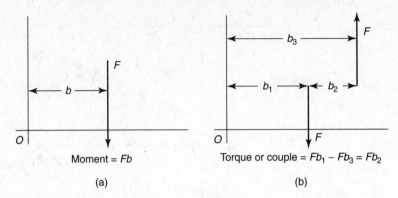

FIGURE 13.2: (a) Definition of moment, (b) definition of torque or couple.

attraction will vary from location to location. Their ratio, however, remains constant and is proportional to the mass, m, as expressed in Eq. (13.2).

As we can see from Eq. (13.2), neither the ratio m/g_c, nor w/g is required for application of Newton's second law. Any ratio F/a, where F is an applied force and a is the resulting acceleration, is just as valid for establishing the necessary value. The ratio w/g is particularly convenient, however, because over the surface of the earth, g is reasonably constant and indeed is often considered a constant in many engineering calculations. As a result, measurement of weight suffices for determining the ratio w/g. The "standard" value of g is 32.1739 ft/s^2, or 9.80665 m/s^2. Rounded values of 32.2 ft/s^2 or 9.81 m/s^2 are commonly used.

Force, in addition to its effect along its line of action, may exert a turning effort relative to any axis other than those intersecting the line of action. Such a turning effect is variously called *torque, moment,* or *couple,* depending on the manner in which it is produced. The term *moment* is applied to conditions such as those illustrated in Fig. 13.2(a), whereas the terms *torque* and *couple* are applied to conditions involving counterbalancing forces, such as those shown in Fig. 13.2(b).

Mass Standards   As stated previously (Section 2.5), the fundamental unit of mass is the kilogram, equal to the mass of the International Prototype Kilogram located at Sévres, France. A gram is defined as a mass equal to one-thousandth of the mass of the International Prototype Kilogram. The commonly used avoirdupois[2] pound is 0.435 592 37 kilogram, as agreed to in 1959 (Section 2.5). Various classifications and tolerances for standard masses are recommended by the National Institute of Standards and Technology and the American Society for Testing and Materials [1, 2].

13.2   MEASURING METHODS

As in other areas of measurement, there are two approaches to the problem of force and weight measurement: (1) direct comparison, and (2) indirect comparison through use of calibrated transducers. Directly comparative methods use some form of beam balance with a null-balance technique. If the beam neither amplifies nor attenuates, the comparison is

[2]From the French meaning "goods of weight."

direct. The simple analytical balance is of this type. Often, however, as in the case of a platform scale, the force is attenuated through a system of levers so that a smaller weight may be used to *balance* the unknown, with the variable in this case being the magnitude of attenuation. This method requires calibration of the system.

   Question: When an equal-arm balance scale is used, are forces or are masses being compared? (Problem 13.2 expands on this query.)

13.3   MECHANICAL WEIGHING SYSTEMS

Mechanical weighing systems originated in Egypt and were probably used as early as 5000 B.C. [3]. The earliest devices were of the cord and *equal-arm* type, traditionally used to symbolize justice. *Unequal-arm* balances were apparently first used in the form shown in Fig. 13.3(a). This device, called a Danish steelyard, was described by Aristotle (384–322 B.C.) in his *Mechanics*. Balance is accomplished by moving the beam through the loop of cord, which acts as the fulcrum point, until balance is obtained. A later unequal-arm balance, the Roman steelyard, which employed fixed pivot points and movable balance weights, is still in use today; see Fig. 13.3(b).

13.3.1   The Analytical Balance

Probably the simplest weight or force measuring system is the ordinary equal-arm beam balance (Fig. 13.4). Basically this device operates on the principle of *moment comparison*. The moment produced by the unknown weight or force is compared with that produced by a known value. When null balance is obtained, the two weights are equal, provided the two arm lengths are identical. A check on arm equivalence may be made easily by simply

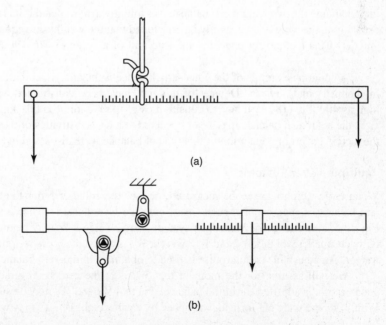

(a)

(b)

FIGURE 13.3: (a) Danish steelyard, (b) Roman steelyard.

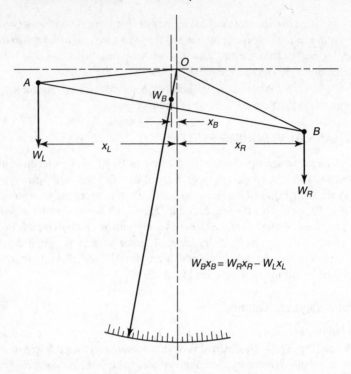

$$W_B x_B = W_R x_R - W_L x_L$$

FIGURE 13.4: Requirement for equilibrium of an analytical balance.

interchanging the two weights. If balance was initially achieved and if it is maintained after exchanging the weights, it can only be concluded that the weights are equal, as are the arm lengths. This method for checking the true null of a system is known as the method of *symmetry*.

A common example of the equal-arm balance is the analytical scale used principally in chemistry and physics. Devices of this type have been constructed with capacities as high as 400 lbm (180 kg), having sensitivities of 0.0002 lbm (0.0001 kg) [4]. In smaller sizes the analytical balance may be constructed to have sensitivities of 0.001 mg. Some of the factors governing operation of this type of balance were discussed in Section 5.10.

13.3.2 Multiple-Lever Systems

When large weights are to be measured, neither the equal-arm nor the simple unequal-arm balance is adequate. In such cases, multiple-lever systems, shown schematically in Fig. 13.5, are often used. With such systems, large weights W may be measured in terms of much smaller weights W_p and W_s. Weight W_p is called the *poise* weight and W_s the *pan weight*. An adjustable counterpoise is used to obtain an initial zero balance.

We will assume for the moment that W_p is at the zero beam graduation, that the counterpoise is adjusted for initial balance, and that W_1 and W_2 may be substituted for W. With W on the scale platform and balanced by a pan weight W_s, we may write the relations

$$T \times b = W_s \times a \qquad (13.3)$$

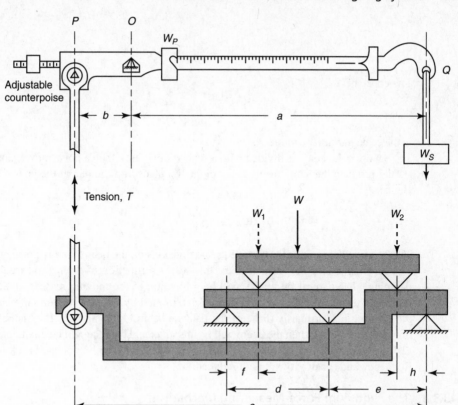

FIGURE 13.5: Multiple-lever system for weighing.

and

$$T \times c = W_1 \frac{f}{d} e + W_2 h \qquad (13.4)$$

Now if we proportion the linkage such that

$$\frac{h}{e} = \frac{f}{d}$$

then

$$T \times c = h(W_1 + W_2) = hW \qquad (13.4a)$$

From this we see that W may be placed anywhere on the platform and that its position relative to the platform knife edges is immaterial.

Solving for T in Eqs. (13.3) and (13.4a) and equating, we obtain

$$\frac{W_s a}{b} = \frac{Wh}{c}$$

or

$$W = \frac{ac}{bh} W_s = R W_s \qquad (13.5)$$

The constant

$$R = \frac{ac}{bh}$$

is the scale *multiplication ratio*.

Now if the beam is divided with a scale of u (lb/in.), then a poise movement of v (in.) should produce the same result as a weight W_p placed on the pan at the end of the beam. Hence,

$$W_p v = uva \quad \text{or} \quad u = \frac{W_p}{a}$$

This relation determines the required scale divisions on the beam for any poise weight W_p.

Dynamic response of a scale of this sort is a function of the natural frequency and damping. The natural frequency will be a function of the moving masses, multiplication ratio, and restoring forces. The latter are determined by the relative vertical placement of the pivot points, primarily those of the balance beam O, P, and Q. If O is below a line drawn from P to Q, then the beam will be unstable, and balance will be unattainable. Pivot O is normally above line PQ, and as the distance above the line is increased, the natural frequency and sensitivity are both reduced.

13.3.3 The Pendulum Force-Measuring Mechanism

Another type of movement comparison device used for measurement of force and weight is shown in Fig. 13.6. This is often referred to as a *pendulum scale*. Basically, the pendulum mechanism is a force-measuring device of the multiple-lever type, with the fixed-length levers replaced by ribbon- or tape-connected sectors. The input, either a direct force or a force proportional to weight and transmitted from a suitable platform, is applied to the load rod. As the load is applied, the sectors rotate about points A, as shown, moving the counter weights outward. This movement increases the counterweight effective moment until the load and balance moments are equalized. Motion of the equalizer bar is converted to indicator movement by a rack and pinion, the sector outlines being proportioned to provide a linear dial scale. This device may be applied to many different force-measuring systems, including dynamometers (Section 13.9).

13.4 ELASTIC TRANSDUCERS

Many force-transducing systems make use of some mechanical elastic member or combination of members. Application of load to the member results in an analogous deflection, usually linear. The deflection is then observed directly and used as a measure of force or load, or a secondary transducer is used to convert the displacement into another form of output, often electrical.

Most force-resisting elastic members adhere to the relation

$$K = \frac{F}{y} \qquad (13.6)$$

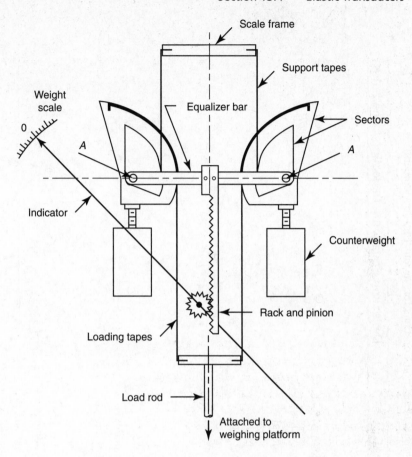

FIGURE 13.6: Essentials of a pendulum scale.

where

F = the applied load,

y = the resulting deflection at the location of F,

K = the deflection constant

To determine the value of the deflection constant of an element, it is necessary to write only the deflection equation, and if the deflection is a linear function of the load, K may be found. Table 13.1 lists representative relations indicating the general form.

Design detail of the detector–transducer element is largely a function of capacity, required sensitivity, and the nature of any secondary transducer and depends on whether the input is static or dynamic. Although it is impossible to discuss all situations, there are several general factors we may consider.

It is normally desirable that the detector–transducer be as sensitive as possible; that is, maximum output per unit input should be obtained. An elastic member would be required that deflects considerably under load, indicating as low a value of K as possible. There are usually conflicting factors, however, with the final design being a compromise. For example, if we were to measure rolling-mill loads by placing cells between the screwdown

TABLE 13.1: Deflection Equations and Deflection Constants for Various Elastic Members

Elastic Element	Deflection Equation	Deflection Constant K
A F = Load L = Length A = Cross-sectional area y = Deflection at load E = Young's modulus	$y = \dfrac{FL}{AE}$	$K = \dfrac{AE}{L}$
B F = Load L = Length E = Young's modulus I = Moment of inertia	$y = \dfrac{1}{48}\dfrac{FL^3}{EI}$	$K = \dfrac{48EI}{L^3}$
C F = Load L = Length E = Young's modulus I = Moment of inertia	$y = \dfrac{1}{3}\dfrac{FL^3}{EI}$	$K = \dfrac{3EI}{L^3}$
D F = Load D_m = Mean coil diameter N = Number of coils E_s = Shear modulus D_w = Wire diameter	$y = \dfrac{8FD_m^3 N}{E_s D_w^4}$	$K = \dfrac{E_s D_w^4}{8 D_m^3 N}$

Elastic Element	Deflection Equation	Deflection Constant K
E F = Load D = Diameter of ring E = Young's modulus I = Moment of inertia of section about centroidal axis of bending section	$y = \dfrac{1}{16}\left(\dfrac{\pi}{2} - \dfrac{4}{\pi}\right)\dfrac{FD^3}{EI}$	$K = \dfrac{16}{(\pi/2) - (4/\pi)}\left(\dfrac{EI}{D^3}\right)$
F K_1 = Deflection constant of member 1 K_2 = Deflection constant of member 2 F = Load	$y = \dfrac{F}{K_1 + K_2}$	$K = K_1 + K_2$
G K_1 = Deflection constant of member 1 K_2 = Deflection constant of member 2 F = Load	$y = F\left(\dfrac{1}{K_1} + \dfrac{1}{K_2}\right)$	$K = \dfrac{1}{(1/K_1) + (1/K_2)}$

and bearing blocks, our application could scarcely tolerate a springy load cell—that is, one that deflected considerably under load. It would be necessary to construct a stiff cell at the expense of elastic sensitivity and then attempt to make up for the loss by using as sensitive a secondary transducer as possible.

Another factor involving sensitivity is response time, or time required to come to equilibrium. This is a function of both damping and natural frequency (see Section 5.11). Fast response corresponds to high natural frequency, and thus a stiff elastic member is needed.

Stress also may be a limiting factor in any loaded member. It is especially important that the stresses remain below the elastic limit, not only in gross section but also at every isolated point. In this respect residual stresses are often of significance. Although load stresses may be well below the elastic limit for the material, it is possible that when they are added to *locked-in* stresses the total may be too great. Even though such a situation occurs only at a single isolated point, hysteresis and nonlinearity will result.

Manufacturing tolerances are yet another factor of importance in the design and application of elastic load elements. Tolerances were discussed in some detail in Section 6.18.

13.4.1   Calibration Adjustment

Various calibration adjustments may be made to account for variation in characteristics of elastic load members. Sometimes a simple check at the time of assembly and the selection of one of several standard scale graduations may suffice. For the coil spring tolerance example (Section 6.18.1) we determined the deflection constant uncertainty from dimensional tolerances to be 9%. At the time of assembly a quick single calibration check and a choice of two faceplates could cut the uncertainty from this source in half. Four plates would reduce it to $\pm2.5\%$. This scheme is often used not only for load-measuring devices but for all varieties of inexpensive instruments employing a scale. It does not provide for calibration adjustment in use, however.

When coil springs are used, means are sometimes provided to adjust the number of effective coils through use of an end connection that may be screwed into or out of the spring, thereby changing the number of active coils and hence the stiffness of the spring. In other cases, the springs may be purposely overdesigned with regard to stress, and the number of coils specified so that in no case may the tolerances add up to give a spring that is too flexible. Then at the time of assembly the springs are buffed on a wheel to obtain the required deflection constant.

If a secondary transducer is used, we may be able to provide for calibration by making adjustments in its characteristics. As an example, we could use a voltage-dividing potentiometer to sense the load deflection of the spring just discussed. We might employ this procedure to provide remote indication or recording. A circuit arrangement could be used in which an adjustable series resistor would be employed to provide calibration for the complete system.

Figure 13.7 illustrates a calibration adjustment scheme that minimizes the need for holding unduly close tolerances. Here the total load is shared between a primary member (a coil spring in this case) and one or more vernier members (the small springs in the figure). The design may call on the vernier members to carry 10% or less of the total load. At the time of assembly the verniers are selected from a range of stiffnesses so as to make the uncertainty of the assembly less than some specified value, say $\pm0.2\%$. Schemes of this sort are adaptable to a wide range of elastic devices.

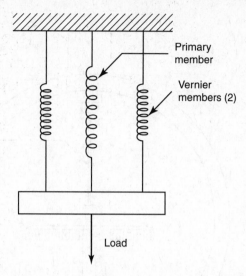

FIGURE 13.7: Approximate calibration method employing paralleling vernier members.

13.4.2   The Proving Ring

The proving ring has long been the *standard* for calibrating materials testing machines and is, in general, the means whereby accurate measurement of large static loads may be obtained. Figure 13.8 shows the construction of a compression-type ring. Capacities generally fall in the range of from 300 to 300,000 lbf (1334 N to 1.334 MN) [5].

Here, again, deflection is used as the measure of applied load, with the deflection measured by means of a precision micrometer. Repeatable micrometer settings are obtained with the aid of a vibrating reed. In use, the reed A is plucked (electrically driven reeds are also available), and the micrometer spindle B is advanced until contact is indicated by the marked damping of the vibration. Although different operators may obtain somewhat different individual readings, consistent differences in readings still will be obtained provided both zero and loaded readings are made by the same person. With 40 to 64 micrometer threads per inch, readings may be made to one or two hundred-thousandths of an inch [5].

The equation given in Table 13.1 for circular rings is derived with the assumption that the radial thickness of the ring is small compared with the radius. Most proving rings are made with a section of appreciable radial thickness. However, Timoshenko [6] shows that use of the thin-ring rather than the thick-ring relations introduces errors of only about 4% for a ratio of section thickness to radius of $\frac{1}{2}$. Increased stiffness on the order of 25% is introduced by the effects of integral bosses [5]. It is therefore apparent that use of the simpler thin-ring equation is normally justified.

Stresses may be calculated from the bending moments M determined by the relation [6]

$$M = \frac{PR}{2}\left(\cos\phi - \frac{2}{\pi}\right) \tag{13.7}$$

Symbols correspond to those shown in Fig. 13.9.

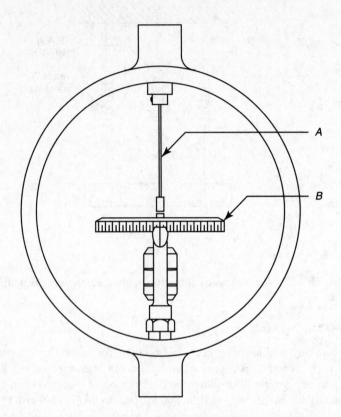

FIGURE 13.8: Compression-type proving ring with vibrating reed.

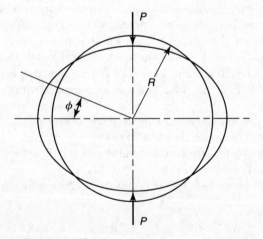

FIGURE 13.9: Ring loaded diametrically in compression.

13.5   STRAIN-GAGE LOAD CELLS

Instead of using total deflection as a measure of load, the strain-gage load cell measures load in terms of unit strain. Resistance strain gages are very suitable for this purpose (see Chapter 12). One of the many possible forms of elastic member is selected, and the gages are mounted to provide maximum output. If the loads to be measured are large, the direct tensile-compressive member may be used. If the loads are small, strain amplification provided by bending may be used to advantage.

Figure 13.10 illustrates the arrangement for a tensile-compressive cell using all four gages sensitive to strain and providing temperature compensation for the gages. The bridge constant (Section 12.8.2) in this case will be $2(1 + \nu)$, where ν is Poisson's ratio for the material. Compression cells of this sort have been used with a capacity of 3 million pounds (13.3 MN) [7]. Simple beam arrangements may also be used, as illustrated in Table 12.4.

Figure 13.11 illustrates proving-ring strain-gage load cells. In Fig. 13.11(a) the bridge output is a function of the bending strains only, the axial components being cancelled in the bridge arrangement. By mounting the gages as shown in Fig. 13.11(b), somewhat greater sensitivity may be obtained because the output includes both the bending and axial components sensed by gages 1 and 4.

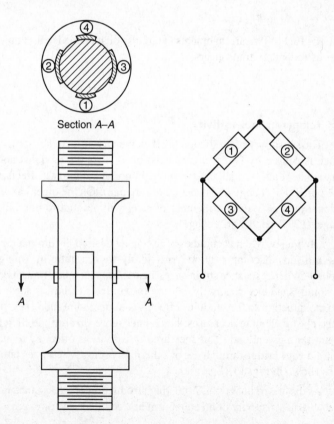

Section *A–A*

FIGURE 13.10: Tension-compression resistance strain-gage load cell.

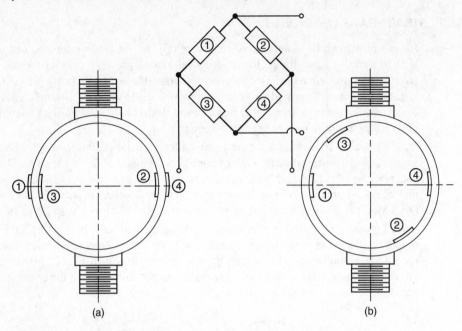

FIGURE 13.11: Two arrangements of circular-shaped load cells employing resistance strain gages as secondary transducers.

Temperature Sensitivity

The sensitivity of elastic load-cell elements is affected by temperature variation. This change is caused by two factors: variation in Young's modulus and altered dimensions. Variation in Young's modulus is the more important of the two effects, amounting to roughly 2.5% per 100°F change (4.5% per 100 K change). On the other hand, the increase in cross-sectional area of a tension member of steel will amount to only about 0.15% per 100°F change (0.27% per 100 K change).

Obviously, when accuracies of $\pm\frac{1}{2}$% are desired, as provided by certain commercial cells, a means of compensation, particularly for variation in Young's modulus, must be supplied. When resistance strain gages are used as secondary transducers, this compensation is accomplished electrically by causing the bridge's electrical sensitivity to change in the opposite direction to the modulus effect. As temperature increases, the deflection constant for the elastic element decreases; it becomes more *springy* and therefore deflects a greater amount for a given load. This increased sensitivity is offset by reducing the sensitivity of the strain gage bridge through use of a thermally sensitive compensating resistance element, R_s, as shown in Fig. 13.12.

As discussed in Section 7.9.6, the introduction of a resistance in an input lead reduces the electrical sensitivity of an equal-arm bridge by the factor expressed as

$$n = \frac{1}{1 + (R_s/R)}$$

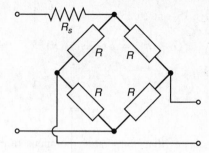

FIGURE 13.12: Schematic diagram of a strain-gage bridge with a compensating resistor.

Requirements for compensation may be analyzed through use of the relation for the initially balanced equal-arm bridge, Eq. (7.17). If we assume

$$2\frac{\Delta R}{R} \ll 4$$

then Eq. (7.17) may be reduced to Eq. (7.18):

$$\frac{\Delta e_o}{e_i} = \frac{k}{4}\frac{\Delta R}{R}$$

This is true, particularly for a *strain-gage bridge* for which $\Delta R/R$ is always small. A bridge constant, k, is included to account for use of more than one active gage. If all four gages are equally active, $k = 4$. For the arrangement shown in Fig. 13.10, $k = 2(1 + \nu)$, where ν is Poisson's ratio. If we account for the compensating resistor, the equation will then read

$$\frac{\Delta e_o}{e_i} = \frac{k}{4}\frac{\Delta R}{R}\left[\frac{1}{1 + (R_s/R)}\right] \tag{13.8}$$

Rewriting Eq. (12.10), we have

$$\varepsilon = \left(\frac{1}{F}\right)\left(\frac{\Delta R}{R}\right)$$

and from the definition of Young's modulus, E [Eq. (12.2)],

$$P = EA\varepsilon$$

we may solve for sensitivity:

$$\frac{\Delta e_o}{P} = \left(\frac{e_i}{4}\right)\left(\frac{FRk}{A}\right)\left[\frac{1}{E(R + R_s)}\right] \tag{13.9}$$

If it is assumed that the gages are arranged for compensation of resistance variation with temperature and that the gage factors F remain unchanged with temperature, and further that any change in the cross-sectional area of the elastic member may be neglected, then complete compensation will be accomplished if the quantity $E(R + R_s)$ remains constant with temperature.

Using Eqs. (6.15) and (6.23), we may write

$$E(R + R_s) = E(1 + c\,\Delta T)[R + R_s(1 + b\,\Delta T)] \tag{13.10}$$

from which we find

$$\frac{R_s}{R} = -\frac{c}{b + c} \tag{13.11}$$

This equation indicates that temperature compensation may possibly be accomplished through proper balancing of the temperature coefficients of Young's modulus, c, and electrical resistivity, b. Because c is usually negative (see Table 6.3), and because the resistances cannot be negative, it follows that

$$b > -c$$

In addition, we may write [see Eq. (6.2)]

$$R_s = \rho\frac{L}{A} = -R\left(\frac{c}{b + c}\right) \tag{13.12}$$

from which

$$L = -\frac{RA}{\rho}\left(\frac{c}{b + c}\right) \tag{13.12a}$$

From these relations, specific requirements for compensation may be derived. After a resistance material, generally in the form of wire, is selected, the required length may be determined through use of Eq. (13.12a).

Although a single resistor would serve, commercial cells normally use two *modulus resistors*, as shown in Fig. 13.13. This technique ensures proper connections regardless of instrumentation and also permits electrical calibration of the gages by shunt resistances as described in Section 12.11. It is necessary, however, to use two calibration resistors, as shown in Fig. 13.14. If each resistor is considered as one-half the total calibration resistance, then the relation given, Eq. (12.15), will remain legitimate.

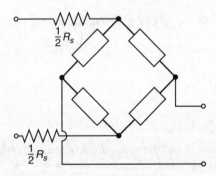

FIGURE 13.13: Strain-gage bridge with two compensating resistors.

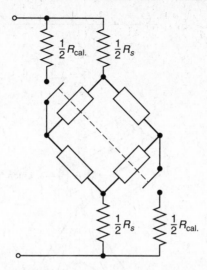

FIGURE 13.14: Schematic diagram of a strain-gage bridge showing how calibration may be accomplished.

13.6 PIEZOELECTRIC LOAD CELLS

Load cells that employ piezoelectric secondary transducers are particularly useful for measuring dynamic loading, especially of an impact or abruptly applied nature. The transducer produces an electrostatic charge (Section 6.14), which is generally conditioned through use of a charge amplifier (Section 7.15.2). Transducer outputs are in terms of coulombs per unit input, with 10 to 20 pC/lbf (2.3 to 4.5 pC/N) being typical. Desirable qualities include wide ranges of working load in a given unit, excellent frequency response, great stiffness, high resolution, and relatively small size. An important limitation is that piezoelectric devices are inherently of a dynamic, rather than a static, nature. Long-term static output stability is not generally practical.

Multiaxis cells are available. When a quartz master crystal is sliced to produce transducer elements, the selection of slicing planes yields elements with different properties. Slices may be taken to produce elements selectively sensitive to tension-compression, shear, or bending (see Section 6.14). By taking advantage of these characteristics, load cells may be designed that provide various combinations of orthogonal load and/or torque outputs.

13.7 BALLISTIC WEIGHING

Figure 13.15 represents a very simplified example of what is termed *ballistic weighing*. Such systems are particularly adaptable to certain production applications [8].

Theoretically, if a mass is suddenly applied to a resisting member having a linear load-deflection characteristic, the dynamic deflection will be exactly twice the final static deflection. This is true so long as damping is absent. This fact may be used as the basis for a weighing system. The basic equation for a system of this type is

$$\frac{m}{g_c}\frac{d^2 y}{dt^2} + Ky = \frac{mg}{g_c} \tag{13.13}$$

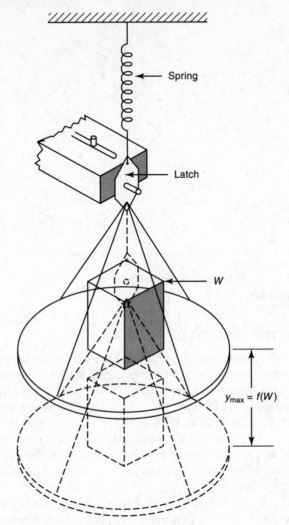

FIGURE 13.15: A ballistic weighing system.

where

m = mass,

K = the deflection constant,

g_c = the dimensional constant,

g = local acceleration due to gravity,

y = deflection,

t = time

A solution is

$$y = \frac{mg}{g_c K}(1 - \cos \omega_n t) \tag{13.13a}$$

for which the maximum value is

$$y_o = \frac{2mg}{g_c K} = 2y_{\text{static}} \qquad (13.14)$$

when $t = \pi\omega_n$ for $\omega_n =$ the undamped natural frequency. The period of oscillation will be

$$T = 2\pi\sqrt{\frac{m}{g_c K}} \qquad (13.15)$$

In operation, the platform is locked (Fig. 13.15) with the spring unstretched; then the weight to be measured is put in place, the system is unlocked, and the maximum excursion is measured. If damping is minimized, the maximum displacement will be linearly proportional to the weight and can be used to measure the weight. Of course, the system is useful only for mass measurement and cannot be used to measure force.

13.8   HYDRAULIC AND PNEUMATIC SYSTEMS

If a force is applied to one side of a piston or diaphragm, and a pressure, either hydraulic or pneumatic, is applied to the other side, some particular value of pressure will be necessary to exactly balance the force. Hydraulic and pneumatic load cells are based on this principle.

For hydraulic systems, conventional piston and cylinder arrangements may be used. However, the friction between piston and cylinder wall and required packings and seals is unpredictable, and thus good accuracy is difficult to obtain.[3] Use of a *floating* piston with a diaphragm-type seal practically eliminates this variable.

Figure 13.16 shows a hydraulic cell in section. This cell is similar to the type used in some materials-testing machines. The piston does not actually contact a cylinder wall in the normal sense, but a thin elastic diaphragm, or bridge ring, of steel is used as the positive seal, which allows small piston movement. Mechanical stops prevent the seal from being overstrained.

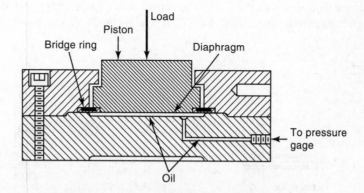

FIGURE 13.16: Section through a hydraulic load cell.

---

[3] An exception to this statement applies to the so-called dead-weight tester (Fig. 14.3). The prescribed procedure is to make certain that the weights, hence piston, are rotating in the cylinder as readings are taken. This practice essentially eliminates seal friction.

When force acts on the piston, the resulting oil pressure is transmitted to some form of pressure-sensing system such as the simple Bourdon gage. If the system is completely filled with fluid, very small transfer or flow will be required. Piston movement may be less than 0.002 in. (50 μm) at full capacity. In this respect, at least, the system will have good dynamic response; however, overall response will be determined very largely by the response of the pressure-sensing element.

Very high capacities and accuracies are possible with cells of this type. Capacities to 5,000,000 lbf (22.2 MN) and accuracies on the order of $\pm\frac{1}{2}\%$ of reading or $\pm\frac{1}{10}\%$ of capacity, whichever is greater, have been attained. Since hydraulic cells are somewhat sensitive to temperature change, provision should be made for adjusting the zero setting. Temperature changes during the measuring process cause errors of about $\frac{1}{4}\%$ per 10°F change (0.45% per 10 K change).

Pneumatic load cells are similar to hydraulic cells in that the applied load is balanced by a pressure acting over a resisting area, with the pressure becoming a measure of the applied load. However, in addition to using air rather than liquid as the pressurized medium, these cells differ from the hydraulic ones in several other important respects.

Pneumatic load cells commonly use diaphragms of a flexible material rather than pistons, and they are designed to regulate the balancing pressure automatically. A typical arrangement is shown in Fig. 13.17. Air pressure is supplied to one side of the diaphragm and allowed to escape through a position-controlling *bleed* valve. The pressure under the diaphragm, therefore, is controlled both by source pressure and bleed valve position. The diaphragm seeks the position that will result in just the proper air pressure to support the load, assuming that the supply pressure is great enough so that its value multiplied by the effective area will at least support the load.

We see that as the load changes magnitude, the measuring diaphragm must change its position slightly. Unless care is used in the design, a nonlinearity may result, the cause of which may be made clear by referring to Fig. 13.18(a). As the diaphragm moves, the portion between the load plate and the fixed housing will alter position as shown. If it is assumed that the diaphragm is of a perfectly flexible material, incapable of transmitting any but tensile forces, then the division of vertical load components transferred to housing and load plate will occur at points A or A', depending on diaphragm position. We see then that

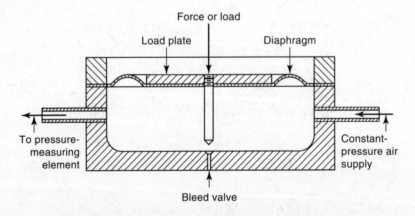

FIGURE 13.17: Section through a pneumatic load cell.

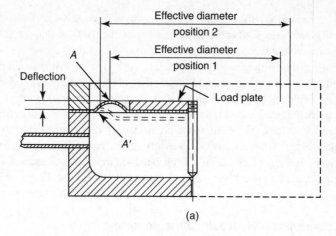

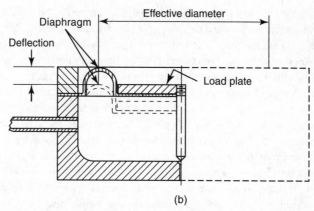

FIGURE 13.18: (a) A section through a diaphragm showing how a change in effective area may take place. (b) When sufficient "roll" is provided, the effective area remains constant.

the effective area will change, depending on the geometry of this portion of the diaphragm. If a complete semicircular roll is provided, as shown in Fig. 13.18(b), this effect will be minimized.

Since simple pneumatic cells may tend to be dynamically unstable, most commercial types provide some form of viscous damper to minimize this tendency. Also, additional chambers and diaphragms may be added to provide for tare adjustment.

Single-unit capacities to 80,000 lbf (356 kN) may be obtained, and by use of parallel units practically any total load or force may be measured. Errors as small as 0.1% of full scale may be expected.

13.9 TORQUE MEASUREMENT

Torque measurement is often associated with determination of mechanical power, either power required to operate a machine or power developed by the machine. In this connection, torque-measuring devices are commonly referred to as *dynamometers*. When so applied, both torque and angular speed must be determined. Another important reason for measuring torque is to obtain load information necessary for stress or deflection analysis.

There are three basic types of torque-measuring apparatus—namely, absorption, driving, and transmission dynamometers. *Absorption dynamometers* dissipate mechanical energy as torque is measured; hence they are particularly useful for measuring power or torque developed by power sources such as engines or electric motors. *Driving dynamometers*, as their name indicates, both measure torque or power and also supply energy to operate the tested devices. They are, therefore, useful in determining performance characteristics of such things as pumps and compressors. *Transmission dynamometers* may be thought of as passive devices placed at an appropriate location within a machine or between machines, simply for the purpose of sensing the torque at that location. They neither add to nor subtract from the transmitted energy or power and are sometimes referred to as *torque meters*.

13.9.1 Mechanical and Hydraulic Dynamometers

Probably the simplest type of absorption dynamometer is the familiar *Prony brake*, which is strictly a mechanical device depending on dry friction for converting the mechanical energy into heat. There are may different forms, two of which are shown in Fig. 13.19.

Another form of dynamometer operating on similar principles is the *water brake*, which uses fluid friction rather than dry friction for dissipating the input energy. Figure 13.20 shows this type of dynamometer in its simplest form. Capacity is a function of two factors, speed and water level. Power absorption is approximately a function of the *cube* of the speed, and the absorption at a given speed may be controlled by adjustment of the water level in the housing. This type of dynamometer may be made in considerably larger capacities than the simple Prony brake because the heat generated may be easily removed by circulating the water into and out of the casing. Trunnion bearings support the dynamometer housing, allowing it freedom to rotate except for restraint imposed by a reaction arm.

In each of the foregoing devices the power-absorbing element tends to rotate with the input shaft of the driving machine. In the case of the Prony brake, the absorbing element is the complete brake assembly, whereas for the water brake it is the housing. In each case such rotation is constrained by a force-measuring device such as some form of scales or load cell, placed at the end of a reaction arm of radius r. By measuring the force at the known radius, the torque T may be computed by the simple relation

$$T = Fr \tag{13.16}$$

If the angular speed of the driver is known, power may be determined from the relation

$$P = 2\pi T\Omega \tag{13.17}$$

where

T = torque,
F = the force measured at radius r,
P = power, and
Ω = angular speed in revolutions per second

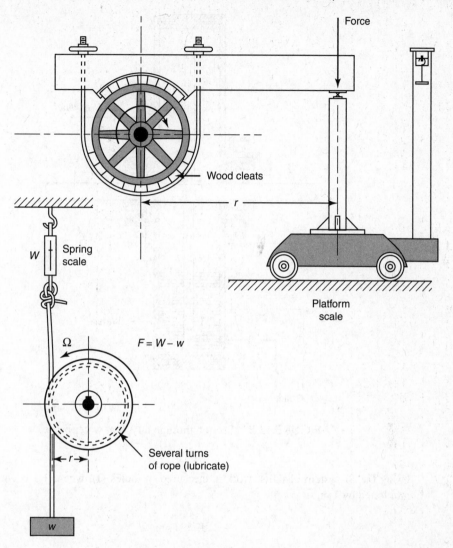

FIGURE 13.19: Two forms of the Prony brake.

At this point it may be wise to carefully consider the units to be used in the above relation-ships. We may rewrite Eq. (13.17) as follows:

$$P = F(2\pi r \Omega) = \text{force} \times \text{distance/time}$$
$$= \text{work/unit time}$$

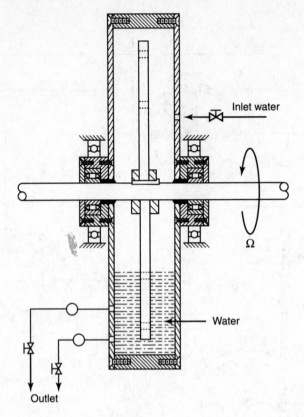

FIGURE 13.20: Section through a typical water brake.

Using the SI system of units, work is measured in joules (J), where 1 J is equal to 1 N multiplied by 1 m, or

$$J = N \cdot m$$

Mechanical power then becomes $N \cdot m/s = J/s = watts$ (W). Checking the units in Eq. (13.17) yields watts. Using the English system of units we find power as determined from Eq. (13.17) to yield units of lbf $\cdot$ ft/s. The English system often goes an additional step by assigning the term *horsepower* (hp) to 550 lbf $\cdot$ ft/s, or

$$hp = \frac{2\pi T \Omega}{550} \qquad (13.18)$$

Conversion from watts to horsepower may be made using the relation

$$W = 745.7 \times hp \qquad (13.19)$$

EXAMPLE 13.1

Calculate the power if $F = 120$ N (or 26.98 lbf), $r = 75$ cm (or 2.46 ft), and $\Omega = 20$ rev/s.

Solution   Using SI units, we have
$$P = 2\pi \cdot 120 \cdot 0.75 \cdot 20 = 11{,}310 \text{ W}$$
Using English units, we have

$$P = 2\pi \cdot 26.98 \cdot 2.46 \cdot 20 = 8340 \text{ lbf} \cdot \text{ft/s}$$
$$= 15.16 \text{ hp}$$

A check on equivalence yields

$$\frac{11{,}310}{15.16} = 746 \text{ W/hp}$$

13.9.2   Electric Dynamometers

Almost any form of rotating electric machine can be used as a driving dynamometer, or as an absorption dynamometer, or as both. As expected, those designed especially for the purpose are most convenient to use. Four possibilities are (1) eddy-current dynamometers; (2) cradled dc dynamometers; (3) dc motors and generators; and (4) ac motors and generators.

Eddy-current dynamometers are strictly of the absorption type. They are incapable of driving a test machine such as a pump or compressor, hence they are only useful for measuring the power from a source such as an internal combustion engine or electric motor.

The eddy-current dynamometer is based on the following principles. When a conducting material moves through a magnetic flux field, voltage is generated, which causes current to flow. If the conductor is a wire forming a part of a complete circuit, current will be caused to flow through that circuit, and with some form of commutating device a form of ac or dc generator may be the result. If the conductor is simply an isolated piece of material, such as a short bar of metal, and not a part of a complete circuit as generally recognized, voltages will still be induced. However, only local currents may flow in practically short-circuit paths within the bar itself. These currents, called eddy currents, become dissipated in the form of heat.

An eddy-current dynamometer consists of a metal disk or wheel that is rotated in the flux of a magnetic field. The field is produced by field elements or coils excited by an external source and attached to the dynamometer housing, which is mounted in trunnion bearings. As the disk turns, eddy currents are generated, and the reaction with the magnetic field tends to rotate the complete housing in the trunnion bearings. Torque is measured in the same manner as for the water brake, and Eqs. (13.16), (13.17), and (13.18) are applicable. Load is controlled by adjusting the field current. As with the water brake, the mechanical energy is converted to heat energy, presenting the problem of satisfactory dissipation. Most eddy-current dynamometers must use water cooling. Particular advantages of this type are the comparatively *small size* for a given capacity and characteristics permitting *good control at low rotating speeds*.

Undoubtedly the most versatile of all types is the *cradled dc dynamometer*, shown in Fig. 13.21. This type of machine is usable both as an absorption and as a driving dynamometer in capacities to 5000 hp (3730 kW). Basically the device is a dc motor

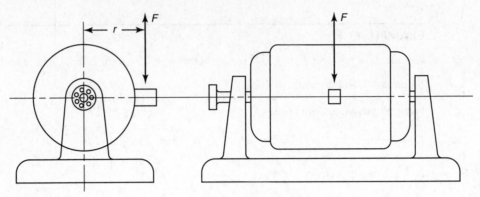

FIGURE 13.21: The general-purpose electric dynamometer.

generator with suitable controls to permit operation in either mode. When used as an absorption dynamometer, it performs as a dc generator and the input mechanical energy is converted to electrical energy, which is dissipated in resistance racks. This latter feature is important, for unlike the eddy-current dynamometer, the heat is dissipated outside of the machine. Cradling in trunnion bearings permits the determination of reaction torque and the direct application of Eqs. (13.16), (13.17), and (13.18).

Provision is made for measuring torque in either direction, depending on the direction of rotation and mode of operation. As a driving dynamometer, the device is used as a dc motor, which presents a problem in certain instances of obtaining an adequate source of dc power for this purpose. Use of either an ac motor-driven dc generator set or a rectified source is required. *Ease of control* and *good performance at low speeds* are features of this type of machine.

Ordinary *electric motors* or *generators* may be adapted for use in dynamometry. This is more feasible when do rather than ac machinery is used. Cradling the motor or generator may be used for either driving or absorbing applications, respectively. By measuring torque reaction and speed, power may be computed. This, of course, requires special effort in designing and fabricating a minimum-friction arrangement. Adjustment of driving speed or absorption load could be provided through control of field current. Load-cell mounting may be used.

Knowledge of motor or generator characteristics versus speed presents another appr-oach. If a dc generator is used as an absorption dynamometer, then

$$\text{Power (absorbed)} = \frac{(e)(i)}{\text{Efficiency}} \tag{13.20}$$

where

$$e = \text{the output voltage, in volts,}$$
$$i = \text{the output current, in amperes, and}$$
$$\text{Efficiency} = \text{the efficiency of the generator}$$

In like manner, Eq. (13.20) holds if a dc motor is used as a *driving* dynamometer, except that e and i are *input* voltage and current, respectively. Both e and i may be measured separately, or a wattmeter may be used and the electric power measured directly.

In many applications, only approximate results may be required, in which case *typical* motor or generator efficiencies supplied by the manufacturer should suffice. For more accurate results, some form of dynamometer would be required to determine the efficiencies for the particular machine to be used. The use of ac motors or generators, while feasible, is considerably more difficult and will not be discussed here. In any case, application of *general-purpose* electrical rotating machinery to dynamometry must be considered special and will not yield as satisfactory results as equipment particularly designed for the purpose.

13.10   TRANSMISSION DYNAMOMETERS

As mentioned earlier, transmission dynamometers may be thought of as passive devices neither appreciably adding to nor subtracting from the energy involved in the test system. Various devices have been used for this purpose, including gear-train arrangements and belt or chain devices.

Any gear box producing a speed change is subjected to a reaction torque equal to the difference between the input and output torques. When the reaction torque of a cradled gear box is measured, a function of either input or output torque may be obtained.

Belt or chain arrangements, in which reaction is a function of the difference between the tight and loose tensions, may also be used. Torque at either main pulley is also a function of the difference between the tight and loose tensions; hence the measured reaction may be calibrated in terms of torque, from which, with speed information, power may be determined. Mechanical losses introduced by arrangements of these types, combined with general awkwardness and cost, make them rather unsatisfactory except for an occasional special application.

More common forms of transmission dynamometers are based on calibrated measurement of unit or total strains in elastic load-carrying members. A popular dynamometer of the elastic type uses bonded strain gages applied to a section of torque-transmitting shaft [9, 10], as shown in Table 12.4. Such a dynamometer, often referred to as a *torque meter*, is used as a coupling between driving and driven machines or between any two portions of a machine. A complete four-arm bridge is used, incorporating modulus gages to minimize temperature sensitivity (Section 13.5). Electrical connections are made through slip rings, with means provided to lift the brushes when they are not in use, thereby minimizing wear. Any of the common strain-gage indicators or recorders are usable to interpret the output. Dynamometers of this type are commercially available in capacities of 100 to 30,000 in. · lbf (12 to 3500 N · m). Accuracies to $\frac{1}{4}$% are claimed.

In most cases resistance strain-gage transducers are most sensitive when bending strains can be used. Figure 13.22 suggests methods whereby torsion may be converted to bending for measurement.

Slip rings are subject to wear and may present annoying maintenance problems when permanent installations are required. For this reason many attempts have been made to devise electrical torque meters that do not require direct electrical connection to the moving shaft. Inductive [11, 12] and capacitive [13] transducers have been used to accomplish this (see Fig. 6.13).

In addition to temperature sensitivity resulting from variation in elastic constants, further variation may be caused in the inductive type by change in magnetic constants with temperature. This may be compensated for by resistors in a manner similar to that used for strain-gage load cells (Section 13.5).

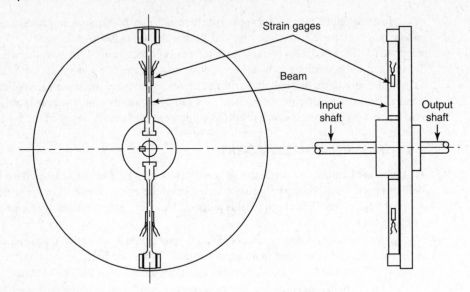

FIGURE 13.22: Transmission dynamometer that employs beams and strain gages for sensing torque.

These types are relatively expensive and cannot be considered general-purpose instruments. However, in permanent installations they provide the advantage of long service without maintenance problems.

SUGGESTED READINGS

ASME PTC 19.7-1980. *Measurement of Shaft Power*. New York: ASME, 1980.

ASME PTC 19.5.1-1964. *Weighing Scales*. New York, 1964.

Butcher, T., L. Crown, R. Suiter, and J. Williams (eds.). *NIST Handbook 44: Specifications, Tolerances, and Other Technical Requirements for Weighing and Measuring Devices*. Gaithersburg, Md.: National Institute of Standards and Technology, 2005.

PROBLEMS

13.1. There are various sources of data on local gravitational accelerations, such as geological surveys, university physics departments, research organizations, and oil, gas or mining companies. Determine the value of gravitational acceleration for your particular locality.

13.2. Consider a simple balance-beam-type scale (Fig. 13.4). Does the scale compare "weights" or does it compare "masses"? Is the scale sensitive to local gravity? Is it as functional on a mountain top as it is at sea level? Would the scale perform its function in gravity-free space?

13.3. A mass of volume V and unit density d is weighed on a sensitive scale. Write a short summary of the problem that may be presented by air buoyancy as it affects measurement accuracy. Include consideration of the type of scale that is used.

13.4. What will 1 kg of water weigh (a) in Ft. Egbert, Alaska? (b) in Key West, Florida? (c) on the moon? (See Appendix D for gravitational data.)

13.5. What will 1 lbm of water weigh in each of the locations listed in Problem 13.4?

13.6. Very often spring scales of the type shown in Fig. 6.24 carry divisions marked in kilograms. Is this practice fundamentally correct? Basically, what does such a device measure when a mass is suspended from it? What is the relationship between mass and weight?

13.7. Assume that a spring of the type referred to in Problem 13.6 is properly calibrated to measure force in newtons. If, on the surface of the moon (gravitational acceleration = 1.67 m/s), a reading of 50 N is obtained when an item is suspended from the scale, what weight would be indicated if the measurement were made under standard conditions on the surface of the earth? What would be the mass of the item in kilograms?

13.8. Assign tolerances to the values given (or determined) in Problems 6.11 and 12.7 and calculate an overall uncertainty to apply to the readout from the beam. (*Note:* Any "electronics" used to evaluate the strain-gage output will also contain uncertainties. Make an estimate for this and include it in the final calculation.) A spreadsheet solution is suggested.

13.9. Referring to Fig. 13.5, show that the scale reading is independent of the location of W on the platform.

13.10. Prepare a spreadsheet template for designing a cantilever-beam-type load cell (see Case C, Table 13.1). Assume a beam with a rectangular cross section.

13.11. Using the template prepared for Problem 13.10, determine the deflection constant for a steel beam 6 in. long, $\frac{3}{8}$ in. wide, and $\frac{1}{8}$ in. thick. Investigate the effect of tolerances on each dimension and on the modulus of elasticity. Assign tolerances on each variable with the aim to control the value of the deflection constant to $\pm 3\%$.

13.12. Referring to Fig. 13.10, show that small transverse and/or angular misalignments of the load relative to the centerline of the cell will not affect the readout.

13.13. A proving-ring-type force transducer is a very reliable device for checking the calibration of material-testing machines. An equation for estimating the deflection constant of the elemental ring, loaded in compression, is given in Table 13.1. If $D = 10 \pm 0.010$ in. (254 ± 0.25 mm), t = the radial thickness of the section= 0.6 ± 0.005 in. (15.24 ± 0.127 mm), w = the axial width of the section = 2 ± 0.015 in. (50.8 ± 0.381 mm), and $E = 30 \times 10^6 \pm 0.5 \times 10^6$ lbf/in.^{2} ($20.68 \times 10^{10} \pm 0.34 \times 10^{10}$ N/m^2), calculate the value of K and its uncertainty, using English units.

13.14. Solve Problem 13.13 using SI units.

13.15. Figure 13.7 shows a scheme for adjusting the calibration of an elastic force measuring system. Prepare a spreadsheet template for the purpose of designing a two-element coil spring arrangement of the type illustrated.

13.16. Using the template devised for Problem 13.15, design a two-element coil spring system to meet the following specifications:

$$K = 100 \text{ lbf/in.} \pm 0.2\%$$

Note that the solution will consist of a primary spring, along with several selectable vernier springs. Each vernier spring should be designed to adjust for a range of primary spring tolerances. The smaller the number of verniers required, the better.

13.17. Review Problems 6.11 and 12.7. Using data from these two problems and from Tables 6.3 and 6.4, select a resistance material and determine the value for a series resistor to provide compensation for temperature-derived variations in Young's modulus for the beam.

13.18. Determine the bridge constant for the arrangement shown in Fig. 13.11(b).

13.19. A torque meter is incorporated in the coupling between an electric motor and a dc generator. If the effect of a small 60-Hz torque component is to be limited to no more than 3% of the readout, what are the limiting natural frequencies for the two mass system? Damping is negligible. (Refer to Section 5.16.2.)

REFERENCES

[1] *Specifications and Tolerances for Field Standard Weights*. National Institute of Standards and Technology Handbook 105-1. Washington, D.C.: U.S. Government Printing Office, 1990.

[2] *Standard Specification for Laboratory Weights And Precision Mass Standards*, ASTM E617-97 (2003). W. Conshohocken, Pa.: ASTM International, 2003.

[3] Weighing machines, *Encyclopedia Britannica*, Encyclopedia Britannica. Inc. Chicago, Ill.: William Benton Publisher, 23:483, 1957.

[4] *Instruments*, 25:1300, Sept. 1952.

[5] Wilson, B. L., D. R. Tate, and G. Borkowski. Proving rings for calibrating testing machines. *NBS Circular C454*. Washington, D.C.: U.S. Government Printing Office, 1946.

[6] Timoshenko, S. *Strength of Materials*, Part II, 2nd ed. New York: Van Nostrand, 1941, p. 88.

[7] High capacity load calibrating devices. *NBS Tech. News Bull*. 37: Sept. 1953.

[8] Bell, R. E., and J. A. Fertle. Electronic weighing on the production line. *Electronics*, 28(6):152.

[9] Ruge, A. C. The bonded wire torquemeter. *SESA Trans*. 1(2):68, 1943.

[10] Rebeske, J. J., Jr. *Investigation of a NACA High-Speed Strain-Gage Torquemeter*, NACA Tech. Note 2003. Jan., 1950.

[11] Langer, B. F. Measurement of torque transmitted by rotating shafts. *J. Appl. Mech*. 67:A.39, March 1945.

[12] Langer, B. F., and K. L. Wommack. The magnetic-coupled torquemeter. *SESA Trans*. 2(2):11, 1944.

[13] Hetenyi, M. *Handbook of Experimental Stress Analysis*. New York: John Wiley, 1950, Chaps. 6 and 7.

C H A P T E R   14

Measurement of Pressure

14.1 INTRODUCTION

Pressure is the normal force exerted by a medium, usually a fluid, on a unit area. In engineering, pressure is most often expressed in pascal (1 Pa = 1 N/m^2) or pounds-force per square inch (lbf/in.^{2}, or psi). Typically, pressure is detected as a differential quantity, that is, as the difference between an unknown pressure and a known reference pressure. Atmospheric pressure is the most common reference, and the resulting pressure difference, known as *gage pressure*, is of obvious importance in determining net loads on pressure vessel and pipe walls. In other cases, the reference pressure is taken to be zero (a complete absence of pressure), and the pressure measured is called *absolute*. In the English system of units, gage and absolute pressure are distinguished by writing psig and psia, respectively. Figure 14.1 illustrates these relationships.

Pressure is often expressed in units of hydrostatic force per unit area at the base of a column of liquid, usually mercury or water. For example, standard atmospheric pressure (101,325 Pa or 14.696 psia) is approximately equal to the pressure exerted at the bottom of a mercury column 760 mm (or 29.92 in.) in height.[1] Therefore, one often finds standard atmospheric pressure specified as 760 mmHg or 29.92 inHg, even though the fundamental unit of pressure is neither millimeters nor inches. Pressure measurement using liquid columns is called *manometry* (see Section 14.4).

An absolute pressure less than atmospheric pressure is often referred to as a *vacuum*. Vacuum is occasionally measured in terms of a negative gage pressure (so that −7 psig = 7 psi vacuum). When the vacuum is nearly complete, however, small variations in

---

[1]The pressure is equal to $\rho g h$, for ρ the liquid density, g the local value of the gravitational body force, and h the column height. Since liquid density varies with temperature, the precise height for 1 atm pressure depends on both local gravity and ambient temperature.

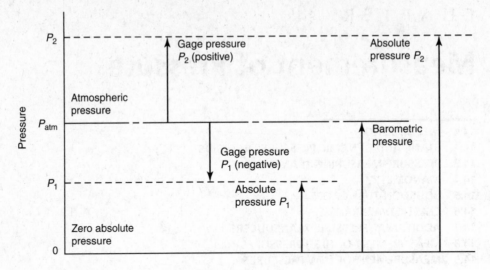

FIGURE 14.1: Relations among absolute, gage, and barometric pressures.

TABLE 14.1: Relation of Various Units of Pressure to the Pascal [1]. H_2O at 4°C; Hg at 0°C.

1 microbar = 0.1 Pa	1 inH_2O = 249.1 Pa
1 μmHg = 0.1333 Pa	1 kPa = 1000 Pa
1 N/m^2 = 1 Pa	1 ftH_2O = 2989 Pa
1 mmH_2O = 9.807 Pa	1 inHg = 3386 Pa
1 mbar = 100 Pa	1 psi = 6895 Pa
1 mmHg = 133.3 Pa	1 bar = 10^5 Pa
1 torr = 133.3 Pa	1 atm = 101325 Pa

atmospheric pressure can produce large errors in the measured gage pressure. Hence, absolute pressure is always used to describe a high vacuum. The low absolute pressures of a high vacuum are sometimes expressed in units of torr (1 torr = 1 mmHg) or micrometers of mercury (μmHg).

High pressures are often written in units of atmospheres (1 atm = 1.01325×10^5 Pa), bar (1 bar = 10^5 Pa), or megapascal (1 MPa = 10^6 Pa). Selected units of pressure measurement are summarized in Table 14.1.

14.2 STATIC AND DYNAMIC PRESSURES IN FLUIDS

When a fluid is at rest, a small pressure sensor in it will read the same *static pressure* at a given position in the fluid no matter how it is oriented. In other words, at any particular point in the fluid, the small surface experiences the same pressure whether it faces upward or downward or left or right. Gravitational force can produce a vertical pressure gradient, causing a higher pressure at lower levels in the fluid, but at any particular level the pressure on the small surface remains independent of its orientation.

When the fluid is in motion, a surface placed in it may experience not only the static

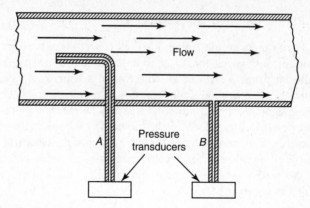

FIGURE 14.2: Impact tube, *A*, and static-pressure tube, *B*. Tube *A* senses the total or stagnation pressure.

pressure, but also a *dynamic pressure*. For example, if the surface is perpendicular to the direction of flow, the fluid must come to rest at the surface. This stagnation of the flow results in the conversion of kinetic energy into an additional pressure on the surface, much like the pressure you feel when standing in the wind. On the other hand, if the surface is parallel to the flow direction, the fluid is not stagnated and flows along the surface without creating any additional pressure. Thus, a pressure transducer's reading in a moving fluid will depend on its orientation.

In Fig. 14.2, two small tubes each sample the pressure in an air duct. Pressure tap *B* senses only the static pressure in the duct. Tube *A*, on the other hand, is aligned so that the flow impacts against its opening, and it senses the *total* or *stagnation pressure*. The static pressure is identical to the pressure one would sense if moving along with the airstream. The stagnation pressure can be defined as that which would be obtained if the stream were brought to rest isentropically. The difference between the stagnation and static pressures results from the motion of the fluid and is called the *velocity pressure* or *dynamic pressure*:

$$\text{Dynamic pressure} = \text{stagnation pressure} - \text{static pressure}$$

As discussed in Section 15.8, this pressure difference can even be exploited for measurement of the fluid's velocity.

We see, therefore, that to obtain and interpret pressure measurements properly, flow conditions must be taken into account. Conversely, to interpret flow measurements properly, the pressure conditions must be considered.

Sound Pressure

Sound waves propagate in an elastic medium as longitudinal pressure variations (along the path of propagation), with pressure fluctuating above and below the static pressure. The instantaneous difference between the pressure at any point and the time-average pressure there is called the *sound pressure*. Because sound pressures are normally relatively small, they are often expressed in units of microbar (1 μbar $= 10^{-1}$ Pa). Measurement of sound pressure is accomplished with microphones and related apparatus, as discussed in Chapter 18.

14.3 PRESSURE-MEASURING TRANSDUCERS

Pressure measurement most often involves converting a pressure difference into a force and then measuring that force. In some cases, the force may be measured directly by comparing it to the weight of an object or of a column of liquid. In other cases, the force may be used to produce a deflection in an elastic member, such as a curved tube or a diaphragm. This deflection, in turn, may be measured either mechanically or by a secondary electrical transducer, such as a strain gage or inductive or capacitive sensor. In other devices, a pressure-induced strain may change the electrical properties of the elastic member itself, as when a piezoelectric material generates a charge in response to a load. Many other methods of pressure measurement have been devised, particularly in connection with vacuum systems.

Pressure, as force per unit area, must ultimately be related back to the standards of length, mass, and time which define force and area. There is no separate standard for pressure. The most common way to connect pressure to the standards is to use a calibrated mass subjected to a known value of gravity to create a force that is supported by the unknown pressure as applied to a carefully measured area. For example, the piston force balance or dead-weight tester (Fig. 14.3) produces a constant pressure that may be used to calibrate other pressure gages. Masses sitting atop a piston are supported by the pressure of a fluid below. If the piston's area is known, the pressure is calculated easily using the known masses and the local value of gravity. These devices are commonly used by standards laboratories for high-accuracy calibration of other pressure sensors. When properly applied, they are accurate to better than 100 parts per million [2]. Similarly high accuracy can be obtained by balancing a pressure force against the hydrostatic pressure at the base of a column of liquid whose height is known; this is called *manometry*.

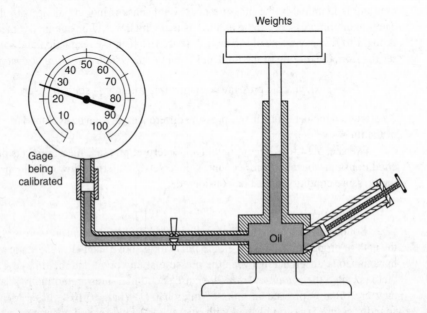

FIGURE 14.3: Dead-weight tester.

When then pressure to be measured varies rapidly in time (a *dynamic* pressure measurement), the transducer used must have sufficiently high frequency response to track the signal. In general, only electromechanical transducers are adequate for dynamic pressure measurements, and then only if the associated elastic members are light enough to respond rapidly to changing conditions. In addition to the response of transducer itself, we must take account of the responsiveness of any connecting tubing or chambers between the transducer and the point at which pressure is to be measured.

As an example, suppose that a diaphragm-type transducer were to be used for measuring the pressure at a specific point on an aircraft skin. In such an application, it may be undesirable to place the diaphragm flush with the aircraft surface. Possibly the size of the diaphragm is too great in comparison with the pressure gradients existing; or perhaps flush mounting would disturb the surface to too great a degree; or it may be necessary to mount the pickup internally to protect it from large temperature variations. In such cases, the pressure would be conducted to the sensing element of the pickup through a passageway, and a small space or cavity would exist over the diaphragm. The passageway and cavity become, in essence, an integral part of the transducer, and the mass, elasticity, and damping properties of the passage and chamber contribute to the overall response of the system. It is obvious that it would be insufficient to consider only the transducer characteristics in assessing the frequency response. This issue is discussed further in Section 14.10.

International conferences on pressure measurement are held periodically by the national standards laboratories of several dozen nations. The proceedings of those conferences should be consulted for detailed information on pressure measurement at the highest levels of accuracy [3, 4].

14.4 MANOMETRY

Manometry refers to the measurement of pressure by comparison to the hydrostatic pressure produced by a column of liquid. The manometer is one of the most elementary measuring devices imaginable. It is simple, inexpensive, and relatively free from error, and yet it may be arranged to almost any degree of sensitivity. Its major disadvantages lie in its pressure ranges and in its poor dynamic response. It is not very practical for measuring pressures greater than, say, 200 kPa (30 psi), and it is incapable of following any but slowly changing pressures.

A simple well-type manometer is shown in Fig. 14.4. A force-equilibrium expression for the net liquid column is

$$(P_{1a}A - P_{2a}A) = Ah\rho \left(\frac{g}{g_c}\right) \tag{14.1}$$

or

$$(P_{1a} - P_{2a}) = P_d = h\rho \left(\frac{g}{g_c}\right) \tag{14.1a}$$

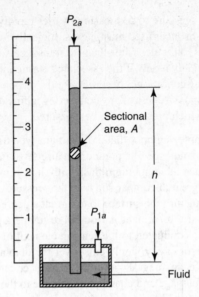

FIGURE 14.4: Well-type manometer.

where

P_{1a} and P_{2a} = the applied absolute pressures,

P_d = the pressure difference or differential pressure,

ρ = the density of the fluid (mass/volume),

h = the net column height, or "head,"

g = the gravitational body force, and

g_c = the dimensional constant (see Table 2.6)

In practice, pressure P_{2a} is commonly atmospheric and

$$(P_{1a} - P_{\text{atm}}) = P_{1g} = h\rho \left(\frac{g}{g_c} \right) \tag{14.2}$$

where

P_{1g} = the gage pressure at point 1

Perhaps it would be wise at this point to make sure we understand the units to be used. In simplified form the preceding equations may be written as

$$P_d = h\rho \left(\frac{g}{g_c} \right) \tag{14.2a}$$

Substituting units in the right-hand side of the equation, we have, for the SI System,

$$(\text{m})(\text{kg/m}^3)(\text{m/s}^2)(\text{N} \cdot \text{s}^2/\text{kg} \cdot \text{m}) = \text{N/m}^2 = \text{Pa}$$

Using the English system of units, we have

$$(ft)(lbm/ft^3)(ft/s^2)(lbf \cdot s^2/lbm \cdot ft) = lbf/ft^2$$

EXAMPLE 14.1

Calculate the pressure at the base of a column of water 1 m (3.281 ft) in height if the local gravity acceleration is 9.75 m/s^2(31.99 ft/s^2) and the temperature is 20°C (68°F).

Solution   From Table D.1 in Appendix D, we find that the density of water at 20°C is 998.2 kg/m^3 (62.32 lbm/ft^3). Using SI units, we have

$$P_{SI} = (1)(998.2)(9.75/1) = 9732 \, Pa \qquad (or \, N/m^2)$$

Using English units, we have

$$P_{Eng} = (3.281)(62.32)(31.99/32.17) = 203.3 \, lbf/ft^2 = 1.412 \, psi$$

Because the fluid density is involved, accurate work will require consideration of temperature variation of density: The manometer possesses a certain amount of temperature sensitivity.

When the pressure P_{1a} is atmospheric and the absolute pressure P_{2a} is made to be zero (as by sealing and evacuating the top of the tube in Fig. 14.4), we obtain the ordinary barometer. In this case the fluid has traditionally been mercury. In recent years, health regulations have strongly discouraged the use of mercury in many settings, so that the traditional mercury barometer has become relatively uncommon.

Figure 14.5 illustrates the function of the U-tube manometer. Pressures are applied to both legs of the U, and the manometer fluid is displaced until force equilibrium is attained. Pressures P_{1a} and P_{2a} are transmitted to the manometer legs through some fluid of density ρ_t, while the manometer fluid has some greater density ρ_m. In general, we see that

$$P_{1a} - P_{2a} = h(\rho_m - \rho_t)\left(\frac{g}{g_c}\right) \qquad (14.3)$$

In many cases the density difference between the two fluids is great enough that the lesser density may be neglected ($\rho_m \gg \rho_t$)—when air is the transmitting fluid and water is the measuring fluid, for instance. In that case, we have what might be called a *simple* U-tube manometer, and Eq. (14.3) reduces to

$$P_{1a} - P_{2a} = h\rho_m\left(\frac{g}{g_c}\right) \qquad (14.3a)$$

EXAMPLE 14.2

Suppose the manometer fluids in Fig. 14.5 are water and mercury. This situation might occur when a manometer is used to measure the differential pressure across a venturi meter (see Section 15.3) through which water is flowing. We will consider both systems of units used in this book along with the following data:

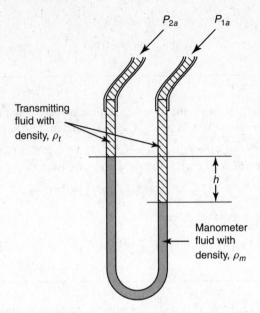

FIGURE 14.5: U-tube manometer.

$h = 10$ in. or $\frac{5}{6}$ ft (0.254 m),

Density of water = 62.38 lbm/ft^3 (999.2 kg/m^3),

Specific gravities of H_2O and Hg = 1 and 13.6, respectively,

Standard gravity applies (32.174 ft/s^2 and 9.80665 m/s^2)

Determine the differential pressure.

Solution In the English system of units,

$$P_{1a} - P_{2a} = \left(\frac{5}{6}\right)(13.6 - 1)(62.38)\left(\frac{32.174}{32.174}\right)$$

$$= 655 \text{ lbf/ft}^2 = 4.55 \text{ psi}$$

For the SI system of units, we have

$$P_{1a} - P_{2a} = (0.254)(13.6 - 1)(999.2)\left(\frac{9.80665}{1}\right)$$

$$= 31,360 \text{ Pa} = 31.4 \text{ kPa}$$

It is left for the reader to show that the two answers represent the same physical quantity and that the unit balance is proper in each case.

In general, a simple U-tube manometer will have a greater pressure range when a more dense measuring fluid is used and a greater sensitivity (change in height per unit change in

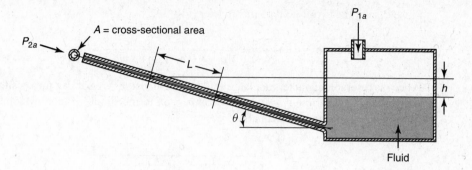

FIGURE 14.6: Inclined-type manometer.

pressure) when a less dense fluid is used. From Eq. (14.3a),

$$\text{Sensitivity} = \frac{h}{\Delta P} = \frac{1}{\rho_m(g/g_c)} \tag{14.3b}$$

Greater sensitivity may also be obtained through a displacement amplification scheme, two of which are shown in Figs. 14.6 and 14.7. For the single inclined leg (Fig. 14.6), in which $h = L \sin \theta$,

$$P_{1a} = \frac{\rho_m(L \sin \theta)g}{g_c} + P_{2a} \tag{14.3c}$$

so that

$$\text{Sensitivity} = \frac{L}{\Delta P} = \frac{1}{\rho_m \sin \theta \, (g/g_c)} \tag{14.3d}$$

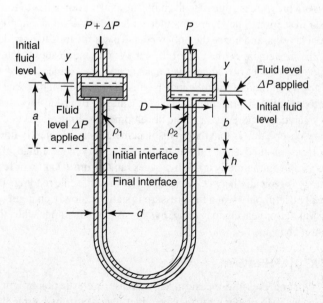

FIGURE 14.7: Two-fluid manometer with reservoirs.

In the case of the two-fluid type manometer (Fig. 14.7),

$$\text{Sensitivity} = \frac{h}{\Delta P} = \frac{1}{[(d/D)^2(\rho_2 + \rho_1) + (\rho_2 - \rho_1)](g/g_c)} \tag{14.4}$$

When the reservoir diameters are large and the fluid densities are similar, the sensitivity can be substantial (a *micromanometer*). In comparison to the simple U-tube manometer, the deflection amplification, M, equals

$$M = \left[\frac{\rho_m}{(d/D)^2(\rho_2 + \rho_1) + (\rho_2 - \rho_1)} \right] \tag{14.4a}$$

where ρ_m = the density of the fluid in the simple manometer and $\rho_2 > \rho_1$. An extensive survey of micromanometers has been given by Brombacher [5].

For high-accuracy measurements, special attention must be given to the measurement of the liquid level in the manometer. The most common means of locating the liquid surface is by direct visual comparison to an adjacent scale. With appropriate lighting and lens arrangements, it is possible to obtain accuracies of several micrometers. For higher resolution, nonvisual techniques are used, including capacitative detection, interferometry, and ultrasound [6].

14.5 BOURDON-TUBE GAGES

Bourdon-tube gages (see Fig. 6.1), like other elastic transducers, operate on the principle that the deflection or deformation accompanying a balance of pressure and elastic forces may be used as a measure of pressure. A tube, normally of oval section, is initially coiled into a circular arc of radius R, as shown in Fig. 14.8. The included angle of the arc is usually less than 360°; however, in some cases, when increased sensitivity is desired, the tube may be formed into a helix of several turns.

As a pressure is applied to the tube, the oval section tends to round out, becoming more circular in section. The inner and outer arc lengths will remain approximately equal to their original lengths, and hence the only recourse is for the tube to uncoil. In the simple pressure gage, the movement of the end of the tube is communicated through linkage and gearing to a pointer whose movement over a scale becomes a measure of pressure (Fig. 6.1). In other forms, the end of the tube may be linked to a position transducer, such as an LVDT, in order to track the displacement. The mechanics of Bourdon-tube action were the subject of many analytical studies prior to the development of finite element computation [7], but the primary observation is that the tip displacement increases nearly linearly with the pressure.

Bourdon-tube gages are available for a very wide range of pressures. Full-scale readings on commercially available gages range from 100 kPa or less to 150 MPa or more. Accuracy is typically between 0.5 and 2% of the full scale reading. Bourdon-tube gages are generally useful only when the pressure is static or slowly changing. Friction and backlash in the linkage or gearing may cause hysteresis in the readings when the direction of pressure variation changes.

14.6 ELASTIC DIAPHRAGMS

Many dynamic pressure-measuring devices use an elastic diaphragm as the primary pressure transducer. Such diaphragms may be either flat or corrugated; the flat type [Fig. 14.9(a)] is

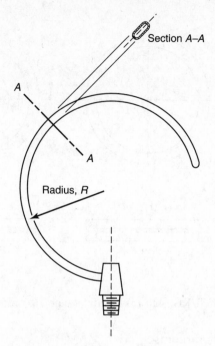

FIGURE 14.8: Basic Bourdon tube.

often used in conjunction with electrical secondary transducers whose sensitivity enables detection of very small diaphragm deflections, whereas the corrugated type [Fig. 14.9(b)] is particularly useful when larger deflections are required, perhaps for driving mechanical linkages.

Diaphragm displacement may be transmitted by mechanical means to some form of indicator, perhaps a pointer and scale as is used in the familiar aneroid barometer. For engineering measurements, particularly when dynamic results are required, diaphragm motion is usually sensed by some form of electrical secondary transducer, whose principle of operation may be resistive, capacitive, inductive, piezoelectric or piezoresistive, among other possibilities.

Diaphragm design for pressure transducers generally involves all the following requirements to some degree:

1. Dimensions and total load must be compatible with physical properties of the material used.
2. Flexibility, and thus sensitivity to pressure change, must provide diaphragm deflections that match the input range of the secondary transducer.
3. Volume of displacement should be minimized to provide reasonable dynamic response.
4. Natural frequency of the diaphragm should be sufficiently high to provide satisfactory frequency response.
5. Output should be linear.

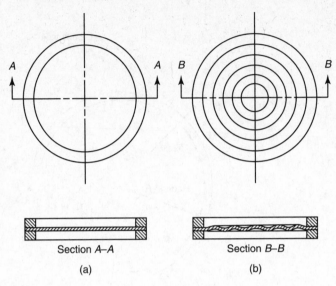

A   *A*   *B*   *B*

Section *A–A*   Section *B–B*

(a)   (b)

FIGURE 14.9: (a) Flat diaphragm, (b) corrugated diaphragm.

14.6.1   Flat Metal Diaphragms

Deflection of flat metal diaphragms is limited either by stress requirements or by deviation from linearity. A general rule is that the maximum deflection that can be tolerated while maintaining a linear pressure–displacement relation is about 30% of the diaphragm thickness.

In certain cases secondary transducers require physical connection with the diaphragm at its center. This is generally true when mechanical linkages are used and is also necessary for certain types of electrical secondary transducers. In addition, auxiliary spring force is sometimes introduced to increase the diaphragm deflection constant. These requirements make necessary some form of boss or reinforcement at the center of the diaphragm face, which reduces a diaphragm flexibility. When a central connection is made, a concentrated force will normally be applied. In general, therefore, the diaphragm may be simultaneously subjected to two deflection forces, the distributed pressure load and a central concentrated force.

An undesirable characteristic of simple flat diaphragms that is often encountered is a nonlinearity referred to as *oil canning*. The term is derived from the action of the bottom of a simple oil can when it is pressed. A slight unintentional dimpling in the assembly of a flat-diaphragm pressure pickup is difficult to eliminate unless special precautions are taken.

Diaphragm design is covered in detail in reference [8].

14.6.2   Corrugated Metal Diaphragms

Corrugated diaphragms are normally used in larger diameters than the flat types. Corrugations permit increased linear deflections and reduced stresses. Since the larger size and deflection reduce the dynamic response of the corrugated diaphragms as compared with the flat type, they are more commonly used in static applications.

Two corrugated diaphragms are often joined at their edges to provide what is referred to as a *pressure capsule*. This is the type commonly used in aneroid barometers.

Similarly, metal bellows are sometimes used as pressure-sensing elements. Bellows are generally useful for pressure ranges from about 3 kPa to 10 MPa full scale. Hysteresis and zero shift are somewhat greater problems with this type of element than with most of the others.

14.6.3 Semiconductor Diaphragms

Diaphragms can be micromachined directly onto silicon chips, producing semiconductor diaphragm pressure sensors (see Fig. 6.16). Semiconductor diaphragms are usually instrumented with directly embedded piezoresistive or capacitative sensors that track their deflection [9]. Bridge circuitry, amplification, temperature compensation, and other signal conditioning can be provided directly on the chip, using standard integrated circuit technologies. Some models are quite inexpensive. In others, the silicon element may be isolated from the measured environment by a steel diaphragm to which it is coupled, providing a fairly rugged transducer. Because these sensors are rather small, with diaphragms of a few square millimeters, their basic frequency response can be very high, exceeding 100 kHz, although packaging requirements may lower the achievable value substantially. A wide range of commercial products now incorporate semiconductor diaphragm pressure sensors, including the engine control systems of most automobiles and battery-powered handheld pressure transducers.

Silicon-diaphragm pressure transducers are sometimes simply called "piezoresistive" transducers. Transducers with full scale ranges of as high as 100 MPa or as little as 150 Pa are on the market.

14.7 ADDITIONAL PRESSURE TRANSDUCERS

A wide variety of pressure transducers are commercially available. Many are based on diaphragm deflection, and most electromechanical transducer principles have been applied to diaphragm-based pressure pickups. Other types of elastic deflection are also used to convert pressure to strain or displacement. The following examples are only representative of the possible variations.

14.7.1 Strain Gages and Flat Diaphragms

An obvious approach is simply to apply strain gages directly to a diaphragm surface and calibrate the measured strain in terms of pressure. One drawback of this method is the small physical area available for mounting the gages; for this reason, gages with short gage lengths or of custom design must be used.

Special spiral grids may used in constructing the strain gage. Grids are mounted in the central area of the diaphragm, with the elements in tension (see Fig. 12.4). Ordinary strain gages may also be used by mounting them as illustrated in Fig. 14.10 [10]. When pressure is applied to the side opposite the gages, the central gage is subject to tension while the outer gage senses compression. The two gages may be used in adjacent bridge arms, thereby adding their individual outputs and simultaneously providing temperature compensation. Some commercial transducers employ four-gage bridges made by depositing thick-film resistors directly onto the diaphragm. For new designs, the effect of gages on diaphragm stiffness and mass should be taken into account.

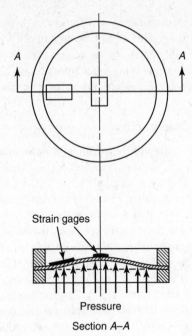

Section *A–A*

FIGURE 14.10: Location of strain gages on flat diaphragm.

Other designs involve connecting a bending beam to the center of the diaphragm by means of a rod. The strain gages are mounted on the beam. The rod serves to provide thermal isolation of the gages from the diaphragm.

Strain-gage based pressure transducers are available in full-scale ranges from roughly 10 kPa to 100 MPa.

14.7.2   Inductive Transducers

Variable inductance (Sections 6.10–6.12) is sometimes used as a form of secondary transduction with a diaphragm. Figure 14.11 illustrates one arrangement of this sort. Flexing of the diaphragm due to applied pressure causes it to move toward one pole piece and away from the other, thereby altering the relative inductances. An inductive bridge circuit may be used, as shown. Variable inductance or variable reluctance transducers are fairly rugged and provide good sensitivity. As a result of the need for ac excitation and the surrounding coils, they are relatively bulky transducers and perhaps best suited for laboratory installations. Inductive transducers are available with full-scale ranges as low as 20 Pa and as high as 70 MPa.

14.7.3   Piezoelectric Transducers

Piezoelectric transducers are typically constructed by placing several quartz columns behind a metal diaphragm with a compressive preloading. Flexion of the diaphragm stresses the quartz columns and creates a piezoelectric charge on them (Section 6.14). By detection of this charge, the pressure is sensed. Because piezoelectric charge dissipates fairly quickly, these devices are best suited for dynamic pressure measurements. The high stiffness of

Differential pressure connections

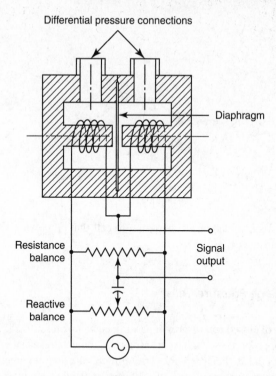

Diaphragm

Resistance
balance

Signal
output

Reactive
balance

FIGURE 14.11: Differential pressure cell with variable inductance secondary transducer.

these transducers gives them very high natural frequencies, with values of 150 kHz being representative. On the low-frequency end, piezoelectric transducers are not usually suitable for cycle periods of more than a few minutes' duration. The sensitivity of these transducers is usually high.

A charge-amplifier is used to convert the piezoelectric charge to voltage (Section 7.15.2), but commercial devices may incorporate this type of signal conditioning within the transducer housing. Piezoelectric pressure transducers have full-scale ranges from 100 kPa to 1 GPa. They have a variety of applications, including measurements in internal combustion engine cylinders, hydraulic systems, and ballistic devices [11]. Low-pressure piezoelectric sensors, capable of resolving 0.1 Pa pressure variations, are used as microphones.

14.7.4 Capacitative Transducers

Capacitative pressure transducers use a flexible diaphragm as one plate of a capacitor (see Fig. 6.14). The change in capacitance caused by the diaphgram displacement is detected using either a capacitance bridge or the change in frequency of an electrical oscillator. Capacitative transducers may have extremely high sensitivity, and they are thus very useful for high-resolution vacuum pressure measurements. Full-scale ranges from about 100 Pa to about 10 MPa are available. Drawbacks of capacitative sensors are temperature sensitivity and long-term drift [12, 13].

Capacitative diaphragms are also the basis of the condenser microphone (see Fig. 18.8).

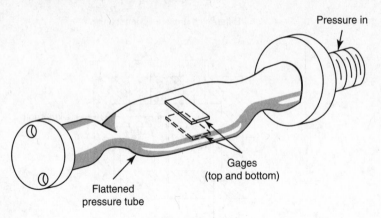

Pressure in

Gages
(top and bottom)

Flattened
pressure tube

FIGURE 14.12: Flattened-tube pressure cell that employs resistance strain gages as secondary transducers.

14.7.5 Strain-Gage Pressure Cells

Any form of closed container will be strained when pressurized. Sensing the resulting strain with an appropriate secondary transducer, such as a resistance strain gage, will provide a measure of the applied pressure. The term *pressure cell* is sometimes applied to this type of pressure-sensing device, and various forms of elastic containers or cells have been devised.

For low pressures, a pinched tube may be used (Fig. 14.12). This arrangement supplies a bending action as the tube tends to round out. Gages may be placed diametrically opposite on the flattened faces, as shown, with two unstressed temperature-compensating gages mounted elsewhere. This arrangement completes the electrical bridge.

Probably the simplest form of strain-gage pressure transducer is a cylindrical tube such as that shown in Fig. 14.13. In this application two active gages mounted in the hoop direction may be used for pressure sensing, along with two temperature-compensating gages mounted in an unstrained location. Temperature-compensating gages are shown mounted on a separate disk fastened to the end of the cell. Design relationships may be found in most mechanical design texts.

The sensitivity of a pair of circumferentially mounted strain gages (Fig. 14.13) with gage factor F is expressed by the relationship[2]

$$\frac{\Delta R}{P_i} = \frac{2FRd^2}{E}\left[\frac{2-\nu}{D^2-d^2}\right] \qquad (14.5)$$

---

[2]Equation (14.5) is based on the Lamé equations for heavy-wall pressure cylinders. See Problem 14.27.

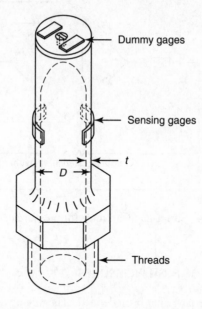

FIGURE 14.13: Cylindrical-type pressure cell.

where

$$\Delta R = \text{the strain-gage resistance change,}$$
$$R = \text{the nominal gage resistance,}$$
$$P_i = \text{the internal pressure,}$$
$$d = \text{the inside diameter of the cylinder,}$$
$$D = \text{the outside diameter of the cylinder,}$$
$$E = \text{Young's modulus,}$$
$$\nu = \text{Poisson's ratio}$$

The bridge constant, 2, appears because two circumferential gages are assumed. If a single strain-sensitive gage is to be used, the sensitivity will be one-half that given by Eq. (14.5). Of course, these relations are true only if elastic conditions are maintained and if the gages are located so as to be unaffected by end restraints.

Improved frequency response may be obtained for a cell of this type by minimizing the internal volume. This may be accomplished by use of a solid "filler" such as a plug, which will reduce the flow into and out of the cell with pressure variation.

Figure 14.14 shows the electrical circuitry used for a transducer of this type. Gage M is a modulus gage, discussed in Section 13.5, used to compensate for variation in Young's modulus with temperature. The calibration and output resistors are adjusted to provide predetermined bridge resistance and calibration.

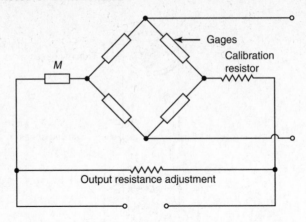

FIGURE 14.14: Strain-gage circuitry for pressure cells employing a modulus gage.

14.8 MEASUREMENT OF HIGH PRESSURES

The high-pressure range has been defined as beginning at about 1 MPa (about 10 atm) and extending upward to the limit of present techniques, which is on the order of 100 GPa (about 10^6 atm) [14]. Various conventional pressure-measuring devices, such as piezoelectric transducers for dynamic measurements and Bourdon-tube gages for static measurements, may be used at pressures as high as 500 MPa to 1 GPa. Bourdon tubes for such pressures are nearly round in section and have a high ratio of wall thickness to diameter. They are, therefore, quite stiff, and the deflection per turn is small. For this reason, high-pressure Bourdon tubes are often made with a number of turns.

14.8.1 Electrical Resistance Pressure Gages

Very high pressures may be measured by electrical resistance gages, which make use of the resistance change brought about by direct application of pressure to the electrical conductor itself. The sensing element consists of a loosely wound coil of relatively fine wire. The length and cross-section of the wire affect its electrical resistance, and both dimensions vary with applied pressure at a rate determined by the bulk modulus of the material. The electrical resistance change may thus be calibrated against the applied pressure.

Figure 14.15 shows a bulk modulus gage in section. In this particular gage, the sensing element does not actually contact the process medium but is separated therefrom by a kerosene-filled bellows. One end of the sensing coil is connected to a central terminal, as shown, while the other end is grounded, thereby completing the necessary electrical circuit.

Although Eq. (12.8) was written with a somewhat different application in mind, it also applies to the situation being discussed. Let us rewrite this relation:

$$\frac{dR}{R} = \frac{dL}{L} - 2\frac{dD}{D} + \frac{d\rho}{\rho} \qquad (14.6)$$

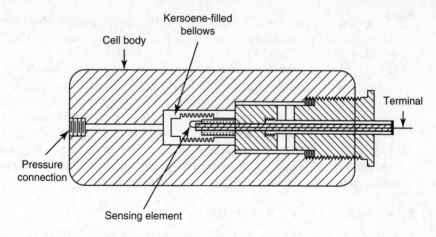

FIGURE 14.15: Section through a bulk-modulus pressure gage.

where

$$R = \text{the electrical resistance,}$$
$$L = \text{the length of the conductor,}$$
$$D = \text{the wire diameter,}$$
$$\rho = \text{the electrical resistivity}$$

The wire will be subject to a biaxial stress condition because the ends, in providing electrical continuity, will generally not be subject to pressure. Using relations of the form expressed by Eqs. (12.2a) and assuming that $\sigma_x = \sigma_y = -P$ and $\sigma_z = 0$, we may write

$$\varepsilon_x = \varepsilon_y = \frac{dD}{D} = -\frac{P}{E}(1 - v) \tag{14.7}$$

and

$$\varepsilon_z = \frac{dL}{L} = \frac{2vP}{E} \tag{14.7a}$$

Combining the above relations gives us

$$\frac{dR}{R} = \frac{2P}{E} + \frac{d\rho}{\rho} \tag{14.8}$$

If the resistivity were independent of pressure, so that $d\rho = 0$, this would yield a linear relationship between resistance change and pressure. In practice, the resistance is better represented by a second-order relationship:

$$\frac{dR}{R} = k_0 P - k_1 P^2 \tag{14.9}$$

where k_0 is known as the pressure coefficient of resistance.

Special alloys are normally used for resistance gages, particularly manganin, which is 84% copper, 12% manganese, and 4% nickel. For manganin, approximate values of the coefficients in Eq. (14.9) are $k_0 = 2.3 \times 10^{-5}$/MPa and $k_1 = 2.5 \times 10^{-10}$/MPa^2. The pressure coefficient k_0 is somewhat sensitive to temperature, which can complicate measurement. In the range to 20 to 50°C, for example, k_0 for manganin increases by 0.02% per degree of temperature increase. A variety of other alloys have been examined for use in electrical resistance pressure gages, although most have either a smaller pressure coefficient than manganin or a greater temperature sensitivity. For example, gold–2.1% chromium has $k_0 = 1 \times 10^{-5}$/MPa and a resistance decrease of 0.08% per degree kelvin in the same temperature range. Despite these somewhat poorer numbers, gold–chromium has found application in high-pressure hydrogen environments [14].

14.9 MEASUREMENT OF LOW PRESSURES

Atmospheric pressure serves as a convenient reference datum, and, in general, pressures below atmospheric may be called low pressures or vacuums. We know, of course, that a *positive* magnitude of absolute pressure exists at all times, even in a vacuum. It is impossible to reach an absolute pressure of zero.

In the older literature, absolute pressure in vacuum systems was often expressed in micrometers of mercury (μmHg), equivalent to 0.133 Pa. Units of torr (equal to 1 mmHg or 133 Pa) were also common. Today's practice is to use the SI unit pascal.

Two basic methods are used for measuring low pressure: (1) *direct* measurement based on a displacement caused by the action of a pressure force, and (2) *indirect* or *inferential* methods wherein pressure is determined through the measurement of certain other pressure-controlled properties, including volume and thermal conductivity. Devices included in the first category encompass most of those discussed in the earlier sections of this chapter, including manometers, Bourdon gages, and the various diaphragm-based electronic transducers. Since these have been discussed in the preceding pages, they need not be discussed further here except to say that their use is generally limited to lowest absolute pressures in the range of 1 to 100 Pa. For measurement of lower pressures, one of the inferential methods is normally required.

14.9.1 The McLeod Gage

Operation of the McLeod gage is based on Boyle's law for the isothermal compression of a gas:

$$P_1 = \frac{P_2 V_2}{V_1} \tag{14.10}$$

where P_1 and P_2 are pressures at initial and final conditions, respectively, and V_1 and V_2 are volumes at corresponding conditions. By compressing a known volume of the low-pressure gas to a higher pressure and measuring the resulting volume and pressure, one can calculate the initial pressure.

Figure 14.16 illustrates the basic construction and operation of one type of McLeod gage. Measurement is made as follows. The unknown pressure source is connected at point A, and the mercury level is adjusted to fill the volume represented by the darker shading. Under these conditions the unknown pressure fills the bulb B and capillary C. Mercury is then forced out of the reservoir D, up into the bulb and reference column E. When the

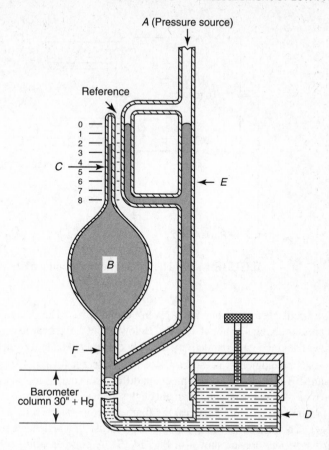

FIGURE 14.16: McLeod vacuum gage.

mercury level reaches the cutoff point F, a *known* volume of gas is trapped in the bulb and capillary. The mercury level is then further raised until it reaches a zero reference point in E. Under these conditions the volume remaining in the capillary is read directly from the scale, and the difference in heights of the two columns is the measure of the trapped pressure. The initial pressure may then be calculated by use of Boyle's law.

The pressure of gases containing vapors cannot normally be measured with a McLeod gage, for the reason that the compression will cause condensation. By use of instruments of different ranges, a total pressure range of from about 0.01 Pa to 10 kPa may be measured with this type of gage.

14.9.2 Thermal Conductivity Gages

The temperature of a given wire through which an electric current is flowing will depend on three factors: the magnitude of the current, the resistivity, and the rate at which the heat generated in the wire by resistive losses is conveyed to the surrounding environment. The latter will be largely dependent on the conductivity of the surrounding media. As the density of a given medium is reduced below a certain level, its conductivity will also decrease and

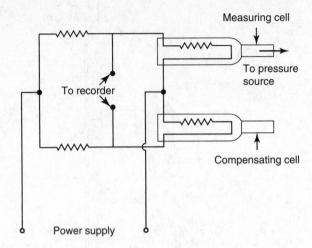

FIGURE 14.17: The Pirani thermal conductivity gage.

the wire will become hotter for a given current flow. The decrease in gas conductivity with pressure begins at about 130 Pa. Below about 1 Pa, heat loss by thermal radiation is dominant, and the gas conductivity effect becomes indetectably small.

This phenomenon is the basis for two different forms of gages for measuring low pressures. Both use a heated filament but differ in the means for measuring the temperature of the wire. A single platinum filament enclosed in a chamber is used by the *Pirani gage*. As the surrounding pressure changes, the filament temperature, and hence its resistance, also changes. The resistance change is measured by use of a resistance bridge that is calibrated in terms of pressure, as shown in Fig. 14.17. A compensating cell is used to minimize variations caused by ambient temperature changes.

A second gage also depending on thermal conductivity is the *thermocouple gage*. In this case the filament temperatures are measured directly by means of thermocouples welded to them. Filaments and thermocouples are arranged in two chambers, as shown schematically in Fig. 14.18. When conditions in both the measuring and reference chambers are the same no thermocouple current will flow. When the pressure in the measuring chamber is altered, changed conductivity will cause a change in temperature, which will then be indicated by a thermocouple current.

In both cases the gages must be calibrated for a definite pressurized medium, for the conductivity is also dependent on gas composition. As noted earlier, gages of these types are useful in the range from about 1 to 100 Pa.

14.9.3 Ionization Gages

For measurement of extremely low pressures—down to 10^{-6} Pa—an *ionization gage* may be used. The maximum pressure for which an ionization gage may be used is about 0.1 Pa (1 μmHg). An ionization cell for pressure measurement is very similar to the old-style triode electronic tube. It possesses a heated filament, a positively biased grid, and a negatively biased plate in an envelope evacuated by the pressure to be measured. The grid draws electrons from the heated filament, and collision between them and gas molecules causes

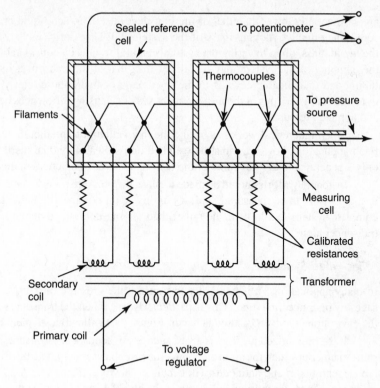

FIGURE 14.18: Thermocouple gage.

ionization of the molecules. The positively charged molecules are then attracted to the plate of the tube, causing a current flow in the external circuit, which is a function of the gas pressure.

Disadvantages of the heated-filament ionization gage are that (1) excessive pressure (0.1 to 0.2 Pa) will cause rapid deterioration of the filament and a short life; and (2) the electron bombardment is a function of filament temperature, and therefore careful control of filament current is required. Another form of ionization gage minimizes these disadvantages by substituting a radioactive source of alpha particles for the heated filament; these gages operate in the much higher-pressure range from 0.01 Pa up to atmospheric pressure [15].

14.10 DYNAMIC CHARACTERISTICS OF PRESSURE-MEASURING SYSTEMS

Basic pressure-measuring transducers are driven, damped, spring-mass systems whose isolated dynamic characteristics are theoretically similar to the generalized systems discussed in Chapter 5. In application, however, the actual dynamic characteristics of the complete pressure-measuring system are usually controlled more by factors extraneous to the basic transducer than by the transducer characteristics alone. In other words, overall dynamic performance is determined less by the transducer than by the manner in which it is inserted into the complete system.

When the transducer is used to measure a dynamic air or gas pressure, the compressibility of the gas in any connecting lines or volumes can introduce natural frequencies that are substantially lower than that of the transducer itself. When liquid pressures are measured, the effective moving mass of the system will necessarily include some portion of the liquid mass, often lowering the system's natural frequency considerably. In both cases, the damping of the system may be greater than that of the transducer itself. Connecting tubing and unavoidable cavities in the pneumatic or hydraulic circuitry thus change the dynamic characteristics of the measurement system, causing differences between measured and applied pressures.

Much theoretical work has been done in an attempt to predict these effects [16–19]. Each application, however, must be weighed on its own individual merits; for this reason only a general summary of some of the factors involved is practical in this discussion.

In general, efforts should be made to place the pressure transducer as close as possible to the pressure to be measured, and, when this is not possible, dynamic calibration of the complete system should be made rather than relying upon the dynamic calibration of the transducer alone.

14.10.1 Gas-Filled Systems

In many applications, it is necessary to transmit the pressure through some form of passageway or connecting tube. This is particularly true when the transducer is bulky or when the environment to be measured is harsh. Figure 14.19 illustrates typical cases.

If the pressurized medium is a gas, such as air, acoustical resonances may occur in the same manner in which the air in an organ pipe resonates. If sympathetic driving frequencies are present, nodes and antinodes will occur, as shown in the figure. A node, characterized by a point of zero air motion, will occur at the blocked end (assuming that the displacement of the pressure-sensing element, such as a diaphragm, is negligible). Maximum pressure variation takes place at this point. Maximum oscillatory motion will occur at the antinodes, and the distance between adjacent nodes and antinodes equals one-fourth the wavelength

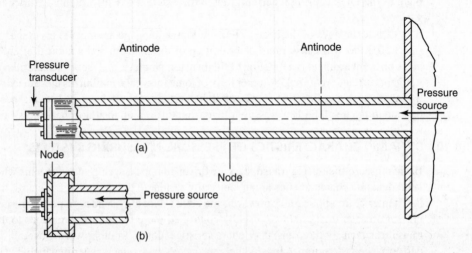

FIGURE 14.19: (a) Gas-filled pressure measuring system, (b) gas-filled pressure-measuring system with cavity.

of the resonating frequency. Theoretical resonant frequencies may be determined from the relation

$$f = \frac{c}{4L}(2n - 1) \tag{14.11}$$

where

$f =$ the resonance frequencies (including both fundamental and
   harmonics), in hertz,

$c =$ the velocity of sound in the pressurized medium, in m/s,

$L =$ the length of the connecting tube, in m,

$n =$ any positive integer (It will be noted from the equation
   that only odd harmonics occur.)

In many cases a cavity is required at the transducer end to adapt the instrument to the tubing, as shown in Fig. 14.19(b). If we assume that the medium is a gas, and that the containing system, including the transducer, is relatively stiff compared with the gas, we have what is known as a Helmholtz resonator. The the mass of gas in the tubing together with the compressibility of that in the cavity form a spring-mass system having an acoustical resonance whose fundamental frequency may be expressed by the relation [20]

$$f = \frac{c}{2\pi} \sqrt{\frac{a}{V(L + \sqrt{\pi a}/2)}} \tag{14.12}$$

where

$a =$ the cross-sectional area of the connecting tube, in m^2,

$V =$ the net internal volume of the cavity, excluding the
   volume of the tube, in m^3

This equation is accurate so long as the tubing volume, aL, is small relative to the cavity volume, V. When the tubing volume is not small, the following approximation may instead be used:

$$f = \frac{c}{2\pi} \sqrt{\frac{a}{V(L + aL/2)}} \tag{14.13}$$

More complete equations for long pneumatic tubes are given by Andersen [19].

14.10.2   Liquid-Filled Systems

When a pressure-measuring system is filled with liquid rather than a gas, a considerably different situation is presented. The liquid becomes a major part of the total moving mass, thereby becoming a significant factor in determining the natural frequency of the system.
   If a single degree of freedom is assumed,

$$f = \frac{1}{2\pi} \sqrt{\frac{k_s}{m}} \tag{14.14}$$

where

$$f = \text{the natural frequency, in hertz,}$$
$$m = \text{the equivalent moving mass, in kg,} = m_1 + m_2,$$
$$m_1 = \text{the mass of moving transducer elements, in kg,}$$
$$m_2 = \text{the equivalent mass of the liquid column, in kg,}$$
$$k_s = \text{the effective transducer stiffness, in N/m}$$

By simplified analysis, White [21] has determined the following approximate relation for the effective mass of the liquid column:

$$m_2 = \frac{4}{3}\rho a L \left(\frac{A}{a}\right)^2 \tag{14.15}$$

where

$$\rho = \text{the fluid density, in kg/m}^3,$$
$$a = \text{the sectional area of the tube, in m}^2,$$
$$L = \text{the length of the tube, in m,}$$
$$A = \text{the effective area of the transducer-sensing element, in m}^2$$

It will be noted that A is the *effective* area, which is not necessarily equal to the actual diaphragm or bellows area but may be defined by the relation

$$A = \frac{\Delta V}{\Delta y} \tag{14.16}$$

where

$$\Delta V = \text{the volume change accompanying sensing-element deflection, in m}^3,$$
$$\Delta y = \text{the significant displacement of the sensing element, in m}$$

Likewise, the transducer stiffness, k_s, may be defined in terms of the pressure change, ΔP (Pa), required to produce the volume change, ΔV:

$$k_s = \frac{\Delta P \cdot A}{\Delta V / A} \tag{14.17}$$

By substitution of Eq. (14.15) into Eq. (14.14), we have

$$f = \frac{1}{2\pi} \sqrt{\frac{k_s}{m_1 + \frac{4}{3}\rho a L (A/a)^2}} \tag{14.18}$$

In many cases the equivalent mass of the liquid, m_2, is of considerably greater magnitude than m_1, and the latter may be ignored without introducing an appreciable discrepancy. By so doing, and writing a in terms of the tube diameter, we get

$$a = \pi D^2 / 4,$$
$$D = \text{tubing I.D., in m,}$$

$$f = \frac{D}{8A} \sqrt{\frac{3k_s}{\pi \rho L}} \tag{14.19}$$

As mentioned before, pressure transducers involve spring-restrained masses in the same manner as do seismic accelerometers, and therefore good frequency response is obtainable only in a frequency range well below the natural frequency of the measuring system itself. For this reason it is desirable that the pressure-measuring system have as high a natural frequency as is consistent with required sensitivity and installation requirements. Inspection of Eq. (14.19) indicates that the diameter of the connecting tube should be as large as practical and that its length should be minimized.

In addition, it has been shown that optimum performance for systems of this general type requires damping in rather definite amounts. White [21] gives the following relation for the damping ratio ξ of a system of the sort being discussed:

$$\xi = \frac{4\pi L\nu(A/a)^2}{\sqrt{k_s m}} \tag{14.20}$$

$$= \frac{4\pi L\nu(A/a)^2}{\sqrt{k_s[m_1 + \frac{4}{3}\rho aL(A/a)^2]}} \tag{14.21}$$

where ν = the viscosity of the fluid. If we ignore m_1 and insert $a = \pi D^2/4$, we may write the equation as

$$\xi = \frac{16\nu A}{D^2}\sqrt{\frac{3L}{\pi k_s \rho}} \tag{14.22}$$

14.11 CALIBRATION METHODS

Static calibration of pressure gages presents no particular problems unless the upper pressure limits are unusually high. The familiar dead-weight tester (Fig. 14.3) may be used to supply accurate reference pressures with which transducer outputs may be compared. Depending upon the specific design, testers of this type are useful to pressures exceeding 1 GPa (150,000 psi).

Although static calibration is desirable, transducers used for dynamic measurement should also receive some form of dynamic calibration, so as to determine the transducer's frequency response. Since pressure transducers may often be modelled as second-order (spring-mass-damper) systems, dynamic calibrations may be used to find the transducer's natural frequency, damping ratio, and static amplitude for use in determining its amplification ratio as a function of frequency, that is, to find the transducer's transfer function (see Sections 5.16 and 5.20) [22, 23].

Dynamic calibration problems consist of (1) obtaining a satisfactory source of pressure, either periodic or pulsed, and (2) reliably determining the true pressure–time relation produced by such a source. These two problems will be discussed in the next few paragraphs.

Dynamic pressure sources may be periodic (steadily oscillating) or aperiodic (transient). Some sources of dynamic pressure are as follows:

I. Periodic pressure sources

(a) Piston and chamber
(b) Rotating valve

    (c) Siren disk

    (d) Acoustic resonator

  II. Aperiodic pressure sources

    (a) Quick-release valve

    (b) Closed combustion bomb

    (c) Shock tube

When a periodic source is used, the frequency of the source may be varied so as to as to measure the transducer's frequency response directly. When aperiodic sources are used, additional calculations will be required.

14.11.1 Periodic Pressure Sources

Most periodic pressure sources are designed for gaseous media, and they are generally limited to relatively low frequencies and amplitudes, particularly if nearly sinusoidal fluctuations are desired. At higher amplitudes or frequencies, the pressure variations tend toward a sawtooth form owing to the nonlinear behavior of compressible gas flows [23, 24].

One source of steady-state periodic calibration pressure is simply an ordinary piston and cylinder arrangement, shown schematically in Fig. 14.20. If the piston stroke is fixed, pressure amplitude may be varied by adjusting the cylinder volume. For a fixed volume, a known repeatable pressure variation can be generated at a single frequency. Amplitude and frequency ranges will depend on the mechanical design; however, peak pressures to 7 MPa and frequencies as high as 100 to 1000 Hz have been reported [22, 25].

A method very similar to this has been used for microphone calibration. In this case required pressure amplitudes are quite small, and instead of the piston being driven with a mechanical linkage, an electromagnetic system is used [26]. Similar approaches have used

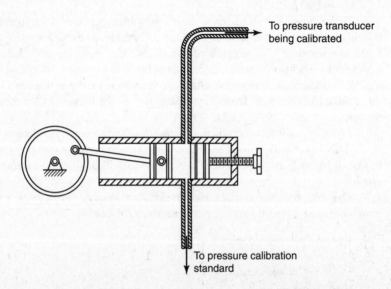

FIGURE 14.20: Schematic diagram of a piston and cylinder periodic pressure source.

vibration test shakers (Section 17.14) as a source of piston motion or of periodic inertial loading, the latter method being adaptable to liquid media [23].

Several pressure supplies have been based upon rotating valves that periodically switch the pressure sent to a transducer between high- and low-pressure supplies. The pressure signal is approximately rectangular in such systems if acoustic resonance can be avoided in the gas-containing volumes. Recent designs have been reported to work at up to 100 kPa and 1 kHz [27, 28].

Figure 14.21 illustrates another method for obtaining a steady-state periodic pressure. A source of this type has been used to 3000 Hz with amplitudes to 7 kPa [29]. A variation of this method uses a motor-driven siren-type disk having a series of holes drilled in it so as alternately to vent a pressure source to atmosphere and then shut it off [30]. Sirens have been used at frequencies to 1 kHz and pressures to 200 kPa [22].

Steady-state sinusoidal pressure generators consisting of an acoustically driven resonant system (see Section 14.10.1) have also been used. Typical arrangements place a loudspeaker at one end of a tube whose opposite end is closed by a movable piston; the piston position is adjusted to give a half-wavelength separation from the loudspeaker. These systems are useable at frequencies up to a few kHz and amplitudes up to a few kPa [17, 23].

All the methods suggested here simply supply sources of pressure variation, but, in themselves, they do not provide means for determining pressure amplitudes or time characteristics. They are useful, therefore, for comparing a pressure transducer having unknown characteristics to one whose performance is known.

14.11.2 Aperiodic Pressure Sources

Periodic sources used to determine dynamic characteristics of pressure transducers are limited by the amplitude and frequency that can be produced. High amplitudes and steady-state frequencies are difficult to obtain simultaneously. For this reason, it is necessary to resort to some form of step change in pressure to determine the transducer's high-amplitude and high-frequency performance. Typically, the transducer's response to a step-changed pressure is recorded, and Laplace transform calculations are then used to determine the

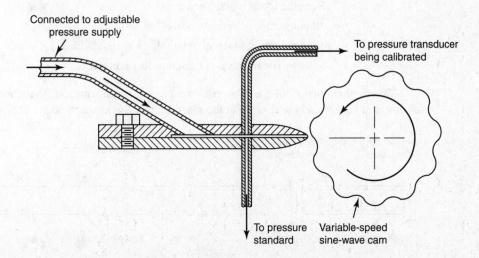

FIGURE 14.21: Jet and cam steady-state pressure source.

transducer's dynamic characteristics.

Various methods are used to produce the necessary pulse. One of the simplest is to use a fast-acting valve between a source of pressure and the transducer. These devices may be configured so that the transducer is held in a high-pressure chamber which is abruptly vented to atmosphere or so that the transducer is kept in a small chamber at low pressure which is abruptly opened to a large chamber at high pressure. Rise times, from 0 to 90% of final pressure, of 1 ms or less have been reported [22–24].

Another source of stepped-pressure is the closed combustion bomb, in which a pressure generator such as a dynamite cap is exploded. Peak pressure is controlled by net internal volume, and pressure steps as high as 5 MPa in 0.3 ms have been obtained [17, 22, 27].

Undoubtedly the *shock tube* provides the nearest thing to a transient pressure standard [24, 31]. Construction of a shock tube is quite simple: It consists of a long tube, closed at both ends, separated into two chambers by a diaphragm, as shown in Fig. 14.22. A pressure differential is built up across the diaphragm, and the diaphragm is burst, either by the pressure differential itself or by means of an externally controlled mechanism or cutter. Rupturing the diaphragm causes a pressure discontinuity, or *shock wave*, to travel at very high speed into the region of lower pressure and a rarefaction wave to travel through the chamber of initially higher pressure. The reduced pressure wave is reflected from the end of the chamber and follows the stepped pressure down the tube at a velocity that is higher because it is added to the velocity already possessed by the gas particles from the pressure step. Figure 14.23 illustrates the sequence of events immediately following the bursting of the diaphragm.

A relationship between pressures and shock-wave velocity may be expressed as follows [32]:

$$\frac{P_1}{P_0} = 1 + \frac{2k}{k+1}(M_0^2 - 1) \tag{14.23}$$

where

P_1 = the intermediate transient pressure,

P_0 = the lower initial pressure,

k = the ratio of specific heats (see Table 15.2),

M_0 = the shock Mach number (shock speed divided by sound speed in region of low pressure)

We see, then, that if the gas properties are known, measurement of the propagation velocity will be sufficient to determine the magnitude of the pressure pulse. Propagation

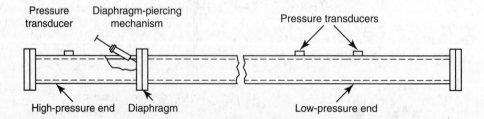

FIGURE 14.22: Basic shock tube.

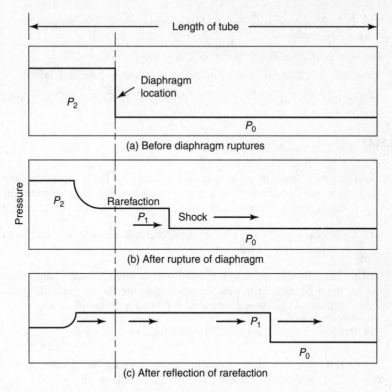

FIGURE 14.23: Pressure sequence in a shock tube before and immediately after diaphragm is ruptured. Abscissa represents longitudinal axis of tube.

velocity may be determined from information supplied by accurately positioned pressure transducers in the wall of the tube. By this means, a known transient pressure pulse may be applied to a pressure transducer or to a complete pressure-measuring system simply by mounting the transducer in the wall of the shock tube. The response characteristics, as determined in this manner, may then be used to calculate the response or transfer function of the device or system over a spectrum of frequencies.

The rise time of a shock tube corresponds to the time for the thin shock wave to pass by and is generally well below 1 microsecond. The test duration is the length of time that the intermediate pressure, P_1, is sustained at the transducer. It is a function of the length of the shock tube and is typically some number of milliseconds. Shock tubes are thus used for calibrations in the frequency range above several hundred hertz. For lower frequency calibrations, a fast-acting valve is preferred [22, 24]. Shock tubes have been used for calibrations at up to 20 MPa.

SUGGESTED READINGS

ASME PTC19.2-1987. Instruments and Apparatus: Part 2. *Pressure Measurement*. New York: American Society of Mechanical Engineers, 1987.

Benedict, R. P. *Fundamentals of Temperature, Pressure and Flow Measurements*, 3rd ed. New York: Wiley-Interscience, 1984.

Fowles, G. *Flow, Level and Pressure Measurement in the Water Industry.* Oxford, U.K.: Butterworth-Heineman, 1993.

Gautschi, G. H. *Piezoelectric Sensorics.* Berlin: Springer-Verlag, 2002.

Kovacs, G. T. A. *Micromachined Transducers Sourcebook.* New York: McGraw-Hill, 1998.

Peggs, G. N. (ed.). *High-Pressure Measurement Techniques.* London: Applied Science Publishers, 1983.

PROBLEMS

14.1. Standard atmospheric pressure is 1.01325×10^5 Pa. What are the equivalents in (a) newtons per square meter, (b) pounds-force per square foot, (c) meters of water, (d) inches of oil, with 0.89 specific gravity, (e) millibars, (f) micrometers, and (g) torr?

14.2. Determine the factors for converting pressure in pascals to "head" in (a) meters of water; (b) centimeters of mercury.

14.3. The following are some commonly encountered pressures (approximate). Convert each to the SI units, Pa or kPa: (a) automobile tire pressure of 32 psig; (b) household water pressure of 120 psia; (c) regulation football pressure of 13 psig.

14.4. First in SI units and then in English units,

  (a) Write expressions relating the height of a fluid column in terms of a reduced gage pressure (a vacuum).
  (b) Under standard conditions of atmosphere and gravity, what is the maximum height to which water may be raised by suction alone?
  (c) Under similar conditions, to what height may a column of mercury be raised?
  (d) On the surface of the moon, what is the height to which water could be raised by suction alone? (See Appendix D for data.)

14.5. The common mercury barometer may be formed by sealing the upper end of a tube (e.g., Fig. 14.4), inverting it and filling it with mercury, then righting the tube into a mercury-filled reservoir. A vacuum is formed over the column and the height of the column is governed primarily by the pressure (air pressure) applied at the base. Under these conditions, is a true zero absolute pressure (a complete vacuum) formed over the column? Investigate the vapor pressure of mercury and determine the degree of error introduced if it is ignored.

14.6. Rewrite Eq. (14.3) in terms of specific gravities.

14.7. Write a few sentences explaining the mechanics of "suction."

14.8. Write a few sentences explaining the operation of a syphon.

14.9. The U-tube-type manometer (Fig. 14.5) uses mercury and water as the manometer and transmitting fluids, respectively. What value of h should be expected at 20°C if the applied differential pressure is 80 kPa (11.6 psig)? (See Appendix D for data.) Use SI units.

14.10. Solve Problem 14.9 using English units.

14.11. A dual-fluid U-tube manometer, Fig. 14.5, located at Fort Egbert, Alaska, displays a pressure, $\Delta h = 5.400$ in. fluid displacement. Under identical conditions, except for location, determine what pressure would be indicated (a) at Key West, Florida, and (b) on the moon. (Refer to Appendix D for data.)

14.12. Figure 14.24 illustrates a manometer installation. Write an expression for determining the static pressure in the conduit in terms of h_1, h_2, and the other pertinent parameters.

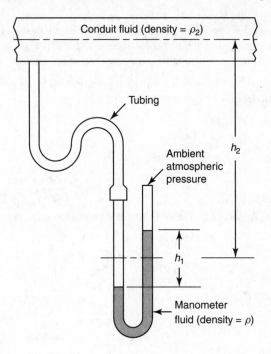

Conduit fluid (density = ρ_2)

Tubing

Ambient atmospheric pressure

h_2

h_1

Manometer fluid (density = ρ)

FIGURE 14.24: Manometer arrangement referred to in Problem 14.12.

14.13. For the conditions shown in Fig. 14.24, if the manometer fluid is Hg and the conduit fluid is H_2O (both at 20°C), $h_1 = 18.4$ cm (7.24 in.), and $h_2 = 0.7$ m (2.30 ft), what pressure exists in the conduit? Solve using SI units.

14.14. Solve Problem 14.13 using English units.

14.15. Express the ratio of sensitivities of an inclined manometer (Fig. 14.6) to that of a simple manometer in terms of the angle θ. For an inclined manometer six times more sensitive than a simple manometer, what should be the angle of inclination?

14.16. Verify Eqs. (14.4) and (14.4a).

14.17. The manometer shown in Fig. 14.7 uses water and carbon tetrachloride as the two fluids (see Appendix D for data). If the area ratio is 0.01, what magnification will result as compared to the simple manometer using (a) water as the fluid and (b) carbon tetrachloride as the fluid?

14.18. Derive an expression for the two-fluid manometer in Fig. 14.7, substituting reservoirs of diameters D_1 and D_2 for the like-sized reservoirs shown in the figure.

14.19. A two-fluid manometer as shown in Fig. 14.7 uses a combination of kerosene (specific gravity, 0.80) and alcohol-diluted water (specific gravity, 0.83). Also, $d = \frac{1}{4}$ in. (6.35 mm) and $D = 2$ in. (50.8 mm). What amplification ratio is obtained with this arrangement as compared to a simple water manometer? What error would be introduced if the ratio of diameters was ignored?

14.20. Confirm Eq. (14.3c).

14.21. Note that Eq. (14.3c) is based on a moving datum—namely, the liquid level in the reservoir. Derive an equation for the differential pressure based on the movement of the liquid in the inclined column only. (Note that a practical solution would be to make provision for adjusting the reservoir level to an index or "zero" line.)

14.22. Figure 14.13 shows a cylindrical pressure cell using two sensing strain gages. If the cylinder may be assumed to be "thin wall," then

$$\sigma_H = \frac{Pd}{2t}$$

and

$$\sigma_L = \frac{Pd}{4t}$$

(See Appendix E, Figure E.10, for symbol meanings.)

For this case, show that

$$\frac{\Delta R}{P} = \frac{FRd}{Et}\left[1 - \frac{\nu}{2}\right]$$

where

$$F = \text{gage factor,}$$
$$R = \text{gage resistance,}$$
$$E = \text{Young's modulus, and}$$
$$\nu = \text{Poisson's ratio}$$

14.23. For a circular diaphragm of the type and loading shown in Fig. 14.9(a), the maximum normal stress occurs in the radial direction at the outer boundary, expressed as follows [33]:

$$\sigma_r = \frac{3}{4}\left(\frac{a}{t}\right)^2 P$$

Greatest linear deflection occurs at the center and is equal to

$$Y_{max} = \left(\frac{3}{16}\right)\left(\frac{Pa^4}{Et^3}\right)(1 - \nu^2)$$

where

$$P = \text{pressure,}$$
$$a = \text{radius,}$$
$$E = \text{Young's modulus, and}$$
$$\nu = \text{Poisson's ratio}$$

For $a = \frac{1}{4}$ in. (6.35 mm), $E = 30 \times 10^6$ psi (20.68×10^7 kPa), and $\nu = 0.3$, and for a design stress of 9×10^4 psi (6.2×10^5 kPa), what maximum deflection may be expected if $P = 300$ psi (2.07×10^3 kPa)?

14.24. Solve Problem 14.23 using SI units and reconcile the two answers.

14.25. A thin-walled, cylindrically sectioned tube of nominal diameter D and wall thickness t is subjected to a pressure P. Circumferential (hoop) and longitudinal stresses may be determined from the relations $\sigma_H = PD/2t$ and $\sigma_L = PD/4t$ (see Example E.2, Appendix E). Using Eqs. (12.2a), show that the corresponding circumferential and longitudinal strains are

$$\varepsilon_H = \left(\frac{PD}{2Et}\right)\left(1 - \frac{1}{2}\nu\right) \text{ and } \varepsilon_L = \left(\frac{PD}{2Et}\right)\left(\frac{1}{2} - \nu\right)$$

14.26. A pressure transducer is constructed from a steel tube having a nominal diameter of 15 mm (0.59 in.) and a wall thickness of 2 mm (0.0787 in.).

(a) If the design stress is limited to 2.75×10^8 Pa (3.99×10^4 psi), what maximum pressure may be applied to the transducer?

(b) For the maximum pressure calculated in part (a), determine the circumferential and longitudinal strains that should be expected. Use $E = 20 \times 10^{10}$ Pa (29×10^6 psi) and $\nu = 0.3$.

14.27. The simple stress relations given in Problem 14.25 assume a uniform stress distribution through the wall of the cylinder. For so-called heavy-wall cylinders, this simplifying assumption leads to error. The following, more complex relations, often referred to as the Lamé equations, must be used:

$$\sigma_H = \frac{P(D^2 + d^2)}{D^2 - d^2} \text{ on the inner surface,}$$

$$\sigma_H = \frac{2Pd^2}{D^2 - d^2} \text{ on the outer surface,}$$

$$\sigma_L = \frac{Pd^2}{D^2 - d^2},$$

$$\sigma_r = -P \text{ on the inner surface}$$

All are principal stresses (see Appendix E).

(a) For a design stress of 2.75×10^8 Pa (3.99×10^4 psi), $d = 2$ cm (0.787 in.), and $D = 5$ cm (1.968 in.), what is the maximum pressure that may be applied?

(b) If the maximum pressure is applied, what circumferential and longitudinal strains should be expected on the outer surface? Use 0.3 for Poisson's ratio and 20×10^{10} Pa for Young's modulus.

14.28. Solve Problem 14.27 using English units.

14.29. Derive Eq. (14.5). Note that this equation is based on the Lamé equations for heavy-wall pressure vessels. See Problem 14.27.

14.30. Confirm Eq. (14.8).

14.31. The speed of sound in an ideal gas, c, may be expressed by the relation [32]

$$c = \sqrt{kRT}$$

where

k = the ratio of specific heats (1.4 for air),

R = the gas constant (287 J/kg·K for air), and

T = absolute temperature (in kelvin)

Using the above equation and Eq. (14.11), determine the change in frequency in percent, corresponding to a temperature change from 10°C to 40°C.

14.32. A pressure-measuring system involves a $\frac{1}{4}$-in.-diameter tube, 24 in. long, connecting a pressure source to a transducer. At the transducer end there is a cylindrical cavity $\frac{1}{2}$ in. in diameter and $\frac{1}{2}$ in. long. Proper performance requires that the frequency of applied pressures be such as to avoid resonance. Calculate the resonance frequency of the system.

14.33. Assume that the 24-in. connecting tube used in Problem 14.32 is reduced to zero length. What will be the resonance frequency? [Use Eq. (14.12), letting $L = 0$.]

14.34. A Helmholtz resonator consists of a spherical cavity to which a circularly sectioned tube is attached. It may be considered as approximating the tube and cavity of Problem 14.32. The resonance frequency of a Helmholtz resonator may be estimated by the relation [32]

$$f = \frac{c}{2\pi} \sqrt{\frac{a}{VL}}$$

The symbols have the same meaning as in Eq. (14.12). Use this equation to estimate the resonance frequency of the system described in Problem 14.32.

REFERENCES

[1] Taylor, B. N. *Guide to the Use of the International System of Units (SI)*. NIST Special Publication 811. Gaithersburg, Md.: National Institute of Standards and Technology, 1995.

[2] Benedict, R. P. *Fundamentals of Temperature, Pressure and Flow Measurements*. 3rd ed. New York: Wiley-Interscience, 1984.

[3] Proceedings of CCM third international conference: Pressure metrology from ultra-high vacuum to very high pressures—10^{-7} to 10^9 Pa. *Metrologia* 36(6), 1999.

[4] Proceedings of CCM second international seminar: Pressure metrology from 1 kPa to 1 GPa. *Metrologia* 30(6), 1993/94.

[5] Brombacher, W. G. *Survey of Micromanometers*. NBS Monograph 114. Washington, D.C.: U.S. Government Printing Office, 1970.

[6] Tilford, C. R. Three and a half centuries later—The modern art of the manometer. *Metrologia* 30:545–552, 1993/94.

[7] Kardos, G. *Bourdon Tubes and Bourdon Tube Gages: An Annotated Bibliography*. New York: American Society of Mechanical Engineers, 1978.

[8] Di Giovanni, M. *Flat and Corrugated Diaphragm Design Handbook*. New York: Marcel Dekker, 1982.

[9] Kovacs, G. T. A. *Micromachined Transducers Sourcebook*. New York: McGraw-Hill, 1998.

[10] Wenk, E. Jr. A diaphragm-type gage for measuring low pressures in fluids. *SESA Proc.* 8(2):90, 1951.

[11] Gautschi, G. H. *Piezoelectric Sensorics*. Berlin: Springer-Verlag, 2002, Chap. 8.

[12] Adams, E. D. High-resolution capacitative pressure gages. *Rev. Sci. Instr.* 64(3):601–611, 1993.

[13] Miiller, A. P. Measurement performance of high-accuracy low-pressure transducers. *Metrologia* 36:617–621, 1998.

[14] Peggs, G. N. (ed.). *High Pressure Measurement Techniques*. London: Applied Science Publishers, 1983.

[15] ASME PTC19.2-1987. Instruments and Apparatus: Part 2. *Pressure Measurement*. New York: American Society of Mechanical Engineers, 1987.

[16] Iberall, A. S. Attenuation of oscillatory pressures in instrument lines. *NBS J. Res.* 45:85, July 1950.

[17] Hylkema, C. G., and R. B. Bowersox. Experimental and mathematical techniques for determining the dynamic response of pressure gages. *ISA Proc.* 8:115, 1953, and *ISA J.* 1:27, February 1954.

[18] Brown, F. T. The transient response of fluid lines. *J. Basic Engr.* 84:547–552, 1962.

[19] Andersen, R. C. *Analysis and Design of Pneumatic Systems*. New York: John Wiley, 1967, Sect. 4.3.

[20] Lord Rayleigh. *The Theory of Sound*, vol. II. 2nd ed. New York: Dover Publications, 1945, p. 188.

[21] White, G. Liquid filled pressure gage systems. *Instrument Notes* 7. Los Angeles: Statham Laboratories, January–February 1949.

[22] Bean, V. E. Dynamic pressure metrology. *Metrologia* 30:737–741, 1993/94.

[23] Hjelmgren, J. *Dynamic Measurement of Pressure—A Literature Survey*. SP Report 2002:34. Borås, Sweden: SP Swedish National Testing and Research Institute, 2002.

[24] Damion, J. P. Means of dynamic calibration for pressure transducers. *Metrologia* 30:743–746, 1993/94.

[25] Taback, I. The response of pressure measuring systems to oscillating pressure. *NACA Tech. Note 1819*: February 1949.

[26] Badmaieff, A. Techniques of microphone calibration. *Audio Eng.* 38: Dec. 1954.

[27] Schweppe, J. L., L. C., Eichberger, D. F. Muster, E. L. Michaels, and G. F. Paskusz. *Methods for the Dynamic Calibration of Pressure Transducers*. National Bureau of Standards Monograph 67. Washington, D.C.: U.S. Department of Commerce, 1963.

[28] Kobota, T., and A. Ooiwa, Square-wave pressure generator using novel rotating valve. *Metrologia* 36:637–640, 1999.

[29] Patterson, J. L. A miniature electrical pressure gage utilizing a stretched flat diaphragm, *NACA Tech. Note 2659*: April 1952.

[30] Meyer, R. D. Dynamic pressure transmitter calibrator. *Rev. Sci. Inst.* 17: 199, May 1946.

[31] Bean, V. E., W. J. Bowers, W. S. Hurst, and G. J. Rosasco, Development of a primary standard for the measurement of dynamic pressure and temperature. *Metrologia* 30:747–750, 1993/94.

[32] Thompson, P. A. *Compressible-Fluid Dynamics*. New York: McGraw-Hill, 1972.

[33] Roark, R. J. *Formulas for Stress and Strain*. New York: McGraw-Hill, 1965, p. 217.

CHAPTER 15

Measurement of Fluid Flow

15.1   INTRODUCTION

Fluid flow encompasses a wide range of situations. The flowing medium may be a liquid, a gas, a granular solid, or any combination thereof. The flow may be laminar or turbulent, steady state or transient. The desired measurement may be the velocity at a point, the rate of flow through a channel, or simply a picture of the entire flow field. Each of these factors affects the selection of an appropriate measurement technique, and many different methods have been developed for the various situations. This chapter, therefore, will present only an outline of some of the more important aspects of the general topic.

The most direct way to measure flow rate is to capture and record the volume or mass that flows during a fixed time interval (a *primary measurement* of flow rate). More often, some other quantity, such as a pressure difference or mechanical response, is used to infer the flow rate (a *secondary measurement*). We may also make a distinction between flowmeters and velocity sensors. *Flowmeters* determine volume or mass flow rates (e.g., liters per minute or kilograms per second) through tubes and channels, whereas *velocity-sensing probes* measure fluid speed (e.g., meters per second) at a point in the flow. Although velocity-sensing probes can be used as building blocks for flowmeters, the converse is rarely true. In addition, *flow-visualization* techniques are sometimes employed to obtain an image of the overall flow field.

A categorization of flow-measurement methods is as follows:

1. Primary or quantity methods

    (a) Weight tanks and so on

    (b) Volume tanks graduated cylinders, bell provers, and so on

2. Flowmeters

    (a) Obstruction meters (responding to pressure differentials)

        i. Venturi meters

        ii. Flow nozzles

        iii. Orifices

        iv. Variable-area meters

    (b) Volume flowmeters (responding to volumetric flow rates)

        i. Turbine and propeller meters

        ii. Electromagnetic flowmeters (liquids only)

        iii. Vortex-shedding meters

        iv. Ultrasonic flowmeters

        v. Positive-displacement meters

    (c) Mass flowmeters (responding to mass flow rates)

        i. Coriolis meters

        ii. Critical flow venturi meters

        iii. Thermal mass flow anemometers

3. Velocity probes

    (a) Pressure probes

        i. Total pressure and Pitot-static tubes

        ii. Direction-sensing probes

    (b) Hot-wire and hot-film anemometers

    (c) Doppler-shift methods

        i. Laser-Doppler anemometer

        ii. Ultrasonic-Doppler anemometer (liquids only)

    (d) Particle-image velocimetry

4. Flow-visualization techniques

    (a) Smoke trails and smoke wires (gases)

    (b) Dye injection, chemical precipitates, particle tracers (liquids)

    (c) Hydrogen bubble technique (liquids)

    (d) Laser-induced fluorescence

    (e) Refractive-index change: interferometry, schlieren, shadowgraph

The preceding outline does not exhaust the list of flow-measuring methods, but it does attempt to include the most common types. Obstruction meters are probably those most often used in industrial practice. Application of some of the methods listed is so obvious that only passing note will be made of them. This is particularly true of *quantity* methods. Weight tanks are especially useful for steady-state calibration of liquid flowmeters, and no particular problems are connected with their use.

15.2   FLOW CHARACTERISTICS

We may measure the flow through a duct or pipe using its mass flow rate, $\dot{m}$, perhaps in kg/s, or its volume flow rate, Q, perhaps expressed in m³/s. These two quantities are related through the density of the fluid, ρ in kg/m³, by

$$\dot{m} = \rho Q \tag{15.1}$$

We may also define the average velocity, V, of the fluid in the duct using the cross-sectional area of the duct, A:

$$V = \frac{\dot{m}}{\rho A} = \frac{Q}{A} \tag{15.2}$$

When fluids move through uniform conduits at very low velocities, the motions of individual particles are generally along lines paralleling the conduit walls. The particle velocity is greatest at the center and zero at the wall, with the velocity distribution as shown in Fig. 15.1(a). Such a flow is called *laminar*.

As the flow rate is increased, a point is reached where the particle motion becomes *turbulent*, showing unsteady, random vortices throughout the pipe. In this case, we think of the time-average velocity distribution, which has the appearance shown in Fig. 15.1(b). The *approximate* velocity at which this change occurs is called the *critical velocity*.

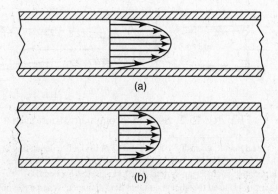

(a)

(b)

FIGURE 15.1: Velocity distribution for (a) laminar flow in a pipe or tube and (b) turbulent flow in a pipe or tube.

Experiments have shown that the critical velocity is a function of several factors that may be put in a dimensionless form called the Reynolds number, Re_D,[1] as follows:

$$\text{Re}_D = \frac{\rho V D}{\mu} \tag{15.3}$$

where

D = the diameter of the pipe (or hydraulic diameter if the pipe
        is not circular),[2]

ρ = the density of the fluid,

V = the average velocity of the fluid,

μ = the dynamic viscosity of the fluid

The critical Reynolds number for pipes is usually between 2100 and 4000. Below this range, the flow will be laminar. Above this range, it will be turbulent.

The volume flow rate, Q, through a pipe or duct is just the integral of the velocity distribution, $V(x, y)$, over the cross-sectional area A:

$$Q = \int_A V(x, y) \, dA \tag{15.4}$$

Flowmeters measure Q and/or V, while velocity probes measure $V(x, y)$. The output of a velocity probe can be integrated to obtain Q.

Changes in fluid velocity or elevation produce changes in pressure. For example, if an incompressible fluid flows from a section of large area at point 1 into a section of smaller area at point 2 (Fig. 15.2), its average velocity must increase, according to Eq. (15.2). The corresponding pressure change is given by Bernoulli's equation for incompressible flow[3] as

$$\frac{P_1 - P_2}{\rho} = \frac{V_2^2 - V_1^2}{2g_c} + \frac{(Z_2 - Z_1)g}{g_c} \tag{15.5}$$

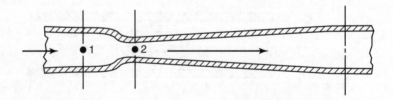

FIGURE 15.2: Section through a restriction in a pipe or tube.

---

[1] The units for dynamic viscosity, μ, are kg/m · s or lbf · s/ft^2, depending on the system of units used. We see that although the form of Eq. (15.3) as written is most common, inclusion of g_c is required to obtain a proper unit balance in the English engineering system. See Example 2.1 for details.

[2] The subscript D is used to indicate nominal pipe diameter. When a Reynolds number is based, for example, on the throat diameter of a venturi or an orifice, a lowercase d is commonly used (e.g., Re_d).

[3] Bernoulli's equation applies to steady lossless flow along a streamline, expressed in terms of local velocity. It may be applied approximately in terms of the average velocity in a duct, so long as the losses are negligible.

where

P = pressure,   N/m^2 (or Pa)   lbf/ft^2
ρ = density,   kg/m^3 lbm/ft^3
V = linear velocity,   m/s   ft/s
Z = elevation,   m   ft
g = acceleration due to gravity,   9.807 m/s^2 32.17 ft/s^2
g_c = dimensional constant   1 kg $\cdot$ m/N $\cdot$ s^2 32.17 lbm $\cdot$ ft/lbf $\cdot$ s^2

As written here, the relationship assumes that there is no mechanical work done on or by the fluid and that there is no heat transferred to or from the fluid as it passes between points 1 and 2. This equation provides the basis for evaluating the operation of flow-measuring devices generally classified as *obstruction meters* and of velocity sensors classified as *pressure probes*.

15.3   OBSTRUCTION METERS

Figure 15.3 shows three common forms of obstruction meters: the venturi tube, the flow nozzle, and the orifice. In each case, the basic meter acts as an obstacle placed in the path of the flowing fluid, causing localized changes in velocity. In conjunction with the velocity change, the pressure will change, as illustrated in the figure. At points of maximum restriction, hence maximum velocity, minimum pressures are found. The difference between this minimum pressure and the upstream pressure is measured so as to determine the velocity.

A certain portion of the pressure drop through an obstruction meter is irrecoverable owing to dissipation of kinetic energy; therefore, the output pressure will always be less than the input pressure. As the figure indicates, the venturi, with its guided reexpansion, is the most efficient. In contrast, losses of about 30%–40% of the differential pressure occur through the orifice meter.

15.3.1   Obstruction Meters for Incompressible Flow

For *incompressible fluids*, with Eq. (15.2),

$$\rho_1 = \rho_2 = \rho \quad \text{and} \quad \dot{m} = \rho A_1 V_1 = \rho A_2 V_2$$

where points 1 and 2 are as indicated in Fig. 15.3. If we let $Z_1 = Z_2$ and substitute $V_1 = (A_2/A_1)V_2$ into Eq. (15.5), we obtain

$$\frac{P_1 - P_2}{\rho} = \frac{V_2^2 - V_1^2}{2g_c} = \frac{V_2^2}{2g_c}\left[1 - \left(\frac{A_2}{A_1}\right)^2\right] \qquad (15.6)$$

which treats the flow as ideal (without any pressure losses). Solving Eq. (15.6) for V_2, we may compute the mass flow rate:

$$\dot{m}_{\text{ideal}} = \rho A_2 V_2 = \left[\frac{A_2}{\sqrt{1 - (A_2/A_1)^2}}\right]\sqrt{2g_c\rho(P_1 - P_2)} \qquad (15.7)$$

For a given meter, A_1 and A_2 are established values, and it is often convenient to calculate a *velocity of approach factor*, E:

$$E = \frac{1}{\sqrt{1 - (A_2/A_1)^2}} \qquad (15.7a)$$

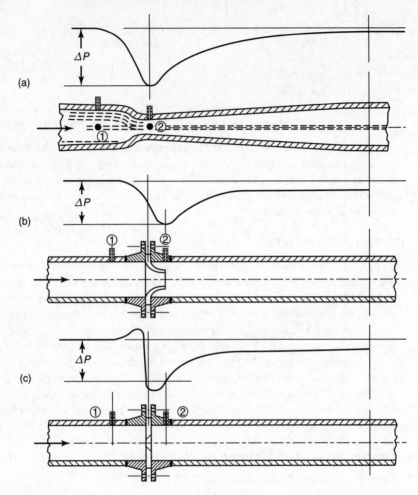

FIGURE 15.3: (a) A venturi tube, (b) a flow nozzle, and (c) an orifice flowmeter.

For circular sections, the area $= \pi(\text{diameter})^2/4$; hence

$$E = \frac{1}{\sqrt{1 - \beta^4}} \qquad (15.7b)$$

where

$$\beta = \frac{d}{D}$$

and

$$d = \text{the smaller diameter},$$
$$D = \text{the larger diameter}$$

To account for losses through the obstruction meter, the *discharge coefficient, C,* is introduced:

$$C = \frac{\dot{m}_{\text{actual}}}{\dot{m}_{\text{ideal}}}$$ (15.7c)

The discharge coefficient, C, is occasionally combined with the velocity of approach factor, E, to define the *flow coefficient, K.*

$$K = CE = \frac{C}{\sqrt{1 - \beta^4}}$$ (15.7d)

The flow coefficient K is used simply as a matter of convenience.

Therefore, we may write

$$\dot{m}_{\text{actual}} = CEA_2\sqrt{2g_c\rho(P_1 - P_2)}$$ (15.8a)

$$= KA_2\sqrt{2g_c\rho(P_1 - P_2)}$$ (15.8b)

and, with $Q_{\text{actual}} = \dot{m}_{\text{actual}}/\rho$,

$$Q_{\text{actual}} = KA_2\sqrt{\frac{2g_c}{\rho}}\ \sqrt{P_1 - P_2}$$ (15.8c)

15.3.2 Venturi Tube Characteristics

An international standard is available for three types of venturi tube flow meters [1]. The standard includes detailed specifications for their construction. Dimensions common to these designs are indicated in Fig. 15.4.

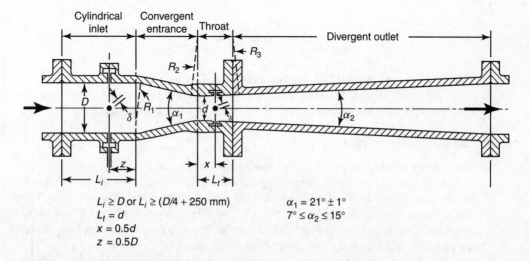

$L_i \geq D$ or $L_i \geq (D/4 + 250\ \text{mm})$ $\alpha_1 = 21° \pm 1°$
$L_t = d$ $7° \leq \alpha_2 \leq 15°$
$x = 0.5d$
$z = 0.5D$

FIGURE 15.4: Recommended proportions of standard venturi tubes. Specifications for the transition radii (R_1, R_2, R_3) and the pressure tap diameters, δ, are given in the ISO standard [1].

TABLE 15.1: Discharge Coefficients for Standard Venturi Tubes [1]

Type	Range	C
As cast	$100 \text{ mm} \leq D \leq 800 \text{ mm}$ $0.3 \leq \beta \leq 0.75$ $2.5 \times 10^5 \leq \text{Re}_D \leq 2 \times 10^6$	$0.984 \pm 0.7\%$
Machined	$50 \text{ mm} \leq D \leq 250 \text{ mm}$ $0.4 \leq \beta \leq 0.75$ $2 \times 10^5 \leq \text{Re}_D \leq 1 \times 10^6$	$0.995 \pm 1\%$
Rough welded sheet iron	$200 \text{ mm} \leq D \leq 1200 \text{ mm}$ $0.4 \leq \beta \leq 0.7$ $2 \times 10^5 \leq \text{Re}_D \leq 2 \times 10^6$	$0.985 \pm 1.5\%$

Standard venturi tubes are high-efficiency devices with discharge coefficients very near the ideal value of $C = 1$. Table 15.1 gives the discharge coefficients for the three standard designs, together with their uncertainties and the specified range of operation for each design. Outside the specified range of operation, the discharge coefficient may differ by a few percent.

15.3.3 Flow-Nozzle Characteristics

Figure 15.5 illustrates two types of flow nozzle, with the dimensions set by the relevant international standard [2]. The approach curve must be proportioned to prevent separation between the flow and the wall, and the parallel section is used to ensure that the flow fills the throat. The discharge coefficient may be calculated to an uncertainty of 2% with the following relationship, as given in the international standard:

$$C = 0.9965 - 6.53\sqrt{\frac{\beta}{\text{Re}_d}} \tag{15.9}$$

This equation applies for $0.2 \leq \beta \leq 0.8$ and $10^4 \leq \text{Re}_D \leq 10^7$.

The usual range of discharge coefficients is shown in Fig. 15.6. Observe that the flow nozzle has greater losses than the venturi, especially at low Reynolds numbers.

15.3.4 Orifice Characteristics

The primary variables in the use of flat-plate orifices are the ratio of orifice to pipe diameter and the pressure tap locations. Figure 15.7 illustrates typical orifice installations. Three tap locations are indicated: (1) flange taps, (2) "$1D$" and "$\frac{1}{2}D$" taps, and (3) corner taps. These are all shown in composite fashion in Fig. 15.7; however, only one set would be used for a given installation. As for the other obstruction meters, an international standard specifies the proper dimensions and positioning of the orifice meter [3].

As fluid flows through an orifice, the necessary transverse velocity components imparted to the fluid as it approaches the obstruction carry through to the downstream side. As a result, the minimum stream section occurs not in the plane of the orifice, but

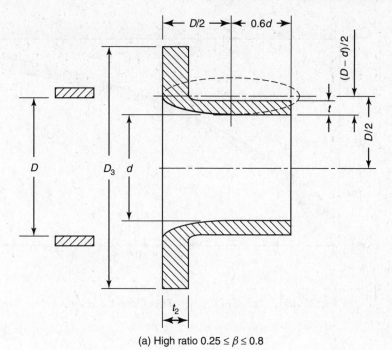

(a) High ratio $0.25 \leq \beta \leq 0.8$

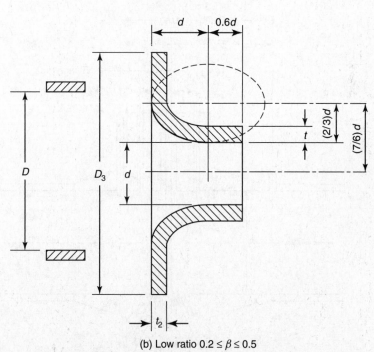

(b) Low ratio $0.2 \leq \beta \leq 0.5$

FIGURE 15.5: Dimensional relations for standard long-radius flow nozzles [2]: (a) high ratio $0.25 \leq \beta \leq 0.8$; (b) low ratio $0.2 \leq \beta \leq 0.5, 3$ mm $\leq t_2 \leq 0.15D$, $t \geq 3$ mm for $D > 65$ mm; else $t \geq 2$ mm.

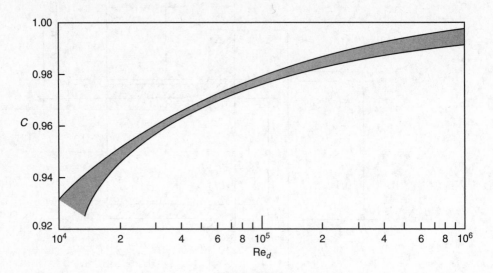

FIGURE 15.6: Range of discharge coefficients for long-radius flow nozzles.

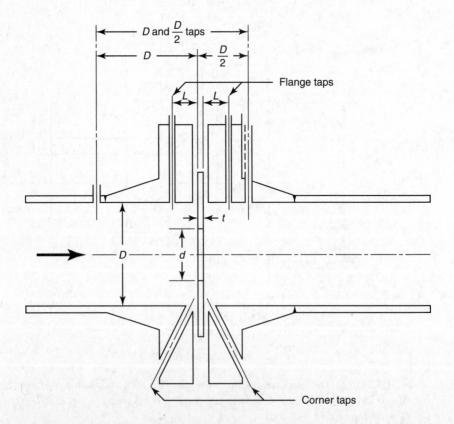

FIGURE 15.7: Locations of pressure taps for use with orifice meters; $L = 25.4$ mm.

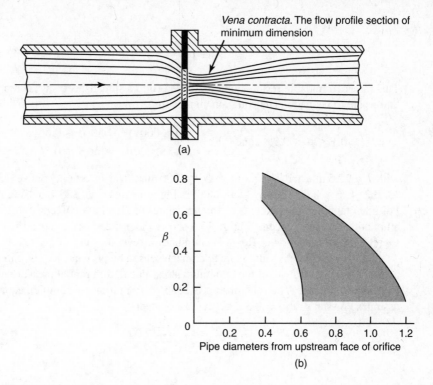

FIGURE 15.8: (a) Diagram illustrating vena contracta location for an orifice; (b) guide for locating vena contracta as measured from orifice face.

somewhat downstream, as shown in Fig. 15.8(a). The term *vena contracta* is applied to the location and conditions of this minimum stream dimension. This is also the location of minimum pressure. A guide for the location of the vena contracta is given in Fig. 15.8(b).

The formation of the vena contracta is sensitive to the shape of the orifice, which must therefore be manufactured with care to obtain accurate measurements. The downstream face of the orifice must be beveled at an angle of approximately 45° if the plate thickness is greater than 2% of the pipe diameter. Similarly, in order to dissipate the effects of pipe bends or valves, substantial lengths of straight pipe may be required upstream of the orifice (many tens of diameters, typically); and sufficient straight length is required downstream to avoid interference with the vena contracta. The standard should be consulted for further details [3].

The discharge coefficient of an orifice meter with corner tappings may be calculated from the following equation [3, 4]:

$$C = 0.5961 + 0.0261\beta^2 - 0.216\beta^8 + 0.000521 \left(\frac{10^6 \beta}{\text{Re}_D} \right)^{0.7}$$

$$+ (0.0188 + 0.0063A)\beta^{3.5} \left(\frac{10^6}{\text{Re}_D} \right)^{0.3} \quad (15.10)$$

in which

$$A = \left(\frac{19,000\beta}{\mathrm{Re}_D} \right)^{0.8} \tag{15.10a}$$

This rather formidable result should be coded (into a spreadsheet, say) if repetitive calculations are to be made. The equation applies for

$$0.1 \leq \beta \leq 0.75 \quad \text{and} \quad \mathrm{Re}_D \geq \begin{cases} 5,000 & \text{for } \beta \leq 0.56 \\ 16,000\beta^2 & \text{for } \beta > 0.56 \end{cases} \tag{15.10b}$$

with $d \geq 12.5$ mm and 71.12 mm $\leq D \leq 1000$ mm. The uncertainty of Eq. (15.10) is 0.5% for $0.2 \leq \beta \leq 0.6$ and $\mathrm{Re}_D \geq 10,000$, rising to as much as 1.25% outside these ranges. For other tapping arrangements or smaller values of D, additional terms enter Eq. (15.10), amounting to a few percent. Figure 15.9 shows the orifice discharge coefficients given by Eq. (15.10) as a function of Reynolds number for several values of β.

An orifice plate is vulnerable to damage caused by pressure surges, entrained debris, and the like. An estimate of the maximum stress due to differential pressure may be found from the following. The relationship is adapted from a rather complex equation [5] and assumes Poisson's ratio = 0.3 (as is typical of steels).

$$\sigma_{\max} = \frac{F D^2 \, \Delta P}{t^2} \tag{15.11}$$

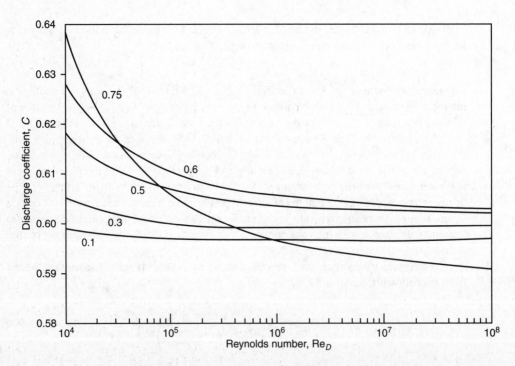

FIGURE 15.9: Range of discharge coefficients for flat-plate orifices with corner tappings as given by Eq. (15.10) for several values of β.

where

σ_{max} = maximum normal stress (radial direction at the clamped edge),
    t = plate thickness,
   ΔP = differential pressure across the plate,
     F = a factor, the value of which may be estimated from the following :

β	0.2	0.3	0.4	0.5	0.6	0.7	0.8
F	0.18	0.17	0.15	0.12	0.09	0.06	0.04

According to the standard, the plate thickness should lie in the range $0.005D \le t \le 0.05D$.

EXAMPLE 15.1

An orifice meter with corner taps is placed into a horizontal 200-mm-diameter (7.874 in.) line that carries 30 L/s (475.5 gallons per min.) of water. If the throat diameter is 120 mm (4.724 in.), what differential pressure may be expected across the pressure taps? The water temperature is 20°C (68°F).

Solve the problem (a) using SI units and (b) using the English engineering system. Use the tables in Appendix D for the properties of water.

Solution   (a) For water at 20°C, the density and dynamic viscosity are

$$\rho = 998.2 \text{ kg/m}^3$$
$$\mu = 10.05 \times 10^{-4} \text{ Pa} \cdot \text{s}$$

The areas of the pipe, A_1, and the orifice, A_2, are

$$A_1 = 0.03142 \text{ m}^2 \quad \text{and} \quad A_2 = 0.01131 \text{ m}^2$$

and the diameter ratio and velocity of approach factor are

$$\beta = \frac{d}{D} = \frac{120}{200} = 0.600$$

$$E = \frac{1}{\sqrt{1 - \beta^4}} = 1.072$$

The velocity in the pipe, V_1, is

$$V_1 = \frac{Q}{A_1} = \frac{30 \times 10^{-3}}{0.03142} = 0.9548 \text{ m/s}$$

The Reynolds number is

$$\text{Re}_D = \frac{\rho V_1 D}{\mu} = \frac{(998.2)(0.9548)(0.200)}{10.05 \times 10^{-4}} = 1.897 \times 10^5$$

Substituting for β and Re_D in Eqs. (15.10a) and (15.10), we find $A = 0.1055$ and

$$C = 0.5961 + 0.0261(0.6)^2 - 0.216(0.6)^8 + 0.000521 \left(\frac{10^6(0.6)}{1.897 \times 10^5} \right)^{0.7}$$

$$+ [0.0188 + 0.0063(0.1055)] (0.6)^{3.5} \left(\frac{10^6}{1.897 \times 10^5} \right)^{0.3} = 0.6084$$

so that $K = CE = (0.6084)(1.072) = 0.6522$. Finally, we may rearrange Eq. (15.8c), remembering that $g_c = 1$ in the SI System:

$$P_1 - P_2 = \left(\frac{Q}{KA_2} \right)^2 \left(\frac{\rho}{2g_c} \right)$$

$$= \left[\frac{30 \times 10^{-3}}{(0.6522)(0.01131)} \right]^2 \left[\frac{998.2}{2(1)} \right]$$

$$= 8255 \text{ Pa} = 8.26 \text{ kPa}$$

(b) For water at 68°F,

$$\rho = 62.32 \text{ lbm/ft}^3 \quad \text{and} \quad \mu = 2.099 \times 10^{-5} \text{ lbf} \cdot \text{s/ft}^2$$

In addition, $Q = 475.5 \text{ gpm} = 1.060 \text{ ft}^3/\text{s}$, $A_1 = 0.3382 \text{ ft}^2$, $A_2 = 0.1217 \text{ ft}^2$, and

$$\beta = \frac{4.724}{7.874} = 0.600$$

$$E = \frac{1}{\sqrt{1 - \beta^4}} = 1.072$$

The velocity in the pipe, V_1, is

$$V_1 = \frac{1.060}{0.3382} = 3.134 \text{ ft/s}$$

The Reynolds number must include g_c in this case (see Example 2.2):

$$Re_D = \frac{\rho V_1 D}{g_c \mu} = \frac{(62.32)(3.134)(7.874/12)}{(32.17)(2.099 \times 10^{-5})} = 1.898 \times 10^5$$

As before, substitution of β and Re_D into Eqs. (15.10a) and (15.10) yields $C = 0.6084$ and $K = 0.6522$. Finally,

$$P_1 - P_2 = [(1.060)/(0.6522)(0.1217)]^2 [(62.32)/2(32.17)]$$

$$= 172.7 \text{ lbf/ft}^2 = 1.203 \text{ lbf/in.}^2$$

See Problem 15.11 for the case of a vertical pipe.

15.3.5 Relative Merits of the Venturi, Flow Nozzle, and Orifice

Good pressure recovery and resistance to abrasion are the primary advantages of the venturi. They are offset, however, by considerably greater cost and space requirements than with the orifice and nozzle. The orifice is inexpensive and may often be installed between existing pipe flanges. However, its pressure recovery is poor, and it is especially susceptible to inaccuracies resulting from wear and abrasion. It is also quite sensitive to upstream flow disturbances, requiring either flow conditioning or significant lengths of straight pipe upstream. The flow nozzle possesses the advantages of the venturi, except that it has lower pressure recovery, and it has the added advantage of shorter physical length. It is expensive compared with the orifice, is relatively difficult to install properly, and is the least accurate of the three meters.

A major disadvantage of these meters is that the pressure drop varies as the square of the flow rate [Eqs. (15.8)]. This means that if these meters are to be used over a wide range of flow rates, pressure-measuring equipment of very wide range will be required. In general, if the pressure range is accommodated, accuracy at low flow rates will be poor: The small pressure readings in that range will be limited by pressure transducer resolution. One solution would be to use two (or more) pressure-measuring systems: one for low flow rates and another for high rates.

15.4 OBSTRUCTION METERS FOR COMPRESSIBLE FLUIDS

When compressible fluids flow through obstruction meters of the types discussed in Section 15.3, the density does not remain constant during the process; that is, $\rho_1 \neq \rho_2$. The usual practice is to base the energy relation, Eq. (15.5), on the density at condition 1 (Fig. 15.2) and to introduce a dimensionless *expansion factor*, Y, into Eq. (15.8a) as follows:

$$\dot{m} = CEA_2Y\sqrt{2g_c\rho_1(P_1 - P_2)} \qquad (15.12)$$

The value of Y is less than or equal to unity, with $Y = 1$ corresponding to incompressible flow.

The expansion factor, Y, may be determined theoretically for gases flowing through nozzles and venturis and experimentally for gases in orifice meters. For nozzles and venturis, Y may be calculated from this isentropic-flow relationship [1]:

$$Y = \left[\left(\frac{P_2}{P_1}\right)^{2/k} \left(\frac{k}{k-1}\right) \left(\frac{1 - (P_2/P_1)^{(k-1)/k}}{1 - (P_2/P_1)}\right) \left(\frac{1 - \beta^4}{1 - \beta^4(P_2/P_1)^{2/k}}\right) \right]^{1/2} \qquad (15.12a)$$

in which

$$k = \frac{\text{specific heat at constant pressure}}{\text{specific heat at constant volume}}$$

Approximate values of the specific heat ratio, k, are given for various gases in Table 15.2. For orifice meters, the following empirical relation should be used [3]:

$$Y = 1 - \left[0.351 + 0.256\beta^4 + 0.93\beta^8\right]\left[1 - \left(\frac{P_2}{P_1}\right)^{1/k}\right] \qquad (15.12b)$$

TABLE 15.2: Approximate Specific Heat Ratios at 20°C

Gas	Specific Heat Ratio, k
Ar, He, Kr, Xe	1.67
Air, CO, H_2, N_2, O_2	1.4
CH_4, CO_2	1.3

The uncertainty of this correlation, in percent, is $3.5(1 - P_2/P_1)$ when $P_2/P_1 \geq 0.75$. In both of these equations, absolute pressures must be used.

Expansion factors, Y, for venturis and nozzles with $k = 1.4$ are shown plotted against pressure ratio in Fig. 15.10(a). Similar values for orifice meters are given in Fig. 15.10(b). From these graphs, we see that compressibility become significant when $(P_1 - P_2)/P_1 \gtrsim 0.01$.

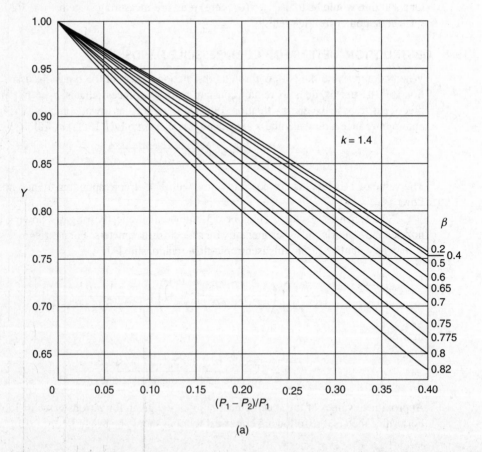

(a)

FIGURE 15.10: Expansion factors for $k = 1.4$: (a) venturis and nozzles [6]; (b) orifice meters.

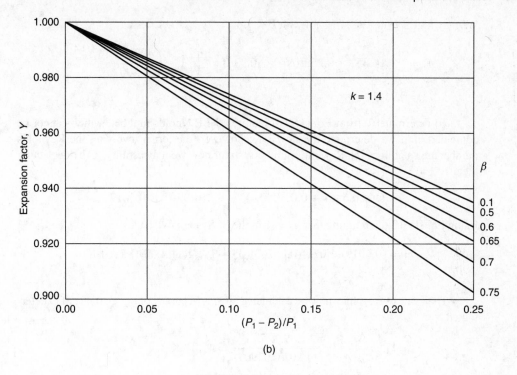

(b)

FIGURE 15.10: *Continued*

EXAMPLE 15.2

An orifice with $\beta = 0.6$ is used to measure flow of 30°C air through a 0.25 m diameter circular-sectioned duct. Estimate the flow rate if the differential pressure between vena contracta taps is 80 kPa and the upstream absolute pressure P_1 is 400 kPa.

Solution   We find the density and dynamic viscosity at 1 atm from Table D.3 in Appendix D. The density, but not the viscosity, must be corrected for the higher pressure in the duct:

$$\rho_1 = \rho_{atmos}\left(\frac{P_1}{P_{atmos}}\right)$$

$$= 1.14[(400 \times 10^3)/(101.325 \times 10^3)]\ \text{kg/m}^3$$

$$= 4.50\ \text{kg/m}^3$$

$$\mu_1 = 1.85 \times 10^{-5}\ \text{Pa} \cdot \text{s}$$

Using the given information, we compute

$$E = \frac{1}{\sqrt{1 - \beta^4}} = 1.072,$$

$$A_1 = 0.04909\ \text{m}^2,\ \text{and}$$

$$A_2 = 0.01767\ \text{m}^2$$

We may find Y from Eq. (15.12b) with $k = 1.4$:

$$Y = 1 - \left[0.351 + 0.256(0.6)^4 + 0.93(0.6)^8\right]\left[1 - \left(\frac{400 - 80}{400}\right)^{1/1.4}\right]$$

$$= 0.9411$$

To determine C from Eq. (15.10) requires the Reynolds number, which cannot be computed until we determine the flow rate; however, the terms involving the Reynolds number are small for high Reynolds numbers, so for now we may estimate C by neglecting them:

$$C \cong 0.5961 + 0.0261(0.6)^2 - 0.216(0.6)^8 = 0.6019$$

Using Eq. (15.12), and noting that $g_c = 1$ in the SI System, we have

$$\dot{m} = (0.6019)(1.072)(0.01767)(0.9411)\sqrt{2(1)(4.50)(80 \times 10^3)}$$

$$= 9.105 \text{ kg/s}$$

The value of Re_D may now be calculated and the accuracy of our estimate of C determined.

$$V_1 = \frac{\dot{m}}{\rho_1 A_1} = \frac{9.105}{(4.50)(0.04909)} = 41.21 \text{ m/s}$$

$$\text{Re}_D = \frac{\rho_1 V_1 D_1}{\mu_1}$$

$$= \frac{(4.50)(41.21)(0.25)}{1.85 \times 10^{-5}}$$

$$= 2.506 \times 10^6$$

Recomputing C using all the terms in Eq. (15.10) yields 0.6045, which is within 0.5% of the value we estimated previously. Greater refinement is not warranted.

15.4.1   Choked Flow and the Critical Flow Venturi Meter

One particular issue necessitates caution when considering the flow of compressible fluids. The pressure difference across a constriction (such as the throat of a nozzle) drives flow through it, as shown in Fig. 15.3. As flow rates are increased, perhaps by decreasing the downstream pressure, the flow velocity through the constriction eventually reaches the speed of sound (Mach number $= 1.0$). When this occurs, downstream pressure changes cannot propagate through the nozzle to affect the upstream flow. Any further decrease in pressure downstream of the constriction has no influence on the mass flow rate. The condition of Mach number equal to unity at a constriction is called *choked flow* [7].

The *critical pressure ratio* at which choked flow occurs may be expressed in terms of the upstream total pressure, P_{t1} and the throat pressure, P_2, as

$$\frac{P_2}{P_{t1}} = \left(\frac{2}{k+1}\right)^{k/(k-1)}$$

(15.13)

(see Section 15.8 for a discussion of total pressure). So long as the pressure ratio P_2/P_{t1} is greater than the value given in Eq. (15.13), the flow may be predicted by Eq. (15.12). However, when the pressure ratio given by Eq. (15.13) is reached, the flow is choked and the mass flow rate cannot be increased by lowering the pressure ratio further. For air, $k = 1.4$, and this ratio is 0.528.

Choked flow conditions can be used for very precise measurement of mass flow rate. Specifically, once sonic flow is reached in the throat of a venturi nozzle, the mass flow rate of a perfect gas is given by

$$\dot{m} = \frac{CA_2\,P_{t1}}{\sqrt{R\,T_{t1}}}\sqrt{k}\left(\frac{2}{k+1}\right)^{(k+1)/2(k-1)} \tag{15.14}$$

for C the discharge coefficient of the nozzle in question, A_2 the throat area, R the gas constant for the particular gas, and T_{t1} the total temperature (see Section 16.10.2). For flow-metering applications, toroidal or cylindrical throat venturi nozzles are used [8]. The discharge coefficients of properly designed nozzles are about 0.99, with a weak dependence on Reynolds number. Additional corrections may be applied to account for the fact that real gases are not perfect.

Critical flow venturi meters have the advantages of very high precision (with uncertainties potentially well below 1%), of a linear relationship between mass flow rate and upstream pressure, and of independence from downstream pressure. They have the disadvantage of relatively limited range. They have primarily been used for laboratory and calibration applications.

15.5 ADDITIONAL FLOWMETERS

The preceding discussion covers flowmeters that directly detect flow-induced pressure differentials. Many other physical phenomena can be adapted to the measurement of flow rate, most of which produce a response proportional to either volume flow rate or mass flow rate. Several of these devices are of high accuracy, especially the better turbine, Coriolis, and positive displacement meters. With the exception of the Coriolis meter, the flowmeters discussed in this section have a linear response to flow rate. This means that their resolution is the same at both high and low flow rates, which is an advantage over the obstruction meters. Table 15.3 summarizes the characteristics of various common flowmeters.

15.5.1 Turbine Meters

The familiar anemometer used by weather stations to measure wind velocity is a simple form of free-steam turbine meter. Somewhat similar rotating-wheel flowmeters have long been used by civil engineers to measure water flow in rivers and streams [10]. Both the cup-type rotors and the propeller types are used for this purpose. In each case the number of turns of the wheel per unit time is counted and used as a measure of the flow rate.

Figure 15.11 illustrates a typical adaptation of these methods to the measurement of flow in tubes and pipes. Rotor motion is sensed by a variable reluctance-type pickup coil. A permanent magnet is encased in the rotor body, and each time a rotor blade passes the pole of the coil, change in permeability of the magnetic circuit produces a voltage pulse which is received by the meter's signal processing electronics. The turbine blades are designed so as to make the rotation rate linearly proportional to the volume flow rate over some

TABLE 15.3: Characteristics of Common Flowmeters [4, 9]. Wider Ranges Are Available for Some Meters If Lower Accuracy Is Acceptable. Specialized Designs May Allow Broader Application of Some Types of Flowmeter

Type	Gas	Liquid	Slurry	Dirty	Accuracy	Range	ΔP	Cost
Differential Pressure Sensing								
Venturi	Y	Y	Y	Y	M	5:1	M	M/H
Flow nozzle	Y	Y		Y	M	5:1	M	M
Orifice	Y	Y			M	5:1	H	L/M
Variable area	Y	Y			VL	10:1	M	L
Volume Flow Rate Sensing								
Turbine	Y	Y			H	10:1(l) 30:1(g)	M	L/M
Electromagnetic		Y	Y	Y	M	100:1	L	M
Vortex	Y	Y		Y	M	10:1	M/H	L/M
Ultrasonic		Y			M	20:1	L	M/H
Positive displacement	Y	Y			H	10:1(l) 80:1(g)	M/H	M/H
Mass Flow Rate Sensing								
Coriolis	Y	Y	Y	Y	H	100:1	L/M	M/H
Critical flow venturi	Y	Y			H	3:1	H	M
Thermal mass flow	Y				L	15:1(l) 50:1(g)	M	M

Y—suitable for this application; H—high, M—medium, L—low, VL—very low; (l)—liquid range, (g)—gas range.
Dirty—suitable for dirty fluids; ΔP—pressure loss through meter.

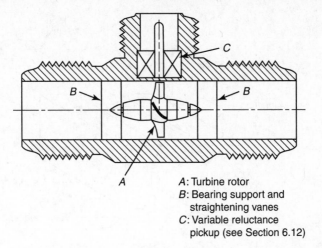

A: Turbine rotor
B: Bearing support and
   straightening vanes
C: Variable reluctance
   pickup (see Section 6.12)

FIGURE 15.11: Turbine flowmeter.

range of flow. Low accuracy models may use the rotor to drive a worm-gear connected to a mechanical counter; these types are sometimes used as water or gas meters.

Turbine meters can be among the most accurate of flow-metering devices. High accuracy models may have uncertainties of 0.25% to 0.5% for liquid flow and 0.25% to 1.5% for gas flow. Bearing wear makes periodic recalibration necessary. Turbine meters lose accuracy at low flow rates, which limits their range of operation. Maximum to minimum flow rates are about 10:1 for liquids and about 30:1 for gases. These meters are less accurate in unsteady flow conditions.

15.5.2 Electromagnetic Flowmeters

Electromagnetic flowmeters are based on Faraday's law of induced voltage for a conductor moving through a magnetic field, expressed by the relation

$$e = SBDV \times 10^{-4} \qquad (15.15)$$

where

$e =$ the induced voltage, in volts,

$S =$ the sensitivity coefficient for the particular meter,

$B =$ the magnetic flux density, in gauss,

$D =$ the length of the conductor, in m,

$V =$ the velocity of the conductor, in m/s

The basic flowmeter arrangement is as shown in Fig. 15.12. The flowing medium is passed through a pipe, a short section of which is subjected to a transverse magnetic flux. The fluid itself acts as the conductor having dimension D equal to pipe diameter and velocity V roughly equal to the average fluid velocity. Fluid motion relative to the field causes a voltage to be induced proportional to the fluid velocity. This emf is detected by electrodes placed in the conduit walls. An alternating magnetic flux is usually used, with

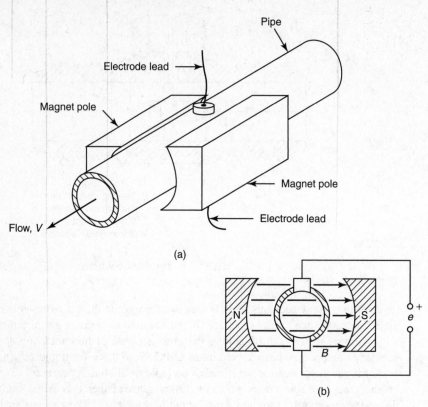

FIGURE 15.12: (a) Schematic arrangement of an electromagnetic flowmeter; (b) section showing electrodes and magnetic field.

both sine-wave and square-wave excitation being common. The output is detected and processed by appropriate circuitry. For a uniform magnetic flux and a uniform velocity, the sensitivity $S = 1$; in practice, neither B nor V is uniform, and the value of S is determined by calibration [11].

For most applications, the fluid need be only slightly electrically conductive (comparable to tap water), and the conduit must be lined with a nonconducting material. The electrodes are placed flush with the inner conduit surfaces and make direct contact with the flowing fluid. With purely sinusoidal excitation, eddy currents and induced voltages may cause drift in the signal at zero flow rate. By the use of either square-wave excitation or signal conditioning circuitry, most modern meters provide accurate accurate output down to zero flow.

Commercial electromagnetic flowmeters have rated accuracies of 0.5% to 2%. They are particularly useful for corrosive liquids and slurries, due to the absence of moving parts, and they are useful for homogeneous gas–liquid flows at low void fraction. The calibration is generally not sensitive to the particular liquid being metered.

15.5.3 Coriolis Flowmeters

Coriolis flowmeters measure the mass flow rate by detecting the Coriolis force on a section of the tube carrying the fluid when the tube is oscillated. This force is proportional to the product of the mass flow rate and the frequency of oscillation. The force is generally detected from the bending or flexion that it causes in the tube.

Coriolis meters can be very accurate, although the accuracy varies over the range of the meter. Values of 0.4% to 1% are typical. The range of flow rates spanned by a specific meter may be 100:1. Coriolis meters are suitable for both gases and liquids, and they are not directly sensitive to viscosity, density, or temperature.

15.5.4 Vortex Shedding Flowmeters

Vortex shedding meters are based on the fact that when a bluff body is placed in a stream, vortices are alternately formed, first to one side of the obstruction and then to the other (Fig. 15.13). The frequency of formation, f in Hz, is a function of flow rate:

$$f = \left(\frac{St}{D}\right) V \tag{15.16}$$

where

$\quad$ St = the Strouhal number,

$\quad$ V = the flow velocity, in m/s,

$\quad$ D = the dimension of the obstruction transverse to the flow direction, in m

The Strouhal number is approximately constant over a broad range of Reynolds numbers, giving a vortex shedding frequency that is proportional to velocity [12]. The value of St may be regarded as a calibration constant for the meter.

Various schemes are used to sense the frequency of vortex formation. The obstructing body may be mounted on an elastic support and the support oscillation sensed by one of a number of means. Heated thermistors downstream, with one to each side of the obstruction, and the flexing of diaphragms have been used to detect the vortices. Another technique makes use of an ultrasonic beam that is amplitude modulated by pulses. In another scheme, a differential piezoelectric pressure transducer is implanted in the rear of the bluff body.

Commercially available vortex shedding meters are accurate to 0.5% to 1.5% of flow rate. They are used for both gas and liquid flows. The calibration is relatively insensitive to viscosity, although the Reynolds number based on pipe diameter should generally be above 2×10^4.

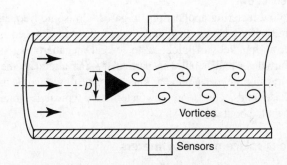

FIGURE 15.13: Vortex shedding caused by bluff body in flow stream.

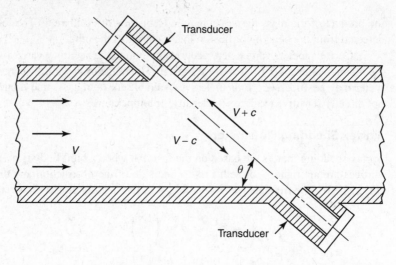

FIGURE 15.14: Ultrasonic flowmeter.

15.5.5 Ultrasonic Flowmeters

Ultrasound refers to acoustic waves at frequencies well above human hearing. Ultrasound can be used for time-of-travel measurements of mean flow velocity. A pair of piezoelectric or magnetostrictive transducers are located on the outside of a conduit a few inches apart (Fig. 15.14). One serves as a sound source and the other serves as the pickup. As the sound wave travels from the source to the receiver, its ordinary velocity in the stationary fluid, c, will be either increased or decreased due to the fluid velocity V. For example, if the sound wave crosses the pipe at an angle θ relative to the flow direction, then the effective velocity of the wave is $c \pm V \cos \theta$, depending on whether the wave moves upstream or downstream. Since a wave travels more slowly in the upstream direction than in the downstream direction, the flow velocity can be determined from the difference in travel time or the relative phase shift between upstream and downstream waves. To obtain both upstream and downstream waves, the function of the two ultrasonic transducers is reversed periodically.

The flow rate is obtained by multiplying the measured velocity V by the pipe's cross-sectional area, as if V were uniform over the cross section. This assumption is appropriate for turbulent flow.

In flow-metering applications in gases, ultrasonic frequencies may range from tens to hundreds of kilohertz. In liquids, where the sound speed is greater, the frequency is higher, from hundreds of kilohertz to several megahertz, so as to keep the wavelengths short. Particles—and bubbles in particular—can attenuate the sound waves, interfering with the transit time measurement; thus, this type of ultrasonic meter is best suited to clean, single-phase fluids. Ultrasonic flowmeters have typical accuracies ranging from a fraction of a percent to about $\pm 5\%$ [11].

15.5.6 Positive Displacement Flowmeters

Positive displacement meters have many forms and variations. Common examples are the water and gas meters used by suppliers to establish charges for services. Basically, displacement meters are rotating hydraulic or pneumatic chambers whose cycles of motion are recorded by some form of counter. The volume displaced on each cycle is known with great accuracy, allowing either the volume flow rate or the total volume passed to be determined. Only such energy from the stream is absorbed as is necessary to overcome the friction in the device, and this is manifested by a pressure drop between inlet and outlet. Many of the configurations used for pumps have been applied to metering. These include reciprocating and oscillating pistons, sliding-vane arrangements, various types of gears and rotors, the nutating (or nodding) disk, helical screw devices, flexing diaphragms, and so forth.

Positive displacement meters have the advantages of high accuracy (from 0.2% to 2%), a very wide range of flow rate (perhaps 100:1 in a given meter), the ability to retain accuracy at very low rates or during on/off flow conditions, and a general insensitivity to the viscosity or velocity profile of the fluid. These meters have the disadvantage of requiring very clean fluids (since particulates may cause wear or jamming), of the potential to create blocked lines (should they jam), and of inducing flow pulsations.

15.5.7 The Variable-Area Meter

The variable-area flowmeter is shown in cross section in Fig. 15.15. This instrument is also known by the trade name *Rotameter*. Two parts are essential, the float and the tapered tube in which the float is free to move. The term *float* is somewhat a misnomer in that it must

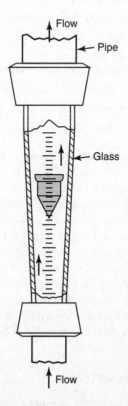

FIGURE 15.15: Variable-area flowmeter.

be heavier than the liquid it displaces. As fluid flows upward through the tube, three forces act on the float: a downward gravity force, and upward pressure, and viscous drag forces.

For a given flow rate, the float assumes a position in the tube where the forces acting on it are in equilibrium. The total drag is dependent on flow rate and the annular area between the float and the tube, and, for a given tube taper, the float's position will be determined by the flow rate alone. A basic equation for the variable-area meter has been developed in the following form [13]:

$$Q = A_w C \left[\frac{2g v_f (\rho_f - \rho_w)}{A_f \rho_w} \right]^{1/2} \tag{15.17}$$

where

$Q =$ the volumetric rate of flow,

$v_f =$ the volume of the float,

$g =$ acceleration due to gravity,

$\rho_f =$ the float density,

$\rho_w =$ the liquid density,

$A_f =$ the area of the float,

$C =$ the discharge coefficient,

$A_w =$ the area of the annular orifice

$\quad = \left(\frac{\pi}{4}\right) [(D + by)^2 - d^2]$,

$D =$ the diameter of the tube when the float is at the zero position,

$b =$ the change in tube diameter per unit change in height,

$d =$ the maximum diameter of the float,

$y =$ the height of the float above zero position

Normally, the values of D, b, and d will be selected to produce an essentially linear variation of A_w with y. Thus, the flow rate is a linear function of the reading.

Certain disadvantages of the variable-area meter are that: the meter must be installed in a vertical position; the float may not be visible when opaque fluids are used; it cannot be used with liquids carrying large percentages of solids in suspension; for high pressures or temperatures, it is expensive; it usually does not provide electronic readout; the calibration is affected by fluid density; and it is not terribly accurate. Advantages include the following: there is a uniform flow scale over the range of the instrument, with the pressure loss fixed at all flow rates; the capacity may be changed with relative ease by changing float and/or tube; many corrosive fluids may be handled without complication; the condition of flow is readily visible; and models are available for very low flow rates. Variable-area meters are typically accurate to no better than 2% of full scale and may have accuracies of only 10% or so.

15.6   CALIBRATION OF FLOWMETERS

Facilities for producing standardized flows are required for flowmeter calibration. Fluid at known rates of flow must be passed through the meter and the rate compared with the meter readout. When the basic flow input is determined through measurement of time and either weight (mass flow) or linear dimensions (volumetric flow), the procedure may be called *primary calibration*. After receiving a primary calibration, a meter may then be used as a *secondary standard* for standardizing other meters through *comparative calibration*. In general, higher accuracy may be obtained for liquid calibrations than for gas calibrations, owing to the low densities and handling difficulties associated with gases [11].

Primary calibration is usually carried out at a constant flow rate by measuring the total flow for a predetermined period of time. Primary calibration in terms of mass is commonly accomplished by means of a *weigh tank*, in which the liquid is collected and weighed (see Section 3.11.2). Although the latter method is normally used only for liquids, with proper facilities it may also be used for calibration with gases. Volumetric displacement of a liquid may be measured in terms of the liquid level in a carefully measured tank or container. For gases, at moderate rates, volume may be determined through use of a *bell prover*, which consists of an inverted bell that creates a gas chamber above a liquid; the lower rim of the bell lies beneath the liquid surface (see Fig. 15.16). Gas is pumped into the chamber through a pipe leading to the flowmeter being calibrated; as the gas flows in, the bell rises further above the liquid, maintaining a constant pressure within the chamber. The bell's displacement provides the measure of volume [14].

Figure 15.17 illustrates a method obviating the requirement for direct mass or volume measurement. A standpipe of known capacity (diameter) is used as a collector. We see that the pressure or head at the base is the analog of the mass or volume as it is collected.

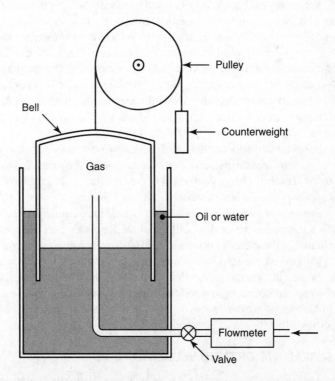

FIGURE 15.16: The bell prover for gas flow calibrations.

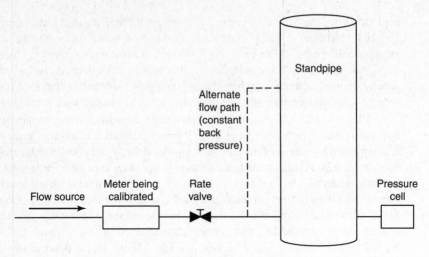

FIGURE 15.17: Standpipe employed for flowmeter calibration.

Calibrations within 0.1% for meters handling 20 to 300,000 kg/h have been reported [15]. The accuracy of this approach will clearly depend upon accurate knowledge of the liquid's density.

Secondary calibration may be accomplished by simply placing a secondary standard in series with the meter to be calibrated and comparing their respective readouts over the desired range of flow rates. High-accuracy meters, such as turbine meters and positive displacement meters, are generally used as secondary standards. It is clear that this procedure requires careful consideration of meter installations, minimizing interactions or other disturbances such as might be caused by nearby line obstructions, like elbows or tees. Flow straighteners may be placed upstream of the standard meter for this purpose. Standard meters are often used in pairs, in order that drift in one of the standard meters will appear as a difference between them.

For obstruction meters that are designed to meet an ISO or other standard, a calibration may be done by measuring the relevant dimensions of the meter and using the corresponding equations for the discharge coefficient or flow coefficient as given in the governing standard. This approach should lead to uncertainties of 1% to 2%.

Changes in the viscosity and density of the fluid being metered will affect the calibration of a meter, insofar as they will affect the Reynolds number and other parameters that may influence the meter's performance. These properties depend not only on the particular fluid, but also on its temperature and pressure. Liquid viscosity is especially sensitive to temperature. In general, calibration should be done using the specific fluid to be metered at the expected operating temperature and pressure. Corrections may be applied to move a calibration to other conditions or fluids, provided that one knows how (or whether) these properties affect the meter's response.

15.7   MEASUREMENTS OF FLUID VELOCITIES

Flow rate is generally proportional to some flow velocity; hence, by measuring the velocity, a measure of flow rate is obtained. Often velocity per se is desired, either the velocity of the

TABLE 15.4: Velocity Measurement Techniques.   Response and Resolution Are Representative Values. Frequency Response in Particular Will Differ Depending Upon Probe Design, Operating Conditions, and Ancillary Hardware; for example, Particle Image Velocimeter (PIV) Resolution Can Reach 0.1 μm when Combined with Microscopy.

Device	Cost	Frequency Response (Hz)	Spatial Resolution (mm)	Reversing Flow?
Pitot tube	Low	< 0.1	5	No
Hot-wire anemometer	Moderate	10,000	1	No
Laser-Doppler anemometer	High	100	0.1	Yes
Ultrasonic-Doppler anemometer	Moderate		10	Yes
Particle image velocimeter	High	30	1	Yes

fluid itself or the velocity *relative* to a fluid. An example of the latter is an aircraft moving through the air [16]. The following sections deal with measurement of the absolute and relative velocities of fluids. The instruments considered measure local velocity—at one point—rather than a spatially averaged velocity. We shall refer to the local velocity as V for simplicity of notation.

Fluid velocities may vary in time, either as a result of unsteadiness in the flow system or as the result of turbulence. In the case of turbulent flow, the velocity fluctuates in time about a mean value. The frequency of these fluctuations may reach some number of kilohertz, depending upon the flow conditions. The velocity to be measured may be either a time-average velocity or an instantaneous velocity. For the latter, of course, we must consider the frequency response of the velocity sensor.

Table 15.4 lists several common techniques for measuring fluid velocity, together with typical frequency response and spatial resolution; the ability to measure a flow that reverses its direction is also noted. We shall discuss these devices in the sections that follow.

15.8   PRESSURE PROBES

A common reason to measure pressure at some point in a fluid is to determine flow conditions at that point. The flowing medium may be gaseous or liquid in a symmetrical conduit or pipe or in a more complex configuration such as a jet engine or compressor.

Point measurement of pressure is accomplished by the use of tubes joining the location in question with some form of pressure transducer. A *pressure probe* is intended insofar as possible to obtain a reliable and interpretable indication of the pressure at the measurement point. Therein lies a difficulty, however, for the mere presence of the probe will alter, to some extent, the pressure being measured.

Many different types of pressure probes are used, with the selection depending on the information required, space available, pressure gradients, and constancy of flow magnitude and direction. Basically, pressure probes measure one or both of two different pressures (Fig. 15.18). We briefly discussed these two pressures, the *static* and *total* pressures, in Section 14.2, and we indicated that the difference is a result of the flow velocity. Specifically,

$$P_t = P_s + P_v \qquad (15.18)$$

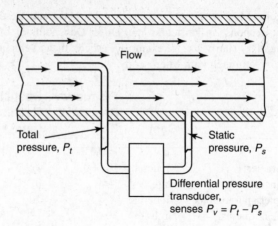

FIGURE 15.18: Total and static pressure probes.

where

P_t = the total pressure (often called the *stagnation* pressure),

P_s = the static pressure, and

P_v = the velocity pressure

15.8.1   Incompressible Fluids

Referring to Eq. (15.5) for incompressible fluids, we may write

$$P_t = P_s + \frac{\rho V^2}{2g_c}$$

where

ρ = the fluid density,

V = the fluid velocity, and

g_c = the dimensional conversion constant for English units ($g_c = 1$ in SI units)

Solving for velocity, we obtain

$$V = \sqrt{\frac{2g_c(P_t - P_s)}{\rho}} = \sqrt{\frac{2g_c(\Delta P)}{\rho}} \qquad (15.19)$$

where $\Delta P = P_t - P_s$. From this we see that velocity may be determined simply by measuring the difference between the total and static pressures.

When velocity is used to measure flow rate, consideration must be given to the velocity distribution across the channel or conduit. A mean may be found by traversing the area to determine the velocity profile, from which the average may be calculated, or a multiplication constant may be determined by calibration (see Problem 15.33).

15.8.2 Compressible Fluids

For high speeds, the compressibility of gases must be accounted for in determining the velocity from pressure measurements. At the probe tip, an isentropic compression of the gas occurs as the pressure changes from P_s to P_t, provided that flow is subsonic. If the flow is supersonic, a shock wave will be produced ahead of the probe. Under subsonic conditions, the equations of isentropic flow [7] lead to the following relationship between velocity and pressure:

$$V = \sqrt{2\left(\frac{k}{k-1}\right)\left(\frac{P_s}{\rho_s}\right)\left[\left(\frac{P_t}{P_s}\right)^{(k-1)/k} - 1\right]g_c}$$

$$= \sqrt{2\left(\frac{k}{k-1}\right)\left(\frac{P_s}{\rho_s}\right)\left[\left(1 + \frac{\Delta P}{P_s}\right)^{(k-1)/k} - 1\right]g_c} \qquad (15.19a)$$

where

k = the ratio of specific heats (see Table 15.2), and

ρ_s = the density of the flowing gas (the *static* density)

This equation should be used for when the flow speed is between 30 and 100% of the speed of sound, that is, for Mach numbers between 0.3 and 1. For supersonic flow, an additional correction is required to account for the shock wave. The speed of sound in an ideal gas is given by $c = \sqrt{kRT}$, where R is the gas constant for the gas in question and T is the absolute temperature.

15.8.3 Total-Pressure Probes

Obtaining a measure of total or stagnation pressure is usually somewhat easier than getting good measure of static pressure, except in cases such as a jet of air issuing into an open room, when a barometer reading provides the static pressure. The simple Pitot tube (named for Henri Pitot) shown at A in Fig. 14.2 is usually adequate for determining stagnation pressure. More often, however, the Pitot tube is combined with static openings, constructed as shown in Fig. 15.19. This is known as a *Pitot-static tube*, or sometimes as a Prandtl-Pitot tube. For steady-flow conditions, a simple differential manometer, often of the inclined type, suffices for pressure measurement, and $P_t - P_s$ is determined directly. When variable conditions exist, some form of pressure transducer, such as one of the diaphragm types, may be used. Of course, care must be exercised in providing adequate response, particularly in the connecting tubing (see Section 14.10).

A major problem in the use of an ordinary Pitot-static tube is to obtain proper alignment of the tube with flow direction. The angle formed between the probe axis and the flow streamline at the pressure opening is called the *yaw angle*. This angle should be zero, but in many situations it may not be constant: The flow may be fixed neither in magnitude nor in direction. In such cases, yaw sensitivity is very important. The Pitot-static tube is particularly sensitive to yaw, as shown in Fig. 15.20. Although sensitivity is influenced by orientation of both stagnation and static openings, the latter probably has the greater effect.

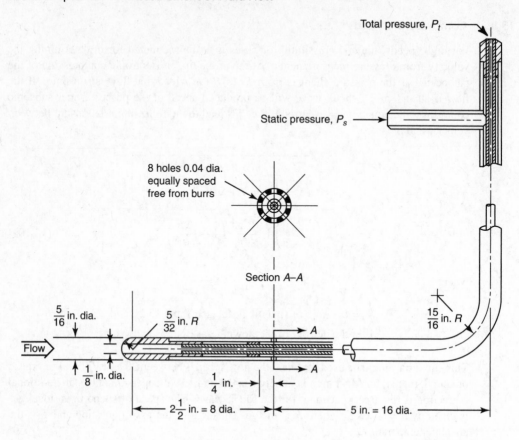

FIGURE 15.19: A Pitot-static tube.

The Kiel tube, designed to measure total or stagnation pressure only (there are no static openings), is shown in Fig. 15.21. It consists of an impact tube surrounded by what is essentially a venturi. The curve demonstrates the striking insensitivity of this type of probe to variations in yaw. Modifications of the Kiel tube make use of a cylindrical duct, beveled at each end, rather than the streamlined venturi. This appears to have little effect on the performance and makes the construction much less expensive.

15.8.4 Static-Pressure Probes

Static-pressure probes have been used in many different forms [17]. Ideally, the simple opening with axis normal to flow direction should be satisfactory. However, slight burrs or yaw introduce appreciable errors. As mentioned previously, in many situations the yaw angle may be continually changing. For these reasons, special static-pressure probes may be used. Figure 15.22 shows several probes of this type and the corresponding yaw sensitivities.

As mentioned earlier, the mere presence of the probe in a pressure-flow situation alters the parameters to be measured. Probes interact with other probes, with their own supports, and with duct or conduit walls. Such interaction is primarily a function of geometry and relative dimensional proportions; it is also a function of Mach number. Much work has been conducted in this area (see, e.g., [18, 19]).

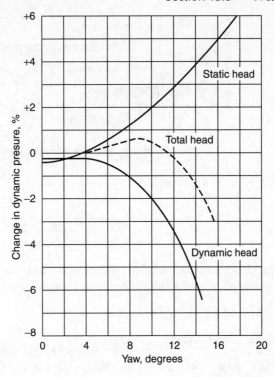

FIGURE 15.20: Yaw sensitivity of a standard Pitot-static tube. (Courtesy: The Airflo Instrument Company, Glastonbury, Connecticut)

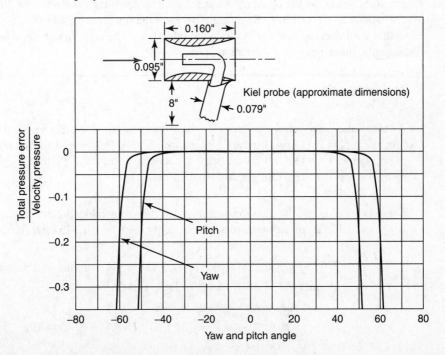

FIGURE 15.21: Kiel-type total-pressure tube and plot of yaw sensitivity. (Courtesy: The Airflo Instrument Company, Glastonbury, Connecticut)

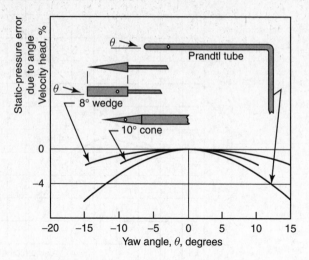

FIGURE 15.22: Angular characteristics of certain static-pressure-sensing elements. (Courtesy: Instrument Society of America, Research Triangle Park, North Carolina)

15.8.5 Direction-Sensing Probes

Figure 15.23 illustrates two forms of direction-sensing or yaw-angle probes. Each of these probes uses two impact tubes. In each case the probe is placed transverse to flow and is rotatable around its axis. The angular position of the probe is then adjusted until the pressures sensed by the openings are equal. When this is the case, the flow direction will correspond to the bisector of the angle between the openings. Probes are also available with a third opening midway between the other two. The additional hole, when properly aligned, senses maximum stagnation pressure.

EXAMPLE 15.3

A Pitot-static tube is used to determine the velocity of air at the center of a pipe. Static pressure is 124 kPa (18 psia), the air temperature is 26.7°C (80°F), and a differential pressure of 96.5 mm of water (3.8 in.) is measured. What is the air velocity? Perform the calculations using: (a) the SI system of units; and (b) the English system.

Solution (a) Using the SI system of units, we find from Table D.2 that the density of air is $\rho_{26.7} = 1.153 \text{ kg/m}^3$ at standard atmospheric pressure of 101.325 kPa. At 124 kPa,

$$\rho_{26.7} = (124/101.325) \times 1.153 = 1.411 \text{ kg/m}^3,$$
$$1 \text{ mm H}_2\text{O} = 9.807 \text{ Pa (from Table 14.1)},$$
$$\Delta P = 96.5 \times 9.807 = 946.4 \text{ Pa},$$
$$P_t = P_s + \Delta P = 124,000 + 946.4 = 124.946 \text{ kPa},$$
$$k = 1.4$$

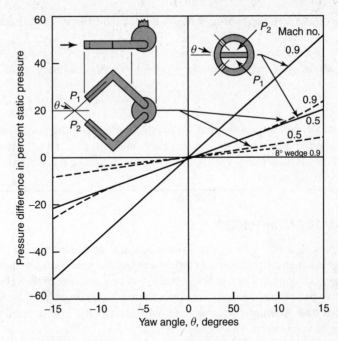

FIGURE 15.23: Special direction-sensing elements and their yaw characteristics. (Courtesy: Instrument Society of America, Research Triangle Park, North Carolina)

Using Eq. (15.19a) with $g_c = 1$, we have

$$V = \sqrt{2(1.4/0.4)(124,000/1.411)[(124,946/124,000)^{(0.4/1.4)} - 1] \times 1}$$
$$= 36.57 \text{ m/s}$$

(b) Using the English engineering system of units, we find from Table D.2 that $\rho_{80} = 0.0735$ lbm/ft^3 at a pressure of 14.7 psia. At 18 psia,

$$\rho_{80} = (0.0735)(18/14.7) = 0.0900 \text{ lbm/ft}^3,$$
$$1 \text{ in. H}_2\text{O} = 5.202 \text{ lbf/ft}^2 \text{ (using results from Table 14.1)},$$
$$\Delta P = 3.8 \times 5.202 = 19.77 \text{ lbf/ft}^2,$$
$$P_s = 18 \times 144 = 2592 \text{ lbf/ft}^2,$$
$$P_t = P_s + \Delta P = 2592 + 19.77 = 2612 \text{ lbf/ft}^2,$$
$$k = 1.4$$

Substituting in Eq. (15.19a) gives us

$$V = \sqrt{2(1.4/0.4)(2592/0.090)[(2612/2592)^{(0.4/1.4)} - 1] \times 32.17}$$
$$= 119 \text{ ft/s (or 36.4 m/s)}$$

Close scrutiny of the arithmetical manipulations required in this example clearly shows that calculation errors of considerable size may easily result from the fact that P_t and

P_s are quite commonly of very nearly the same magnitudes: Calculation of the ratio must be quite precise.

To determine the effect of neglecting compressibility, we may substitute the values for part (a) of this example directly into Eq. (15.19), obtaining

$$V = \sqrt{2 \times 1 \times (946.4)/1.411} = 36.63 \text{ m/s}$$

Essentially the same answer is obtained. In this case, the flow is nearly incompressible because the Mach number is low: The sound speed in air at this temperature is about 347 m/s. At higher flow speeds, the agreement would be much worse. (Problem 15.31 also bears on this matter.)

15.9 THERMAL ANEMOMETRY

A heated object in a moving stream loses heat at a rate that increases with the fluid velocity. If the object is electrically heated at a known power, it will reach a temperature determined by the rate of cooling. Thus, its temperature will be a measure of the velocity. Conversely, the heating power may be controlled by a feedback system to hold the temperature constant. In that case, heating power is a measure of velocity. These relations are the basis for *thermal anemometry*.

The most commonly used thermal velocity probes are the *hot-wire* and *hot-film* *anemometers*. The hot-wire anemometer consists of a fine wire supported by two larger-diameter prongs; an electric current heats the wire to a temperature well above the fluid temperature (Fig. 15.24). Typically, the wire is 4 to 10 μm in diameter, is 1 mm in length, and is made of platinum or tungsten. These fine wires are extremely fragile, so hot-wire probes are used only in clean gas flows. In liquids, or in rugged gas-flow applications, the

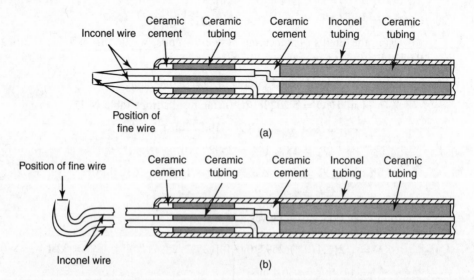

FIGURE 15.24: Two forms of hot-wire anemometer probes: (a) wire mounted normal to probe axis, (b) wire mounted parallel to probe axis.

hot-film probe is used instead. Here, a quartz fiber is suspended between the prongs, and a platinum film coated onto the fiber surface provides the electrically heated element. The fiber, with a diameter of 25 to 150 μm, has much greater mechanical strength than fine wire.

Hot-wire and hot-film probes are most often operated at constant temperature using a feedback-controlled bridge (Fig. 15.25). The probe forms one leg of a voltage-sensitive deflection bridge (Section 7.9). Current flowing through the bridge provides heating power to the wire. The resistance of the wire is a function of temperature (Section 16.4.1), and any increase in flow velocity tends to lower the wire's temperature, reducing its resistance and causing bridge imbalance. The voltage imbalance drives a feedback amplifier, which increases the voltage and current supplied to the bridge; the added current increases the heating power, thus raising the wire temperature and resistance and restoring bridge balance. The voltage supplied to the bridge also serves as the circuit output. High-quality bridges are available commercially; alternatively, acceptable bridges can be built at minimal cost [20].

The relation between flow speed and bridge output is obtained by equating the electrical power to the heat loss. The heating power in watts is

$$\text{Electrical power} = \frac{e_w^2}{R_w} = \left(\frac{R_w}{R_o + R_w} e_o \right)^2 \frac{1}{R_w}$$

$$= e_o^2 \left(\frac{R_w}{(R_o + R_w)^2} \right)$$

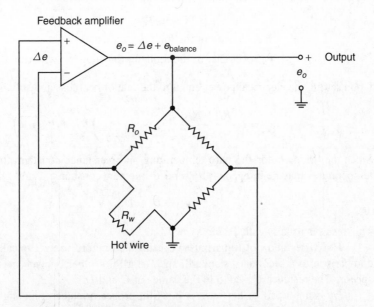

FIGURE 15.25: Constant-temperature-anemometer bridge circuit.

where

$$e_w = \text{the voltage across the wire (V)},$$
$$R_w = \text{the wire resistance (V)},$$
$$R_o = \text{the upper leg's resistance } (\Omega), \text{ and}$$
$$e_o = \text{the output voltage } (\Omega)$$

Heat loss from the wire is mainly by convection to the fluid: Thermal radiation is negligible, and conduction to the supporting prongs may be accounted for by calibration. The rate of heat loss in watts is given by

$$\text{Rate of heat loss} = A_w h (T_w - T_f)$$

where

$$A_w = \text{the wire surface area } (\text{m}^2),$$
$$T_w = \text{the wire temperature (K)},$$
$$T_f = \text{the fluid temperature (K), and}$$
$$h = \text{convective heat transfer coefficient } (\text{W/m}^2\text{K})$$

The heat transfer coefficient for small-diameter wires is given by

$$h = A + B\sqrt{\rho V}$$

where

$$A \text{ and } B = \text{constants that depend on the wire diameter, the fluid,}$$
$$\text{and the temperatures;}$$
$$\rho = \text{density of the fluid } (\text{kg/m}^3); \text{ and}$$
$$V = \text{velocity of the fluid approaching the wire (m/s)}$$

Upon setting the electrical power equal to the rate of heat loss and solving, we obtain

$$e_o^2 = \left(\frac{(R_o + R_w)^2}{R_w} (T_w - T_f) A_w \right) \left(A + B\sqrt{\rho V} \right) \tag{15.20}$$

Since the bridge holds the wire temperature and resistance constant, the resistances and temperatures may be lumped with A and B into new constants, C and D:

$$e_o^2 = C + D\sqrt{\rho V} \tag{15.20a}$$

This result is usually called *King's law* [21].

Hot wires and hot films must be calibrated before use. Typically, a side-by-side comparison to a secondary standard, such as a Pitot tube, is made over a range of flow speeds. The results are used to fit the values of C and D.

By using a pair of hot-wire or hot-film probes oriented at an angle to one another, two components of the flow velocity vector may be measured. Since the wires are typically oriented at $\pm 45°$ to the probe body, such an arrangement is called an *X-wire probe*.

Significant complications arise if the fluid temperature varies. Small changes in T_f may be corrected for by placing a fine-wire resistance thermometer (a *cold-wire* probe) adjacent to the hot wire, so as to measure the temperature difference in Eq. (15.20). Larger temperature changes necessitate a further modification of Eq. (15.20), in which A and B become functions of temperature [22, 23].

The primary value of the hot-wire anemometer lies in its high frequency response and excellent spatial resolution. The frequency response of a hot wire can easily reach 10 kHz; if loss of spatial resolution [24] is unimportant, hot wires can be applied at frequencies several times higher. On the other hand, owing to their relatively high cost and inherent fragility, hot wires are usually justifiable only when their fast response is essential to the measurement at hand. The most common situation requiring high-frequency response is the measurement of the fluctuating velocity in turbulent flows.

Thermal flowmeters are also available. These devices measure flow rates in tubes or pipes. In one form, a section of a pipe wall is electrically heated; the resulting increase in fluid temperature, downstream, is proportional to the mass-flow rate in the pipe. Some automotive fuel-injection systems use a type of thermal flowmeter as part of the engine control system. A hot-film sensor is located in the intake manifold; this sensor is of relatively rugged construction, lowering its frequency response in favor of improved reliability. The hot-film signal identifies ρV; when the latter is multiplied by the manifold cross section, the total mass-flow rate is obtained. A temperature sensor is incorporated to compensate for changes in environmental conditions. The measured mass-flow rate enables a microprocessor to set the fuel injectors for the correct fuel mixture.

Hot-wire and hot-film anemometers have been studied extensively, and a large body of literature is available on their performance and use [23, 25].

15.10   DOPPLER-SHIFT MEASUREMENTS

When light or sound waves of a given frequency are scattered off of particles in a moving fluid, they undergo a frequency change or *Doppler shift*. The Doppler shift of the scattered waves is proportional to the speed of the scattering particle. Thus, by measuring the frequency difference between the scattered and unscattered waves, the particle or flow speed may be found. The wave sources most often used in fluid velocimetry are laser light and ultrasound.

The Doppler shift is responsible for the familiar change in pitch as a moving source of sound passes, such as that heard in a siren or car horn. The size of the Doppler shift in a wave scattered from a moving particle depends on the particle's direction relative to the incident wave, as well as the observer's position (Fig. 15.26). A calculation shows that the frequency shift observed is [26]:

$$\Delta f = \left(\frac{2V}{\lambda}\right) \cos \beta \sin \left(\frac{\alpha}{2}\right) \qquad (15.21)$$

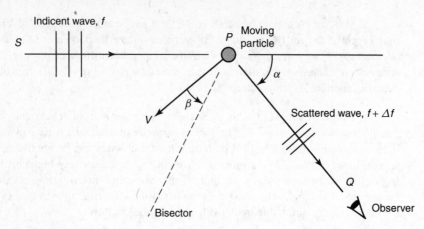

FIGURE 15.26: When an incident wave of frequency f is scattered from a particle moving at speed V, the observer sees a scattered wave of frequency $f + \Delta f$, where Δf is the Doppler frequency shift (Eq. 15.21).

where

Δf = the Doppler frequency shift (Hz),

V = the particle velocity (m/s),

λ = the wavelength of the original wave before scattering (m),

β = the angle between the velocity vector and the bisector of the angle SPQ,

α = the angle between the observer and the axis of the incoming wave

For the purpose of flow measurement, the important aspect of the Doppler shift is its proportionality to the particle velocity. If we can measure the shift, we can find the particle speed; and if the particle moves with the flow, this speed should equal the fluid speed.

Because laser light and ultrasonic waves have relatively high frequencies, the Doppler shift is only a small fraction of the original wave's frequency. For example, the fractional frequency change in scattered laser light may be only $1/10^8$ (see Example 4.3). The Doppler frequency is usually resolved by *heterodyning* the scattered wave with an unshifted reference wave to produce a measurable beat frequency (see Section 4.4.1).

15.10.1 Laser-Doppler Anemometry

The original *laser-Doppler anemometer* (LDA) used separate scattering and reference beams to create an optical heterodyne at a photodetector [Fig. 15.27(a)]. The light scattered by particles in the flow interfered with the light from the reference beam to produce beats at a frequency of one-half the Doppler shift [see Eq. (4.9)]. Unfortunately, these systems were fairly difficult to align because the intensity of the reference beam must nearly equal that of the scattered light in order to achieve an acceptable heterodyne [compare Fig. 4.7(a) to Fig. 4.7(b)]. Consequently, reference beam systems have largely given way to the *differential Doppler* approach shown in Fig. 15.27(b).

The differential Doppler system splits the laser into two equal intensity beams, which are focused into an intersection point. A particle passing through the intersection scatters

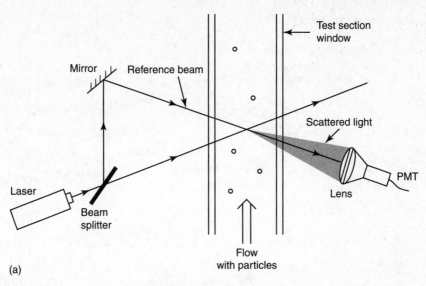

(a)

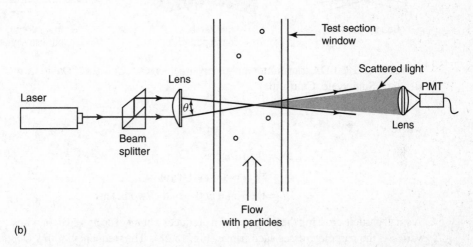

(b)

FIGURE 15.27: Laser-Doppler optical systems: (a) reference-beam arrangement; (b) differential-Doppler arrangement.

light from *both* beams, and this light is collected by a photomultiplier tube (PMT). The Doppler shift for each beam is equal and opposite, by virtue of their different angles, but the intensities of the two scattered waves are now identical. The resulting beat frequency at the detector is equal in magnitude to the Doppler shift of the beams. Figure 15.28 shows a commercial LDA transmitter/receiver arrangement.

The signal detected by a differential LDA may also be interpreted in terms of the light and dark *interference fringes* produced at the beam crossover point (see Fig. 15.28). The distance between these fringes may be shown to be

$$\delta = \frac{\lambda}{2\sin(\theta/2)} \tag{15.22}$$

Laser velocimetry optics

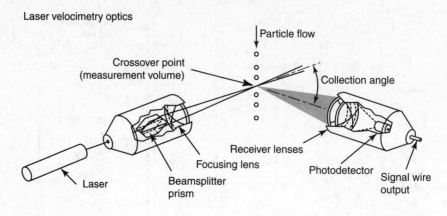

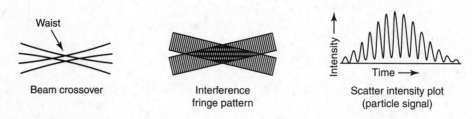

Beam crossover

Interference
fringe pattern

Scatter intensity plot
(particle signal)

FIGURE 15.28: LDA transmitter and receiver packages. (Courtesy: David Carr, Aerometrics Inc., Sunnyvale, California)

where

$$\delta = \text{the fringe spacing (m)},$$
$$\lambda = \text{the laser wavelength (m)},$$
$$\theta = \text{the angle between the two beams}$$

A small particle crossing the fringe pattern produces a burst of scattered light whose intensity varies as the particle crosses each fringe (Fig. 15.28). The frequency of this *Doppler burst* is just the particle velocity divided by the fringe spacing:

$$f_D = \frac{V_x}{\delta} = \left(\frac{2V_x}{\lambda}\right) \sin\left(\frac{\theta}{2}\right) \tag{15.23}$$

where

$$f_D = \text{the Doppler-burst frequency (Hz)},$$
$$V_x = \text{the particle velocity in the direction normal to the fringes (m/s)}$$

Note that the Doppler burst frequency depends only on the velocity component *normal* to the plane of the fringes. Also note that the burst frequency is independent of the position of the photomultiplier tube.

    The PMT signal is processed digitally to find the frequency, f_D, and thus the particle velocity, V_x. One common scheme identifies the Doppler frequency by calculating the

Fourier transform of the burst signal (Section 4.6), and others use digital time-correlation techniques [27].

The beam intersection volume can be quite small, depending on the focusing optics. These *probe volumes* are elliptical, with a major axis of 0.1 to 1 mm and fringe spacings measured in micrometers. The scattering particles are usually seeded into the flow. For liquids, natural impurities may provide acceptable seed particles; if not, adding small polystyrene spheres or even a little milk will work. In gases, an aerosol of nonvolatile oil can be used. For good signal quality, the scattering particles should generally be of diameter smaller than the fringe spacing; seed particle diameters of about 1 μm are common for gas flows.

Particles in the flow cross the probe volume at random time intervals, and, as a result, the data consist of a sequence of individual velocity measurements that must usually be analyzed statistically. An additional consequence is that the frequency response of LDA systems depends primarily on the rate at which particles cross the intersection point, rather than on the optical or electronic configuration. Frequency responses can reach hundreds of hertz with good seeding conditions.

The actual intensity of the scattered light depends upon the angle from which the scattering particle is viewed as well as the ratio of the laser wavelength to particle diameter and the particle's index of refraction [28]. The strongest signals are obtained when the scattering particle's diameter is several times the wavelength and when the probe volume is viewed using a small collection angle (see Fig. 15.28). However, the collection angle may be increased up to 180° if space is limited and weaker signals can be tolerated.

Finally, it should be noted that ordinary LDA cannot distinguish the flow direction: Positive and negative values of V_x will produce the same Doppler shift. This difficulty is overcome by applying an additional, known *frequency shift* to one of the laser beams. This causes the fringe pattern to move at the shift frequency, so that a stationary particle would scatter light at the shift frequency. For a moving particle, the shift frequency is increased or decreased by the amount of the Doppler shift. The measured burst frequency no longer passes through zero when the flow direction changes; instead, it becomes larger or smaller relative to the shift frequency. With frequency shift, LDA becomes one of the few velocimeters capable of measuring flow reversals. Furthermore, if additional pairs of laser beams are used, LDA can measure two or three components of velocity simultaneously.

Commercial LDA systems are very expensive, and their use is justified only when local and nonintrusive measurements are absolutely imperative. If intrusion can be tolerated and if flow reversal or flowborne particulates are not an issue, thermal anemometers are a more economical alternative.

15.10.2   Ultrasonic-Doppler Anemometry

Ultrasonic waves in the range of tens of kilohertz to several megahertz have been applied to Doppler flow measurements in fluids. The ultrasonic transmitter and receiver may be piezoelectric and are often designed to be clamped to the outside of a pipe.

Like LDA, ultrasonic-Doppler anemometry requires that particles be present in the flow. In industrial settings, fluids are often clean, and deliberate seeding may very undesirable. As a result, ultrasonic-Doppler systems have found greatest application to the measurement of slurries and dirty liquids which already include particulates. Ultrasonic-Doppler flowmeters are accurate no better than 2%, and they can be quite susceptible to misinstallation errors [11].

Because ultrasonic beams tend to be relatively large (measured in centimeters, perhaps), the spatial resolution of ultrasonic Doppler systems is poor. One consequence is that ultrasonic-Doppler lacks laser Doppler's ability to measure the fluctuating velocities of turbulent flow, which are generally of much smaller scale.

15.11 FLOW VISUALIZATION

Many flows are so complex that the designer may have difficulty predicting their form. An experimental visualization of the flow field then becomes an integral part of the design process. For example, an understanding of the flow past an automobile is essential to determining drag and improving fuel efficiency, particularly for the flow toward the rear of the car and in its wake. Normally, a scale model of the proposed auto body design is tested in a wind tunnel. Smoke trails introduced upstream of the car are used to make the flow visible as it passes the model. The designer can then identify regions of separated flow and adjust the body contour to reduce the drag.

The advantage, of course, is that flow visualization can illustrate an entire flow field, whereas velocity probes yield information at only a single point. Regions of separation, recirculation, and pressure loss may be identified without detailed (and more expensive) velocity measurements or calculations.

Techniques of flow visualization are widely varied. Some methods introduce a visible material, such as particles or a dye, into the flow. In other cases, density variations (and thus refractive index variations) in the fluid itself may be rendered visible. The following lists a few of the common techniques. Further discussion may be found in the Suggested Readings for this chapter.

1. *Smoke wire visualization*: A thin steel wire (~0.1 mm) is coated with oil and placed in an air flow. An electric pulse resistively heats the wire, causing the oil to form smoke. The thin line of smoke is carried with the flow, showing the fluid pathlines.

2. *Hydrogen bubble visualization*: A very fine wire, often platinum of 25 to 100 μm diameter, is placed in a water flow. A second, flat electrode is placed nearby in the flow. When dc power of about 100 V is applied to the wire, a current passes through the water, causing electrolysis at the wire surface. Tiny hydrogen bubbles are created. These bubbles are too small to experience much buoyancy, and they instead follow the water as a visible marker [29].

3. *Particulate tracer visualization*: Reflective or colored particles make the flow pattern visible. In liquids, particles of near-liquid density are preferred; diameters of 25 to 200 μm are typical. Examples are polystyrene spheres, hollow glass spheres, fish scales, and aluminum or magnesium flakes. In gases, very-small-diameter particles must be used, since larger particles tend to settle out of the flow. Smokes of hydrocarbon oils or titanium dioxide (which are composed of particles 0.01 to 0.5 μm diameter) have been used, as have oil droplets, various hollow spheres, and even helium-filled soap bubbles [30, 31].

4. *Dye injection*: A colored dye is bled into a liquid flow through a small hole or holes in the surface of a test object. The dye track shows the path taken by the liquid as it passes the object.

5. *Chemical indicators*: A chemical change is produced, often electrolytically, to cause a solution to change color or to cause formation of a fine colloidal precipitate. For

example, a hydrogen-bubble electrical arrangement may alternatively be used to cause a thymol-blue solution to change color from orange to blue.

6. *Laser-induced fluorescence*: Fluorescence is the tendency of some molecules to absorb light of one color (or frequency) and reemit light of a different color (a lower frequency). Fluorescent dyes are added to water, as in regular dye injection, but a now thin sheet of laser light is used to excite the dye in a specific plane of the flow. The resulting fluorescence provides visualization of the flow in that plane alone. Typical dyes for use with a blue-green argon-ion laser include rhodamine (fluoresces dark red or yellow) and fluoresceine (fluoresces green). Laser-induced fluorescence can be adapted to gas flows as well [32].

7. *Refractive-index-change visualizations*: A fluid's refractive index changes with its density (or temperature). Variations in refractive index will deflect or phase-shift light passing through a fluid; and with an appropriate optical arrangement, these effects can be made visible. For example, the abrupt density change at a shock wave deflects light and can be made to appear as a thin shadow in a photograph (a *shadowgraph*). Smaller density changes in compressible flows or temperature gradients in buoyancy-driven flows may also be visualized (and quantitatively measured) using refractive-index methods such as the *schlieren* technique and *holographic interferometry* [33].

Digital image processing (see Section 8.12) is a useful adjunct to many of the preceding methods. For example, a time series of digitized images can be processed to trace the time history of particle motions, yielding a whole-field velocity measurement. This technique is known as *particle-image velocimetry*, or PIV [31, 34]. PIV systems are commercially available.

SUGGESTED READINGS

H.-E. Albrecht, M. Borys, N. Damaschke, and C. Tropea. *Laser Doppler and Phase Doppler Measurement Techniques*. Berlin: Springer-Verlag, 2003.

Baker, R. C. *An Introductory Guide to Flow Measurement (ASME Edition)*. New York: ASME Press, 2003.

Baker, R. C. *Flow Measurement Handbook*. Cambridge, U.K.: Cambridge University Press, 2000.

Bruun, H. H. *Hot-Wire Anemometry: Principles and Signal Analysis*. Oxford, U.K.: Oxford University Press, 1995.

Goldstein, R. J. *Fluid Mechanics Measurements*. 2nd ed. Washington, D.C.: Taylor and Francis, 1996.

Lynnworth, L. C. *Ultrasonic Measurements for Process Control: Theory, Techniques, and Applications*. San Diego: Academic Press, 1989.

Merzkirch, W. *Flow Visualization*. 2nd ed. Orlando, Fla.: Academic Press, 1987.

Miller, R. W. *Flow Measurement Engineering Handbook*. 3rd ed. New York: McGraw-Hill, 1996.

Raffel, M., C. E. Willert, and J. Kompenhans. *Particle Image Velocimetry*. Berlin: Springer-Verlag, 1998.

Yang, W.-J. (ed.) *Handbook of Flow Visualization*. 2nd ed. Washington, D.C.: Taylor and Francis, 2001.

PROBLEMS

15.1. Water at 30°C (86°F) flows through a 10-cm (3.94-in.) I.D. pipe at an average velocity of 6 m/s (19.68 ft/s). Calculate the value of Re_D, using SI units. Check for proper unit balance. Obtain required data from Appendix D.

15.2. Solve Problem 15.1 using English units and check for unit balance. (*Note*: The same numerical result should be obtained in both cases.)

15.3. If Problem 15.1 had specified 4-in., Schedule 120 commercial steel pipe, what would be the resulting Re_D? (Check any general engineering handbook for the significance of pipe "schedule" numbers.)

15.4. Check Eq. (15.5) for a balance of units using (a) the SI system of units and (b) the English system.

15.5. An 8-in. ID (20.32-cm) pipe is connected by means of a reducer to a 6-in. ID pipe. Kerosene (see Appendix D for data) flows through the system at 2000 gal/min (7.57 m³/ min). What pressure differential in inches of Hg should exist between the larger and the smaller pipes? What differential pressure should be found if the fluid is changed to water at the same flow rate?

15.6. Solve Problem 15.5 using SI units and reconcile the answers.

15.7. A section of horizontally oriented pipe gradually tapers from a diameter of 16 cm (6.3 in.) to 8 cm (3.15 in.) over a length of 3 m (9.84 ft). Oil having a specific gravity of 0.85 flows at 0.05 m³/s (1.77 ft³/s). Assuming no energy loss, what pressure differential should exist across the tapered section? Check unit balance.

15.8. Solve Problem 15.7 using English units.

15.9. If the conduit described in Problem 15.7 is oriented into a vertical position with the larger diameter at the lowest point, what differential pressure across the section should be found? Solve (a) using SI units and (b) using English units. (c) Check the two answers for equivalency. (d) Would there be a difference if the smaller diameter is at the lowest position?

15.10. Show that Eq. (15.8c) may be written as follows:

$$Q = KA_2\sqrt{2gh}$$

where h = the differential pressure across the meter, measured in the "head" of the flowing fluid. Check the unit balance.

15.11. When an obstruction meter is placed in a vertical run of pipe (as opposed to a horizontal run), what precautions must be made in measuring the differential pressure?

15.12. Prepare a spreadsheet template for solving venturi tube problems.

15.13. A machined venturi meter is placed in a horizontal run of 8-in., schedule 40 commercial pipe [I.D. = 7.981 in. (202.7 mm)] for the purpose of metering medium heating oil as it is pumped into a large storage tank. The throat diameter of the venturi is 6.00 in. (152.4 mm). If the differential pressure is held to 2.54 psi (17.5 kPa) for one-half hour, how many gallons (liters) of oil should have been pumped? The oil temperature is 60°F (15.6°C) and its specific gravity is 0.86.

15.14. Solve Problem 15.13 using SI units.

15.15. Water at 15°C and 650 kPa flows through a 15 × 10 cm (15-cm pipe and 10-cm throat diameter) as-cast venturi tube. A differential pressure of 25 kPa is measured. Calculate the flow rate (a) in kg/min and (b) in m³/h.

15.16. A rough-welded venturi meter with a 40-cm (15.75-in.) diameter throat is used to meter 15°C (59°F) air in a 60-cm (23.62-in.) duct. If the differential pressure is measured to be 84 mm (3.31 in.) of water and the upstream pressure (absolute) is 125 kPa (18.13 psi), what is the flow rate in kg/s? In m³/s?

15.17. Solve Problem 15.16 using English units.

15.18. Prepare a spreadsheet template to be used for solving corner-tapped orifice meter problems.

15.19. Use the spreadsheet template prepared in answer to Problem 15.18 to check the calculations in Example 15.1(a).

15.20. A corner-tapped orifice meter is used to measure the flow of kerosene in a 75-mm (2.95-in.) diameter line. If $\beta = 0.4$, what differential pressure may be expected for a flow rate of 1.3 m³/min (47.67 ft³/min) when the temperature of the fluid is 10°C (50°F)?

15.21. Solve Problem 15.20 using English units.

15.22. A corner-tapped orifice meter is used to measure the flow of 60°F water in a 4-in. I.D. pipe. Prepare a plot of differential pressure readout (inches of H_2O) versus β over a range $0.2 < \beta < 0.75$ if the flow rate is fixed at 230 gal/min. (*Note*: A spreadsheet solution is recommended.)

15.23. A corner-tapped orifice meter is to be used in a 4-in. ID pipe carrying water whose temperature may vary between 40°F and 120°F. The range of flow rate is from 150 to 600 gal/min. The differential pressure across the orifice will, among other things, depend on the value of β. To minimize losses, it is desirable to use as large an orifice diameter as is feasible.

     Investigate and specify the type of differential pressure sensor to be used. Considering sensitivity limits, determine the largest practical value of β that will satisfy your specifications. Write a statement explaining the selections you have made. It is suggested that a spreadsheet be used.

15.24. A orifice meter has a nominal throat diameter of 2.75 in. If the orifice is calibrated at 60°F, plot the percent error introduced caused by temperature change over a range of 35°F to 120°F for (a) a steel plate, and (b) a brass plate. See Table 6.3 for coefficients.

15.25. If the steel plate used for a $\frac{3}{16}$-in.-thick orifice plate in a 5-in.-diameter pipe has a yield strength of 45,000 lbf/in.² and $\beta = 0.5$, estimate the differential pressure above which distortion may be expected.

15.26. Direct secondary calibration is used to determine the flow coefficient for an orifice to meter the flow of nitrogen. For an orifice inlet pressure of 25 psia (172 kPa), the flow rate determined by the primary meter is 9 lbm/min (19.8 kg/m) at 68°F (20°C). If the differential pressure across the orifice being calibrated is 3.1 in. (7.87 cm) of water, the conduit diameter is 4 in. (10.16 cm), and $\beta = 0.5$, what is the value of $(K \times Y)$? (Use $\rho = 0.0726$ lbm/ft³ at 68°F and standard pressure.) Solve using English units.

15.27. Solve Problem 15.26 using SI units.

15.28. Using Eq. (15.13), plot P_2/P_{t1} versus k over a range of $k = 1$ to 1.4.

15.29. A Pitot-static tube is used to measure the velocity of 20°C (68°F) water flowing in an open channel. If a differential pressure of 6 cm (2.36 in.) of H_2O is measured, what is the corresponding flow velocity? Check the result by using English units and comparing your answer to the SI result.

15.30. Using the data given in Problem 15.29, what would be the result if the temperature of the flowing water were 50°C, the water in the manometer were at 5°C, and the measurements are made at (a) Key West, Florida, and (b) Ft. Egbert, Alaska? (*Note*: See Appendix D for data and overlook "significant figures" by carrying the results to the degree required to show a difference.)

15.31. A Pitot-static tube is used to measure the velocity of an aircraft. If the air temperature and pressure are 5°C (41°F) and 90 kPa (13.2 psia), respectively, what is the aircraft velocity in km/h if the differential pressure is 450 mm (17.5 in.) of water? (a) Solve using Eq. (15.19), then (b) using Eq. (15.19a).

15.32. Solve Problem 15.31 using English units.

15.33. The velocity profile for turbulent flow in a smooth pipe is sometimes given as [35]

$$\frac{V(r)}{V_{\text{center}}} = \left[1 - \left(\frac{2r}{D} \right) \right]^{1/n}$$

where D = the pipe diameter, r = the radial coordinate from the center of the pipe, and n ranges in value from about 6 to 10, depending on the Reynolds number. For $n = 8$, determine the value of r at which a Pitot tube should be placed to provide the velocity V_{av}, such that $Q = A V_{\text{av}}$.

15.34. Like a Pitot-static tube, a hot-wire probe may also suffer from angular misalignment errors. Figure 15.29 defines three possible angles that the probe may have relative to its desired alignment. Describe separately what error is incurred when each angle is nonzero. Give qualitative answers and assume that the angles remain less than 45°.

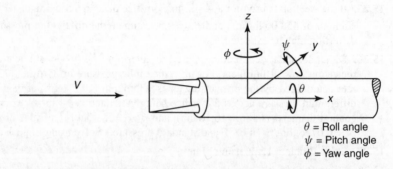

θ = Roll angle
ψ = Pitch angle
ϕ = Yaw angle

FIGURE 15.29: Angular orientation of a hot-wire probe.

REFERENCES

[1] ISO 5167-4:2003 *Measurement of fluid flow by means of differential devices inserted in a circular cross-section running full—Part 4: Venturi tubes.* Geneva: International Organization for Standardization, 2003.

[2] ISO 5167-3:2003 *Measurement of fluid flow by means of differential devices inserted in a circular cross-section running full—Part 3: Nozzles and Venturi nozzles.* Geneva: International Organization for Standardization, 2003.

[3] ISO 5167-2:2003 *Measurement of fluid flow by means of differential devices inserted in a circular cross-section running full—Part 2: Orifice plates.* Geneva: International Organization for Standardization, 2003.

[4] Baker, R. C. *An Introductory Guide to Flow Measurement (ASME Edition).* New York: ASME Press, 2003, App. 4.3.

[5] Roark, R. J. *Formulas for Stress and Strain.* 4th ed. New York: McGraw-Hill, 1965, p. 221, Case 17.

[6] ASME PTC 19.5, Application—Pt. II of Fluid Meters: Interim Supplement on Instruments and Apparatus, 1972, p. 232.

[7] Thompson, P. A. *Compressible-Fluid Dynamics.* New York: McGraw-Hill, 1972.

[8] ISO 9300:1995 *Measurement of gas flow by means of critical flow Venturi nozzles.* Geneva: International Organization for Standardization, 1995.

[9] Miller, R. W. *Flow Measurement Engineering Handbook.* 3rd ed. New York: McGraw-Hill, 1996.

[10] Nagler, F. A. Use of current meters for precise measurement of flow. *ASME Trans.* 57:59, 1935.

[11] Baker, R. C. *Flow Measurement Handbook.* Cambridge, U.K.: Cambridge University Press, 2000.

[12] Lienhard IV, J. H., and J. H. Lienhard V. *A Heat Transfer Textbook.* 3rd ed. Cambridge, Mass.: Phlogiston Press, 2003, Section 7.6.

[13] Schoenborn, E. M., and A. P. Colburn. The flow mechanism and performance of the rotameter. *Trans. AIChE* 35(3):359, 1939.

[14] Wright, J. D., and G. E. Mattingly. *NIST Calibration Services for Gas Flow Meters: Piston Provers and Bell Prover Facilities.* Gaithersburg, Md.: National Institute of Standards and Technology, Special Publication 250-98, 1998.

[15] Jarret, F. H. Standpipes simplify flowmeter calibration. *Control Eng.* 1:37, December 1954.

[16] Gracey, W. Measurement of aircraft speed and altitude. *NASA Reference Pub. 1046*: 1980.

[17] Gracey, W. Measurement of static pressure on aircraft. *NACA Tech. Note 4184*: November 1957.

[18] Gettelman, C. C., and L. N. Krause. Considerations entering into the selection of probes for pressure measurement in jet engines. *ISA Proc.* 7:134, 1952.

[19] Krause, L. N., and C. C. Gettelman. Effect of interaction among probes, supports, duct walls and jet boundaries on pressure measurements in ducts and jets. *ISA Proc.* 7:138, 1952.

[20] Itsweire, E. C., and K. N. Helland. A high-performance low-cost constant-temperature hot-wire anemometer. *J. Phys. E: Sci. Instrum.* 16:549–553, 1983.

[21] King, L. V. On the convection of heat from small cylinders in a stream of fluid, with applications to hot-wire anemometry. *Phil. Trans. Roy. Soc. (London)* 214, 14, Ser. A:373–432, 1914.

[22] Lienhard V, J. H. *The decay of turbulence in thermally stratified flow.* Doctoral dissertation, University of California, San Diego, 1988, Chapter 3.

[23] Bruun, H. H. *Hot-Wire Anemometry: Principles and Signal Analysis.* Oxford, U.K.: Oxford University Press, 1995.

[24] Wyngaard, J. C. Measurement of small-scale turbulence structure with hot wires. *J. Phys. E: Sci. Instrum.* 1:1105–1108, 1968.

[25] Freymuth, P. *Bibliography of Thermal Anemometry.* 2nd ed. St. Paul, Minn.: TSI, Inc., 1993.

[26] Drain, L. E. *The Laser Doppler Technique.* New York: John Wiley, 1980, Chapter 3.

[27] H.-E. Albrecht, M. Borys, N. Damaschke, and C. Tropea. *Laser Doppler and Phase Doppler Measurement Techniques.* Berlin: Springer-Verlag, 2003.

[28] Van de Hulst, H. C. *Light Scattering by Small Particles.* New York: Dover Publications, 1981.

[29] Geller, E. W. An electrochemical method of visualizing the boundary layer. *J. Aero. Sci.* 22:869–870, 1955.

[30] Merzkirch, W. *Flow Visualization.* 2nd ed. Orlando, Fla.: Academic Press, 1987, p. 46.

[31] Raffel, M., C. E. Willert, and J. Kompenhans. *Particle Image Velocimetry.* Berlin: Springer-Verlag, 1998, pp. 13–22.

[32] Hansen, R. K., and J. M. Seitzman. Planar Fluorescence Imaging in Gases, in Yang, W.-J. (ed.) *Handbook of Flow Visualization.* 2nd ed. Washington, D.C.: Taylor and Francis, 2001.

[33] Merzkirch, W. *Flow Visualization.* 2nd ed. Orlando, Fla.: Academic Press, 1987, Chap. 3.

[34] Adrian, R. J. Particle-imaging techniques for experimental fluid mechanics. *Annu. Rev. Fluid Mech.* 23:261–304, 1991.

[35] Fox, R. W., and A. T. McDonald. *Introduction to Fluid Mechanics.* 5th ed. New York: John Wiley, 1998, pp. 353–354.

C H A P T E R 16

Temperature Measurements

16.1 INTRODUCTION

Temperature change is usually measured by observing the change in a temperature-dependent physical property. Unlike the direct comparison of other fundamental physical quantities to a calibrated standard (as mass is measured by comparison to the International Prototype Kilogram), direct comparison of an unknown temperature to a reference temperature is relatively difficult. In formal thermodynamics, this comparison is made by connecting a Carnot engine between two systems at different temperatures. In practical thermometry, temperature is instead gauged by its effect on quantities such as volume, pressure, electrical resistance, or radiated energy.

Section 2.7 described the International Practical Temperature scale (ITS-90). This scale *assigns* values of temperature to a few highly reproducible states of matter, such as certain freezing points and triple points. These defined reference temperatures then provide calibration points for various special thermometers, which are used to interpolate between the reference points. The interpolating thermometers undergo a change in some other physical property, such as pressure or electrical resistance, as their temperature changes, and the value of that property is used to infer the corresponding temperature. In a sense, temperature itself is never directly sampled in practical thermometry.

In this chapter, we survey a selection of common temperature-sensing techniques. These thermometers are based on changes in a broad range of physical properties, among which are the following:

1. Changes in physical dimensions

   (a) Liquid-in-glass thermometers
   (b) Bimetallic elements

2. Changes in gas pressure or vapor pressure

   (a) Constant-volume gas thermometers
   (b) Pressure thermometers (gas, vapor, and liquid filled)

3. Changes in electrical properties

   (a) Resistance thermometers (RTD, PRT)
   (b) Thermistors
   (c) Thermocouples
   (d) Semiconductor-junction sensors

4. Changes in emitted thermal radiation

   (a) Thermal and photon sensors
   (b) Total-radiation pyrometers
   (c) Optical and two-color pyrometers
   (d) Infrared pyrometers

5. Changes in chemical phase

   (a) Fusible indicators
   (b) Liquid crystals
   (c) Temperature-reference (fixed-point) cells

Of these methods, electrical sensors are perhaps the most broadly used, particularly when automatic or remote recording is desired or when temperature sensors are incorporated into control systems. Bimetallic elements are used in various low-accuracy, low-cost applications. Radiant sensors are used for noncontact temperature sensing, either in high-temperature applications like combustors or for infrared sensing at lower temperatures; since radiant sensors are optical in nature, they are also adaptable to whole-field temperature measurement—so-called thermal imaging. The familiar liquid-in-glass thermometer continues to appear in both laboratory and household situations, primarily because of its ease of use and low cost. Changes in chemical phase are somewhat less often applied in engineering work.

Table 16.1 outlines approximate ranges and uncertainties of various temperature-measuring devices. The values listed in the table are only approximate, and many untabulated factors will cause deviations from the values listed. Among those factors are the precise form of electrical signal conditioning employed, the influence of manufacturer's and/or laboratory calibration techniques, and dedicated efforts to extend the operating range of a particular sensor.

TABLE 16.1: Characteristics of Various Temperature-Measuring Elements and Devices (Data from Various Sources)

Type	Useful Range*	Limits of Uncertainty*	Comments
Liquid in Glass			
Mercury filled	−37 to 320°C −35 to 600°F	0.1°C 0.2°F	Low cost. Remote reading not practical.
Pressurized mercury	−37 to 650°C −35 to 1200°F	0.1°C 0.2°F	Lower limit of mercury-filled thermometers determined by freezing point of mercury.
Alcohol	−75 to 130°C −100 to 200°F	0.5°C 1°F	Upper limit determined by boiling point.
Bimetal	−65 to 430°C −80 to 800°F	0.5 to 12°C 1 to 20°F	Rugged. Inexpensive.
Pressure Systems			
Gas (laboratory)	−270 to 100°C −450 to 212°F	0.002 to 0.2°C 0.005 to 0.5°F	Very accurate. Quite fragile. Not easily used. Used as an interpolating standard for ITS-90 (see Section 2.7).
Gas (industrial)	−270 to 760°C −450 to 1400°F	0.5 to 2% of full scale	Bourdon pressure gage used for readout. Rugged, with wide range.
Liquid (except mercury)	−90 to 370°C −125 to 700°F	1°C 2°F	Relative elevations of readout and sensing bulb are critical. Smallest bulb. Up to 3 m (10 ft) capillary.
Liquid (mercury)	−37 to 630°C −35 to 1200°F	0.5 to 2% of full scale	Same as above.
Vapor pressure	−75 to 340°C −100 to 650°F	0.5 to 2% of full scale	Fast response. Nonlinear. Lowest cost.

TABLE 16.1: *continued*

Type	Useful Range*	Limits of Uncertainty*	Comments
Thermocouples			
General	−270 to 2300°C −454 to 4200°F	±1 to 2°C ±2 to 4°F	Extreme ranges. Usually inexpensive. Small size.
Type B: Pt/30% Rh (+) vs. Pt/6% Rh (−)	870 to 1700°C 1600 to 3100°F	±0.5%[‡]	Not for reducing atmosphere. Use nonmetallic sheath. Preferred to types S and R above 1200°C.
Type E: Chromel[†] (+) vs. Constantan[†] (−)	−250 to 900°C −420 to 1650°F	±0.5%[‡]	Highest output of common thermocouples. Not for reducing or vacuum atmosphere. Best common type for cryogenics.
Type J: Fe (+) vs. Constantan[†] (−)	0 to 760°C 32 to 1400°F	±0.75%[‡]	Most atmospheres. Popular and inexpensive. Calibrations less consistent.
Type K: Chromel[†] (+) vs. Alumel[†] (−)	−250 to 1260°C −420 to 2300°F	±0.75%[‡]	Oxidizing or inert atmosphere. Less oxidation than types E, J, and T. Most linear type. Popular.
Type N: Nicrosil[†] (+) vs. Nisil[†] (−)	−250 to 1260°C −420 to 2300°F	±0.75%[‡]	Higher stability and oxidation resistance than types E, J, K, and T above 1000°C. Oxidizing or inert atmosphere.
Type R: Pt/13% Rh (+) vs. Pt (−)	−50 to 1480°C −50 to 2700°F	±0.25%[‡]	Not for reducing atmospheres. Use nonmetallic sheath.
Type S: Pt/10% Rh (+) vs. Pt (−)	−50 to 1480°C −50 to 2700°F	±0.25%[‡]	Not for reducing atmosphere. Most stable type. Use nonmetallic sheath.

TABLE 16.1: *continued*

Type	Useful Range*	Limits of Uncertainty*	Comments
Type T: Cu (+) vs. Constantan[†] (−)	−200 to 370°C −330 to 700°F	±0.75%[‡]	All atmospheres. Stable. The high thermal conductivity may cause errors.
W/5% Re (+) vs. W/26% Re (−)	400 to 2310°C 750 to 4200°F	±1.0%[‡]	No standards. Not for oxidizing or reducing atmospheres. Highest temperature limit of all thermocouples.
Resistance			
Platinum	−260 to 980°C −435 to 1800°F	0.02 to 0.2°C 0.04 to 0.4°F	High repeatability. Linear. Used as an interpolating device for ITS-90 (see Section 2.7). Sensor can be used as far as 1500 m (5000 ft) from readout.
Nickel	−180 to 320°C −300 to 600°F	—	High repeatability. Nonlinear. Produces greater resistance change per degree than does Pt. Sensor can be as far as 1500 m (5000 ft) from readout.
Thermistor (metal oxide)	−100 to 315°C −150 to 600°F	0.3°C 0.5°F	Negative temperature coefficient. Highly nonlinear. Less stable than metal types.
Thermistor (doped germanium)	−273 to −173°C −459 to −280°F	0.03°C 0.05°F	High repeatability. Nonlinear. Negative temperature coefficient. Cryogenic sensor.

TABLE 16.1: *continued*

Type	Useful Range*	Limits of Uncertainty*	Comments
Thermistor (carbon-glass)	−272 to 50°C −458 to 125°F	0.05°C 0.1°F	High repeatability. Nonlinear. Negative temperature coefficient. Cryogenic sensor.
Semiconductor Junction Diode (silicon, GaAlAs)	−272 to 50°C −457 to 125°F	0.05°C 0.1°F	Nonlinear. High accuracy requires calibration. Cryogenic sensor.
Linear integrated circuit	−50 to 150°C −60 to 300°F	0.5°C 1°F	Inexpensive. Linear. Easily integrated into electronics. Limited temperature range.
Pyrometers Total radiation	200 to 2000°C 400 to 3600°F	0.75 to 2%[§]	Rugged. Least complex type.
Spectral band	250 to 3000°C 500 to 5400°F	0.5 to 1%[§]	More sensitive than total radiation type.
Disappearing filament	800 to 4200°C 1500 to 7600°F	0.5 to 2%[§]	Requires manual manipulation by operator.
Infrared	−75 to 2500°C −100 to 4500°F	0.5 to 2%[§]	Particularly useful for low temperatures.

[*]Approximate values. Actual values depend on many factors such as sheathing and insulation, physical size of sensor or thermocouple wire gage, purity of materials, calibration employed, etc. Types such as thermocouples and resistance thermometers require additional signal-conditioning apparatus; values given are for sensors only. Unsuitable ranges of certain thermocouple types are omitted.

[†]Trade names of alloys.

[‡]In higher ranges. Percentages refer to temperature in °C.

[§]For measurement of blackbodies with $\varepsilon = 1$. For surfaces of lower or poorly known emissivity, the error may be much larger.

16.2   USE OF THERMAL EXPANSION

16.2.1   Liquid-in-Glass Thermometers

The ordinary thermometer is an example of the liquid-in-glass type (Fig. 16.1). Its essential elements are a relatively large bulb at the lower end, a capillary tube with scale, and a liquid filling both the bulb and a portion of the capillary. In addition, an expansion chamber is generally incorporated at the upper end to serve as a safety reservoir when the intended temperature range is exceeded.

As the temperature is raised, the greater expansion of the liquid compared with that of the glass causes it to rise in the capillary or stem of the thermometer, and the height of rise is used as a measure of the temperature. The volume enclosed in the stem above the liquid may either contain a vacuum or be filled with air or another gas. For the higher temperature ranges, an inert gas at a carefully controlled initial pressure is introduced in this volume, thereby raising the boiling point of the liquid and increasing the total useful range. In addition, it is claimed that such pressure minimizes the potential for column separation.

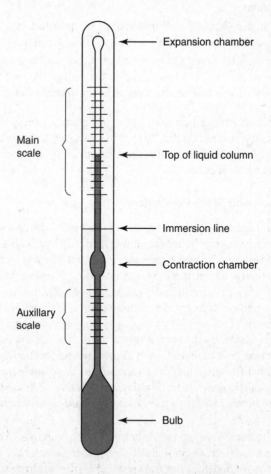

FIGURE 16.1: Liquid-in-glass thermometer.

High-grade liquid-in-glass thermometers may include several additional features. An *immersion line* may be inscribed on the thermometer to indicate the depth to which it should be submerged into the measured enviroment. A contraction chamber may be provided to shorten the overall length of capillary needed and to prevent bubbles from being formed into the bulb when the thermometer is cooled. Finally, an auxillary scale may be provided for checking calibration points outside the main range of the thermometer, such as 0°C or 100°C.

Several desirable properties for the liquid used in a glass thermometer are as follows:

1. The temperature–dimension relationship should be linear, permitting a linear instrument scale.

2. The liquid should have as large a coefficient of expansion as possible. For this reason, alcohol is better than mercury. Its larger expansion makes possible larger capillary bores and hence provides easier reading.

3. The liquid should accommodate a reasonable temperature range without change of phase. Mercury is limited at the low-temperature end by its freezing point, $-37.97°F$ (or $-38.87°C$), and spirits are limited at the high-temperature end by their boiling points.

4. The liquid should be clearly visible when drawn into a fine thread. Mercury is obviously acceptable in this regard, whereas alcohol is usable only if dye is added.

5. Preferably, the liquid should not adhere to the capillary walls. When rapid temperature drops occur, any film remaining on the wall of the tube will cause a reading that is too low. In this respect, mercury is better than alcohol.

Within its temperature range, mercury is undoubtedly the best liquid for liquid-in-glass thermometers and is generally used in the higher-grade instruments. Alcohol is usually satisfactory. Other liquids are also used, primarily for the purpose of extending the useful ranges to lower temperatures.

16.2.2 Calibration and Stem Correction

High-grade liquid-in-glass thermometers are made with the scale etched directly on the thermometer stem, thereby making it mechanically impossible to shift the scale relative to the stem. The care with which the scale is laid out depends on the intended accuracy of the instrument (and to a large extent governs its cost). The process of establishing *bench marks* from which a scale is determined is known as *pointing*, and two or more *marks* or *points* are required. In spite of contrary intentions, a particular thermometer will exhibit some degree of nonlinearity. This may be caused by nonlinear temperature–dimension characteristics of liquid or glass or by the nonuniformity of the bore of the column. In the simplest case, two points may be established, such as the freezing and boiling points of water, and equal divisions used to interpolate (and extrapolate) the complete scale. For a more accurate scale, additional points—sometimes as many as five—are used. Calibration points for this purpose are obtained through use of known phase-equilibrium temperatures, as discussed in Section 16.12.

Greatest sensitivity to temperature is at the bulb, where the largest volume of liquid is contained; however, all portions of a glass thermometer are temperature-sensitive. With temperature variation, the stem and upper bulb will also change dimensions, thereby altering the available liquid space and hence the thermometer reading. Thus, if maximum accuracy is to be attained, it is necessary to prescribe *how* a glass thermometer is to be subjected to the temperature. Greatest control is obtained when the thermometer's total measuring length (to within a few divisions of the top of the liquid column) is immersed in a uniform temperature bath. Often this is not possible, especially when the medium is liquid. A common practice, therefore, is to calibrate the thermometer for a given partial immersion, with the proper depth of immersion indicated by a line scribed around the stem (the immersion line). This technique does not ensure absolute uniformity because the upper portion of the stem is still exposed to the ambient temperature. Thermometer accuracy is prescribed only for the specified partial immersion and a specified ambient temperature. Thermometers are called *total immersion thermometers* in the former case and *partial immersion thermometers* in the latter case.

  If some part of the stem is at a condition different from that used for calibration, an *estimate* of the correct reading may be obtained from the following relation [1, 2]:

$$T = T_{obs} + kn\,(T_1 - T_2) \tag{16.1}$$

where

T = the correct temperature,

T_{obs} = the observed temperature reading,

T_1 = stem temperature specified for calibration of a partial immersion thermometer or the bath temperature of a total immersion thermometer,

T_2 = the temperature of emergent stem (this may be determined by attaching a second thermometer to the stem of the main thermometer or approximated by the ambient temperature surrounding the emergent stem),

k = the differential expansion coefficient between liquid and glass (for mercury thermometers, commonly used values are 0.00009 for the Fahrenheit scale and 0.00016 for the Celsius scale),

n = number of scale degrees equivalent to the length of the emergent stem

  The value n is determined as follows: For a total immersion thermometer, n should be the number of scale degrees between the point of emergence and the top of the liquid column. For the partial immersion thermometer, n should be the number of scale degrees that fit in the distance between the scribed calibration immersion line and the top of the liquid column.

  Another factor influencing liquid-in-glass thermometer calibration is a variation in the applied pressure, particularly in pressure applied to the bulb. The resulting elastic deformation causes displacement of the column and hence an incorrect reading. Normal variation in atmospheric pressure is not usually of importance, except for the most precise work. However, if the thermometer is subjected to system pressures of higher values, considerable error may be introduced.

16.2.3   Bimetal Temperature-Sensing Elements

When two metal strips having different thermal expansion coefficients are brazed together, a change in temperature will cause a free deflection of the assembly [3]. Such bimetal strips form the basis for control devices such as the analog home heating system thermostat. They are also used to some extent for temperature measurement. In the latter case, the sensing strip is commonly wrapped into a helical form, similar to a simple helical spring. As the temperature changes, the free end of the helix rotates (the diameter of the helix either increasing or decreasing due to the differential action). The rotational motion is directly indicated by the movement of a pointer over a circular scale.

Thermometers with bimetallic temperature-sensitive elements are often used because of their ruggedness, their ease of reading, their low cost, and the convenience of their particular form. (See Problem 16.4.)

16.3   PRESSURE THERMOMETERS

Figure 16.2 illustrates a simple *constant-volume* gas thermometer. Gas, usually hydrogen or helium, is contained in bulb A. A mercury column, B, is adjusted so that reference point C is maintained. In this manner, a constant volume of gas is held in the bulb and adjoining capillary. Mercury column h is a measure of the gas pressure and can be calibrated in terms of temperature.

In this form the apparatus is fragile, difficult to use, and restricted to the laboratory. It does, however, illustrate the working principle of a group of practical instruments called *pressure thermometers*.

Figure 16.3 shows the essentials of the practical *pressure thermometer*. The necessary parts are bulb A, tube B, pressure-sensing gage C, and some sort of filling medium. Pressure thermometers are called *liquid filled, gas filled*, or *vapor filled*, depending on whether the filling medium is completely liquid, completely gaseous, or a combination of a liquid and its vapor. A primary advantage of these thermometers is that they can provide sufficient force output to permit the direct driving of recording and controlling devices. The pressure-type temperature-sensing system is usually less costly than other systems. Tubes as long as 100 m (330 ft) may be used successfully.

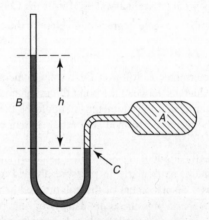

FIGURE 16.2: Sketch illustrating the essentials of a constant-volume gas thermometer.

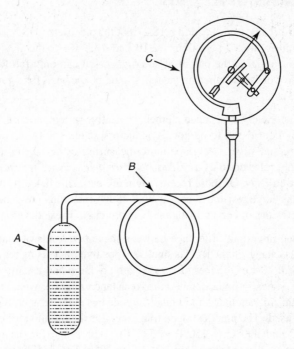

FIGURE 16.3:   Schematic diagram showing the operation of a practical pressure thermometer.

Expansion (or contraction) of bulb A and the contained fluid or gas, caused by temperature change, alters the volume and pressure in the system. In the case of the liquid-filled system, the sensing device C acts primarily as a differential volume indicator, with the volume increment serving as an analog of temperature. For the gas- or vapor-filled systems, the sensing device serves primarily as a pressure indicator, with the pressure providing the measure of temperature. In both cases, of course, both pressure and volume change.

Ideally the tube or capillary should serve simply as a connecting link between the bulb and the indicator. When liquid- or gas-filled systems are used, the tube and its filling are also temperature sensitive, and any difference from calibration conditions along the tube introduces output error. This error is reduced by increasing the ratio of bulb volume to tube volume. Unfortunately, increasing bulb size reduces the time response of a system, which may introduce problems of another nature. On the other hand, reducing tube size, within reason, does not degrade response particularly because, in any case, flow rate is negligible. Another source of error that should not be overlooked is any pressure gradient resulting from difference in elevation of bulb and indicator not accounted for by calibration.

Temperature along the tube is not a factor for vapor-pressure systems, however, so long as a free liquid surface exists in the bulb. In this case, Dalton's law of vapors applies, which states that if both phases (liquid and vapor) are present, only one pressure is possible for a given temperature. This is an important advantage of the vapor-pressure system. In many cases, though, the tube in this type of system will be filled with liquid, and hence the system is susceptible to error caused by elevation difference.

16.4 THERMORESISTIVE ELEMENTS

We have already seen (Section 6.18.2) that the electrical resistance of most materials varies with temperature. In Sections 12.10 and 13.5 we found this to supply a troublesome extraneous input to the output of strain gages. It can only follow that this relation, which proves so worrisome when unwanted, should be the basis for a good method of temperature measurement.

Historically, resistance elements sensitive to temperature were first made of metals generally considered to be good conductors of electricity. Examples are nickel, copper, platinum, and silver. A temperature-measuring device using an element of this type is commonly referred to as a *resistance thermometer*, or a *resistance temperature detector*, abbreviated *RTD*. Of more recent origin are elements made from semiconducting materials having large resistance coefficients. Such materials are usually some combination of metallic oxides of cobalt, manganese, and nickel. These devices are called *thermistors*.

One important difference between these two kinds of material is that, whereas the resistance change in the RTD is small and positive (increasing temperature causes increased resistance), that of the thermistor is relatively large and usually negative. In addition, the RTD has a nearly linear temperature-resistance relation, whereas that of the thermistor is nonlinear. Still another important difference lies in the temperature ranges over which each may be used. The practical operating range for the thermistor lies between approximately $-100°C$ and $300°C$ ($-150°F$ to $575°F$). The range for the resistance thermometer is much greater, being from about $-260°C$ to $1000°C$ ($-435°F$ to $1800°F$). Finally, the metal resistance elements are more time-stable than the semiconductor oxides; hence they provide better reproducibility with lower hysteresis.

16.4.1 Resistance Thermometers

Evidence of the importance and accuracy of the resistance thermometer may be obtained by recalling that the International Temperature Scale of 1990 specifies a platinum resistance thermometer as the interpolation standard over the range from $-259.35°C$ to $961.78°C$ (see Section 2.7).

Certain properties are desirable in a material used for resistance thermometer elements. The material should have a resistivity permitting fabrication in convenient sizes without excessive bulk, which would degrade time response. In addition, its thermal coefficient of resistivity should be high and as constant as possible, thereby providing an approximately linear output of reasonable magnitude. The material should be corrosion resistant and should not undergo phase changes in the temperature range of interest. Finally, it should be available in a condition providing reproducible and consistent results. In regard to this last requirement, it has been found that to produce precision resistance thermometers, great care must be exercised in minimizing residual strains, which requires careful heat treatment subsequent to forming.

As is generally the case in such matters, no material is universally acceptable for resistance-thermometer elements. Platinum is undoubtedly the material most commonly used, although others such as nickel, copper, tungsten, silver, and iron have also been employed. The specific choice normally depends upon which compromises may be accepted.

The temperature-resistance relation of an RTD must be determined experimentally. For most metals, the result can be accurately represented as

$$R(T) = R_0 \left[1 + A(T - T_0) + B(T - T_0)^2 \right] \tag{16.2}$$

where

$$R(T) = \text{the resistance at temperature } T,$$
$$R_0 = \text{the resistance at a reference temperature } T_0,$$
$$A \text{ and } B = \text{temperature coefficients of resistance depending on material}$$

Over a limited temperature interval (perhaps 50 K for platinum), a linear approximation to the resistance variation may be quite acceptable,

$$R(T) = R_0[1 + A(T - T_0)] \tag{16.2a}$$

but for the highest accuracy, a polynomial fit is required [4].

The resistance element is most often a metal wire wrapped around an electrically insulating support of glass, ceramic, or mica. The latter may have a variety of configurations, ranging from a simple flat strip, as shown in Fig. 16.4, to intricate "bird-cage" arrangements [5]. The mounted element is then provided with a protective enclosure. When permanent installations are made and when additional protection from corrosion or mechanical abuse is required, a *well* or *socket* may be used, such as shown in Fig. 16.5.

More recently, thin films of metal-glass slurry have been used as resistance elements. These films are deposited onto a ceramic substrate and laser trimmed. Film RTDs are less expensive than the wire RTDs and have a larger resistance for a given size; however, they are also somewhat less stable [6]. Resistance elements similar in construction to foil strain gages are available as well. The resistance grid is deposited onto a supporting film, such as Kapton, which may then be cemented to a surface. These sensors are generally designed to have low strain sensitivity and high temperature sensitivity.

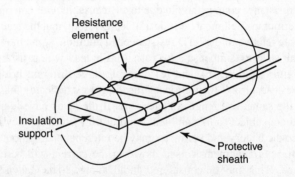

FIGURE 16.4: Section illustrating the construction of a simple RTD.

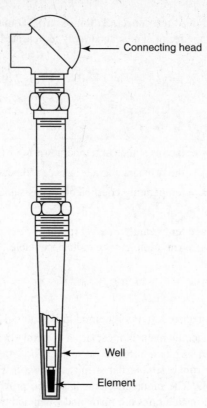

Connecting head

Well

Element

FIGURE 16.5: Installation assembly for an industrial-type resistance thermometer.

Table 16.2 describes characteristics of several typical commercially available resistance thermometers.

16.4.2   Instrumentation for Resistance Thermometry

Some form of electrical bridge is normally used to measure the resistance change in the RTD. However, particular attention must be given to the manner in which the thermometer is connected into the bridge. Leads of some length appropriate to the situation are required, and any resistance change therein due to any cause, including temperature, may be credited to the thermometer element. It is desirable, therefore, that the lead resistance be kept as low as possible relative to the RTD resistance. In addition, some modification may be made to the circuit so as to compensate for variations in lead-wire resistance.

Figures 16.6(a), (b), and (c) illustrate three different bridge arrangements used to minimize lead error. Inspection of the diagrams indicates that arms AD and DC each contain the same lead lengths. Therefore, if the leads have identical properties to begin with and are subject to like ambient conditions, the effects they introduce will cancel. In each case the battery and voltmeter may be interchanged without affecting balance. When the Siemens arrangement is used, however, no current will be carried by the center lead at balance. This may be considered an advantage. The Callender arrangement is useful when thermometers are used in both arms AD and DC to provide an output proportional

TABLE 16.2: Typical Properties of Resistance-Thermometer Elements

Type of Element	Case Material	Temperature Range, °C (°F)	Resistance, Ω	Temperature coefficient, A, Ω/(Ω · °C) (approx.)	Limits of Error*, K	Response,† s
Platinum (laboratory)	Pyrex glass	−190 to 540 (−310 to 1000)	25 at 0°C	0.00385	±0.01	
Platinum (industrial)	Stainless steel	−200 to 125 (−325 to 260)‡ −18 to 540 (0 to 1000)§	100 at 0°C 100 at 0°C	0.00385	±1 ±2	10 to 30 10 to 30
Platinum (film)	Ceramic coating	−50 to 600 (−60 to 1100)	100 at 0°C	0.00385	±0.3	~1
Rhodium-iron	Alumina and glass	−272 to 200 (−458 to 390)	27 at 0°C	0.0037	±0.04	
Copper	Brass	−75 to 120 (−100 to 250)	10 at 25°C	0.0038	±0.5	20 to 60
Nickel	Brass	0 to 120 (32 to 250)	100 at 20°C	0.0067	±0.3	20 to 60

*Typical values.
†Time required to detect 90% of any temperature change in water moving at 30 cm/s. The lower value is for the thermometer case only, whereas the higher value is for the thermometer in a protective well. Actual values vary considerably with sensor packaging and flow conditions. Response in gases will be much slower than in liquids; response will be faster at higher flow speeds.
‡Low range.
§High range.

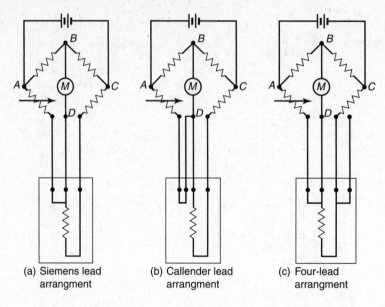

(a) Siemens lead arrangement

(b) Callender lead arrangement

(c) Four-lead arrangement

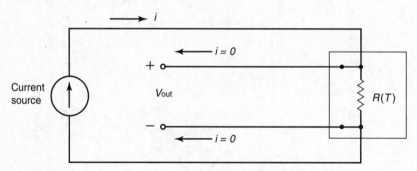

(d) Four-wire constant-current circuit

FIGURE 16.6: Four methods for compensating for lead resistance.

to temperature differential between the two thermometers. The four-lead arrangement is used in the same way as the one with three leads. Provision is made, however, for using any combination of three, thereby permitting checking for unequal lead resistance. By averaging readings, more accurate results are possible. Some form of this arrangement is used where highest accuracies are desired.

The general practice is to use the bridge in the null-balance form, but the deflection bridge may also be used (see Section 7.9). In general, the null-balance arrangement is limited to measurement of static or slowly changing temperatures, whereas the deflection bridge is used for more rapidly changing inputs. Dynamic changes are most conveniently recorded rather than simply indicated, and for this purpose either the self-balancing or the deflection types may be used, depending on time rate of temperature change.

When a resistance bridge is used for measurement, current will necessarily flow through each bridge arm. An error may, therefore, be introduced by $i^2 R$ heating of the resistance thermometer. For resistance thermometers, such an error will in general be small because the gross effects in individual arms will be largely balanced by similar effects in the other arms. An estimate of the overall error resulting from ohmic heating may be had by making readings at different current values and extrapolating to zero current.

Figure 16.6(d) shows a four-wire constant-current circuit for RTD resistance measurement. In this case, the current source holds i constant, irrespective of changes in either lead or sensor resistance. The output voltage, V_{out}, is read with a high-input-impedance meter, so that no current is drawn through the output leads and no voltage drop occurs along them. Thus, the output voltage is a linear function of sensor resistance, $V_{out} = iR(T)$, and it is independent of the lead resistance. However, because this circuit is essentially a ballast-type circuit, it lacks a bridge circuit's sensitivity to small resistance changes (Section 7.8). Also, ohmic heating effects are still present.

16.4.3 Thermistors

The thermistor is a thermally sensitive variable resistor made of a ceramic-like semiconducting material. Unlike metal resistance thermometers, thermistors generally respond to an increase of temperature with a decrease in resistance. This happens because increasing temperature usually makes more charge carriers available in a semiconductor (see Section 6.15.1). Figure 16.7 shows typical temperature-resistance relations for thermistors in relation to that of a typical RTD.

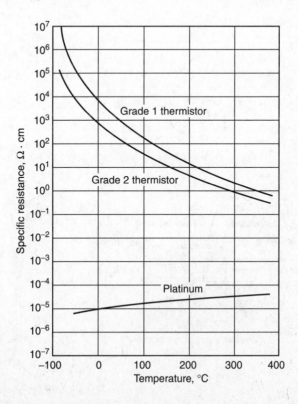

FIGURE 16.7: Typical thermistor temperature-resistance relations.

Thermistors are often composed of oxides of manganese, nickel, and cobalt in formulations having resistivities of 100 to 450,000 $\Omega \cdot$ cm. In cryogenic applications, doped germanium and carbon-impregnated glass are used. Thermistors are available in various forms, such as shown in Fig. 16.8. Some types are packaged for specific applications, such as air and water temperature sensors for automobile engines or as surface mountable chips for printed circuit boards. Table 16.3 lists some typical properties of commercially available thermistors.

The temperature-resistance function for a thermistor is given by the relationship

$$R = R_0 \exp\left[\beta\left(\frac{1}{T} - \frac{1}{T_0}\right)\right] \tag{16.3}$$

where

R = the resistance at any temperature T, in K,

R_0 = the resistance at reference temperature T_0, in K,

β = a constant, in K

The constant β depends on the thermistor material; for metal-oxide thermistors, β is typically in the range of 1000 to 5000 K.

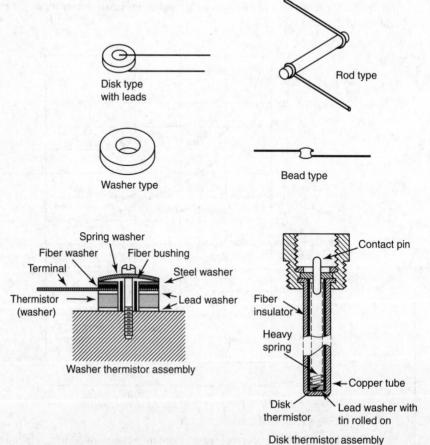

FIGURE 16.8: Various thermistor forms commercially available.

TABLE 16.3: Representative Metal-Oxide Thermistor Specifications. A Wide Range of Resistances Is Available in Any Specific Shape

Type (see Fig. 16.8)	Resistance			Approximate Maximum Continuous Temperature, °C
	At 0°C	At 25°C	At 50°C	
Bead	165 kΩ	60 kΩ	25 kΩ	—
Glass-coated bead	8.8 kΩ	3.1 kΩ	1.3 kΩ	300
Washer	28.3 Ω	10 Ω	4.1 Ω	150
Washer	3270 Ω	1000 Ω	360 Ω	150
Rod	103 kΩ	31.5 kΩ	11.3 kΩ	150
Rod	327 kΩ	100 kΩ	36 kΩ	150
Disk	283 Ω	100 Ω	40.7 Ω	125

When a thermistor is used in an electrical circuit, current normally flows through it, resulting in ohmic heating. The temperature of the thermistor is then raised, by an amount depending on the resistance to heat dissipation. For a given configuration and a given ambient temperature, a specific thermistor temperature will be obtained together with a specific electrical resistance. Through proper application of thermal analysis and electrical circuit analysis, thermistors may thus be used for both measurement and control of temperature. In addition, they are quite useful for compensating electrical circuitry for changing ambient temperature—largely because the *decreasing* electric resistance of the thermistor is in contrast to *increasing* resistance of other most electrical components when temperature rises. Also, thermistors can facilitate time-delay actions through proper balancing of electrical and heat transfer conditions.

Figure 16.9 illustrates typical thermistor self-heating response characteristics. Of course, the environment (the heat transfer condition) is a major factor in an actual application. Thermistors can be quite small (a few millimeters in diameter), so their response to changes in ambient temperature may potentially be very rapid.

The inherently high sensitivity possessed by thermistors permits the use of very simple electrical circuitry for temperature measurement. Ordinary ohmmeters may be used within the limits of accuracy of the meter itself. More often one of the various forms of resistance bridge is used (Section 7.9), either in the null-balance form or as a deflection bridge. Simple ballast circuits (Section 7.6) are also usable. In some cases, special *linearizing* circuits are used to obtain an output voltage that varies linearly with temperature.

Through use of the thermistor's temperature-resistance characteristics alone, or in conjunction with controlled heat transfer, thermistors have been used for measurement of many quantities, including pressure, liquid level, and power. They are also used for temperature control, timing (through use of their delay characteristics in combination with relays), overload protectors, warning devices, and so on.

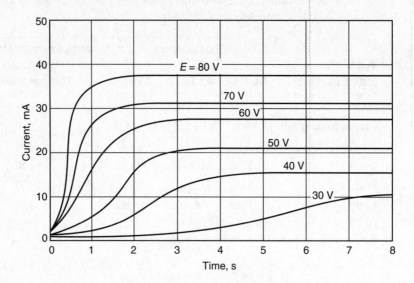

FIGURE 16.9: Typical current–time relations for thermistors.

Thermistors can also be made to have large *positive* temperature coefficients. These *PTC sensors* can be made from semiconductor oxides having barium titanate as the main component; they can also be made with heavily doped silicon. PTC sensors show an enormous increase of resistance with increasing temperature; this resistance change can be tailored to occur abruptly at a given temperature, which makes PTC thermistors useful as temperature-controlled switching elements. In conjunction with ohmic self-heating, they are also applied as current-limiting devices.

16.5 THERMOCOUPLES

In 1821, T. J. Seebeck discovered that an electric potential occurs when two different metals are joined into a loop and the two junctions are held at different temperatures [7]. Subsequently, this potential was shown to be caused by an electromotive force present in any conductor experiencing a temperature gradient (Fig. 16.10). This *Seebeck emf* is a voltage difference between the two ends of the conductor that depends on the temperature difference of the ends and a material property called the *Seebeck coefficient, σ*:

$$\mathcal{E}(T_2) - \mathcal{E}(T_1) = \int_{T_1}^{T_2} \sigma(T)\, dT \qquad (16.4)$$

If the ends of the wire have the same temperature, no emf occurs—even if the middle of the wire is hotter or colder.

When wires of two different materials, A and B, are connected as shown in Fig. 16.11, the emf that occurs depends on the temperatures of free ends of the two wires and the temperature of the junction between the two wires. In particular, if the two free ends have a temperature T_{ref} and the junction has a temperature T_m, the voltage difference between

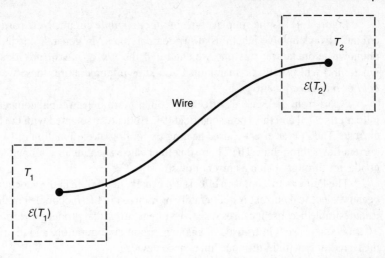

FIGURE 16.10: Seebeck emf between the ends of a wire of varying temperature.

the free ends is

$$E = \int_{T_{\text{ref}}}^{T_m} \sigma_A \, dT \; + \; \int_{T_m}^{T_{\text{ref}}} \sigma_B \, dT$$
$$= [\mathcal{E}_A(T_m) - \mathcal{E}_A(T_{\text{ref}})] \; + \; [\mathcal{E}_B(T_{\text{ref}}) - \mathcal{E}_B(T_m)]$$
$$= [\mathcal{E}_A(T_m) - \mathcal{E}_B(T_m)] \; - \; [\mathcal{E}_A(T_{\text{ref}}) - \mathcal{E}_B(T_{\text{ref}})]$$

If we define the *relative Seebeck emf* of materials A and B as $\mathcal{E}_{AB}(T) \equiv \mathcal{E}_A(T) - \mathcal{E}_B(T)$, then

$$E = \mathcal{E}_{AB}(T_m) - \mathcal{E}_{AB}(T_{\text{ref}}) \tag{16.5}$$

For any given pair of materials, the relative Seebeck emf can be measured and tabulated as a function of temperature. Moreover, if the composition of the wires is carefully controlled, this emf is highly reproducible and provides a reliable means of temperature measurement. For example, in the circuit of Fig. 16.11, if the temperature T_{ref} is known, a measurement of E allows determination of the unknown temperature T_m using tabled values of $\mathcal{E}_{AB}(T)$ and Eq. (16.5).

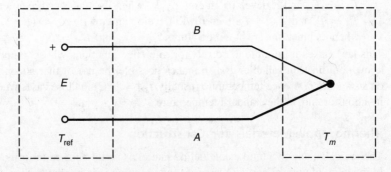

FIGURE 16.11: Net voltage when different conductors are connected.

Figure 16.11 is the simplest type of *thermocouple* circuit. Note particularly that such circuits always involve junctions at *two* temperatures. In general, one junction senses the unknown temperature; this one we shall call the *hot* or *measuring* junction. The other junction(s) will usually be maintained at a known temperature; these we shall refer to as *cold* or *reference* junctions.

Apart from the Seebeck effect, two other thermoelectric phenomena are known, the Peltier effect [8] and the Thomson effect [9]. Both are associated with the flow of electrical current. The Peltier effect causes heating or cooling at the junction of two metals when a current flows through it. The Thomson effect causes heating or cooling of a nonisothermal conductor through which a current flows.

The Peltier and Thomson effects are usually negligible in thermocouple thermometry because electrical current is deliberately minimized in thermocouple circuits. Current flow is undesirable because it causes resistive voltage drops that produce errors in the circuit emf. Thermocouples should be used in an open circuit configuration, and their emf's should be measured only with high-input-impedance devices.

16.5.1   Application Laws for Thermocouples

The following laws for thermocouple behavior can be proven by integrating the Seebeck emf around the circuits described [10].

> **Law of intermediate metals.** Insertion of an intermediate metal into a thermocouple circuit will not affect the net emf, provided that the two junctions introduced by the third metal are at identical temperatures.

This law follows directly from Eq. (16.4), since a wire whose ends have the same temperature ($T_2 = T_1$) produces no net emf. Applications of the law are shown in Fig. 16.12. In part (a) of the figure, a third metal C is introduced within the circuit. If the two new junctions r and s have the same temperature, wire C will create no additional potential and the net voltage E of the circuit will be unchanged. Part (b) of the figure shows that the third metal may even be placed in at the measuring junction, so long as the junctions r and s are both at temperature T_m. This makes possible the use of joining materials, such as soft or hard solder, in fabricating the thermocouple junctions. In addition, a thermocouple may be embedded directly into the surface or interior of either a conductor or nonconductor without altering the thermocouple's usefulness.

The following law may be proven using Eq. (16.5).

> **Law of intermediate temperatures.** If a simple thermocouple circuit (Fig. 16.11) develops an emf E_{T_2,T_1} when its junctions are at temperatures $T_m = T_2$ and $T_{\text{ref}} = T_1$, and it develops an emf E_{T_3,T_2} when its junctions are at temperatures $T_m = T_3$ and $T_{\text{ref}} = T_2$, then it will develop an emf $E_{T_3,T_1} = (E_{T_3,T_2} + E_{T_2,T_1})$ when its junctions are at temperatures $T_m = T_3$ and $T_{\text{ref}} = T_1$.

This law makes it possible to correct for reference junctions whose temperatures may be known but not controllable. It also makes possible the use of thermocouple tables based on a standard reference temperature (usually $T_{\text{ref}} = 0°C$) in those cases when the reference junctions are not at the standard temperature.

16.5.2   Thermocouple Materials and Construction

Theoretically, any two unlike conducting materials could be used to form a thermocouple. In practice, of course, certain materials and combinations are better than others, owing to

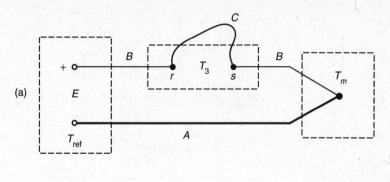

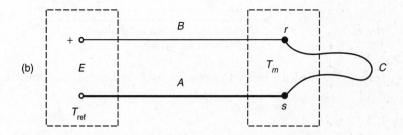

FIGURE 16.12: Diagrams illustrating the law of intermediate metals.

higher or more smoothly varying voltage, better resistance to oxidation, higher temperature limits, or other desirable attributes. Table 16.1 lists the most common types and some of their characteristics. The letter types are ANSI standard thermocouples.

Thermocouple wire is available in spools or as part of a complete probe assembly. When using spooled wire, the thermocouples may be prepared by twisting the two wires together and brazing, or preferably welding, the junction (Fig. 16.13). The wire may be sold with a Teflon jacket, or, for higher temperatures, a glass-fiber jacket may be used. The jacket of the negative thermoelement is color-coded red for all thermocouple types. When bare wires are used, electrical separation of the leads can be maintained with ceramic separating elements (Fig. 16.14).

Thermocouples are often housed in a closed-ended tube, either to form a more rugged probe or as protection against high temperatures or corrosive environments. Probe sheaths are commonly made from stainless steel, inconel, or a hard-fired ceramic. The thermocouple may be separated from the sheath by a mineral-oxide powder or a ceramic insulator. Many different probe configurations are commercially available. Figure 16.5 shows a section through one typical protective tube.

In certain instances, it is desirable to know precisely where within a thermocouple junction the indicated temperature occurs. This is especially important when either the size of the junction or the temperature gradient through it is large. This *effective location* is indicated in Figs. 16.13 and 16.14 as the point j.

Usually, a heavier wire size is needed when the operating temperature of the thermocouple increases. This helps minimize the effect of corrosion on the thermocouple voltage. ASTM (American Society for Testing and Materials) provides standards for wire gauge

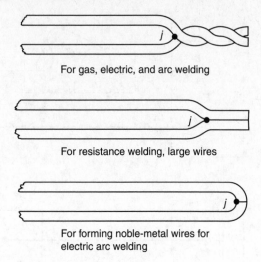

For gas, electric, and arc welding

For resistance welding, large wires

For forming noble-metal wires for
electric arc welding

FIGURE 16.13: Common forms of thermocouple construction.

as a function of operating temperature [10]. As the wire size increases, however, so does
the mass of the measuring junction; hence, the time required to change the junction's temperature increases. Protective tubes also reduce thermocouple response markedly. Some
compromise between time response and durability is usually required.

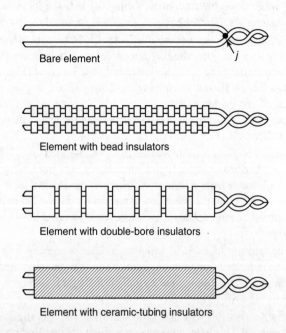

Bare element

Element with bead insulators

Element with double-bore insulators

Element with ceramic-tubing insulators

FIGURE 16.14: Methods of separating thermocouple leads.

16.5.3 Values of the Thermocouple EMF

The magnitude of the thermoelectric emf is actually quite small, generally a value in millivolts. The output varies among the different types, always increasing with the temperature, as shown in Fig. 16.15. A few numerical values of voltage are given in Table 16.4. More detailed data for type K thermocouples is given in Table 16.5. In each case, the voltage E is obtained from Eq. (16.5) with $T_{\text{ref}} = 0°\text{C}$.

Full tables of thermocouple emf for the letter-designated types are published by the National Institute of Standards and Technology [11]. Those tables list the emf to five or six digits at increments of $1°\text{C}$ over the full temperature range for each thermocouple type; they also provide polynomial functions for the voltage as a function of temperature. The tables are adopted as standards by ANSI and ASTM.

For purposes of data reduction or computer data processing, some of the NIST polynomials are given in Tables 16.6 and 16.7. Table 16.6 lists the *exact* NIST reference functions for E as a function of T_m. Table 16.7 lists NIST's *approximate* inverse functions (T_m as a function of E) and their maximum errors. Polynomials for other thermocouple types and temperature ranges are given in reference [11].

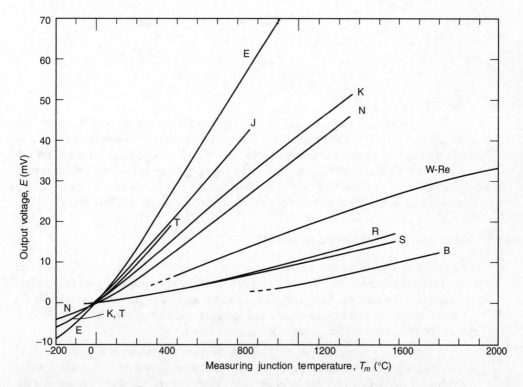

FIGURE 16.15: Thermocouple voltage versus temperature for reference junctions at $0°\text{C}$. Voltages are shown only for the recommended range of use.

TABLE 16.4: Thermocouple Voltage E in Millivolts versus Temperature T_m for Reference Junctions at $T_{ref} = 0°C$. Values Are Limited to the Recommended Range of Use [11]

	Thermocouple Type				
Temperature °C (°F)	Chromel vs. Constantan E	Iron vs. Constantan J	Chromel vs. Alumel K	Pt/10% Rh vs. Platinum S	Copper vs. Constantan T
−200 (−328)	−8.825		−5.891		−5.603
−150 (−238)	−7.279		−4.913		−4.648
−100 (−148)	−5.237		−3.554		−3.379
−50 (−58)	−2.787		−1.889	−0.236	−1.819
0 (32)	0.000	0.000	0.000	0.000	0.000
50 (122)	3.048	2.585	2.023	0.299	2.036
100 (212)	6.319	5.269	4.096	0.646	4.279
150 (302)	9.789	8.010	6.138	1.029	6.704
200 (392)	13.421	10.779	8.139	1.441	9.288
300 (572)	21.036	16.327	12.209	2.323	14.862
400 (752)	28.946	21.848	16.397	3.259	
600 (1112)	45.093	33.102	24.906	5.239	
800 (1472)	61.017		33.275	7.345	
1000 (1832)			41.276	9.587	
1200 (2192)			48.838	11.951	
1400 (2552)				14.373	

At a given temperature, type E thermocouples have the highest output voltage, but even this is less than 70 millivolts. The temperature sensitivity of thermocouples is also relatively low. For example, Table 16.4 shows the type E voltage to increase from 6.319 to 13.421 mV between 100°C and 200°C. The average change per degree Celsius is only 71 μV! Because of these factors, thermocouples require accurate and sensitive voltage measurement, and in practice cannot resolve temperature changes less than about 0.1°C.

16.5.4 Measurement of Thermocouple EMF

Historically, thermocouple emf was measured using a voltage-balancing potentiometer and a fixed voltage reference, the standard cell [10]. Today, a high-quality digital voltmeter is sufficient for all but the most exacting measurements. In fact, digital voltmeter circuitry is often combined with a microprocessor and LED to produce a thermocouple thermometer that directly displays the measuring-junction temperature.

A very simple thermocouple measuring arrangement is shown in Fig. 16.16. Here, the measuring junction senses a temperature T_m. The reference junctions are located where the thermocouple wires meet the input terminals of the voltmeter. If we assume that the meter's two input terminals have the same temperature, T_{ref}, this circuit is identical to that in Fig. 16.11 and the voltage detected will be $E = \mathcal{E}_{AB}(T_m) - \mathcal{E}_{AB}(T_{ref})$.

TABLE 16.5: Voltage E in Millivolts versus Temperature T_m for Type K Thermocouples Having Reference Junctions at $T_{ref} = 0°C$ [11]

°C	Type K				
	0	5	10	15	20
−200	−5.891	−5.813	−5.730	−5.642	−5.550
−175	−5.454	−5.354	−5.250	−5.141	−5.029
−150	−4.913	−4.793	−4.669	−4.542	−4.411
−125	−4.276	−4.138	−3.997	−3.852	−3.705
−100	−3.554	−3.400	−3.243	−3.083	−2.920
−75	−2.755	−2.587	−2.416	−2.243	−2.067
−50	−1.889	−1.709	−1.527	−1.343	−1.156
−25	−0.968	−0.778	−0.586	−0.392	−0.197
0	0.000	0.198	0.397	0.597	0.798
25	1.000	1.203	1.407	1.612	1.817
50	2.023	2.230	2.437	2.644	2.851
75	3.059	3.267	3.474	3.682	3.889
100	4.096	4.303	4.509	4.715	4.920
125	5.124	5.328	5.532	5.735	5.937
150	6.138	6.340	6.540	6.741	6.941
175	7.140	7.340	7.540	7.739	7.939
200	8.139	8.338	8.539	8.739	8.940
225	9.141	9.343	9.545	9.747	9.950
250	10.153	10.357	10.561	10.766	10.971
275	11.176	11.382	11.588	11.795	12.002
300	12.209	12.416	12.624	12.832	13.040
325	13.248	13.457	13.665	13.875	14.084
350	14.293	14.503	14.713	14.923	15.133
375	15.343	15.554	15.764	15.975	16.186

To make use of the voltage reading, T_{ref} must be measured using a temperature sensor located at the meter's input terminals. In digital thermocouple thermometers, this second sensor might be an integrated-circuit temperature sensor (Section 16.6) or a calibrated thermistor (Section 16.4.3) coupled to the microprocessor. In old potentiometers, it was simply a good liquid-in-glass thermometer. The reader may well ask why the thermocouple is needed at all—why not just use the second sensor measure T_m? General answers to this question include the very broad temperature range of thermocouples, their ruggedness, their small size, and their potentially fast time response.

TABLE 16.6: Polynomial Expansion for Thermocouple Output E in Millivolts as a Function of Measuring Junction Temperature T_m in Degrees Celsius: $E = a_0 + a_1 T_m + a_2 T_m^2 + \cdots + a_n T_m^n$. For Type K, add $\alpha_0 \exp\{\alpha_1 (T_m - 126.9686)^2\}$ to the polynomial. Reference junctions at 0°C [11]

Type	E		J	K	T
Temperature Range	−270°C to 0°C	0°C to 1000°C	−210°C to 760°C	0°C to 1372°C	0°C to 400°C
a_0	0.000 000 000 0	0.000 000 000 0	0.000 000 000 0	$-1.760\,041\,368\,6 \times 10^{-2}$	0.000 000 000 0
a_1	$5.866\,550\,870\,8 \times 10^{-2}$	$5.866\,550\,871\,0 \times 10^{-2}$	$5.038\,118\,781\,5 \times 10^{-2}$	$3.892\,120\,497\,5 \times 10^{-2}$	$3.874\,810\,636\,4 \times 10^{-2}$
a_2	$4.541\,097\,712\,4 \times 10^{-5}$	$4.503\,227\,558\,2 \times 10^{-5}$	$3.047\,583\,693\,0 \times 10^{-5}$	$1.855\,877\,003\,2 \times 10^{-5}$	$3.329\,222\,788\,0 \times 10^{-5}$
a_3	$-7.799\,804\,868\,6 \times 10^{-7}$	$2.890\,840\,721\,2 \times 10^{-8}$	$-8.568\,106\,572\,0 \times 10^{-8}$	$-9.945\,759\,287\,4 \times 10^{-8}$	$2.061\,824\,340\,4 \times 10^{-7}$
a_4	$-2.580\,016\,084\,3 \times 10^{-8}$	$-3.305\,689\,665\,2 \times 10^{-10}$	$1.322\,819\,529\,5 \times 10^{-10}$	$3.184\,094\,571\,9 \times 10^{-10}$	$-2.188\,225\,684\,6 \times 10^{-9}$
a_5	$-5.945\,258\,305\,7 \times 10^{-10}$	$6.502\,440\,327\,0 \times 10^{-13}$	$-1.705\,295\,833\,7 \times 10^{-13}$	$-5.607\,284\,488\,9 \times 10^{-13}$	$1.099\,688\,092\,8 \times 10^{-11}$
a_6	$-9.321\,405\,866\,7 \times 10^{-12}$	$-1.919\,749\,550\,4 \times 10^{-16}$	$2.094\,809\,069\,7 \times 10^{-16}$	$5.607\,505\,905\,9 \times 10^{-16}$	$-3.081\,575\,877\,2 \times 10^{-14}$
a_7	$-1.028\,760\,553\,4 \times 10^{-13}$	$-1.253\,660\,049\,7 \times 10^{-18}$	$-1.253\,839\,533\,6 \times 10^{-19}$	$-3.202\,072\,000\,3 \times 10^{-19}$	$4.547\,913\,529\,0 \times 10^{-17}$
a_8	$-8.037\,012\,362\,1 \times 10^{-16}$	$2.148\,921\,756\,9 \times 10^{-21}$	$1.563\,172\,569\,7 \times 10^{-23}$	$9.715\,114\,715\,2 \times 10^{-23}$	$-2.751\,290\,167\,3 \times 10^{-20}$
a_9	$-4.397\,949\,739\,1 \times 10^{-18}$	$-1.438\,804\,178\,2 \times 10^{-24}$		$-1.210\,472\,127\,5 \times 10^{-26}$	
a_{10}	$-1.641\,477\,635\,5 \times 10^{-20}$	$3.596\,089\,948\,1 \times 10^{-28}$			
a_{11}	$-3.967\,361\,951\,6 \times 10^{-23}$			$\alpha_0 = 1.185\,97\,6 \times 10^{-1}$	
a_{12}	$-5.582\,732\,872\,1 \times 10^{-26}$			$\alpha_1 = -1.183\,43\,2 \times 10^{-4}$	
a_{13}	$-3.465\,784\,201\,3 \times 10^{-29}$				

TABLE 16.7: Polynomial Expansion of Measuring Junction Temperature T_m in Degrees Celsius as a Function of Thermocouple Output E in Millivolts: $T_m = c_0 + c_1 E + c_2 E^2 + \cdots + c_n E^n$. Reference Junctions at 0°C [11]

Type	E		K		
Temperature Range	−200°C to 0°C	0°C to 1000°C	−200°C to 0°C	0°C to 500°C	500°C to 1372°C
EMF Range	−8.825 mV to 0.0 mV	0.0 mV to 76.373 mV	−5.891 mV to 0.0 mV	0.0 mV to 20.644 mV	20.644 mV to 54.886 mV
c_0	0.000 000 0	0.000 000 0	0.000 000 0	0.000 000 0	$-1.318\ 058 \times 10^2$
c_1	$1.697\ 728\ 8 \times 10^1$	$1.705\ 703\ 5 \times 10^1$	$2.517\ 346\ 2 \times 10^1$	$2.508\ 355 \times 10^1$	$4.830\ 222 \times 10^1$
c_2	$-4.351\ 497\ 0 \times 10^{-1}$	$-2.330\ 175\ 9 \times 10^{-1}$	$-1.166\ 287\ 8$	$7.860\ 106 \times 10^{-2}$	$-1.646\ 031$
c_3	$-1.585\ 969\ 7 \times 10^{-1}$	$6.543\ 558\ 5 \times 10^{-3}$	$-1.083\ 363\ 8$	$-2.503\ 131 \times 10^{-1}$	$5.464\ 731 \times 10^{-2}$
c_4	$-9.250\ 287\ 1 \times 10^{-2}$	$-7.356\ 274\ 9 \times 10^{-5}$	$-8.977\ 354\ 0 \times 10^{-1}$	$8.315\ 270 \times 10^{-2}$	$-9.650\ 715 \times 10^{-4}$
c_5	$-2.608\ 431\ 4 \times 10^{-2}$	$-1.789\ 600\ 1 \times 10^{-6}$	$-3.734\ 237\ 7 \times 10^{-1}$	$-1.228\ 034 \times 10^{-2}$	$8.802\ 193 \times 10^{-6}$
c_6	$-4.136\ 019\ 9 \times 10^{-3}$	$8.403\ 616\ 5 \times 10^{-8}$	$-8.663\ 264\ 3 \times 10^{-2}$	$9.804\ 036 \times 10^{-4}$	$-3.110\ 810 \times 10^{-8}$
c_7	$-3.403\ 403\ 0 \times 10^{-4}$	$-1.373\ 587\ 9 \times 10^{-9}$	$-1.045\ 059\ 8 \times 10^{-2}$	$-4.413\ 030 \times 10^{-5}$	
c_8	$-1.156\ 489\ 0 \times 10^{-5}$	$1.062\ 982\ 3 \times 10^{-11}$	$-5.192\ 057\ 7 \times 10^{-4}$	$1.057\ 734 \times 10^{-6}$	
c_9		$-3.244\ 708\ 7 \times 10^{-14}$		$-1.052\ 755 \times 10^{-8}$	
Maximum Error	±0.03°C	±0.02°C	±0.04°C	±0.05°C	±0.06°C

TABLE 16.7: *continued*

Type	J	N	S	T	
Temperature Range	0°C to 760°C	600°C to 1300°C	250°C to 1200°C	−200°C to 0°C	0°C to 400°C
EMF Range	0.0 mV to 42.919 mV	20.613 mV to 47.513 mV	1.874 mV to 11.950 mV	−5.603 mV to 0.0 mV	0.0 mV to 20.872 mV
c_0	0.000 000	$1.972\ 485 \times 10^1$	$1.291\ 507\ 177 \times 10^1$	0.000 000 0	0.000 000
c_1	$1.978\ 425 \times 10^1$	$3.300\ 943 \times 10^1$	$1.466\ 298\ 863 \times 10^2$	$2.594\ 919\ 2 \times 10^1$	$2.592\ 800 \times 10^1$
c_2	$-2.001\ 204 \times 10^{-1}$	$-3.915\ 159 \times 10^{-1}$	$-1.534\ 713\ 402 \times 10^1$	$-2.131\ 696\ 7 \times 10^{-1}$	$-7.602\ 961 \times 10^{-1}$
c_3	$1.036\ 969 \times 10^{-2}$	$9.855\ 391 \times 10^{-3}$	$3.145\ 945\ 973$	$7.901\ 869\ 2 \times 10^{-1}$	$4.637\ 791 \times 10^{-2}$
c_4	$-2.549\ 687 \times 10^{-4}$	$-1.274\ 371 \times 10^{-4}$	$-4.163\ 257\ 839 \times 10^{-1}$	$4.252\ 777\ 7 \times 10^{-1}$	$-2.165\ 394 \times 10^{-3}$
c_5	$3.585\ 153 \times 10^{-6}$	$7.767\ 022 \times 10^{-7}$	$3.187\ 963\ 771 \times 10^{-2}$	$1.330\ 447\ 3 \times 10^{-1}$	$6.048\ 144 \times 10^{-5}$
c_6	$-5.344\ 285 \times 10^{-8}$		$-1.291\ 637\ 500 \times 10^{-3}$	$2.024\ 144\ 6 \times 10^{-2}$	$-7.293\ 422 \times 10^{-7}$
c_7	$5.099\ 890 \times 10^{-10}$		$2.183\ 475\ 087 \times 10^{-5}$	$1.266\ 817\ 1 \times 10^{-3}$	
c_8			$-1.447\ 379\ 511 \times 10^{-7}$		
c_9			$8.211\ 272\ 125 \times 10^{-9}$		
Maximum Error	±0.04°C	±0.04°C	±0.01°C	±0.04°C	±0.03°C

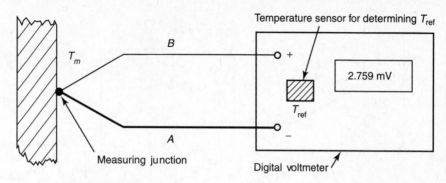

FIGURE 16.16: Simple arrangement for measuring thermocouple emf.

EXAMPLE 16.1

Suppose that the arrangement shown in Fig. 16.16 uses a type K thermocouple (chromel-alumel) and has a reference junction temperature of 20°C. If the voltmeter reads 2.759 mV, what is the measuring junction temperature?

Solution   Because the readout is for reference junctions at 20°C while the TC tables are referenced to 0°C, we must use the law of intermediate temperatures to correct the displayed emf, as follows:

$$E_{T_m,0} = E_{T_m,20} + E_{20,0}$$

Here, $E_{T_m,0}$ and $E_{T_m,20}$ are the emf's for the unknown temperature referenced to 0 and 20°C, respectively, and $E_{20,0}$ is the emf for 20°C referenced to 0°C.

From Table 16.5, we read $E_{20,0} = 0.798$ mV; hence,

$$E_{T_m,0} = 2.759 + 0.798 = 3.557 \text{ mV}$$

Interpolation of the tabulated values yields $T_m = 87°C$.

To illustrate the use of the polynomial relationships in Tables 16.6 and 16.7, let us check the value obtained in Example 16.1. What temperature does the equation give for a type K thermocouple with reference junctions at 0°C and an emf of 3.557 mV?

$$\begin{aligned} T_m &= 0.000 + (2.508355 \times 10^1) \times (3.557) + (7.860106 \times 10^{-2}) \times (3.557)^2 + \cdots °C \\ &= 86.96°C \end{aligned}$$

Thermocouple wire is relatively expensive compared to most common materials, such as ordinary copper. To help reduce costs, thermocouple circuits are often constructed using cheaper lead wires between the thermocouple and the recording instrument. An arrangement of this type is shown in Fig. 16.17. As before, we shall assume that the input terminals of the voltmeter are at the same temperature, T_{meter}. The reference junctions of this circuit are where the thermocouple wires meet the copper leads. In order for the lead wire arrangement to work, these two junctions must also be at a common temperature, T_{ref}.

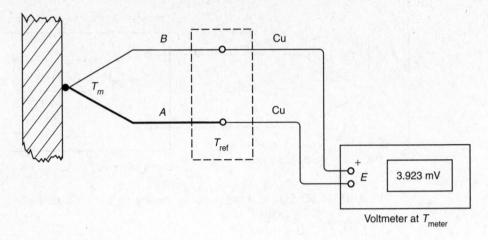

FIGURE 16.17: Extension wires used in a thermocouple circuit.

The output of the circuit is easy to calculate by summing the Seebeck emf's of each wire:

$$E = [\mathcal{E}_{Cu}(T_{ref}) - \mathcal{E}_{Cu}(T_{meter})] + [\mathcal{E}_A(T_m) - \mathcal{E}_A(T_{ref})]$$
$$+ [\mathcal{E}_B(T_{ref}) - \mathcal{E}_B(T_m)] + [\mathcal{E}_{Cu}(T_{meter}) - \mathcal{E}_{Cu}(T_{ref})]$$
$$= \mathcal{E}_{AB}(T_m) - \mathcal{E}_{AB}(T_{ref}) \tag{16.6}$$

In other words, because the two copper leads have the same endpoint temperatures (T_{ref} and T_{meter}), their Seebeck emf's cancel. The circuit's output voltage is identical to that for the circuits in Figs. 16.11 and 16.16.

For laboratory work, the reference junction temperature is usually carefully controlled. A very common arrangement places the reference junctions in an ice-water mixture held in a Dewar flask, as shown for two cases in Fig. 16.18. The ice bath remains at 0°C (32°F) while the ice melts. The voltage of both circuits shown is

$$E = \mathcal{E}_{FeCn}(T_m) - \mathcal{E}_{FeCn}(0°C)$$

Ice bath references can be very accurate when they are used carefully; however, if the ice at bottom of the Dewar melts, the remaining water can warm to as much as 4°C while ice still floats at the top of the Dewar. This may create a significant error. (See Section 16.12 for further discussion of ice baths.)

EXAMPLE 16.2

During an experiment, a thermocouple circuit has its reference junctions in an ice bath and its measuring junction in a duct carrying warm air at 40°C. The experiment lasts several hours and much of the ice in the Dewar melts, leaving the reference junctions in 4°C water at the bottom of the Dewar. If the experimenter nevertheless assumes that the junctions are at 0°C, what will be the error in his measurement?

Solution   The actual output voltage will be $E_{40,4}$. According to the law of intermediate temperatures, $E_{40,0} = E_{40,4} + E_{4,0}$. Thus,

$$E_{40,4} = E_{40,0} - E_{4,0} = 1.612 - 0.158 = 1.454 \text{ mV}$$

from Table 16.5. The experimenter is assuming this voltage represents

$$E_{T_m,0} = 1.454 \text{ mV}$$

which corresponds to $T_m = 36.1°C$. Thus, he will report a temperature of 36.1°C rather than the correct temperature of 40°C. The melting of the ice bath has caused a −3.9°C error.

EXAMPLE 16.3

A Dewar flask contains liquid nitrogen. Because the upper surface of the nitrogen is in contact with warmer air, the nitrogen tends to remain near its boiling point temperature of 77 K. A warm copper block is dropped into the Dewar, where it cools. The block has a type K thermocouple embedded at its center, and the ends of the thermocouple connect to copper lead wires outside the block in the nitrogen (Fig. 16.19). How does this circuit work, and what is the block temperature when the output of the thermocouple circuit is 0.579 mV?

Solution   The reference junctions are those where the copper wire joins the thermocouple wire. The liquid nitrogen acts as a fixed-temperature bath holding the reference junctions at $77 \text{ K} = -196°C$ while the copper block cools.

To find the temperature for an output of 0.579 mV, we must correct for reference junctions that are not at 0°C, as we did in previous examples:

$$E_{T_m,-196} = E_{T_m,0} + E_{0,-196} = E_{T_m,0} - E_{-196,0}$$

Note that the reference and measuring junctions were interchanged in the last term, changing its sign.[1] From Table 16.5, we find $E_{-196,0} = -5.829$ mV. Thus,

$$E_{T_m,0} = 0.579 + (-5.829) = -5.250 \text{ mV}$$

and from the table we read $T_m = -165°C$.

16.5.5   Electronic Instruments

As previously mentioned, digital circuits can be used to ˙
mometers. These devices connect directly to the ˙ˊ
external reference junction control. A micropre
couple wires to a displayed temperature, using, for ˎ
equations. The reference junction voltage, at the thei.
be accounted for.

---

[1] Specifically, $E_{0,-196} = \mathcal{E}_{CrAl}(0) - \mathcal{E}_{CrAl}(-196) = -[\mathcal{E}_{CrAl}(-196)$

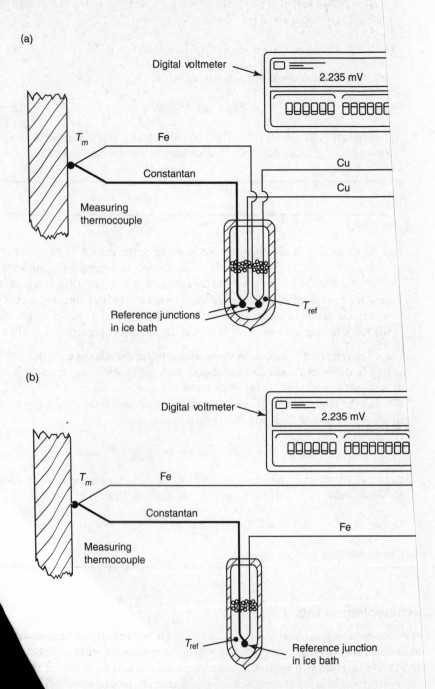

(a)

Digital voltmeter

2.235 mV

T_m

Fe

Constantan

Cu

Cu

Measuring
thermocouple

Reference junctions
in ice bath

T_{ref}

(b)

Digital voltmeter

2.235 mV

T_m

Fe

Constantan

Fe

Measuring
thermocouple

T_{ref}

Reference junction
in ice bath

FIGURE 16.18: Systems using an ice bath to fix the reference junction tempe

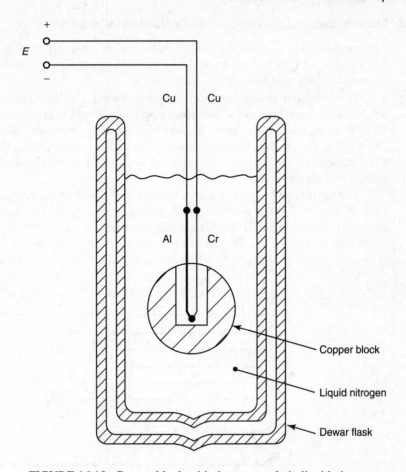

FIGURE 16.19: Copper block with thermocouple in liquid nitrogen.

Reference junction compensation in electronic thermometers may be accomplished in several ways. A calibrated temperature sensor may be incorporated at the input terminals to the thermometer and linked to the microprocessor. Alternatively, an electronic circuit may be used to mimic an icepoint reference junction. Such *electronic icepoints* produce a stable emf equivalent to that of a 0°C junction for the particular thermocouple type [12]. Integrated circuits including cold-junction compensation and amplification are also available at low cost.

Typical specifications for a digital thermocouple thermometer are as follows:

Range	−100 to 1000°C
Resolution	0.1 or 1°C
Response time	Less than 2 s
Input impedance	100 MΩ
Selectable display scale	°C or °F

These thermometers are available as handheld, panel-mounted, or benchtop units. They are often restricted to a single wire type, owing to the incorporation of specific temperature–emf and ice-point relationships.

16.5.6 Thermopiles and Thermocouples Connected in Parallel

Thermocouples may be connected electrically in series or in parallel, as shown in Fig. 16.20. When connected in series, the combination is called a *thermopile*, whereas the parallel-connected arrangement has no particular name.

The total output from *n* thermocouples connected to form a thermopile [Fig. 16.20(a)] will be equal to the sums of the individual emf's, and if the thermocouples are identical, the total output will equal *n* times the emf of a single thermocouple. For example, the thermopile in Fig. 16.20(a) has five thermocouples and an output of $5[\mathcal{E}_{AB}(T_1) - \mathcal{E}_{AB}(T_2)]$. The purpose of using a thermopile rather than a single thermocouple, of course, is to obtain a more sensitive element.

The thermocouples in a thermopile should usually be clustered together as closely as possible, so as to measure the temperature at only a single point. On the other hand, the junctions must remain electrically separated in order to avoid short-circuiting the individual thermocouple emf's. The fabrication of compact thermopiles has been greatly improved by modern circuit-manufacturing techniques that deposit fine-featured metal films onto

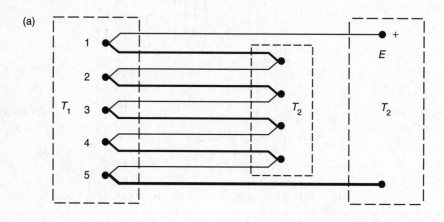

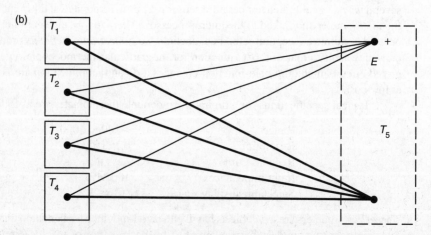

FIGURE 16.20: (a) Series-connected thermocouples forming a thermopile; (b) parallel-connected thermocouples.

suitable substrates. Such thin-film thermopiles are used in heat flux gauges (Section 16.11) and pyrometers (Section 16.8).

Parallel connection provides an averaging of the thermocouples' temperatures, which is advantageous in certain cases. This arrangement is *not* usually referred to as a thermopile. Similarly, a series-connected thermopile with junctions spread over an area can provide a spatially averaged temperature.

16.6   SEMICONDUCTOR-JUNCTION TEMPERATURE SENSORS

The junction between differently doped regions of a semiconductor has a voltage-current curve that depends strongly on temperature (see Section 6.15). This dependence has been harnessed in two types of temperature sensors: diode sensors and monolithic integrated-circuit sensors [Fig. 16.21(a)]. Like many semiconductor sensors, these devices have maximum operating temperatures of 100 to 150°C. Both types can be small, having dimensions of a few millimeters.

Semiconductor diode sensors, when properly calibrated, are the more accurate. The diode is powered with a fixed forward current of about 10 μA, and the resulting forward voltage is measured with a four-wire constant-current circuit [the diode replaces the resistor in Fig. 16.6(d)]. The diode's forward voltage is a decreasing function of temperature, known from the calibration [Fig. 16.21(b)]. Diode sensors can be accurate to about 50 mK for temperatures between 1.4 K and 300 K [13]. Typical temperature-sensing diodes are made from either silicon or gallium-aluminum-arsenide, and they are often applied in cryogenic temperature measurements. Precision diodes are relatively expensive.

Monolithic integrated-circuit devices use silicon transistors to generate an output current proportional to absolute temperature. A modest voltage (4 to 30 V) is applied to the sensor and the current through the circuit is monitored with an ammeter [Fig. 16.21(c)]. One such sensor is the Analog Devices AD590, which produces a current in microamperes numerically equal to the absolute temperature in kelvin (e.g., 298 μA at 298 K or 25°C). Because the device is a current source, its susceptibility to voltage noise and lead-wire errors is minimal. IC temperature sensors are inexpensive (a few dollars). They are applied as sensors for control circuits, as temperature-compensation elements in precision electronics, and even as electronic icepoints for thermocouple circuits. Accuracy is about 0.5°C.

16.7   THE LINEAR QUARTZ THERMOMETER

The relationship between temperature and the resonating frequency of a quartz crystal has long been recognized. In general, the relationship is nonlinear, and for many applications very considerable effort has been expended in attempts to minimize the frequency drift caused by temperature variation. Hammond [14] discovered a new crystal orientation called the "LC" or "linear cut," which provides a temperature–frequency relationship of 1000 Hz/°C with a deviation from the best straight line of less than 0.05% over a range of −40°C to 230°C (−40°F to 446°F). This linearity may be compared with a value of 0.55% for the platinum-resistance thermometer.

The nominal resonator frequency is 28 MHz, and the sensor output is compared to a reference frequency of 28.208 MHz supplied by a reference oscillator. The frequency difference is detected, converted to pulses, and passed to an electronic counter, which provides a digital display of the temperature magnitude. Various probes are available, all with time constants of 1 s. Resolution is dependent on repetitive readout rate, with a value of 0.0001°C attainable in 10 s. Readouts as fast as four per second may be obtained. Absolute

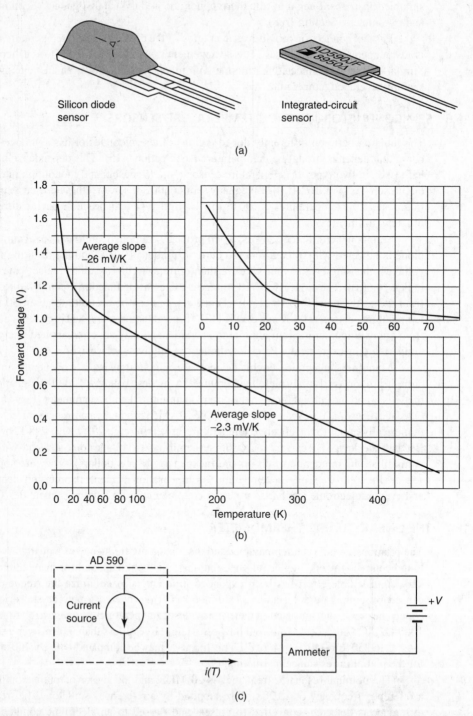

FIGURE 16.21: (a) Semiconductor junction sensors. (Courtesy: Lake Shore Cryotronics, Inc., and Analog Devices, Inc.); (b) diode forward voltage versus temperature (Lake Shore Standard Curve 10); (c) typical AD590 measuring circuit.

accuracy is rated at ±0.040°C. Remote sensing to 3000 m is possible.

16.8 PYROMETRY

The term *pyrometry* is derived from the Greek words *pyros*, meaning "fire," and *metron*, meaning "to measure." Literally, the term means general temperature measurement. However, in engineering usage, the word has historically referred to the measurement of temperatures in the range extending upward from about 500°C (≈1000°F). Although certain thermocouples and resistance-type thermometers can be used above 500°C, pyrometry normally implies thermal-radiation measurement of temperature without contacting the object being measured.

Electromagnetic radiation extends over a wide range of wavelengths (or frequencies), as illustrated in Fig. 16.22. Pyrometry is based on sampling the energies in certain bandwidths of this spectrum. At any given wavelength, a body radiates energy of an intensity that depends on the body's temperature. By evaluating the emitted energy at known wavelengths, the temperature of the body can be found.

Pyrometers are essentially photodetectors designed specifically for temperature measurement. Like ordinary photodetectors (Section 6.16), pyrometers are of two general types: *thermal detectors* and *photon detectors*. Thermal detectors are based on the temperature rise produced when the energy radiated from a body is focused onto a target, heating it. The target temperature may be sensed with a thermopile, a pyroelectric element (Section 6.14),

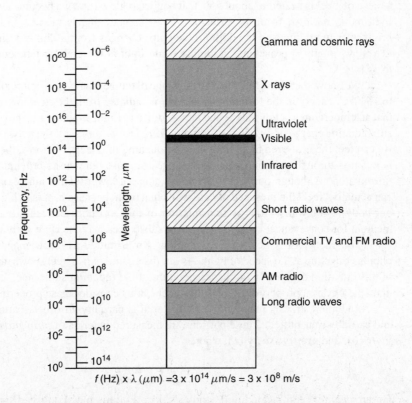

$$f \text{ (Hz)} \times \lambda \text{ } (\mu m) = 3 \times 10^{14} \text{ } \mu m/s = 3 \times 10^8 \text{ m/s}$$

FIGURE 16.22: The electromagnetic radiation spectrum.

or a thermistor or RTD. Photon detectors usually use semiconductors of either the photocon-ductive or photodiode type. In those devices, the sensor responds directly to the intensity of radiated light by a corresponding change in its resistance or in its junction current or voltage.

Pyrometers may also be classified by the set of wavelengths measured. A *total-radiation pyrometer* absorbs energy at all wavelengths or, at least, over a very broad range of wavelengths. A *spectral-band pyrometer* (or *optical* pyrometer) measures radiated energy over a narrow band of wavelengths; the band will often be narrow enough to be considered a single wavelength. A *wide-band pyrometer* uses a broader range of wavelengths, usually in order to obtain a stronger signal. The *infrared pyrometer* is a type of wide-band pyrometer used for measurements near room temperature, where radiation is weak and mainly on infrared wavelengths. The *two-color pyrometer* compares the radiated energy at two specific wavelengths in order to determine the temperature.

Because infrared pyrometry can be used at or below room temperature, it overturns the traditional perception of radiation pyrometry as a strictly high-temperature technique. In fact, the Greek meaning of pyrometry, mentioned before, is no longer so inappropriate. Irre-spective of the temperature range, however, radiation pyrometers retain the distinguishing feature of finding an object's temperature without directly contacting it.

16.8.1   Radiation Pyrometry Theory

All bodies at temperatures above absolute zero radiate energy. Not only do they radiate or emit energy, but they also receive and absorb it from other sources. We all know that when a piece of steel is heated to about 550°C it begins to glow (i.e., we become aware of visible light being *radiated* from its surface). As the temperature is raised, the light becomes brighter or more intense. In addition, the color changes from a dull red, through orange to yellow, finally approaching an almost white light at the melting temperature (1430°C to 1540°C).

We know, therefore, that through the range of temperatures from approximately 550°C to 1540°C, energy in the form of *visible light* is radiated from the steel. We can also sense that at temperatures below 550°C and almost down to room temperature, the piece of steel is still radiating energy or *heat* in the form of *infrared radiation*, for if the mass is large enough we can feel the heat even though we are not touching the steel. We know, then, that energy is radiated through certain temperature ranges because our senses provide the necessary information. Although our senses are not as acute at lower temperatures, on occasion we can actually "feel" the presence of cold walls in a room because heat is being radiated *from* our body *to* the walls. Energy transmission of this sort does not require an intervening medium for conveyance; in fact, intervening substances may interfere with transmission.

The energy of which we are speaking is transmitted as electromagnetic waves or photons traveling at the speed of light. As our discussion of hot steel shows, the wavelength of this radiation depends upon the temperature of the radiating substance. It also depends upon the physical properties of the substance. Let us consider these properties further.

Radiation striking the surface of a material is partially absorbed, partially reflected, and partially transmitted. These portions are measured in terms of *absorptivity* (α), *reflectivity* (ρ), and *transmissivity* (τ), where

$$\alpha + \rho + \tau = 1 \tag{16.7}$$

For an ideal reflector, a condition approached by a highly metal polished surface, $\rho \rightarrow 1$.

In many cases, gases represent substances of high transmissivity, for which $\tau \to 1$; for opaque materials, on the other hand, $\tau = 0$. A small opening into a large cavity approaches an ideal absorber, or *black body*, for which $\alpha \to 1$; this is because a photon entering the cavity is very unlikely to be reflected back out the opening.

A body in radiative equilibrium with its surroundings emits as much energy as it absorbs. It follows, therefore, that a good absorber is also a good radiator, and it may be concluded that the *ideal radiator* is one for which the value of α is equal to unity. In other words, a black body is both a perfect absorber and a perfect emitter of radiation. When we refer to emitted radiation as distinguished from absorption, the term *emissivity* (ε) is used rather than absorptivity (α). However, the two are directly related by *Kirchhoff's law*,

$$\varepsilon = \alpha$$

In general, each of the properties α, ρ, τ, and ε is a function of wavelength, temperature, and the angle with which radiation approaches or leaves the surface. Fortunately, for angles within about 50° of the normal to the surface, the angular dependence is weak enough to be ignored. Pyrometers are generally used within this range of angles. For opaque bodies, with $\tau = 0$, Kirchhoff's law and Eq. 16.7 show that α and ρ can be determined if ε is known. Table 16.8 lists values of emissivities for certain materials, averaged over *all* wavelengths.[2]

As we mentioned earlier, the radiated *color* changes with increasing temperature. Change in color, of course, corresponds to change in wavelength, and the wavelength of *maximum* radiation decreases with an increase in temperature. A decrease in wavelength shifts the color from the reds toward the yellows. Steel at 540°C has a deep red color. At 815°C the color is a bright red, and at 1200°C the color appears white. The corresponding radiant energy *maximums* occur at wavelengths of 3.5, 2.6, and 1.9 μm, respectively.

If we should heat an ideal radiator to various temperatures and determine the relative intensities at each wavelength, we would obtain the energy-distribution curves shown in Fig. 16.23. Not only is the radiation intensity of a higher-temperature body increased, but the wavelength of maximum emission is also shifted toward shorter wavelengths. The intensity distribution, or *spectrum*, for an ideal radiator (black body) may be expressed as follows [15]:

$$E_{\lambda,b} = \frac{C_1}{\lambda^5 (e^{C_2/\lambda T} - 1)} \tag{16.8}$$

---

[2]Kirchhoff's law is strictly valid only at any particular wavelength. When α is averaged over all wavelengths, the value may differ from the wavelength-averaged ε if the body is absorbing radiation from a much hotter or colder body [15]. In the parlance of radiation theory, values at a single wavelength are called *spectral values*, whereas those averaged over all wavelengths are called *total values*.

TABLE 16.8: Total Emissivity for Certain Surfaces [15]

Surface	Temperature, °C	Emissivity
Polished silver	200	0.01–0.04
Polished aluminum	200–600	0.04–0.06
Platinum wire	40–1370	0.04–0.19
Heavily oxided aluminum	90–540	0.20–0.33
Rusted iron	40	0.61–0.85
Rolled sheet steel	40	0.66
Roofing paper	40	0.91
Plaster	40–260	0.92
Rough red brick	40	0.93
Rough concrete	40	0.94
Smooth glass	40	0.94
Water, ≥ 0.1 mm deep	40	0.96
Black body	—	1.00

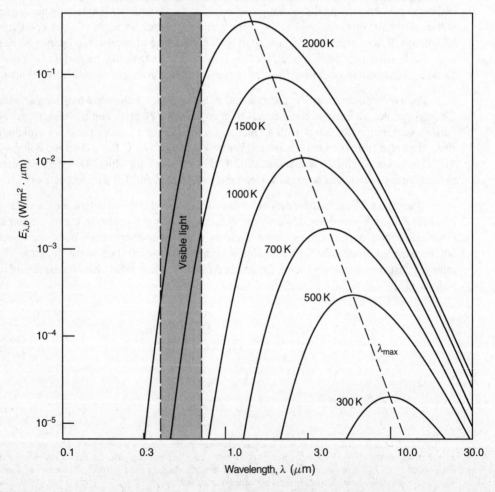

FIGURE 16.23: Radiation intensity as a function of wavelength and temperature for an ideal radiator, or black body, with $\varepsilon = 1$.

where

$E_{\lambda,b}$ = the energy emitted by a blackbody at wavelength λ,
  in W/m$^2 \cdot \mu$m,

T = the absolute temperature, in K,

λ = the wavelength, in μm,

$C_1 = 374.18$ MW $\cdot \mu$m^4/m^2,

$C_2 = 14388 \; \mu$m $\cdot$ K

The wavelength of peak intensity for a particular temperature is given by the *Wien displacement law*:

$$\lambda_{\max} T = 2897.8 \; \mu\text{m} \cdot \text{K} \qquad (16.8a)$$

For a nonideal body, the intensity distribution must be multiplied by the value of the emissivity, $\varepsilon(\lambda)$, that is appropriate to the wavelength considered

$$E_\lambda = \varepsilon(\lambda) E_{\lambda,b} = \frac{\varepsilon(\lambda) C_1}{\lambda^5 (e^{C_2/\lambda T} - 1)} \qquad (16.8b)$$

where E_λ denotes the spectrum of a nonideal body.

These relations are the basis for spectral-band and two-color pyrometers. Optical filters are used to eliminate all but the wavelengths of interest, whose intensities are then measured. The body's temperature is calculated from the measured intensity, using Eq. (16.8b) and the value of $\varepsilon(\lambda)$. Note that uncertainty in $\varepsilon(\lambda)$ leads directly to uncertainty in the measured temperature.

For total-radiation pyrometers, no filters are used and radiation from all wavelengths is sampled. Thus, the detector receives energy from the source at a rate proportional to the total radiant energy, q, emitted by the surface, as given by the *Stefan-Boltzmann law*:

$$q = \int_0^\infty E_\lambda \, d\lambda = \int_0^\infty \varepsilon(\lambda) E_{\lambda,b} \, d\lambda = \varepsilon \sigma T^4 \qquad (16.9)$$

where

q = radiant heat flux emitted by the source, in W/m^2,

$\varepsilon(\lambda)$ = the emissivity of source at wavelength λ,

ε = the emissivity of source averaged over all wavelengths,

T = absolute temperature of source, in K,

σ = the Stefan-Boltzmann constant, 5.6704×10^{-8} W/m^{2}K^4

The radiation q is focused onto the detector, whose temperature rise is measured. With a knowledge of the source's emissivity, ε, the source temperature can be calculated from Eq. (16.9). A calibration test is required to establish the relationship between the detector's temperature rise and q.

Particular attention must be given to the optical system of a radiation pyrometer, and appropriate optical glasses must be selected to pass the necessary range of wavelengths. Pyrex glass may be used for the range of 0.3 to 2.7 μm, fused silica for 0.3 to 3.8 μm, and calcium fluoride for 0.3 to 10 μm. Thus, while Pyrex glass may be used for high-temperature measurement based on short wavelengths, it is practically opaque to the long-wavelength radiation of low-temperature sources, say, those below 550°C. By choosing calcium fluoride and adding appropriate filters, a radiation pyrometer may instead be made to sense only longer infrared wavelengths (2 to 10 μm, for instance).

Although radiation pyrometers may theoretically be used at any reasonable distance from a target, some practical limitations should be mentioned. First, the size of target will largely determine the degree of temperature averaging, and in general, the greater the distance from the source, the greater the averaging. Second, the nature of the intervening atmosphere will have a decided effect on the pyrometer reading. If smoke or dust is present, or if certain gases or solids, even though they may appear to be transparent, are in the path, considerable energy absorption may occur. This problem will be particularly troublesome if such absorbents are not constant but vary with time. Third, heat radiated from surrounding bodies may be reflected from the measured object into the pyrometer, particularly if the measured body has a low emissivity. For example, if the steel discussed previously is located inside a furnace, radiation from the walls of the furnace may be reflected from the steel to the pyrometer. In other cases, reflected sunlight may cause errors. For these reasons, minimum practical distance is advisable, along with careful selection of pyrometer sighting arrangements.

For all pyrometers, calibration is essential to accounting for the effects of the optical system, the detector response, and, when it is unknown, the source emissivity.

16.8.2   Total-Radiation Pyrometry

Figure 16.24 shows, in simplified form, a thermal-detector total-radiation pyrometer. Essential parts of the device consist of some light-directing means, shown here as baffles but which are more often lenses, and an approximate blackbody receiver with means for sensing temperature. Although the sensing element may be any of the types discussed earlier in this chapter, it is generally some form of thermopile or pyroelectric sensor; occasionally, a thermistor or gas-pressure thermometer is used. A balance is quickly established between the energy absorbed by the receiver and that dissipated by conduction through the leads and radiation emitted to the surroundings. The receiver equilibrium temperature then becomes the measure of source temperature, with the scale established by calibration.

Figure 16.25 shows a sectional view of a commercially available pyrometer. Although total-radiation pyrometry is primarily used for temperatures above 550°C, the pyrometer shown is selected to illustrate an instrument sensitive to very low-level radiation (50°C to 375°C). The arrangement, however, is typical of general radiation pyrometry practice. A lens-and-mirror system is used to focus the radiant energy onto a thermopile. The thermocouple reference temperature is supplied by maintaining the assembly at constant temperature through use of a heater controlled by a resistance thermometer. In many cases compensation is obtained through use of temperature-compensating resistors in the electrical circuit.

As previously noted, a knowledge of the radiating body's emissivity is required in order to find the temperature. Handbook data may be used if accurate values are available

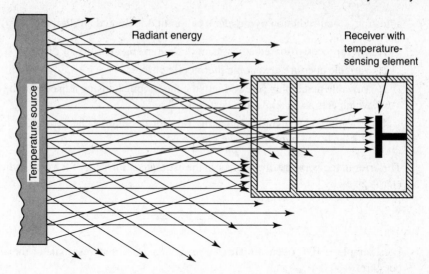

FIGURE 16.24: A simplified form of total-radiation pyrometer.

for the particular surface being measured. Alternatively, a calibration may be done by comparing the pyrometer readout with that of some standardized device, such as a thermocouple, attached to the radiating source. Often a single-point calibration suffices. Mechanisms for

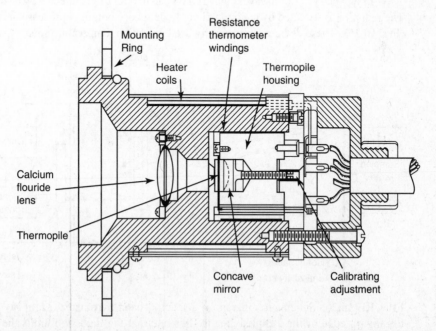

FIGURE 16.25: Section through a commercially available low-temperature, total radiation pyrometer (Courtesy: Honeywell, Inc., Process Control Division, Ft. Washington, Pennsylvania).

adjusting a total-radiation pyrometer's calibration include the following:

1. Microprocessor-based adjustment of the sensing circuitry or display;
2. Variable aperture area at the thermopile or lens;
3. Movable metal plug screwed into the thermopile housing adjacent to the hot junction which acts as an adjustible heat sink;
4. Movable concave mirror reflecting varying amounts of energy back to the thermopile of a lens-type pyrometer (see Fig. 16.25).

The error in temperature due to error in the estimated emissivity for a total radiation pyrometer is given by [16]

$$\frac{\Delta T}{T} = -\frac{1}{4}\frac{\Delta \varepsilon}{\varepsilon} \tag{16.10}$$

For example, a 10% overestimate of ε leads to a 2.5% underestimate of the absolute temperature.

16.8.3 Spectral-Band Pyrometry

Spectral-band, or optical, pyrometers measure radiant intensity at only one or two specific wavelengths, which are isolated by use of appropriate filters. The intensity is found either by using the output of a calibrated thermal or photon detector or by visual comparison to a calibrated source. The temperature dependence of the intensity distribution, Eq. (16.8), provides the necessary relation between measured intensity and temperature.

In typical systems, a wavelength of 0.65 μm to 0.85 μm is selected by an optical filter. The radiation is detected by a silicon photodiode whose output is processed electronically (Fig. 16.26). These devices are generally used with high temperature sources, above about

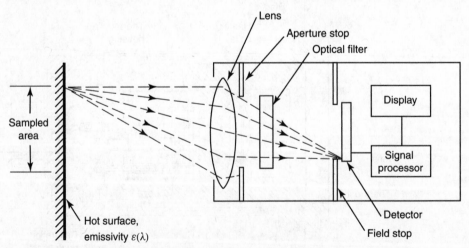

FIGURE 16.26: Schematic diagram of a spectral-band pyrometer. Light rays are shown leaving on edge of the sampled area to illustrate that the field stop limits the image size whereas the aperture stop limits the amount of light collected. Similar rays may be drawn from any point in the sampled area.

500°C. The source emissivity must be known at the wavelength used, and the appropriate value of $\varepsilon(\lambda)$ is input to the pyrometer's microprocessor.

These devices are less sensitive to uncertainties in the emissivity than total radiation pyrometers. The relationship of temperature error and emissivity error is [16]

$$\frac{\Delta T}{T} = -\frac{\lambda T}{C_2}\frac{\Delta \varepsilon}{\varepsilon} \tag{16.11}$$

for $C_2/\lambda T > 2$. Since typical values of $C_2/\lambda T$ are 10 or more, a 10% overestimate of ε leads to 1% or smaller underestimate of the absolute temperature. In addition, the radiation emitted at a single, short wavelength varies more rapidly with temperature than the total radiation, making spectral-band pyrometers more sensitive to temperature change than total-radiation pyrometers.

The *disappearing filament pyrometer* is an example of the visual comparison type (Fig. 16.27). The intensity of an electrically heated filament is varied to match the source intensity at a particular wavelength. In use, the pyrometer is sighted at the unknown temperature source at a distance such that the objective lens focuses the source in the plane of the lamp filament. The eyepiece is then adjusted so that the filament and the source appear superimposed and in focus to the observer. In general, the filament will appear either hotter than or colder than the unknown source, as shown in Fig. 16.28. When the battery current is adjusted, the filament (or any prescribed portion such as the tip) may be made to disappear, as indicated in Fig. 16.28(c). The current indicated by the milliammeter to obtain this condition may then be used as the temperature readout. A red filter is generally used to obtain approximately monochromatic conditions, and an absorption filter is used so that the filament may be operated at reduced intensity, thereby prolonging its life. The filaments are typically tungsten and can be calibrated to high accuracy. The disappearing filament pyrometer has, however, largely been replaced by electronic spectral-band pyrometers.

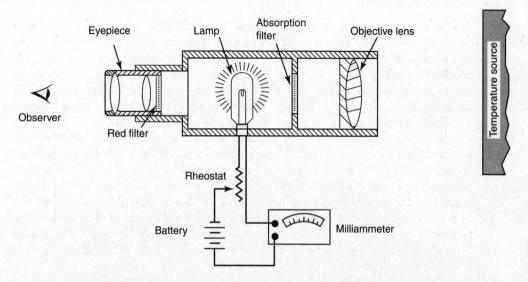

FIGURE 16.27: Schematic diagram of a disappearing filament pyrometer.

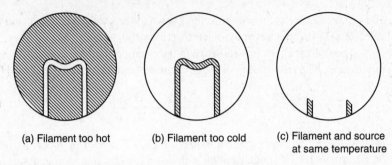

(a) Filament too hot   (b) Filament too cold   (c) Filament and source at same temperature

FIGURE 16.28: Appearance of filament when (a) filament temperature is too high, (b) filament temperature is too low, and (c) filament temperature is correct.

Two-color pyrometry is an adaptation of spectral-band pyrometry that minimizes the influence of the emissivity. Specifically, two-color pyrometers measure the source intensity at two adjacent wavelengths, λ_1 and λ_2. If the wavelengths are close and the emissivity is not too rapidly varying, then the emissivity will be nearly the same for each wavelength. Hence, the ratio of measured intensities depends only on temperature:

$$\frac{\varepsilon(\lambda_1)E_{\lambda_1}}{\varepsilon(\lambda_2)E_{\lambda_2}} \approx \frac{E_{\lambda_1}}{E_{\lambda_2}} = \left(\frac{\lambda_2}{\lambda_1}\right)^5 \frac{(e^{C_2/\lambda_2 T} - 1)}{(e^{C_2/\lambda_1 T} - 1)}$$

Two-color pyrometery is particularly useful for situations in which the source emissivity varies in time, such as during the processing of steel and aluminum [17]. The two-color pyrometer can also be used with small objects that do not fill the instrument's field of view, since only the ratio of intensities is required. Color-ratio pyrometry is the defining temperature standard for ITS-90 in the range above 1234.93 K (Section 2.7) [4].

16.8.4 Infrared Pyrometry

The infrared region begins at a wavelength of about 0.75 μm—where the visible region ends—and extends upward to wavelengths of about 1000 μm. *Infrared pyrometry* is simply an adaptation of spectral-band pyrometry to sensing a range of infrared wavelengths. The benefit of infrared sensing is found in Wien's displacement law (Fig. 16.23), which shows that the peak radiant intensity of low-temperature bodies occurs in the infrared. For example, a body at 25°C (298 K) radiates at a peak wavelength of 9.7 μm. Thus, infrared detection is essential to radiant measurements of near-room-temperature objects.

Common infrared detectors use optical filters to isolate some portion of the interval between 2 and 14 μm. This interval corresponds to peak radiant temperatures between about 200 and 1400 K. However, this interval is also dictated by the need to avoid infrared absorption by air and water vapor between the source and the detector: atmospheric absorption is at a minimum in several bands between 2 and 5 μm and in the interval from 8 to 14 μm. Most commercial infrared pyrometers are centered on one of these transmitting bandwidths. Specialized devices, such as satellite-based far-infrared detectors, may respond to wavelengths up to 100 μm; however, detectors for such long wavelengths generally must be operated at cryogenic temperatures.

Detectors for infrared pyrometers are often thermopile heat flux sensors of the Gardon gage type, as described in Section 16.11. Pyroelectric materials are also quite useful as broadband IR (infrared) sensors. These materials have an intrinsic electrostatic polarization that decreases with an increase in temperature. Pyroelectric elements, then, act as thermal detectors that respond to a change in temperature by developing a charge, in much the same way that piezoelectric materials respond to strain. Pyroelectric materials include ferroelectric crystals, such as triglisine sulfate (TGS) and lithium tantalate ($LiTaO_3$), and some organic polymer films, such as polyvinylidene fluoride (PVDF) [18].

Like piezoelectric charge, pyroelectric charge dissipates in time. To obtain measurements of steady temperatures, a rotating "chopper" may be used to interrupt the radiation periodically, producing a square-wave signal whose peak amplitude is related to the target temperature [19]. Electrical signal conditioning requirements are otherwise similar to those of piezoelectric devices (Section 7.15.2). Apart from their use in temperature measurement, pyroelectric sensors are commonly used in motion sensors for light switches and burglar alarms.

Focusing optics for infrared applications are complicated by the poor infrared transmissivity of ordinary glasses. Infrared lenses and windows are sometimes made from calcium fluoride or germanium, which have high IR transmissivity. Mirrors are usually of the first-surface type, having the reflective coating on top rather than beneath a layer of glass [19]. When low temperature levels are to be detected, background radiation from the pyrometer's own mirrors, lenses, and windows may be comparable to that from the source. Here, rotating choppers are again useful in periodically removing the source radiation, so that the background signal may be identified and subtracted.

As with other pyrometers, IR pyrometers depend upon a knowledge of the emissivity. Equation (16.11) may be used to estimate uncertainties for these devices. Errors in temperature due to errors in emissivity tend to be larger for infrared pyrometers, owing to the longer wavelengths involved.

16.8.5   Thermal Imaging

One important application of infrared pyrometry is to whole-field temperature measurement, called *thermal imaging* or *infrared thermography*. Infrared thermography is used in medical imaging, in testing buildings for heat leakage, in satellite surveys, in nighttime surveillance, and in measuring temperature distributions in electronic equipment. Many devices operate by scanning the image across a single, cooled photon detector. More recent designs take advantage of the CCD (charge-coupled diode) array technology developed for digital video cameras. Here, a two-dimensional array of detectors is positioned behind a camera lens to record the thermal image of the field viewed. At present, arrays of photoconductive sensors are usually employed. Surprisingly, photon detectors (such as silicon diodes) are not currently used in IR arrays; this is because the candidate materials either lack broadband sensitivity above 1 μm (like silicon) or are difficult to fabricate as detector arrays.

Uncertainties in the emissivity can be a particular problem for thermal imaging systems because the emissivity may vary across the field of view.

16.9   OTHER METHODS OF TEMPERATURE INDICATION

One method of temperature measurement given in the introduction to this chapter has not been referred to in the intervening pages. It is the application of changes chemical state or phase. Several techniques based on this principle should be mentioned [20].

Seger cones have long been used in the ceramic industry as a means of checking temperatures. These devices are simply small cones made of an oxide and glass. When a predetermined temperature is reached, the tip of the cone softens and curls over, thereby providing the indication that the temperature has been reached. Seger cones are made in a standard series covering a range from 600°C to 2000°C.

Somewhat similar temperature-level indicators are available in the forms of crayonlike sticks, lacquer, and pill-like pellets. Each may be calibrated at temperature intervals through a range of about 50°C to 1100°C. The crayon or lacquer is stroked or brushed on the part whose temperature is to be indicated. After the lacquer dries, it and the crayon marks appear dull and chalky. When the calibration temperature is reached, the marks become liquid and shiny. The pellets are used in a similar manner, except that they simply melt and assume a shiny liquidlike appearance as the stated temperature is reached. By using crayons, lacquer, or pellets covering various temperatures within a range, the maximum temperature attained during a test may be rather closely determined.

Liquid crystals are perhaps the most colorful of temperature indicators. The liquid crystal is a meso-phase state of certain organic compounds that shares properties of both liquids and crystals. As temperature increases past a threshold, liquid crystals successively scatter reds, yellows, greens, blues, and violets until an upper threshold temperature is reached. By changing the crystal composition, the entire color change can made occur over an interval of as much as 50°C or as little as 1°C. Liquid crystals are useful from roughly −30°C to 120°C; they may resolve temperature changes as small as 0.1°C.

Liquid crystals are available commercially in various temperature ranges and packagings. They may be encapsulated into tiny pellets which are suspended into a liquid slurry for later use. They may also be coated in a thin film over a blackened plastic or paper sheet that makes the scattered light more visible; the sheet is then affixed to a surface whose temperature distribution is to be observed. High-accuracy measurements using liquid crystals usually employ a digital camera to record the color pattern on a surface. The digital image is analyzed on a computer to determine the temperature distribution.

16.10 TEMPERATURE MEASUREMENT ERRORS

The number of potential sources of error associated with temperature measurement is unlimited. However, several are significant enough to warrant special note. These will be discussed in the next several pages.

16.10.1 Errors Associated with Convection, Radiation, and Conduction

Any temperature-measuring element senses temperature because heat is transferred between the surroundings and the element until some kind of equilibrium condition is reached.

As an example, consider the use of a thermocouple to measure the gas temperature in a furnace. When a thermocouple probe is inserted through the wall of a furnace (assume it to be gas-or coal fired), heat is transferred to it from the immersing gases by convection. Heat also reaches the probe through radiation from the furnace walls and from incandescent solids such as a fuel bed or those carried along by the swirling gases. Finally, heat will flow from the probe by conduction through any connecting leads or supports. The temperature indicated by the probe therefore will be a function of all these environmental factors, and consideration must be given to their effects to interpret or control the results intelligently.

Convection Effects

When the temperature of a gas or liquid is to be measured, we rely on heat flow by convection to bring the probe to the fluid temperature. The rate of heat transfer between the fluid and the probe is described by the following relationship:

$$Q = hA(T_f - T_t), \tag{16.12}$$

where

$$Q = \text{the heat transferred, in W,}$$

$$h = \text{the heat transfer coefficient, in W/m}^2\text{K,}$$

$$A = \text{the surface area of the probe, in m}^2,$$

$$T_f = \text{the fluid temperature, in K, and}$$

$$T_t = \text{the probe temperature, in K}$$

The heat transfer coefficient, h, may be predicted using using correlations or equations appropriate to the particular situation at hand [15]. However, several factors should be noted. Heat transfer coefficients are generally higher in flowing fluids than in fluids at rest, and they rise as the fluid velocity increases. Heat transfer coefficients in liquids are typically one to two orders of magnitude greater than in gases, other things being the same. In addition, heat transfer coefficients are generally higher for small objects than for larger ones. Representative ranges of h are as follows, for gases and nonmetallic liquids:

Gases at rest	5–25 W/m^{2}K
Flowing gases	10–200 W/m^{2}K
Liquids at rest	50–2000 W/m^{2}K
Flowing liquids	100–10,000 W/m^{2}K

EXAMPLE 16.4

A 5-mm-diameter thermistor sits in a 4 m/s air flow. It shows a resistance of 40 kΩ, corresponding to a temperature of 30.0°C. If the sensing current is 1 mA and the heat transfer coefficient is $h = 110$ W/m^{2}K, what is the air temperature?

Solution The electrical power dissipated in the thermistor must be removed by convection. Hence,

$$Q = hA(T_{\text{therm}} - T_{\text{air}}) = i^2 R$$

so that

$$T_{\text{air}} = T_{\text{therm}} - \frac{i^2 R}{h(\pi D^2)}$$

$$= 30.0 - \frac{(10^{-3})^2(40 \times 10^3)}{(110)\pi(0.005)^2}$$

$$= 25.4°C$$

The error, 4.6 K, may be reduced by lowering the current to the thermistor.

Radiation Effects

Radiation between the probe and any source or sink of different temperature is a function of the difference in the fourth powers of the *absolute* temperatures. It is generally true, therefore, that radiation becomes an increasingly important source of temperature error as the temperatures increase. As discussed in Section 16.8.1, radiant-heat transfer is also a function of the emissivities of the objects involved. For this reason, a bright, shiny probe is less affected by thermal radiation than is one tarnished or covered with soot.

Calculation of radiation heat exchange generally requires consideration of the geometric configuration of the objects involved. One simple case is particularly important, however: that of a small object exchanging radiation with a large isothermal enclosure [15]. In this case, we may estimate the radiation heat transfer as follows:

$$Q = \varepsilon A \sigma (T_t^4 - T_e^4) \tag{16.13}$$

where

Q = the heat transferred from the probe, in W,

ε = the emissivity of the probe,

A = the surface area of the probe, in m^2,

σ = the Stefan-Boltzmann constant, 5.6704×10^{-8} W/m^{2}K^4,

T_e = the temperature of the enclosure, in K, and

T_t = the probe temperature, in K

This equation will apply when the surface area of the probe is small compared to the surface area of its enclosure.

EXAMPLE 16.5

A thermocouple is located within a large exhaust duct. The thermocouple has an emissivity of $\varepsilon = 0.5$. The walls of the duct have a temperature of 400 K. The thermocouple reads 470 K. If the heat transfer coefficient between the air and the thermocouple is $h = 80$ W/m^{2}K, what is the air temperature?

Solution   The thermocouple gains heat from the air heat by convection and loses heat by radiation to the the walls. Thus,

$$Q_{\text{conv}} = Q_{\text{rad}}$$
$$hA(T_{\text{air}} - T_{\text{tc}}) = \varepsilon A \sigma (T_{\text{tc}}^4 - T_{\text{wall}}^4)$$

Hence,

$$
\begin{aligned}
T_{\text{air}} &= \frac{\varepsilon \sigma}{h}(T_{\text{tc}}^4 - T_{\text{wall}}^4) + T_{\text{tc}} \\
&= \frac{0.5(5.6704 \times 10^{-8})}{80}[(470)^4 - (400)^4)] + 470 \\
&= 478.2 \text{ K}
\end{aligned}
$$

Radiation error may be largely eliminated through the introduction of *radiation shielding*. This consists of placing reflective barriers around the probe, which prevent the probe from "seeing" the radiant source or sink, as the case may be. For low-temperature work, such shields may simply be made of sheet metal appropriately formed to provide the necessary protection. Aluminum is a common choice, because its very low emissivity sharply limits radiation heat transfer. At higher temperatures, metal or ceramic sleeves or tubes may be used. In applications where gas temperatures are desired, however, care must be exercised in placing radiation shields so as not to cause stagnation of flow around the probe. As pointed out earlier, desirable convection transfer is a function of gas velocity.

Consideration of these factors led quite naturally to the development of an aspirated high-temperature probe known as the *high-velocity thermocouple* (HVT) [21]. Figure 16.29 illustrates an aspirated probe with several types of tips. Gas is induced through the end, over the temperature-sensing element, and either is exhausted to the exterior or, if it will not alter process or measurement functions, may be returned to the source. A renewable shield provides radiation protection for the element, and through use of aspiration, convective transfer is enhanced. Gas mass-flux past the element should be not less than 20 kg/m^2s for maximum effectiveness.

When a single shield is used, as shown in Fig. 16.29(b), the shield temperature is largely controlled by convective transfer from the aspirated gas through it. Its exterior, however, is subject to thermal-radiation effects, and thus its equilibrium temperature, and hence that of the sensing element, will still be somewhat influenced by radiation. Maximum shielding may be obtained through use of multiple shields, as shown in the lower two sections of Fig. 16.29(b). Thermocouples using multiple shielding are known as *multiple high-velocity thermocouples* (MHVTs) [21]. The effectiveness of both the HVT and MHVT relative to a bare thermocouple is illustrated in Fig. 16.30.

Our discussion of radiation effects has been centered largely on the high-temperature application of thermocouples. It should of course be clear that the principles involved apply to *any* temperature-measuring system or situation. Radiation may introduce errors at low temperatures as well as at high ones, and it will present similar problems to all types of sensing elements. When the fluids are liquid rather than gaseous, the problem is essentially eliminated because most liquids, and to some degree water vapor in air, act as strong absorbers of thermal radiation.

Conduction Effects

In general, if heat is conducted through the region of a temperature sensor, temperature gradients will be present. These gradients may lead to differences between the sensor's reading and the temperature that is desired.

Consider a temperature-measuring probe that is placed into a warm fluid. This probe will require a mechanical support, and, in general some connection must be made to an external readout or recording device. These supports and connections provide paths through which heat may be conducted to cooler walls or surroundings. Ideally, convection between the fluid and the probe will hold the probe at the fluid's temperature. But if heat is steadily conducted out of the probe by the supports, the sensing element may remain at a temperature somewhat below the fluid. In the language of heat transfer, such a probe is said to behave as a heat conducting "fin" [15].

Such fin effects can be minimized by introducing a nonconductive separator into the probe's supporting structure (a so-called *thermal break*) or by having a sufficiently long

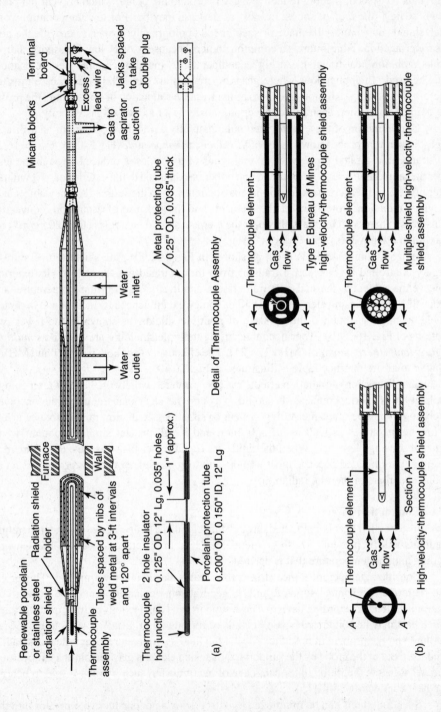

Micarta blocks

Terminal board

Excess lead wire

Jacks spaced to take double plug

Gas to aspirator suction

Renewable porcelain or stainless steel radiation shield

Radiation shield holder

Tubes spaced by nibs of weld metal at 3-ft intervals and 120° apart

Thermocouple assembly

Water intlet

Water outlet

Furnace Wall

Thermocouple hot junction

Metal protecting tube 0.25" OD, 0.035" thick

2 hole insulator 0.125" OD, 12" Lg, 0.035" holes

1" (approx.)

Porcelain protection tube 0.200" OD, 0.150" ID, 12" Lg

Detail of Thermocouple Assembly

(a)

Thermocouple element

Gas flow

Type E Bureau of Mines high-velocity-thermocouple shield assembly

A — A

Thermocouple element

Gas flow

Multiple-shield high-velocity-thermocouple shield assembly

A — A

Thermocouple element

Gas flow

Section A–A

High-velocity-thermocouple shield assembly

A — A

(b)

FIGURE 16.29: (a) Section through an aspirated high-velocity thermocouple, HVT; (b) various tips used on high-velocity thermocouples. (Courtesy: Babcock & Wilcox, Barberton, Ohio)

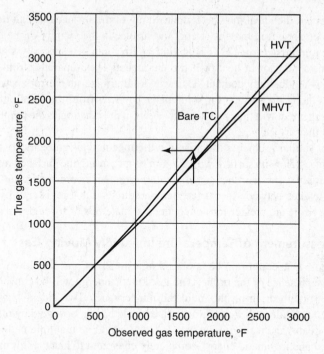

FIGURE 16.30: Graphical representation of the effectiveness of the high-velocity thermocouple (Courtesy: Babcock & Wilcox, Barberton, Ohio)

probe. Assuming that the sensing element of the probe (a thermocouple junction, say) is at the tip of the probe and that the base of the probe is attached directly to a wall, fin conduction errors will be miminal if the length of the probe satifies the following relationship:

$$L \geq 5\sqrt{\frac{kA_c}{hP}} \tag{16.14}$$

where

L = the probe length from sensor to wall, in m,

h = the heat transfer coefficient between probe and fluid, in W/m²K,

P = the perimeter of the probe (e.g., π times its diameter), in m,

A_c = the cross-sectional area of the probe, in m², and

k = the thermal conductivity of the probe, in W/m · K

In the case of a probe whose cross section includes several different materials, such as a stainless steel jacket with a powered filler inside, the product kA_c should be computed for each part of the cross section, and the results added to together to get an effective value of kA_c for the probe. For example, in the case just described, $(kA_c)_{\text{effective}} = (kA_c)_{\text{steel}} + (kA_c)_{\text{filler}}$. The perimeter is based only on the outside surface of the probe, in contact with the fluid.

    Conduction effects are also of concern for surface temperature measurements. If a thermocouple is affixed to a hot surface and the lead wires of the thermocouple extend into cooler air, then heat may be conducted out of the surface through the wires of the thermo-couple. The result may be to cool the surface in the vicinity of the thermocouple, creating a measurement error [22]. This kind of error can be especially sigificant if the surface is a poor conductor of heat or if the thermocouple is separated from the surface by a layer of low-conductivity cement. One way to minimize such errors is to run the thermocouple wires along the surface for some distance away from the location of measurement. The cooling effect will thus be moved away from the junction to the point where the leads depart from the surface.

Similar problems can arise if a thermocouple is embedded within an object at some distance below its surface with the aim of measuring the surface temperature. If heat flows through the surface, then a temperature gradient is present in the object, which leads to a difference between the surface temperature and that registered by the embedded sensor. Such errors are greater for nonconductive materials or for high rates of heat transfer.

16.10.2   Measurement of Temperature in Rapidly Moving Gas

When a temperature probe is placed in a stream of gas, the flow will be partially stopped by the presence of the probe. The lost kinetic energy will be converted to heat, which will have some bearing on the indicated temperature. Two "ideal" states may be defined for such a condition. A *static* or true state would be that observed by instruments moving with the stream, and a *stagnation* or *total* state would be that obtained if the gas were brought to rest and its kinetic energy completely converted to heat, resulting in a temperature rise. A fixed probe inserted into the moving stream will indicate conditions lying between the two states. For exhaust gases from internal combustion engines, we find that temperature differences between the two states may be as great as 200°C [23].

An expression relating total and static temperature for a moving gas may be written as follows [24]:

$$T_t - T_s = \frac{V^2}{2c_p} \tag{16.15}$$

This relation may also be written

$$\frac{T_t}{T_s} = 1 + \frac{1}{2}(k-1)M^2 \tag{16.15a}$$

where

T_t = the total or stagnation temperature, in K,

T_s = the static or true temperature, in K,

V = the velocity of flow, in m/s,

c_p = the specific heat at constant pressure, in J/kg · K,

k = the ratio of specific heats,

M = the Mach number, V/(sound speed)

The effectiveness of a probe in bringing about kinetic energy conversion is described by the ratio

$$r = \frac{T_i - T_s}{T_t - T_s} \tag{16.16}$$

where

$$T_i = \text{the temperature indicated by the probe, in K,}$$
$$r = \text{the } recovery\ factor, \text{which is proportional}$$
$$\text{to the energy conversion}$$

If $r = 1$, the probe would measure the stagnation temperature, and if $r = 0$, it would measure the static or true temperature. Experiment has shown that for a given instrument used under adiabatic conditions, the recovery factor is essentially a constant and is a function of the probe configuration. It changes little with composition, temperature, pressure, or velocity of the flowing gas [24]. In practice, however, heat losses due to thermal radiation and heat conduction will cause some probes to become more sensitive to variations in flow speed [25].

Combining Eqs. (16.15) and (16.16), we obtain

$$T_s = T_i - \frac{rV^2}{2c_p} \tag{16.17}$$

or

$$T_t = T_i + \frac{(1-r)V^2}{2c_p} \tag{16.17a}$$

The recovery factor, r, for a given probe may be determined experimentally [23, 25]. However, this approach does not generally provide sufficient information to determine either the true or the stagnation temperature. Inspection of Eqs. (16.17) and (16.17a) indicates that in addition to knowing the indicated temperature T_i and the recovery factor r, we must know the stream velocity and the specific heat of the gas. When these values are known, the relations yield the desired temperatures directly. In many cases, however, it is particularly difficult to determine the flow velocity, and further theoretical consideration of the situation is required.

For sonic velocities, with $M = 1$,

$$T_t = \phi T_i \tag{16.18}$$

where

$$\phi = \frac{k+1}{2+r(k-1)} \tag{16.19}$$

One solution to the problem of temperature measurement in high-velocity gases has been to make the measurement at Mach 1, through use of an instrument called a *sonic-flow pyrometer*. Such a device is shown in Fig. 16.31. The basic instrument comprises a temperature-sensing element (thermocouple) located at the throat of a nozzle. Gas whose temperature is to be measured is aspirated (or pressurized by the process) through the nozzle to produce critical or sonic velocity at the nozzle throat. Under these conditions, Eqs. (16.18) and (16.19) apply, and in this manner determination of flow velocity need not be made. It is still necessary to know the ratio of specific heats, but these can usually be determined or estimated with sufficient accuracy.

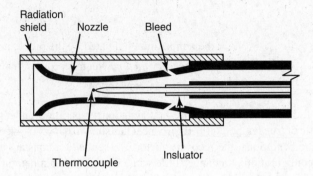

FIGURE 16.31: Schematic diagram of a sonic-flow pyrometer. (Courtesy: National Institute of Standards and Technology [24])

16.10.3 Temperature Element Response

An ideal temperature transducer would faithfully respond to fluctuating inputs regardless of the time rate of temperature change; however, the ideal is not realized in practice. A time lag exists between cause and effect, and the system seldom, if ever, actually indicates true temperature input. Figure 16.32 illustrates dramatically the magnitude of errors that may result from poor response.

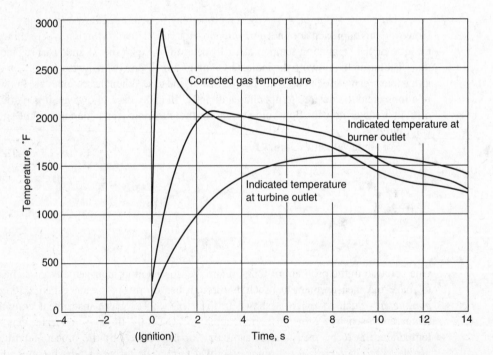

FIGURE 16.32: Temperature-time record made from two thermocouples of different size and location during the starting cycle of a large jet engine. (Courtesy: Instrument Society of America [26])

The time lag is determined by the particular heat transfer circumstances that apply, and the complexity of the situation depends to a large extent on the relative importance of the convective, conductive, and radiative components. If we assume that radiation and conduction are minimized by design and application, we may equate the energy absorbed by the probe per unit time to the rate of heat transfer by convection [15]:

$$mc\left(\frac{dT_p}{dt}\right) = hA(T_g - T_p) \qquad (16.20)$$

or

$$\tau\left(\frac{dT_p}{dt}\right) + T_p = T_g \qquad (16.20a)$$

where

T_p = the temperature of the probe, in K,

T_g = the temperature of the surrounding fluid, in K,

c = the specific heat capacity of the probe, in J/kg $\cdot$ K,

m = the mass of the probe, in kg,

t = the time, in s,

h = the convective heat-transfer coefficient, in W/m$^2 \cdot$ K,

A = the surface area of the probe exposed to fluid, in m^2,

$\tau = mc/hA$ = the time constant, in s

If the probe is initially at a temperature T_{p0} when it is put in contact with the fluid, we may write Eq. (16.20a) as follows:

$$\int_0^t dt = \tau \int_{T_{p0}}^T \frac{dT_p}{T_g - T_p} \qquad (16.20b)$$

Solving gives us

$$T_p = T_g - (T_g - T_{p0})e^{-t/\tau}$$

If we let

$$\Delta T_p = T_p - T_{p0}$$

then

$$\Delta T_p = (T_g - T_{p0})(1 - e^{-t/\tau}) \qquad (16.21)$$

This relation describes the response when a probe at temperature T_{p0} is suddenly exposed to a fluid temperature T_g. This would be approximated if the probe were quickly inserted through the wall of a furnace or immersed suddenly in a liquid bath.

The quantity τ will be recognized as the *time constant* or *characteristic* time for the probe, or the time in seconds required for 63.2% of the maximum possible change $T_g - T_{p0}$

(see Section 5.15.1). Obviously, τ should be as small as possible, and inspection shows, as should be expected, that this condition corresponds to low mass, low heat capacity, high transfer coefficient, and large area. Probes with low time constant provide fast response, and vice versa.

In response to a steady sinusoidal variation in fluid temperature of angular frequency ω, the probe temperature will oscillate with reduced amplitude and will lag in phase and time (see Example 5.4 in Section 5.15.2). To obtain faithful response, we would want $\omega\tau \ll 1$.

It must be remembered that the time constant for a given probe is not determined by the probe alone. The convective heat transfer coefficient is also dependent on the character of the fluid flow. For this reason, a given probe may show different time constants when subjected to different conditions. For example, a particular probe will usually respond much faster in a liquid flow than in a gas flow.

Although practical probe response characteristics may, in many cases, be closely approximated by the application of Eq. (16.21), in many other cases the response is complicated by the presence of other objects that absorb or transfer heat. The corresponding temperature response may be characterized by multiple time constants.

For example, the case of the common thermometer in a well or a thermocouple or resistance thermometer in a protective sheath (Fig. 16.5) may be better approximated by a two-time-constant model [27, 28]. Both probe and jacket will have characteristic time constants. Let us analyze this situation as follows. We will assume that a probe-jacket assembly (Fig. 16.33) at temperature T_1 is suddenly inserted into a medium at temperature T_2. In the manner of Eq. (16.20), we may write two relationships, as follows:

$$m_j c_j \frac{dT_j}{dt} = h_j A_j (T_2 - T_j) - h_p A_p (T_j - T_p) \tag{16.22}$$

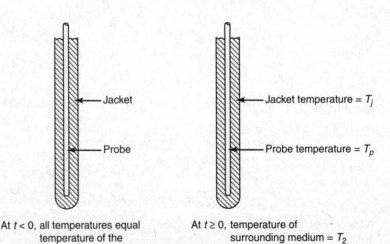

At $t < 0$, all temperatures equal temperature of the surrounding medium = T_1

At $t \geq 0$, temperature of surrounding medium = T_2

Jacket temperature = T_j

Probe temperature = T_p

FIGURE 16.33: Temperature probe in jacket subjected to a step change in temperature.

and

$$m_p c_p \frac{dT_p}{dt} = h_p A_p (T_j - T_p) \tag{16.22a}$$

where subscripts j and p refer to the protective jacket and the probe, respectively. The relationships may be rewritten as

$$\tau_j \frac{dT_j}{dt} = T_2 - T_j - \frac{h_p A_p}{h_j A_j} (T_j - T_p) \tag{16.23}$$

and

$$\tau_p \frac{dT_p}{dt} = T_j - T_p \tag{16.23a}$$

Simplification may be obtained if we assume that the last term in Eq. (16.23) may be neglected. This assumption will be legitimate if

$$\frac{h_p A_p}{h_j A_j} \ll 1$$

When this assumption is made, Eqs. (16.23) and (16.23a) may be combined, to yield

$$\tau_j \tau_p \frac{d^2 T_p}{dt^2} + (\tau_j + \tau_p) \frac{dT_p}{dt} + T_p = T_2 \tag{16.24}$$

The solution to this equation is

$$\frac{T_2 - T_p}{T_2 - T_1} = \frac{\Delta T}{\Delta T_{\max}} = \left(\frac{\zeta}{\zeta - 1} \right) e^{-t/\zeta \tau_p} - \left(\frac{1}{\zeta - 1} \right) e^{-t/\tau_p} \tag{16.25}$$

where

$\Delta T = $ the momentary difference between the actual and indicated temperatures,

$\Delta T_{\max} = $ the difference between the temperature of the medium and the probe

temperature at $t = 0$,

$\zeta = \dfrac{\tau_j}{\tau_p}$

Characteristics for various values of ζ are shown in Fig. 16.34. It is seen that for $\zeta = 0$, Eq. (16.25) reverts to Eq. (16.21). In addition, as the time constant for the jacket is increased, the overall lag is increased, as one would suspect it should be.

16.10.4 Compensation for Temperature Element Response

Time lag in electrical temperature-sensing elements, such as thermocouples and resistance thermometers, may be compensated approximately by use of digital signal processing techniques or by the introduction of appropriate electrical networks. The digital approach involves taking a Fourier transform of a measured time-series of temperature and performing an appropriate convolution to correct for the time-lag [29, 30]. This method is

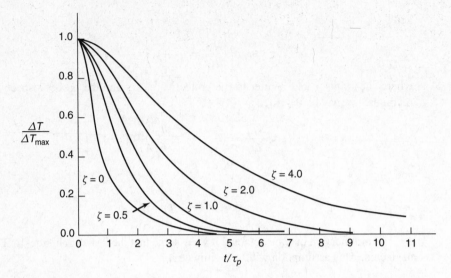

FIGURE 16.34: Two-time constant problem: Plot of $\Delta T/\Delta T_{\max}$ versus t/τ_p for various ratios of $\zeta = \tau_j/\tau_p$.

very appropriate to measurements in turbulent fluid flow, where high frequency response is needed and for which digital signal processing is almost universally applied.

The electrical technique involves selecting a type of filter (Section 7.17) whose electrical-time characteristics complement those of the sensing element [31, 32]. Figure 16.35 illustrates a simple form of such a compensator. In the example illustrated, thermocouple response drops off with increased input frequency (as shown in terms of multiples of time-constant reciprocals). By proper choice of resistors and capacitance, satisfactory combined response may be extended approximately 100 times.

Both approaches require that the sensor's time constant be accurately known under the conditions in which it is used. In general, however, the best practice is to minimize the sensor's time constant so that compensation will not be required.

16.11   MEASUREMENT OF HEAT FLUX

Heat flux is the rate of heat flow per unit area. The common units are W/m² or Btu/h · ft². We can write an expression for heat flux as follows:

$$q = -k\frac{dT}{dx} \tag{16.26}$$

where

$q =$ heat flux, in W/m²,

$k =$ the thermal conductivity of the material, in W/m · K,

$T =$ temperature, in K,

$x =$ material dimension in the direction of heat flow, in m

Knowledge of heat flux rather than temperature is of particular value in designing systems to *avoid* excessive temperatures. Examples might involve supersonic aircraft, gas

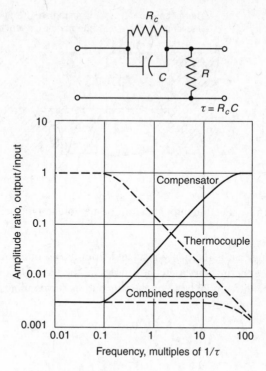

FIGURE 16.35: Curves illustrating compensating action of a simple RC network. (Courtesy: Instrument Society of America)

turbine blades, combustor walls in rocket motors, and similar situations where heat loads are of concern.

Heat flux gages are of several forms [33], of which three have particular importance: the slug type (Fig. 16.36); the foil or membrane type (Fig. 16.37), which is also known as the Gardon gage; and the thin-film layered type (Fig. 16.38).

As shown in Fig. 16.36, the essentials of the slug-type meter include a concentrated

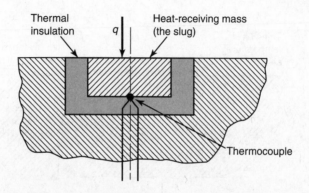

FIGURE 16.36: Section through a slug-type heat flux sensor.

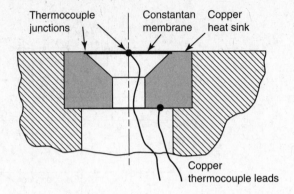

FIGURE 16.37: Section through a foil- or membrane-type heat flux sensor (Gardon gage).

mass or slug that is thermally insulated from its surroundings and a temperature sensor, commonly a thermocouple. As heat flows in, the thermal isolation of the slug produces a temperature differential between the slug and its surroundings. The governing relation is

$$q A_{\text{top}} = mc \frac{dT}{dt} + Q_{\text{loss}} \qquad (16.27)$$

where

A_{top} = the surface area of the top of the slug, in m^2,

m = the mass of the slug, in kg,

c = the specific heat capacity of the slug, in J/kg · K,

T = the slug temperature, in K,

t = the time, in s,

Q_{loss} = the rate of heat loss through the thermal insulation, in W

Slug temperature is measured by the sensor and, through calibration, its derivative is the analog of flux. The slug should have a high thermal conductivity so that its temperature

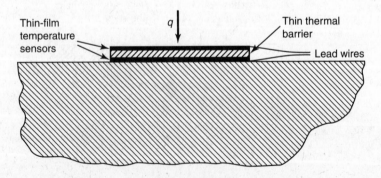

FIGURE 16.38: Thin-film layered heat-flux gage (vertical scale exaggerated).

will be uniform, and Q_{loss} must be minimized. A primary disadvantage of the gage is that it is not useful for steady-state conditions. This gage is also called a *thermal capacitance calorimeter*.

Construction of the Gardon gage [34], or *asymptotic calorimeter*, is shown in Fig. 16.37. It consists of an embedded copper heat sink, a thin membrane of constantan, and an integral thermocouple. The nature of the construction provides two copper-constantan thermocouple junctions, one at the center of the membrane and the other at the interface between the membrane and the heat sink. Thermocouple output, therefore, is a function of the temperature differential between the center and the periphery of the membrane. This, in turn, is a function of the rate of heat flow from the membrane into the sink. The governing relationship is

$$q = 4\left(\frac{tk}{R^2}\right)\Delta T$$
$$= Ce \tag{16.28}$$

where

> $t =$ the membrane thickness, in m,
>
> $k =$ the thermal conductivity of the membrane material, in W/m · K,
>
> $R =$ the membrane radius, in m,
>
> $\Delta T =$ the temperature difference between the center and the edge, in K
>
> $C =$ a calibration constant, in W/m^2 · K · mV,
>
> $e =$ the output of the thermocouple, in mV

With microfabrication technology, it is possible to create a large number of thermocouples, connected as a thermopile, on a membrane just a few millimeters in diameter. Gages of this type can be packaged as a TO-can (Fig. 7.21). They are very widely used as sensors in infrared heat detectors (see Section 16.8).

The layered gage is shown in Fig. 16.38. Here temperature sensors are attached to the upper and lower surfaces of a thin, thermally resisting layer. The heat flux is obtained directly from the measured temperature difference by approximating Eq. (16.26):

$$q = -k\frac{dT}{dx} \approx k\frac{\Delta T}{\delta} \tag{16.29}$$

where

> $k =$ the thermal conductivity of the resisting film, in W/m · K,
>
> $\Delta T =$ the temperature difference between the upper and lower
>
> surface, in K,
>
> $\delta =$ the thickness of the resisting film, in m

The temperature sensors are usually either RTDs or thermocouples. Because the temperature difference across the thin barrier is very small, a differential thermopile may be used to obtain ΔT. As many as 100 or more thermocouple junctions are connected in series, with successive junctions located on opposite sides of the film.

Layered gages became capable of high-frequency response only with the advent of microfabrication technology, which made it possible to deposit thin-film temperature sensors (0.1 to 0.5 μm thick) onto relatively thin (1 to 75 μm) thermal barriers. In some applications, the sensors are deposited onto each side of a Kapton film [35]; in other designs, the sensors and the thermal barrier are sequentially sputtered onto a supporting ceramic substrate [36]. This technology is continuing to evolve rapidly.

Calibration of heat flux meters can involve radiant, conductive, and convective heat transfer and generally depends upon both the type of gage and its specific application [33]. One common approach uses thermal radiation from a black body source at known temperature to heat the gage with a known heat flux [37].

16.12   CALIBRATION OF TEMPERATURE-MEASURING DEVICES

As noted in Section 1.7, if a measurement is to be meaningful, the measuring procedure and apparatus must be provable. This statement is true for all areas of measurement, but for some reason the impression seems prevalent that it is less true for temperature-measuring systems than for other systems. For example, it is generally thought that the only limitation in the use of thermocouple tables is in satisfying the requirement for metal combination indicated in the table heading. Mercury-in-glass scale divisions and resistance-thermometer characteristics are commonly accepted without question. And it is assumed that once proved, the calibrations will hold indefinitely.

Of course, we know that these ideas are incorrect. Thermocouple output is very dependent on purity of elementary metals and consistency and homogeneity of alloys. Alloys of supposedly like characteristics but manufactured by different companies may have temperature–emf relations sufficiently at variance to require different tables. In addition, aging with use will alter thermocouple outputs. Resistance-thermometer stability is very dependent on freedom from residual strains in the element, and comparative results from like elements require very careful use and control of the metallurgy of the materials.

Any calibration of a temperature measuring system must be traceable to the International Temperature Scale of 1990 (ITS-90). Direct calibration on this scale uses the fixed physical conditions (melting points, triple points, etc.) and interpolating equations described in Section 2.7. Such calibrations are done almost exclusively at standards laboratories. Thermometers so calibrated are primary standard thermometers, and other thermometers are calibrated by some type of comparison to them.

In general practice, thermometers are calibrated by putting the thermometer into a known-temperature enclosure and reading its output. The enclosure temperature, in turn, is known either because another calibrated thermometer is in it or because the enclosure is held at a known of freezing or melting point temperature.

Known-temperature enclosures are of two general types, one of which, the fixed-point enclosure, has already been mentioned. A fixed-point enclosure might house a quantity of a material whose freezing point is known; often, this is a material whose temperature is assigned by ITS-90 (see Table 2.5). Glass is used for the container material for the lower temperatures and graphite is used for the higher temperatures. Integral heating coils are employed. To use the fixed-point cell, the temperature sensor to be calibrated is placed in a well extending into the center of the container. The heater is then turned on, and the temperature carried above the melting point of the reference substance and held until melting is completed. The cell is then permitted to cool, and when the freezing point is reached, the temperature stabilizes and remains constant at the specified value as long as liquid and solid are both present. This period may persist from several minutes to several days, depending on the particular cell. Accuracies of approximately 0.1 K may be easily attained and, if great care is exercised, accuracies of better than 1 mK may be reached [4].

The other class of calibration enclosures are the variable-temperature enclosures, which include furnaces for high-temperature work, stirred liquid baths for use from about -50 to $600°C$, and cryostats for very-low-temperature work [38]. These are more commonly used than fixed-point cells, for reasons of lower cost and greater flexibility. With a variable-temperature enclosure, it is possible to obtain calibration data at a number of adjacent temperatures, which may then be used to adjust the coefficients of an appropriate output-to-temperature equation [e.g., Eq. (16.2).]

The simplest possible calibration enclosure is the icepoint bath, which consists of crushed ice and water held in an insulated container, such as a Dewar flask or an expanded-polystyrene box. The ice-point provides a convenient way to check the calibration of any temperature sensor whose range includes $0°C$. Properly done, an icebath can produce a temperature of $0°C$ to an accuracy of better than 10 mK. The most important guidelines for making an icepoint are the following [17].

- Clean water should be used to make the ice and for the water that is added. Distilled water is best. Salts or organic contaminants should be carefully avoided, as these may depress the freezing point temperature. For low-accuracy work, tap water may be sufficient, although it should have an electrical resistivity above $0.5 \ M\Omega \cdot m$.

- The ice should be shaved or crushed to pieces no larger than 5 mm diameter.

- The ice should be tightly packed into the container and then filled with water. The mixture must make good contact with the sensor.

- Liquid that accumulates at the bottom of the container should be siphoned off periodically.

In addition to the primary fixed points established by ITS-90 and the icepoint, numerous secondary fixed points have been tabulated (see, for example, [39]). Some examples include the sublimation point of carbon dioxide or dry ice ($-78.5°C$), the triple point of n-docosane ($43.9°C$), the boiling point of water ($100°C$), and the freezing point of lead ($327.5°C$). Fixed-point cells based on both primary and secondary materials are commercially available.

Calibration of liquid-in-glass thermometers is discussed in detail in [40].

SUGGESTED READINGS

ASME 19.3-1974 (R1998) *Temperature Measurement*. New York: American Society of Mechanical Engineers, 1998.

ASTM. *Manual on the Use of Thermocouples in Temperature Measurement*. 4th ed. ASTM manual series MNL 12. Philadelphia: American Society for Testing and Materials, 1993.

Bentley, R. E. (ed.) *Handbook of Temperature Measurement*, Vols. 1–3. Singapore: Springer-Verlag, 1998.

Burns, G. W., and M. G. Scroger. *Temperature-Electromotive Force Reference Functions and Tables for Letter-Designated Thermocouple Types Based on the ITS-90*. NIST Monograph 175. Washington, D.C.: U.S. Department of Commerce, National Institute of Standards and Technology, 1993. (Supercedes NBS Monograph 125)

Goldstein, R. J., P. H. Chen, and H. D. Chang. Measurement of Temperature and Heat Transfer, Chapter 16 of Rohsenow, W. M., J. P. Hartnett, and Y. I. Cho (eds.), *Handbook of Heat Transfer*. 3rd ed. New York: McGraw-Hill, 1998.

Lawton, B., and G. Klingenberg. *Transient Temperature in Engineering and Science*. New York: Oxford University Press, 1996.

Lienhard IV, J. H., and J. H. Lienhard V. *A Heat Transfer Textbook*. 3rd ed. Cambridge, Mass.: Phlogiston Press, 2003.

Mangum, B. W., and G. T. Furukawa. *Guidelines for Realizing the International Temperature Scale of 1990 (ITS-90)*. NIST Technical Note 1265. Washington, D.C.: U.S. Department of Commerce, National Institute of Standards and Technology, 1990.

McGee, T. D. *Principles and Methods of Temperature Measurement*. New York: John Wiley, 1988.

Michalski, L., K. Eckersdorf, and J. McGhee. *Temperature Measurement*. New York: John Wiley, 1991.

Nicholas, J. V., and D. R. White. *Traceable Temperatures: An Introduction to Temperature Measurement and Calibration*. 2nd ed. New York: John Wiley, 2001.

Preston-Thomas, H. The International Temperature Scale of 1990 (ITS-90). *Metrologia* 27:3–10, 1990 (with corrections in *Metrologia* 27:107, 1990).

PROBLEMS

16.1. At what temperature readings do the Celsius and Fahrenheit scales coincide?

16.2. The temperature indicated by a "total immersion" mercury-in-glass thermometer is 70°C (158°F). Actual immersion is to the 5°C (41°F) mark. What correction should be applied to account for the partial immersion? Assume ambient temperature is 20°C (68°F).

16.3. The uncertainty of a thermometer is stated to be ±1% of "full scale." If the thermometer range is −20°C to 120°C, plot the uncertainty as a percentage of the reading over the thermometer's range.

16.4. The following relation [3] may be used to determine the radius of curvature, r, of a bimetal strip that is initially flat at temperature T_0:

$$r = \frac{t\{3(1+m)^2 + (1+mn)[m^2 + (1/mn)]\}}{6(\alpha_2 - \alpha_1)(T - T_0)(1+m)^2}$$

where

$$t = \text{the combined thickness of the two strips,}$$

$$m = \text{the ratio of thicknesses of low- to high-expansion components,}$$

$$n = \text{the ratio of Young's modulus values of low-to high-expansion components,}$$

$$\alpha_1 \text{ and } \alpha_2 = \text{coefficients of linear expansion, with } \alpha_1 < \alpha_2,$$

$$T = \text{the temperature, in °C or °F, depending on the units for } \alpha_1 \text{ and } \alpha_2$$

(a) Devise a spreadsheet template to solve for r, using the preceding equation.

(b) If 12-cm-long by 1-mm-thick strips of phosphor-bronze and Invar are brazed together to form a bimetal temperature sensor, determine the deflection of the free end per degree change in temperature. Recall that for a beam in bending, $1/r = d^2y/dx^2$. See Table 6.3 for material properties. Let $T_0 = 20°C$ and $T = 100°C$.

16.5. Search the literature for the range of values for the constants A and B in Eq. (16.2) for commonly used resistance thermometer materials. (See the Suggested Readings for this chapter.)

16.6. The element of a resistance thermometer is constructed of a 50-cm (19.7-in.) length of 0.03-mm (0.0012-in.) nickel wire. What will be the nominal resistance of the element? (See Table 12.1 for resistivity.) If we assume that the temperature coefficient of resistivity is constant over the common range of ambient temperatures, what will be the change in resistance of the element per degree C? Per degree F?

16.7. If platinum is substituted for nickel in Problem 16.6, what are the calculated values?

16.8. The circuit shown in Fig. 16.6(a) is used with a platinum resistance thermometer having a resistance of 1200 Ω at 200°C. Also, $R_{AB} = R_{BC} = 8000\ \Omega$ and $R_{DC} = 6800\ \Omega$. Using the data for platinum listed in Problem 3.40, plot the bridge output voltage (assume a high-impedance readout device) versus temperature over a range of 0°C to 500°C. Refer to Sections 7.9 through 7.9.3 for bridge circuit relationships.

16.9. The circuit shown in Fig. 16.39 is used to drive a platinum resistance thermometer, $R(T)$, whose temperature response is given by Eq. (16.2a). If the temperature varies over a range of $\Delta T = \pm 10$ K, show that the current i through the sensor is approximately constant and that the output voltage is approximately given by $e_o = (A \cdot e_i / 2101)\,\Delta T$, where A is the temperature coefficient from Eq. (16.2a).

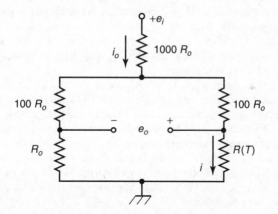

FIGURE 16.39: Circuit for Problem 16.9.

16.10. Investigate the techniques used by the various automobile manufacturers for measuring and displaying engine temperatures. What accuracies do you think are obtained by the various systems?

16.11. Devise a simple thermistor calibration facility consisting of a variable-temperature environment, an accurate resistance-measuring means that avoids ohmic heating of the element, and a reliable temperature-measuring system to be used as the "standard." Calibrate several thermistors and evaluate their degree of adherence to Eqs. (16.3). (Avoid the problems implied in Problem 16.15.)

16.12. Prepare spreadsheet templates for Eq. (16.3) to solve

(a) for R when T, T_0, R_0, and β are given;
(b) for β when T, T_0, R, and R_0 are given.

16.13. The following are data for the calibration of a thermistor.

Temperature, °F	Resistance, kΩ
78	3.16
76	3.23
72.5	3.89
68	4.24
65	4.47
61	4.76
58	5.31
54	5.77
50.5	6.37
47.5	6.80

For each line of data, calculate the value of β, using Eq. (16.3), and T_0 and R_0 corresponding to the values of 68°F (20°C). Some spread in the results will be found; however, use the average of the calculated values as the magnitude of β. (Use of the spreadsheets prepared in answer to the previous problem is recommended.)

16.14. Write Eq. (16.3) using the value of β found in answer to Problem 16.13, and plot the result over the range of data. Spot-check several points.

16.15. A small insulated box is constructed for the purpose of obtaining temperature calibration data for thermistors. Provision is made for mounting a thermistor within the box and bringing suitable leads out for connection to a commercial Wheatstone bridge. The bulb of a standardized mercury-in-glass thermometer is inserted into the box for the purpose of determining reference temperatures. A small heating element (a miniature soldering iron tip) is used as a heat source.

After the heater is turned on, thermistor resistances and thermometer readings are periodically made as the temperature rises from ambient to a maximum. The heater is then turned off and further data are taken as the temperature falls.

It is quickly noted, however, that there is a very considerable discrepancy in the "heating" resistance–temperature relationship compared with the corresponding "cooling" data. Why should this have been expected? Criticize the design of the arrangement described above when used for the stated purpose. How would you make a *simple* laboratory setup for obtaining reasonably accurate calibration data for a thermistor over a temperature range of, say, 80°F to 400°F?

16.16. Prove the law of intermediate metals for the situation shown in Fig. 16.12(a) using Eq. (16.4).

16.17. Prove the law of intermediate temperatures using Eq. (16.5).

16.18. Show that the output of the circuit in Fig. 16.18(b) is $E = \mathcal{E}_{\text{FeCn}}(T_m) - \mathcal{E}_{\text{FeCn}}(0°\text{C})$.

16.19. An ice-bath reference junction is used with a copper-constantan thermocouple. For four different conditions, millivolt outputs are read as follows: $-4.334, 0.00, 8.133$, and 11.130. What are the respective junction temperatures (a) in degrees C and (b) in degrees F?

16.20. Chromel-alumel thermocouples are used for measuring the temperatures at various points in an air conditioning unit. A reference junction temperature of 22.8°C is recorded. If the following emf outputs are supplied by the various couples, determine the corresponding temperatures: $-1.689, -1.108, -0.113$, and 3.185 mV.

16.21. A K-type thermocouple circuit (Fig. 16.40) has its reference junction in liquid nitrogen at 1 atm pressure (77 K). The output voltage is measured to be 1.340 mV.
(a) What is the temperature of the measuring junction?

(b) What would the output voltage be if the measuring junction were now placed in room temperature air (20°C)?

(c) What would the output voltage be if the reference junction were placed in room temperature air (20°C) while the measuring junction remained at the temperature of part **(a)**?

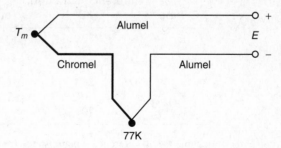

FIGURE 16.40: Circuit for Problem 16.21.

16.22. Using the equation for E versus T for type K thermocouples given in Table 16.6, spot-check at least five points in Table 16.5 to satisfy yourself that the tabulated and calculated values agree.

16.23. Select a thermocouple type and write a computer program or spreadsheet that calculates measuring junction temperature as a function of circuit emf, assuming that the reference junction temperature is 0°C. If you will be performing experiments with thermocouples, select the thermocouple type used in your lab. Take the appropriate equation from Table 16.7.

16.24. Write a spreadsheet (or other computer program) that calculates thermocouple outputs using the equations in Tables 16.6 and 16.7.

(a) Write the software to find temperature as a function of emf for each type in Table 16.7.

(b) Write the software to find emf as a function of temperature for each type in Table 16.6. Spot-check your results against the data in Table 16.4.

16.25. Use appropriate equations from Table 16.6 to calculate the emf's that are expected for the following situations, assuming a circuit such as that shown in Fig. 16.16. If you wrote a thermocouple spreadsheet, you may use it in your calculations.

TC Type	Temperature at the Reference Junction, °C	Temperature at the Measuring Junction, °C
E	0	500
E	20	750
J	0	650
K	80	1150
T	100	315

16.26. Use appropriate equations from Tables 16.6 and 16.7 to calculate the temperature at the measuring junction for each of the following situations. Assume a circuit such as that shown in Fig. 16.16. If you wrote a thermocouple spreadsheet, you may use it in your calculations.

TC Type	Temperature at the Reference Junction, °C	emf, mV
E	0	60.63
E	90	-4.36
J	15	16.30
J	280	-11.78
K	0	29.79
K	700	-8.14

16.27. A type N thermocouple has its reference junctions at 0°C. Plot the measuring junction temperature as a function of emf over the range 0.0 mV to 45.0 mV, using equations from Table 16.7. If you have written a thermocouple spreadsheet, you may use it to do the job.

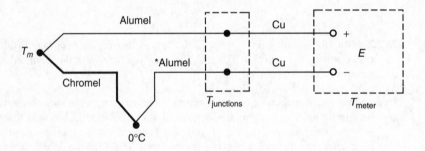

FIGURE 16.41: Circuit for Problem 16.28.

16.28. Consider the thermocouple circuit shown in Fig. 16.41. Copper extension wires are connected to the thermocouple circuit. Suppose that the extension wire junctions are also at 0°C. Show that the output voltage is unchanged when the second piece of alumel (marked *) is eliminated and the copper lead is directly connected to the chromel wire.

16.29. The temperature difference between two points on a heat exchanger is desired. The measuring and reference junctions of a chromel-alumel thermocouple are embedded within the inlet and outlet tubes, 1 and 2, respectively, and an emf of 0.381 mV is read. Why does this reading provide insufficient data to determine the differential temperature accurately? What additional information must be obtained before the answer may be found?

16.30. Show that the output voltage of the thermopile in Fig. 16.20(a) is $5[\mathcal{E}_{AB}(T_1) - \mathcal{E}_{AB}(T_2)]$.

16.31. A thermopile is constructed from three type K thermocouples (Fig. 16.42). The measuring junctions each have different temperatures (T_1, T_2, and T_3), and the reference junctions are all at 0°C.

(a) What is the output emf, E, of the thermopile when $T_1 = 20°C$, $T_2 = 25°C$, and $T_3 = 30°C$?

(b) If the user of the thermopile were to assume that all three temperatures were equal, what temperature would he or she calculate? How does this compare to the actual average of the three temperatures?

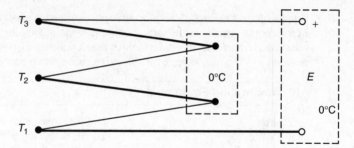

FIGURE 16.42: Circuit for Problem 16.31.

(**c**) If instead, $T_1 = 800°C$, $T_2 = 1000°C$, and $T_3 = 1200°C$, what temperature would the user calculate? How does this compare to the true average temperature?

16.32. The sensitivity of a measuring device is the rate of change of its output with respect to its input (see Section 6.1). For a thermocouple circuit, then, the sensitivity is dE/dT_m.
(**a**) Estimate the sensitivities of type E, K, and S thermocouples near 100°C. Assume an icepoint reference junction, and use the data in Table 16.4 or 16.5.
(**b**) If you are measuring thermocouple emf with a voltmeter that has a resolution 0.1 mV, what is the smallest temperature change that you can detect with type E, type K, and type S thermocouples, respectively?
(**c**) To detect a change of 0.1°C using a type K thermocouple, what is the minimum resolution that the voltmeter must have?
(**d**) Estimate the sensitivity of the thermopile in Fig. 16.20(a), assuming that type K thermocouples are used, that $T_2 = 0°C$, and that T_1 is near 100°C. To detect a change of 0.1°C with this thermopile, what minimum resolution must the voltmeter have?

16.33. Thermocouple wire is generally manufactured to meet ASTM tolerances that vary by wire type. For commercial-grade type E thermocouples between 0°C and 900°C, this tolerance is ±1.7°C or ±0.5% (whichever is greater). The tolerance is a bound on the bias error of the thermocouple's temperature as calculated from the measured emf using the NIST standards. The actual size of the bias error for any specific thermocouple is unknown. To measure temperature to an accuracy better than the standard tolerance, the thermocouple must first be calibrated.

(**a**) A type E thermocouple probe using an electronic icepoint is calibrated by comparison to a platinum resistance thermometer (PRT). Both sensors are placed into a well-stirred temperature-controlled liquid bath. The following readings are obtained.

PRT Temperature, °C	Thermocouple emf, mV
25.0	1.526
35.1	2.146
50.2	3.091
63.5	3.941
74.9	4.681

Use these data to find an equation for the thermocouple's temperature as a function of its emf.

(**b**) Suppose that the calibrated thermocouple is now used to measure a temperature. The thermocouple circuit has the same electronic icepoint but now uses a lower-quality voltmeter. Discuss the uncertainty in that measurement and estimate its size (at 95% confidence). You may assume the following uncertainties in the equipment used.

Calibration Equipment

PRT	$\pm0.05°C$ (bias, 95%)
Voltmeter	±0.001 mV (bias, 95%)
Liquid bath	temperature fluctuations are small and random

Thermocouple Equipment

Electronic ice point	$\pm0.25°C$ (bias, 95%)
Voltmeter	±0.005 mV (precision, 95%)

16.34. To eliminate high-frequency electrical noise, a low-pass filter is connected to a type K thermocouple circuit, as shown in Fig. 16.43. The output is displayed on an oscilloscope having 1 MΩ input impedance. The thermocouple circuit's output is steady (i.e., a dc voltage), and the filter's cut-off frequency is known to be appropriate.

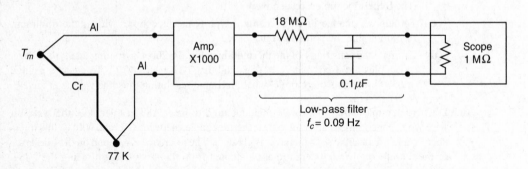

FIGURE 16.43: Circuit for Problem 16.34.

When the voltages are read from the scope and converted to temperatures, the results are clearly wrong: The apparent value of T_m is close to 77 K, when it should be much higher. Explain why, and describe a better filter.

16.35. Plot the error in measured temperature as a function of temperature for the following types of pyrometers, assuming a 10% error in emissivity: a total radiation pyrometer; a spectral-band pryometer operating at 0.9 μm; and an infrared pyrometer sensitive to wavelengths from 8 to 14 μm. Use a range of 200°C to 2000°C.

16.36. Referring to Eq. (16.15a), plot the ratio of static to total temperatures versus Mach number over the range of $M = 0$ to 3 and for $k = 1.3, 1.4,$ and 1.5.

16.37. Referring to Eqs. (16.18) and (16.19), plot the ratio of indicated to static temperatures versus the recovery factor over the range of $r = 0$ to 1, for $k = 1.3$, 1.4, and 1.5.

16.38. If you did not work Problems 5.3 and 5.14, do so now.

16.39. The following temperature–time data were recorded after a probe was placed in a warm environment:

Time, s	Temperature, °C
0	20
4	83
8	123
12	152
20	182
30	194
40	201
50	203

(**a**) Plot the data points.

(**b**) From the plot, determine a time constant for the system.

(**c**) Write a response equation assuming first-order process.

(**d**) Calculate sufficient points to plot the theoretical curve.

(**e**) Decide, on the basis of the plot, whether or not the process may be considered a single time-constant first-order type.

16.40. A "two time-constant" temperature transducer has time constants in the ratio $\zeta = 4/1$, where $\tau_p = 1.5$ s. If the transducer, initially at a temperature of 80°C, is suddenly immersed in a 500°C environment, what will be the temperature indicated after 3 s?

16.41. If the transducer in Problem 16.40 is initially at 500°C and is suddenly immersed in an 80°C environment, what temperature will be indicated after 3 s?

16.42. The performance of a temperature-measuring system approximates that dictated by two time-constant theory: $\tau_p = 10$ s and $\tau_j = 25$ s. If the system is subjected to the temperature input shown in Fig. 16.44, we see that the probe will not have sufficient time to produce a readout approximating T_{max}. If, however, at the end of the 18-s pulse, the system indicates 135°C, what must be the value of T_{max}?

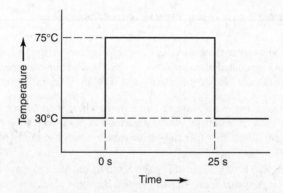

FIGURE 16.44: Temperature–time relationship for Problem 16.42.

16.43. The behavior of a temperature-measuring system approximates two time-constant theory with $\tau_p = 6$ s and $\tau_j = 14$ s. If the system experiences a perturbation as shown in Fig. 16.45, what will be the indicated temperature at $t = 15$ s? At $t = 25$ s? At 25 s the temperatures of the probe and the jacket will not be the same; however, on the assumption that both are at probe temperature, estimate the indicated temperature at 60 s. Will the calculated value be too high or too low? State your reasoning in answering the last question.

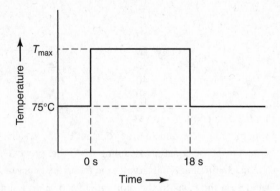

FIGURE 16.45: Temperature–time relationship for Problem 16.43.

16.44. We wish to have both a continuous record and an instantaneous readout of energy flow rate from heated water passing through a pipe. Temperature, pressure, and rate of flow each vary over a range of values.

(**a**) Analyze the problem and prepare a block diagram of the various measurements and functional problems that must be solved.

(**b**) Insofar as you can, detail the steps to a solution.

(*Note:* Figure 8.23 may help in providing a starting point.)

16.45. A thin-film layered heat-flux gage consists of a Kapton film with a single thin-foil thermocouple junction coated on either side (Fig. 16.38). The film has a thickness of 25 μm (± 1 μm at 95% confidence) and a conductivity of 0.20 W/m $\cdot$ K (± 0.01 W/m $\cdot$ K at 95% confidence).

(**a**) If the thermocouples operate in a range where their output is 40 μV/°C, and if the thermocouple voltage can be measured to ± 1 μV (at 95% confidence), what is the smallest heat flux that can be measured to an uncertainty of no more than 10% (at 95% confidence)?

(**b**) Describe how the heat-flux gage could be modified to measure a heat flux one-tenth as large as that found in part (**a**) with the same percentage uncertainty.

(**c**) Thermocouple junctions may have systematic error of up to 2°C. What should be done to ensure accurate heat flux readings with this gage?

REFERENCES

[1] Wise, J. A. *NIST Measurement Services: Liquid-in-Glass Thermometer Calibration Service*. NIST Special Publication 250-23. Washington, D.C.: U.S. Department of Commerce, National Institute of Standards and Technology, 1989.

[2] Horrigan, E. C. Liquid-expansion thermometers, Chapter 5 in Bentley, R. E. (ed.), *Resistance and Liquid-in-Glass Thermometry*. Singapore: Springer-Verlag, 1998. (*Handbook of Temperature Measurement*, Vol. 2)

[3] Eskin, S. G., and J. R. Fritze. Thermostatic bimetals. *ASME Trans*. 62(5):433–442, 1940.

[4] Mangum, B. W., and G. T. Furukawa. *Guidelines for Realizing the International Temperature Scale of 1990 (ITS-90)*. NIST Technical Note 1265. Washington, D.C.: U.S. Department of Commerce, National Institute of Standards and Technology, 1990.

[5] Riddle, J. L., G. T. Furukawa, and H. H. Plumb. *Platinum Resistance Thermometry*. NBS Monograph 126, Washington, D.C.: U.S. Government Printing Office, April 1973.

[6] *The Temperature Handbook*. Stamford, Conn.: Omega Engineering, Inc., 1989.

[7] Seebeck, T. J. *Evidence of the Thermal Current of the Combination Bi-Cu by Its Action on Magnetic Needle*. Berlin: Abt. d. Königl, Akad. d. Wiss. 1822–23, p. 265.

[8] Peltier, M. Investigation of the heat developed by electric currents in homogeneous materials and at the junction of two different conductors. *Ann. Chim. Phys.* 56:371, 1834.

[9] Thomson, W. Theory of thermoelectricity in crystals. *Trans. Edinburgh Soc.* 21:153, 1847. Also in *Math. Phys. Papers* 1:232, 266, 1882.

[10] ASTM. *Manual on the Use of Thermocouples in Temperature Measurement*. 4th ed. ASTM manual series MNL 12. Philadelphia: American Society for Testing and Materials, 1993.

[11] Burns, G. W., and M. G. Scroger. *Temperature-Electromotive Force Reference Functions and Tables for Letter-Designated Thermocouple Types Based on the ITS-90*. NIST Monograph 175. Washington, D.C.: U.S. Department of Commerce, National Institute of Standards and Technology, 1993.

[12] Muth, S., Jr. Reference junctions. *Instr. Control Systems* 40:133–134, May 1967.

[13] *Lake Shore Product Catalog*. Westerville, Ohio: Lake Shore Cryotronics, Inc., 1991.

[14] Hammond, D. L., and A. Benjaminson. Linear quartz thermometer. *Instr. Control Systems*, 38(10):115, 1965.

[15] Lienhard IV, J. H., and J. H. Lienhard V. *A Heat Transfer Textbook*. 3rd ed. Cambridge, Mass.: Phlogiston Press, 2003.

[16] Ballico, M. J. Radiation Thermometry, Chapter 4 in Bentley, R. E. (ed.), *Temperature and Humidity Measurement*. Singapore: Springer-Verlag, 1998. (*Handbook of Temperature Measurement*, Vol. 1)

[17] Nicholas, J. V., and D. R. White. *Traceable Temperatures: An Introduction to Temperature Measurement and Calibration*. 2nd ed. New York: John Wiley, 2001.

[18] Keyes, R. J. (ed.) *Optical and Infrared Detectors*. 2nd ed. New York: Springer-Verlag, 1980.

[19] *Handbook of Infrared Radiation Measurement*. Stamford, Conn.: Barnes Engineering Company, 1983.

[20] Childs, P. R. N. Advances in Temperature Measurement, in J. P. Hartnett, T. F. Irvine, Y. I. Cho, and G. A. Greene (eds.), *Advances in Heat Transfer*, Vol. 36. San Diego: Academic Press, 2002.

[21] *Steam, Its Generation and Use*. 38th ed. New York: The Babcock and Wilcox Company, 1975.

[22] Tszeng, T. C., and V. Saraf. A study of fin effects in the measurement of temperature using surface mounted thermocouples. *J. Heat Transfer*, 125(5):926–935, Oct. 2003.

[23] Hottel, H. C., and A. Kalitinsky. Temperature measurement in high-velocity air streams. *J. Appl. Mech.*, 67:A25, March 1945.

[24] Lalos, G. T., A sonic-flow pyrometer for measuring gas temperatures. *NBS J. Res.* 47(3):179, Sept. 1951.

[25] Benedict, R. P. *Fundamentals of Temperature, Pressure, and Flow Measurements*. 3rd ed. New York: John Wiley, 1984.

[26] Moffat, R. J. How to specify thermocouple response. *ISA J.*, 4:219, June 1957.

[27] Linahan, T. C.. The dynamic response of industrial thermometers in wells. *ASME Trans.* 78(4):759–763, May 1956.

[28] Coon, G. A. Responses of temperature-sensing-element analogs. *ASME Trans.* 79(8):1857–1868, Nov. 1957.

[29] Fralick, G. C., and L. J. Forney. Frequency response of a supported thermocouple wire: Effects of axial conduction. *Rev. Sci. Instrum.*, 64(11):3236–3244, Nov. 1993.

[30] Hopkins, K. H., J. C. LaRue, and G. E. Samuelson. Effect of variable time constants on compensated thermocouple measurements. *NATO Advanced Study Institute on Instrumentation for Combustion and Flow in Engines*. Dordrecht: Kluwer, 1988.

[31] Shepard, C. E., and I. Warshawsky. Electrical techniques for compensation of thermal time lag of thermocouples and resistance thermometer elements. *NACA Tech. Note 2703*: May 1952.

[32] Shepard, C. E., and I. Warshawsky. Electrical techniques for time lag compensation of thermocouples used in jet engine gas temperature measurements. *ISA J.*, 1:119, Nov. 1953.

[33] Diller, T. E. Advances in Heat Flux Measurements, in J. P. Hartnett, T. F. Irvine, and Y. I. Cho (eds.), *Advances in Heat Transfer*, Vol. 23. New York: Academic Press, 1993.

[34] Gardon, R. An instrument for the direct measurement of intense thermal radiation. *Rev. Sci. Instr.* 24(5):366–370, May 1953.

[35] Epstein, A. H., G. R. Guenette, R. J. G. Norton, and C. Yuzhang. High-frequency response heat-flux gauge. *Rev. Sci. Instr.* 57(4):639–649, April 1986.

[36] Hager, J. M., S. Simmons, D. Smith, S. Onishi, L. W. Langley, and T. E. Diller. *Experimental Performance of a Heat Flux Microsensor.* New York: ASME Paper No. 90-GT-256, 1990.

[37] Pitts, W. M., and Burgess, S. R. *Collected Reports and Publications by the National Institute of Standards and Technology on Heat Flux Gage Calibration and Usage.* NIST Special Publication 971. Washington, D.C.: U.S. Department of Commerce, National Institute of Standards and Technology, 2001.

[38] Horrigan, E. C. Calibration Enclosures, Chapter 8 in Bentley, R. E. (ed.) *Resistance and Liquid-in-Glass Thermometry.* Singapore: Springer-Verlag, 1998. (*Handbook of Temperature Measurement*, Vol. 2)

[39] Crovini, L., R. E. Bedford, and A. Moser. Extended list of secondary reference points. *Metrologia* 13:197–206, 1977.

[40] Horrigan, E. C. Calibration and Use of Liquid-in-Glass Thermometers, Chapter 6 in Bentley, R. E. (ed.), *Resistance and Liquid-in-Glass Thermometry.* Singapore: Springer-Verlag, 1998. (*Handbook of Temperature Measurement*, Vol. 2)

CHAPTER 17

Measurement of Motion

17.1 INTRODUCTION

Mechanical motion may be defined in terms of various parameters as listed in Table 17.1. One or more of the values may be constant with time, periodically varying, or changing in a complex manner. Measurement of static displacement was discussed in detail in Chapter 11. Very broadly, if the displacement-time variation is of a generally continuous form with some degree of repetitive nature, it is thought of as being a *vibration*. On the other hand, if the action is of a single-event form, a transient, with the motion generally decaying or damping out before further dynamic action takes place, then it may be referred to as *shock*. Obviously, shock action may be repetitive and in any case the displacement-time relationships will normally contain vibratory characteristics. To be so termed, however, shock must in general possess the property of being discontinuous. Additionally, steep wavefronts are often associated with shock action, although this is not a necessary characteristic.

In any event, both mechanical shock and mechanical vibration involve the parameters of frequency, amplitude, and waveform. Basic measurement normally consists of applying the necessary instrumentation to obtain a time-based record of displacement, velocity, or acceleration. Subsequent analysis can then provide such additional information as the frequencies and amplitudes of harmonic components and derivable displacement-time relationships that were not directly measured.

TABLE 17.1: Motion Parameters

Motion Parameter	Defining Relationships	
	For Linear Motion	**For Angular Motion**
Displacement	$s = f(t)$	$\theta = g(t)$
Velocity	$v = \dfrac{ds}{dt}$	$\omega = \dfrac{d\theta}{dt}$
Acceleration	$a = \dfrac{dv}{dt} = \dfrac{d^2 s}{dt^2}$	$\alpha = \dfrac{d\omega}{dt} = \dfrac{d^2\theta}{dt^2}$
Jerk	$\dfrac{da}{dt}$	$\dfrac{d\alpha}{dt}$

In many respects instrumentation used for vibration measurements are directly applicable to shock measurement. On the other hand, testing procedures and methods are quite different. The fundamental aspects of acceleration velocity and displacement measurements can be determined through examination of the most basic device for measuring these quantities, the seismic transducer.

17.2   VIBROMETERS AND ACCELEROMETERS

Current nomenclature applies the term *vibration pickup* or *vibrometer* to detector-transducers yielding an output, usually a voltage, that is proportional to either *displacement* or *velocity*. Whether displacement or velocity is sensed is determined primarily by the secondary transducing element. For example, if a differential transformer (Sections 6.11 and 11.13) or a voltage-dividing potentiometer (Sections 6.6 and 7.7) is used, the output will be proportional to a displacement. On the other hand, if a variable-reluctance element (Section 6.12) is used, the output will be a function of velocity.

The term *accelerometer* is applied to those pickups whose outputs are functions of *acceleration*. There is a basic difference in design and application between vibration pickups and accelerometers.

17.3   ELEMENTARY VIBROMETERS AND VIBRATION DETECTORS

In spite of the tremendous advances made in vibration-measuring instrumentation, one of the most sensitive vibration detectors is the human touch. Tests conducted by a company specializing in balancing machines determined that the average person can detect, by means of his or her fingertips, sinusoidal vibrations having amplitudes as low as 0.3 μm (12 μin.) [1]. When the vibrating member was tightly gripped, the average minimum detectable amplitude was only slightly greater than 0.025 μm (1 μin.). In both cases, by fingertip touch and by gripping, greatest sensitivity occurred at a frequency of about 300 Hz.

When amplitudes of motion are greater than, say, $\frac{1}{32}$ in. or about 1 mm, a simple and useful tool is the *vibrating wedge*, shown in Fig. 17.1(a). This is simply a wedge of paper or other thin material of contrasting tone, often black, attached to the surface of the vibrating

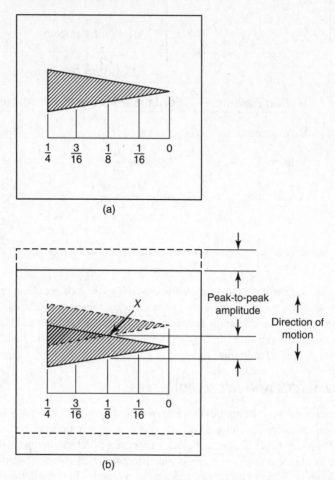

FIGURE 17.1: Vibrating-wedge amplitude indicator: (a) stationary wedge, (b) extreme positions of wedge.

member. The axis of symmetry of the wedge is placed at right angles to the motion. As the member vibrates, the wedge successively assumes two extreme positions, as shown in Fig. 17.1(b). The resulting double image is well defined, with the center portion remaining the color of the wedge and the remainder of the images a compromise between dark and light. By observing the location of the point where the images overlap, marked X, one can obtain a measure of the amplitude. At this point the width of the wedge is equal to the double amplitude of the motion. This device does not yield any information as to the waveform of the motion. A simple apparatus for measuring frequency involves a small cantilever beam whose resonance frequency may be varied by changing its effective length. In use, the instrument case is held against the member whose frequency is to be measured, and the beam length is slowly adjusted, searching for the length of beam at which resonance will occur. When this condition is found, the end of the beam whips back and forth with considerable amplitude. The device is quite sensitive, with the accuracy limited only by the resolution of the scale.

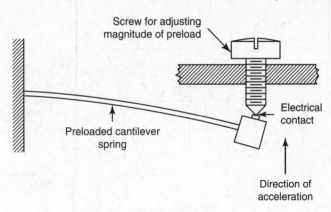

FIGURE 17.2: Preloaded spring-type accelerometer.

17.4 ELEMENTARY ACCELEROMETERS

Probably the most elementary acceleration-*measuring* device is the acceleration-level indicator. There are different forms of this instrument, but they are all of the yes-or-no variety, indicating that a predetermined level of acceleration has or has not been reached. Figure 17.2 is a schematic diagram of one such instrument, which makes use of a preloaded electrical contact [2]. In theory, when the effect of the inertia forces acting on the spring and mass exceed the preload setting, contact will be broken, and this action may then be used to trip some form of indicator. Rather elaborate forms of this arrangement have been devised. A second acceleration level indicator is of the *one-shot* type. Acceleration level is determined by whether or not a tension member fractures. Strictly brittle materials should be used for the tension member; otherwise cold working caused by previous acceleration history will change the physical properties and hence the calibration. Since such materials do not exist, this limitation is an important one. Each of the approaches described can be considered as providing only rough indications, whose primary value lies in their simplicity.

17.5 THE SEISMIC INSTRUMENT

Vibration pickups and accelerometers are usually of the "seismic mass" form illustrated schematically in Fig. 17.3. A spring-supported mass is mounted in a suitable housing, with a sensing element provided to detect the relative motion between the mass and the housing. As we will see later, damping may also be provided. In the figure this is represented by a dashpot mounted between the mass and the housing.

Basically, the action of the seismic instrument is a function of acceleration through the inertia of the mass. The output, however, is determined by the *relative* motion between the mass and the housing. This principle results in two varieties of seismic mass instruments, the *vibrometer* and the *accelerometer*. Several of the more commonly used types of vibration pickups employ a variable-reluctance transducer, in which the relative motion between a coil and the flux field from a permanent magnet is used. In this case, the instrument is velocity sensitive because the output is proportional to the rate at which the lines of flux are cut.

By proper selection of natural frequency and damping, it is possible to design the seismic instrument so that the relative displacement between mass and housing is a function of acceleration. The output from such an instrument could therefore be calibrated in terms of

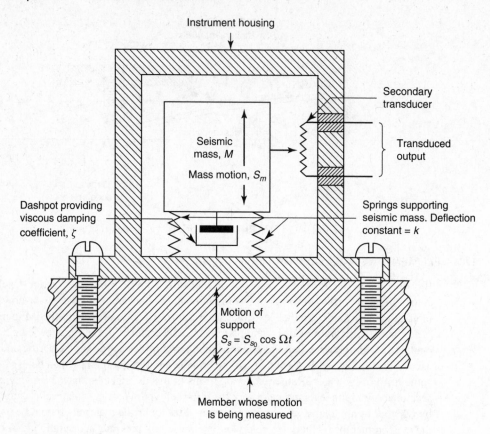

FIGURE 17.3: Seismic type of motion-measuring instrument.

acceleration, and the instrument would be an accelerometer. The fundamental requirements for the two types of instrument will be developed in the following sections.

17.6   GENERAL THEORY OF THE SEISMIC INSTRUMENT

Figure 17.3 shows a one-degree-of-freedom system with viscous damping, excited by a harmonic motion supplied to the support. Special note should be taken of the fact that *simple harmonic excitation* is assumed, which, strictly speaking, restricts the relations to be developed to a rather limited case. As we will see, however, much can be learned about seismic instruments by studying this special case. Let

M = the mass of the seismic element,

g_c = the dimensional constant,

k = the deflection constant for the spring support,

ζ = the damping coefficient,

S_m = the absolute displacement of mass M, measured from the static equilibrium condition,

S_{m_0} = the displacement amplitude of mass M,

S_s = the absolute displacement of the supporting member

$\quad = S_{s_0} \cos \Omega t$,

$S_r = S_m - S_s$

$\quad$ = the relative displacement between the mass and the support, which is the

$\qquad$ displacement that the secondary transducer will detect,

S_{r_0} = the relative displacement amplitude between the mass and the supporting

$\qquad$ member,

t = any instant of time from $t = 0$,

Ω = the exciting frequency,

ω_n = the undamped natural frequency of the system = $\sqrt{kg_c/M}$,

ϕ = the phase angle

Applying Newton's second law to the free body of mass M, we find that the differential equation for the motion of the mass will be

$$\frac{M}{g_c} \frac{d^2 S_m}{dt^2} + \zeta \frac{d S_r}{dt} + k S_r = 0 \tag{17.1}$$

Each term represents a force. The first is the inertia force, the second is the damping force, and the third is the spring force. Substituting,

$$S_m = S_s + S_r \tag{17.2}$$

we get

$$\frac{M}{g_c} \frac{d^2 S_r}{dt^2} + \zeta \frac{d S_r}{dt} + k S_r = -\frac{M}{g_c} \frac{d^2 S_s}{dt^2} \tag{17.3}$$

However,

$$S_s = S_{s_0} \cos \Omega t$$

Then

$$\frac{M}{g_c} \frac{d^2 S_r}{dt^2} + \zeta \frac{d S_r}{dt} + k S_r = \frac{M}{g_c} S_{s_0} \Omega^2 \cos \Omega t \tag{17.4}$$

This equation is a linear differential equation of the second order with constant coefficients and is very similar to Eq. (5.24). Therefore, by comparison, the solution may be written as

$$S_r = e^{-t/\tau} \left[A \cos \sqrt{1 - \xi^2} \, \omega_n t + B \sin \sqrt{1 - \xi^2} \, \omega_n t \right]$$

$$+ \frac{(M/kg_c) S_{s_0} \Omega^2 \cos(\Omega t - \phi)}{\sqrt{\left[1 - (\Omega/\omega_n)^2\right]^2 + \left[2\xi(\Omega/\omega_n)\right]^2}} \tag{17.5}$$

where ξ = the ratio of the damping coefficient to the critical damping coefficient and

$$\phi = \tan^{-1}\left[\frac{2\xi(\Omega/\omega_n)}{1 - (\Omega/\omega_n)^2}\right] \tag{17.6}$$

The first term on the right-hand side of Eq. (17.5) provides the transient component and the second term the steady-state component. If we assume a time interval that is several time constants in length, the transient term may be neglected. We may then write

$$S_{r_0} = \frac{S_{s_0}(\Omega/\omega_n)^2}{\sqrt{\left[1 - (\Omega/\omega_n)^2\right]^2 + [2\xi(\Omega/\omega_n)]^2}} \tag{17.7}$$

17.6.1   The Vibration Pickup

Let us now consider just what we have in Eq. (17.7) by recalling that S_{r_0} is the relative displacement amplitude between the seismic mass M and the support and that S_{s_0} is the displacement amplitude of the instrument housing and hence of the supporting member to which it is attached. We should also recall that the instrument output will be some function of S_r, the relative displacement, and not a direct function of the quantity we wish to measure, S_s.

An inspection of Eq. (17.7) forces us to the conclusion that if the relative amplitude is to be a direct linear function of the support amplitude, it will be necessary that the total coefficient of S_{s_0} be a constant. For a given instrument, the undamped natural frequency and damping will be *built in*; hence the only variable will be the forcing frequency, Ω. Let us see, then, how the function behaves by plotting the ratio S_{r_0}/S_{s_0} versus Ω/ω_n. This can be done for various damping ratios, thereby obtaining a family of curves. Figure 17.4 is the result. Inspection of the curves shows that for values of Ω/ω_n considerably greater than 1.0, the amplitude ratio is indeed near unity, which is as desired. It may also be observed that the value of the damping ratio is not important for high values of Ω/ω_n. However, in the region near a frequency ratio of 1.0, the amplitude ratio varies considerably and is quite dependent on damping. Below $\Omega/\omega_n = 1.0$, the ratios of amplitude break widely from unity. It may also be observed by inspection of Fig. 17.4 that for certain damping ratios, the amplitude ratio does not stray very far from unity, *even in the vicinity of resonance*.

We may conclude from our inspection that *damping on the order of 65% to 70% of critical is desirable* if the instrument is to be used in the frequency region just above resonance. We also see that, in any case, damping of a general-purpose instrument is a compromise, and inherent errors resulting from the principle of operation will be present. To these would be added errors that may be introduced by the secondary transducer and the second- and third-stage instrumentation.

As an example, let us check the discrepancy for the following conditions:

$\quad \xi$ = the damping ratio = 0.68,

$\quad S_{s_0}$ = 0.015 in.,

$\quad f_n$ = the natural undamped frequency of the instrument = 4.75 Hz,

$\quad f_e$ = the exciting frequency = 7 Hz

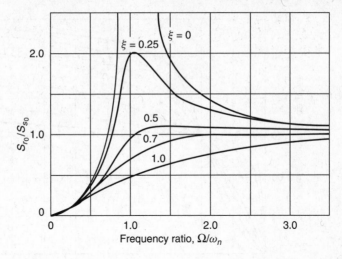

FIGURE 17.4: Response of a seismic instrument to harmonic displacement.

Then

$$\frac{f_e}{f_n} = \frac{7}{4.75} = 1.474, \qquad \left(\frac{f_e}{f_n}\right)^2 = 2.17$$

Using Eq. (17.7), we get

$$S_{r0} = \frac{2.17 \times 0.015}{\sqrt{[1 - (2.17)]^2 + (2 \times 0.68 \times 1.474)^2}}$$

$$= 0.01404,$$

$$\text{Inherent error} = \left(\frac{0.01404}{0.015} - 1\right) 100 = -6.38\%$$

17.6.2 Phase Shift in the Seismic Vibrometer

Let us now turn our attention to the phase relation between relative amplitude and support amplitude. Naturally it would be very desirable to have a zero phase relation for all frequencies. A plot of Eq. (17.6) is shown in Fig. 17.5. This indicates that for *zero damping* the seismic mass moves exactly in phase with the support (but not with the same amplitude), so long as the forcing frequency is below resonance. Above resonance the mass motion is completely out of phase (180°) with the support motion. At resonance there is a sudden shift in phase. For other damping values, a similar shift takes place, except there is a gradual change with frequency ratio.

A simple experiment verifies this phase-shift relation. A crude support-excited seismic mass can be constructed by tying together five or six rubber bands in series to form a long, soft *spring* and attaching a mass of, say, 0.25 kg or $\frac{1}{2}$ lb to one end while holding the other end in the hand. When the hand is moved up and down, a relative motion is obtained and the natural frequency of the system can easily be found, which is the frequency that provides greatest amplitudes with least effort. Now try moving the hand up and down at a frequency considerably below the natural frequency. It will be observed that the mass

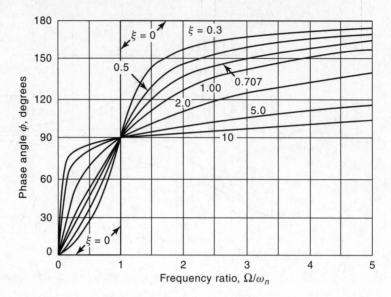

FIGURE 17.5: Relations among phase angle, frequency ratio, and damping for a seismic instrument.

moves up and down at nearly the same time as the hand. The motion of the seismic mass is approximately *in phase* with the motion of the supporting member, the hand. Now move the hand up and down at a frequency considerably above the natural frequency. It will be observed that the mass now moves downward as the hand moves up, and the weight moves up as the hand moves down. The motions are *out of phase*.

Our observations indicate that from the standpoint of phase shift, a "best" solution would be to design an instrument with zero damping (if that were possible). However, the amplitude relation near resonance would then be in serious error.

Perhaps any amplitude and phase-shift effects near resonance, such as we have been discussing, could be accounted for by a calibration! It would seem at first glance that such a possibility would be good, and indeed it would be feasible if *single-frequency* harmonic motions were always encountered. In fact, if simple sinusoidal motion were always to be measured, phase shift would not be of consequence. We would not care particularly whether the peak relative motion coincided exactly with the peak support motion so long as the waveform and measured amplitudes were correct. The difficulty arises when the input is a complex waveform, made up of the fundamental and many other harmonics, with each harmonic simultaneously experiencing a different phase shift.

As we saw in Sections 5.4 and 5.5, if certain harmonic terms in a complex waveform shift relative to the remaining terms, the shape of the resulting wave is distorted, and an incorrect output results. On the other hand, bodily shifts of all harmonics without *relative* changes preserve the true shape, and in most applications no problem results.

We find that there are three possible ways in which distortion from phase shift may be minimized. First, if there is no lag for any of the terms, there will be no distortion. Second, if all components lag by 180°, their relative values remain unchanged. And finally, if the shifts are in proportion to the harmonic orders (i.e., there is a linear shift with frequency), correct relative relations will be retained.

Zero shift requires no further comment other than to suggest that it rarely, if ever, exists. When a 180° shift takes place, all sine and cosine terms will simply have their signs reversed, and their relative magnitudes will remain unaffected.

A phase shift linear with frequency would be of the type in which the first harmonic lagged by, say, ϕ degrees, the second by 2ϕ, the third by 3ϕ, and so on. Let us consider this situation by means of the following relation:

$$f(t) = A_0 + A_1 \cos \omega t + A_2 \cos 2\omega t + A_3 \cos 3\omega t + \cdots \qquad (17.8)$$

Linear phase shifts would alter this equation to read

$$\begin{aligned} f(t) &= A_0 + A_1 \cos(\omega t - \phi) + A_2 \cos(2\omega t - 2\phi) \\ &\quad + A_3 \cos(3\omega t - 3\phi) + \cdots \\ &= A_0 + A_1 \cos \beta + A_2 \cos 2\beta + A_3 \cos 3\beta + \cdots \qquad (17.8a) \end{aligned}$$

where $\beta = \omega t - \phi$. We see, then, that the whole relation is retarded uniformly and that each term retains the same relative harmonic relationship with the other terms. Therefore, there will be no phase distortion. As we shall see, the vibration pickup approximates the second situation (i.e., 180° phase shift), whereas the accelerometer is of the linear phase-shift type.

Figure 17.5 shows that in the frequency region above resonance, used by a seismic-type displacement or velocity pickup, phase shift approaches 180° as the frequency ratio is increased. The swiftness with which it does so, however, is determined by the damping. For zero damping, the change is immediate as the exciting frequency passes through the instrument's resonant frequency. At higher damping rates, the approach to 180° shift is considerably reduced. We see, therefore, that damping requirements for good amplitude and phase response in this frequency area *are in conflict*, and some degree of compromise is required. In general, however, amplitude response is more of a problem than phase response, and commercial instruments are often designed with 60% to 70% of critical damping, although in some cases the damping is kept to a minimum. In any case, the greater the frequency ratio above unity, the more accurately will the relative motion to which the vibrometer responds represent the desired motion.

17.6.3 General Rule for Vibrometers

We may say, therefore, that in order for a vibration pickup of the seismic mass type to yield satisfactory motion information, use of the instrument must be restricted to input forcing frequencies above its own undamped natural frequency. Hence the lower the instrument's undamped natural frequency, the greater its range. In addition, in the frequency region immediately above resonance, compromised amplitude and phase response must be accepted. Of course, the displacement range that can be accommodated is limited by the design of the particular instrument. In general, the vibrometers of larger physical size permit measurement of larger displacement amplitudes. However, as size is increased, so too is the loading on the signal source.

17.7 THE SEISMIC ACCELEROMETER

We now turn our attention to a similar type of seismic instrument: the accelerometer. Basically, the construction of the accelerometer is the same as that of the vibrometer (Fig. 17.3), except that its design parameters are adjusted so that its output is proportional to the applied acceleration.

Let us rewrite Eq. (17.7) as follows:

$$S_{r0} = \frac{S_{s0}\Omega^2}{\omega_n^2 \sqrt{[1 - (\Omega/\omega_n)^2]^2 + [2\xi\Omega/\omega_n]^2}} \tag{17.9}$$

or

$$S_{r0} = \frac{a_{s0}}{\omega_n^2 \sqrt{[1 - (\Omega/\omega_n)^2]^2 + [2\xi\Omega/\omega_n]^2}} \tag{17.10}$$

in which a_{s0} is the acceleration amplitude of the supporting member.

Inspection of Eq. (17.10) clarifies the problem of properly designing and using an accelerometer. In order that the relative displacement between the supporting member and the seismic mass may be used as a measure of the support acceleration, the radical in the equation should be constant. The term ω_n^2 in the denominator is fixed for a given instrument and does not change with application. Hence, if the radical is a constant, the relative displacement will be directly proportional to the acceleration. Let

$$K = \frac{1}{\sqrt{[1 - (\Omega/\omega_n)^2]^2 + [2\xi\Omega/\omega_n]^2}} \tag{17.11}$$

By plotting K versus Ω/ω_n for various damping ratios, we obtain Fig. 17.6. Inspection of the plot indicates that the only possibility of maintaining a reasonably constant amplitude ratio as the forcing frequency changes is over a range of frequency ratio between 0.0 and about 0.40 and for a damping ratio of around 0.7. The extent of the usable range depends on the magnitude of error that may be tolerated.

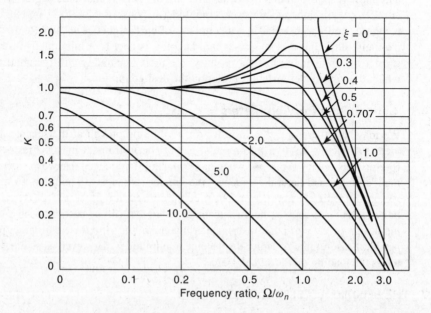

FIGURE 17.6: Response of a seismic instrument to sinusoidal acceleration.

17.7.1   Phase Lag in the Accelerometer

Referring again to Fig. 17.5 and to the limited accelerometer operating range just indicated—that is, $\Omega/\omega_n = 0$ to about 0.4 and $\xi = 0.70$—we see that the phase changes very nearly linearly with frequency. This relationship is fortunate, for as we have seen, it results in good phase response.

17.7.2   Practical Design of Seismic Instruments

We may now say that in order for a seismic instrument to provide satisfactory acceleration data, it must be used at forcing frequencies *below* approximately 40% of its own undamped natural frequency and the instrument damping should be on the order of 70% of critical damping.

It may be observed that both vibration pickups and accelerometers may use about the same damping; however, the range of usefulness of the two instruments lies on opposite sides of their undamped natural frequencies. The vibration pickup is made to a low undamped natural frequency, which means that it uses a "soft" sprung mass. On the other hand, the accelerometer must be used well below its own undamped natural frequency; therefore, it uses a "stiff" sprung mass. This makes the accelerometer an inherently less sensitive but more rugged instrument than the vibration pickup.

Figure 17.3 may be used to represent either a vibrometer or an accelerometer. As developed in Sections 17.6 and 17.7, the basic readout for a seismic instrument is the relative motion between the mass and the supporting structure. To sense this motion we require a relative-motion secondary transducer. (The mass-spring combination forms the primary transducer.) Although a voltage-dividing potentiometer is shown in the figure, at one time or another essentially all the appropriate transducing principles discussed in this book have been used for this purpose. A list includes variable-reluctance and variable-inductance devices, both bonded and unbonded strain gages, piezoresistive and piezoelectric sensors, variable-capacitance transducers, and some quite uncommon devices not mentioned in the preceding discussion.

Most of the devices listed above are displacement sensors. Variable reluctance is an exception. In this case the output is a function of velocity: the *rate* at which the magnetic lines of flux are cut. Sensitivity of this type is therefore in volts per unit velocity rather than volts per unit displacement. Variable-reluctance transducers have been very successfully used as vibrometers. There are two basic designs: (1) A permanent magnet forms a part of the seismic mass, which moves relative to pickup coils anchored to the case; or (2) the magnet is fixed to the housing and the coil forms a part of the seismic mass. Neither has a marked advantage, although it is obvious that for the moving coil, the electrical circuit becomes more critical.

Undoubtedly the most popular type of accelerometer makes use of a piezoelectric element in some form as shown in Fig. 17.7. Polycrystalline ceramics including barium titanate, lead zirconate, lead titanate, and lead metaniobate are among the piezoelectric materials that have been used [3]. Various design arrangements, as shown in the figure, are also used; the type depends on the characteristics desired, such as frequency range and sensitivity.

Important advantages enjoyed by the piezoelectric type are high sensitivity, extreme compactness, and ruggedness. Although the damping ratio is relatively low (0.002 to 0.25), the useful linear frequency ranges that may be attained are still large because of the high undamped natural frequencies inherent in the design (up to 100,000 Hz).

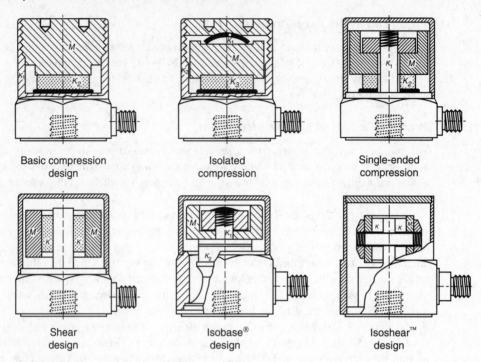

Basic compression design

Isolated compression

Single-ended compression

Shear design

Isobase® design

Isoshear™ design

FIGURE 17.7: Typical piezoelectric-type accelerometer designs. (Courtesy: Endevco Corp., San Juan Capistrano, California)

The output impedance of a piezoelectric device is high and presents certain problems associated with proper matching, noise, and connecting-cable motion and length. Either an impedance-transforming amplifier (Section 7.14) or a charge amplifier (Section 7.15.2) is normally required for proper signal conditioning. Each device has both advantages and disadvantages. Greatest effectiveness is attained when the instrumentation is located near the accelerometer. In fact, modern IC technology has made it possible, in some instances, to incorporate the amplifier circuitry within the accelerometer housing. Proper selection of instrumentation will, therefore, depend on application.

Silicon micromachining, or MEMS technology, has been applied to make low-cost semiconductor accelerometers. These may operate on piezoresistive principles such as described in Section 6.15.1. Integrated-circuit technologies allow the signal conditioning to be built in, so that these accelerometers are relatively easy to integrate into other products.

17.8 CALIBRATION

To be useful as amplitude-measuring instruments, both vibration pickups and accelerometers must be calibrated by determining the units of output signal (usually voltage) per unit of input (displacement, velocity, or acceleration). For the accelerometer, volts per g could be determined. Also, the calibration should indicate how such "constants" vary over the useful frequency range.

There are two basic approaches to the calibration of seismic-type transducers: (1) by absolute methods (based directly on the physical concepts of mass, length, and time), and (2) by comparative techniques.

The latter approach uses a "standard" against which the subject transducer is compared. It is clear that the standard must have highly reliable characteristics whose own calibration is not questioned. "Identical" motions are then imposed on subject and standard and the two outputs compared. Although this method would appear to be quite simple and is undoubtedly the one most commonly used, there are many pitfalls to be avoided. Error-free results depend on a number of factors [4, 5].

1. The impressed motions must *indeed* be identical.
2. Readout apparatus associated with the standard should preferably be and remain a part of the standard, and the *entire* system should have traceable calibration.
3. Associated readout apparatus in both circuits must have identical responses.
4. The standard must have long-term reliability.

None of these requirements is easily achieved.

The following two sections discuss some of the more fundamental methods applied to calibration of seismic-type transducers.

17.9   CALIBRATION OF VIBROMETERS

Vibration pickups are often calibrated by subjecting them to steady-state harmonic motion of known amplitude and frequency. The output of the pickup is then a sinusoidal voltage that is measured either by a reliable voltmeter or a cathode-ray oscilloscope. The primary problem, of course, is in obtaining a harmonic motion of *known* amplitude and frequency.

Electromechanical exciters are commonly used [6]. Devices of this sort are described in detail in Section 17.14. Exciters of this type are capable of producing usable amplitudes at frequencies to several thousand cycles per second.

17.10   CALIBRATION OF ACCELEROMETERS

Accelerometer calibration methods may be classified as follows:

1. Static

   (a) Plus or minus 1 g turnover method
   (b) Centrifuge method

2. Steady-state periodic

   (a) Using a sinusoidal shaker or exciter for back-to-back accelerometer calibration

17.10.1   Static Calibration

Plus or Minus 1 g Turnover Method

Low-range accelerometers may be given a 2 g step calibration by simply rotating the sensitive axis from one vertical position 180° through to the other vertical position (i.e., by simply turning the accelerometer upside down). This method is positive but is, of course, limited in the magnitude of acceleration that may be applied. A simple fixture is described in reference [7]. Of course, for precise calibration, the value of local gravity acceleration must be used.

Centrifuge Method
Practically unlimited values of static acceleration may be determined by a centrifuge or rotating table. The normal component of acceleration toward the center of rotation is expressed by the relation

$$a_n = r(2\pi f)^2 \qquad (17.12)$$

where

a_n = the acceleration of the seismic mass,

r = the radius of rotation measured from the center of the table to the center of gravity of the seismic mass,

f = table rotation speed, rev/s

It is assumed here that the axis of rotation is vertical. One of the disadvantages of this method, though not serious, is that of making electrical connections to the instrument.

17.10.2   Steady-State Periodic Calibration

Back-to-Back Accelerometer Calibration
To calibrate a vibration accelerometer is to determine its sensitivity (in mV/g or pC/g) at various frequencies of interest. The Instrument Society of America–approved back-to-back comparison method is probably the most convenient and least expensive technique.

The back-to-back calibration involves coupling the test accelerometer directly to a NIST traceable double-ended calibration standard accelerometer and driving the pair with a vibration exciter at various g levels. If a NIST traceable double ended calibration standard accelerometer is not readily (or financially) available, a secondary accelerometer may be substituted.

The accelerometers are connected back to back as shown in Fig. 17.8. The vibration exciter frequency and amplitude can be adjusted and the output of the "calibration standard" and test accelerometer can be compared. When accelerometers (or vibration pickups) are calibrated by this method, the mass of the test unit relative to the test standard must be considered. If test unit mass is more than 30% of the test standard mass, then mass loading compensation must be considered [8].

17.11   RESPONSE OF THE SEISMIC INSTRUMENT TO TRANSIENTS

Our discussion of seismic instruments to this point has been largely in terms of simple harmonic motion. How will these instruments respond to complex waveforms and transients? As we saw in Section 4.4 and developed further in Section 5.16, complex waveforms can be analyzed as a series of simple sinusoidal components in appropriate amplitude and phase relationships. It would seem then that a seismic instrument capable of responding faithfully to a range of individual harmonic inputs should also respond faithfully to complex inputs

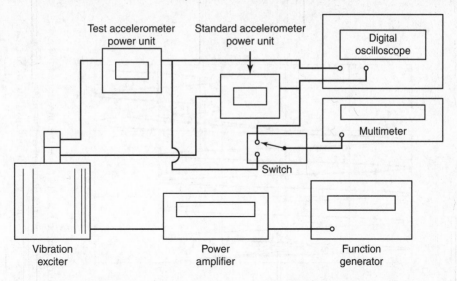

FIGURE 17.8: Back-to-back accelerometer calibration.

made up of frequency components within that range. Of course, if such an assumption is legitimate, the inherent nature of the accelerometer places it initially in a much more restricted area of operation than the vibration pickup. Whereas the accelerometer is limited to a frequency range up to roughly 40% of its own natural frequency, the only frequency restriction on the vibration pickup is that it be operated above its own natural frequency. In comparing the relative merits of the vibrometer and the accelerometer, we must not forget, however, that in terms of phase response the advantage is definitely on the side of the accelerometer (Sections 17.6.2 and 17.7.1). This is an important advantage.

By constructing the vibration pickup with a lightly sprung seismic mass, we can satisfactorily cover almost any frequency range. For the accelerometer, frequency components above the approximate 40% value may be unavoidable. In general, however, higher-frequency inputs are attenuated. Figure 17.9 shows the theoretical accelerometer response to square-wave pulses [9, 10]. Results are shown for different damping ratios and undamped natural frequencies.

First, these curves confirm our previous conclusion that a damping ratio of about 0.7 is near optimum. They also show that insofar as mass response is concerned, use of an instrument with as high an undamped natural frequency as possible is desirable. This is not surprising, because our previous investigation has pointed to this conclusion. A high undamped natural frequency, however, requires a stiff suspension. The result is that an extra burden is placed on the secondary transducer because of the small relative motions between mass and instrument housing. Hence, what is gained in response may be immediately lost in resolution.

The situation emphasizes even more the fact that accelerometer selection must be based on compromise. It means that the accelerometer cannot be selected entirely on its own merits but that the whole system must be considered and then the accelerometer selected with the highest undamped natural frequency consistent with satisfactory overall response.

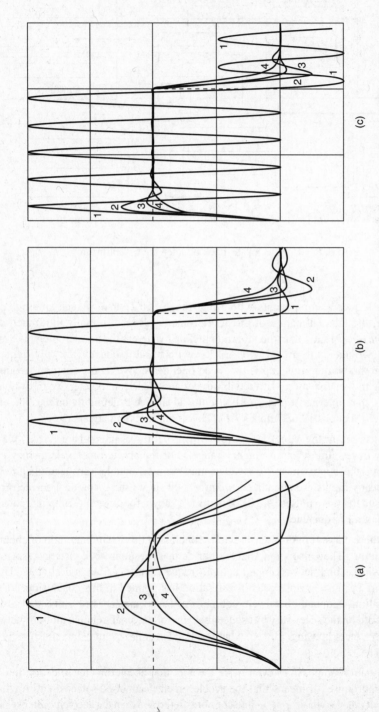

(a)

(b)

(c)

FIGURE 17.9: Response to a square pulse of acceleration of an accelerometer (a), whose natural period is 1.014 times the duration of the pulse, (b) whose natural period is 0.334 times the duration of the pulse, (c) whose natural period is 0.203 times the duration of the pulse. (1) For zero damping. (2) For damping ratio = 0.4. (3) For damping ratio = 0.7. (4) For damping ratio = 1.0 (Courtesy: National Institute of Standards and Technology)

17.12   MEASUREMENT OF VELOCITY BY THE LASER VELOCITY TRANSDUCER

Figure 11.6 shows a simplified form of laser displacement sensor. If the movable reflector is attached to a vibrating surface $[\delta = \delta(t)]$, the back-reflected beam is combined with the initially split beam, causing a number of successive dark fringes to be seen by the photodetector. The number of fringes per unit of time represents the surface velocity. The velocity sensed by this transducer is the velocity component of the movable reflector along the direction of the laser beam.

In practice the movable reflector can be a retroreflective tape, which can easily be attached to most surfaces. The operating, or standoff, distance of this device is usually 1.0 m or less. Since this is a noncontacting-type velocity sensor, it can be used for the velocity measurement of structures where the application of seismic-type transducers would greatly alter the structure mass.

Typical uses of this transducer are as follows:

1. Velocity survey of a hot surface such as a combustion engine manifold
2. Velocity survey of a vibrating membrane
3. Orbit analysis of rotating shafts in rotating machinery
4. Measurement of velocities of machine elements where attachment of seismic transducers is impossible

17.13   VIBRATION AND SHOCK TESTING

Vibration and shock-test systems are particularly important in relation to numerous R & D contracts. Many specifications require that equipment perform satisfactorily at definite levels of steady-state or transient dynamic conditions. Such testing requires the use of special test facilities, often unique for the test at hand but involving principles common to all.

Numerous items for civilian consumption require dynamic testing as part of their development. All types of vibration-isolating methods require testing to determine their effectiveness. Certain material fatigue testing uses vibration test methods. Specific examples of items subjected to dynamic tests include many automobile parts, such as car radios, clocks, headlamps, radiators, ignition components, and larger parts like fenders and body panels. Also, many aircraft components and other items for use by the armed services must meet definite vibration and shock specifications. Missile components are subjected to extremely severe dynamic conditions of both mechanical and acoustic origin.

It might be assumed that dynamic testing should exactly simulate field conditions. However, this situation is not always necessary or even desirable. First, field conditions themselves are often nonrepetitive; situations at one time are not duplicated at another time. Conditions and requirements today differ from those of yesterday. Hence, to define a set of *normal* operating conditions is often difficult if not impossible. Dynamic testing, on the other hand, may be used to pinpoint particular areas of weakness under accurately controlled and measurable conditions. For example, factors such as accurately determined resonance frequencies, destructive amplitude-frequency combinations, and so on, may be uncovered in the development stage of a design. With such information the design engineer then may judge whether corrective measures are required, or perhaps determine that such conditions lie outside operating ranges and are therefore unimportant. Another factor making dynamic testing attractive is that accelerated testing is possible. Field testing, in many cases, would require inordinate lengths of time.

Our discussion of dynamic testing is divided into two parts: vibration testing and shock testing.

17.14   VIBRATIONAL EXCITER SYSTEMS

In order to submit a test item to a specified vibration, a source of motion is required. Devices used for supplying vibrational excitation are usually referred to simply as *shakers* or *exciters*. In most cases, simple harmonic motion is provided, but systems supplying complex waveforms are also available.

There are various forms of shakers, the variation depending on the source of driving force. In general, the primary source of motion may be electromagnetic, mechanical, piezoelectric, or hydraulic-pneumatic or, in certain cases, acoustical. Each is subject to inherent limitations, which usually dictate the choice.

17.14.1   Electromagnetic Systems

A section through a small electromagnetic exciter is shown in Fig. 17.10. This consists of a field coil, which supplies a fixed magnetic flux across the air gap h, and a driver coil supplied from a variable-frequency source. Permanent magnets are also sometimes used for the fixed field. Support of the driving coil is by means of flexure springs, which permit the coil to reciprocate when driven by the force interaction between the two magnetic fields. We see that the electromagnetic driving head is very similar to the field and voice coil arrangement in the ordinary radio loudspeaker.

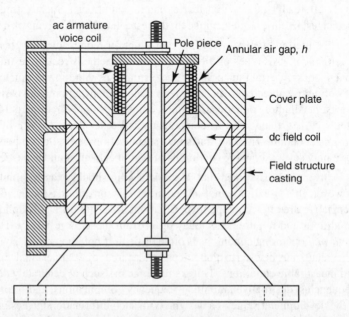

FIGURE 17.10: Sectional view showing internal construction of an electromagnetic shaker head.

An electromagnetic shaker is rated according to its vector force capacity, which in turn is limited by the current-carrying ability of the voice coil. Temperature limitations of the insulation basically determine the shaker force capacity. The driving force is commonly simple harmonic (complex waveforms are also used) and may be thought of as a rotating vector in the manner of harmonic displacements discussed in Section 4.2. The force used for the rating is the vector force exerted between the voice and field coils.

Rated force, however, is never completely available for driving the test item. It is the force developed within the system, from which must be subtracted the force required by the moving portion of the shaker system proper. It may be expressed as

$$F_n = F_t - F_a \tag{17.13}$$

where

F_n = the net usable force available to shake the test item,

F_t = the manufacturer's rated capacity, or total force provided by the magnetic interaction of the voice and field coils,

F_a = the force required to accelerate the moving parts of the shaker system, including the voice coil, table, and appropriate portions of the voice coil flexure beams

In practice, it is often convenient to think in terms of the total vector force, F_t, and to simply add the weight of the shaker's moving parts to that of the test item and any required accessories such as mounting brackets and the like. Table 17.2 lists the specifications for typical commercially available electromagnetic shaker systems.

17.14.2 Mechanical-Type Exciters

There are two basic types of mechanical shakers: the directly driven system and the inertia system. The directly driven shaker consists simply of a test table that is caused to reciprocate

TABLE 17.2: Specifications for Typical Electromagnetic Exciter Systems

Maximum Rated Force, lbf	Frequency Range, Hz	Maximum Double Amplitude, in.	Weight of Moving Armature, lbm
50	0–5,000	$\frac{1}{2}$	$\frac{3}{4}$
100	0–10,000	1	$1\frac{1}{4}$
300	0–5,000	1	$2\frac{3}{4}$
1,500	0–4,000	$1\frac{1}{4}$	20
3,200	5–3,000	$\frac{1}{2}$	25
20,000	5–3,000	1	95
40,000	2–2,000	1	250

by some form of mechanical linkage. Crank and connecting rod mechanisms, Scotch yokes, or cams may be used for this purpose.

Another mechanical type uses counterrotating masses to apply the driving force. Force adjustment is provided by relative offset of the weights and the counterrotation cancels shaking forces in one direction, say the x-direction, while supplementing the y-force. Frequency is controlled by a variable-speed motor.

There are two primary advantages in such inertia systems. In the first place, high force capacities are not difficult to obtain, and second, the shaking amplitude of the system remains unchanged by frequency cycling. Therefore, if a system is set up to provide a 0.05-in. amplitude at 20 Hz, changing the frequency to 50 Hz will not alter the amplitude. The reason will be understood if it is remembered that both the *available* exciting force and the *required* accelerating force are harmonic functions of the *square* of the exciting frequency; hence as the requirement changes with frequency, so too does the available force.

17.14.3 Hydraulic and Pneumatic Systems

Important disadvantages of the electromagnetic and mechanical shaker systems are limited load capacity and limited frequency, respectively. As a result, the search for other sources of controllable excitation has led to investigation in the areas of hydraulics and pneumatics.

Figure 17.11 illustrates, in block form, a hydraulic system used for vibration testing [11]. In this arrangement an electrically actuated servo valve operates a main control valve, in turn regulating flow to each end of a main driving cylinder. Large capacities (to 500,000 lbf or 2.2 MN) and relatively high frequencies (to 400 Hz), with amplitudes as great as 18 in. (460 mm), have been attained. Of course, the maximum values cannot be attained simultaneously. As would be expected, a primary problem in designing a satisfactory system of this sort has been in developing valving with sufficient capacity and response to operate at the required speeds.

17.14.4 Relative Merits and Limitations of Each System

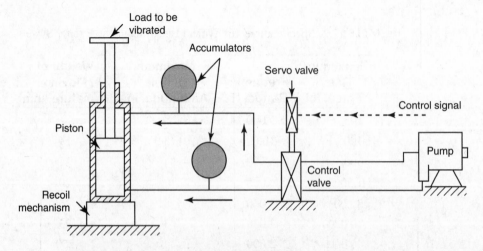

FIGURE 17.11: Block diagram of a hydraulically operated shaker.

Frequency Range

The upper frequency ranges are available only through use of the electromagnetic shaker. In general, the larger the force capacity of the electromagnetic exciter, the lower its upper frequency will be. However, even the 40,000-lbf (180-kN) shaker listed in Table 17.2 boasts an upper useful frequency of 2000 Hz. To attain this value with a mechanical exciter would require speeds of 120,000 rpm. The maximum frequency available from the smaller mechanical units is limited to approximately 120 Hz (7200 rpm) and for the larger machines to 60 Hz (3600 rpm). Hydraulic units are presently limited to about 2000 Hz.

Force Limitations

Electromagnetic shakers have been built with maximum vector force ratings of 40,000 lbf (180 kN). Variable-frequency power sources for shakers of this type and size are very expensive. Within the frequency limitations of mechanical and hydraulic systems, corresponding or higher force capacities may be obtained at lower costs by hydraulic shakers. Careful design of mechanical and hydraulic types is required, however, or maintenance costs become an important factor. Mechanical shakers are particularly susceptible to bearing and gear failures, whereas valve and packing problems are inherent in the hydraulic ones.

Maximum Excursion

One inch, or slightly more, may be considered the upper limit of peak-to-peak displacement for the electromagnetic exciter. Mechanical types may provide displacements as great as 5 or 6 in.; however, total excursions as great as 18 in. have been provided by the hydraulic-type exciter.

Magnetic Fields

Because the electromagnetic shaker requires a relatively intense fixed magnetic field, special precautions are sometimes required in testing certain items such as solenoids or relays, or any device in which induced voltages may be a problem. Although the flux is rather completely restricted to the magnetic field structure, relatively high stray flux is nevertheless present in the immediate vicinity of the shaker. Operation of items sensitive to magnetic fields may therefore be affected. Degaussing coils are sometimes used around the table to reduce flux level.

Nonsinusoidal Excitation

Shaker head motions may be sinusoidal or complex, periodic or completely random. Although sinusoidal motion is by far the most common, other waveforms and random motions are sometimes specified [12]. In this area, the electromagnetic shaker enjoys almost exclusive franchise. Although the hydraulic type may produce nonharmonic motion, precise control of a complex waveform is not easy. Here again, future development of valving may alter the situation.

The voice coil of the ordinary loudspeaker normally produces a complex random motion, depending on the sound to be reproduced. Complex random shaker head motions are obtained in essentially the same manner. Instead of using a fixed-frequency harmonic oscillator as the signal source, either a strictly random or a predetermined random signal source is used. Electronic *noise* sources are available, or a record of the motion of the actual end use of the device may be recorded on magnetic tape and used as the signal source for driving the shaker. As an example, electronic gear may be subjected to combat-vehicle motions by first recording the output of motion transducers and then using the record to drive a shaker. In this manner, controlled repetition of an identical program is possible.

17.15 VIBRATION TEST METHODS

Two basic methods are used in applying a sinusoidal force to the test item: the *brute-force method* and the *resonance method* [13]. In the first case the item is attached or mounted on the shaker table, and the shaker supplies sufficient force to literally drive the item back and forth through its motion. The second method makes use of a mechanical spring-mass fixture having the desired natural frequency. The test item is mounted as a part of the system that is excited by the shaker. The shaker simply supplies the energy dissipated by damping.

17.15.1 The Brute-Force Method

Brute-force testing requires that the exciter supply all the accelerating force to drive the item through the prescribed motion. Such motion is generally sinusoidal, although complex waveforms may be used. The problems inherent in an arrangement of this sort are shown by the following example.

Suppose a vibration test specification calls for sinusoidally shaking a 10-kg test item at 100 Hz with a displacement amplitude of 2 mm (double amplitude = 4 mm). What force amplitude will be required?

Maximum force will correspond to maximum acceleration, and maximum acceleration may be calculated as follows:

$$\text{Circular frequency} = \Omega = 2\pi \times 100 = 628 \text{ rad/s},$$

$$\text{Maximum acceleration} = \text{the displacement amplitude} \times (\Omega)^2$$

$$= \left(\frac{2}{1000}\right)(628)^2 = 789 \text{ m/s}^2,$$

$$\text{Maximum force} = \frac{ma}{g_c} = \frac{(10 \times 789)}{1 \ (\text{kg} \cdot \text{m/s}^2)(\text{N} \cdot \text{s}^2/\text{kg} \cdot \text{m})}$$

$$= 7890 \text{ N, or about } 1770 \text{ lbf}$$

This, of course, is the force amplitude required to shake the test item only. If support fixtures are required, they too must be shaken along with the moving coil of the shaker itself. Suppose these items (the fixture and voice-coil assembly) have a mass of 5 kg; then an additional vector force of 3945 N is required. The rated capacity of the shaker must therefore be a minimum of about 12,000 N.

17.15.2 The Resonance Method

The resonant system uses some form of spring-mass fixture to which the test item is attached. Figure 17.12 illustrates a small experimental setup whose characteristics will be described. The test item weighed 7.7 lbm and the test specifications required a sinusoidal vibration at 50 Hz with an amplitude of $\pm\frac{1}{8}$ in. A spring-supported table was designed, as shown in the

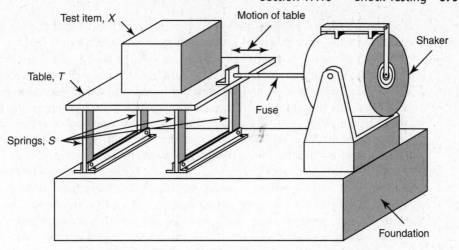

Test item, *X*

Motion of table

Shaker

Table, *T*

Fuse

Springs, *S*

Foundation

FIGURE 17.12: Exciter-driven table for horizontal motion.

figure. As initially tested, the resonance frequency was 58 Hz. Addition of a small mass fine-tuned the frequency to 50 Hz. The final weight distribution was as follows:

Test item	7.7
Table with mounting accessories	4.75
Moving weight of exciter	0.35
$\frac{1}{3}$ weight of leaf springs	0.50
Total	13.3 lbm

As a test of the maximum capacity of the system at 50 Hz, it was found that the 5-lbf shaker could actually move the table with test load through an amplitude of ± 0.17 in. at 50 Hz. The force required to accomplish this may be calculated as follows:

$$\text{Maximum acceleration} = S_0 \Omega^2 = (0.17)(2\pi \times 50)^2 = 16,750 \text{ in./s}^2$$

$$\text{Necessary accelerating force} = \frac{ma}{g_c} = \left(\frac{16,750}{386}\right) \times 13.3 = 577 \text{ lbf}$$

Obviously the 5-lbf shaker did not supply the force: The necessary accelerating forces were supplied almost in their entirety by the springs.

It will be readily realized that a resonant system of this type is limited to *one frequency*. Although a limited range of application might be designed into such a system through use of adjustable springs and masses, in general the system must be designed for the problem at hand and for that only.

Other forms of mechanically resonant systems include vertical spring-mass arrangements, the free-free beam [14, 15], tuning-fork systems [14], and so on.

17.16 SHOCK TESTING

Mechanical engineers are called on to design machinery to operate at higher and higher speeds. As speed goes up, accelerations increase, for the most part, not in direct proportion, but as the square of the speed. Both the magnitude of acceleration and acceleration gradients are increased. Resulting body loads often become much greater than applied loading, therefore becoming very significant factors in the design. The complexity of many problems has led to an area of investigation generally referred to as *shock testing* [16].

Actually, shock testing is only one of two phases of a broader classification that might better be called *acceleration testing.* Acceleration testing includes any test wherein acceleration loading is of primary significance. Included would be tests involving static or relatively slowly changing accelerations of any magnitude. Shock testing, on the other hand, is usually thought of as involving acceleration transients of moderate to high magnitude. In both cases the basic problem is to determine the ability of the test item to continue functioning properly either during or after application of such loading.

The more passive type of acceleration testing involves constant or relatively slowly changing accelerations, which, however, may be of high magnitude. It involves the use of centrifuges, rocket sleds, maneuvering aircraft, and the like, for the purpose of testing the capabilities of system components, including the human body, to withstand sustained or slowly changing high-level accelerations. Such tests are usually of a specialized nature, generally applied to the study of performance in high-speed aircraft and missiles. Therefore, we shall only note this phase in passing and shall devote our primary attention to the first type of acceleration testing—namely, shock testing.

Most military apparatus must satisfactorily pass specified shock tests before acceptance. Equipment aboard ship, for example, is subject to shock from the ship's own armament, noncontact mine explosions, and the like. Aircraft equipment must withstand sharp maneuvering and landing loads, and artillery and communication equipment is subject to severe handling in crossing rough terrain. In addition, many items of industrial and civilian application are also subject to shock, often simply caused by normal handling during distribution, such as railroad-car humping, mail chuting, and so on.

As a result, shock testing has become accepted as a necessary step in determining the usefulness of many items. It is becoming generally recognized, however, that to be meaningful, considerably more than magnitude of acceleration must be considered. In addition to magnitude, the rate and duration of application, along with the dynamic characteristics of the test item, must all be studied in setting up a useful shock test.

In general terms, shock testing may be divided into two broad categories: *low energy* and *high energy* (Fig. 17.13). Low-energy testing corresponds to the application of high accelerations over short time intervals. The terms *sharp, intense, violent,* and *abrupt* might be applied. However, the resulting velocities (hence energies) may not be great. On the other hand, high-energy shock is applied for lengths of time permitting the buildup (or deterioration) of relatively high velocities. Acceleration magnitudes accompanying high-energy shock are commonly relatively low. This type of shock might be referred to as *impulsive, dynamic,* and so on. Sometimes the latter is referred to as *energy loading* as opposed to *impact loading.* The severity of a shock test is very subjective: For certain items, high-energy shock is more severe; for others, the opposite may be true.

17.17 SHOCK RIGS

Several different methods are used for producing the necessary motion for shock testing. The approach generally taken is to store the required energy in some form of potential energy until needed, then to release it at a rate supplying the desired acceleration–time relation. Methods for doing this include the use of compressed air or hydraulic fluid, loaded springs, and the acceleration of gravity. The latter is the most commonly used.

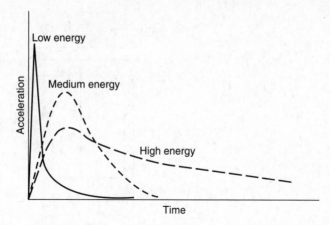

FIGURE 17.13: Typical characteristics of mechanical shock-producing machines of low, medium, and high energy.

17.17.1   Air Gun Shock-Producing Devices

Basically, the air gun system uses a piston, which moves within a tube or barrel under the action of high-pressure air applied to one face of the piston. Energy is stored by pressurizing the air in an accumulator. The high pressure is applied to the piston while it is restrained by a mechanical latching mechanism. When released, the piston with test item attached is sharply accelerated. Air trapped in the downstream portion of the cylinder serves to decelerate the piston, finally bringing it to rest. Machines based on this principle of operation have been made with energy capacities of more than 10^6 ft · lbf (1.4×10^6 N · m) [17].

17.17.2   Spring-Loaded Test Rigs

As the name indicates, these machines use some form of mechanical spring for storing the energy required for acceleration. One machine designed to provide vertical accelerations uses helical tension springs attached at one end to a test carriage and at the other end to anchors that may be moved to put various initial tensions in the springs. With the springs initially tensioned, the test carriage (with test item) is released by a mechanical triggering mechanism and is accelerated suddenly upward. After the carriage has traveled a predetermined distance, the carriage ends of the spring strike stationary hooks. The spring's working stroke is thereby limited; however, the carriage continues upward until stopped by gravity.

17.17.3   A Hydraulic-Pneumatic Rig

The essentials for one hydraulic-pneumatic machine are illustrated in Fig. 17.14. In operation, an initial pneumatic pressure is introduced to chamber B. This pressure acts over the larger piston area. Hydraulic pressure is then increased in chamber A until the hydraulic pressure over the small area overcomes the pneumatic pressure acting over the larger area. At the instant the piston is lifted, the hydraulic pressure is suddenly applied to the larger area, producing a sharp impulsive-type upward motion. The form of the pulse is controlled by the shape of the metering pin [18, 19].

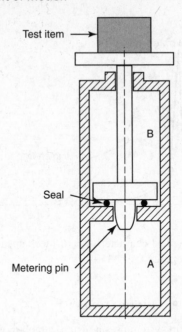

FIGURE 17.14: A form of hydropneumatic shock-producing device.

17.17.4 Gravity Rigs

There are two commonly used gravity-type shock rigs: the drop type and the hammer type. The hammer type is often referred to as a high-*g* machine and normally provides higher values of acceleration than the drop machine does.

Basically the drop machine consists of a platform to which the test item is attached, an elevating system for raising the platform, a releasing device that allows the platform to drop, and an impact pad or arrester against which the platform strikes. Guides are provided for controlling the fall (Fig. 17.15).

Acceleration–time relations are adjusted by controlling the height of drop and type of arrester pad. Pad selection is of great importance in determining the exact shock characteristics. If the pad is very rigid, for example, an acceleration pulse of very short duration results. On the other hand, a more flexible pad provides a longer time base. *Magnitude* of peak acceleration is controlled by adjusting the height of drop.

Rubber pads shaped to provide the desired acceleration–time pulse may be used. Sand pits have also been used in conjunction with variously shaped impacting surfaces. As another example, the test platform may be equipped with shaped pins or punches that strike and penetrate blocks of lead: The shape of the pins is designed to provide the desired pulse form.

Figure 17.16 illustrates the operation of a hammer-type shock-producing device that has been used to study the effects of head injury resulting from an impact. Strain propagation throughout the skull structure resulting from controlled frontal, rear, or side impacts was evaluated. In addition, extensive studies have been conducted on pressure wave propagation through simulated brain tissue [20].

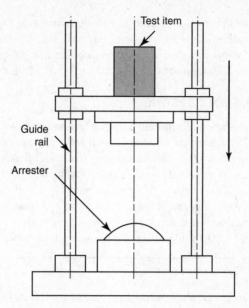

FIGURE 17.15: A drop-type mechanical shock-producing machine.

17.17.5   Relative Merits and Limitations of Each Shock Rig

Each of the shock-testing machines discussed in this section possessed certain distinctive characteristics. The air gun type produces what may be called a *high-energy* shock. Generally speaking, high energy is synonymous with *high velocity*, and to reach a high velocity, considerable displacement of the test item is required. High velocity can be acquired only by relatively large accelerations, or relatively long time intervals, or a combination of the two. In either case, the test item will be displaced a considerable distance.

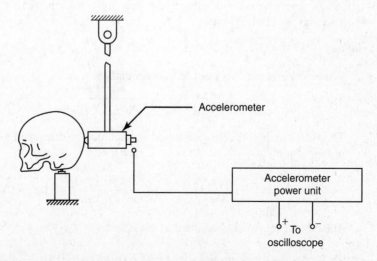

FIGURE 17.16: An application of a pendulum-type shock-loading device to a research project.

On the other hand, the drop and hammer machines are of the low-energy category. High acceleration levels are possible, but only for short time intervals. This results in comparatively low test-item velocities and hence low energies. The hydraulic-pneumatic machine would be classified as a medium-energy machine.

17.18 PRACTICAL SHOCK TESTING

Some of the parameters pertinent to shock testing are illustrated in the following example. Figure 17.17(a) illustrates an arrangement for a laboratory experiment in dynamic-stress analysis. Pendulum OA is caused to swing freely and strike a steel cantilever beam BC, on the end of which is mounted a small mass. A small plastic disk is placed at the point of impact to promote inelastic impact. Strain gages on each side of the beam are placed at D. Gage output is appropriately amplified and fed to an oscilloscope for readout. Figure 17.17(b) shows the oscilloscope trace obtained when the pendulum swings from rest through an arc equivalent to $h = 0.18$ in. Calibration-based readout values are shown in the figure.

EXAMPLE 17.1

Applying theoretical relationships, compare analytical and experimental values of (a) maximum strain, (b) time of contact between pendulum tup and beam, and (c) the period of free vibration of the beam. Pertinent dimensions and masses are shown in the sketch. Equivalent mass, m_1, of the pendulum is taken as the mass of the tup plus one-third of the mass of the arm. Equivalent mass, m_2, is taken as the sum of the small mass at B plus one-third of the mass of the beam.

Note that the initial potential energy of the pendulum will be completely converted into kinetic energy at impact (assuming negligible losses). On the basis of the inelastic conditions between masses m_1 and m_2, the law of conservation of momentum will apply, from which the impacting energy loss may be evaluated. The remaining energy must be absorbed by the beam.

Time of contact may be estimated using the assumption that the strain-time relationship is a half sine wave corresponding to the free vibration of the beam as though both m_1 and m_2 are rigidly attached to it.

Solution

1. The velocity of initial contact of hammer and beam is

$$V_1 = \sqrt{2gh} = \sqrt{2 \times 386 \times 0.18} = 11.8 \text{ in./s}$$

The velocity immediately after initial contact may be calculated using conservation of momentum, or

$$V_2 = \left(\frac{m_1}{m_1 + m_2} \right) V_1 = \left(\frac{1.96}{1.96 + 0.168} \right) \times 11.8 = 10.9 \text{ in./s}$$

Therefore, energy to be absorbed by the beam is

$$U_2 = \frac{\frac{1}{2}(m_1 + m_2)V_2^2}{g_c} = \frac{\frac{1}{2}(1.96 + 0.168) \times (10.9)^2}{386} = 0.327 \text{ in.} \cdot \text{lbf}$$

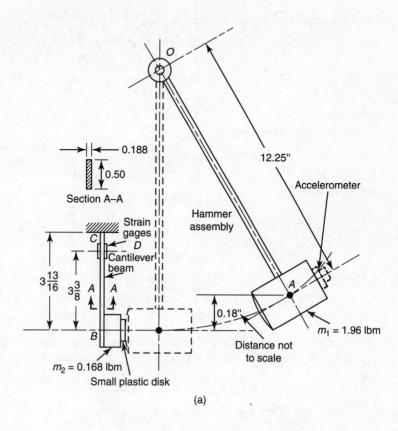

(a)

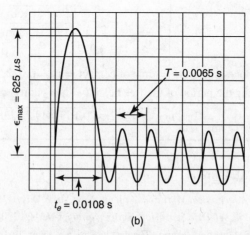

(b)

FIGURE 17.17: (a) Laboratory setup for demonstrating analysis of mechanical shock; (b) typical experimental result obtained from apparatus shown in (a).

For the beam,

$$I = \frac{bd^3}{12} = \frac{0.5 \times (0.188)^3}{12} = 2.769 \times 10^{-4} \text{ in.}^4$$

Work required to deflect the beam is

$$U_2 = \int_0^L \frac{M_1^2 \, dx}{2EI} = \frac{F^2 L^3}{6EI}$$

where

$$M_1 = Fx \qquad (0 \leq x \leq L)$$

Substituting and solving for F_{max},

$$F_{max} = \sqrt{\frac{6EIU_2}{L^3}} = \sqrt{\frac{6 \times 30 \times 10^6 \times 2.769 \times 0.327 \times 10^{-4}}{(3.8125)^3}}$$

$$= 17.15 \text{ lbf}$$

At the strain gages, $M_2 = 3.375 \times F_{max} = 57.88 \text{ in.} \cdot \text{lbf}$

$$\varepsilon = \frac{M_2 c}{EI} \qquad \text{(where } c = 0.188/2)$$

$$= \frac{57.88 \times 0.094}{30 \times 10^6 \times 2.769 \times 10^{-4}}$$

$$= 655 \text{ } \mu\text{-strain}$$

2. To estimate the time of contact between pendulum tup and beam, assume that for one-half cycle of vibration the tup is rigidly attached to the end of the beam; therefore,

$$\text{Period of vibration} = 2\pi \sqrt{\frac{m_1 + m_2}{kg_c}}$$

$$= 2\pi \sqrt{\frac{1.96 + 0.168}{449.7 \times 386}}$$

$$= 0.022 \text{ s}$$

Time for one-half cycle = time of contact = $t_e = 0.022/2 = 0.011$ s.

3. Calculating the period of the free beam (with m_2 at its end),

$$T = 2\pi \sqrt{m_2/kg_c} = 2\pi \sqrt{0.168/449.7 \times 386} = 0.0062 \text{ s}$$

SUGGESTED READINGS

Den Hartog, J. P. *Mechanical Vibrations*. 4th ed. New York: Dover, 1985.

Gautschi, G. H. *Piezoelectric Sensorics*. Berlin: Springer-Verlag, 2002.

McConnell, K. G. *Vibration Testing: Theory and Practice*. New York: John Wiley, 1995.

Weaver, W. Jr., S. P. Timoshenko, and D. H. Young. *Vibration Problems in Engineering*. 5th ed. New York: John Wiley, 1990.

PROBLEMS

17.1. A simple frequency meter is described in the final paragraph of Section 17.3. The undamped first-mode frequency of a uniformly sectioned cantilever beam may be calculated from the expression

$$f = 3.52\sqrt{\frac{EIg_c}{mL^4}} \text{ Hz}$$

where

$E =$ Young's modulus for the material of the beam,
$m =$ the mass of the beam per unit length,
$I =$ the area moment of inertia of the beam section,
$L =$ the length of the beam

For a steel wire of $1\frac{1}{2}$-mm (0.0039-in.) diameter, plot the resonance frequencies over a range of $L = 10$ to 25 cm (3.94 to 9.84 in.). (Use density of steel = 7900 kg/m^3.)

17.2. Referring to Problem 17.1, we may design an instrument of wider range by providing a series of small masses to be attached to the outer end of the beam. In this case,

$$f = \frac{1}{2\pi}\sqrt{\frac{3EIg_c}{(M_1 + M_B)L^3}} \text{ Hz}$$

where

$M_1 =$ the mass of the attachment,
$M_B =$ one-third the mass of the beam

Design a system to cover the range of $50 < f < 2000$ Hz.

17.3. Obtain an inexpensive crystal-type phonograph pickup designed for replaceable phonograph needles. The least expensive will be satisfactory. Place a small overhanging mass in the stylus chuck, connect the pickup output to an oscilloscope, and use it to investigate various sources of mechanical vibration. (*Note*: The device should be quite useful as a "frequency" pickup; however, it will not be adequate for meaningful amplitude readout.)

17.4. The waveform from a mechanical vibration is sensed by a velocity-sensitive vibrometer. The oscilloscope trace indicates that the motion is essentially simple harmonic. A 1-kHz oscillator is used for time calibration, and 4 cycles of the vibration are found to correspond to 24 cycles from the oscillator. Calibrated vibrometer output indicates a velocity amplitude ($\frac{1}{2}$ peak-to-peak) of 3.8 mm/s. Determine (a) the displacement amplitude in mm and (b) the acceleration amplitude in standard g's.

17.5. A vibrometer is used to measure the time-dependent displacement of a machine vibrating with the motion

$$y = 0.5\sin(3\pi t) + 0.8\sin(10\pi t)$$

where y is in cm and t is in s. If the vibrometer has an undamped natural frequency of 1 Hz and a critical damping ratio of 0.65, determine the vibrometer time-dependent output and explain any discrepancies between the machine vibration and the vibrometer readings.

17.6. An accelerometer is used for measuring the amplitude of a mechanical vibration. The following data are obtained:

> Waveform: simple sinusoidal,
>
> Period of vibration = 0.0023 s,
>
> Output voltage from accelerometer = 0.213 V rms,
>
> Accelerometer calibration = 0.187 V/standard g

What vibrational displacement (amplitude) is sensed by the accelerometer? Express your answer (a) in millimeters; (b) in inches.

17.7. An accelerometer is designed to have a maximum practical inherent error of 4% for measurements having frequencies in the range of 0 to 10,000 Hz. If the damping constant is 50 N · s/m, determine the spring constant and suspended mass.

17.8. A seismic-type vibrometer is characteristically a relatively fragile instrument, whereas an accelerometer is relatively rugged. Explain why this is so.

17.9. Refer to most any introductory vibrations text and review the material on viscous damping including that on the logarithmic decrement. Use such equipment as may be available to obtain a displacement-time record for a decaying damped free vibration, such as that shown in Fig. 17.18(a). The vibrating element may be a simple cantilever beam with concentrated end mass, as shown in Fig. 17.18(b), and the like. The phono-pickup vibration sensor of Problem 17.3 should suffice as a transducer and an oscilloscope or oscillograph as the terminating device. Using data as defined in the figure, determine the damped frequency of vibration and the viscous damping coefficient

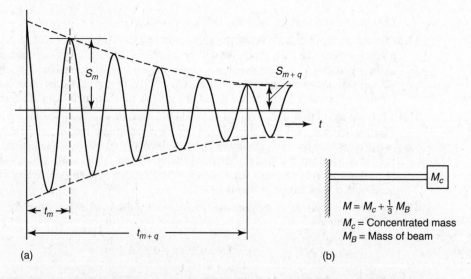

(a)

$M = M_c + \frac{1}{3} M_B$
M_c = Concentrated mass
M_B = Mass of beam

(b)

FIGURE 17.18: (a) Displacement-time record and (b) equipment setup for Problem 17.9.

$$c = \frac{M \omega_{nd} \delta}{\pi g_c}$$

where

ω_{nd} = the circular frequency of free vibration,

δ = the logarithmic decrement

$$= \frac{1}{q} \ln \left(\frac{S_m}{S_{m+q}} \right)$$

17.10. An electromagnetic-type sinusoidal vibration exciter having a rated force capacity of 25 N (5.62 lbf) is to be used to excite a test item weighing 3 kg (6.61 lbm). If the moving parts of the shaker have a mass of 0.75 kg (1.65 lbm) and the amplitude of the vibration is 0.15 mm (0.0059 in.) (double amplitude = 0.30 mm), determine the maximum excitation frequency that can be applied.

17.11. At full rated load of 5000 N, a load cell deflects 0.10 mm. If it is used to measure the thrust of a small (200-kg) jet engine, what maximum frequency component of the thrust may be accurately measured if the inherent error is limited to 2%? Damping is negligible. What is the basis that you use for determining the limiting frequency?

REFERENCES

[1] Fibikar, R. J. Touch and vibration sensitivity, *Prod. Eng.* 27: November 1956, p. 177.

[2] Hudson, D. E., and O. D. Terrell. A preloaded spring accelerometer for shock and impact measurements. *SESA Proc.* 9(1):1, 1951.

[3] Pennington, D. *Piezoelectric Accelerometer Manual*. Pasadena, Calif.: Endevco Corporation, 1965.

[4] Kistler, W. P. Precision calibration of accelerometers for shock and vibration. *Test Eng.* 16, May 1966.

[5] Edelman, S. Additional thoughts on precision calibration of accelerometers. *Test Eng.* 17, November 1966.

[6] Lewis, R. C. Electro-dynamic calibration for vibration pickups. *Prod. Eng.* 22:September 1951.

[7] Easily made device calibrates accelerometer. *NBS Tech. News Bull.* 94, June 1966.

[8] www.dytran.com/graphics/a9.pdf

[9] Levy, S., and W. D. Kroll. Response of accelerometers to transient accelerations. *NBS J. Res.* 45:4, October 1950.

[10] Welch, W. P. A proposed new shock-measuring instrument. *SESA Proc.* 5(1):39, 1947.

[11] Adler, J. A. Hydraulic shakers. *Test Eng.* April 1963.

[12] Crandall, S. H. *Random Vibration*. New York: John Wiley, 1959.

[13] Unholtz, K. Factors to consider in setting up vibration test specifications, *Mach. Des.* 28:6, March 22, 1956.

[14] Wozney, G. P. Resonant vibration fatigue testing. *Exp. Mech.* January 1962.

[15] Application and design formulae for free-free resonant beams. *MB Vibration Notebook*. New Haven, Conn.: MB Manufacturing Co., March 1955.

[16] Lazarus, M. Shock testing: A design guide. *Mach. Des.* October 12, 1967.

[17] Armstrong, J. H. Shock-testing technology at the Naval Ordnance Laboratory. *SESA Proc.* 6(1):55, 1948.

[18] Brown, J. Selection factors for mechanical buffers. *Prod. Eng.* 21:156, November 1950.

[19] Brown, J. Further principles of buffer design, *Prod. Eng.* 21:125, December 1950.

[20] Marangoni, R. D., C. A. Saez, D. A. Weyel, and R. A. Polosky. Impact stresses in human head-neck model. *J. Eng. Mech. Div., ASCE* 104(EM1):1978.

C H A P T E R 18

Acoustical Measurements

18.1 INTRODUCTION

Sound may be described on the basis of two considerably different points of view: (a) from the standpoint of the physical phenomenon itself, or (b) in terms of the "psychoacoustical" effect sensed through the human process of hearing. It is very important that these basically different aspects be kept continually in mind. To measure the particular physical parameters associated with a specific sound, either of simple or complex waveform, is a much simpler assignment than to attempt to evaluate the effects of the parameters as sensed by human hearing.

Occasionally, the reasons for measuring a sound may not be associated with hearing. For example, sound pressure variations accompanying high-thrust rocket motor or jet engine operation may be of sufficient magnitude to endanger the structural integrity of the missile or aircraft [1]. Structural fatigue failures have been induced by sound excitation. In such cases measurement of the parameters that are involved does not directly include the psychoacoustical relationship. But in the great majority of cases the effect of the measured sound *is* directly related to human hearing, and this added complication is therefore unavoidable.

Noise may be defined as unwanted sound. Noise affects human activities in many ways. Excessive noise may make communication by direct speech difficult or impossible. Noise may be a factor in marketing appliances or other equipment. Prolonged ambient noise levels may eventually cause permanent damage to hearing or, of course, they may simply impair efficiency of workers because of the annoyance factor. All these aspects of sound are unavoidably coupled with human hearing. Seldom are mechanical engineers concerned with the production of *pleasing* sounds. Almost always they are concerned with noise, its abatement, and its control.

Physically, airborne sound is, within a certain range of frequencies, a periodic variation in air pressure about the atmospheric mean. The air particles oscillate along the direction of propagation, and for this reason the waveform is said to be longitudinal. For a single tone or frequency (as opposed to a sound of complex form), the oscillation is simple harmonic and may be expressed as [2]

$$S = S_0 \cos\left[\frac{2\pi}{\lambda}(x \pm ct)\right] \tag{18.1}$$

where

S = the displacement of a particle of the transmitting medium,

S_0 = the displacement amplitude,

λ = the wavelength = $\dfrac{c}{f}$,

x = the distance from some origin (e.g., the source), in the direction of propagation,

c = the velocity of propagation,

t = time,

f = frequency

For right-running wave in a gaseous medium [2],

$$p = -B\frac{\partial S}{\partial x} = -B\frac{2\pi}{\lambda}S_0 \sin\left[\frac{2\pi}{\lambda}(ct - x)\right] \tag{18.2}$$

where

p = the pressure variation about an ambient pressure, P_{amb}, and

B = the adiabatic bulk modulus

18.2 CHARACTERIZATION OF SOUND (NOISE)

In its simplest form, sound is a *pure tone*; that is, it is of one frequency. Such a source is extremely difficult to produce. In addition to the inherent purity of the signal source and its coupling to the air, it requires the elimination of all reflections from surrounding objects. Such reflections or reverberations produce standing waves that at the point of observation can produce distortions caused by interaction between the directly incident wave and the returning reflections. *Free-field* conditions may be approximated in an *anechoic (no-echo) chamber*.[1] The walls of such a chamber are lined with sound-absorbing materials formed in wedges such that small reflections as may result from the initial encounter with the wall are directed again and again into the absorbent materials until essentially all the energy has been absorbed. In such an environment, sound simply travels outward and away from the

---

[1] An imperfect but inexpensive substitute for an anechoic chamber is an open field in as quiet a location as possible. Preferably the site should be on a slight hill or hummock.

source, with no return. Even so, the mere placement of a transducer into the space will cause unwanted distortions.

Random noise, rather than a pure tone, is by far the more common subject of investigation. Random noise is produced from a number of discrete sources whose outputs combine to form the whole. The sounds reaching the transducer (or the ear) are commonly from more than the initiating sources alone. The basic originating sources of sound energy may be called *primary* sources. In mechanical engineering these sources typically result from interacting machine parts such as gear teeth or bearings; from vibrating members of a wide variety, such as housing panels, shafts, and supports; or from hydraulic, pneumatic, or combustion sources. Sound waves traveling outward from a primary source are intercepted by surrounding objects such as ceilings, walls, and other machinery. A part of such incident waves is absorbed and the remainder is reflected (Fig. 18.1). Each point of reflection then becomes a *secondary source*, "heard" by any pickup device and thereby becoming confused with the initial or primary sounds. Combinations of primary and secondary sounds from *different* reflecting sources produce *standing waves*, resulting in reinforcements and nulls throughout the environment. In the most common situation the primary sounds are not pure tones but possess certain randomnesses in both amplitudes and frequencies. As a result, the standing waves are not necessarily fixed in space but may be thought of as dancing about throughout the environment. Beats may result (Sections 4.4 and 10.6). It is obvious, then, that evaluation of the parameters associated with a given sound (e.g., from an internal combustion engine or an air compressor) becomes very difficult indeed. Under normal circumstances, separation of subject and environment becomes relatively impossible. In fact, one may question the real value of such a separation. Although anechoic-chamber testing may help in understanding or treating certain specific noise sources, induction of meaningful, true-to-life results must include interaction between both the primary and the secondary sources. The environment is an unavoidable adjunct to any analysis.

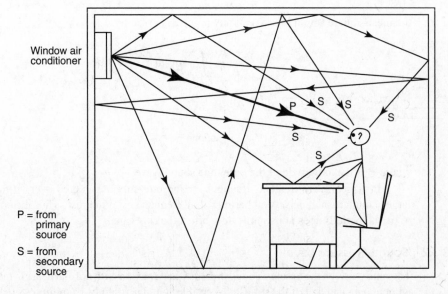

FIGURE 18.1: Sketch illustrating primary and secondary sources of sound (noise).

18.3 BASIC ACOUSTICAL PARAMETERS

18.3.1 Sound Pressure

In the presence of a sound wave, the instantaneous difference in air pressure and the average air pressure at a point is called *sound pressure*. The unit of measurement is the pascal (Pa), or newtons per square meter, or, more commonly, the micronewton per square meter (μN/m^2). The microbar (μbar) is also used. It is equal to one dyne per square centimeter.[2]

As noted previously, sound pressure is the difference between instantaneous absolute pressure and the ambient pressure. We now define the mean-square sound pressure as

$$p_{\text{rms}}^2 = \frac{1}{T} \int_0^T p^2 \, dt \tag{18.3}$$

where

$$p = p(t) = \text{instantaneous sound pressure at a point (Pa)},$$
$$p_{\text{rms}} = \text{root-mean-square sound pressure (Pa)},$$
$$T = \text{averaging measurement time (s)}$$

For a pure-tone sound wave described by

$$p = P \sin \omega \left(\frac{x}{c} - t \right) \tag{18.3a}$$

where

$$\omega = 2\pi f$$

the mean-square sound pressure at any point x is given by

$$p_{\text{rms}}^2 = \frac{P^2}{T} \int_0^T \sin^2 \omega \left(\frac{x}{c} - t \right) dt \tag{18.3b}$$

If we choose $T = 1/f = 2\pi/\omega$,

$$p_{\text{rms}} = \frac{P}{\sqrt{2}} \tag{18.3c}$$

where P is the amplitude of the pure tone sound wave.

In typical measurement situations, sound pressure is averaged over a time interval large enough to include several periods of all frequencies of interest. Thus the contribution of partial periods does not significantly affect the rms value.

18.3.2 Sound Pressure Level

The ratio between the greatest sound pressure that a person with normal hearing may tolerate without pain and that of the softest discernible sound is roughly 10 million to 1 (Fig. 18.2). This tremendous range suggests the use of some form of logarithmic scale. Recall that the

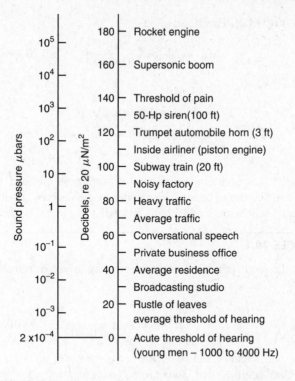

FIGURE 18.2: Typical sound pressures and sound pressure sources.

decibel (Section 7.12) provides such a scale and it is on this basis that most sound or noise measurements are made. Basically, the decibel is a measure of *power ratio*:

$$dB = 10 \ \log_{10} \left(\frac{\text{power}_1}{\text{power}_2} \right) \tag{18.4}$$

Because sound power is proportional to the *square* of sound pressure [3], Eq. (18.4) may be written as

$$dB = 10 \ \log_{10} \frac{p_1^2}{p_0^2} = 20 \ \log_{10} \left(\frac{p_1}{p_0} \right) \tag{18.5}$$

We see that the decibel is not an absolute quantity but a comparative one, which, however, can be used *in the manner of* an absolute quantity if referred to some generally accepted base. This is done and the *rms* value,

$$p_0 = 0.00002 \ \text{N/m}^2 = 20 \ \mu\text{N/m}^2$$

is widely accepted as the standard reference for sound pressure level. This value takes on added significance when we note that it corresponds closely to the acute threshold of

---

[2] $1 \ \mu\text{N/m}^2 = 0.00001 \ \mu\text{bar}$.

hearing (Fig. 18.2). Therefore,

$$\text{SPL} = 20 \ \log\left(\frac{p}{0.00002}\right) \qquad \text{re: } 20 \ \mu\text{N/m}^2, \qquad (18.6)$$

where

$$\text{SPL} = \text{the sound pressure level, in dB}$$
$$p = \text{the rms pressure from a sound source, in N/m}^2$$

Note: Inasmuch as all sound level measurements use base 10 logarithms, the reference will be omitted throughout the remainder of the chapter. In addition, the appended "re: $20 \ \mu\text{N/m}^2$" will be omitted with the understanding that this is the standard of reference.

EXAMPLE 18.1

What is the sound pressure level corresponding to an rms sound pressure of $1 \ \text{N/m}^2$?

Solution

$$\text{SPL} = 20 \ \log\left(\frac{1}{0.00002}\right) \approx 94 \ \text{dB}$$

Special attention should be directed to the use of the word *level*. Various terms yet to be discussed use the word: *sound level, loudness level, noise level,* and so on. Use of the word *level* implies a logarithmic scale of measurement expressed in decibels. Remembering this fact should help in keeping the units straight.

18.3.3 Power, Intensity, and Power Level

As suggested by Eq. (18.4), sound involves energy whose magnitude can be expressed in terms of power. The common unit is the watt, W. As the power radiates outward from an ideal point source, it will be spread continually over larger and larger areas of space. *Sound intensity* at any location is expressed in terms of watts per unit area. For a plane or spherical wave, the intensity I in the direction of propagation is [3] (see Fig. 18.3)

$$I = \frac{(p_{\text{rms}})^2}{\rho_0 c} = \frac{W}{A} \qquad (18.7)$$

where

$$\rho_0 = \text{the average mass density of the medium,}$$
$$c = \text{the speed of sound in the medium,}$$
$$W = \text{acoustic power in watts,}$$
$$A = \text{area}$$

A further discussion of the application of vector sound intensity measurement is given in Section 18.7.3.

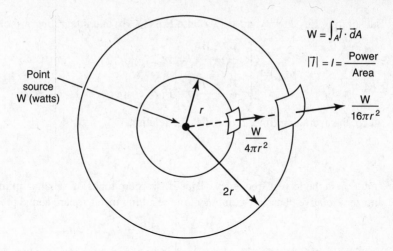

$$W = \int_A \vec{I} \cdot \vec{d}A$$

$$|\vec{I}| = I = \frac{\text{Power}}{\text{Area}}$$

$$\frac{W}{16\pi r^2}$$

$$\frac{W}{4\pi r^2}$$

FIGURE 18.3: Relation between sound power and intensity for a spherical wave.

Sound power level (PWL) is expressed in decibels and therefore must be given in terms of a reference level that is usually taken as 10^{-12} W. Power level is therefore defined as

$$\text{PWL} = 10 \ \log(W/10^{-12}) \ \text{dB} \qquad \text{re: } 10^{-12} \ \text{W} \qquad (18.8)$$

There is no instrument for measuring power level directly. However, the quantity can be calculated from sound pressure level measurements [4].

18.3.4 Combination of Sound Pressure Levels

When two pure-tone sounds occur at the same time, the combined effect depends on the sound pressure amplitudes, frequencies, and phase relationship at the receiver. Consider two sound waves at a point in space described by

$$p_1 = P_1 \cos(\omega_1 t + \phi_1), \qquad (18.9)$$
$$p_2 = P_2 \cos(\omega_2 t + \phi_2) \qquad (18.9a)$$

where

$$p = \text{instantaneous sound pressure,}$$
$$P = \text{sound pressure amplitude,}$$
$$\omega = 2\pi f = \text{circular frequency,}$$
$$\phi = \text{phase angle}$$

The instantaneous sound pressure due to the two waves is the sum of the two instantaneous sound pressures. The mean-square sound pressure of the combined pure tones is given by

$$p_{\text{rms}}^2 = \frac{1}{T} \int_0^T (p_1 + p_2)^2 \ dt \qquad (18.9b)$$

Substituting Eqs. (18.9) and (18.9a) into Eq. (18.9b) and integrating, we obtain [5]

$$
p_{rms}^2 = \begin{cases} \dfrac{P_1^2 + P_2^2}{2} = p_{rms1}^2 + p_{rms2}^2 & \omega_1 \neq \omega_2 \\[2ex] \dfrac{P_1^2 + P_2^2}{2} + P_1 P_2 \cos(\phi_1 - \phi_2) & \omega_1 = \omega_2 \end{cases} \tag{18.9c}
$$

where the averaging time, T, satisfies the relation

$$
T \gg \frac{1}{f_{min}}
$$

and f_{min} is the lowest frequency. Thus if two pure tones of equal amplitudes, frequency, and zero relative phase are combined, the resulting mean-square sound pressure is

$$
p_{rms}^2 = P_1^2 + P_1^2 \cos 0 = 2P_1^2 = 4p_{rms1}^2
$$

and the resulting increase in sound pressure level over the sound pressure level of one pure tone is

$$
\mathrm{SPL}_{comb} - \mathrm{SPL}_1 = 20 \; \log_{10}\left(\frac{2p_{rms1}}{p_0}\right) - 20 \; \log_{10}\frac{p_{rms1}}{p_0}
$$
$$
= 20 \; \log_{10} 2 = 6.02
$$

which is about a 6-dB increase.

If we now determine the effect of adding two pure tones of equal sound pressure magnitudes but different frequencies, the first expression of Eq. (18.9c) should be used. Thus the difference between the combined sound pressure level and the sound pressure level of one pure tone is

$$
\mathrm{SPL}_{comb} - \mathrm{SPL}_1 = 20 \; \log_{10}\frac{\sqrt{2}p_{rms1}}{p_0} - 20 \; \log_{10}\frac{p_{rms1}}{p_0}
$$
$$
= 20 \; \log_{10}\sqrt{2} = 3.01
$$

or about 3 dB. Most industrial and community noise problems that we experience represent a combination of many *uncorrelated* sources. There may be random variations in both amplitude and frequency. For *uncorrelated* noise sources, the first expression in Eq. (18.9c) applies. Thus the total mean-square sound pressure is simply the sum of the component source mean-square pressures. In order to simplify the combination of *uncorrelated* noise sources, Fig. 18.4 can be used to add or subtract noise sources.

EXAMPLE 18.2

A business machine is added to an office. The original ambient SPL was 68 dB. After the machine was added, the sound pressure level rose to 72 dB. What SPL was contributed by the machine?

Solution   Using Fig. 18.4, the difference between the total and the smaller levels is 4-dB. The 4 dB vertical line intersects the difference-between-decibel-levels line at 1.8 dB. Thus the machine decibel level is $68 + 1.8 = 69.8$ dB.

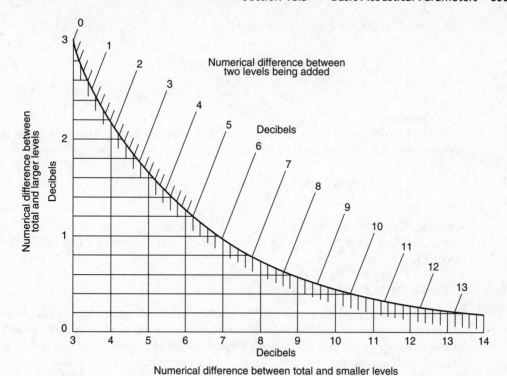

FIGURE 18.4: Diagram for adding or subtracting decibels. (Courtesy: GenRad Inc., Concord, Massachusetts)

18.3.5   Attenuation with Distance

In a *lossless, free space* there is a 6-dB decrease in sound pressure level (SPL) for each doubling of distance [6]. This may be shown as follows. From Eq. (18.5), we have

$$\text{SPL}_1 = 20 \ \log \left(\frac{p_1}{p_0} \right) \quad \text{and} \quad \text{SPL}_2 = 20 \ \log \left(\frac{p_2}{p_0} \right)$$

therefore

$$\text{SPL}_2 - \text{SPL}_1 = 20 \left[\log(p_2) - \log(p_1) \right] = 20 \ \log \left(\frac{p_2}{p_1} \right) \tag{18.10}$$

From Eq. (18.7) we see that for a spherical wave

$$\frac{p^2}{\rho c} = \frac{W}{(4\pi r^2)}$$

hence

$$\frac{p_2}{p_1} = \frac{r_1}{r_2} \tag{18.11}$$

Combining Eqs. (18.10) and (18.11) gives us

$$SPL_2 - SPL_1 = 20 \ \log \left(\frac{r_1}{r_2} \right) \qquad (18.12)$$

Question

In a free field, what will be the difference in sound pressure levels between points A and B if point B is twice as far from the source as is A?

$$SPL_A - SPL_B = 20 \ \log \frac{r_B}{r_A} = 20 \ \log 2 = 6.02$$

Caution   The relation expressed by Eq. (18.12) holds only for free-field conditions. In certain instances the relation may be used to confirm the existence of a free field. (See Section 18.8.)

18.4 PSYCHOACOUSTIC RELATIONSHIPS

As mentioned earlier, in only a relatively few situations is human hearing disassociated from sound measurements. Measuring systems and techniques are therefore unavoidably greatly influenced by the physiological and psychological makeup of the human ear as a transducer and the brain as the final evaluator. In terms of input magnitudes and frequencies, the human hearing system is quite nonlinear. Figure 18.5 shows average thresholds of hearing and tolerances for young persons. It will be noted that the greatest sensitivity occurs at about 4000 Hz and that a considerably greater SPL is required for equal reception at both

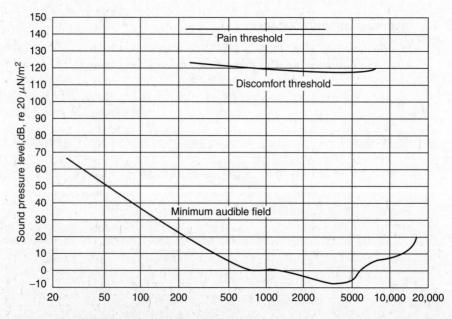

FIGURE 18.5: Thresholds of hearing and tolerance for young people with good hearing. (Courtesy: GenRad Inc., Concord, Massachusetts)

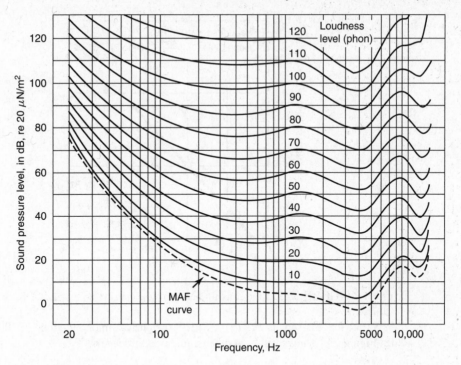

FIGURE 18.6: Free-field equal-loudness contours for pure tones. (Courtesy: GenRad Inc., Concord, Massachusetts)

lower and higher frequencies. Figure 18.6 shows the free-field *equal-loudness* contours for pure *tones* as determined by Robinson and Dadson at the National Physical Laboratories, Teddington, England.

Loudness is a measure of relative sound magnitudes or strengths *as judged by the listener*. It is a subjective quantity depending on both the physical waveform emanating from the source and on the *average* of many human hearing systems (persons) as receptors. The quantity is measured in *loudness level* and the unit is called the *phon* (pronounced "fon," to rhyme with "up*on*"). The loudness level in phons is numerically equal to the sound pressure level in dB at the frequency of 1000 Hz. Note that for each contour in Fig. 18.6, the loudness level and the sound pressure level are equal at only one point, namely, at 1000 Hz. It is important to keep in mind that the equal-loudness contours of Fig. 18.6 are based on *pure tones*. Loudness levels of complex sounds are considered in Section 18.8.

To further understand the curves in Fig. 18.6, first note that the SPL of 30 dB at 1000 Hz corresponds to 30 phons. On average, for a person to sense a loudness of 30 phons at 100 Hz would require an SPL of 44 dB; at 9000 Hz, 40 dB; and so on.

It is clear that loudness level in terms of the phon is a logarithmic quantity. Although this is useful, still another measure of strength is employed. *Loudness* (note the absence of the word *level*) is measured with a linear unit, the *sone* (rhymes with zone). One sone is the loudness of a 1000-Hz tone with an SPL of 40 dB (note that this also corresponds to 40 phons). A tone that sounds *n* times as loud has a loudness of *n* sones, and so on.

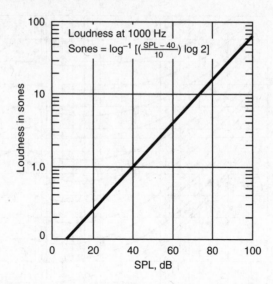

FIGURE 18.7: Relationship between loudness in sones and SPL.

We see that the sone is tied to the SPL at the one common pure tone of 1000 Hz and 40 dB. Figure 18.7 shows values at other sound pressure levels.

18.5   SOUND-MEASURING APPARATUS AND TECHNIQUES

Measurement of the parameters associated with sound use a basic system made up of a detector–transducer (the microphone), intermediate modifying devices (amplifiers and filtering systems), and readout means (a meter, oscilloscope, or recording apparatus). Most sound-measuring systems are used to obtain psychoacoustically related information. It is therefore necessary to build into the apparatus nonlinearities approximating those of the average human ear. Elaborate filtering networks also provide the basis for analyzers, devices for separating and identifying the various frequency components or ranges of components forming a complex sound.

18.5.1   Microphones

Most microphones incorporate a thin diaphragm as the primary transducer, which is moved by the air acting against it. The mechanical movement of the diaphragm is converted to an electrical output by means of some form of secondary transducer that provides an analogous electrical signal.

Common microphones may be classified on the basis of the secondary transducer, as follows:

1. Capacitor or condenser
2. Crystal
3. Electrodynamic (moving coil or ribbon)

The *capacitor* or *condenser microphone* is probably the most respected microphone for sound measurement purposes. It is arranged with a diaphragm forming one plate of an

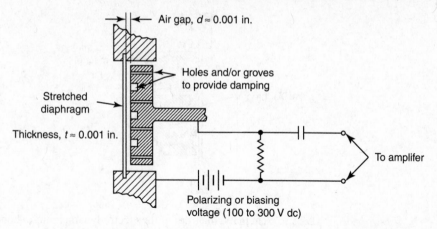

FIGURE 18.8: Schematic representation of the condenser-type microphone.

air-dielectric capacitor (Fig. 18.8). Movement of the diaphragm caused by impingement of sound pressure results in an output voltage [7]:

$$E(t) = E_{\text{bias}} \frac{d'(t)}{d_0} \tag{18.13}$$

where

$$E(t) = \text{the output voltage},$$
$$E_{\text{bias}} = \text{the polarizing voltage},$$
$$d_0 = \text{the original separation of the plates},$$
$$d'(t) = \text{the change in plate separation caused by sound pressure fluctuations}$$

The capacitive microphone is widely used as the primary transducer for sound measurement purposes.

The *electret microphone* is a special form of the condenser type. Whereas the common condenser type requires an external polarizing voltage, the electret type is self-polarizing. The diaphragm is constructed of a plastic sheet that has a conductive coating on one side. The coating serves as one side of the capacitor.

The *crystal microphone* uses a piezoelectric-type element (Section 6.14), generally activated by bending. For greatest sensitivity, a cantilevered element is mechanically linked to the diaphragm. Other constructions use direct contact between diaphragm and element, either by cementing (element placed in bending) or by direct bearing (element in compression). Crystal microphones are extensively used for serious sound measurement.

The *electrodynamic microphone* uses the principle of the moving conductor in a magnetic field. The field is commonly provided by a permanent magnet, thereby placing the transducer in the variable-reluctance category (Section 6.12). As the diaphragm is moved, the voltage induced is proportional to the *velocity* of the coil relative to the magnetic field, thereby providing an analogous electrical output. Two different constructions are used,

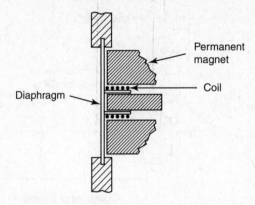

FIGURE 18.9: Schematic diagram of the electrodynamic microphone.

the *moving coil* (Fig. 18.9) and the *ribbon* type. The inductive member of the latter type consists of a single element in the form of a ribbon that serves the dual purpose of "coil" and diaphragm.

Microphone Selection Factors

An ideal microphone used for measurement would have the following characteristics:

1. Flat frequency response over the audible range
2. Nondirectivity
3. Predictable, repeatable sensitivity over the complete dynamic range
4. At the lowest sound level to be measured, output signal that is several times the system's internal noise level
5. Minimum dimensions and weight
6. Output that is unaffected by all environmental conditions except sound pressure

The capacitor-type microphone undoubtedly enjoys the top position for sound measurement use, although the crystal type runs a close second. As with all measurement, the presence of the sensor (microphone) unavoidably alters (loads) the signal to be measured. In this application the microphone should be as physically small as possible. It is obvious, however, that size, particularly diaphragm diameter, must have an important influence on both sensitivity and response. Microphones are therefore available in a range of sizes, and final selection must be based on a balance of the requirements for the specific application. Table 18.1 summarizes microphone characteristics.

18.5.2 The Sound Level Meter

The basic sound level meter is a measuring system that senses the input sound pressure and provides a meter readout yielding a measure of the sound magnitude. The sound may be wideband, it may have random frequency distribution, or it may contain discrete tones. Each of these factors will, of course, affect the readout.

Generally the system includes weighting networks (filters) that roughly match the instrument's response to that of human hearing. The readout, therefore, includes a psychoacoustical factor and provides a number ranking the sound magnitude in terms of the ability of the human measuring system.

TABLE 18.1: Summary of Microphone Characteristics

Type of Microphone	Principle of Operation	Relative Impedance	Linearity	Advantages	Disadvantages
Capacitor	Capacitive	Very high	Excellent	Stable: holds calibration. Low sensitivity to vibration. Wide range.	Sensitive to temperature and pressure variations. Relatively fragile. Requires high polarizing voltage. Requires impedance-coupling device near microphone.
Crystal	Piezoelectric	High	Good to excellent	Self-generating. May be hermetically sealed. Relatively rugged. Relatively inexpensive.	Requires impedance-matching device. Relatively sensitive to vibration.
Electrodynamic	Reluctive	Low	Good	Self-generating.	Physically large.

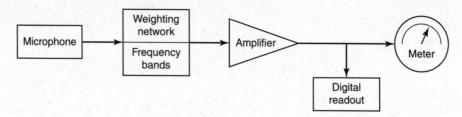

FIGURE 18.10: Block diagram of a typical sound level meter, or sound level recorder.

A block diagram of the basic sound level meter is shown in Fig. 18.10, and Fig. 18.11 shows a commercially available system. When used in conjunction with a readout, the system is commonly referred to as a *sound level recorder*.

Figure 18.12 displays the internationally standardized weighting characteristics selectable by panel switch. We can see that the filter responses selectively discriminate against low and high frequencies, much as the human ear does. It is customary to use characteristics A for sound levels below 55 dB, B for sounds between 55 and 85 dB, and C for levels above 85 dB, as shown in Fig. 18.12. Certain broad generalities as to the frequency makeup of a sound may be made by taking separate readings with each network.

Calibration of sound level meters may be accomplished using a precision acoustic calibrator, such as shown in Fig. 18.13. Calibrators produce a specified sound pressure level (e.g., 94 dB or 114 dB) at a given frequency (e.g., 250 Hz or 1 kHz). Further discussion of acoustic calibration is given in Section 18.9.

18.5.3 Frequency Spectrum Analysis

Although determination of a value of sound pressure or sound level provides a measure of sound intensity, it yields no indication of frequency distribution. For noise abatement purposes, for example, it is very desirable to know the predominant frequencies involved. This can often point directly to the prominent noise sources.

Determination of intensities versus frequency is referred to as *spectrum analysis* and is accomplished through the use of band-pass filters (Section 7.16). Various combinations of filters may be used, determined by their relative band-pass widths. Probably the most commonly used are the "full-octave" filters having center frequencies as follows: 31.5, 63, 125, 250, 500, 1000, 2000, 4000, 8000, and 16,000 Hz. Figure 18.14 depicts typical frequency characteristics of the filters.

In addition to full-octave spectrum division, $\frac{1}{2}$, $\frac{1}{3}$, $\frac{1}{10}$, and other fractional octave divisions are also used. Although the ear may be capable of distinguishing pure tones in the presence of other tones, it does tend to integrate complex sounds over roughly $\frac{1}{3}$-octave intervals [8], thus lending some additional importance to $\frac{1}{3}$-octave analyzers. Table 18.2 lists the octave and $\frac{1}{3}$-octave band center frequencies and lower and upper band limits.

Simple analyzers are used by taking a separate reading for each pass band. It is seen therefore that an appreciable period of time is required to scan the range. Not only does the bulk of data increase with reduced bandwidth, but the constancy of the sound source also becomes of greater importance as the necessary time for measurement increases. A partial solution to the latter problem is to use a tape recorder for sampling and then analyze

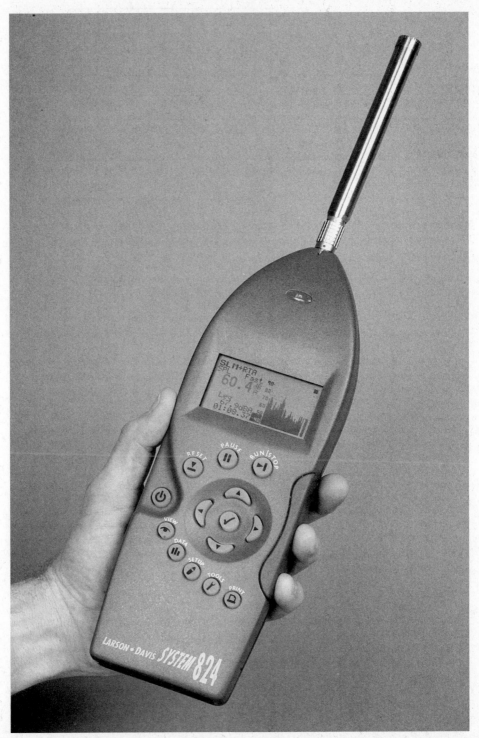

FIGURE 18.11: Sound level meter, Model 824 (Courtesy: Larson-Davis, Inc.).

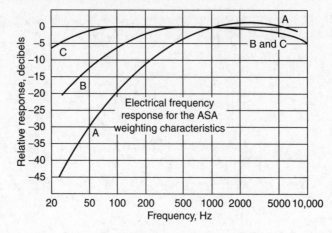

FIGURE 18.12: Standard weighting characteristics for sound level meters from the Acoustical Society of America (ASA).

FIGURE 18.13: Precision acoustic calibrator. (Courtesy: Larson-Davis, Inc.)

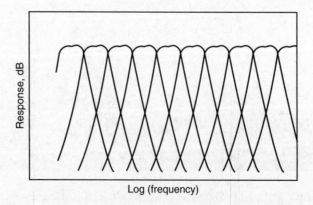

FIGURE 18.14: Typical overlapping filter characteristics of band-type sound analyzers.

TABLE 18.2: Octave Band and $\frac{1}{3}$-Octave Band Limits

Octave Band Center Frequency	Lower Limit	Upper Limit
31.5 Hz	22.4 Hz	45 Hz
63	45	90
125	90	180
250	180	355
500	355	710
1 kHz	710	1.4 kHz
2	1.4 kHz	2.8
4	2.8	5.6
8	5.6	11.2
16	11.2	22.4

$\frac{1}{3}$-Octave Band Center Frequency	Lower Limit	Upper Limit
20.0 Hz	18.0 Hz	22.4 Hz
25.0	22.4	28.0
31.5	28.0	35.5
40	35.5	45
50	45	56
63	56	71
80	71	90
100	90	112
125	112	140
160	140	180
200	180	224
250	224	280
315	280	355
400	355	450
500	450	560
630	560	710
800	710	900
1,000	900	1,120
1,250	1,120	1,400
1,600	1,400	1,800
2,000	1,800	2,240
2,500	2,240	2,800
3,150	2,800	3,550
4,000	3,550	4,500
5,000	4,500	5,600
6,300	5,600	7,100
8,000	7,100	9,000
10,000	9,000	11,200
12,500	11,200	14,000
16,000	14,000	18,000
20,000	18,000	22,400

the recorded sound at leisure. It is obvious that to be useful for faithfully recording a sound source, the tape recorder must be of highest quality. Commonly available recorders for speech and music are not usable; their frequency response requirements are purposely made nonlinear over the required range of frequencies.

18.5.4   The Discrete Fourier Transform

Application of the DFT is of particular usefulness in acoustical work. Recall (Section 4.6) that this calculation produces an amplitude-versus-frequency display. It provides a very convenient means for determining the contributions of the various harmonic components making up a complex noise signal.

18.6   APPLIED SPECTRUM ANALYSIS

Harmonic or Fourier analysis of complex waveforms is discussed in Section 4.6 and is further mentioned in Appendix B. From those discussions it should be clear that even for relatively simple waveshapes, the number of required numerical manipulations may easily make the procedure prohibitive from a time-benefit standpoint. This limitation becomes especially important when the variety of nonrepetitive conditions needing analysis taxes the capacity of even the larger computers. For this reason an accelerated procedure referred to as the *fast Fourier transform*, or *FFT*, has been developed. If N represents the number of harmonic coefficients to be determined, ordinary harmonic analysis requires roughly N^2 separate computations, whereas the FFT requires approximately $N\log_2 N$—a marked reduction. It has been found that many waveforms encountered in acoustics, mechanical vibrations, and most electrical quantities permit practical application of the FFT.

Figure 18.15 is a schematic diagram of an FFT analyzing system. As shown, the signal is sensed by a microphone, vibration pickup, and so on and is then conditioned by amplification or filtering, as may be required. A sample of the conditioned input is then taken over a time interval, T. Then the time base is divided into an integral number of equal increments. The number is usually a power of 2.[3] For *each time increment*, Δt, the A/D converter then outputs an amplitude in digital form for analysis by the logic section. The function of the logic section is to transform (convert) the sample from the time domain to the frequency domain for final display or numerical listing.

The output from the FFT calculation usually consists of the amplitude of a sine and cosine pair for each integral multiple of the fundamental frequency.[4] The number of pairs obtained is just equal to the number of original data points. Each sine and cosine pair can be considered as a vector with real and imaginary parts, which can be combined into a vector with absolute magnitude and a phase angle (see Section 4.4). For many acoustical analyses the phase angle is ignored and the vector magnitude at each integral multiple of the fundamental frequency is used as the result of the analysis.

The squared value of each vector magnitude is sometimes called the *autospectral value* and the set of the squared values, the *power spectrum*, even though these values may not represent power at all.

The sampling time interval, T, is often referred to as the *data window duration*. The

---

[3]Note that $2^8 = 256_{10}$ is commonly used. Counting zero as the initial value, this corresponds to 0 to 255_{10} or 0 to FF_{16}, or 1 byte of information.

[4]The fundamental frequency here is directly related to the time duration of the waveform being analyzed [see Eq. (4.21)].

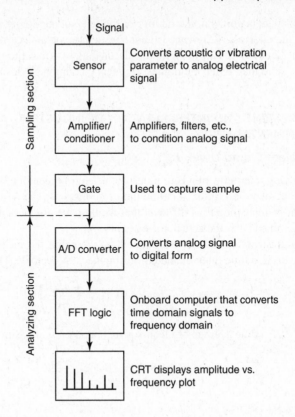

FIGURE 18.15: Block diagram of an FFT analyzing system.

experimenter sets this value by adjustment of the gating time, hence the sample's length. When obviously repetitive waveforms are encountered, the logical value is clear: the period of the cycle. But when the input is complex and nonrepetitive, the selection of a proper sampling time interval is in doubt. In such a case many primary sources of signal plus secondary sources may be present. In mechanical applications these sources may consist of gear and bearing noises, elastic resonances, combustion sounds, hydraulic noises, and the like. In the case of acoustical measurements, the environment contributes also. In many situations there are identifiable frequency sources such as rpm's, rate of gear tooth meshings, and so on. Because of its analyzing speed, the FFT system becomes extremely useful in cases of this sort. A variety of sampling time intervals may be tried and, by comparison of readouts, a "best" result selected. In certain instances, it may be decided that no significant, constant harmonic relationships are present. Of course, this result also represents pertinent information.

Some final considerations in the proper use of the FFT analysis include (1) the "shape" of the data window chosen in addition to its time duration, and (2) the A/D converter's sampling frequency rate in relation to the highest frequency component desired from the analysis. With regard to (1), the shape of the data window indicates how the data points will be weighted. For example, a rectangular window indicates that all data points will be treated as equally important, whereas a "raised cosine" or "Hanning" window gives greater emphasis to data points in the interior of the interval and lesser importance to those at the ends of the interval. (Reference [9] contains a more detailed discussion of data windows.) For consideration of (2), the simple rule of thumb is that the data sampling rate, $1/\Delta t$, must be more than twice the highest frequency component desired from the analysis (see Section 4.7.2).

18.7   MEASUREMENT AND INTERPRETATION OF INDUSTRIAL AND ENVIRONMENTAL NOISE

18.7.1   Equivalent Sound Level, L_{eq}

For studying long-term trends in industrial or environmental noise, it is convenient to use a single number descriptor to define the noise history. The descriptor most often used is L_{eq}, that is, the continuous dB level that would have produced the same sound energy in the same time (T) as the actual noise history.

Equivalent sound level is obtained by averaging the mean-squared sound pressure over the desired time interval and converting back to decibels. Thus

$$L_{eq} = 10 \, \log_{10} \left(\frac{\overline{p_{rms}^2}}{p_0^2} \right) \tag{18.14}$$

where $\overline{p_{rms}^2}$ = time average of mean-square sound pressure. Since

$$SPL = 10 \, \log_{10} \frac{p_{rms}^2}{p_0^2}$$

and

$$\frac{p_{rms}^2}{p_0^2} = 10^{SPL/10} \tag{18.15}$$

we can write Eq. (18.14) as

$$L_{eq} = 10 \, \log_{10} \left(\frac{1}{T} \int_0^T 10^{SPL/10} \, dt \right) \tag{18.16}$$

where

$$L_{eq} = \text{equivalent sound level, in dB,}$$
$$SPL = \text{rms sound pressure level,}$$
$$T = \text{averaging time, which is specified}$$

Note that equivalent sound level is an energy average and will, in general, differ from the arithmetic average of the sound levels and the median level. High SPL readings tend to dominate the equivalent sound level. Readings of 20 dB or more below the peak level make a small contribution to the equivalent sound level. Most integrating sound level meters, such as the one shown in Fig. 18.11, calculate the L_{eq} electronically for any user-specified averaging time. Typical averaging times can range from minutes to days, depending upon the specific situation [5]. Figure 18.16 illustrates the difference between a time varying SPL reading and its L_{eq} value. Although the L_{eq} value may be determined for any weighted scale, it is most commonly used with the A-weighted scale.

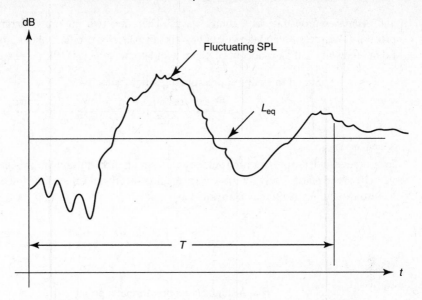

FIGURE 18.16: Comparison of SPL and L_{eq} values.

18.7.2   Sound Exposure Level (SEL)

When noise is of a transient nature, such as an automobile passing by an observer, an airplane passing overhead, or impact noise caused by a forging operation, the usual acoustic measurement of SPL is not only difficult to obtain but also is more difficult to interpret. Some sound level meters have a peak-hold capability, whereby the maximum SPL value measured may be captured and stored. The question arises as to how important this value is, especially when comparing one measurement with a second in which some noise attenuation fixes have been performed.

For this situation the L_{eq} measurement as described in Section 18.7.1 is performed, except that the averaging time T is taken as 1.0 s. The sound exposure level is thus defined as

$$ \text{SEL} = 10 \, \log_{10} \left(\int_0^T 10^{\text{SPL}/10} \, dt \right) \tag{18.17} $$

where T is 1 s. If we compare Eq. (18.16) with Eq. (18.17), we observe that

$$ \text{SEL} = L_{eq} + 10 \, \log_{10} T \tag{18.18} $$

and the SEL and L_{eq} value differ significantly only for large values of averaging time T. Thus the SEL reading, since it is a measure of acoustic energy, can be used to compare unrelated noise events because it is normalized to 1.0 s.

18.7.3   Sound Intensity Measurement

When a particle of air is displaced from its mean position, there is a temporary increase in pressure. The pressure increase acts in two ways: to restore the particle to its original position and to pass on the disturbance to the next particle. The cycles of pressure increases and decreases propagate as a sound wave. There are two important parameters in this process. The pressure and the velocity of the air particles oscillate about a fixed position. Sound intensity is the product of pressure and particle velocity [10].

$$\text{Sound intensity} = \text{pressure} \times \text{particle velocity}$$

$$= \frac{\text{force}}{\text{area}} \cdot \frac{\text{distance}}{\text{time}} = \frac{\text{energy}}{\text{area} \times \text{time}} = \frac{\text{power}}{\text{area}}$$

In an active field, the pressure and velocity vary simultaneously and the pressure and particle velocity are in phase. Only for this case is the time-averaged value of the intensity not equal to zero. Sound intensity may be defined as

$$I_r = \frac{1}{T} \int_0^T p u_r \, dt \qquad (18.19)$$

where

$$p = \text{instantaneous pressure at a point,}$$
$$u_r = \text{air-particle velocity in the } r\text{-direction}$$

The air-particle velocity at a point can be expressed in terms of the pressure gradient at a point [5]:

$$u_r = -\frac{1}{\rho} \int_{-\infty}^t \frac{\partial p}{\partial r} dt \approx -\frac{1}{\rho} \int_0^t \frac{(p_B - p_A)}{\Delta r} dt$$

where the partial derivative has been replaced by a finite-difference approximation. If the pressure, p, is replaced by the average pressure, then the sound intensity determined by Eq. (18.19) becomes

$$I_r = -\frac{1}{T} \int_0^T \frac{(p_A + p_B)}{2\rho} \left(\int_0^t \frac{(p_B - p_A)}{\Delta r} dt \right) dt \qquad (18.20)$$

where

$$\rho = \text{mass density of air,}$$
$$\Delta r = \text{spacing between points } A \text{ and } B,$$
$$p_A \text{ and } p_B = \text{instantaneous pressures at points } A \text{ and } B, \text{ respectively,}$$
$$T = \text{averaging time}$$

Sound intensity is determined by a two-microphone method, where the two microphones are placed face to face and separated by a distance Δr by a spacer, as shown in Fig. 18.17(a). Figure 18.17(b) shows the directivity characteristic for the sound intensity measuring system. Note that when the angle θ is 90°, the sound intensity component is zero, since there is no difference in the pressure signals being measured. This feature makes this measurement useful in locating the sources of noise in complex sound fields.

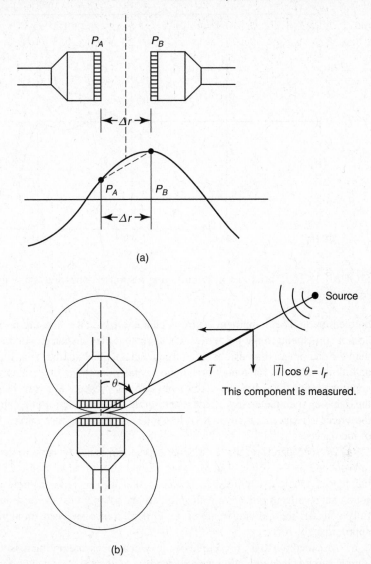

FIGURE 18.17: Sound intensity measuring systems: (a) microphone spacing and pressure gradient approximation, (b) sound intensity component measured.

A sound intensity analyzing system consists of a two-microphone probe system and an analyzer. The microphone probe system measures the two pressures p_A and p_B and the analyzer does the integration to find the sound intensity. A typical commercial system is the Bruel and Kjaer Type 3360 Sound Intensity Analysis System.

18.8   NOTES ON SOME PRACTICAL ASPECTS OF SOUND MEASUREMENTS

With all measurements, the act of measuring disrupts the process being evaluated. The sound pressure existing at the microphone diaphragm is *not* the sound pressure that would exist at that location were the microphone not present. *True* free-field conditions cannot

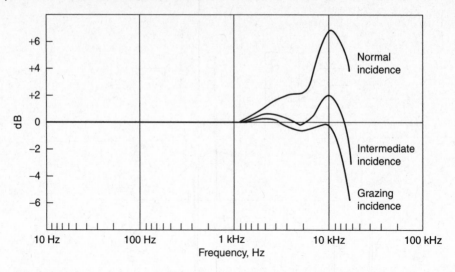

FIGURE 18.18: Variations in microphone response depending on angle of incidence of sound wave.

be measured. The diaphragm stiffness characteristics, the housing properties, and so on do not correspond to the properties of the air that they displace. For wavelengths smaller than the diaphragm dimensions (high frequencies), the diaphragm has the characteristics of an infinite wall, and for sounds arriving perpendicular to the diaphragm face, the pressures approach twice the value were the diaphragm not present. For wavelengths several times the diaphragm diameter, this effect is negligible. In addition, for other angles of incidence[5] the effect is reduced. Figure 18.18 illustrates typical response curves for various angles of incidence.

We see, therefore, that as a general practice, it is advisable to avoid "pointing" the microphone at the sound source. One should use an angle approaching 90° (grazing) incidence. One must remember, however, that in a reverberant field there are multiple sound sources from reflective walls, and so on, so one must exercise judgment and carefully consider the individual situation. One reference suggests an angle of incidence of approximately 70°.

An exception to the preceding occurs when one is attempting to isolate discrete sound inputs, perhaps in conjunction with a spectrum analyzer or band analyzer. In such a case it may be desirable to point the microphone to suspected areas and *also* to place the microphone much closer to the source than would otherwise be done.

At what distance from the source should the microphone be placed when a general measurement is being made? Figure 18.19 shows a condition typical of many sound fields, produced by such sources as machinery: sources made up of many individual noisemakers such as gears, bearings, housings, and so on. The cross-sectional areas in the figure depict distances over which *transverse* movement of the microphone produces variations in SPL readings for a presumably constant input. In the near-distance interval the discrete sources may be identifiable. In the far distances the field is reverberant, and reflective sources cause SPL variations.

---

[5]Zero degrees incidence occurs when the microphone axis is parallel to the direction of sound propagation (i.e., the wavefront is parallel to the microphone diaphragm).

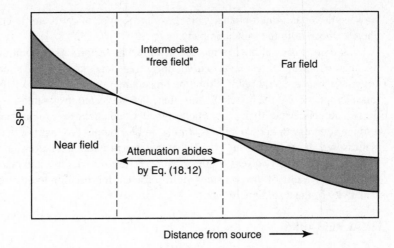

FIGURE 18.19: Typical SPL variations dependent upon microphone placement.

In the free-field interval, transverse movement of the microphone should make little or no difference in readings. In addition, movement toward or away from the source should provide readings approximating those predicted by Eq. (18.12). This latter fact may be helpful in establishing that free-field conditions do or do not prevail. It is important to note that in many cases the free-field interval is not present: Near- and far-field conditions overlap.

18.9   CALIBRATION METHODS

As with most electromechanical measuring systems, sound measurement involves a detector–transducer (the microphone) followed by intermediate electrical/electronic signal conditioning and some sort of readout stage. Calibration of the complete system involves introducing a known sound pressure variation and comparing the known with the readout. In practice such a calibration may be made; more commonly, the various component stages may be calibrated separately. For example, sound level meters often provide a simple check of the electrical and readout stages by substituting a carefully controlled and known voltage for the microphone output and adjusting the readout to a predetermined value for the particular instrument. This procedure ignores the microphone altogether; however, it does provide a convenient method for checking the remainder of the system. Acoustic calibrators (see Fig. 18.13) can be used to calibrate the entire meter, including the microphone.

Comparative methods may also be used whereby a microphone or a complete system is directly compared with a "standard." Of course, as with all comparative methods, it is imperative that great care be exercised to ensure that a true comparison is made. It is necessary that both the standard and the test microphones "hear" the identical sound. Our introductory discussion at the beginning of this chapter should indicate the difficulties in achieving this result.

Standard sound sources involve mechanical loudspeaker-type drivers for producing the necessary pressure fluctuations. One system uses two battery-driven pistons moving in opposition in a cylinder at 250 Hz. A precalibrated pressure level of 124 dB $\pm$ 0.2 dB is produced. A somewhat similar system uses a small, rugged loudspeaker as the driver. In both cases proper calibration is maintained only if the design coupling cavity is employed between driver and microphone. This restricts this type of calibrator to compatible microphones: those from the same company.

A more basic calibration method is known as *reciprocity* calibration. In addition to the microphone under test, a *reversible*, linear transducer and a sound source with a proper coupling cavity are required. Calibration procedure is divided into two steps. First, both the test microphone and the reversible transducer are subjected to a common sound source and the two outputs; hence their ratio is determined. Second, the reversible transducer is used as a sound source with known input (current), and the output (voltage) of the test microphone is measured. It can be shown [11, 12] that this step provides a relationship that is a function of the product of the two sensitivities: that of the test microphone and that of the reversible transducer. Results of the two steps yield sufficient information to determine the absolute sensitivity of the tested microphone.

18.10   FINAL REMARKS

Sound is a complex physical quantity. Its evaluation through measurement becomes doubly difficult in comparison with evaluation of other engineering quantities because of the necessary involvement of human hearing. It must, therefore, be recognized that the foregoing sections serve merely as an introduction to the subject. The student is directed to the appended list of Suggested Readings for further study.

Little distinction has been made throughout this chapter between sound and noise. The latter has been considered merely as "undesirable" sound. Noise, however, is becoming an increasingly important problem to the engineer due to the controls that are set up by ordinance and statute as well as company–union agreements and court decisions intended to protect the employee. As a result, engineers are called on to design "quiet" machinery and processes. Engineers are increasingly criticized for their part in "silence pollution." (We do not use the common term *noise pollution* because it is not noise that is polluted.) It is quite necessary, therefore, that they be knowledgeable not only in the field of noise measurement, but also in the theory and art of noise abatement.

SUGGESTED READINGS

Bell, L. H. and D. H. Bell. *Industrial Noise Control: Fundamentals and Applications*. 2nd ed. New York: Marcel Dekker, 1993.

Beranek, L. L. *Acoustical Measurements*. New York: American Institute of Physics, 1988.

Beranek, L. L., and I. L. Vér (eds.). *Noise and Vibration Control Engineering: Principles and Applications*. New York: Wiley-Interscience, 1992.

Lord, H. W., W. S. Gatley, and H. A. Evensen. *Noise Control for Engineers*. Malabar, Fla.: Krieger, 1987.

Rayleigh, J. W. S. *The Theory of Sound*, vols. 1 and 2. New York: Dover, 1976.

PROBLEMS

18.1. Prepare a spreadsheet template for adding sound pressure levels from multiple sources. Provide for up to eight entries. Check the template against the results for the examples in Section 18.3.4.

18.2. Pumps are being selected for a fluids-handling system that is being designed. The upper capacity of the system is 10,000 gal/min. Among many parameters considered is that of noise. Combinations from a range of six pumps are being considered. The pump capacities along with the individual sound pressure levels that they generate are listed as follows:

Capacity (gal/min)	SPL (dB)
2000	89
2500	91
4000	93
5000	95
6000	96
7500	98

Determine the combination of pumps that will yield the lowest combined sound pressure level in dB (*Note*: The advantages of spreadsheet templates become obvious when working this and the following problem.)

18.3. The results from Problem 18.2 may be further refined by recognizing that more involved piping will be required as the number of pumps increases, thereby resulting in an added sound source. Use the following estimates of pipe noise to the various combinations of pumps and recalculate the total sound pressure levels.

Number of Pumps	Add the Following dB
2	80
3	83
4	86
5	89

18.4. The noise level at the factory property line caused by 12 identical air compressors running simultaneously is 60 dB. If the maximum SPL permitted at this location is 57 dB, how many compressors may be run simultaneously?

18.5. Calculate the SPL and intensity at a distance of 10 m from a uniformly radiating source of 2.0 W.

18.6. A sound pressure level meter used to measure the SPL of an engine without a muffler gave a reading of 120 dB. After attaching the muffler, the same meter gave a reading of 90 dB. Determine (a) the rms sound pressure before the muffler was attached, and (b) the percentage reduction in rms sound pressure amplitude when the muffler was used.

18.7. The SPL of a single rocket engine on its test stand and at a distance of $\frac{1}{2}$ mi. is 108 dB. What SPL, at the same distance, would result if a cluster of five such engines were tested together? If two of the five were shut down, what drop in SPL would be expected?

18.8. Data from two lawnmower manufacturers regarding the noise produced by each as measured on the factory floor resulted in the following:

Mower	SPL (A)	Distance Directly Above Mower
A	100 dB	30 ft
B	85 dB	50 ft

Estimate the SPL (A) when these mowers are used in grass cutting at a distance of 100 ft at an ambient temperature of 25°C.

18.9. The following data were obtained from an octave band frequency analysis of a steam boiler based on "flat," or "linear," frequency weighting.

(**a**) Construct a decibel-versus-frequency histogram of the data.

(**b**) Superimpose on this histogram the following data corrected for A-weighting.

Octave Band Center Frequency, Hz	SPL, dB
31.5	79
63.0	77
125.0	76
250.0	81
500.0	80
1,000.0	82
2,000.0	82
4,000.0	72
8,000.0	67
16,000.0	60

18.10. In order to determine whether free-field conditions are approximated, an engineer finds the SPL at 10 m from the source to be 89 dB. What must be the SPL at 7.5 m in order for free-field conditions to exist?

18.11. Figure 18.20 illustrates a simple experimental setup that may be used to demonstrate a variety of basic principles discussed throughout this book. A small transistor-type loud-speaker (available from most electronic parts dealers) is cemented to a length of plastic pipe [Fig. 18.20(a)]. A signal generator is used to drive the speaker with a sinusoidal

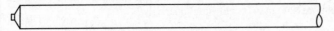

(a) Small transistor-type loudspeaker
cemented to a length of plastic pipe

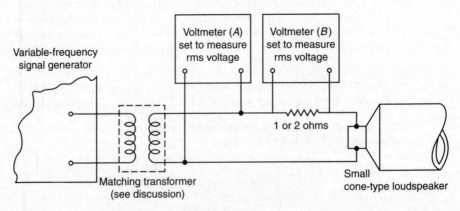

(b) Electrical circuitry

FIGURE 18.20: Setup for Problem 18.11.

waveform [Fig. 18.20(b)]. The tube-speaker combination possesses two fundamental resonance frequencies, that of the "organ pipe" (Section 14.10.1) and the acoustical/electrical resonance frequency of the speaker element. Voltmeter B, along with the 1 or 2 Ω resistor, is used to monitor the current to the speaker, and voltmeter A monitors the voltage. If voltage measured by A is held constant as a range of frequencies is swept by the signal generator, a sharp drop in current will be found at the frequency corresponding to resonance of the "organ pipe." This is the frequency at which oscillation may be maintained with minimum *power* expenditure. If the measuring system is sensitive enough, other resonances may also be determined, that of the speaker itself and the higher-mode frequencies of the pipe. The minimum-power principle is often useful in vibration testing employing shaker-driven systems (Chapter 17). Increased power transfer may be had by insertion of a matching transformer (Section 7.21). The turns ratio is not too critical; however, many signal generators have a 600-Ω output impedance and the input impedance of the speaker may be about 16 Ω. Care must be exercised to prevent overdriving the speaker and destroying its voice coil.

REFERENCES

[1] Skilling, D. C. Acoustical testing at Northrup Aircraft. *SESA Proc.* 16(2):121, 1959.

[2] Randall, R. H. *An Introduction to Acoustics*. Reading, Mass.: Addison-Wesley, 1951.

[3] Beranek, L. L. *Acoustics*. New York: McGraw-Hill, 1954, p. 12.

[4] Peterson, A. P. G. *Handbook of Noise Measurement*. 9th ed. Concord, Mass.: GenRad, Inc., 1980.

[5] Wilson, C. E. *Noise Control*. New York: Harper & Row, 1989.

[6] Beranek, L. L. (ed.). *Noise Reduction*. New York: McGraw-Hill, 1960, p. 186.

[7] Keast, D. N. *Measurements in Mechanical Dynamics*. New York: McGraw-Hill, 1967.

[8] Ranz, J. R. Noise measurement methods, *Mach. Des.* November 10, 1966.

[9] Peterson, A. P. G. *Handbook of Noise Measurement*. 9th ed. Concord, Mass.: GenRad, Inc., 1980, p. 133.

[10] *Sound Intensity*, Technical Note. Brüel and Kjaer, 1986. 2850 Naerum Denmark.

[11] Various papers, Handbook 77, Vol. II, *Precision Measurement and Calibration, Heat and Mechanics*, National Bureau of Standards, 1961.

[12] *American National Standard Method for Calibration of Microphones*, S1.10–1966. New York: ANSI, 1966.

PART THREE
APPENDICES

APPENDIX A

Standards and Conversion Equations

Standards

Gravitation acceleration
9.80665 m/s^2
32.174 ft/s^2

Standard atmospheric pressure
101325 Pa
14.696 psia

Dimensional constants
$g_c = 1 \text{ kg} \cdot \text{m/N} \cdot \text{s}^2$
$g_c = 32.174 \text{ lbm} \cdot \text{ft/lbf} \cdot \text{s}^2$

Conversion Equations

Throughout the listing that follows, the computer printout convention for decimal place is used; for example, $\text{E} + 03 = 10^3 = 1000$. Likewise, $\text{E} - 02 = 10^{-2} = 0.01$.

The relationships given are in the form of conversion factors which are used to obtain the pertinent SI unit (see Table A.1). Consider, for example, the conversion from inches to meters, which is given in the listing as follows:

$$\text{m} = 2.540 \, \text{E} - 02 \, \times \, \text{in.}$$

If we wish to find the number of meters corresponding to 36.00 in. (1 English yard), we would multiply the number of inches by the conversion factor:

$$\text{length in m} = \frac{2.540 \times 10^{-2} \text{ m}}{\text{in.}} \times 36.00 \text{ in.} = 0.9144 \text{ m}$$

that is, 36.00 in. or 1 English yard, is equal to 0.9144 m.

On the other hand, if we wish to find the number of inches that are the equivalent of 5.00 m, we would divide by the conversion factor:

$$\text{length in inches} = 5.00 \text{ m} \times \frac{\text{in.}}{2.540 \times 10^{-2} \text{ m}} = 197 \text{ in.}$$

That is, 5.00 m and 197 in. represent approximately the identical length.

TABLE A.1: Conversion Factors

	SI			English or Other
Acceleration	m/s^2	=	$3.048\,000$ E $-$ 01 $\times$	ft/s^2
	m/s^2	=	$2.540\,000$ E $-$ 02 $\times$	$in./s^2$
Area	m^2	=	$6.451\,600$ E $-$ 04 $\times$	$in.^2$
	m^2	=	$9.290\,304$ E $-$ 02 $\times$	ft^2
Density (mass/vol)	kg/m^3	=	$1.601\,846$ E $+$ 01 $\times$	lbm/ft^3
	kg/m^3	=	$2.767\,990$ E $+$ 04 $\times$	$lbm/in.^3$
Energy (work)	J	=	$1.055\,056$ E $+$ 03 $\times$	Btu*
	J	=	$1.355\,818$ E $+$ 00 $\times$	ft $\cdot$ lbf
	J	=	$4.186\,800$ E $+$ 00 $\times$	calorie*
	J	=	$3.600\,000$ E $+$ 03 $\times$	W $\cdot$ h
Flow rate (vol/time)	m^3/s	=	$4.719\,474$ E $-$ 04 $\times$	ft^3/min (cfm)
	m^3/s	=	$6.309\,020$ E $-$ 05 $\times$	U.S. liq gal/min (gpm)
Force	N	=	$1.000\,000$ E $-$ 05 $\times$	dyne
	N	=	$4.448\,222$ E $+$ 00 $\times$	lbf
Length	m	=	$1.000\,000$ E $-$ 10 $\times$	angstrom
	m	=	$1.000\,000$ E $-$ 06 $\times$	micrometer (μm)
	m	=	$2.540\,000$ E $-$ 08 $\times$	microinch (μin.)
	m	=	$2.540\,000$ E $-$ 02 $\times$	in.
	m	=	$3.048\,000$ E $-$ 01 $\times$	ft
	km	=	$1.609\,344$ E $+$ 00 $\times$	statute mile
Mass	kg	=	$4.535\,924$ E $-$ 01 $\times$	lbm
	kg	=	$1.459\,390$ E $+$ 01 $\times$	slug
Moment (torque)	N $\cdot$ m	=	$1.129\,848$ E $-$ 01 $\times$	lbf $\cdot$ in.
	N $\cdot$ m	=	$1.355\,818$ E $+$ 00 $\times$	lbf $\cdot$ ft
Power	W	=	$2.930\,711$ E $-$ 01 $\times$	Btu/h
	W	=	$2.259\,697$ E $-$ 02 $\times$	ft $\cdot$ lbf/ min
	W	=	$7.456\,999$ E $+$ 02 $\times$	hp (550 ft $\cdot$ lbf/s)
Pressure (stress)	Pa	=	$3.386\,38$ E $+$ 03 $\times$	inHg (0°C)
	Pa	=	$2.490\,82$ E $+$ 02 $\times$	inH_2O(4°C)
	Pa	=	$4.788\,026$ E $+$ 01 $\times$	lbf/ft^2(psf)
	Pa	=	$9.806\,38$ E $+$ 00 $\times$	mmH_2O (4°C)
	Pa	=	$1.333\,22$ E $+$ 02 $\times$	mmHg (0°C)
	Pa	=	$6.894\,757$ E $+$ 03 $\times$	$lbf/in.^2$(psi)
	Pa	=	$1.000\,000$ E $+$ 05 $\times$	bar
	Pa	=	$1.013\,250$ E $+$ 05 $\times$	atmosphere

* These are the International Table values of the Btu and the calorie. Other definitions are generally within 0.1%.

TABLE A.1: *continued*

	SI			English or Other
Stress (see Pressure)				
Temperature	K	=	°C + 273.15	
	K	=	(°F + 459.67)/1.8	
	K	=	°R/1.8	
	°C	=	(°F − 32)/1.8	
Torque (see Moment)				
Velocity	m/s	=	3.048 000 E − 01	× ft/s
	m/s	=	4.470 400 E − 01	× mph
Viscosity	Pa · s	=	4.788 026 E + 01	× lbf · s/ft^2
	Pa · s	=	1.000 000 E − 03	× centipoise
Volume	m^3	=	3.785 412 E − 03	× U.S. liq gal
	m^3	=	1.638 706 E − 05	× in.^{3}
	m^3	=	2.831 685 E − 02	× ft^3
	m^3	=	1.000 000 E − 03	× liter
Volume/time (see Flow rate)				

Some Additional Conversion Factors

ft · lbf	=	7.782 E + 02	×	Btu
gallons	=	4.329 E − 03	×	in.^{3}
mph	=	6.214 E − 01	×	km/h
knots	=	5.396 E − 01	×	km/h
miles	=	6.214 E − 04	×	m
°R	=	459.67	+	°F

APPENDIX B

Theoretical Basis for Fourier Analysis

The theoretical basis for the harmonic-analysis procedure may be described as follows: Any single-valued function $y(x)$ that is continuous (except for a finite number of finite discontinuities) in the interval $-\pi$ to π, and which has only a finite number of maxima and minima in that interval, can be represented by a series in the form

$$
\begin{aligned}
y(x) = {} & \frac{A_0}{2} + A_1 \cos x + A_2 \cos 2x + \cdots + A_n \cos nx + \cdots \\
& + B_1 \sin x + B_2 \sin 2x + \cdots + B_n \sin nx + \cdots
\end{aligned}
\tag{B.1}
$$

If each term in Eq. (B.1) is multiplied by dx and integrated over any interval of 2π length, all sine and cosine terms will drop out, leaving

$$
\int_a^{2\pi+a} y(x)\, dx = \int_a^{2\pi+a} \frac{A_0}{2}\, dx = A_0 \pi
$$

or

$$
A_0 = \frac{1}{\pi} \int_a^{2\pi+a} y(x)\, dx
\tag{B.2}
$$

The factor A_n may be determined if we multiply both sides of Eq. (B.1) by $\cos mx\, dx$ and integrate each term over the interval of 2π. In general, there are the following terms:

$$
\int_a^{2\pi+a} \sin nx \cos mx\, dx = 0
$$

and

$$
\int_a^{2\pi+a} \cos nx \cos mx\, dx = 0, \qquad \text{except for } m = n
$$

For the special case $m = n$,

$$
\int_a^{2\pi+a} \cos^2 nx\, dx = \frac{1}{2n} [nx + \sin nx \cos nx]_{-\pi}^{\pi}
$$
$$
= \pi
$$

hence

$$
\int_a^{2\pi+a} y(x) \cos nx\, dx = A_n \pi
$$

or

$$A_n = \frac{1}{\pi} \int_a^{2\pi+a} y(x)\cos nx \, dx \qquad (B.3)$$

[*Note*: For $n = 0$, Eq. (B.3) reduces to Eq. (B.2).]

In like manner, if we multiply both sides of Eq. (B.1) by $\sin mx \, dx$ and integrate term by term over the interval 2π, we may obtain

$$B_n = \frac{1}{\pi} \int_a^{2\pi+a} y(x)\sin nx \, dx \qquad (B.4)$$

CALCULATION OF FOURIER COEFFICIENTS FOR SPECIAL PERIODIC WAVEFORMS

Square Wave

Considering the square wave shown in Fig. 4.9(a), we have for one full period

$$y(x) = A \qquad (0 \le x \le \pi)$$
$$y(x) = -A \quad (\pi \le x \le 2\pi) \qquad (B.5)$$

where $x = \omega t$. Applying Eq. (B.2) to Eq. (B.5), where we choose $a = 0$ for convenience, we have

$$A_0 = \frac{1}{\pi} \int_0^\pi A \, dx - \frac{1}{\pi} \int_\pi^{2\pi} A \, dx,$$
$$A_0 = 0 \qquad (B.6)$$

From Eq. (B.3) we have

$$A_n = \frac{1}{\pi} \int_0^\pi A \cos nx \, dx - \frac{1}{\pi} \int_\pi^{2\pi} A \cos nx \, dx$$

or

$$A_n = 0 \qquad (B.7)$$

Similarly, from Eq. (B.4),

$$B_n = \frac{1}{\pi} \int_0^\pi A \sin nx \, dx - \frac{1}{\pi} \int_\pi^{2\pi} A \sin nx \, dx$$

or

$$B_n = \frac{2A}{n\pi} [1 - \cos n\pi]$$

and, finally,

$$B_n = \begin{cases} \dfrac{4A}{n\pi} & \text{for } n \text{ odd,} \\ 0 & \text{for } n \text{ even} \end{cases} \qquad (B.8)$$

Substituting the results from Eqs. (B.6), (B.7), and (B.8) into Eq. (B.1), we obtain

$$y(x) = y(\omega t) = \frac{4A}{\pi} \sum_{n=1}^{\infty} \frac{\sin(2n-1)\omega t}{(2n-1)} \qquad (B.9)$$

Sawtooth Wave

For the sawtooth wave shown in Fig. 4.9(d), we have

$$
y(x) = \begin{cases} \dfrac{A}{\pi}x & (0 \leq x \leq \pi), \\[2mm] 2A - \dfrac{A}{\pi}x & (\pi \leq x \leq 2\pi) \end{cases} \tag{B.10}
$$

Applying Eqs. (B.2) and (B.3),

$$
A_0 = \frac{1}{\pi} \int_0^\pi \frac{Ax}{\pi}\, dx + \frac{1}{\pi} \int_\pi^{2\pi} \left(2A - \frac{A}{\pi}x\right) dx = A \tag{B.11}
$$

and

$$
A_n = \frac{1}{\pi} \int_0^\pi \frac{Ax}{\pi} \cos x\, dx + \frac{1}{\pi} \int_\pi^{2\pi} \left(2A - \frac{A}{\pi}x\right) \cos nx\, dx
$$

or

$$
A_n = \frac{2A}{n^2\pi^2} [\cos n\pi - 1]
$$

Considering various integer values for n, we obtain

$$
A_n = \begin{cases} 0 & \text{for } n \text{ even,} \\[2mm] -\dfrac{4A}{n^2\pi^2} & \text{for } n \text{ odd} \end{cases} \tag{B.12}
$$

Finally, from Eq. (B.4),

$$
B_n = \frac{1}{\pi} \int_0^\pi \frac{Ax}{\pi} \sin nx\, dx + \frac{1}{\pi} \int_\pi^{2\pi} \left(2A - \frac{Ax}{\pi}\right) \sin nx\, dx
$$

and therefore

$$
B_n = 0 \tag{B.13}
$$

Using the results of Eqs. (B.11), (B.12), and (B.13) in Eq. (B.1), the Fourier series for the sawtooth wave becomes

$$
y(x) = y(\omega t) = \frac{A}{2} - \frac{4A}{\pi^2} \sum_{n=1}^{\infty} \frac{\cos(2n-1)\omega t}{(2n-1)^2} \tag{B.14}
$$

Fourier series for some other waveforms are given in Table 4.1.

DISCRETE FOURIER ANALYSIS

When an analog waveform is sampled at N points separated by time increments of Δt, the apparent waveform has a period $T = N \Delta t$ and a fundamental frequency $\Delta f \equiv 1/T$, and it is defined at the sampling times $t_r = r \Delta t$, for $r = 1, \ldots, N$. In a manner analogous to the continuous Fourier series, a discretely sampled waveform can be represented as a finite sum of frequency components having frequencies $0, \Delta f, 2\,\Delta f, \ldots, (N/2)\Delta f$:

$$
y(t_r) = \frac{A_0}{2} + \sum_{n=1}^{N/2-1} [A_n \cos(2\pi n\,\Delta f\, t_r) + B_n \sin(2\pi n\,\Delta f\, t_r)]
$$
$$
+ \frac{A_{N/2}}{2} \cos\left(2\pi \frac{N}{2}\Delta f\, t_r\right) \tag{B.15}
$$

for N an even number. Only a finite number of frequencies are present in this sum, because frequencies greater than $(N/2)\,\Delta f$ (the *Nyquist frequency*; Section 4.7.2) vary too rapidly to be resolved with the points taken.

The coefficients of the discrete Fourier series are found in essentially the same way as those of the continuous Fourier series, except that we must sum rather than integrate. Hence, we multiply Eq. (B.15) by the mth frequency component and sum over time:

$$
\sum_{r=1}^{N} y(t_r)\cos(2\pi m\,\Delta f\, t_r) = \frac{A_0}{2} \sum_{r=1}^{N} \cos(2\pi m\,\Delta f\, t_r)
$$
$$
+ \sum_{n=1}^{N/2-1}\sum_{r=1}^{N} \cos(2\pi m\,\Delta f\, t_r)
$$
$$
\times [A_n \cos(2\pi n\,\Delta f\, t_r) + B_n \sin(2\pi n\,\Delta f t_r)]
$$
$$
+ \frac{A_{N/2}}{2} \sum_{r=1}^{N} \cos(2\pi m\,\Delta f\, t_r) \cos\left(2\pi \frac{N}{2}\Delta f\, t_r\right) \tag{B.16}
$$

It turns out that

$$
\sum_{r=1}^{N} \cos(2\pi m\,\Delta f\, t_r) \cos(2\pi n\,\Delta f\, t_r) = \begin{cases} 0 & \text{for } m \neq n \\ \dfrac{N}{2} & \text{for } m = n \\ N & \text{for } m = n = 0 \text{ or } m = n = \dfrac{N}{2} \end{cases}
$$

Likewise,

$$
\sum_{r=1}^{N} \cos(2\pi m\,\Delta f\, t_r) \sin(2\pi n\,\Delta f\, t_r) \equiv 0
$$

Thus, Eq. (B.16) reduces to the statement that

$$
\sum_{r=1}^{N} y(t_r) \cos(2\pi m\,\Delta f\, t_r) = \frac{N}{2} A_m \qquad m = 0, 1, \ldots, \frac{N}{2}
$$

A similar calculation fixes B_n. From this, the harmonic coefficients of the discrete Fourier series are

$$A_n = \frac{2}{N} \sum_{r=1}^{N} y(r \, \Delta t) \cos\left(\frac{2\pi r n}{N}\right) \qquad n = 0, 1, \ldots, \frac{N}{2} \qquad \text{(B.17)}$$

$$B_n = \frac{2}{N} \sum_{r=1}^{N} y(r \, \Delta t) \sin\left(\frac{2\pi r n}{N}\right) \qquad n = 1, 2, \ldots, \frac{N}{2} - 1 \qquad \text{(B.18)}$$

APPENDIX C

Number Systems

In general, a number N_b written in the base b system has a magnitude given by

$$N_b = \sum a_n b^n = \cdots + a_2 b^2 + a_1 b^1 + a_0 b^0 + a_{-1} b^{-1} + a_{-2} b^{-2} \cdots \qquad \text{(C.1)}$$

where

b = base or radix of the particular number system,
   = number of distinct character types used to express a quantity,
N_b = a number in the system,
a_n = an integer coefficient that may range from 0 to $b - 1$,
n = positional value or power of b for each coefficient

The meaning of this equation is best illustrated by example, several of which follow.

The Decimal System

The decimal system is based on ten digits, 0 through 9: the base or radix is 10. Hence

$$N_{10} = \sum a_n 10^n \qquad \text{(C.2)}$$

For example,

$$524.3_{10} = (5)(10)^2 + (2)(10)^1 + (4)(10)^0 + (3)(10)^{-1}$$

The digits 5, 2, 4, and 3 are the coefficients corresponding to the respective *positions* or powers 2, 1, 0, and -1. In the decimal system the coefficients may have integral values ranging from 0 through 9, inclusive.

The Binary System

The binary system employs two digits only, namely, 0 and 1. The decimal value corresponding to a binary number N_2 is given by the sum

$$N_2 = \sum a_n 2^n \qquad \text{(C.3)}$$

For example, the representation 1101.1_2 is interpreted in the decimal system as

$$(1)(2)^3 + (1)(2)^2 + (0)(2)^1 + (1)(2)^0 + (1)(2)^{-1}$$

$$= 8 + 4 + 0 + 1 + \frac{1}{2} = 13.5_{10}$$

This example demonstrates the procedure for *converting* a binary number to the equivalent decimal number. The two numbers 1101.1_2 and 13.5_{10} each represent an identical quantity.

The Octal System

This system is based on eight digits, 0 through 7. The decimal value corresponding to an octal number N_8 is given by

$$N_8 = \sum a_n 8^n \tag{C.4}$$

The coefficients a_n do not include the decimal digits 8 and 9.

For example, 375.3_8 is expressed in the decimal system as

$$375.3_8 = (3)(8)^2 + (7)(8)^1 + (5)(8)^0 + (3)(8)^{-1}$$
$$= 192_{10} + 56_{10} + 5_{10} + 0.375_{10} = 253.375_{10}$$

The Hexadecimal System

This system employs 16 as the base or radix. Additional digital symbols are required to represent values above 9, and letters have been chosen for this purpose. Thus, the digits of the hexadecimal system are 0, 1, 2, 3, 4, 5, 6, 7, 8, 9, A, B, C, D, E, and F. These 16 characters correspond to the base-10 numbers 0 through 15, respectively. The decimal value of a hexadecimal number N_{16} is again obtained by summation:

$$N_{16} = \sum a_n (16)^n \tag{C.5}$$

For example,

$$D8B.2_{16} = (D)_{16}(16)_{10}^2 + (8)_{16}(16)_{10}^1 + (B)_{16}(16)_{10}^0 + (2)_{16}(16)_{10}^{-1}$$
$$= (13)_{10}(16)_{10}^2 + (8)_{10}(16)_{10}^1 + (11)_{10}(16)_{10}^0 + (2)_{10}(16)_{10}^{-1}$$
$$= 3467.125_{10}$$

Octal and Hexadecimal Formatted Binary

A binary number may be arranged (grouped) in ways that make it easy to convert it to either octal or hexadecimal equivalents.

Consider 110111001100_2, which we will rewrite as

$$110\ 111\ 001\ 100_2$$

If we consider each subgroup as a single octal digit, we obtain

$$6\ 7\ 1\ 4 \quad \text{or} \quad 6714_8$$

This works because the binary numbers 000_2 through 111_2 are equivalent to the octal numbers 0_8 through 7_8, which span the radix of the octal system.

In like manner, we can rearrange the same binary number into subgroups of four, as follows:

$$1101\ 1100\ 1100_2$$

We see that this is equal to DCC_{16}. This approach works because the binary numbers 0000_2 through 1111_2 are equivalent to the hexadecimal numbers 0_{16} through F_{16}.

We can confirm the legitimacy of all of this by converting each number, the binary, the octal, and the hexadecimal to the equivalent decimal number, which we find is equal to 3532_{10}.

Why Need We Be Concerned with the Various Number Systems?

1. The decimal system is common in everyday usage, but it is not a convenient system around which to build or use a computer. A computer would have to distinguish among 10 different states, as opposed to only two when binary is used. As discussed in Chapter 8, digital circuitry generally relies on on/off or high/low voltage states that lend themselves to a two-state counting system.
2. Although binary requires a much longer string of symbols to define a given magnitude, the advantage of the simple Yes/No operation more than offsets the use of longer numbers. Machine language employs binary arithmetic.
3. *Why octal or hexadecimal*? The fundamental computer language is binary, but for a human that system would be extremely awkward and error prone due to the long strings of symbols representing each number. Hexadecimal and/or octal can be thought of as simply a crutch used by humans to communicate with the computer in the computer's own tongue—machine or assembly language.

CONVERTING A BASE 10 NUMBER TO ONE OF A DIFFERENT RADIX

In the preceding paragraphs, we established procedures for converting numbers of various bases to equivalent magnitudes of base 10. We shall now demonstrate a method[1] for performing the reverse: converting a base 10 number to equivalent binary, octal, and hexadecimal numbers. We will use Tables C.1, C.2, and C.3.

TABLE C.1: Decimal Values of 2^n

n		2^n
	etc.	
−1	↑	0.5
0		1
1		2
2		4
3		8
4		16
5		32
6		64
7		128
8		256
9		512
10		1024
11		2048
12	↓	4096
	etc.	

[1]There are other methods in addition to the one demonstrated.

TABLE C.2: Decimal Values of $(a)(8)^n$

Coefficients	Powers, n						
a	6	5	4	3	2	1	0
1	262 144	32 768	4 096	512	64	8	1
2	524 288	65 536	8 192	1 024	128	16	2
3	786 432	98 304	12 288	1 536	192	24	3
4	1 048 576	131 072	16 384	2 048	256	32	4
5	1 310 720	163 840	20 480	2 560	320	40	5
6	1 572 864	196 608	24 576	3 072	384	48	6
7	1 835 008	229 376	28 672	3 584	448	56	7

TABLE C.3: Decimal Values of $(a)(16)^n$

Coefficients	Powers, n						
a	6	5	4	3	2	1	0
1	16 777 216	1 048 576	65 536	4 096	256	16	1
2	33 554 432	2 097 152	131 072	8 192	512	32	2
3	50 331 648	3 145 728	196 608	12 288	768	48	3
4	67 108 864	4 194 304	262 144	16 384	1 024	64	4
5	83 886 080	5 242 880	327 680	20 480	1 280	80	5
6	100 663 296	6 291 456	393 216	24 576	1 536	96	6
7	117 440 512	7 340 032	458 752	28 672	1 792	112	7
8	134 217 728	8 388 608	524 288	32 768	2 048	128	8
9	150 994 944	9 437 184	589 824	36 864	2 304	144	9
A	167 772 160	10 485 760	655 360	40 960	2 560	160	10
B	184 549 376	11 534 336	720 896	45 056	2 816	176	11
C	201 326 592	12 582 912	786 432	49 152	3 072	192	12
D	218 103 808	13 631 488	851 968	53 248	3 328	208	13
E	234 881 024	14 680 064	917 504	57 344	3 584	224	14
F	251 658 240	15 728 640	983 040	61 440	3 840	240	15

Decimal to Binary

Let's convert the number 713_{10} to the equivalent binary number. We will do this by successively subtracting the largest values of 2^n that each remainder will permit. The procedure is demonstrated as follows (refer to Table C.1):

$$
\begin{array}{rl}
713_{10} & \\
\underline{-512} & = 2^9 \\
201_{10} & \\
\underline{-128} & = 2^7 \\
73_{10} & \\
\underline{-64} & = 2^6 \\
9_{10} & \\
\underline{-8} & = 2^3 \\
1_{10} & \\
\underline{-1} & = 2^0 \\
0 &
\end{array}
$$

Recalling that the powers of the radix correspond to the positional orders, we may write:

$$1011001001_2 = 713_{10}$$

Decimal to Octal

Let us now employ Table C.2 to convert the number 713_{10} to the equivalent octal number. A procedure similar to the one used for the previous example may be used. In this case, however, we select the "largest" components in terms of both powers of the radix, 8, and also the required coefficients.

$$
\begin{array}{rl}
713_{10} & \\
\underline{-512} & = 1 \times 8^3 \\
201_{10} & \\
\underline{-192} & = 3 \times 8^2 \\
9_{10} & \\
\underline{-8} & = 1 \times 8^1 \\
1_{10} & \\
\underline{-1} & = 1 \times 8^0 \\
0 &
\end{array}
$$

From this we determine $713_{10} = 1311_8$.

Decimal to Hexadecimal

In a similar manner, convert 713_{10} to the equivalent hexadecimal number using Table C.3.

$$
\begin{array}{rl}
713_{10} & \\
\underline{-512} & = 2 \times 16^2 \\
201_{10} & \\
\underline{-192} & = C \times 16^1 \\
9_{10} & \\
\underline{-9} & = 9 \times 16^0 \\
0 &
\end{array}
$$

From this calculation we may write $713_{10} = 2C9_{16}$.

APPENDIX D

Some Useful Data

TABLE D.1: Properties of Water: SI System

Temperature degrees C (F)	Absolute Viscosity Pa · s	Density kg/m^3
5 (41)	15.19×10^{-4}	1000
10 (50)	13.08	999.7
15 (59)	11.40	999.1
20 (68)	10.05	998.2
25 (77)	8.937	997.0
30 (86)	8.007	995.7
35 (95)	7.225	994.1
40 (104)	6.560	992.2
45 (113)	5.988	990.3
50 (122)	5.494	988.1

TABLE D.2: Properties of Water: English System

Temperature degrees F (C)	Absolute Viscosity lbf · s/ft^2	Density lbm/ft^3
40 (4.44)	3.23×10^{-5}	62.42
50 (10.0)	2.72	62.41
60 (15.56)	2.33	62.37
70 (21.11)	2.02	62.30
80 (26.67)	1.77	62.22
90 (32.22)	1.58	62.11
100 (37.78)	1.43	61.99
110 (43.33)	1.30	61.86
120 (48.89)	1.15	61.71

TABLE D.3: Properties of Dry Air at Atmospheric Pressure: SI System

Temperature degrees C (F)	Absolute Viscosity* Pa · s	Density† kg/m³
0 (32)	1.68×10^{-5}	1.26
10 (50)	1.73	1.22
20 (68)	1.80	1.18
30 (86)	1.85	1.14
40 (104)	1.91	1.10
50 (122)	1.97	1.07
60 (140)	2.03	1.04
70 (158)	2.09	1.00
80 (176)	2.15	0.97
90 (194)	2.22	0.94
100 (212)	2.28	0.924

*Over the range from atmospheric pressure to about 7000 kPa ($\approx$ 1000 psia) the viscosity of dry air increases at a rate of approximately 1% for each 700 kPa (100 psi) increase in pressure.

†For pressures other than atmospheric, use $\rho/\rho_{atmos} = P/P_{atmos}$.

TABLE D.4: Properties of Dry Air at Atmospheric Pressure: English System

Temperature degrees F (C)	Absolute Viscosity* lbf · s/ft²	Density† lbm/ft³
40 (4.44)	0.362×10^{-6}	0.0794
50 (10.0)	0.368	0.0779
60 (15.6)	0.374	0.0764
70 (21.1)	0.379	0.0749
80 (26.7)	0.385	0.0735
90 (32.2)	0.390	0.0722
100 (37.8)	0.396	0.0709
110 (43.3)	0.401	0.0697
120 (48.9)	0.407	0.0685

*Over the range from atmospheric pressure to about 7000 kPa ($\approx$ 1000 psia) the viscosity of dry air increases at a rate of approximately 1% for each 700 kPa (100 psi) increase in pressure.

†For pressures other than atmospheric, use $\rho/\rho_{atmos} = P/P_{atmos}$.

TABLE D.5: Some Values of Gravitational Acceleration

	m/s^2	ft/s^2
Standard	9.806 65	32.174
Location		
Ft. Egbert, Alaska	9.821 83	32.224
Key West, Florida	9.789 70	32.118
Batavia, Java	9.781 78	32.092
Karajak Glacier, Greenland	9.825 34	32.235
Pittsburgh, Pennsylvania	9.801 05	32.156
Latitude 40°26'40"		
Longitude 79°57'13"W		
Elevation 908.35 ft		
Moon	1.67	5.48
Planet Mercury	3.92	12.86
Planet Jupiter	26.46	87.07

TABLE D.6: Specific Gravities* of Selected Materials

Material	Specific Gravity
Mercury	13.596 @ 0°C
	13.546 @ 20°C
	13.690 @ −38.8°C (liquid at freezing point)
	14.193 @ −38.8°C (solid at freezing point)
Gasoline	0.66 to 0.69
Kerosene	0.82
Seawater	1.025
Oil (Meriam Red, a common manometer oil)	0.823
Carbon tetrachloride	1.60
Tetrabromo-ethane	2.96
Ethyl alcohol/water mixture at 20°C	
% alcohol by weight	
0	1.00
20	0.97
40	0.94
60	0.89
80	0.84
100	0.79

*Specific gravity is the ratio of the mass of a body to that of an equal volume of water at 4°C or at some other specified temperature.

TABLE D.7: Some Additional Material Properties

Temperature (°F)	Medium Heating Oil		Heavy Heating Oil	
	Specific Gravity	Viscosity (lbf · s/ft^2)	Specific Gravity	Viscosity (lbf · s/ft^2)
40	0.865	10.82×10^{-5}	0.918	789×10^{-5}
60	0.858	7.85	0.912	390
80	0.851	6.03	0.905	200
100	0.843	4.59	0.899	109

TABLE D.8: Kerosene Viscosity

Temperature, °C	Viscosity, Pa · s
0	28.7×10^{-4}
20	19.2
40	13.4
60	9.6
80	7.7
100	6.7

TABLE D.9: Young's Modulus

For steel	$E \approx 29.5 \times 10^6$ lbf/in.^{2}	$\approx 20.3 \times 10^{10}$ Pa
For aluminum	$E \approx 10 \times 10^6$ lbf/in.^{2}	$\approx 6.9 \times 10^{10}$ Pa

A P P E N D I X E

Stress and Strain Relationships

E.1   THE GENERAL PLANE STRESS SITUATION

Suppose an element dx wide by dy high is selected from a general plane stress situation in equilibrium, as shown in Fig. E.1. Assume that the element is of uniform thickness, t, normal to the paper.

From strength of materials it will be remembered that, for equilibrium, τ_{xy} must be equal to τ_{yx}. We will therefore employ the symbol τ_{xy} for both. We will also assume that not only must the complete element be in equilibrium, but so also must be all its parts. Therefore, if the element is bisected by a diagonal ds long, each half of the element must also be in equilibrium.

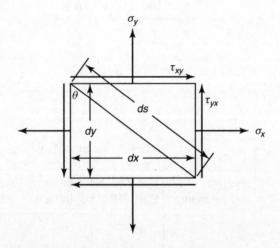

FIGURE E.1: Element subject to plane stresses.

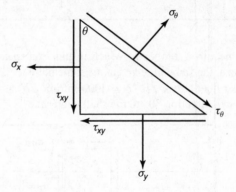

FIGURE E.2: Element used to define positive stress directions.

As shown in Fig. E.2, there must be a normal stress σ_θ and a shear stress τ_θ acting on the diagonal area. If the various stresses shown on the partial element, Fig. E.2, are multiplied by the areas over which they act, forces are obtained as shown in Fig. E.3. Let all the directions be considered positive, as shown. We note that $dy/ds = \cos\theta$ and $dx/ds = \sin\theta$. Summing forces normal to the diagonal plane and solving for σ_θ we obtain

$$\sigma_\theta = \sigma_x \cos^2\theta + \sigma_y \sin^2\theta + 2\tau_{xy} \sin\theta \cos\theta$$

A more convenient form of this equation may be had in terms of double angles. By substituting trigonometric equivalents,

$$\sigma_\theta = \frac{1}{2}(\sigma_x + \sigma_y) + \frac{1}{2}(\sigma_x - \sigma_y)\cos 2\theta + \tau_{xy}\sin 2\theta \qquad (E.1)$$

Using this equation, the stress on any plane may be determined if values of σ_x, σ_y, and τ_{xy} are known.

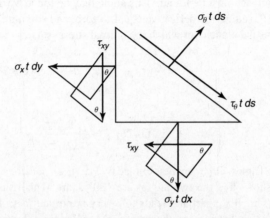

FIGURE E.3: Element illustrating the requirement for force equilibrium.

EXAMPLE E.1

A shaft is subject to a torque, T, which results in a shear stress, $Tc/J = 9500$ lbf/in.^{2} (Fig. E.4), and at the same time and at the same point a bending moment due to gear loads causes an outer fiber stress, $Mc/I = 4000$ lbf/in.^{2}. What will be the normal stress on the outer surface in a direction 30° to the shaft centerline?

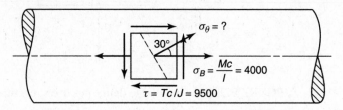

FIGURE E.4: Stresses acting on an element on the outer surface of a shaft subject to torsion and bending.

Solution

$$\sigma_{30°} = \frac{1}{2}(\sigma_x + \sigma_y) + \frac{1}{2}(\sigma_x - \sigma_y)\cos 2\theta + \tau_{xy}\sin 2\theta$$

$$= \frac{1}{2}(4000 + 0) + \frac{1}{2}(4000 - 0)\cos 60° + 9500\sin 60°$$

$$= 11,240 \text{ lbf/in.}^2$$

Of course, the normal stress, $11,240$ lbf/in.^{2}, is not necessarily the maximum normal stress, because the angle 30° was chosen at random; undoubtedly some other angle may result in a larger normal stress.

E.2   DIRECTION AND MAGNITUDES OF PRINCIPAL STRESSES

To calculate the maximum normal stress, the particular angle θ_1 determining the plane over which it will act must be found. This angle may be found by differentiating Eq. (E.1) with respect to θ, setting the derivative equal to zero, and solving for the angle θ_1. This should also give us the plane over which the normal stress is a minimum.

$$\frac{d\sigma_\theta}{d\theta} = -(\sigma_x - \sigma_y)\sin 2\theta + 2\tau_{xy}\cos 2\theta = 0$$

or

$$\tan 2\theta_{1,2} = \frac{\pm 2\tau_{xy}}{\pm(\sigma_x - \sigma_y)} \tag{E.2}$$

Two angles, $2\theta_{1,2}$, are determined by Eq. (E.2), and consideration of the trigonometry involved shows that the two angles are 180° apart. This result means, then, that the two angles $\theta_{1,2}$ are 90° apart and leads to a very important fact: *The planes of maximum and minimum normal stress are always at right angles to each other.*

The maximum and minimum normal stresses are called the *principal stresses*, and the planes over which they act are called the *principal planes*. We have just found, therefore, that the principal planes are at right angles to each other. If we know the direction of the maximum normal stress, we automatically know the direction of the minimum normal stress.

Now we would like to find an expression for the principal stresses. From Eq. (E.2) we may write

$$\sin 2\theta_{1,2} = \frac{2\tau_{xy}}{\sqrt{(\sigma_x - \sigma_y)^2 + (2\tau_{xy})^2}},$$

$$\cos 2\theta_{1,2} = \frac{(\sigma_x - \sigma_y)}{\sqrt{(\sigma_x - \sigma_y)^2 + (2\tau_{xy})^2}} \tag{E.3}$$

Substituting these values in Eq. (E.1) gives us the principal stresses, which we shall designate σ_1 and σ_2, and

$$\sigma_{\theta \text{ max}} = \sigma_1 = \frac{1}{2}(\sigma_x + \sigma_y) + \frac{1}{2}\sqrt{(\sigma_x - \sigma_y)^2 + (2\tau_{xy})^2},$$

$$\sigma_{\theta \text{ min}} = \sigma_2 = \frac{1}{2}(\sigma_x + \sigma_y) - \frac{1}{2}\sqrt{(\sigma_x - \sigma_y)^2 + (2\tau_{xy})^2} \tag{E.4}$$

EXAMPLE E.2

Referring to Example E.1, determine the magnitudes of the principal stresses and the positions of the principal planes relative to the shaft centerline.

Solution   Using Eq. (E.4),

$$\sigma_1 = \frac{1}{2}(4000 + 0) + \frac{1}{2}\sqrt{(4000 + 0)^2 + (2 \times 9500)^2}$$

$$= 11{,}708 \text{ lbf/in.}^2$$

and

$$\sigma_2 = \frac{1}{2}(4000 + 0) - \frac{1}{2}\sqrt{(4000 + 0)^2 + (2 \times 9500)^2}$$

$$= -7708 \text{ lbf/in.}^2$$

From Eq. (E.2),

$$\tan 2\theta = \frac{2 \times 9500}{4000 - 0} = \frac{19{,}000}{4000} = 4.75,$$

$$2\theta = 78.1°,$$

$$\theta = 39.05°$$

The orientation is as shown in Fig. E.5.

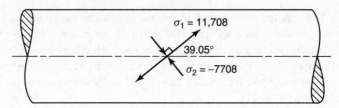

FIGURE E.5: The principal stresses corresponding to the situation shown in Fig. E.4.

E.3 VARIATION IN SHEAR STRESS WITH DIRECTION

Following the same procedure used for normal stresses, and again referring to Fig. E.3, if the forces parallel to the diagonal plane are summed, the following shear relations are obtained. The equation for the shear stress on any plane in terms of σ_x, σ_y, and θ is

$$\tau_\theta = \frac{1}{2}(\sigma_x - \sigma_y)\sin 2\theta - \tau_{xy}\cos 2\theta \qquad (E.5)$$

The angle determining the planes over which the shear stresses are maximum and minimum may be determined by the relation

$$\tan 2\theta_s = \frac{\mp(\sigma_x - \sigma_y)}{\pm 2\tau_{xy}} \qquad (E.6)$$

By substituting the angles determined by Eq. (E.6) in Eq. (E.5), relations for maximum and minimum shear stress may be obtained.

$$\text{Maximum shear stress} = \tau_{\theta\,\max} = \frac{1}{2}\sqrt{(\sigma_x - \sigma_y)^2 + (2\tau_{xy})^2},$$

$$\text{Minimum shear stress} = \tau_{\theta\,\min} = -\frac{1}{2}\sqrt{(\sigma_x - \sigma_y)^2 + (2\tau_{xy})^2} \qquad (E.7)$$

We must be careful, however, because the shear stress extremes given by Eq. (E.7) account for only the x- and y-directions. Consideration of the three-dimensional condition often shows the greatest shear stress to occur on yet another plane. See Section E.6.

E.4 SHEAR STRESS ON PRINCIPAL PLANES

Equations (E.7) allow us to determine the shear stress on any plane defined by θ. Therefore, let us substitute the expressions for $\theta_{1,2}$, Eq. (E.2), in Eq. (E.5) and thereby determine the shear stresses acting over the principal planes. Doing this gives

$$\tau_{1,2} = \frac{-\frac{1}{2}(\sigma_x - \sigma_y)(2\tau_{xy}) + \tau_{xy}(\sigma_x - \sigma_y)}{\sqrt{(\sigma_x - \sigma_y)^2 + (2\tau_{xy})^2}}$$

$$= 0$$

This proves a very important fact about any plane stress situation: *The shear stresses on the principal planes are zero.* This in itself often provides the necessary clue to determine the orientation of the principal planes by inspection. In any case where it can be said, "there can be no shear on this plane," then the fact we have just established tells us that the plane we are referring to is a principal plane. Or, often just as important, if it can be said that shear stresses *do* exist on a plane, we know the plane *cannot* be one of the principal planes.

In strain-gage applications, knowing the directions of the principal planes at the point of interest provides a very decided advantage. With this information, gages may be aligned in the principal directions, and usually only two gages are required. More important, however, is the fact that the calculations become much simpler and less time consuming (Section 12.15.2).

E.5   GENERAL STRESS EQUATIONS IN TERMS OF PRINCIPAL STRESSES

Checking back on our original assumptions in Section E.1, we see that we assumed a simple element subject to two orthogonal normal stresses, σ_x and σ_y, and shear stresses, τ_{xy}. Using the information since developed—namely, that the principal stresses are at right angles to each other and that the shear stresses are zero on the principal planes—we may now rewrite certain of our equations in terms of principal stresses σ_1 and σ_2.

By selecting just the right element orientation (i.e., aligning it with the principal planes), our basic element could be made to appear as it does in Fig. E.6. The stresses σ_1 and σ_2 are orthogonal, and we know also that there will be no shear on the planes over which they act. Therefore, any of our equations written so far may be modified by substituting σ_1 and σ_2 for σ_x and σ_y, respectively, and making the shear stress equal to zero.

Whereas substitution in Eqs. (E.1) and (E.5) yields particularly useful relations, substitution in most of the others simply confirms our definitions.

Substituting in Eqs. (E.1) and (E.5) gives

$$\sigma_\theta = \frac{1}{2}(\sigma_1 + \sigma_2) + \frac{1}{2}(\sigma_1 - \sigma_2)\cos 2\theta, \tag{E.8}$$

$$\tau_\theta = \frac{1}{2}(\sigma_1 - \sigma_2)\sin 2\theta \tag{E.8a}$$

These equations are particularly useful in helping us visualize the overall stress condition as shown in the following section.

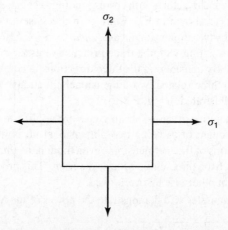

FIGURE E.6: An element subject to principal stresses.

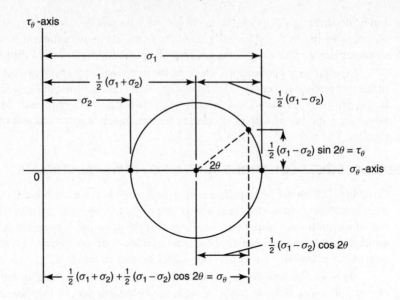

FIGURE E.7: Mohr's circle for plane stresses.

E.6 MOHR'S CIRCLE FOR STRESS

Let us establish a coordinate system with σ_θ plotted as the abscissa and τ_θ as the ordinate (Fig. E.7). The shear stress corresponding to the principal stresses is zero; hence σ_1 and σ_2 will be plotted along the σ_θ-axis.

If a circle is drawn passing through σ_1 and σ_2 points and having its center on the σ_θ-axis, the construction shown in Fig. E.7 will result. It will be noted that for any point on the circle the distance along the abscissa represents σ_θ, and the ordinate distance represents τ_θ. This construction, which is very useful in helping to visualize stress situations, is known as Mohr's stress circle.

At this point we should consider the third or z-direction. In general, three orthogonal stresses σ_x, σ_y, and σ_z, along with corresponding shear stresses τ_{xy}, τ_{yz}, and σ_{zx}, will occur on an element, as shown in Fig. E.8(a). In this case a third principal plane exists, over which, as for the two dimensional case, *shear is zero*. Also, it can be shown[1] that the three principal planes, along with the three principal stresses σ_1, σ_2, and σ_3, are at right angles to one another. By considering the three directions in combinations of two, we may reduce the problem to three related two-dimensional situations. The resulting combined Mohr's diagrams are illustrated in Fig. E.8(b).

In the majority of cases, strain gages are applied to free, unloaded surfaces and the condition is thought of as being two-dimensional. It is wise, however, to consider every condition in terms of three dimensions, even though the third stress may be zero, and to plot or merely sketch the three-circle Mohr's diagram. This procedure often reveals a maximum shear that might otherwise be overlooked.

A few examples will demonstrate the power of the Mohr diagram.

---

[1] For example, in most intermediate solid mechanics books.

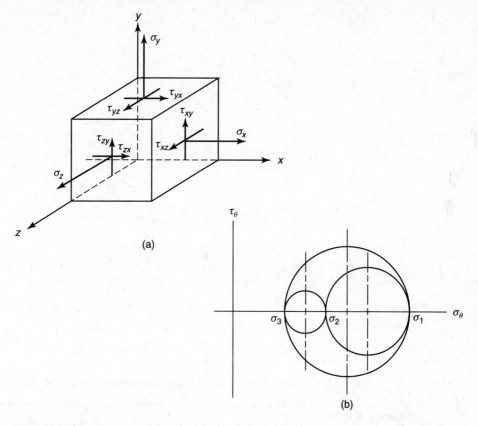

FIGURE E.8: (a) An element subjected to normal and shear stresses on three orthogonal planes; (b) Mohr's circles for stress for the element shown in (a).

EXAMPLE E.3

Figure E.9(a) shows a simple tension member. We know there is no shear stress on a transverse section; hence we know that this must be a principal plane. Since the other principal plane must be normal to the first principal plane and hence be aligned with the axis of the specimen, the normal stress on this plane must be zero. Therefore,

$$\sigma_1 = \frac{F}{A} \quad \text{and} \quad \sigma_2 = \sigma_3 = 0$$

Plotting Mohr's circle for this situation gives us Fig. E.9(b). One of the Mohr circles degenerates to a point in this case.

By inspection we see that the maximum shear stress is equal to $F/2A$ at an angle $2\theta = 90°$ or $\theta = 45°$ measured relative to the axis of the specimen. This confirms our previous knowledge of the stress condition for this simple situation.

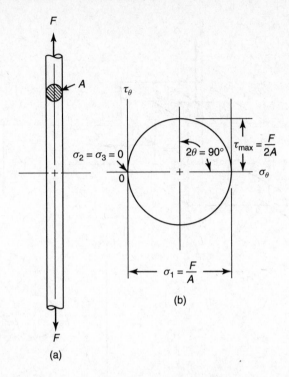

(b)

(a)

FIGURE E.9: Mohr's circle of stresses for a simple tension member.

EXAMPLE E.4

Figure E.10(a) shows a thin-walled cylindrical pressure vessel. From elementary theory, the hoop, longitudinal, and normal stresses may be calculated by the following relations:

$$\sigma_H = \frac{PD}{2t}, \quad \sigma_L = \frac{PD}{4t}, \quad \text{and} \quad \sigma_N = 0$$

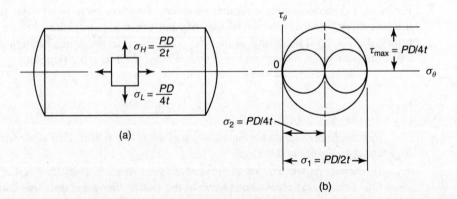

(a)

(b)

FIGURE E.10: Mohr's circle of stresses for the free surface of a cylindrical pressure vessel; P = pressure, D = shell diameter, and t = shell-wall thickness.

Consideration of the nature of the stress field makes it difficult to imagine shear stresses on planes parallel to the hoop, longitudinal, or normal directions. Assuming this to be correct, then the circumferential, the longitudinal, and the normal directions must be the principal directions and σ_H, σ_L, and σ_N must be the principal stresses. Mohr's circle for this situation is shown in Fig. E.10(b). The maximum shear is seen to be $PD/4t$, over a plane inclined 45° to the circumferential and normal directions.

EXAMPLE E.5

Figure E.11(a) shows a shaft in simple torsion. From courses in strength of materials we know that the shear stress on the outer fiber acting on a plane normal to the shaft centerline is equal to Tc/J. The fact that shear stress exists on this plane eliminates it from consideration as a principal plane. Since principal planes must be normal to each other, we immediately think of the one other symmetrical possibility—the two planes inclined 45° to the shaft centerline. Careful consideration of the stresses that may exist on these two planes leads us to conclude that tension would exist on one and compression on the other. The wringing of a wet towel is often used as an example of this situation. In the one 45° direction, the threads of the towel are obviously in tension, while in the other 45° direction, compression is employed to squeeze out the water.

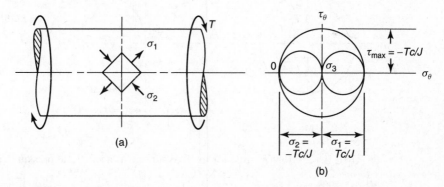

FIGURE E.11: Mohr's circle for a shaft subject to "pure" torsion; T = torque, J = polar moment of inertia of section, and C = distance from neutral axis to fiber of interest.

Because of symmetry, we are led to the conclusion that the magnitudes of the two stresses are equal. Plotting equal tensile and compressive principal stresses, using Mohr's circle construction, gives us Fig. E.11(b).

The third principal direction is normal to the shaft surface and we see that $\sigma_N = 0$. Although the preceding discussion can hardly be considered rigorous proof, Fig. E.11(b) does represent the actual stress situation for a shaft subject to pure torsion. We know that maximum shear stress is equal to Tc/J. Therefore, inspection shows us that the principal stresses σ_1 and σ_2 must also have the sample magnitude, Tc/J, tension and compression, respectively.

EXAMPLE E.6

Figure E.12(a) shows a thin-walled spherical pressure vessel, for which elementary theory shows that the stress in the wall abides by the following relation:

$$\sigma = \frac{PD}{4t}$$

In this case it is difficult to see how direction has significance. At any point on the outside of the shell, the normal stresses must be equal in all directions, simply because of symmetry. We must therefore conclude that

$$\sigma_1 = \sigma_2 = \frac{PD}{4t} \quad \text{and} \quad \sigma_3 = \sigma_N = 0$$

Mohr's diagram for this condition is shown in Fig. E.12(b), from which it is seen that $\tau_{\text{max}} = PD/8t$.

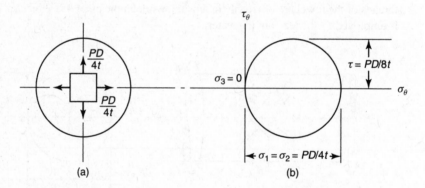

FIGURE E.12: Mohr's circle for the free surface of a spherical pressure vessel.

From the preceding discussion and consideration of Mohr's circle construction, we may now make the following general observations:

1. A stress state involving shear without normal stress is impossible.
2. Maximum shear stress always occurs on planes oriented 45° to the principal stresses and is equal to one-half the algebraic difference of the principal stresses.
3. The shear stresses on any mutually perpendicular planes are of equal magnitude.
4. The sum of the normal stresses on any mutually perpendicular planes is a constant.
5. The maximum ratio of shear stress to principal stress occurs when the principal stresses are of equal magnitude but opposite sign.

E.7   STRAIN AT A POINT

Through use of Hooke's law and the stress relations developed in the preceding pages, the following relations for strain at a point may be derived:

$$\varepsilon_\theta = \frac{1}{2}(\varepsilon_x + \varepsilon_y) + \frac{1}{2}(\varepsilon_x - \varepsilon_y)\cos 2\theta + \frac{\gamma_{xy}}{2}\sin 2\theta, \tag{E.9}$$

$$\frac{\gamma_\theta}{2} = \frac{1}{2}(\varepsilon_x - \varepsilon_y)\sin 2\theta - \frac{\gamma_{xy}}{2}\cos 2\theta \tag{E.9a}$$

Comparison of the above two equations with Eqs. (E.1) and (E.5), respectively, indicates that with a minor exception (the shear strains γ are divided by 2, whereas their counterparts are not), the stress and the strain relations at a point are functionally alike. It follows, therefore, that we can draw a Mohr's diagram for strain, provided the ordinate is made $\gamma_\theta/2$. This is sometimes useful in treating strain-rosette data.

EXAMPLE E.7

Power piping is subject to a combination of loading whose complexity will serve as an interesting example of a combined stress situation. In addition to pressure loading, differential expansion between the hot and cold conditions may superimpose bending, torsional, and axial loading.

Of course, the primary problem involved in piping design is the determination of the loading brought about by pipe expansion, end movements, and movement-limiting stops. In the simple situations good estimates of these loads may be determined analytically, through the use of computer programs.

In this example it will be assumed that such preliminary work has been finished and the critically stressed location found. The remaining problem, then, is to combine the stress components and to determine the net stress condition. The problem is as follows:

Pipe Data (14-in. Schedule 100)

Outside diameter = 14 in.,
Wall thickness = 0.937 in.,
Inside diameter = 12.125 in.,
Bending moment of inertia = 825 in.^{4},
Bending section modulus = 117.9 in.^{3},
Torsional moment of inertia = 1650 in.^{4},
Torsional section modulus = 235.8 in.^{3},
Cross-sectional area = 38.47 in.^{2},
Young's modulus = 23×10^6 lbf/in.^{2},
Poisson's ratio = 0.29

Loading Data

Internal pressure = 620 lbf/in.^{2},
Bending moment = $700,000$ in. $\cdot$ lbf,
Torsional moment = $480,000$ in. $\cdot$ lbf,
Axial load = $35,000$ lbf tension

Problem. For the outer surface of the pipe, calculate (a) the maximum shear stress, (b) the principal stresses, (c) the direction of the stress σ_1 relative to the axis of the pipe, (d) the axial and circumferential unit strains, and (e) the principal strains. Also, (f) sketch Mohr's diagrams for stress and for strain.

Solution   The stress components are found as follows:

$$\text{The ratio } \frac{\text{I.D.}}{\text{O.D.}} = \frac{12.125}{14} = 0.87$$

which is the range usually termed *thin wall*. Hence,

$$\sigma_H = \text{hoop stress} = \frac{PD}{2t} = \frac{620 \times 12.125}{2 \times 0.937} = 4011 \text{ lbf/in.}^2,$$

$$\sigma_L = \text{longitudinal stress} = \frac{PD}{4t} = \frac{1}{2}\sigma_H = 2005 \text{ lbf/in.}^2,$$

$$\sigma_B = \text{bending stress} = \frac{Mc}{I} = \frac{700,000}{117.9} = \pm5937 \text{ lbf/in.}^2,$$

$$\tau = \text{torsional stress} = \frac{Tc}{J} = \frac{480,000}{235.8} = 2035 \text{ lbf/in.}^2,$$

$$\sigma_A = \text{axial stress} = \frac{F}{A} = \frac{35,000}{38.47} = 910 \text{ lbf/in.}^2$$

These conditions are illustrated in Fig. E.13.

1. Using Eq. (E.7), we obtain

$$\tau = \frac{1}{2}\sqrt{(8852 - 4011)^2 + (2 \times 2035)^2} = 3162 \text{ lbf/in.}^2$$

From Fig. E.14 we see, however, that the true maximum shear stress is

$$\tau_{max} = \left(\frac{9593}{2}\right) = 4796 \text{ lbf/in.}^2$$

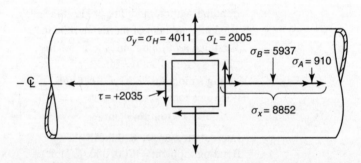

FIGURE E.13: Axial and hoop stresses acting in the pipe of Example E.7.

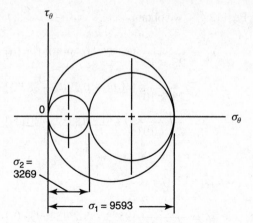

FIGURE E.14: Principal stresses and principal stress directions for the pipe in Example E.7.

2. Using Eq. (E.4) (see also Fig. E.15), we get

$$\sigma_1 = \frac{1}{2}(8852 + 4011) + 3162 = 9593 \text{ lbf/in.}^2,$$

$$\sigma_2 = \frac{1}{2}(8852 + 4011) - 3162 = 3269 \text{ lbf/in.}^2$$

Also,

$$\sigma_3 = \sigma_N = 0$$

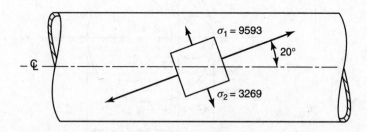

FIGURE E.15: Principal stress element for Example E.7.

3. Using Eq. (E.2), we obtain

$$\tan 2\theta_{\sigma_1} = \frac{2 \times 2035}{(8852 - 4011)} = \frac{4072}{4841} = 0.8407,$$

$$2\theta_{\sigma_1} = 40°4', \qquad \theta_{\sigma_1} = 20°2'$$

4. Using Eq. (12.40), we obtain

$$\varepsilon_x = \frac{1}{23 \times 10^6}[8852 - 0.29(4011 + 0)] = 334 \ \mu\text{-strain},$$

$$\varepsilon_y = \frac{1}{23 \times 10^6}[4011 - 0.29(0 + 8852)] = 63 \ \mu\text{-strain},$$

$$\varepsilon_z = \frac{1}{23 \times 10^6}[0 - 0.29(8852 + 4011)] = -162 \ \mu\text{-strain}$$

5. Again, from Eq. (12.4),

$$\varepsilon_1 = \frac{1}{23 \times 10^6}[9593 - 0.29(3269 + 0)] = 376 \ \mu\text{-strain},$$

$$\varepsilon_2 = \frac{1}{23 \times 10^6}[3269 - 0.29(0 + 9539)] = 22 \ \mu\text{-strain},$$

$$\varepsilon_3 = \frac{1}{23 \times 10^6}[0 - 0.29(9593 + 3269)] = -162 \ \mu\text{-strain}$$

6. Mohr's diagrams for stress and for strain are shown in Figs. E.14 and E.16, respectively.

[*Student assignment*: Modify the preceding calculations and diagrams for conditions on the inner pipe surface ($\sigma_3 = \sigma_N = -620$ lbf/in.^{2}).]

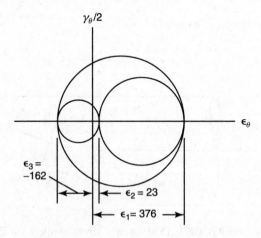

FIGURE E.16: Mohr's strain diagram (outer surface) for Example E.7.

APPENDIX F

Statistical Tests of Least Squares Fits

The method of least squares is described in Section 3.14.1. Here, we consider some statistical tests of least squares results. Recall that least squares is essentially a method of averaging out the *y precision error* in data that satisfy an underlying straight-line relationship, $y = a + bx$.

As in Section 3.14.1, for the various measured values of x_i, y_i is the experimentally determined ordinate and $y(x_i) = a + bx_i$ is the corresponding value calculated from the fitted line; n is the number of experimental observations used. Least squares minimizes the sum, S^2, of squared vertical deviations from the fitted line:

$$S^2 = \sum_{i=1}^{n} [y_i - y(x_i)]^2$$

to obtain

$$a = \frac{\sum y_i \sum x_i^2 - \sum x_i \sum x_i y_i}{n \sum x_i^2 - \left(\sum x_i\right)^2} \tag{F.1}$$

and

$$b = \frac{n \sum x_i y_i - \sum x_i \sum y_i}{n \sum x_i^2 - \left(\sum x_i\right)^2} \tag{F.2}$$

Our objective in this appendix is to characterize the quality of the least squares fit.

The simplest test of the fit, the correlation coefficient r^2, is described in Section 3.14.1. An algebraic form of r^2 convenient for hand calculation is

$$r^2 = b \cdot \frac{n \sum x_i y_i - \sum x_i \sum y_i}{n \sum y_i^2 - \left(\sum y_i\right)^2} \tag{F.3}$$

When $r^2 \to 1$, the precision error goes to zero, yielding a "perfect" fit. However, as noted previously, one usually obtains $|r| > 0.9$ whenever the data appear to fall on a straight line; the correlation coefficient is not a very sensitive indicator of the precision of the data.

Instead, we may consider the *standard error* of the y-data about the fit, $s_{y/x}$:

$$s_{y/x} = \left(\frac{1}{(n-2)} \sum_{i=1}^{n} [y_i - y(x_i)]^2\right)^{1/2} = \sqrt{\frac{S^2}{n-2}} \tag{F.4}$$

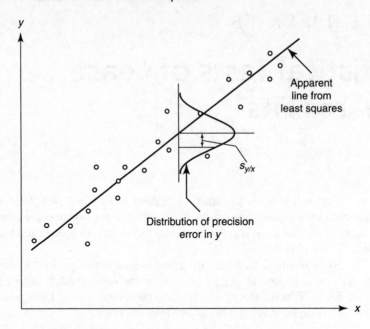

FIGURE F.1: Precision error in least squares fitting.

This quantity approximates the standard deviation of the precision error in y_i (Fig. F.1); if the precision error is small, so will be $s_{y/x}$.

We can make $s_{y/x}$ more revealing by introducing *total* squared variation of the data set (which includes both the precision error and the straight-line variation of y with x),

$$S_{yy}^2 = \sum_{i=1}^{n} (y_i - y_m)^2$$

where y_m is the mean measured y_i:

$$y_m = \frac{1}{n} \sum_{i=1}^{n} y_i$$

Letting $s_{yy}^2 = S_{yy}^2 / (n-1)$ be the *mean* total variation and performing some algebra, we find

$$\frac{s_{y/x}}{s_{yy}} = \left(\frac{n-1}{n-2} \right)^{1/2} \left(1 - r^2 \right)^{1/2} \tag{F.5}$$

The fraction on the left is the ratio of the standard deviation of the data about the line to the mean total variation of the data; thus $(1 - r^2)^{1/2}$ is a better gauge of the precision error in the data than is r itself [1]. For pocket-calculator work, the following formula is useful in calculating $s_{y/x}$ from Eq. (F.5):

$$s_{yy}^2 = \frac{\sum_{i=1}^{n} y_i^2 - \frac{1}{n} \left(\sum_{i=1}^{n} y_i \right)^2}{n-1} \tag{F.6}$$

Most calculators provide the various sums as part of the least squares calculation.

When least squares is justifiable, confidence intervals for the slope and intercept can be calculated under the assumption that the precision error in y_i satisfies the normal distribution [1, 2]. In this case, the true slope lies within the $c\%$ confidence interval

$$b \pm t_{\alpha/2,\nu} \frac{s_{y/x}}{S_{xx}} \qquad (c\%) \tag{F.7}$$

and the true intercept is within

$$a \pm t_{\alpha/2,\nu}\, s_{y/x} \sqrt{\frac{1}{n} + \frac{x_m^2}{S_{xx}^2}} \qquad (c\%) \tag{F.8}$$

where $t_{\alpha/2,\nu}$ is the t-statistic with $\nu = n - 2$ degrees of freedom at an $\alpha = (1 - c)$ level of significance, and S_{xx}^2 is

$$S_{xx}^2 = \sum_{i=1}^{n} (x_i - x_m)^2 = \sum_{i=1}^{n} x_i^2 - \frac{1}{n} \left(\sum_{i=1}^{n} x_i \right)^2 \tag{F.9}$$

for $x_m = (1/n) \sum x_i$. Like other statistical confidence intervals, those just given may become unacceptably broad when only a few data points have been fitted; an eyeball estimate of the uncertainty may again yield a more realistic result.

Finally, statistical tests for outliers can also be applied, under the assumption of normally distributed precision error in y_i. For example, if $s_{y/x}$ approximates the vertical standard deviation of the data, then only one point in 370 will have more than $3s_{y/x}$ vertical deviation from the line; for a small data set, points beyond $3s_{y/x}$ may be considered so unlikely as to be discarded.

EXAMPLE F.1

In Example 3.16, a least squares fit was used to find the stiffness of a cantilever beam. Quantify the vertical precision error in the data, and use a statistical confidence interval to estimate the uncertainty in the beam stiffness.

Solution   In Example 3.16, we found

$$n = 9$$
$$\sum x = 1801$$
$$\sum x^2 = 5.109 \times 10^5$$
$$\sum y = 33.50$$
$$\sum y^2 = 179.3$$
$$\sum xy = 9959$$

The data and the line-fit are shown in Fig. 3.25.

From Eq. (F.5), we calculate $s_{y/x}/s_{yy} = 0.0970$, which indicates that the data's vertical standard deviation about the fitted line is only 9.7% of the mean vertical variation of the data. Thus most of the vertical variation of the data is explained by the straight-line relation between y and x rather than precision error.

The confidence interval for the slope is calculated from Eq. (F.7) (at 95% certainty, say). With Eq. (F.6) and (F.9) and Table 3.5, we obtain

$$s_{yy} = 2.612$$
$$S_{xx} = 387.8$$
$$s_{y/x} = 0.2534$$
$$t_{0.025,7} = 2.365$$

from which the uncertainty in the slope is

$$\pm t_{\alpha/2,\nu}\frac{s_{y/x}}{S_{xx}} = \pm 0.001543 \qquad (95\%)$$

Hence, with 95% certainty, $0.0175 < b < 0.0205$ and so 478 N/m $< K <$ 560 N/m. The 95% uncertainty in K is about $\pm 8\%$.

The statistically determined uncertainty is about the same as that obtained from the eyeball estimated confidence limits on the fitted line (Fig. 3.25). Which approach is easier to use?

REFERENCES

[1] McClintock, F. A. *Statistical Estimation: Linear Regression and the Single Variable.* Research Memo 274, Fatigue and Plasticity Laboratory, February 14, 1987. Cambridge: Massachusetts Institute of Technology.

[2] Miller, I. R., J. E. Freund, and R. Johnson. *Probability and Statistics for Engineers.* 4th ed. Englewood Cliffs, N.J.: Prentice Hall, 1990.

Answers to Selected Problems

Chapter 2

2.2 (b) 310.93 K

2.3 (b) 1, 680.23°R

2.5 0.01124 lbf

2.8 1 gal/min $= 63.09$ cm^3/s

Chapter 3

3.1 183.2 to 216.8

3.3 (a) 10.2 k$\Omega \pm 0.0863$ kΩ

3.4 2.24%

3.7 7.45%

3.10 1.256%

3.12 $\approx \pm 10\%$

3.15 ≈ 65 marbles

3.18 ≈ 93

3.19 Packing is significant

3.20 Significant difference at 99% confidence level

3.23 $1.239 < \mu < 1.279$

3.24 No significant difference

3.29 No difference

3.30 Coin not fair

3.32 Normally distributed at 95% confidence level

3.35 Significant variation from standard

3.38 Force $= 28.8 + 30.7 \times$ Deflection

Chapter 4

4.1 (a) $15/2\pi$ rad; (b) $y(t) = 100 + 109.77 \cos(15t - 1.046)$

4.8 $V(t) \approx 6.7 \cos(2026.8t) + 2.9 \sin(3040.3t) - 0.9 \cos(4053.7t)$

4.12 (d) $V(t) \approx 10 \sin \omega t + 4 \cos \omega t - 2 \sin 6 \omega t$

$\omega \approx 2\pi/.036$ rad/s

4.13 (a) $f_{Nyq} = 2048$ Hz (b) 2 Hz (c) 1000 Hz, 1500 Hz, 2000 Hz

4.16 $f_5 > 4000$ Hz, $N = 8000$ pts. sampled

Chapter 5

5.3 $1.57°F$

5.5 $f \approx 160$ Hz

5.6   35 Hz

5.11 $\tau = V_0/m°$

5.16 $f \approx 1{,}124$ Hz

5.19   Error $= 27.8\%$; $\phi \approx 24°$

5.21   **(a)** 2.0 s **(b)** $e = 0.78E$ volts

Chapter 6

6.4 $U_K/K \simeq \pm 3\%$

6.11 $U_K/K \approx \pm 5.7\%$

6.13 $K = 498$ N/m

Chapter 7

7.4 $e_0/e_i = (kR_T/R_B)\,[kR_T/R_M + kR_T/R_B + 1]^{-1}$

7.6 $R_2 = 63\ \Omega$

7.9 $R_1 = 141.9\ \Omega$

7.13   For 100 N, $\Delta e_0 = e_0 = 0.24$ V

7.29 $R_3 = 1\ \text{k}\Omega$

7.33 $e_{\text{rms}} = 6.0$ V

7.34   **(a)** $f_c = 4.82$ kHz

Chapter 8

8.1   **(b)** 10

8.5   **(a)** $\varepsilon_V/\Delta V_{fs} = \frac{1}{2^8}$

Chapter 9

9.4 $0.584\ E_o$

9.5   0.50

9.6   10.0

Chapter 12

12.1 $\nu = 0.3$

12.2   **(a)** $\sigma = 12{,}750$ psi

12.8   **(a)** $\sigma_a = 9300$ psi

12.11 $F = 1.68$

12.20 $\Delta e_o/e_i = (F/4)(2\varepsilon_T)$

12.23 $P_{\text{max}} = 53{,}800$ watts

Chapter 13

13.5 $F = 0.17$ lbf

13.7 $F_e = 293$ N

13.13 $U_K/K = 3.11\%$

Chapter 14

14.9 $h = 0.651$ m

14.13 $P_2 - P_1 = 16.65$ kPa

14.15 $O = 9.6°$

14.17 $M_{H_2O} = 1.67$

14.23 $y_{max} = 3.4 \times 10^{-3}$ in.

14.27 $\varepsilon_H = 322\ \mu$ strain; $\varepsilon_L = 76\ \mu$ strain

14.33 $f = 3400$ Hz

14.34 $f = 310$ Hz

Chapter 15

15.3 $Re_D = 6.9 \times 10^5$

15.7 $\Delta P = 39.35$ kPa

15.25 $\Delta P = 530$ psi

15.29 $V = 1.085$ m/s

Chapter 16

16.2 $T = 70.5°C$

16.6 $\Delta R = 0.33\ \Omega/°C$

16.39 $T_p = 179°C$

16.40 $T_p = 400°C$

16.41 $T_{max} = 276°F$

Chapter 17

17.4 $a_o = 0.41$ g

17.7 $k = 13.4 \times 10^6$ N/m

Chapter 18

18.2   Five 2000 gpm pumps

18.5 $I = W/A = 0.0016$ W/m^2

18.7   **(a)** $SPL_5 = 115$ dB; **(b)** $SPL_3 = 112.8$ dB, Decrease = 2.2 dB

18.10 $SPL = 91.5$ dB

Index